PRECALCULUS MATHEMATICS
WITH
ELEMENTARY FUNCTIONS

PRECALCULUS MATHEMATICS WITH ELEMENTARY FUNCTIONS

Lawrence P. Runyan
Shoreline Community College
Seattle, Washington

ALLYN AND BACON, INC.
Boston · London · Sydney · Toronto

Copyright © 1977 by Allyn and Bacon, Inc.,
470 Atlantic Avenue, Boston, Massachusetts 02210.

All rights reserved. Printed in the United States of America. No part of the material protected by this copyright notice may be reproduced or utilized in any form or by any means, electronic or mechanical, including photocopying, recording, or by any information storage and retrieval system, without written permission from the copyright owner.

Library of Congress Cataloging in Publication Data

Runyan, Lawrence P
  Precalculus mathematics with elementary functions.

	1. Mathematics—1961-		I. Title.
QA39.2.R86		512'.1	76-30356

ISBN 0-205-05573-7

To My Wife

Preface xi
Suggested Course Outlines xiii

PART I Basics

1. **ALGEBRA REVIEW** 3

    1.1 The Real Number System 3
    1.2 Order 6
    1.3 Inequalities and Absolute Value 9
    1.4 Laws of Exponents 12
    1.5 Complex Numbers 15
    1.6 Quadratic Equations 19
    1.7 Quadratic in Form 23
    1.8 Equations Involving Fractions 24
    1.9 Linear and Quadratic Inequalities 26
    1.10 Radical Expressions 30
    1.11 The Cartesian Coordinate System 31
    1.12 Sequences and Series (Optional) 36

2. **SET THEORY AND LOGIC** 42

    2.1 Introduction 42
    2.2 Sets and Subsets 42
    2.3 Unions and Intersections 43
    2.4 Complement 44
    2.5 DeMorgan's Laws 44
    2.6 The Cartesian Product 46
    2.7 Logic—A Proposition and Its Truth Value 47
    2.8 Connectives and Truth Tables 47
    2.9 Negations 49
    2.10 Conditional Propositions 52
    2.11 Logic and Mathematical Proofs 54
    2.12 Mathematical Induction 56
    2.13 Historical Comments on Sets and Logic 60

PART II Elementary Analysis of Functions

3. **FUNCTIONS AND RELATIONS** 65

    3.1 Introduction 65
    3.2 Functions 66
    3.3 Creating New Functions by Reflections 70
    3.4 Symmetry—Even and Odd Functions (Optional) 76

vii

CONTENTS

- 3.5 Translations, Compressions and Expansions 78
- 3.6 Inverses 86
- 3.7 Operations on Functions 91

4. LINEAR FUNCTIONS 99
- 4.1 The Slope of a Straight Line 99
- 4.2 The Distance Formula 104
- 4.3 Graphing and Curve Fitting for Linear Functions 107
- 4.4 Linear Functions and Their Inverses 110

5. QUADRATIC FUNCTIONS 112
- 5.1 Introduction 112
- 5.2 Graphing Quadratic Functions 112
- 5.3 Inverses of Quadratic Functions (Optional) 116
- 5.4 Maximums and Minimums—An Application 118
- 5.5 Curve Fitting (Optional) 122

6. THEORY OF EQUATIONS 126
- 6.1 Introduction to Polynomials 126
- 6.2 Synthetic Division 126
- 6.3 Rational Zeros of Polynomial Functions 129
- 6.4 Graphing Polynomial Functions 134
- 6.5 Irrational Zeros of Polynomial Functions 139
- 6.6 Complex Zeros of Polynomial Functions (Optional) 141
- 6.7 Historical Comments on Polynomials 143

7. QUOTIENTS OF POLYNOMIALS: RATIONAL FUNCTIONS 145
- 7.1 Introduction 145
- 7.2 Limits, Continuity, and Asymptotes 146
- 7.3 Introduction to Graphing Rational Functions 151
- 7.4 Further Graphing of Rational Functions 154
- 7.5 Partial Fractions (Optional) 159

8. EXPONENTIAL AND LOGARITHMIC FUNCTIONS 160
- 8.1 Introduction 160
- 8.2 Exponential Functions 160
- 8.3 A New Base and Some Applications for Exponential Functions 163
- 8.4 Logarithmic Functions 169
- 8.5 Equations Involving Logarithms 174
- 8.6 Computations with Logarithms (Optional) 177
- 8.7 Applications of Logarithms 180
- 8.8 More Applications Involving Logarithms 184
- 8.9 Historical Comments on Logarithms (Optional) 189

9. THE CIRCULAR FUNCTIONS 190
- 9.1 Introduction 190
- 9.2 The Radian 190
- 9.3 The Cosine and Sine Functions 193
- 9.4 Graphing the Cosine and Sine Functions 199
- 9.5 More Circular Functions 205
- 9.6 The Inverses of the Circular Functions 212

10. TRIGONOMETRY 217
- 10.1 Introduction 217
- 10.2 Right Triangle Trigonometry 217
- 10.3 The Fundamental Identities and the Reduction Formulas 222
- 10.4 Basic Equations and Identities 225
- 10.5 Historical Comments on Trigonometry 227

11. TRIGONOMETRIC EQUATIONS AND IDENTITIES 228
- 11.1 Trigonometric Equations 228
- 11.2 The Addition Formulas 231
- 11.3 The Half-Angle and Double-Angle Formulas 234

12. APPLICATIONS OF TRIGONOMETRY 238
- 12.1 Introduction 238
- 12.2 Right-Triangle Applications 238
- 12.3 Azimuth, Bearing, and Heading 243
- 12.4 Applications of the Law of Sines and the Law of Cosines 246
- 12.5 Introduction to Polar Coordinates 252
- 12.6 Properties of Polar Coordinates 258
- 12.7 Complex Numbers in Polar Form 265
- 12.8 The Trigonometric Functions of Small Angles (Optional) 271
- 12.9 Simple Harmonic Motion (Optional) 275

PART III Algebra

13. **THE CONIC SECTIONS** 281
    - 13.1 Introduction 281
    - 13.2 The Circle 281
    - 13.3 The Parabola 284
    - 13.4 The Ellipse 288
    - 13.5 The Hyperbola 292
    - 13.6 Rotations and the Conic Sections (Optional) 297
    - 13.7 Solid Analytic Geometry (Optional) 301

14. **MATRICES AND DETERMINANTS** 309
    - 14.1 Introduction 309
    - 14.2 Matrices 309
    - 14.3 Matrix Multiplication 312
    - 14.4 Determinants 314
    - 14.5 Properties of Determinants 317
    - 14.6 The Inverse of a Matrix 319
    - 14.7 Vectors 322
    - 14.8 Historical Comments on Vectors (Optional) 326

15. **SYSTEMS OF EQUATIONS AND INEQUALITIES** 328
    - 15.1 Linear Equations in Two Unknowns 328
    - 15.2 Cramer's Rule for Two Equations with Two Unknowns 331
    - 15.3 Three-Dimensional Space 334
    - 15.4 Matrices and Linear Systems (Optional) 337
    - 15.5 Non-Unique Solutions and Gaussian Reduction 338
    - 15.6 Echelon Form (Optional) 342
    - 15.7 System of Nonlinear Equations 344
    - 15.8 Systems of Inequalities 347
    - 15.9 Linear Programming 350

16. **COUNTING, PROBABILITY, AND THE BINOMIAL THEOREM** 356
    - 16.1 Counting 356
    - 16.2 Permutations 358
    - 16.3 Combinations 360
    - 16.4 Probability 362
    - 16.5 Theorems on Probability 366
    - 16.6 Finite Probability Distributions 371
    - 16.7 The Binomial Theorem 373

Bibliography 379
APPENDIX I A Chronology 383
APPENDIX II Factorial Notation 386
APPENDIX III Summation Notation 387
APPENDIX IV Conversion Factors 389
APPENDIX V The Meaning of Limit 390
APPENDIX VI Tables 392
- Exponential Functions 393
- Logarithms Base 10 394
- Logarithms Base e 396
- Trigonometric Functions 397

Index I–1

Preface

PURPOSE

This text covers all of the precalculus mathematics most students will need to prepare for either a short course in calculus or the traditional calculus sequence. Because of the inclusion of a review chapter (Chapter 1) and the choice of many basic problems in each set of exercises, this text is especially well suited for students who have been away from mathematics for some time. The completeness of the coverage and independence of the chapters allows the instructor considerable flexibility in choosing those topics most appropriate for courses in elementary functions, college algebra and trigonometry, or precalculus. Several suggested course outlines follow this preface.

APPROACH

There is a general trend towards more rigor as the text progresses and as the students mature mathematically. A "discovery approach" is used in the examples once the necessary background for such investigations has been covered.

The examples in each section are chosen to illustrate the concepts as simply as possible. In some cases the reader is encouraged to supply basic details in the discussion of new ideas.

The geometric interpretation of algebraic models is emphasized throughout.

Historical comments are included to help the reader gain another perspective of precalculus mathematics.

EXERCISES

The problem sets have been carefully grouped two ways. First they are grouped as A, B, and C. The A problems are remedial or very basic, fundamental

PREFACE

exercises; the B problems cover the traditional core material; and the C problems are applications varying in complexity from easy to rather challenging. Secondly, most of the exercises have two or three parts that are grouped in order that students can grasp comparisons between problems. This reinforces the fact that important concepts, such as the solutions to $|x - 4| = 2$ and $|4 - x| = 2$, are the same, or that $x^2 - 4x + 4$ and $x^2 - 6x + 9$ are fundamentally similar.

Many exercises are designed to take advantage of hand-held calculators. These are not simply long arithmetic problems but rather ones intended to extend the reader's knowledge beyond the limits imposed by requiring reasonably simple calculations. Such exercises are indicated by a large dot (•).

The exercises marked with a star (★) have complete solutions in the answer section at the end of the book. Many other exercises have explanations that are also included in the answer section.

ACKNOWLEDGMENTS

I would like to thank my colleagues and their students who class tested the first drafts: J. Andersen, R. Bell, B. Johnson, C. Main, and F. Prydz. I also want to thank H. Hubbard for reviewing my first draft and my colleagues in other departments for their help with the applications, including: R. Love, R. Petersen, D. Rosenquist, and R. Vaughan. The constructive criticism and encouragement I received are greatly appreciated. I want to thank the reviewers of the later drafts, including: Karl Folley, University of Detroit, Dan C. Farris, State University of New York at Cortland, and Allen Christian, Eastfield College. Also special thanks to Dale Comstock, Central Washington State College, and Richard Semmler of Northern Virginia Community College for their careful readings. Last but not least, to my typists, B. Grayson and P. Steiner—special thanks for your patient work.

Lawrence P. Runyan

Suggested Course Outlines

Many of the sections may or may not be included depending upon the pace of the course and each instructor's personal preference. Many may also be assigned as readings, particularly the sections on applications and historical comments.

One Quarter Elementary Functions
Chapters 1 (omit 1.12), 3 (omit 3.4), 4, 5 (omit 5.3, 5.4, 5.5), 6 (omit 6.5, 6.6, 6.7), 7 (omit 7.6), 8 (omit 8.3, 8.6, 8.7, 8.8, 8.9), 9, 10 (omit 10.5), 11, 15 (15.1, 15.3, 15.8 only), 16 (16.7 only).

One Semester Elementary Functions
Chapters 1, 2 (2.12 only), 3, 4, 5 (omit 5.3), 6 (omit 6.6, 6.7), 7 (omit 7.6), 8 (omit 8.8, 8.9), 9, 10 (omit 10.5), 11, 12 (12.1, 12.2, 12.4, 12.5 only), 13 (omit 13.6, 13.7), 14 (omit 14.5, 14.6, 14.7, 14.8), 15 (omit 15.5, 15.6, 15.9), 16 (omit 16.5, 16.6).

A Five Credit College Algebra and a Three Credit Trigonometry
   College Algebra (five quarter credits) Chapters 1 (omit 1.12), 2 (2.1–2.5 and 2.12 only), 3 (omit 3.4), 4, 5 (omit 5.3, 5.5), 6 (omit 6.5, 6.6), 7 (omit 7.6), 8 (omit 8.6, 8.8), 13 (omit 13.6, 13.7), 14 (omit 14.5, 14.7), 15 (omit 15.6, 15.7, 15.9), 16 (16.7 only).
   Trigonometry (three quarter credits)
Chapters 9, 10, 11, 12 (omit 12.6, 12.8, 12.9), 13 (13.6 only).

A Two Semester Precalculus Sequence
Include all of the text with selected sections from the following optional sections: 3.4, 5.3, 5.5, 6.5, 6.6, 7.6, 8.8, 12.6, 12.8, 12.9, 13.6, 14.7, 15.7, 15.9, and 16.6.

SCIENTIFIC ILLUSTRATORS, a textbook production service located in Champaign, Illinois, prepared the illustrations for this book. The service is owned and operated by George E. Morris, who has written two textbooks (*Technical Illustrating* and *Engineering— A Decision-Making Process*), holds degrees in Mechanical Engineering from the University of Illinois, and teaches college mathematics part time. The artist for this book was Jeffrey J. Mellander.

PRECALCULUS MATHEMATICS
WITH
ELEMENTARY FUNCTIONS

PART I

Basics

CHAPTER 1

Algebra Review

This chapter is a review of the basic algebraic skills essential to the study of the precalculus mathematics contained in the remainder of the text.

1.1 THE REAL NUMBER SYSTEM

First, let's discuss the composition of our number system and some of its fundamental properties. We call this system the **real number** system. There are two major types of real numbers, **rational** and **irrational** numbers. Below are some rational numbers

$$3, \quad \tfrac{1}{3}, \quad 1.5, \quad 1.333\cdots, \quad -\tfrac{19}{5}$$

and some irrational ones

$$\sqrt{3}, \quad \pi, \quad \sqrt[3]{5}, \quad 1.010110111\cdots$$

Following are two different ways of expressing rational numbers. Any real number that cannot be expressed in these forms is irrational.[1]

1. A rational number has a terminating or repeating decimal representation.
2. A rational number can be written in the form p/q, where p and q are relatively prime integers and $q \neq 0$.

Two numbers are relatively prime if they have no common factors other than 1. A number like $\tfrac{15}{25}$ is a rational number even though 15 and 25 are not relative prime (they have a common factor of 5) because

$$\frac{15}{25} = \frac{3}{5}$$

where 3 and 5 are relatively prime.

Below are some rational numbers written in both forms.

$$\tfrac{2}{5} = 0.4, \qquad \tfrac{2}{3} = 0.666\cdots, \qquad 0.12 = \tfrac{3}{25}$$

The rational numbers are broken down into two smaller classifications. The first is the **integers**, which consist of zero, the natural numbers

$$1, 2, 3, 4, 5, \ldots$$

CHAPTER 1   Algebra Review

TABLE 1.1. *The Real Number Field*

PROPERTY	ADDITION	MULTIPLICATION
Closure	The sum of any two real numbers is also a real number.	The product of any two real numbers is also a real number.
Commutative[3] (Law of Order)	$a + b = b + a$ for all real numbers a and b.	$a \cdot b = b \cdot a$ for all real numbers a and b.
Associative (Law of Grouping)	$a + (b + c) = (a + b) + c$ for all real numbers a, b, and c.	$a \cdot (b \cdot c) = (a \cdot b) \cdot c$ for all real numbers a, b, and c.
Identity	$a + 0 = a$ for all real numbers a. 0 is called the additive identity. It is unique for the system of real numbers.	$a \cdot 1 = a$ for all real numbers a. 1 is called the multiplicative identity. It is unique for the system of real numbers.
Inverses	$a + (-a) = 0$ for all real numbers a. This additive inverse of a, $(-a)$, is unique for each a.	$a \cdot 1/a = 1$ for all real numbers $a \neq 0$. This multiplicative inverse of a, $1/a$, is unique for each a.
Distributive (Law of the Common Factor)	$a \cdot (b + c) = ab + ac$ for all real numbers a, b, and c.	

and their negatives

$$\ldots, -3, -2, -1$$

The second classification, **fractions**, includes all rational numbers that are not equivalent to some integer. Thus, $\frac{3}{4}$ is a fraction, but $\frac{12}{3}$ is not since it is simply another name for the integer 4.

The following chart summarizes the real number system.

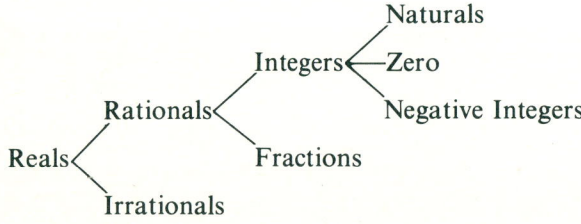

Now let's examine the properties of the real numbers under the usual operations of addition and multiplication. These properties, called "field properties," are listed in Table 1.1.[2]

The distributive property requires special attention since it describes how the operations of addition and multiplication interact. We can show how the distributive property explains why the operations are not defined independently and are really closely related. Consider the sum $3 + 3$ and the product $3 \cdot 2$. By the distributive law we have

$$3 + 3 = 3 \cdot 1 + 3 \cdot 1$$
$$= 3(1 + 1)$$
$$= 3 \cdot 2$$

There are two properties of equality that are fundamental to studies of the field properties and subsequent algebras. They are stated below for real numbers a, b, and c.

PROPERTIES OF EQUALITY.
 I. If $a = b$, then $a + c = b + c$. The Additive Law of Equality.
 II. If $a = b$, then $ac = bc$. The Multiplicative Law for Equality.

EXERCISE 1.1

A. FUNDAMENTALS

 1. Name the rational numbers in the following lists.
 (a) $-\frac{3}{4}, 2, 0, \sqrt{2}, \pi$ (b) $1.4, 1.33\cdots, \sqrt[3]{8}, \sqrt[3]{7}$

SECTION 1.1   The Real Number System

2. Name the integers in the following lists.
(a) $1, \frac{1}{2}, 0, -\frac{3}{5}, \sqrt{2}$ (b) $-2, \frac{3}{5}, \frac{9}{3}, -4.0, -\sqrt{4}$

Name the field property illustrated by each of the following equations.

3. (a) $3 + (x + 4) = 3 + (4 + x)$
   (b) $3 + (4 + x) = (3 + 4) + x = 7 + x$
4. (a) $2x + 3x = (2 + 3)x = 5x$
   (b) $5x + 3x = 8x$
5. (a) $2 + (5 + x) = 7 + x$
   (b) $2 + (x + 5) = 7 + x$
6. (a) $4(x + \frac{1}{4}) = 4x + 1$
   (b) $2(x + 1) + (-2) = (2x + 2) + (-2)$
                      $= 2x + [2 + (-2)]$
                      $= 2x + 0 = 2x$
7. (a) $[(2 + 3) + x] + 3x = (2 + 3) + (x + 3x)$
   (b) $3 + [(2 + x) + 4x] = (3 + 2) + (x + 4x)$

Supply the field property that justifies each of the indicated steps in the solutions of the following problems.

8. $(x + 2)(x + 3) = (x + 2) \cdot x + (x + 2) \cdot 3$
                           (a) _____
   $= (x^2 + 2x) + (x \cdot 3 + 6)$
                        (b) _____
   $= (x^2 + 2x) + (3x + 6)$
                        (c) _____
   $= x^2 + (2x + 3x) + 6$
                        (d) _____
   $= x^2 + (2 + 3)x + 6$
                        (e) _____
   $= x^2 + 5x + 6$
                        Definition of $2 + 3$

9. If $3x + 5 = 8$, then $x = 1$.
   $3x + 5 = 8$
   $(3x + 5) + (-5) = 8 + (-5)$
                  The additive law of equality
   $3x + [5 + (-5)] = 3$
                  (a) _____
                  additive inverse
   $3x + (0) = 3$
   $3x = 3$ (b) _____
   $\frac{1}{3}(3x) = \frac{1}{3} \cdot 3$ (c) _____
   $(\frac{1}{3} \cdot 3)x = \frac{1}{3} \cdot 3$ (d) _____

$1 \cdot x = 1$ (e) _____
$x = 1$ *multiplicative identity*

10. If $3(x + 2) = 2x + 8$, then $x = 2$.
    $3(x + 2) = 2x + 8$
    $3x + 6 = 2x + 8$
            (a) _____
    $(3x + 6) + (-6) = (2x + 8) + (-6)$
            additive law of equality
    $3x + [6 + (-6)] = 2x + [8 + (-6)]$
            (b) _____
    $3x + 0 = 2x + 2$
            additive inverses
    $3x = 2x + 2$
            (c) _____
    $(-2x) + 3x = (-2x) + (2x + 2)$
            (d) _____
    $(-2 + 3)x = -2x + (2x + 2)$
            (e) _____
    $1 \cdot x = (-2x + 2x) + 2$
            (f) _____
    $x = (-2 + 2)x + 2$
            distributive
    $x = 0 \cdot x + 2$
            (g) _____
    $x = 2$
            additive identity

B. ESSENTIALS

Use the field properties to justify the solution of the following equations. State the reason for each step as in Exercises 8, 9, and 10.

★11. If $x + 2 = 4$, then $x = 2$.
12. If $2x = 6$, then $x = 3$.
13. If $2x + 2 = 10$, then $x = 4$.
14. If $x + (-1) = 5$, then $x = 6$.
15. If $\frac{1}{3} \cdot x = 3$, then $x = 9$.
16. If $\frac{1}{2}x + (-2) = 3$, then $x = 10$.
★17. If $2x + 1 = x + 2$, then $x = 1$.
18. If $2(x + 1) = x + 3$, then $x = 1$.
★19. If $2(x + 3) = 3x + (-4)$, then $x = 10$.

C. APPLICATIONS

Prove the following theorems for the real number system.

★20. If $a \cdot b = 0$, then $a = 0$ or $b = 0$.

21. The additive identity is unique.

22. The multiplicative identity is unique.

23. The additive inverse of each real number is unique.

★24. The multiplicative inverse of each real number is unique.

25. The rational numbers under addition and multiplication satisfy all the field properties listed in Table 1.1. The rational number system is a field.

26. Under addition and multiplication some systems do not satisfy all the field properties. Illustrate with examples those properties that are *not* satisfied by
(a) natural numbers
(b) integers
(c) irrational numbers

27. The definition of a real number given by Dedekind in 1876 is still being used. He described a real number as a cut (a Dedekind cut) as follows. To define $\sqrt{2}$, divide the rational number line into two parts, L and G, where L contains all rationals whose square

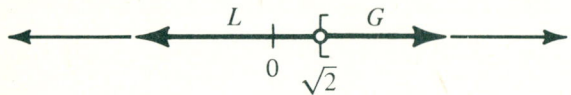

is less than 2 and G contains those rationals whose square is greater than or equal to 2. The cut is the largest number smaller than or equal to all the numbers in G *and* greater than all those in L. If this number is in G, it is rational. If not, it is irrational. Define $\sqrt{11}$ as a Dedekind cut.

1.2 ORDER

Have you noticed that in any pair of unequal real numbers, one of them is always smaller? This is another distinguishing property of the real number system called *order*. Order creates three classes of

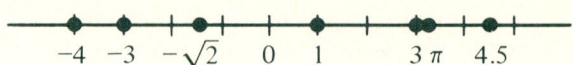

FIGURE 1.1. The Real Number Line.

real numbers: positive, negative, and zero. These classes are depicted on the real number line in Figure 1.1, which shows positive numbers increasing to the right of zero and negative numbers decreasing to the left of zero. Notice that if a real number a is found to the left of another real number b on the real number line, then a is less than b. This is written symbolically as

$$a < b$$

There are two common, equivalent definitions of $a < b$ (a less than b) for real numbers a and b.

DEFINITION 1.1.
$a < b$ means $b - a$ is positive.

DEFINITION 1.2.
$a < b$ means there exists a positive number c such that $a + c = b$.

If a is greater than b, written $a > b$, we simply realize that b is smaller than a and use the above definitions. Therefore, we have Definition 1.3.

DEFINITION 1.3.
$a > b$ means $b < a$.

There are other properties of this ordered system of real numbers that relate to the addition and multiplication of signed numbers (positive and negative). The four basic properties of order are listed next.

> **THE ORDER PROPERTIES.**
> For all real numbers a, b, and c.
>   I. Either $a = b$, $a < b$, or $a > b$. Trichotomy Law
> II. If $a < b$ and $b < c$, then $a < c$. Transitive Law
> III. If $a < b$, then $a + c < b + c$. Additive Law
> IV. If $a < b$ and $c > 0$, then $ac < bc$. Multiplicative Law

If we know only that a is not greater than b, from the trichotomy law we know that either $a = b$ or $a < b$, which is written $a \leq b$.

> **DEFINITION 1.4.**
> $a \leq b$ means $a < b$ or $a = b$.

Consider Theorems I and II. Their proofs illustrate how the order properties and the definitions are used in proving theorems on order.

This first theorem states that if a is negative, then its additive inverse, $-a$, is positive.

> **THEOREM I.**
> For any real number a, if $a < 0$, then $-a > 0$.

Proof: If $a < 0$, then, by Property III,
$$-a + a < -a + 0$$
which implies that
$$0 < -a$$
By Definition 1.3 we have
$$-a > 0$$

> **THEOREM II.**
> If a is any real number, then $a^2 > 0$.

Proof:

CASE I. If $a > 0$, then, by Property IV,
$$a \cdot a \geq a \cdot 0$$
and
$$a^2 \geq 0$$

CASE II. If $a > 0$, then, by Theorem I,
$$-a > 0$$
Now, using the result of Case I (since $-a$ is positive), we have
$$(-a)^2 > 0 \quad \text{and} \quad a^2 > 0$$

The following multiplicative property for inequalities is essential in the fact that it differs from the familiar multiplicative property for equalities. It reminds us that whenever we multiply both sides of an inequality, such as
$$5 < 7$$
by a negative number, say -2, then the resulting inequality,
$$-10 > -14$$
has the opposite order from the original, since "5 is less than 7" becomes "-10 is greater than -14." This property is formally stated in Theorem III.

> **THEOREM III.**
> If $a < b$ and c is negative ($c < 0$), then $ac > bc$.

Now, study each of the examples and fill in the blanks with "$<$" or "$>$" to formulate the main theorems on order.

EXAMPLE I. Since $2 \cdot 5 \_\_\_ 0$ and $(-2)(-5) \_\_\_ 0$, we can formulate Theorems IV and V.

CHAPTER 1   Algebra Review

THEOREM IV.
If $a > 0$ and $b > 0$, then $a \cdot b$ ____ 0.

THEOREM V.
If $a < 0$ and $b < 0$, then $a \cdot b$ ____ 0.

EXAMPLE II. Since $2(-5)$ ____ 0, we can complete Theorem VI.

THEOREM VI.
If $a > 0$ and b ____ 0, then $a \cdot b$ ____ 0.

EXAMPLE III. Since $3 + 4$ ____ 0 and $(-2) + (-5)$ ____ 0, we have Theorems VII and VIII.

THEOREM VII.
If $a > 0$ and $b > 0$, then $a + b$ ____ 0.

THEOREM VIII.
If $a < 0$ and $b < 0$, then $a + b$ ____ 0.

EXAMPLE IV. Since $5 + (-2)$ ____ 0 and $3 + (-7)$ ____ 0, we can develop Theorem IX.

THEOREM IX.
If $0 < a < b$, then $b + (-a)$ ____ 0 and $a + (-b)$ ____ 0.

The procedure you followed to complete Theorems IV through IX is part of the scientific method used by mathematicians in creating new theorems. The first step is to consider appropriate examples. Then a generalization, the hypothesized theorem, is formulated. After trying enough examples to be reasonably convinced of the likelihood that the theorem is valid, the mathematician constructs a proof based upon established definitions and properties and previously proven theorems. A proof is necessary to completely establish the validity of a theorem, since it may not be possible to consider all possible examples.

Now, let's complete the process for Theorem V by constructing its proof. We will leave the proofs of the other theorems as exercises. This theorem states that the product of two negative numbers is always positive.

THEOREM V. For all real numbers a and b, if $a < 0$ and $b < 0$, then $a \cdot b > 0$.

Proof: If $a < 0$, then, by Theorem I,
$$-a > 0$$
By Property IV,
$$b < 0$$
becomes
$$(-a)b < -a \cdot 0$$
The inequality
$$-ab < 0$$
implies, by Theorem I,
$$ab > 0$$

EXERCISE 1.2

A. FUNDAMENTALS

1. (a) $(-2)(-3)$ (b) $-(2)(-3)$
    (c) $-(-2)(3)$

2. (a) $7 + (-9)$ (b) $7 - 9$ (c) $-9 + 7$

3. (a) $3 - (-8)$ (b) $3 + (-1)(-8)$
    (c) $3 + 8$

4. (a) $(-2)(-5 - 4)$
    (b) $(-2)(-5) + (-2)(-4)$

5. $(-3)(2 - 4)$

6. $(-3)(-4) - 3(2)$

7. $2(-3) + 4(-3)$

B. ESSENTIALS

Prove the following for all real numbers a and b.

8. If $a \leq b$ and $b \leq a$, then $a = b$.
9. ★(a) If $a < b$ and $c < 0$, then $ac > bc$.
   (b) If $a < 0$ and $b > 0$, then $a \cdot b < 0$.
10. (a) If $a < 0$ and $b < 0$, then $a + b < 0$.
    (b) If $a < b$ and $b < 0$, then $a + b < 0$.
    (c) If $a < 0 < b$, then $(-a) + b > 0$.
11. ★(a) If $a \cdot b > 0$ and $a < 0$, then $b < 0$.
    (b) If $a \cdot b < 0$ and $b < 0$, then $a > 0$.
12. (a) If x is a real number, then $x(2 - x) \leq 1$.
    (b) If x and y are real numbers, then $x^2 + y^2 \geq 2xy$.

1.3 INEQUALITIES AND ABSOLUTE VALUE

To order the real numbers, we often arrange them on the real number line shown in Figure 1.2.

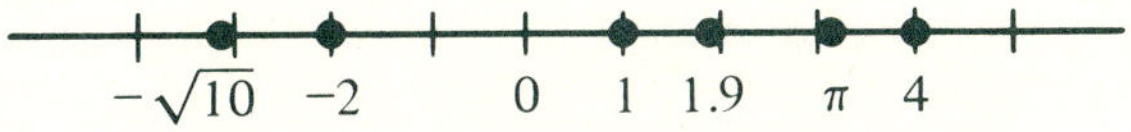

FIGURE 1.2. *The Real Number Line.*

In such orderings the distance any number is away from 0 on the real number line is so important that it has been given a name. This distance is called the **absolute value** of a real number. Its symbol is two vertical lines, $|\ \ |$. Then $|-4| = 4$, $|-1| = 1$, $|3| = 3$, and $|5| = 5$, since -4 is 4 units away from zero, etc. Notice that $|3| = |-3|$, since they are the same distance from the origin. Absolute value, since it measures distances, is *always* positive or zero. Definition 1.5 is a formal definition of absolute value.

DEFINITION 1.5.
$$|x| = \begin{cases} x, & \text{if } x \text{ is positive } (x > 0) \\ 0, & \text{if } x \text{ is } 0 \ (x = 0) \\ -x, & \text{if } x \text{ is negative } (x < 0) \end{cases}$$

Remember that if x is negative, $-x$ is positive, so that the absolute value of x is always nonnegative.

Often we encounter questions such as, "For which values of x is $|x - 4| = 7$?" Both $x = 11$ and $x = -3$ satisfy the equation $|x - 4| = 7$. By examining the geometric relationship between these values and $x = 4$, we can understand what the question is really asking. See Figure 1.3.

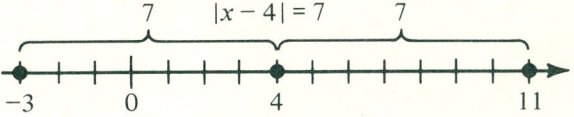

FIGURE 1.3.

In the equation $11 - 4 = 7$, we see that the difference between the two numbers 11 and 4 yields the distance between them. For the equation $(-3) - 4 = -7$, the absolute value of the difference, $|(-3) - 4| = |-7|$, yields the distance between the numbers -3 and 4, which is also 7. Hence, $x = -3$ and $x = 11$ are the only two numbers that are located 7 units away from $x = 4$.

Then, solving the equation

$$|x - 7| = 1$$

means finding the two numbers 1 unit away from $x = 7$. These are $x = 6$ and $x = 8$, the only solutions of the equation. The equation

$$|x + 7| = 2$$

is equivalent to

$$|x - (-7)| = 2$$

which requires the values of x to represent numbers 2 units away from -7. These are $x = $ ____?____ and $x = $ ____?____, since

$$|(-9) - (-7)| = |-9 + 7| = |-2| = 2$$

and

$$|(-5) - (-7)| = |-5 + 7| = |2| = 2$$

Often we encounter questions such as, "For which values of x is $|x| < 3$?" This is asking for the numbers that are less than 3 units away from the origin, 0. The answer then is all values of x between -3 and 3. Therefore,

if $|x| < 3$, then $-3 < x < 3$

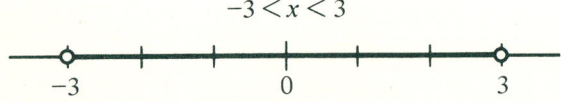

FIGURE 1.4.

The question, "For what values of x is $|x| > 5$?" is asking for all numbers more than 5 units away from the origin. The correct values of x are those greater than 5 or less than -5. Therefore,

if $|x| > 5$, then $x < -5$ or $x > 5$

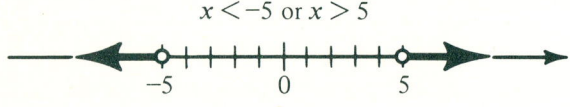

FIGURE 1.5.

These results are summarized in Theorems X and XI, where $K > 0$.

> THEOREM X.
> $|x| < K$ is equivalent to $-K < x < K$.

> THEOREM XI.
> $|x| > K$ is equivalent to $x < -K$ or $x > K$.

We often encounter slightly more complicated problems, for example, trying to find the values of x that satisfy the inequalities

(a) $|x - 3| < 5$ and (b) $|x - 1| > 3$

Inequality (a) is asking for all numbers less than 5 units away from 3. By examining Figure 1.6, we

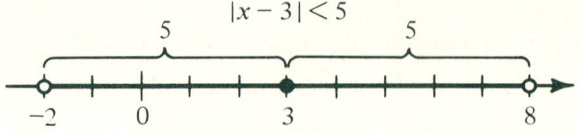

FIGURE 1.6.

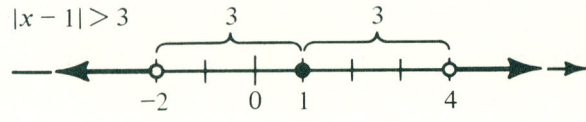

FIGURE 1.7.

can see that they are those numbers between -2 and $+8$, or

if $|x - 3| < 5$, then $-2 < x < 8$

Inequality (b) is asking for all numbers more than 3 units away from 1, as shown in Figure 1.7. Therefore, the values of x must be either less than -2 or greater than 4. Therefore,

if $|x - 1| > 3$, then $x < -2$ or $x > 4$

The algebraic solution of these inequalities involves using either Theorem X or XI to eliminate the absolute value symbol. By Theorem X,

$$|x - 3| < 5$$

is equivalent to

$$-5 < x - 3 < 5$$

We solve this by adding 3 to all three places, the -5, the $x - 3$, and the $+5$:

$$-2 < x < 8$$

We must add 3 to all three places because $-5 < x - 3 < 5$ means $-5 < x - 3$ and $x - 3 < 5$. Solving these separately yields $-2 < x$ and $x < 8$, which are combined to get the solution, $-2 < x < 8$. See Figure 1.8.

By Theorem XI, $|x - 1| > 3$ is equivalent to $x - 1 < -3$ or $x - 1 > 3$, which are each solved for x by adding 1 to both sides of each inequality, yielding $x < -2$ or $x > 4$. See Figure 1.9.

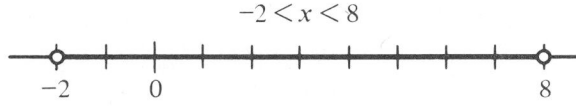

FIGURE 1.8.

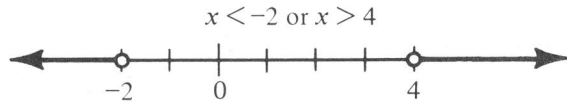

FIGURE 1.9.

EXERCISE 1.3

A. FUNDAMENTALS

Solve for the values of the variables that satisfy the following inequalities. Graph the answers on the real number line.

1. (a) $x + 2 < 3$ (b) $2x + 4 < 8$
   (c) $3x - 1 < x + 5$

2. (a) $3x - 5 \geq 7$ (b) $4x - 2 \geq x + 2$
   (c) $2(x + 3) \geq x - 7$

3. (a) $-2x \leq 8$ (b) $-x/5 \leq 2$
   (c) $2x \leq 3x - 4$

4. (a) $2(3 - y) > 8$
   (b) $3(2 - 4y) < 6 - 10y$
   (c) $4(y + 7) > 7(y - 8)$

5. (a) $3(x - 5) \leq 2(x + 1)$
   (b) $3(x + 7) \leq 4(x + 2)$
   (c) $3(4 - x) \leq 2(x - 3)$

6. Explain the geometric interpretation (using the real number line) of the following inequalities.

   (a) $|x| < 4$ (b) $|x| \geq 3$

7. Explain the geometric interpretation of the following inequalities.

   (a) $|x - 3| \leq 4$ (b) $|x - 2| \geq 3$
   (c) $|x + 1| \leq 2$

B. ESSENTIALS

Solve the following and graph the solutions.

8. (a) $|t| = 7$ (b) $|t| = 8$
   (c) $|t| = -2$

9. (a) $|x| > 4$ (b) $|x| \geq 4$
   (c) $|x| \geq 0$

10. (a) $|x| = 3$ (b) $|x - 1| = 3$
    (c) $|x - (-1)| = 3$

11. (a) $|x| < 3$ (b) $|x - 1| < 3$
    (c) $|x - (-1)| < 3$

12. (a) $|x| > 4$ (b) $|x - 2| > 4$
    (c) $|x + 2| > 4$

13. (a) $|1 - x| < 4$ (b) $|x - 1| < 4$

14. (a) $|3 - 2x| > 7$ (b) $|2x - 3| > 7$

15. (a) $|3x - 6| = 9$ (b) $3|x - 2| = 9$

16. (a) $|2x + 8| > 12$ (b) $2|x + 4| > 12$

17. (a) $|2 - 4x| > 3$ (b) $2|2x - 1| > 3$

Use the following definition to find the solutions of the equations below.

DEFINITION 1.6.
For all real numbers x, $|x| = \sqrt{x^2}$.

---

18. (a) $|x| = 3$ (b) $x^2 = 9$
19. (a) $|x| = 4$ ★(b) $(x - 1)^2 = 16$
20. (a) $(2x - 3)^2 = 1$ (b) $(2x + 3)^2 = 1$
21. (a) $(1 - 2x)^2 = 4$ (b) $(2x - 1)^2 = 4$
22. (a) $|x - 3| = 2$ (b) $x^2 - 6x + 9 = 4$
23. (a) $(2x + 1)^2 = 25$
    (b) $4x^2 + 4x + 1 = 25$

Use the Theorems XII and XIII to solve the inequalities given below. x and y are real numbers.

THEOREM XII.
If $|x| \leq |y|$, then $x^2 \leq y^2$.

CHAPTER 1   Algebra Review

> THEOREM XIII.
> If $|x| \geq |y|$, then $x^2 \geq y^2$.

24. ★(a) $|x+1| \leq |x|$ (b) $|x-2| \leq |x|$
25. (a) $|x+2| \leq |x-1|$
    (b) $|x-3| \leq |x+4|$
26. (a) $|x+3| \geq |x|$ (b) $|x| \geq |x+4|$
27. (a) $|x+2| \geq |x+1|$
    (b) $|x-3| \geq |x+2|$
28. (a) $|x+4| < |x-3|$
    (b) $|x+2| \geq |x-3|$
29. $|\sqrt{x^2} - x| > 1$

C. APPLICATIONS

30. Without using Theorems XII and XIII, prove that if $|x+2| \leq |x|$, then $x \leq -1$. (HINT: you will need to use the definition of absolute value for the following four cases: $x+2 \geq 0$ and $x \geq 0$, $x+2 \geq 0$ and $x \leq 0$, $x+2 \leq 0$ and $x \geq 0$, and $x+2 \leq 0$ and $x \leq 0$.)

31. Use the triangle inequality to prove the following for all real numbers a and b.
$$||a| - |b|| \leq |a-b|$$

32. Prove Theorem XII.
33. Prove Theorem XIII.

1.4 LAWS OF EXPONENTS

Do you recall that 2^3 means that 2 is used as a factor three times, as in
$$2^3 = 2 \cdot 2 \cdot 2 = 8$$
The 3 is the **exponent** and the 2 is the **base** in the expression 2^3. In any case, by carefully considering the following examples, you may be able to formulate the first three theorems on exponents. We will begin with integral exponents and positive real bases. Fill in the blanks as you go.

EXAMPLE I.
(a) $2^3 \cdot 2^2 = (2 \cdot 2 \cdot 2) \cdot (2 \cdot 2) = \underline{\quad ? \quad}$
(b) $x^4 \cdot x^3 = (\underline{\quad ? \quad})(\underline{\quad ? \quad}) = \underline{\quad ? \quad}$

To multiply exponential expressions that have the same base, we $\underline{\quad ? \quad}$ their exponents.

EXAMPLE II.
(a) $\dfrac{2^5}{2^3} = \dfrac{2 \cdot 2 \cdot 2 \cdot 2 \cdot 2}{2 \cdot 2 \cdot 2} = 2 \cdot 2 = \underline{\quad ? \quad}$
(b) $\dfrac{x^4}{x^3} = \dfrac{\quad ? \quad}{\quad ? \quad} = \underline{\quad ? \quad}$

To divide exponential expressions that have the same base, we $\underline{\quad ? \quad}$ their exponents.

EXAMPLE III.
(a) $(2^2)^3 = (2^2)(2^2)(2^2) = 2^{2+2+2} = \underline{\quad ? \quad}$
(b) $(x^3)^4 = \underline{\quad ? \quad} = \underline{\quad ? \quad}$

To raise an exponential expression to a power, we $\underline{\quad ? \quad}$ the exponents.

Your answers to the examples should have been 2^5, x^7, 2^2, x^1, 2^6, and x^{12}. The blanks should have been filled with "add," "subtract," and "multiply."

In order to keep the first three theorems from producing inconsistent results, we must define bases raised to the zeroth power, negative exponents, and fractional exponents in specific ways. Study the following examples to see how this is done.

EXAMPLE IV. We know that
$$\frac{2}{2} = \underline{\quad ? \quad}$$
But, since we subtract exponents when we divide such expressions, we also have
$$\frac{2}{2} = \frac{2^1}{2^1} = 2^{1-1} = \underline{\quad ? \quad}$$

SECTION 1.4 Laws of Exponents

Therefore, we must define 2^0 to equal 1. Any base (except 0) raised to the zeroth power must equal ___?___. The expression 0^0 is undefined.

EXAMPLE V. Since $3^0 = 1$, we have

$$\frac{1}{3} = \frac{3^0}{3^1}$$

which, by subtracting exponents, becomes

$$\frac{1}{3} = \frac{3^0}{3^1} = 3^{0-1} = \underline{\quad ? \quad}$$

Therefore,

$$3^{-1} = \underline{\quad ? \quad}$$

Negative exponents must be defined to mean reciprocals.

EXAMPLE VI. Consider the fractional exponent in the expression $5^{1/2}$. Since we multiply exponents when we raise to a power, we have

$$(5^{1/2})^2 = 5^{(1/2) \cdot 2} = \underline{\quad ? \quad}$$

But, we know that

$$(\sqrt{5})^2 = \underline{\quad ? \quad}$$

Therefore, $5^{1/2} = \sqrt{5}$ and such fractional exponents must be defined to mean roots.

In a similar manner, the fractional exponents $1/3, 1/4, \cdots$, mean cube root, fourth root, etc. Therefore,

$$8^{1/3} = 2, \quad 16^{1/4} = 2, \quad 81^{1/4} = 3$$

This last definition leads us to consider more general rational exponents. Consider the next examples that combine fractional, negative, and zero exponents.

EXAMPLE VII.
(a) $4^{3/2} = (4^{1/2})^3 = (\sqrt{4})^3 = (\underline{\;?\;})^3 = \underline{\;?\;}$
(b) $8^{2/3} = (8^{1/3})^2 = (\sqrt[3]{8})^2 = (\underline{\;?\;})^2 = \underline{\;?\;}$

(c) $4^{3/2} = (4^3)^{1/2} = \sqrt{4^3} = \underline{\;?\;} = \underline{\;?\;}$
(d) $8^{2/3} = (8^2)^{1/3} = \sqrt[3]{8^2} = \sqrt[3]{64} = \underline{\;?\;}$

The answer to (a) and (c) is 8 since they are the same problem worked two different ways. The answer to (b) and (d) is 4.

EXAMPLE VIII.

(a) $\left(\dfrac{4}{16}\right)^{1/2} = \sqrt{\dfrac{4}{16}} = \dfrac{\sqrt{4}}{\sqrt{16}} = \dfrac{?}{?}$

(b) $\left(\dfrac{4^{1/2}}{27^{1/3}}\right)^2 = \left(\dfrac{\sqrt{4}}{\sqrt[3]{27}}\right)^2 = \left(\dfrac{?}{?}\right)^2 = \dfrac{(?)^2}{(?)^2}$

$$= \dfrac{?}{?}$$

(c) $\left(\dfrac{x^3 y}{xy^2}\right)^2 = \dfrac{x^{3 \cdot 2} y^{1 \cdot 2}}{x^{1 \cdot 2} y^{2 \cdot 2}} = \dfrac{?}{?} = \dfrac{x^?}{y^?}$

(d) $\left(\dfrac{x^3 y}{xy^2}\right)^2 = \left(\dfrac{x^2}{y}\right)^2 = \dfrac{?}{?}$

Your answers should be $\dfrac{1}{2}, \dfrac{4}{9}, \dfrac{x^4}{y^2}$, and $\dfrac{x^4}{y^2}$.

EXAMPLE IX.

(a) $\dfrac{x^2 y^0 z^7}{x^4 y z^2} = \dfrac{1 \cdot z^{7-2}}{x^{4-2} y} = \dfrac{?}{?}$

(b) $\left(\dfrac{x^{1/2} y^{3/4}}{2 z^0 y^{1/4}}\right)^2 = \left(\dfrac{x^{1/2} y^?}{2}\right)^2 = \dfrac{(x^{1/2})^2 (?)^2}{2^2}$

$$= \dfrac{?}{?}$$

The answers are $\dfrac{z^5}{x^2 y}$ and $\dfrac{xy}{4}$.

Frequently it is advisable to work with expressions that do not contain roots in their denominators. The process of clearing roots from a denominator is called, "rationalizing the denominator." The following examples illustrate the process for rationalizing denominators.

CHAPTER 1 Algebra Review

EXAMPLE X.

(a) $\dfrac{1}{\sqrt{2}} + \dfrac{3}{\sqrt{3}} \approx \dfrac{1}{1.4142} + \dfrac{3}{1.7320}$

which, by *long* division is approximately 2.4392. But, if we rationalize the denominators first, we have a much easier process.

$$\dfrac{1}{\sqrt{2}} \cdot \dfrac{\sqrt{2}}{\sqrt{2}} + \dfrac{3}{\sqrt{3}} \cdot \dfrac{\sqrt{3}}{\sqrt{3}} = \dfrac{\sqrt{2}}{2} + \dfrac{3\sqrt{3}}{3}$$

$$= \dfrac{\sqrt{2}}{2} + \sqrt{3}$$

$$\approx \dfrac{1.4142}{2} + 1.7320$$

$$= 2.4392$$

(b) $\dfrac{10}{\sqrt{5}} \cdot \dfrac{?}{?} = \dfrac{?\sqrt{5}}{?} = \underline{\ \ ?\ \ } \approx 4.472$

Before doing example (c), multiply the following two irrational numbers. They are often called *conjugates*. You should get 1 for the product:

$(2 + \sqrt{3})(2 - \sqrt{3}) = 2(2 - \sqrt{3})$
$\qquad\qquad\qquad\quad + \sqrt{3}(2 - \sqrt{3}) = \underline{\ \ ?\ \ }$

(c) $\dfrac{1}{2 - \sqrt{3}} \cdot \dfrac{2 + \sqrt{3}}{2 + \sqrt{3}} = \dfrac{2 + \sqrt{3}}{4 - \sqrt{9}} = \dfrac{?}{?}$

$$\approx -3.1462$$

Now we are ready to summarize the theorems and definitions of exponents that are often referred to as *the laws of exponents*.

DEFINITION 1.7.
If $a > 0$ and m and n are natural numbers, then

(i) $a^m = a \cdot a \cdot \cdots \cdot a \qquad m$ times
(ii) $a^{-m} = 1/a^m$
(iii) $a^{1/m} = \sqrt[m]{a}$
(iv) $a^{m/n} = \begin{cases} \sqrt[n]{a^m} \\ (\sqrt[n]{a})^m \end{cases}$
(v) $a^0 = 1$

THEOREM XIV.
If $a, b > 0$, and m and n are rational numbers, then

(i) $a^m \cdot a^n = a^{m+n}$

(ii) $\dfrac{a^m}{a^n} = a^{m-n}$

(iii) $(a^m)^n = a^{m \cdot n}$

(iv) $(a \cdot b)^m = a^m b^m$

(v) $\left(\dfrac{a}{b}\right)^m = \dfrac{a^m}{b^m}$

EXERCISE 1.4

A. FUNDAMENTALS
Simplify the following. Remember that x and y must be greater than zero.

1. (a) $x^2 \cdot x^3$ (b) $x^5 \cdot x^7 \cdot x^3$
   (c) $x^{5/2} \cdot x^{5/2}$

2. (a) $x^{3/2} \cdot x^{3/2}$ (b) $(x^{1/2} \cdot x^{1/2})^3$
   (c) $(x^3 \cdot x^3)^{1/2}$

3. (a) $x^2 \cdot x^2 \cdot x^2$ (b) $(x^2)^3$ (c) $(x^3)^2$

4. (a) $x^{2/3} \cdot x^{1/3}$ (b) $(x^2 \cdot x)^{1/3}$
   (c) $(x^{1/3})^2 \cdot x^{1/3}$

5. (a) $\dfrac{x^2}{y^3} \cdot \dfrac{x^2}{y^3}$ (b) $\left(\dfrac{x^2}{y^3}\right)^2$ (c) $\dfrac{(x^2)^2}{(y^3)^2}$

6. (a) $\dfrac{(x^2)^{1/2}}{(y^2)^{1/2}}$, $x, y > 0$ (b) $\left(\dfrac{x^2}{y^2}\right)^{1/2}$
   (c) $\sqrt{\dfrac{x^2}{y^2}}$

7. (a) $\dfrac{4^{1/2}(x^6)^{1/2}}{(x^4)^{1/2}}$ (b) $\left(\dfrac{4x^6}{x^4}\right)^{1/2}$
   (c) $\sqrt{\dfrac{4x^6}{x^4}}$

8. (a) $\left(\dfrac{8}{27}\right)^{1/3}$ (b) $\left[\left(\dfrac{4}{9}\right)^{1/2}\right]^3$

(c) $\left(\dfrac{2^2}{3^2}\right)^{3/2}$

Rationalize the denominators of the following expressions.

9. (a) $\dfrac{1}{\sqrt{6}}$ (b) $\dfrac{12}{\sqrt{6}}$

10. $\dfrac{1}{\sqrt{3}} - \dfrac{2}{\sqrt{5}}$ 11. $\dfrac{1-\sqrt{3}}{\sqrt{3}}$

12. (a) $\dfrac{2}{\sqrt{3}-\sqrt{2}}$ (b) $\dfrac{\sqrt{5}}{1-\sqrt{5}}$

B. ESSENTIALS

Simplify the following, assuming that all bases are positive.

13. (a) $\dfrac{x^3}{x}$ (b) $\dfrac{x^{-1}}{x}$ (c) $\dfrac{x^{-1}}{x^{-2}}$

14. (a) $\dfrac{x^3 y^2}{xy^3}$ (b) $\dfrac{x^{-1} y^2}{xy^{-1}}$ (c) $\dfrac{x^{-1} y^{-2}}{x^{-2} y^{-1}}$

15. (a) $\dfrac{8x^3 y^2 z}{2xy^3 z^0}$ (b) $\dfrac{5x^{-1} y^2 z^2}{10xy^{-1}}$

★(c) $\dfrac{2x^{-2} y^{-1} z^{-3}}{8x^{-1} y^{-2} z}$

16. (a) $\dfrac{a^{-3} b^2 c^{-1}}{a^2 b^{-1} c^{-2}}$ (b) $\left(\dfrac{a^3 b^{-2} c}{a^{-2} bc^2}\right)^{-1}$

(c) $\dfrac{(a^3 b^{-2} c)^{-1}}{(a^{-2} bc^2)^{-1}}$

17. (a) $\left[\left(\dfrac{x^2 y^4}{z^6}\right)^{1/2}\right]^3$ (b) $\left(\dfrac{x^4 y^6}{z^2}\right)^{3/2}$

(c) $\left(\dfrac{x^{-2} y^2}{x^2 y^{-4}}\right)$

18. (a) $x^{1/2}(x^{3/2} + x^{1/2})$ (b) $\dfrac{x^{3/2} + x^{1/2}}{x^{1/2}}$

(c) $(\sqrt{x^3} + \sqrt{x})\sqrt{x^3}$

19. $\dfrac{x^{-1/2} - y^{-1/2}}{x^{-1/2} y^{-1/2}}$ 20. $\dfrac{1}{\sqrt{x}} + \dfrac{1}{x}$

★21. $(x+1)^{1/2} - x(x+1)^{-1/2}$

22. $(2x-3)^{1/2} - 4x^2(2x-3)^{-3/2}$

1.5 COMPLEX NUMBERS

There is another important system of numbers that contains all real numbers and more. Consider the following equation:

$$x^2 + 1 = 0$$

Solving for x yields

$$x = \pm\sqrt{-1}$$

which involves the square root of a negative number. Since no real number multiplied times itself can be negative, the square roots of negative numbers do not exist in the system of real numbers. This is why the imaginary numbers were created. Notice that we are always able to find $\sqrt{-1}$ in any imaginary number.

$$\sqrt{-16} = \sqrt{4^2}\sqrt{-1} = 4\sqrt{-1}$$
$$\sqrt{-8} = \sqrt{8}\sqrt{-1} = 2\sqrt{2}\sqrt{-1}$$
$$-\sqrt{-16} = -\sqrt{16}\sqrt{-1} = -4\sqrt{-1}$$

In each case the numbers above are simplified to the form $b\sqrt{-1}$, where b is a real number. Every imaginary number can be written in the form bi by simply defining i as follows:

$$i^2 = -1 \quad \text{or} \quad i = \sqrt{-1}$$

Notice that -1 has two roots, $+\sqrt{-1}$ and $-\sqrt{-1}$, but the positive one is chosen for i.

Complex numbers[4] are all possible combinations of real and imaginary numbers. They can be expressed as $a + b\sqrt{-1}$.

CHAPTER 1   Algebra Review

> **DEFINITION 1.8.**
> A complex number can be written in the form
> $$a + bi, \quad i = \sqrt{-1}$$
> where a and b are real numbers. a is called the *real part* and bi the *imaginary part* of the complex number.

The complex numbers include all real numbers since they can all be expressed in the form $a + bi$ by letting $b = 0$. For example,

$$3 = 3 + 0i$$
$$0 = 0 + 0i$$
$$-2 = -2 + 0i$$

The complex numbers also contain all imaginary numbers. For example,

$$\sqrt{-4} = 2i$$

Do you see how we would write $2i$ in the form $a + bi$?

Before discussing addition and multiplication of two complex numbers, we must define what it means for two complex numbers to be equal. Two complex numbers are equal if and only if their real and imaginary components are equal. Hence,

$$a + bi = x + yi$$

means

$$a = x \quad \text{and} \quad b = y$$

Now let's discuss the addition and multiplication of complex numbers. The definitions are made in the most natural way, since they must apply to the addition and multiplication of all the complex numbers that are simply real numbers. How would you add the two complex numbers $3 + 4i$ and $2 + 7i$? If you added the real parts of each $(3 + 2)$ and then added the imaginary parts of each $(4i + 7i)$, getting $5 + 11i$, you are right!

> **DEFINITION 1.9.**
> If $a + bi$ and $c + di$ are complex numbers, then their sum is
> $$(a + bi) + (c + di) = (a + c) + (b + d)i$$

EXAMPLE I.
(a) $(3 + 4i) + (2 - 8i) = 5 - 4i$
(b) $(2 - 3i) + (2 - 5i) = 4 - 8i$

Stop and multiply the two complex numbers $3 + 4i$ and $2 + 3i$, remembering that $i^2 = -1$. You should get $-6 + 17i$.

> **DEFINITION 1.10.**
> If $a + bi$ and $c + di$ are complex numbers, then their product is
> $$(a + bi) \cdot (c + di) = ac + adi + bci + bdi^2$$
> $$= (ac - bd) + (ad + bc)i$$

EXAMPLE II.
(a) $(3 + 4i)(2 - 8i) = 6 - 24i + 8i - 32i^2$
$$= (6 + 32) + (-24 + 8)i$$
$$= 38 - 16i$$
(b) $(\sqrt{2} - 3i) \cdot (2 - 5i) = 2\sqrt{2} - 5\sqrt{2}i - 6i + 15i^2$
$$= (2\sqrt{2} - 15)$$
$$+ (-5\sqrt{2} - 6)i$$

Notice that the answers were always written in the form $a + bi$, illustrating the fact that the complex numbers are closed under addition and multiplication. That is, the sum, or product, of two complex numbers is always a complex number.

All of the other field properties also apply to complex numbers, but the order properties do not. In any ordered field (such as the ordered field of real numbers) all four order properties (Section 1.2) must be satisfied. To show that the field of complex numbers is not ordered, we will show that attempted orderings will lead to a contradiction.

The trichotomy law states that any number is either positive, negative, or zero. Since $i \neq 0$, let's see what happens if we assume it is positive. If

$$i > 0$$

then multiplying both sides by the positive number i yields

$$i^2 > 0$$

This is a contradiction, since $i^2 = -1$ and $-1 < 0$. Now show that assuming i is negative results in the same contradiction. Remember that multiplying both sides of an inequality such as $i < 0$ by a negative number reverses the order of the inequality.

The extension of the field properties from the system of real numbers (Table 1.1) to the system of complex numbers is quite natural except for that of multiplicative inverses. It seems natural to suspect that

$$(a + bi) \cdot \frac{1}{a + bi} = 1, \quad a \text{ and } b \neq 0$$

so the multiplicative inverse of $a + bi$ is $\frac{1}{a + bi}$. In order to make this natural conclusion valid, we have to prove that $\frac{1}{a + bi}$ is a complex number. To accomplish this, we must show that $\frac{1}{a + bi}$ can be written in the form $A + Bi$. We will multiply $\frac{1}{a + bi}$ by the quotient $\frac{a - bi}{a - bi}$ (which equals 1), assuming that the definition of multiplication of complex numbers extends naturally to multiplying quotients of complex numbers. First, multiply these two complex numbers:

$$(a + bi) \cdot (a - bi) = \underline{\quad ? \quad}$$

Now,

$$\frac{1}{a + bi} = \frac{1}{a + bi} \cdot \frac{a - bi}{a - bi} = \frac{a - bi}{a^2 + b^2}$$

Therefore,

$$\frac{1}{a + bi} = \frac{a}{a^2 + b^2} + \frac{-b}{a^2 + b^2} i$$

which, since it is written in the form $A + Bi$, is a complex number.

Finding the multiplicative inverse of a nonzero complex number actually defines division for the complexes, since

$$\frac{3 + 2i}{2 - 3i} = (3 + 2i) \cdot \frac{1}{2 - 3i}$$

Division by $2 - 3i$ is simply multiplication of the complex number by its multiplicative inverse. To find the multiplicative inverse $a + bi$ of $2 - 3i$, we set their product equal to 1. Then,

$$(a + bi)(2 - 3i) = 1$$

and

$$2a + 3b + (2b - 3a)i = 1 + 0i$$

Then, by definition of equality for complex numbers,

$$2a + 3b = 1 \quad \text{and} \quad 2b - 3a = 0$$

which, by solving the second for a and substituting the result into the first, yields

$$a = \frac{2b}{3} \quad \text{and} \quad 2\left(\frac{2b}{3}\right) + 3b = 1$$

Then, we find that

$$a = \frac{2}{13} \quad \text{and} \quad b = \frac{3}{13}$$

Stop here and verify that

$$(2 - 3i) \cdot \left(\frac{2}{13} + \frac{3}{13} i\right) = 1$$

Now we can complete the original division problem:

$$\frac{3 + 2i}{2 - 3i} = (3 + 2i)\left(\frac{1}{2 - 3i}\right)$$

$$= (3 + 2i)\left(\frac{2}{13} + \frac{3}{13} i\right) = \underline{\quad ? \quad}$$

CHAPTER 1 Algebra Review

This process can be shortened by multiplying by $\dfrac{2+3i}{2+3i}$, as follows:

$$\frac{3+2i}{2-3i} \cdot \frac{2+3i}{2+3i} = \frac{(6-6)+(9+4)i}{4+9} = 0+i = i$$

We call $2+3i$ the complex conjugate of $2-3i$. In general, we have Definition 1.11.

> **DEFINITION 1.11.**
> The complex conjugate $\bar{z}$ of $z = a+bi$ is $\bar{z} = a-bi$.

Notice that $z \cdot \bar{z} = a^2 + b^2$.

We can represent complex numbers geometrically, but two axes are required for one complex number.

Several complex numbers are plotted in Figure 1.10 by using a horizontal axis for the real part of each complex number, and a vertical axis for the imaginary part.

The complex conjugate of $z = 3+4i$, $\bar{z} = 3-4i$, is located on the opposite side of the real axis.

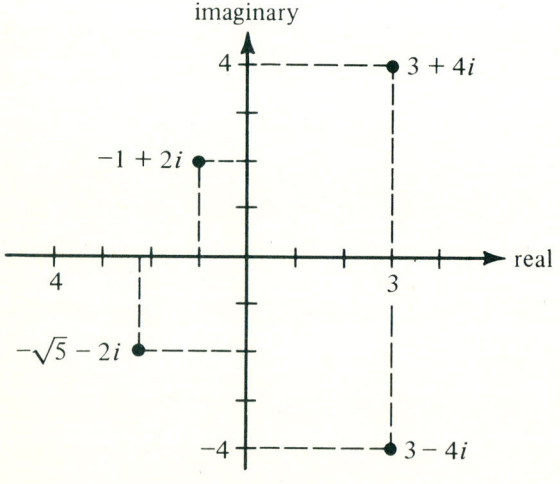

FIGURE 1.10.

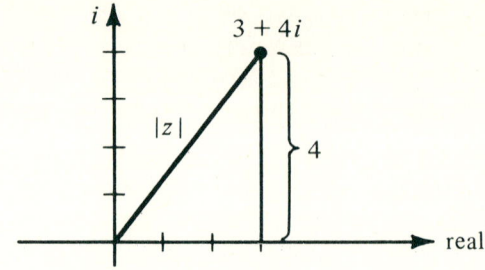

FIGURE 1.11.

By using the Pythagorean Theorem, we can find the distance that $z = 3+4i$ is from the origin. This distance is called the *modulus* or *absolute value* of z and is written $|z|$. See Figure 1.11. In Figure 1.11,

$$|z|^2 = 3^2 + 4^2 \quad \text{or} \quad |z| = \sqrt{9+16} = 5$$

In general, $z = a+bi$ yields

$$|z| = \sqrt{a^2 + b^2}$$

Multiply $z = 3+4i$ and its conjugate and relate this product, $z \cdot \bar{z}$, to the modulus of z, $|z|$.

If you express a real number, say -5, in the form $a+bi$ (that is, $-5 = -5+0i$), and then find its modulus, you will have a common alternative definition of the absolute value of a real number. Try it and see.

EXERCISE 1.5

A. FUNDAMENTALS
Express each of the following in the form $a+bi$.

1. (a) $(3+4i)+(2+3i)$
   (b) $(2-5i)+(-3+i)$
   (c) $(\sqrt{2}+3i)+(1+2i)$

2. (a) $(2+3i) \cdot (2-3i)$
   (b) $(-2-3i) \cdot (-2+3i)$
   (c) $(\sqrt{2}-i) \cdot (\sqrt{2}+i)$

3. (a) $(2+2i)(4+6i)$
   (b) $(-3-3i)(6+9i)$
   (c) $(4+4i)(8+12i)$

4. (a) $(1 + 8i) \div (2 + 3i)$
   (b) $(7 - 4i) \div (-2 - i)$

5. (a) $\dfrac{3 + 4i}{3 - 4i}$ (b) $\dfrac{5 - 2i}{5 + 2i}$

6. (a) $\dfrac{(2 + 3i) + (5 - 2i)}{2 - 2i}$
   (b) $\dfrac{(3 - 2i)(2 - 8i)}{4 - 6i}$

7. Plot each of the following complex numbers, then find the modulus for each.
(a) $3 + 4i$ (b) $6 + 8i$ (c) $12 - 5i$
(d) $-5 - 12i$ (e) 9 (f) $-3i$

B. ESSENTIALS

8. Find the additive and multiplicative inverses for each of the following complex numbers:
(a) $8 - 3i$ (b) $2 + i$

9. Compare $|z_1 + z_2|$ and $|z_1| + |z_2|$ for $z_1 = 2 + 3i$ and $z_2 = 4 + 5i$.

10. Simplify:
(a) $(3i)^3$ (b) $(2i)^5$ (c) $(3i)^4$ ★(d) i^{27}
(e) i^{53}

★11. Compare $|2z|$ and $2|z|$ for $z = 6 - 8i$ and for for $z = -3 + 4i$.

C. APPLICATIONS
Prove or disprove the following equations.

12. (a) If $x = -1$, then $x^3 = -1$.
    ★(b) If $x = \dfrac{1 + \sqrt{3}\,i}{2}$, then $x^3 = -1$.
    (c) If $x = \dfrac{1 - \sqrt{3}\,i}{2}$, then $x^3 = -1$.

13. (a) If $x = 9i$, then $x^2 = -81$.
    (b) If $x = -9i$, then $x^2 = -81$.

14. If $x = 1 + i$ and $y = 1 - i$, then $x^4 = y^4$.

★15. $\sqrt{i}$ is a complex number. (HINT: Show that $\sqrt{i}$ can be written in the form $a + bi$ by letting $\sqrt{i} = a + bi$ and squaring both sides.) Assume that $(\sqrt{i})^2 = i$.

1.6 QUADRATIC EQUATIONS

Now we shall review three methods used to solve equations of the form

$$ax^2 + bx + c = 0, \qquad a \neq 0$$

which are called **quadratic equations**. If a quadratic equation is factorable into the product of two linear terms (which usually involve only integers), then we can apply the following theorem to find the solutions. The theorem states that the product of two numbers can equal zero only if at least one of the numbers is itself zero.

> **THEOREM XV.**
> For any real numbers a and b, if $a \cdot b = 0$, then either $a = 0$ or $b = 0$.

EXAMPLE I. Solve for x if
$$x^2 - 2x - 3 = 0$$

By factoring, we have
$$(x - 3)(x + 1) = 0$$

which, by Theorem XV, yields the solutions

$$\begin{array}{lll} x - 3 = 0 & \text{or} & x + 1 = 0 \\ x = 3 & \text{or} & x = -1 \end{array}$$

If a quadratic equation does not easily factor, then we either use the quadratic formula or a technique called "completing the square" to solve the equation. Completing the square involves transforming a quadratic equation such as

$$x^2 - 4x - 1 = 0$$

into the form

$$(x - 2)^2 = 5$$

where there is a perfect-square quadratic on the left.

It is then solved by simply taking the square root of both sides of the equation, yielding

$$x - 2 = +\sqrt{5} \quad \text{or} \quad x - 2 = -\sqrt{5}$$

We then add 2 to both sides to get the final solution

$$x = 2 + \sqrt{5} \quad \text{or} \quad x = 2 - \sqrt{5}$$

The two irrational solutions, $x = 2 + \sqrt{5}$ and $x = 2 - \sqrt{5}$, equal approximately 2.2 and -0.2, respectively.

Carefully examine the following product to determine the relationship between the middle term and the last term:

$$(x + a)^2 = x^2 + \mathbf{2ax} + \mathbf{a^2}$$

Dividing the coefficient of the middle term in half and then squaring that answer results in the third term. Then, to make $x^2 + 6x + \underline{\ ?\ }$ a perfect square, the last term must be the quantity 6 divided by 2 and then squared, or $(\frac{6}{2})^2 = 3^2 = 9$. Then

$$x^2 + 6x + 9 = (x + 3)^2$$

is a perfect-square quadratic. To solve an equation by completing the square, proceed as follows.

$$x^2 + 6x + 1 = 0$$

implies

$$x^2 + 6x = -1$$

Then, adding 9 to both sides completes the square on the left:

$$x^2 + 6x + 9 = -1 + 9$$

Therefore,

$$(x + 3)^2 = 8$$

and

$$x + 3 = \pm\sqrt{8} = \pm 2\sqrt{2}$$

Hence

$$x = -3 \pm 2\sqrt{2}$$

and our solutions are $x = -3 + 2\sqrt{2} \approx -0.17$ and $x = -3 - 2\sqrt{2} \approx -5.8$.

For $2x^2 - 3x - 1 = 0$, we must first divide both sides of the equation by 2, yielding

$$x^2 - \frac{3}{2}x = \frac{1}{2}$$

The relationship between the coefficient of the x-term and the constant term is based upon $(x + a)^2$ equaling $x^2 + 2ax + a^2$, which has a coefficient of 1 in front of the x^2 term. Then, half of $(-\frac{3}{2})$ squared is $(-\frac{3}{4})^2$, or $\frac{9}{16}$. Adding $\frac{9}{16}$ to both sides yields

$$x^2 - \frac{3}{2}x + \frac{9}{16} = \frac{1}{2} + \frac{9}{16}$$

and

$$\left(x - \frac{3}{4}\right)^2 = \frac{17}{16}$$

Then, taking the square root of both sides yields

$$x - \frac{3}{4} = \pm\sqrt{\frac{17}{16}} = \pm\frac{\sqrt{17}}{4}$$

and

$$x = \frac{3}{4} \pm \frac{\sqrt{17}}{4} = \frac{3 \pm \sqrt{17}}{4}$$

The quadratic formula[5] is easier to use, and since it expresses the final result of completing the square directly, it works for all quadratic equations. The formula is stated in Theorem XVI and its derivation follows.

THEOREM XVI.
If $ax^2 + bx + c = 0$, where $a \neq 0$, then

$$x = \frac{-b \pm \sqrt{b^2 - 4ac}}{2a}$$

Proof: We will complete the square on the general equation, $ax^2 + bx + c = 0$, to derive the quadratic formula.

$$ax^2 + bx + c = 0$$

becomes, by dividing through by a,

$$x^2 + \frac{b}{a}x + \frac{c}{a} = 0$$

and

$$x^2 + \frac{b}{a}x = -\frac{c}{a}$$

To complete the square, we divide $\frac{b}{a}$ by 2, yielding $\frac{b}{2a}$, which is then squared. Since $\left(\frac{b}{2a}\right)^2 = \frac{b^2}{4a^2}$, we have

$$x^2 + \frac{b}{a}x + \frac{b^2}{4a^2} = \frac{b^2}{4a^2} - \frac{c}{a}$$

$$\left(x + \frac{b}{2a}\right)^2 = \frac{b^2}{4a^2} - \frac{4ac}{4a^2} = \frac{b^2 - 4ac}{4a^2}$$

$$x + \frac{b}{2a} = \pm\sqrt{\frac{b^2 - 4ac}{4a^2}} = \pm\frac{\sqrt{b^2 - 4ac}}{2a}$$

$$x = -\frac{b}{2a} \pm \frac{\sqrt{b^2 - 4ac}}{2a}$$

and, finally,

$$x = \frac{-b \pm \sqrt{b^2 - 4ac}}{2a}$$

Now, to solve an equation such as

$$2x^2 - 3x - 4 = 0$$

we simply notice that $a = 2$, $b = -3$, and $c = -4$ and substitute these values into the quadratic formula. Hence,

$$x = \frac{-b \pm \sqrt{b^2 - 4ac}}{2a}$$

$$= \frac{-(-3) \pm \sqrt{(-3)^2 - 4(2)(-4)}}{2(2)}$$

or

$$x = \frac{3 \pm \sqrt{9 + 32}}{4} = \frac{3 \pm \sqrt{41}}{4}$$

By examining $b^2 - 4ac$ we can determine something about the nature of the solutions of a quadratic equation. $b^2 - 4ac$ is called the *discriminant*[6] of a quadratic equation. It is the number under the square root symbol in the quadratic formula. The various possibilities are summarized below.

If
$$ax^2 + bx + c = 0, \qquad a \neq 0$$
then
$$x = \frac{-b \pm \sqrt{b^2 - 4ac}}{2a}$$
will yield

(i) two real solutions if $b^2 - 4ac > 0$,
(ii) one real solution (or two equal solutions) if $b^2 - 4ac = 0$.
(iii) no real solutions if $b^2 - 4ac < 0$.

(The two solutions will be complex numbers.)

EXERCISE 1.6

A. FUNDAMENTALS
Solve the following quadratic equations for x by factoring and then applying Theorem XV.

1. (a) $x^2 - 4x + 4 = 0$
   (b) $x^2 - 6x + 9 = 0$
   (c) $x^2 - 10x + 25 = 0$

2. (a) $x^2 + 8x + 16 = 0$
   (b) $x^2 + 2x + 1 = 0$
   (c) $x^2 + 12x + 36 = 0$

3. (a) $4x^2 - 12x + 9 = 0$
   (b) $4x^2 + 4x + 1 = 0$
   (c) $9x^2 - 12x + 4 = 0$

4. (a) $x^2 - 9 = 0$ (b) $x^2 - 16 = 0$
   (c) $x^2 - 1 = 0$

CHAPTER 1 Algebra Review

5. (a) $x^2 - 25 = 0$ (b) $4x^2 - 9 = 0$
   (c) $9x^2 - 16 = 0$

6. (a) $x^2 - 8x + 7 = 0$
   (b) $x^2 - 11x + 10 = 0$
   (c) $x^2 - 12x + 11 = 0$

7. (a) $x^2 + 3x + 2 = 0$
   (b) $x^2 + 7x + 6 = 0$
   (c) $x^2 + 11x + 10 = 0$

8. (a) $x^2 - 2x - 3 = 0$
   (b) $x^2 - 5x - 6 = 0$
   (c) $x^2 - 8x - 9 = 0$

9. (a) $x^2 + 11x - 12 = 0$
   (b) $x^2 + 2x - 3 = 0$
   (c) $x^2 + 5x - 6 = 0$

10. (a) $x^2 + 7x + 12 = 0$
    (b) $x^2 + 8x + 12 = 0$
    (c) $x^2 + 5x + 6 = 0$

11. (a) $x^2 - 7x + 12 = 0$
    (b) $x^2 - 8x + 12 = 0$
    (c) $x^2 - 5x + 6 = 0$

12. (a) $6x^2 + x - 2 = 0$
    (b) $8x^2 - 10x - 3 = 0$
    (c) $6x^2 - 17x + 12 = 0$

Use the technique called *completing the square* to solve for x in the following quadratic equations.

13. (a) $x^2 = 4$ (b) $(x - 1)^2 = 4$
    (c) $(x - 4)^2 = 9$

14. (a) $(x - 1)^2 = 5$ (b) $x^2 - 2x + 1 = 3$
    (c) $x^2 - 4x = 2$

15. (a) $x^2 - 6x + 6 = 0$
    (b) $x^2 - 3x - 1 = 0$
    ★(c) $2x^2 - 4x - 1 = 0$

Use the quadratic formula to solve for x in the following quadratic equations.

16. $2x^2 - 3x - 5 = 0$ 17. $3x^2 - x - 1 = 0$
18. $2x^2 + x - 3 = 0$ 19. $4x^2 - x - 2 = 0$
20. $6x^2 - 2x - 4 = 0$ 21. $3x^2 + 3x + 3 = 0$

B. ESSENTIALS

Without solving the following equations, determine whether the solutions will contain two real roots, one real root, or no real roots (two complex roots).

★22. $2x^2 - x - 1 = 0$ 23. $4x^2 - 12x + 9 = 0$

24. $3x^2 - 4x + 2 = 0$ 25. $3x^2 - x - 1 = 0$

★26. $x^2 + 16 = 0$

For the following quadratic equations, find the values of K that will yield (i) two real solutions, (ii) no real solutions, or (iii) one real solution.

★27. $3x^2 - 2x + K = 0$

28. $2x^2 - x + K = 1$

29. $2Kx^2 - 3x - 1 = 0$

30. $x - 3Kx^2 = 2$

31. Find the value of a such that the following equation has only one solution, $x = 4$.

$$4x^2 - 32x + a = 0$$

32. Find the value of a such that the following equation has reciprocal real roots.

$$ax^2 - 10x + 3 = 0$$

33. Find the value of a such that the solutions of the following equation differ by two.

$$x^2 - 8x + a = 0$$

34. What must be the values of b and c for the following equation to have only one solution?

$$x^2 + bx + c = 0$$

35. Find the values of the following expressions, where r_1 and r_2 are the roots of the quadratic equation $ax^2 + bx + c = 0$.

(a) $r_1 + r_2$ (b) $r_1 \cdot r_2$ (c) $r_1^2 + r_2^2$

(d) $\dfrac{r_1 + r_2}{2}$

36. The Babylonians interpreted the quadratic equation $x^2 - 10x + 16 = 0$ as defining areas. In $x^2 + 16 = 10x$, 16 refers to the excess area that equates the areas of an $x \cdot x$ square and a $10 \cdot x$ rectangle. The solution as Euclid would present it and as the Babylonians probably solved it is as follows.

(a) Draw $AB = 10$, bisected at C and $CO = 4$.
(b) Use $r = \frac{10}{2} = 5$ as a radius to scribe an arc on CB at D using O as the center.

Explain why the measure of DB is the solution, x. (HINT: The interpretation involves areas. We still call x^2 the square of x.)

37. The Greeks divided quadratic equations into three types, $x^2 + bx = c^2$, $x^2 = bx + c^2$, and $x^2 + c^2 = bx$, where $b, c > 0$. Why?

1.7 QUADRATIC IN FORM

The technique of quadratic equations may be used to solve some equations that are not quadratic equations. These are equations having only two nonconstant terms, such as

$$x - 3\sqrt{x} + 2 = 0$$
$$x^4 - 2x^2 - 3 = 0$$
$$(x + 1)^6 - 3(x + 1)^3 - 40 = 0$$

where the first term's variable is the square of the variable in the second term. A simple substitution will result in a quadratic equation. For these three equations, let $u = x$, $u = x^2$, and $u = (x + 1)^3$, successively. Then we have $u^2 = x$, $u^2 = x^4$, and $u^2 = (x + 1)^6$, respectively. The three equations become the following quadratic equations:

$$u^2 - 3u + 2 = 0$$
$$u^2 - 2u - 3 = 0$$
$$u^2 - 3u - 40 = 0$$

We call this procedure "a change in variable," and the result is called "a quadratic equation in u."

Let's now examine the complete solution for each equation, one equation at a time.

EXAMPLE I. Solve the following equation for x:

$$x - 3\sqrt{x} + 2 = 0$$

Solution: Let $u = \sqrt{x}$. Then $u^2 = x$, and

$$u^2 - 3u + 2 = 0$$
$$(u - 2)(u - 1) = 0$$
$$u - 2 = 0 \quad \text{or} \quad u - 1 = 0$$
$$u = 2 \quad \text{or} \quad u = 1$$

But $x = u^2$ means that

$$x = 4 \quad \text{or} \quad x = 1$$

EXAMPLE II. Solve the following equation:

$$x^4 - 2x^2 - 3 = 0$$

Solution: Let $u = x^2$. Then $u^2 = x^4$, and

$$u^2 - 2u - 3 = 0$$
$$(u - 3)(u + 1) = 0$$
$$u - 3 = 0 \quad \text{or} \quad u + 1 = 0$$
$$u = 3 \quad \text{or} \quad u = -1$$

Since $u = x^2$, we have $x = \pm\sqrt{u}$. Hence $x = \pm\sqrt{3}$. Since $u = -1$ will not yield any additional *real* solutions, it is usually discarded. If the complex numbers were allowed as answers, then our solutions would include $x = \pm\sqrt{-1} = \pm i$.

EXAMPLE III. Solve the following equation:

$$(x + 1)^6 - 3(x + 1)^3 - 40 = 0$$

Solution: Let $u = (x + 1)^3$. Then $u^2 = (x + 1)^6$, and

$$u^2 - 3u - 40 = 0$$
$$(u - 8)(u + 5) = 0$$
$$u - 8 = 0 \quad \text{or} \quad u + 5 = 0$$
$$u = 8 \quad \text{or} \quad u = -5$$

But $u = (x + 1)^3$ implies that

$$x + 1 = \sqrt[3]{u}$$

or

$$x = -1 + \sqrt[3]{u}$$

Then $u = 8$ implies that

$$x = -1 + \sqrt[3]{8} = -1 + 2 = +1$$

and $u = -5$ yields
$$x = -1 + \sqrt[3]{-5} = -1 - \sqrt[3]{5}$$

EXERCISE 1.7

A. FUNDAMENTALS

Determine which of the following equations are "quadratic in form" and give the substitution necessary to obtain a "quadratic equation in u" whenever possible.

1. $x^4 - 2x^2 - 1 = 0$ 2. $x + 3\sqrt{x} + 2 = 0$
3. $x^{2/3} + 2x^{1/3} - 1 = 0$
4. $(x + 1)^2 + 3(x + 1) - 2 = 0$
5. $x^2 = 1 - 2\sqrt{x}$ 6. $2x = 3 - x^4$

Solve the following "quadratic in form" equations.

7. (a) $x^4 + x^2 - 2 = 0$ (b) $x^4 - 8x^2 = 9$
   (c) $x^4 = 14x^2 + 32$
8. (a) $x + \sqrt{x} - 2 = 0$ (b) $x - \sqrt{x} = 2$
   (c) $x = 4\sqrt{x} - 3$
9. ★(a) $x^{2/3} + x^{1/3} = 2$
   (b) $x - 5x^{1/2} + 6 = 0$
   (c) $x^{1/2} + 3x^{1/4} + 2 = 0$
10. (a) $(x + 1)^2 - 2(x + 1) - 3 = 0$
    (b) $(x - 3)^2 + 3x - 7 = 0$
    (c) $x - 1 = 2\sqrt{x - 1} + 3$

B. ESSENTIALS

Solve the following "quadratic in form" equations.

★11. $\dfrac{4}{x} + \dfrac{3}{\sqrt{x}} = 1$ 12. $x(\sqrt{x} + x^2) = 2$

13. $\dfrac{1}{x} + \dfrac{3}{\sqrt{x}} = 4$ 14. $x - 2 = 2\sqrt{x + 1}$

1.8 EQUATIONS INVOLVING FRACTIONS

The easiest way to solve any equation involving fractions is to eliminate the denominators from the fractions first. This is accomplished by finding the Least Common Multiple (LCM) of all the denominators involved, then multiplying both sides of the equation by this common denominator.

EXAMPLE I. Solve the following equation for x:
$$\frac{3x}{4} - \frac{2}{3} = \frac{1}{6}$$

Solution: The LCM for 3, 4, and 6 is 12. Therefore, we multiply both sides of the equation by 12 to clear the fractions. Then
$$12\left(\frac{3x}{4} - \frac{2}{3}\right) = 12 \cdot \frac{1}{6}$$
$$12 \cdot \frac{3x}{4} - 12 \cdot \frac{2}{3} = 2$$
$$3(3x) - 4(2) = 2$$
$$9x - 8 = 2$$
$$9x = 10$$
$$x = \frac{10}{9}$$

Do you know how the least common multiple of 3, 4, and 6 is found to be 12? Remember that it requires factoring each term completely, then multiplying all the factors together using each factor most number of times it occurs as a factor in any one number. Since $3 = 1 \cdot 3$, $4 = 2 \cdot 2$, and $6 = 2 \cdot 3$, their LCM is $2 \cdot 2 \cdot 3$ or 12. Do you see why we used two 2's and one 3 to get the LCM?

Now, equations involving variables in the denominator are solved similarly, except that the variable cannot assume values that will cause division by zero.

EXAMPLE II. Solve the following equation for x:
$$\frac{x}{3x + 3} - \frac{1}{8} = \frac{3}{x + 1}$$

Solution: The denominators factor as $3x + 3 = 3(x + 1)$, $8 = 2 \cdot 2 \cdot 2$, and $x + 1$, yielding an LCM of $2 \cdot 2 \cdot 2 \cdot 3(x + 1)$ or $24(x + 1)$. We now multiply both sides of the equation by this LCM to eliminate all fractions. Note that $x \neq -1$. Why? Then

$$24(x + 1)\left(\frac{x}{3x + 3} - \frac{1}{8}\right) = 24(x + 1) \cdot \frac{3}{x + 1}$$

$$8(x) - 3(x + 1)(1) = 24(3)$$

$$5x - 3 = 72$$

$$5x = 75$$

$$x = 15$$

It is easy to verify that the answer we obtained is indeed correct by substituting it back into the original equation as a check. For example, if $x = 15$, then

$$\frac{x}{3x + 3} - \frac{1}{8} = \frac{15}{3(15) + 3} - \frac{1}{8}$$

$$= \frac{15}{48} - \frac{1}{8} = \frac{5}{16} - \frac{2}{16} = \frac{3}{16}$$

and

$$\frac{3}{x + 1} = \frac{3}{15 + 1} = \frac{3}{16}$$

Then, by the transitive law of equality, if $x = 15$,

$$\frac{x}{3x + 3} - \frac{1}{8} = \frac{3}{x + 1}$$

The above procedure may result in a quadratic equation to solve, as in the next example.

EXAMPLE III. Solve the following equation for x:

$$\frac{3x}{x^2 - 3x - 10} - \frac{x}{x - 5} = \frac{3}{2x + 4}$$

Solution: Factoring the denominators yields

$$\frac{3x}{(x - 5)(x + 2)} - \frac{x}{x - 5} = \frac{3}{2(x + 2)}$$

and the LCM is $2(x - 5)(x + 2)$. Therefore,

$$2(x - 5)(x + 2)\frac{3x}{(x - 5)(x + 2)}$$

$$- 2(x - 5)(x + 2)\frac{x}{x - 5}$$

$$= 2(x - 5)(x + 2)\frac{3}{2(x + 2)}$$

$$2(3x) - 2(x + 2) \cdot x = (x - 5) \cdot 3$$

$$6x - 2x^2 - 4x = 3x - 15$$

$$2x^2 + x - 15 = 0$$

$$(2x - 5)(x + 3) = 0$$

$$2x - 5 = 0 \quad \text{or} \quad x + 3 = 0$$

$$x = \tfrac{5}{2} \quad \text{or} \quad x = -3$$

Let's check these answers. If $x = \tfrac{5}{2}$, then

$$\frac{3x}{x^2 - 3x - 10} - \frac{x}{x - 5} = \frac{15/2}{-45/4} - \frac{5/2}{-5/2}$$

$$= -\frac{2}{3} + 1 = \frac{1}{3}$$

and

$$\frac{3}{2x + 4} = \frac{3}{9} = \frac{1}{3}$$

If $x = -3$, then

$$\frac{3x}{x^2 - 3x - 10} - \frac{x}{x - 5} = \frac{?}{?} - \frac{?}{?} = ?$$

and

$$\frac{3}{2x + 4} = \frac{3}{?} = ?$$

EXERCISE 1.8

A. FUNDAMENTALS
Find the LCM for each of the following.

1. (a) 2, 6, 12   (b) 3, 9, 27
   (c) 5, 10, 20

CHAPTER 1   Algebra Review

2. (a) 3, 6, 9   (b) 4, 6, 14   (c) 6, 12, 15
3. (a) 3, 7, 11   (b) 2, 5, 9   (c) 8, 9, 25
4. (a) 30, 70, 180 (b) 105, 175, 525
   (c) 150, 180, 225
5. (a) $2x + 4, x^2 + 2x$
   (b) $3x - 9, 2x^2 - 6x$
6. (a) $x^2 - 4, 3x^2 - 9, 2x^3 - 8x$
   (b) $2x - 6, x^2 + 4x + 3, x^2 - 2x - 3$

Solve the following equations for x.

7. (a) $\dfrac{3x}{4} = \dfrac{2x}{4}$ (b) $\dfrac{x-1}{2} = \dfrac{3}{2}$
   (c) $\dfrac{2x-3}{5} = \dfrac{x-1}{5}$

8. (a) $\dfrac{x}{3} - \dfrac{2x}{3} = \dfrac{5x+1}{3}$
   (b) $\dfrac{x-1}{5} = \dfrac{2x}{5} - \dfrac{3}{5}$
   (c) $\dfrac{2x-1}{15} = \dfrac{3x}{15} - \dfrac{9}{15}$

9. (a) $\dfrac{x+3}{x} = \dfrac{x-3}{2}$ (b) $\dfrac{x}{x-1} = \dfrac{x}{3}$
   (c) $\dfrac{3}{x-2} = \dfrac{x+2}{x-2}$

B. ESSENTIALS

Solve the following equations.

10. ★(a) $\dfrac{2}{3x-6} - \dfrac{5}{x-2} = -\dfrac{1}{3}$
    (b) $\dfrac{7}{2x+6} - \dfrac{3}{x+3} = \dfrac{1}{2}$

11. (a) $\dfrac{4}{z+4} - \dfrac{z}{2z+8} = \dfrac{1}{3z+12}$
    (b) $\dfrac{3}{12-6z} - \dfrac{1}{z-2} = \dfrac{z}{8z-16}$

12. (a) $\dfrac{2x+1}{x-3} + \dfrac{1}{x+3} = \dfrac{x^2}{x^2-9}$
    (b) $\dfrac{x-3}{x+4} - \dfrac{1}{4-x} = \dfrac{2x^2}{x^2-16}$

13. $\dfrac{3}{6t+4} + \dfrac{2}{3t+2} = \dfrac{1}{2}$

14. $\dfrac{5s^2+6}{s^2-4} + \dfrac{3}{2-s} + \dfrac{7}{s+2} = 5$

15. $\dfrac{1}{y+3} + \dfrac{y+4}{y^2+4y+3} = \dfrac{1}{y+1}$

16. $\dfrac{x}{3x^2-8x+4} + \dfrac{3}{3x^2+7x-6} = \dfrac{1}{x^2+x-6}$

17. $\dfrac{2x}{4x^2-16x+15} - \dfrac{1}{2x^2-7x+5} = \dfrac{-3}{2x^2-5x+3}$

18. ★(a) $x^2 + x = \dfrac{42}{x^2+x} - 1$

    (HINT: Let $u = x^2 + x$.)

    (b) $\dfrac{x^2-6}{x} = 6 - \dfrac{5x}{x^2-6}$

1.9   LINEAR AND QUADRATIC INEQUALITIES

This section is concerned with the study of two classes of inequalities: *linear inequalities* such as

$$Ax + By + C \leq 0$$

where x and y occur only to the first power, and *quadratic inequalities* such as

$$ax^2 + bx + c \leq 0$$

where $a \neq 0$.

Solving linear inequalities is similar to solving linear equations, except when we multiply or divide both sides of the inequality by a negative number. At this step we reverse the order of the inequality (less than becomes greater than and greater than becomes less than).

EXAMPLE I.   Solve $2x - 3 < 5$ for x.

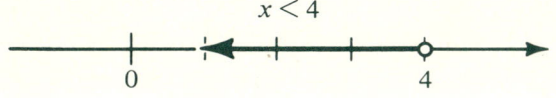

FIGURE 1.12.

Solution: If
$$2x - 3 < 5$$
then
$$2x < 8$$
and
$$x < 4$$

This is illustrated on the real number line in Figure 1.12.

EXAMPLE II. Solve $3 - 2x \leq 15$ for x.

Solution: If
$$3 - 2x \leq 15$$
then
$$-2x \leq 12$$
and
$$x \geq -6$$

This is graphed in Figure 1.13.

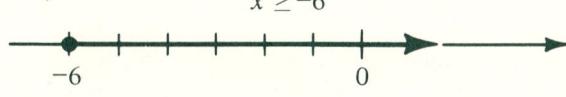

FIGURE 1.13.

Graphically, we will note the difference between a strict inequality, such as $x < 4$ and $x \geq -6$, which includes the possibility of equality, by using a circle at $x = 4$ in the former and a solid dot at $x = -6$ in the latter.

The solution of quadratic inequalities is slightly more complicated. One way to solve an inequality such as $x^2 + 2x - 8 > 0$ is to consider all possible combinations of its factors, $x - 2$ and $x + 4$, that will have a positive product. The product will be positive whenever the factors $x - 2$ and $x + 4$ have the same sign, either both positive or both negative. Symbolically, we write
$$(x - 2)(x + 4) > 0$$
is equivalent to

$x - 2 > 0$ and $x + 4 > 0$ (both positive)

or

$x - 2 < 0$ and $x + 4 < 0$ (both negative)

CASE I. Requiring that both factors be positive yields
$$x - 2 > 0 \quad \text{and} \quad x + 4 > 0$$
which implies
$$x > 2 \quad \text{and} \quad x > -4$$
This yields the solution
$$x > 2$$
since any value of x greater than 2 is also greater than -4.

CASE II. Requiring that both factors be negative yields
$$x - 2 < 0 \quad \text{and} \quad x + 4 < 0$$
which implies
$$x < 2 \quad \text{and} \quad x < -4$$
This yields the solution
$$x < -4$$
since any number less than -4 is also less than 2.

If $x > 2$, then the factors $x - 2$ and $x + 4$ are both positive, and their product $(x - 2)(x + 4) = x^2 + 2x - 8$ will be positive. If $x < -4$, then the factors are both negative and their product, $x^2 + 2x - 8$, is positive. Therefore, the values of x that satisfy $x^2 + 2x - 8 > 0$ are
$$x > 2 \quad \text{or} \quad x < -4$$
The answer is graphed in Figure 1.14 (page 28).

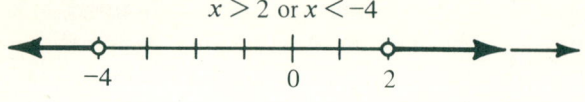

FIGURE 1.14.

The following sketch of the curve $y = x^2 + 2x - 8$ will help us interpret the original inequality and its resulting solution. We will review the graphing of these equations in Section 1.11.

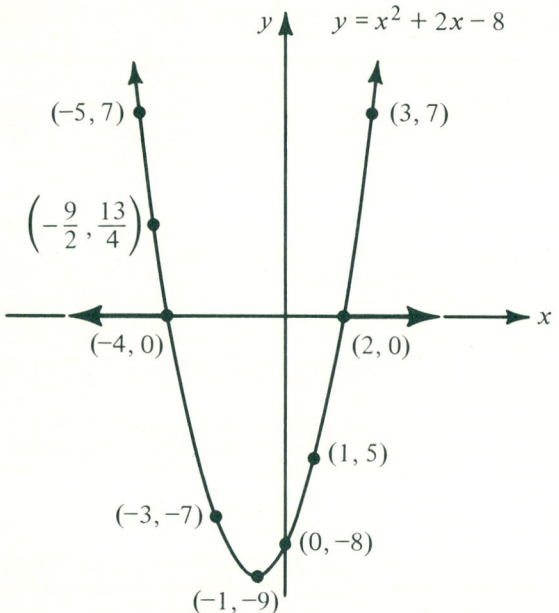

FIGURE 1.15.

Notice that the values of y corresponding to $x > 2$ or $x < -4$ are all positive. Since $y = x^2 + 2x - 8$, choosing $x > 2$ or $x < -4$ will result in a positive value for $x^2 + 2x - 8$, which is y.

Can you use the graph of $y = x^2 + 2x - 8$ to find the values of x that generate negative values for $y = x^2 + 2x - 8$? These values of x between -4 and 2 represent the solution of the inequality $x^2 + 2x - 8 < 0$.

With a little practice, we can use this geometric information to help us solve quadratic inequalities.

Notice that the two values of x for which $y = x^2 + 2x - 8 = 0$, $x = 2$ and $x = -4$, are the boundary points for the solution sets for both inequalities, $x^2 + 2x - 8 > 0$ and $x^2 + 2x - 8 < 0$. To decide whether or not the solution to a given quadratic inequality lies between the boundary points found by solving the corresponding equation, simply test the inequality with any sample point (usually $x = 0$). If the sample point satisfies the original inequality, then all the points in that region will comprise the desired solution set. If the equation does not factor, use the quadratic formula to solve for the boundary points of the solution. See Exercise 1.9, Problem 10, for cases in which there are no real solutions for the corresponding quadratic equation.

Now we will use an example to illustrate this geometric method of solution for inequalities.

EXAMPLE III. Find the values of x satisfying the following equation:

$$\frac{x+1}{x-3} < 0$$

Solution: First, multiply both sides by the *positive* number $(x - 3)^2$:

$$(x-3)^2 \frac{x+1}{x-3} < 0 \cdot (x-3)^2$$

After simplification, this becomes

$$(x-3)(x+1) < 0$$

Since $(x - 3)(x + 1) = 0$ for $x = 3$ or $x = -1$, the boundary points for the solution set are -1 and 3. If we let $x = 0$, then the resulting inequality

$$(0-3)(0+1) < 0$$
$$-3 < 0$$

is true. Then, since $x = 0$ lies between the boundary points, the solution consists of all points between $x = -1$ and $x = 3$, excluding -1 and 3. See Figure 1.16.

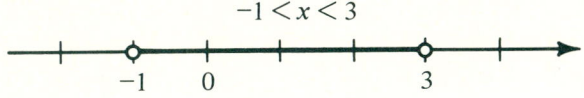

FIGURE 1.16.

EXERCISE 1.9

A. FUNDAMENTALS

Solve the following linear inequalities. Remember that multiplying or dividing by a negative number reverses the inequality. Graph the solutions on the real number line.

1. (a) $2x \leq 4$ (b) $x + 3 \leq 7$
   (c) $2x + 3 \leq 7$
2. (a) $-x \leq 2$ (b) $-2x \leq 6$
   (c) $4 - 3x \leq 10$
3. (a) $2x - 1 > 5$ (b) $2x - 1 = 5$
   (c) $2x - 1 < 5$
4. (a) $-x > 4$ (b) $4 - x > 10$
   (c) $5 - 8x > -11$
5. (a) $2x + 1 \leq x + 3$ (b) $2(x + 1) < x - 3$
   (c) $3(x - 1) < 2(x - 7)$
6. (a) $-4 > x$ (b) $2x + 3 < 3x$
   (c) $5x + 7 < 8(x - 1)$
7. (a) $4 - 2x \leq 7(1 - x)$
   (b) $3 - 8x \geq 2(3 - 2x)$
   (c) $x(x - 3) < x(x - 2)$

Solve the following nonlinear inequalities and graph the solutions on the real number line. Simply determine the boundary points. Then, by testing $x = 0$ (or any other value of x) in the original inequality, decide if the solution is between the two boundaries or outside of them.

8. (a) $(x - 2)(x + 2) \leq 0$
   (b) $(x - 2)(x + 2) < 0$
   (c) $(x - 2)(x + 2) > 0$
9. (a) $(x - 1)(x + 2) < 0$
   (b) $(x - 1)(x + 2) \geq 0$
10. (a) $(x - 4)^2 \geq 0$ (b) $x^2 - 8x + 20 \geq 0$
    (c) $-x^2 + 8x - 20 \geq 0$
11. (a) $x^2 - 4x - 5 < 0$ (b) $x^2 - 3x < 4$
    (c) $x(x - 1) < 6$
12. (a) $x(x + 1)(x - 1) \leq 0$
    (b) $(x + 1)(x - 1)(x + 3) > 0$
    (c) $(x + 2)(x + 3)(x - 4) \geq 0$

(HINT: Here there are three boundaries for the solution set. Test a real number from each resulting region.)

B. ESSENTIALS

Solve the following by first changing the given inequality into a quadratic inequality, then proceed as in Problems 8–12. Remember to multiply only by expressions whose signs you *know* are either positive or negative.

13. $\dfrac{x - 2}{x - 1} \leq 0$

14. $\dfrac{3}{x + 1} \geq 1$

15. $\dfrac{x - 10}{x + 3} \geq 2$

16. (a) $x^2 \leq 9$ ★(b) $4x^2 \leq 9$
17. (a) $(x - 1)^2 - 9 \leq 0$ (b) $(x + 2)^2 \leq 4$

Consider the possible cases to determine the solutions for the following quadratic inequalities. Remember: for greater than (or equal to) zero, both factors must have the same sign; for less than (or equal to) zero, the factors must have opposite signs.

18. (a) $(x - 3)(x + 5) \leq 0$
    (b) $(x - 3)(x + 5) > 0$
19. (a) $(2x - 3)(3x + 1) < 0$
    (b) $(2x - 3)(3x + 1) \geq 0$
20. $x^2 - 3x - 4 \leq 0$
21. $x^2 - 7x + 12 < 0$
22. $x(6x - 1) \geq 2$
23. $6(x^2 - 2) > x$
★24. $x(6x + 7) \geq 3$

Consider cases to solve the following problems. Since the sign rules for division are identical to those for multiplication, these cases will be the same as Problems 18–24.

25. $\dfrac{x-1}{x+4} \leq 0$

26. $\dfrac{2x-3}{x+5} > 0$

27. $\dfrac{x^2 + 4x + 3}{x^2 - 3x - 4} \leq 0$

28. $\left|\dfrac{x}{x - \sqrt{2}}\right| \leq 2$

Prove the following:

★29. If $x^2 < a^2$, then $-a < x < a$.

30. If $x^2 > a^2$, then $x < -a$ or $x > a$.

31. If $x > 0$ and $y > 0$, then $x^2 < y^2$ is equivalent to $x < y$.

(HINT: you must prove two theorems: (i) if $x^2 < y^2$, then $x < y$; and (ii) if $x < y$, then $x^2 < y^2$.)

1.10 RADICAL EXPRESSIONS

Radical expressions are those that involve roots, usually square roots. The radical expression

$$\sqrt{x} = 3$$

yields

$$(\sqrt{x})^2 = 3^2$$
$$x = 9$$

Since $(\sqrt{x})^2 = x$ whenever $x \geq 0$, squaring both sides eliminates the radical, thereby allowing us to proceed as usual to find the solution.

After squaring both sides, it is not unusual to find a solution for the resulting equation that does not satisfy the original one. These invalid solutions are called *extraneous roots*. We can see how this occurs by working the next two examples.

EXAMPLE I. Solve $\sqrt{x + 17} = x - 3$ for x.

Solution: If

$$\sqrt{x + 17} = x - 3$$

then

$$(\sqrt{x + 17})^2 = (x - 3)^2$$
$$x + 17 = x^2 - 6x + 9$$
$$x^2 - 7x - 8 = 0$$
$$(x - 8)(x + 1) = 0$$
$$x = 8 \quad \text{or} \quad x = -1$$

But only $x = 8$ satisfies the original equation, because $\sqrt{16} = +4$, not ± 4.

EXAMPLE II. Solve $\sqrt{x + 17} = 3 - x$.

Solution: If

$$\sqrt{x + 17} = 3 - x$$

then

$$(\sqrt{x + 17})^2 = (3 - x)^2$$
$$x + 17 = 9 - 6x + x^2$$
$$x^2 - 7x - 8 = 0$$
$$(x - 8)(x + 1) = 0$$
$$x = 8 \quad \text{or} \quad x = -1$$

But only $x = -1$ satisfies the original equation.

The difference between the two examples is found on the right side, where $x - 3$ and $3 - x$ are negatives of each other. What do you notice about the third lines in each solution? This is why we often get extraneous roots when we square both sides of an equation. The sign difference between $x - 3$ and $3 - x$ was obliterated in the process. This means we *must* check our answers whenever we square both sides of an equation.

By being careful how we arrange our terms before squaring, we are often able to minimize the amount of work required to complete the solution. This is illustrated by the next two examples.

EXAMPLE III. Solve $\sqrt{x + 3} - \sqrt{2x + 1} = 0$ for x.

Solution: If
$$\sqrt{x+3} - \sqrt{2x+1} = 0$$
then, by squaring both sides, we get
$$(\sqrt{x+3} - \sqrt{2x+1})^2 = 0^2$$
and
$$(\sqrt{x+3})^2 - 2\sqrt{x+3}\sqrt{2x+1} + (-\sqrt{2x+1})^2 = 0 \cdots$$

There is a much better way to proceed. If we add $\sqrt{2x+1}$ to both sides of the equation and then square, we get
$$\sqrt{x+3} = \sqrt{2x+1}$$
$$(\sqrt{x+3})^2 = (\sqrt{2x+1})^2$$
which implies that
$$x + 3 = 2x + 1$$
and
$$x = 2$$

If we check $x = 2$, we see that
$$\sqrt{x+3} = \sqrt{5}$$
and
$$\sqrt{2x+1} = \sqrt{5}$$
Therefore, if $x = 2$,
$$\sqrt{x+3} = \sqrt{2x+1}$$

EXERCISE 1.10

A. FUNDAMENTALS
Solve the following equations and then check your solutions to eliminate all extraneous roots.

1. (a) $\sqrt{x} = 4$ (b) $\sqrt{x} = -4$
2. (a) $\sqrt{x+2} = 8$ (b) $\sqrt{x+2} = -8$
3. (a) $\sqrt{x} = 4$ (b) $\sqrt{-x} = 4$
4. (a) $\sqrt{x} = x - 2$ (b) $\sqrt{x} = 2 - x$
5. (a) $\sqrt{x+3} = x - 3$
   (b) $\sqrt{x+3} = 3 - x$
6. (a) $\sqrt{7-6x} = 2x - 1$
   (b) $\sqrt{2x+19} = 3x - 4$
7. (a) $\sqrt{-3x-5} = x + 1$
   (b) $\sqrt{5x+14} = x + 4$

B. ESSENTIALS
Solving the following equations may require more than one squaring. Before you square both sides, be sure to write the equations in the form that will yield the simplest result. Check your answers.

8. (a) $\sqrt{x-3} + \sqrt{x+5} = 4$
   (b) $\sqrt{x+5} = 4 - \sqrt{x-3}$
9. $\sqrt{2x-1} - \sqrt{2-x} = 0$
10. $\sqrt{t} + \sqrt{t+3} = 3$
11. $\sqrt{13-3x} = 2 - \sqrt{x+5}$
★12. $\sqrt{4x-3} - \sqrt{6-2x} = 3$
13. $\sqrt{75x+214} = 2 + 3\sqrt{7x+18}$
14. $\sqrt{y+1} - \sqrt{2y-3} - \sqrt{3y-2} = 0$
15. $\dfrac{2x - 3\sqrt{x}}{3} = \dfrac{3}{2x + 3\sqrt{x}}$
16. $2 - \sqrt{x} = \dfrac{1}{2 + \sqrt{x}}$

1.11 THE CARTESIAN COORDINATE SYSTEM

The extension of the one-dimension geometric representation of the real numbers—the real number line—to two dimensions is quite natural. The two-dimensional picture depicts all possible pairs of real numbers (x, y), called *ordered pairs*. The first number is called the x-component and the second the y-component. This two-dimensional geometric model is called the Cartesian coordinate system.[7]

The Cartesian coordinate system in Figure 1.17 depicts two perpendicular real number lines intersecting at the origin $(0, 0)$. The quadrants are named and a few typical ordered pairs are shown.

CHAPTER 1 Algebra Review

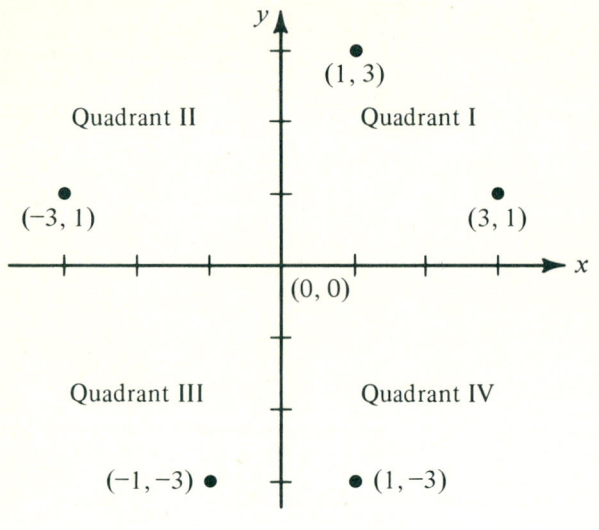

FIGURE 1.17.

This coordinate system often allows us to create geometric models of algebraic expressions and is therefore essential to further mathematical study.

The solutions of equations such as $y = x$, $y = 4$, and $x = -3$ generate infinite numbers of ordered pairs whose geometric models are pictured in Figure 1.18. Only a few of the points are labeled, but all the points on the lines satisfy the corresponding equations and all pairs of values of x and y that satisfy one of the equations are points on the corresponding line.

Inequalities such as $x \geq 0$, $y \leq 0$, and $y \leq x$ can be interpreted geometrically as half of the xy-plane bounded respectively by the lines $x = 0$, $y = 0$, and $y = x$, shown in Figure 1.19. A few sample points are labeled in the shaded regions corresponding to each inequality.

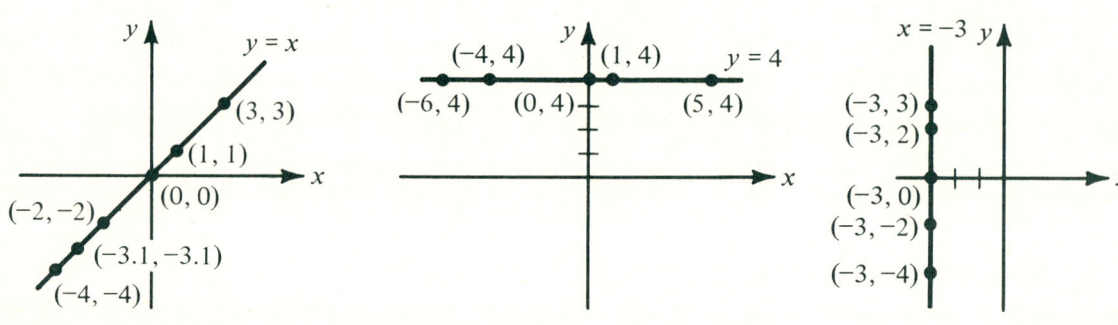

FIGURE 1.18.

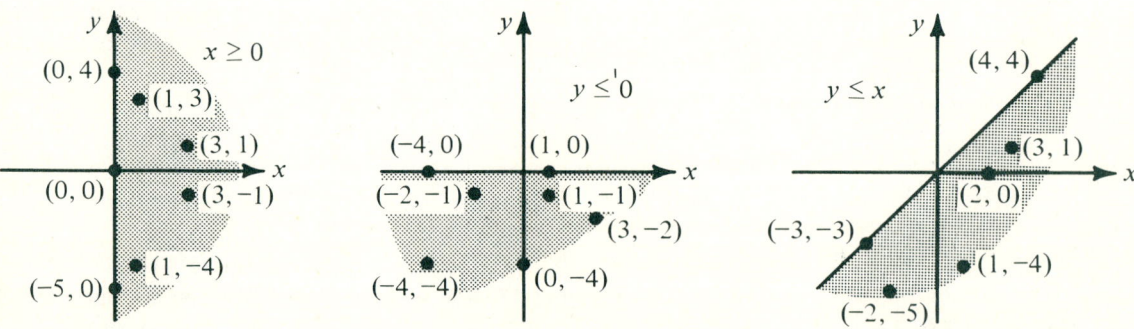

FIGURE 1.19.

SECTION 1.11   The Cartesian Coordinate System

Notice that, in the first part of the figure, all of the points (and any other point in the shaded region) have first coordinates (x-coordinates) greater than or equal to zero. The second part has points whose y-coordinates are negative or zero, and the third part shows points with y-coordinates smaller than or equal to their x-coordinates.

There are a few graphs of fundamental importance in elementary mathematics. Several are associated with equations of the form

$$y = x^n$$

where $n = 1, 2, 3, \ldots$. By choosing values of x at random and substituting them into a given equation, we can find corresponding values of y and generate the following tables of ordered pairs.

Supply the missing entries.

x	$y = x$	x	$y = x^2$	x	$y = x^3$	x	$y = x^4$
-1	-1	-2	___	-2	-8	-2	16
0	___	-1	1	-1	___	-1	1
1	___	0	___	0	0	0	___
2	2	1	___	1	1	1	___
3	___	2	4	2	___	2	___

Draw the corresponding graphs in Figure 1.20 after plotting the rest of the points from the tables.

The differences between the four graphs can be seen more clearly if we plot more points between $x = -1$ and $x = 1$. See Figure 1.21 (page 34).

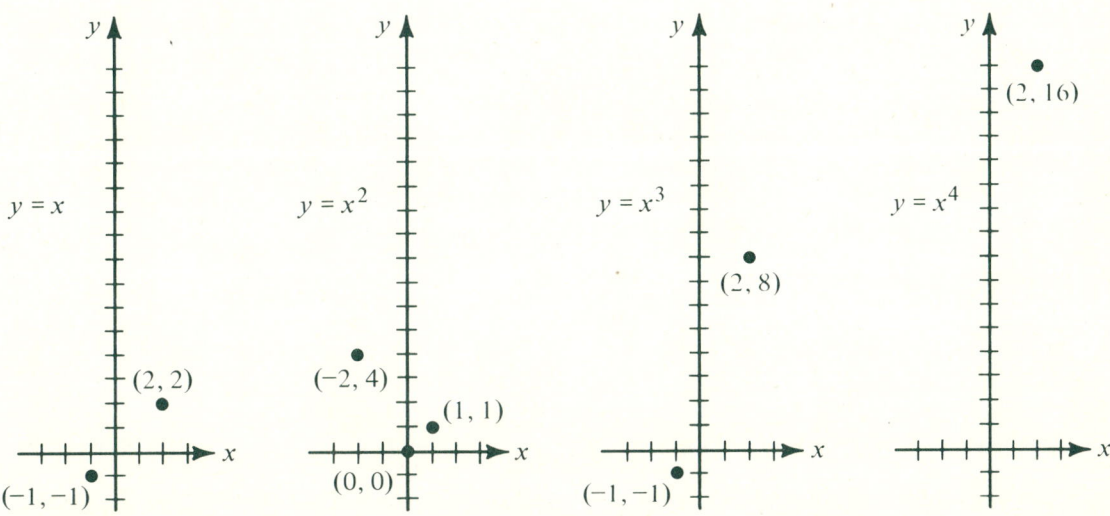

FIGURE 1.20.

Notice that the size of the exponent determines the rate at which the curve rises or falls. For $0 < x < 1$, the higher the exponent the slower the graph rises. But for $x > 1$, the higher the exponent the faster the curve rises. This is seen from the fact that beyond $x = 1$, $y = x^4$ increases more rapidly with increasing values of x than $y = x^2$. What can you determine about the size of the exponent relative to the behavior of the graphs for $-1 < x < 0$ and for $x < -1$?

Now let's graph another common equation, $y = \sqrt{x}$. Notice that since the square root of x will yield real numbers only if x is greater than or equal to zero, our graphing will be restricted to

CHAPTER 1 Algebra Review

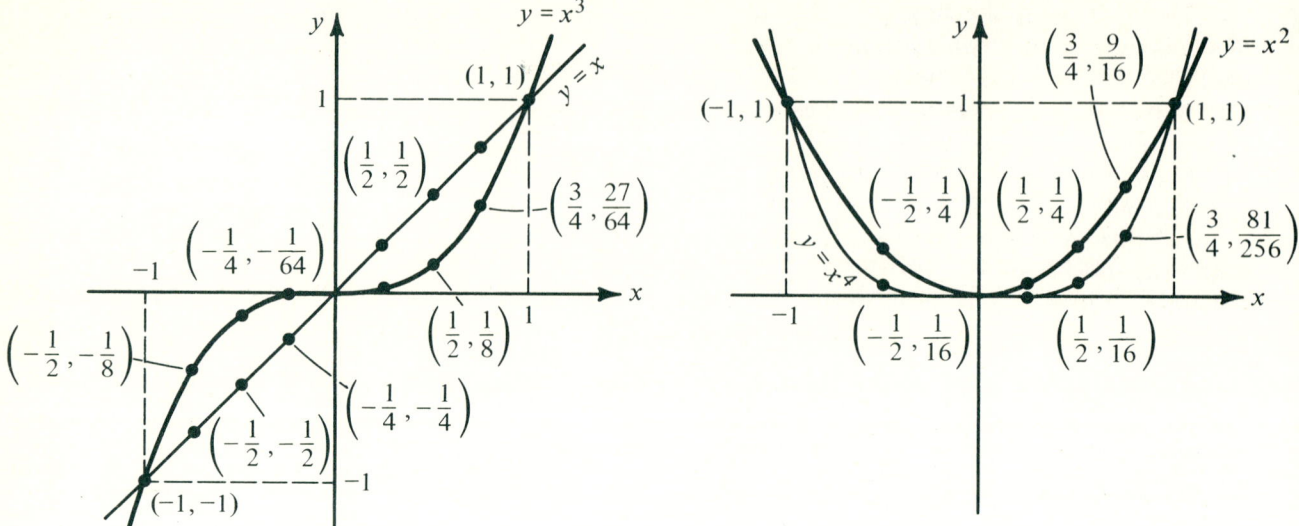

FIGURE 1.21.

that portion of the plane satisfying $x \geq 0$. Since $\sqrt{x}$ means the principle or positive square root of x, the y-values will be greater than or equal to zero.

Complete the table, then plot the points, then draw the graph.

x	$y = \sqrt{x}$
0	——
2	1.4
3	1.7
4	——
5	2.2
9	——

Now, let's graph $y = |x|$. Since the absolute value is always positive or zero, our graph will be in that portion of the xy-plane for which $y \geq 0$. Complete the following table of values, then draw the graph.

x	$y = |x|$
−2	——
−1	1
0	——
1	——
2	2
3	——
4	——

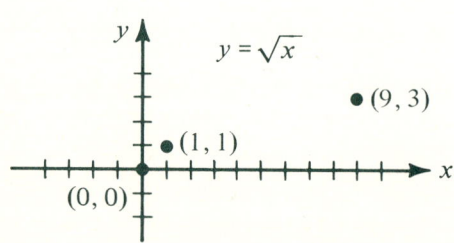

FIGURE 1.22.

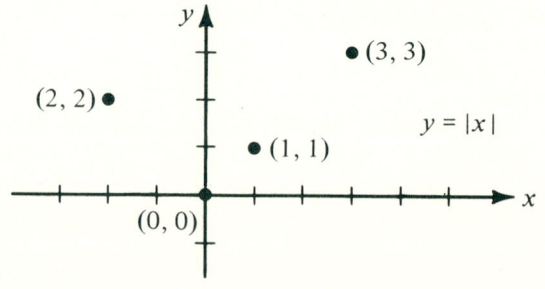

FIGURE 1.23.

34

There is one last graph that is appropriate at this point, that of the "greatest integer" or "step" equation $y = [x]$. (Notice that brackets are used for "step equations.") For each value of x, y takes on the value of the greatest integer less than or equal to x. For example, since the greatest integer less than or equal to $\frac{3}{2}$ is 1, $x = \frac{3}{2}$ yields $y = 1$. Complete the following evaluations of $y = [x]$, then draw its graph. Notice that all values of x between -3 and -2, including -3 but excluding -2, are paired with -3, the greatest integer *less than* these real numbers.

x	$y = [x]$
-3	-3
-2.5	
-2.1	
-2	-2
-1.5	
-1	
-0.9	-1
0	0
0.9	0
2	

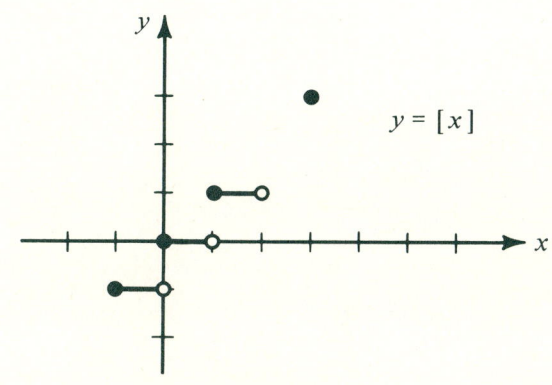

FIGURE 1.24.

EXERCISE 1.11

A. FUNDAMENTALS

Sketch the graphs of the following equations by substituting allowable values for x, finding the resulting value of y, plotting these resulting points, and then connecting them with a smooth curve.

1. (a) $y = x$ (b) $y = 2$ (c) $y = x + 2$
2. (a) $y = x$ (b) $y = -3$
   (c) $y = x - 3$
3. (a) $y = x^2$ (b) $y = 3$
   (c) $y = x^2 + 3$
4. (a) $y = x^2$ (b) $y = -4$
   (c) $y = x^2 - 4$
5. (a) $y = x^3$ (b) $y = 4$
   (c) $y = x^3 + 4$
6. (a) $y = \sqrt{x}$ (b) $y = -3$
   (c) $y = \sqrt{x} - 3$
★7. (a) $y = x$ (b) $y = x^2$
   (c) $y = x^2 + x$
8. (a) $y = |x|$ (b) $y = x$
   (c) $y = |x| + x$
9. (a) $y = |x|$ (b) $y = -x$
   (c) $y = |x| - x$

B. ESSENTIALS

10. (a) $y = x^2$ (b) $y = x^3$
    (c) $y = x^3 + x^2$
11. (a) $y = x^2$ (b) $y = x^2 + 4$
    (c) $y = x^2 - 4$
12. (a) $y = \sqrt{x}$ (b) $y = \sqrt{x} + 4$
    (c) $y = \sqrt{x} - 4$
13. $y = \begin{cases} x & \text{if } x \geq 0 \\ x^2 & \text{if } x < 0 \end{cases}$
★14. $y = \begin{cases} [x] & \text{if } x > 0 \\ |x| & \text{if } x \leq 0 \end{cases}$
15. $y = \begin{cases} x^2 & \text{if } x \geq 2 \\ x + 2 & \text{if } x < 2 \end{cases}$
16. $y = \sqrt{|x|}$
17. $y = |x^3|$
18. (a) $y = [2x]$ (b) $y = 2[x]$
19. (a) $y = [\tfrac{1}{2}x]$ (b) $y = \tfrac{1}{2}[x]$

CHAPTER 1   Algebra Review

20. (a) $y = [x + 2]$ (b) $y = [x - 2]$
21. $|xy| = 1$
• 22. (a) $y = [x^3]$ (b) $y = [x^4]$
• 23. $y = |x| \cdot \sqrt{x}$
• 24. $y = \sqrt{x^3}$
• 25. (a) $y = x^2 + x$ (b) $y = x^3 + x^2$
    (c) $y = x^4 + x^3$
• 26. $y = x^4 + x^3 + x^2 + x + 1$

C. APPLICATIONS

★27. Explain the geometric significance of the value of the discriminant for each of the following by examining its graph.

(a) $y = x^2$ (b) $y = x^2 + 1$
(c) $y = x^2 - 1$

28. Draw a graph that depicts the cost of a package of hamburger weighing x kilograms if the cost is 63 cents per kilogram.

29. Draw a graph depicting the area of a circle of radius r for any real number r.

30. Draw a graph that will show the postage required on a letter of x ounces.

31. Draw a graph of your daily bank balance over a period of two months. (You may wish to use fictitious amounts.)

32. Draw a graph showing the amount of federal income tax paid by single persons with adjusted gross incomes of $0.00 to $20,000.00.

1.12   SEQUENCES AND SERIES (OPTIONAL)

A **sequence** of n real numbers $a_1, a_2, a_3, \ldots, a_n$ will be represented as $\{a_j\}_{i=1}^{n}$ and a particular subsequence, for example, the first four terms, as $\{a_i\}_{i=1}^{4}$.

A **series** is the sum of the terms of a sequence. We will write the series $S = a_1 + a_2 + \cdots + a_n$, using sigma notation (see Appendix III), as

$$S = \sum_{i=1}^{n} a_i$$

Let's begin this study with two special types of sequences, arithmetic and geometric.

Arithmetic sequences are ones such as the following. Supply the missing terms.

2, 4, __?__ , 8, __?__ , __?__ , 14, 16
3, 7, __?__ , 15, __?__ , __?__ , __?__ , 31
1, $\frac{3}{2}$, __?__ , $\frac{5}{2}$, __?__ , __?__ , 4

For each of the sequences, the difference between any two consecutive terms is the same. These common differences, d, for the three previous sequences are 2, 4, and $\frac{1}{2}$, respectively. Notice that any term can be found by adding d to the preceding term. That is, for the *fourth terms* of each of the above sequences,

$8 = \underline{6} + d, \quad d = 2$
$15 = \underline{11} + d, \quad d = 4$
$\frac{5}{2} = \underline{2} + d, \quad d = \frac{1}{2}$

In general, we define an arithmetic sequence as follows.

DEFINITION 1.12.
An arithmetic sequence $a_1, a_2, \cdots, a_n$ has a common difference d such that

$a_{i+1} = a_i + d$ for $i = 1$ to $n - 1$.

If we create the series corresponding to the first arithmetic sequence above, we have

$S = \underline{2} + 4 + 6 + 8 + 10 + 12 + 14 + \underline{16}$

where we can write this sum as $S = 4 \cdot 18$, since

2 + 16 = _____
4 + 14 = _____
6 + 12 = _____
8 + 10 = _____

36

Notice that we were able to form 4 sums totaling 18 because there were 8 terms in the series. If there are an even number of terms, the sum will always be half the number of terms times the sum of the first and last terms,

$$S = \frac{n}{2}(a_1 + a_n)$$

Consider the third sequence above, whose seven-term series is

$$S = 1 + \tfrac{3}{2} + 2 + \tfrac{5}{2} + 3 + \tfrac{7}{2} + 4$$

Notice that the above formula yields

$$S = \frac{n}{2}(a_1 + a_n) = \frac{7}{2}(1 + 4) = \frac{7 \cdot 5}{5} = \frac{35}{2}$$

and the indicated grouped sums also yield

$$S = 5 + 5 + 5 + \frac{5}{2} = \frac{35}{2}$$

If we rewrite this as the following seven-term series,

$$S = \left(\tfrac{5}{2}+\tfrac{5}{2}\right) + \left(\tfrac{5}{2}+\tfrac{5}{2}\right) + \left(\tfrac{5}{2}+\tfrac{5}{2}\right) + \tfrac{5}{2} = 7\left(\tfrac{5}{2}\right)$$

we have written

$$S = \frac{n}{2}(a_1 + a_n)$$

as

$$S = n\left(\frac{a_1 + a_n}{2}\right) = 7\left(\frac{1+4}{2}\right)$$

Thus, S equals the sum of 7 terms, each equaling $\frac{1}{2}(a_1 + a_n) = \frac{5}{2}$, the average of the first and last term. Then S is the **sum** of the first n terms of a **sequence** and the **value** of its corresponding **series**.

SECTION 1.12   Sequences and Series (Optional)

> **THEOREM XVII.**
> The value, S, of the first n terms of an arithmetic series
> $$S = a_1 + a_2 + \cdots + a_n = \sum_{i=1}^{n} a_i$$
> is
> $$S = n\left(\frac{a_1 + a_n}{2}\right)$$

See Section 2.12 for the proof.

Now, let's examine geometric sequences and series. Supply the missing terms for the following geometric sequences.

$$2, 4, \underline{\ ?\ }, 16, \underline{\ ?\ }, 64$$
$$8, 4, \underline{\ ?\ }, \underline{\ ?\ }, \tfrac{1}{2}, \underline{\ ?\ }, \tfrac{1}{8}$$
$$8, 6, \underline{\ ?\ }, \tfrac{27}{8}, \underline{\ ?\ }, \underline{\ ?\ }$$

In each of these and any other geometric sequences, the ratio, r, of any two consecutive terms is the same. The *fourth terms* in each sequence can, therefore, be found by multiplying the preceding term by r:

$$16 = r \cdot 8, \qquad r = 2$$
$$1 = r \cdot 2, \qquad r = \tfrac{1}{2}$$
$$\tfrac{27}{8} = r \cdot \tfrac{9}{2}, \qquad r = \tfrac{3}{4}$$

Write the *fifth term* as r times the fourth for each sequence:

$$\underline{\ \ \ \ } = r \cdot 16, \qquad r = 2$$
$$\tfrac{1}{2} = r \cdot \underline{\ \ }, \qquad r = \tfrac{1}{2}$$
$$\underline{\ \ \ \ } = r \cdot \underline{\ \ }, \qquad r = \tfrac{3}{4}$$

This leads to the Definition 1.13.

> **DEFINITION 1.13.**
> A geometric sequence $a_1, a_2, \ldots, a_n$ has a common ratio r such that
> $$a_{i+1} = r \cdot a_i \text{ for } i = 1 \text{ to } n - 1$$

37

CHAPTER 1   Algebra Review

EXAMPLE I. Classify the following sequences as geometric or arithmetic and give the common ratio, r, or the common difference, d, as appropriate.

(a) 5, 2, −1, −4, −7 _____ _____
(b) 1, 4, 16, 64 _____ _____
(c) 16, 8, 4, 2, 1 _____ _____
(d) $\frac{11}{4}, \frac{52}{2}, \frac{9}{4}, 2$ _____ _____

You should have found (a) and (d) to be arithmetic and (b) and (c) to be geometric with (a) $d = -3$, (b) $r = 4$, (c) $r = \frac{1}{2}$, and (d) $d = \frac{1}{4}$.

Here is a clever ploy that we use to find the value of a geometric series. Consider

$$S = 2 + 4 + 8 + 16 + 32 + 64$$

Write $r \cdot S$ when $r = 2$:

$$r \cdot S = 2(2 + 4 + \cdots + 64)$$
$$= 4 + 8 + 16 + 32 + 64 + 128$$

Now, subtracting $r \cdot S$ and S, we have

$$r \cdot S - S = (4 + 8 + 16 + 32 + 64 + 128)$$
$$- (2 + 4 + 8 + 16 + 32 + 64)$$
$$2 \cdot S - S = 128 - 2$$

and

$$S = 126$$

Try it for the geometric sequence 16, 8, 4, 2, 1, $\frac{1}{2}$, where $r = \frac{1}{2}$, by completing the following series.

$$S = 16 + \underline{\quad} + \underline{\quad} + \underline{\quad} + \underline{\quad} + \frac{1}{2}$$
$$r \cdot S = \underline{\quad} \cdot S = 8 + \underline{\quad} + \underline{\quad} + \underline{\quad}$$
$$+ \underline{\quad} + \underline{\quad} + \frac{1}{4}$$

Therefore,

$$r \cdot S - S = (8 + 4 + 2 + 1 + \tfrac{1}{2} + \tfrac{1}{4})$$
$$- (16 + 8 + 4 + 2 + 1 + \tfrac{1}{2})$$

and

$$(r - 1)S = \underline{\quad\quad ? \quad\quad}$$

Then, $r = \frac{1}{2}$ yields

$$\left(\underline{\quad ? \quad}\right) \cdot S = \frac{-63}{4}$$

Now, dividing both sides by $-\frac{1}{2}$ the above coefficient of S yields the value of the series,

$$S = \frac{63}{2}$$

The proof of Theorem XVIII is completed in much the same way.

THEOREM XVIII.
The value of the first n terms of the geometric series

$$S = a_1 + a_2 + \cdots + a_n = \sum_{i=1}^{n} a_i$$

is

$$S = \frac{a_1 - ra_n}{1 - r} = \frac{a_1 - a_1 r^n}{1 - r}$$

The derivation of the first formula is very similar to our examples. The only catch is noting that $a_{i+1} - a_i \cdot r = 0$, since $a_{i+1} = a_i \cdot r$. Using this formula, we can show that

$$a_n = a_1 \cdot r^{n-1}$$

since

$$a_1 = a_1 \cdot r^0$$
$$a_2 = a_1 \cdot r^1$$
$$a_3 = a_2 \cdot r^1 = (a_1 \cdot r^1) \cdot r^1 = a_1 \cdot r^2, \text{ etc.}$$

Then, a simple substitution for a_n in the first expression yields the second:

$$\frac{a_1 - ra_n}{1 - r} = \frac{a_1 - r(a_1 r^{n-1})}{1 - r} = \frac{a_1 - a_1 r^n}{1 - r}$$

There are many other types of sequences and series in addition to arithmetic and geometric.[8]

The "*p*-series," for example, are of the form

$$\sum_{i=1}^{n} i^p$$

where the power, *p*, is usually a natural number. We will prove the following formulas for the "*p*-two" and "*p*-three" series in Chapter 2:

$$\sum_{i=1}^{n} i^2 = 1^2 + 2^2 + 3^2 + \cdots + n^2$$

$$= \frac{n(n+1)(2n+1)}{6}$$

$$\sum_{i=1}^{n} i^3 = 1^3 + 2^3 + \cdots + n^3 = \frac{n^2(n+1)^2}{4}$$

Some irrational numbers can be approximated by series. Usually the accuracy increases when the number of terms of the series increases. Following are two "series approximations" for the irrational numbers π and e, where $\pi \approx 3.14159$ and $e \approx 2.47183$. Factorial notation, $n!$, is defined as the product, $n! = n(n-1) \cdots 3 \cdot 2 \cdot 1$.

$$\pi \approx 4 \left(1 - \frac{1}{3} + \frac{1}{5} - \frac{1}{7} + \frac{1}{9} - + \cdots + \frac{(-1)^{n+1}}{2n-1} \right)$$

$$e \approx 1 + 1 + \frac{1}{2!} + \frac{1}{3!} + \frac{1}{4!} + \cdots + \frac{1}{n!}$$

If you have access to a calculator, add the first few terms of the above series and compare your results with the given approximate values. How many terms does it take to get five-place accuracy?

EXERCISE 1.12

A. FUNDAMENTALS

Decide whether each of the following series is arithmetic or geometric. Then find *d* or *r* and the sum *S*.

1. $3 + 4 + 5 + 6 + 7 + 8 = \sum_{i=3}^{8} i$

2. $2 + 4 + 8 + 16 + 32 = \sum_{i=1}^{5} 2^i$

3. (a) $\sum_{i=1}^{10} 2i = 2 + 4 + \cdots + 20$

   (b) $\sum_{i=1}^{10} (2i + 1) = 3 + 5 + \cdots + 21$

4. (a) $4 + 8 + 12 + \cdots + 48 = \sum_{i=1}^{12} 4i$

   (b) $4 + 16 + 64 + \cdots + 1{,}024 = \sum_{i=1}^{5} 4^i$

5. (a) $\sum_{i=1}^{3} 3(2i)$ (b) $\sum_{i=1}^{3} 2(3^i)$

6. (a) $\sum_{i=1}^{5} (3 + 2i)$ (b) $\sum_{i=1}^{5} 3 \cdot 2i$

B. ESSENTIALS

Use Theorems XVII and XVIII to find the values of the following series.

★7. (a) $\sum_{i=1}^{100} 2i$ (b) $\sum_{i=1}^{100} (2i - 1)$

8. (a) $\sum_{i=1}^{5} 3 \cdot 2^i$ (b) $\sum_{i=1}^{5} 3 \cdot 2^{-i}$

9. (a) $\sum_{i=1}^{10} (2i + 2^i)$ (b) $\sum_{i=1}^{10} (3^i - 3i)$

10. Use the formulas for the *p*-two and *p*-three series to evaluate the following.

(a) $\sum_{i=1}^{10} i^2$ (b) $\sum_{i=1}^{10} i^3$ ★c $\sum_{i=1}^{10} (2i^3 - 3i^2)$

C. APPLICATIONS

The sums of an infinite number of positive terms of some geometric sequences will be finite. To understand this, we must form *partial sums* for a series, where the *n*th partial sums, S_n, is simply the sum of the first *n* terms. Then, we create the sequence of these partial sums. If this sequence gets closer to some finite number S as we increase the number of terms used, we call S the value of the infinite series.

• **11.** Consider

$$S = \sum_{i=1}^{\infty} 2^{-1} = \tfrac{1}{2} + \tfrac{1}{4} + \tfrac{1}{8} + \cdots$$

The first partial sum, S_1, is $\tfrac{1}{2}$; the second, S_2, is $\tfrac{1}{2} + \tfrac{1}{4} = \tfrac{3}{4}$; the third, S_3, is $\tfrac{1}{2} + \tfrac{1}{4} + \tfrac{1}{8}$, or 0.895. $S_4 = S_3 + \tfrac{1}{16} = 0.9375$, $S_5 = 0.96875$, $S_6 = 0.984$.

The sequence of partial sums,

$S_1, S_2, S_3, S_4, S_5, \ldots = 0.5, 0.75, 0.875,$
$\qquad\qquad\qquad\qquad 0.9375, 0.96875, 0.984, \ldots$

is approaching 1.
(a) Calculate $S_6, S_7, S_8,$ and S_9 to indicate that the sequence of partial sums is approaching 1.
(b) Graph the points (1, .5), (2, .75), etc., to illustrate the geometric interpretation of this sequence.

12. Find the first 5 partial sums of the following. For those that seem to be approaching a specific value, name the value.

(a) $\displaystyle\sum_{i=1}^{\infty} 3^{-i}$ (b) $\displaystyle\sum_{i=1}^{\infty} (\tfrac{1}{2})^{2i}$

(c) $\displaystyle\sum_{i=1}^{\infty} (1 + \tfrac{1}{2})^i$ (d) $\displaystyle\sum_{i=1}^{\infty} 2^i$

• **13.** Find the first 20 partial sums of each of the infinite series in Problem 12. Guess the value.

14. The value of an infinite geometric series with $|r| < 1$ is given by

$$S = \frac{a_1}{1 - r}$$

Use this formula to find the value of the series in Problem 12.

15. Suppose a ball is dropped from a height of 1 meter. If the ball continually (forever) bounces back half as far as it just fell, how far will the ball travel?

16. Drop the ball in Problem 15 from 10 meters. How far does it travel?

17. Find the values of the following infinite series.

(a) $\tfrac{3}{10} + \tfrac{3}{100} + \tfrac{3}{1000} + \cdots = \displaystyle\sum_{i=1}^{\infty} \frac{3}{10^i}$

(b) $\displaystyle\sum_{i=1}^{\infty} 9 \cdot 10^{-i}$

ENDNOTES

1. In 1876 (see Exercise 1.1, Problem 27) Dedekind extended the rational numbers by defining the irrationals. This completed the real numbers.

2. Dedekind established the concept of a number field in 1879 by defining the fields of real and complex numbers.

3. The nineteenth century British school of mathematicians worked with the commutative, associative, and distributive laws of algebra.

4. The complex numbers were hinted at by Cardan, in 1545, who said that $\sqrt{-9}$ is neither $+3$ or -3 but of "hideous nature." He was baffled by imaginaries. He solved the problem of dividing 10 into two parts whose product is 40. (Can you?) Although he solved the problem, he did not understand the solution because he regarded solutions of equations as lengths of line segments.
In 1637 Descartes first called these nonreal solutions "imaginary," distinguishing between the real and imaginary roots of an equation.
In 1748 Euler used the letter "i" to stand for $\sqrt{-1}$.
Gauss called the numbers of the form $a + bi$ "complex," and it was his systematic use of complex numbers that finally led mathematicians to understand their relevance.
It is significant to note that neither Leibniz nor Newton, two of the greatest seventeenth century mathematicians, had any understanding of complex numbers. Newton thought they were insignificant because they made impossible solutions seem possible. Leibniz said, "The Divine Spirit found a sublime outlet in that wonder of analysis, that potent of the ideal world, that amphibian between being and not-being, which we call the imaginary root of negative unity."
The graphic representation of complex numbers

was introduced around 1813 by Jean-Robert Argand and is still referred to as an Argand diagram.

5. By 2000 B.C. the Babylonians were using a form of the quadratic formula to solve the problem of finding a number that when added to its reciprocal yielded a fixed value, $x + 1/x = a$. The equations were formulated and solved by geometric procedures (see Exercise 1.6, Problem 36).

6. Newton discovered the relationship between the solutions of a quadratic equation and its discriminant.

7. Viète, de Fermat, and Descartes influenced the development of analytic geometry, the marriage of plane geometry and algebra. Viète used algebra to solve geometric equations. De Fermat used an axis system with only positive coordinates to draw parts of the graphs of some equations. In 1637, Descartes, after whom the Cartesian coordinate system is named, used a complete coordinate system. Descartes was the first to state that each algebraic equation in x and y has a unique geometric interpretation, its "curve." Thus mathematicians were no longer restricted to curves that could be constructed with a compass and straightedge.

Descartes described the construction of a curve as finding the unknown lengths, y, from the known ones, x, using the equation. Then, by stating that for each known length, x, the unknown length, y, can be constructed, he presented the idea of graphing.

8. Finite sequences and series are ancient, but infinite series baffled mathematicians until relatively modern times. Zeno's paradoxes of motion such as the following:

> Motion does not exist since any motion must arrive at the halfway point before the end.

were not understood, because infinite series were not understood. How could the sum of an infinite number of positive numbers be finite?

Fibonacci came upon his series,

$$0, 1, 1, 2, 3, 5, 8, 13, \cdots$$

while studying the progeny of rabbits in about 1225.

Brook Taylor developed the single most powerful method for expanding a function in an infinite series in 1712.

Jakob and Johann Bernoulli did a great deal of work with series in the seventeenth and eighteenth centuries, Jakob, for example, proved that the harmonic series, $1 + \frac{1}{2} + \frac{1}{3} + \frac{1}{4} + \cdots$, is infinite (divergent).

Fourier used intuition to create his famous trigonometric series in 1822. It was used as a basis for much future work.

CHAPTER 2

Set Theory and Logic

2.1 INTRODUCTION

Set theory and logic are used extensively in the terminology and structure of mathematical language and thought. These concepts help clarify precise meaning and simplify notation throughout most areas of mathematics. It is therefore essential to have at least a basic understanding of these two concepts in order to communicate in modern mathematical terms. The two areas are studied together because there is a close, natural correspondence between them, which will become apparent as we near the end of this chapter.

2.2 SETS AND SUBSETS

For our purposes, a set can be thought of as *any well-defined collection of objects*. By well-defined, we mean that we know *exactly* what things are contained in the set. Sets will usually be defined with capital letters. Braces are used when naming the contents of a set. Its contents are called **elements**. If A and B are sets containing the numbers 1, 2, 3 and 1, 2, respectively, we would write

$$A = \{1, 2, 3\} \quad \text{and} \quad B = \{1, 2\}$$

We say, "3 is an element of A," and we write

$$3 \in A$$

Set A can be written in many equivalent ways, some of which are

$$A = \{1, 2, 3\} = \{3, 1, 2\} = \{1, 1, 2, 3, 3\}$$

The listing of the elements of a set only names the elements. It does not count how many of each are present nor does it give any ordering of the elements. That is why the three above sets are considered equal.

Do you notice the relationship that exists between sets $A = \{1, 2, 3\}$ and $B = \{1, 2\}$? This relationship is called "set inclusion," and we say that B is a **subset** of A.

SECTION 2.3 Unions and Intersections

> **DEFINITION 2.1.**
> The set B is a subset of set A if every element in B is also in A. We write
> $$B \subseteq A$$

Below is a list of most of the subsets of $A = \{1, 2, 3\}$.

$$\{1\}, \{2\}, \{3\}, \{1, 2\}, \{1, 3\}, \{2, 3\}$$

There are only two more,

$$\{\ \} \quad \text{and} \quad \{1, 2, 3\}$$

which are called **trivial subsets** since the whole set (the one containing all the elements) and the **empty set** (the one containing no elements) are always subsets of any set. The other subsets are called **proper subsets** of A, but it is seldom important to make the distinction. The empty set, symbolized by $\emptyset$, is not readily accepted as a subset of every set. The argument asserting that $\emptyset$ is a subset of every set is called a vacuous argument: "Every element in $\emptyset$ is also in A because there are no elements in $\emptyset$."

There is another important set which is the opposite of the empty set, $\emptyset$. It is called the **universal set**, U. This set contains every element being considered, and for our work will usually consist of all real numbers.

2.3 UNIONS AND INTERSECTIONS

There are two fundamental operations on sets that we will now pursue. Consider the following sets: $A = \{1, 2, 3\}$, $B = \{2, 3, 4\}$, $C = \{1, 2, 3, 4\}$, and $D = \{2, 3\}$. Do you see how set C is related to sets A and B? We say that C is the **union** of A and B.

> **DEFINITION 2.2.**
> The set A union B consists of the elements found in set A or B. We write
> $$A \cup B$$

The union of two sets is therefore the combination of all the elements of the sets. That is, if an element is in either set A or set B or both A and B, then it is in their union $A \cup B$.

Do you see how set D is related to sets A and B? We say that D is the **intersection** of sets A and B.

> **DEFINITION 2.3.**
> The set A intersection B consists of all of the elements in both A and B. We write
> $$A \cap B$$

The intersection therefore consists of the elements the sets have in common.

Consider the two sets $X = \{1, 2\}$ and $Y = \{3, 4\}$. What is true about their intersection? We say the sets are **disjoint**.

> **DEFINITION 2.4.**
> Two sets A and B are said to be *disjoint* if they have no elements in common. We write
> $$A \cap B = \emptyset$$

Some examples of Venn diagrams are given in Figure 2.1. Figure 2.1 depicts the three previous definitions.

CHAPTER 2   Set Theory and Logic

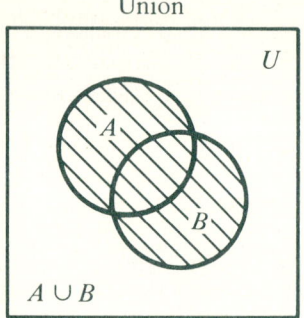
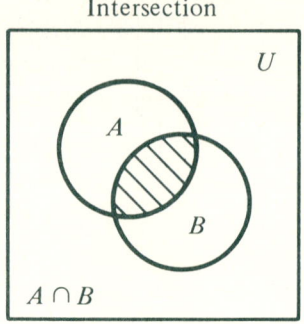
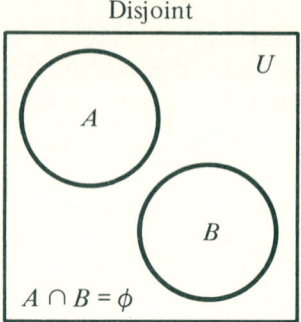

FIGURE 2.1.

2.4 COMPLEMENT

Whenever we define a set A, we have also defined another set called the **complement of** A.

> DEFINITION 2.5.
> The *complement* of set A consists of all of the elements in the universal set that are not in A. We write A' for the complement of A.

If $U = \{0, 1, \ldots, 9\}$ and $A = \{0, 1, 3, 5, 7, 9\}$, then $A' = \{2, 4, 6, 8\}$. Notice that $A \cap A' = \emptyset$ and $A \cup A' = U$. Figure 2.2 shows a Venn diagram of a set A and its complement A'.

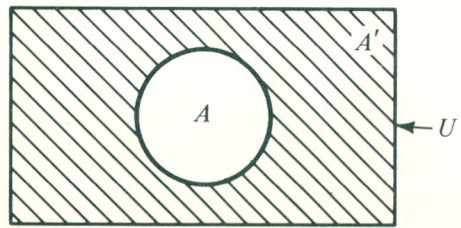

FIGURE 2.2.

2.5 DEMORGAN'S LAWS

Let's now examine the complement of the union of two sets, then the complement of the intersection of two sets. To sketch a Venn diagram of $(A \cup B)'$ we must first find $A \cup B$, then find the complement of that result. By carefully examining the two Venn diagrams in Figure 2.3, you might determine one of the important relationships discovered by the French mathematician DeMorgan. Stop and try.

Did you see that the area with single cross-hatching on the Venn diagram for $(A \cup B)'$ is the same as the area with double cross-hatching on the right? The double cross-hatched area on the right is the intersection of A' and B'. There are two equations illustrating the complements of the union and the intersection of two sets. These rules for complementation, called **DeMorgan's Laws**, follow:

> DEMORGAN'S LAWS.
> $$(A \cup B)' = A' \cap B'$$
> and
> $$(A \cap B)' = A' \cup B'$$

The second rule will be shown in Problem 23.

SECTION 2.5   Demorgan's Laws

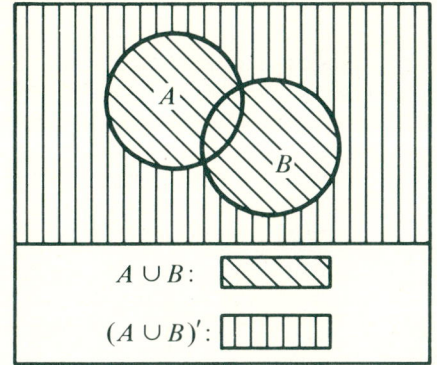 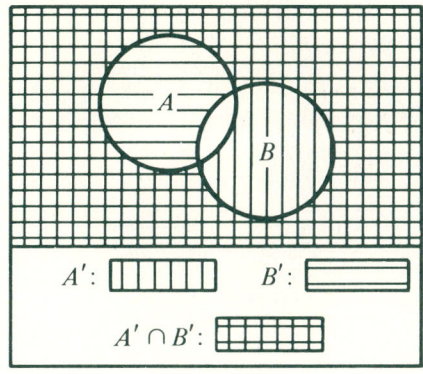

FIGURE 2.3.

EXERCISE 2.5

A. FUNDAMENTALS
Let $U = \{0, 1, 2, 3, 4, 5, 6, 7, 8, 9\}$, $A = \{0, 2, 4, 6, 8\}$, $B = \{1, 3, 5\}$, $C = \{2, 4, 6\}$, $D = \{6, 7, 8\}$, and $E = \{1, 3, 5, 7\}$. Find the following sets. Look for related results.

1. $A \cup B$ 2. $A \cap B$ 3. $C \cap D$
4. $B \cap D$ 5. $A \cup C$ 6. $A \cup E$
7. (a) A' (b) E' (c) $A' \cup E'$
   (d) $(A \cap E)'$
8. (a) $(A \cap B)'$ (b) $(A \cup B)'$
   (c) $A' \cap B'$ (d) $A \cup B'$
9. (a) $D' \cup E'$ (b) $D' \cap E'$
   (c) $(D \cup E)'$ (d) $(D \cap E)'$
10. (a) $A \cap (C \cup D)$ (b) $(A \cap C) \cup D$
    (c) $(A \cap C) \cup (A \cap D)$
    (d) $(A \cup D) \cap (C \cup D)$
11. (a) $A \cup (C \cap D)$ (b) $(A \cup C) \cap D$
    (c) $(A \cup C) \cap (A \cup D)$
12. (a) $B \cap (D \cap E)$ (b) $(B \cap D) \cap E$
13. (a) $(A \cup B) \cup D$ (b) $A \cup (B \cup D)$

Decide whether the following are true or false. Some depict only proper or improper use of the symbols of set theory.

14. $A \in U$ 15. $B \subseteq E$ 16. $A \subseteq A$
17. $2 \in A$ 18. $1 \subseteq B$ 19. $\emptyset \subseteq B$
20. $\{0\} \subseteq C$ 21. $A \cap E = \emptyset$
22. $A \cap E = \{\emptyset\}$

B. ESSENTIALS
Use Venn diagrams to illustrate the following relationships.

23. DeMorgan's Law: $(A \cap B)' = A' \cup B'$
24. The Distributive Laws:
    (a) $A \cap (B \cup C) = (A \cap B) \cup (A \cap C)$
    ★(b) $A \cup (B \cap C) = (A \cup B) \cap (A \cup C)$
25. The Associative Laws:
    (a) $A \cap (B \cap C) = (A \cap B) \cap C$
    (b) $A \cup (B \cup C) = (A \cup B) \cup C$

Algebraic solutions are often depicted as solution sets using the notation

$$\{x \mid x \in \text{reals}\}$$

which is read, "the set of all x, such that x is an element of the reals." Using the sets $U = \{x \mid x \in \text{reals}\}$, $A = \{x \mid x \in \text{integers}\}$, $B = \{x \mid x \in \text{rationals}\}$, $C = \{x \mid x \in \text{naturals}\}$, and $D = \{0\}$, find the following sets. You may wish to refer to the diagram in Section 1.1.

26. (a) $B \cup U$ (b) $B \cap U$
27. (a) $A \cap B$ (b) $A \cup B$
28. (a) $B \cup B'$ (b) $C \cap B'$
29. (a) $D \cap C$ (b) $A \cap C$

45

CHAPTER 2   Set Theory and Logic

Write the solutions of the following equations using the set notation above.

30. $2x - 3 < 5$ 31. $1 - 3x \geq 7$

32. $(x - 2)(x + 3) < 0$

33. $x^2 - 4x - 5 \geq 0$

Use the algebraic properties of sets listed in Problems 23–25 to help prove the following.

34. ★(a) $A \cap (A' \cup B) = A \cap B$
    (b) $A \cup (B \cap A') = A \cup B$

★35. $A \cup (A \cup B)' = A \cup B'$

36. $[(B \cup A')' \cup A']' = A \cap B$

37. $A \cap A = A$ (HINT: Begin with $A \cap U = A$ and $U = A \cup A'$.)

38. $A \cup A = A$

2.6 THE CARTESIAN PRODUCT

There is another way of combining two sets to form a new set called the **Cartesian product** or **cross-product**. The **Cartesian product** is the basis for most geometric discussions in this text, because it is a set of ordered pairs, such as $\{(1, 2), (2, 1)\}$, where, since $(1, 2) \neq (2, 1)$, the order of the components of each ordered pair is important. Consider $A = \{a, b\}$ and $B = \{1, 2, 3\}$. If we list the set of all ordered pairs that have an element from set A as the first part of the pair, and an element from set B as the second, we have the Cartesian product of A and B. We say, "A cross B," where

$$A \times B = \{(a, 1), (a, 2), (a, 3), (b, 1), (b, 2), (b, 3)\}$$

Notice that the Cartesian product is generally not commutative, since

$$B \times A = \{(1, a), (1, b), (2, a), (2, b)\}$$

and

$$(1, a) \neq (a, 1)$$

means

$$A \times B \neq B \times A$$

DEFINITION 2.6.
The Cartesian product of two sets A and B is the new set,

$$A \times B = \{(a, b) | a \in A \text{ and } b \in B\}$$

The Cartesian coordinate system is a representation of the cross product of the real numbers with themselves $R \times R$. Any graph on the Cartesian coordinate system represents simply a specific subset of $R \times R$.

EXERCISE 2.6

A. FUNDAMENTALS
Let $A = \{a, b, c\}$ and $B = \{p, d, q\}$. Find the following sets.

1. $A \times B$

2. $B \times A$

3. The largest subset of $A \times B$ that does not contain any of the elements of A more than once. There is more than one correct answer.

4. All the subsets of $A \times B$ containing 3 ordered pairs that depict a pairing of each element of A with exactly one element of B, and each element of B with exactly one element of A. This correspondence is called "one to one."

5. (a) If A contains 3 elements and B contains 3 elements, as above, how many elements are contained in $A \times B$?
    (b) Repeat (a) if A contains 2 and B contains 3.
    (c) Repeat (a) if A contains 3 and B contains 2.

B. ESSENTIALS
Solve the following.

6. If A contains m elements and B contains n, how many elements are contained in $A \times B$?

★7. Let $A = \{0, 1\}$ and $B = \{2\}$. Name all subsets of $A \times B$ that do not have any elements of A present more than once.

★8. From the list in Problem 7, list all subsets which also represent a one-to-one pairing between sets A and B.

9. Repeat Problems 7 and 8 for $B \times A$.

10. Repeat Problem 7 for $A = \{0, 1\}$ and $B = \{2, 3, 4\}$.

C. APPLICATIONS
Prove the following.

★11. $(A \times B) \cup (A \times C) = A \times (B \cup C)$

12. $(A \times B) \cap (A \times C) = A \times (B \cap C)$

2.7 LOGIC—A PROPOSITION AND ITS TRUTH VALUE

Mathematical logic is based upon Aristotelian logic, and has two restrictive characteristics. First, the *law of the excluded middle* requires each statement to be either true or false. There is no "maybe" in Aristotelian logic. Secondly, only declarative statements whose **truth value** can be determined are allowed as **propositions** of this logical system. Below are some sentences that can be classified as to whether or not they qualify as *propositions* in our logical system. Decide whether they qualify as propositions by being definitely true or definitely false declarative sentences.

a. Henry Kissinger received the Nobel Peace Prize in 1973.
b. The awarding of the Nobel Peace Prize to Henry Kissinger was appropriate.
c. Has the Vietnam War ended?
d. Help solve the energy crisis!
e. The United States government controls all of our energy sources.

Statements (a) and (e) are the only ones that qualify as propositions. Proposition (a) is true and (e) is false. Statement (b) is not a proposition because its truth value cannot be decisively determined.

Statements (c) and (d) are not propositions because they are not declarative sentences.

Propositions are referred to by using the letters $p, q, r, s, \ldots$. Generally, the letters p and q are used.

2.8 CONNECTIVES AND TRUTH TABLES

Now let's use the connectives "and," "or," and "implies" to build propositions. For example, if p and q are defined as:

p: It is raining.

q: It is cloudy

Then

(i) "p and q" is: It is raining *and* it is cloudy.
(ii) "p or q" is: It is raining *or* it is cloudy.
(iii) "p implies q" is: It is raining *implies* that it is cloudy.

The names and symbolic representation of these new propositions are:

(i) p and q, $p \wedge q$, **conjunction**.
(ii) p or q, $p \vee q$, **disjunction**.
(iii) p implies q, $p \Rightarrow q$, **implication**.

Once we have constructed the three new propositions above from the simpler propositions, p and q, it is essential to determine which combinations of truth values for p and q result in the new propositions being true, and which do not.

In order for a conjunction, p and q, to be true, both p must be true and q must be true. Therefore, (i) above would be true only if it were both cloudy and raining. If it were a sunny day with no rain (p false and q false), (i) would be false. If it were either cloudy but not raining (p false, q true), or raining but not cloudy (p true, q false), (i) would again be false. These combinations are summarized in the following *truth table*, which defines the truth values of the proposition.

The Conjunction

p	q	p AND q ($p \wedge q$)
T	T	T
T	F	F
F	T	F
F	F	F

The disjunction, p or q, is false only if both p and q are false. Therefore, proposition (ii) above is true if it is cloudy and raining (p true, q true), if it is raining but not cloudy (p true, q false), or if it is not raining but it is cloudy (p false, q true). It is false only when it is neither raining nor cloudy (p false, q false). This is summarized in the following truth table.

The Disjunction

p	q	p OR q ($p \vee q$)
T	T	T
T	F	T
F	T	T
F	F	F

In the English language there are two uses of "or." The inclusive "or" (either p or q or both) is the "or" used in the field of logic. The exclusive "or" (either one or the other, but not both) is not used in logic.

The implication "p implies q" is the most important proposition in mathematics. It usually appears in the equivalent form "if p, then q," and is used in most theorems in mathematics. *An implication can only be false if the hypothesis, p, is true but the conclusion, q, is not*, since it says, "if p is true, then q is true." Therefore, the only combination that would make (iii) above false would be: it is raining, but not cloudy. This is unlikely since we would generally regard the implication as a true statement. This means *only* that, if it is raining, it also *must* be cloudy (if p is true, then q must be true also). These results are summarized in the following truth table.

The Implication

p	q	p IMPLIES q ($p \Rightarrow q$)
T	T	T
T	F	F
F	T	T
F	F	T

The last two lines in the truth table for an implication deserve special attention. Notice that in each case, the hypothesis, p, is false yet the implication is classified as true. To understand this, it is helpful to remember that a truth table shows the truth value of the implication for various combinations of truth values of the hypothesis and conclusion.

Consider the third line for the following true implication.

$p \Rightarrow q$: If it is raining, then it is cloudy.

The fact that it is possible for it not to be raining (p false) and still be cloudy (q true) does not cause the implication to be false. A possible combination for the fourth line in the truth table is p false (it is not raining) and q false (it is not cloudy). The implication is true because the combination p true (it is raining) and q false (it is not cloudy) is not reasonable.

Consider the following false implication.

$p \Rightarrow q$: If my car is black, then all cars are black.

Several combinations of truth values of p and q are possible, but since p can be true and q false (my car is black, but not all cars are black), the implication is false.

An implication is a conditional statement. It simply requires q to be true whenever p is true (if p, then q). Nothing else is involved.

The next combination of propositions to be considered involves both the implication and the conjunction. *If two propositions, p and q, say exactly the same thing, then we say that p is equivalent to q*, and write

$$p \Leftrightarrow q$$

where $p \Leftrightarrow q$ means $(p \Rightarrow q) \wedge (q \Rightarrow p)$. It is common to say, "$p$ if and only if q," or

$$p \text{ iff } q$$

where $p \Rightarrow q$ is (p only if q) and $q \Rightarrow p$ is (p if q). The truth table for equivalence follows.

Equivalence

p	p	p IS EQUIVALENT TO q ($p \Leftrightarrow q$)
T	T	T
T	F	F
F	T	F
F	F	T

Notice that the truth value of $p \Leftrightarrow q$ is true whenever p and q have the same truth value.

EXERCISE 2.8

A. FUNDAMENTALS

1. Which of the following statements are propositions?
   (a) Love is beautiful.
   (b) Do you enjoy reading?
   (c) Math is your favorite subject.
   (d) Read this book carefully!
   (e) Bob Hope is the world's greatest golfer.

2. Translate the following propositions into symbols using only p, q, r, $\wedge$, $\vee$, and $\Rightarrow$.
   (a) John is a student and a teacher.
   (b) Harry likes Mary or Jane.
   (c) If you like peanuts, you'll love Skippy.
   ★(d) All politicians are liars.
   (e) If the dew point is 28 and the temperature is 50, then dew will form.
   (f) If it rains or snows, we don't go to school.

3. State the conditions that must exist in order for the following propositions to be true:
   (a) Richard is honest and humble.
   (b) John is a tennis and a volleyball player.

4. State the conditions that must exist in order for the following propositions to be false:
   (a) Henry is a statesman or a politician.
   (b) Richard is a victim or he is a criminal.
   (c) If you are a politician, then you are dishonest.
   (d) If you want to be heard, then you must vote.

B. ESSENTIALS
Write truth tables for the following propositions:

5. (a) $p \Rightarrow q$ (b) $p \wedge q$
   (c) $(p \wedge q) \Rightarrow q$

6. (a) $p \vee q$ (b) $(p \vee q) \Rightarrow (p \wedge q)$
   (c) $(p \vee q) \Leftrightarrow (p \wedge q)$

★7. Associative Law:
   (a) $p \wedge (q \wedge r)$ (b) $(p \wedge q) \wedge r$

8. Distributive Law:
   (a) $p \wedge (q \vee r)$
   (b) $(p \wedge q) \vee (p \wedge r)$

★9. Distributive Law:
   (a) $p \vee (q \wedge r)$ (b) $(p \vee q) \wedge (p \vee r)$

10. Associative Law:
    (a) $p \vee (q \vee r)$ (b) $(p \vee q) \vee r$

2.9 NEGATIONS

The negation of a proposition is a proposition with exactly the opposite meaning and truth value. Then, if p were defined as

p: John is an excellent student.

then its **negation**, not p, written $\sim p$, would be

$\sim p$: John is not an excellent student.

If we continue negating, we have

$\sim(\sim p)$: It is not true that John is not an excellent student.

which is saying that John is an excellent student. Therefore,

$$\sim(\sim p) \Leftrightarrow p$$

CHAPTER 2   Set Theory and Logic

Try writing the negations of the following propositions:

$p \wedge q$: John is a man and a male chauvinist.

$p \vee q$: Mary is beautiful or a feminist.

other than

$\sim(p \wedge q)$: It is not true that John is a man and a male chauvinist.

or

$\sim(p \vee q)$: It is not true that Mary is beautiful or a feminist.

If you negated the propositions as,

$(\sim p) \vee (\sim q)$: John is not a man or not a male chauvinist.

and

$(\sim p) \wedge (\sim q)$: Mary is not beautiful and not a feminist.

then you were right. These rules for negating conjunctions and disjunctions, called **DeMorgan's Laws**, are summarized below. Notice that these are analogous to DeMorgan's laws for sets given in Section 2.5.

DeMorgan's Laws.

$$\sim(p \wedge q) \Leftrightarrow \sim p \vee \sim q$$

and

$$\sim(p \vee q) \Leftrightarrow \sim p \wedge \sim q$$

Notice if we say that it is not true that p is true and q is true $\sim(p \wedge q)$, then either p is false or q is false (or both). If we say that it is not true that p or q is true $[\sim(p \vee q)]$, then neither is true. Both p and q are false.

We can prove DeMorgan's Laws by showing that the truth table for $\sim(p \wedge q)$ is identical to that for $\sim p \vee \sim q$. The details of one of the proofs is outlined in the following truth table. The numbers at the bottom indicate the order in which the truth table was completed. Notice that the truth values of the equivalence relation are all true. *Any logical statement that is true under every combination of possible truth values is called a tautology.*

p	q	$\sim p$	$\sim q$	$p \wedge q$	$\sim(p \wedge q)$	$\Leftrightarrow$	$\sim p \vee \sim q$
T	T	F	F	T	F	T	F
T	F	F	T	F	T	T	T
F	T	T	F	F	T	T	T
F	F	T	T	F	T	T	T
3		4	1		2	6	5

See if you can construct a similar proof for DeMorgan's other law, $\sim(p \vee q) \Leftrightarrow \sim p \wedge q$.

Now let's investigate the negation of the implication. Completing the following truth table will show us three implications that are not the negation of $p \Rightarrow q$ because their truth values are not F, T, F, F. Fill in the missing truth values.

p	q	$p \Rightarrow q$	$\sim p \Rightarrow \sim q$	$q \Rightarrow p$	$\sim q \Rightarrow \sim p$	
T	T					1. *implication*
T	F	F				2. *inverse*
F	T					3. *converse*
F	F					4. *contrapositive*
		1	2	3	4	

You should have found the three remaining false entries (one in each column) to be on the third, third, and second lines, respectively. The truth tables for an implication and its contrapositive are identical. We have our first theorem on logic.

Theorem I.

$$(p \Rightarrow q) \Leftrightarrow (\sim q \Rightarrow \sim p)$$

Now, complete the following truth table and you will discover the negation of $p \Rightarrow q$. This will seem more reasonable if you remember that the negation of $p \Rightarrow q$ must be true for only one combination of truth values of p and q, since $p \Rightarrow q$ is only false once.

p	q	$p \Rightarrow q$	$\sim p \vee q$	$\sim(\sim p \vee q)$	$p \wedge \sim q$
T	T	T			
T	F	F			
F	T	T			
F	F	T			

You should have found the following results.

THEOREM II.
$$(p \Rightarrow q) \Leftrightarrow (\sim p \vee q)$$

THEOREM III.
$$\sim(p \Rightarrow q) \Leftrightarrow \sim(\sim p \vee q) \Leftrightarrow p \wedge \sim q$$

The first and second entries have identical truth tables, as do the third and fourth. The truth values of $p \Rightarrow q$ are *exactly* the opposite of the truth values of $p \wedge \sim q$, proving that they are negations of each other.

EXERCISE 2.9

A. FUNDAMENTALS
Negate the following propositions carefully.

1. John is a student and a fireman.
2. Mary is a doctor and a mother.
3. Henry is a teacher but not a researcher.
4. Joyce is an author but not a poet.
5. If it rains, it pours.
6. If you don't try, then you'll pass.

7. (a) Translate the above six propositions into symbols.
   (b) Translate the negations of the above six propositions into symbols.

8. Write the truth table for the tautology $p \vee \sim p$.

9. What conditions must occur in order to make the following sentence false? Whether you like this book or not, you must study it. [HINT: It is $(p \vee \sim p) \Rightarrow q$.]

10. Write the converse and the contrapositives for each of the following implications. Remember that an implication is logically equivalent to its contrapositive.
    (a) If the humidity is 100%, then it is not raining.
    (b) If the dew point is 28° and the temperature is 30°, then no dew is forming.

B. ESSENTIALS
Use truth tables to prove the following tautologies.

★11. $\sim(p \vee q) \Leftrightarrow \sim p \wedge \sim q$ (DeMorgan's law)

12. $\sim(\sim p) \Leftrightarrow p$ (double negation)

★13. $[(p \Rightarrow q) \wedge p] \Rightarrow q$ (*modus ponens*)

★14. $[(p \Rightarrow q) \wedge \sim q] \Rightarrow \sim p$ (*modus tollens*)

15. $\sim(p \Rightarrow q) \Leftrightarrow p \wedge \sim q$

16. $[(p \Rightarrow q) \wedge (q \Rightarrow r)] \Rightarrow (p \Rightarrow r)$
    (Transitive law)

Translate the following arguments into symbolic form. Then, using the logical tautologies discussed in the text and in problems 11–16, determine whether or not the following are valid arguments.

★17. If I do not skip class, then I will pass; if I sleep in class, I will not pass. Therefore, if I sleep in class, I will skip class.
(HINT: Symbolically, the argument is $[(\sim p \Rightarrow q) \wedge (r \Rightarrow \sim q)] \Rightarrow (r \Rightarrow p)$.)

18. The relative humidity is high or it is not raining. It is raining or the temperature is rising. Therefore, if the temperature drops, then the relative humidity is high.

★19. It is not true that the temperature is above freezing and it is snowing. If it is raining, then the temperature is above freezing. Therefore, if it is not raining, then it is snowing.

20. If your home is dry, then plants won't grow there. Either your home is dry or plants grow there. Therefore, a dry home is equivalent to plants not growing.

★21. The temperature is 38° or dew is not forming. If dew does not form then the temperature does not rise. The wind is not blowing or the temperature rises. Therefore, if the temperature is not 38°, then the wind is not blowing.

22. Explain why the following common form for a "check" does not represent a valid argument.

$$x = -1 \wedge x^2 - 3x - 4 = 0 \Rightarrow$$
$$(-1)^2 - 3(-1) - 4 = 0 \Rightarrow$$
$$1 + 3 - 4 = 0 \Rightarrow$$
$$0 = 0$$

2.10 CONDITIONAL PROPOSITIONS

Since most mathematical statements, equations, inequalities, theorems, etc., contain variables, their truth values cannot be determined until a value is assigned to the variable(s) involved. Such statements are called **conditional propositions** and the collection of values of the variable that make the statements true is called its **truth set**. For the conditional proposition

$$p(x): x \text{ is an odd integer}$$

read "p of x", $x = 3$ is contained in the truth set, since

$$p(3): 3 \text{ is an odd integer}$$

is true. $x = 4$ is not, however, since

$$p(4): 4 \text{ is an odd integer}$$

is false.

By writing the truth set for $p(x)$ as T_p, we can readily see the natural correspondence between set theory and logic.

Let

$$p(x): 3x - 4 \leq 5$$

and

$$q(x): 2x - 3 \leq 7$$

Then their respective truth sets are

$$T_p = \{x \mid x \leq 3\}$$

and

$$T_q = \{x \mid x \leq 5\}$$

since

$$3x - 4 \leq 5 \Leftrightarrow x \leq 3$$

and

$$2x - 3 \leq 7 \Leftrightarrow x \leq 5$$

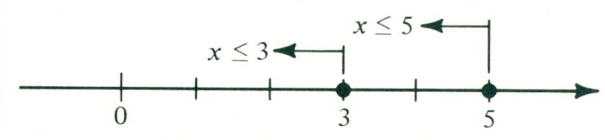

FIGURE 2.4.

Do you see that $p(x) \Rightarrow q(x)$? What is the relationship between T_p and T_q? This is summarized below:

$$p(x) \Rightarrow q(x) \Leftrightarrow T_p \subseteq T_q$$

The relationship between the truth sets for equivalent propositions is the natural extension of that for implications:

$$p(x) \Leftrightarrow q(x) \Leftrightarrow T_p = T_q$$

Now to examine conjunctions, disjunctions, and negations, let

$$p(x): (x + 4)(x - 1) \leq 0$$

and

$$q(x): (x + 2)(x - 3) \leq 0$$

Then

$$T_p = \{x \mid -4 \leq x \leq 1\}$$

and

$$T_q = \{x \mid -2 \leq x \leq 3\}$$

What are the values of x that make both $p(x)$ and $q(x)$ true? This set is $T_{p \wedge q}$. How is it related to T_p and T_q?

Can you find the connection between $T_{p \wedge q}$ and T_p and T_q? If you cannot, examine the following using the above definitions of $p(x)$ and $q(x)$.

THEOREM IV.

$$T_{p \wedge q} = T_p \cap T_q$$

THEOREM V.

$$T_{p \vee q} = T_p \cup T_q$$

For negations, notice that

$$\sim p(x): (x+4)(x-1) < 0$$

and

$$T_{\sim p} = \{x \mid x < -4 \text{ or } x > 1\}$$

Did you see that the truth set for the negation of $p(x)$ is the complement of the truth set of $p(x)$? Then, formally, we have the following theorem.

THEOREM VI.

$$T_{\sim p} = (T_p)'$$

In general, the true value of a conditional proposition cannot be determined until a value of the variable is chosen. Often in mathematics it is necessary to change an ambiguous conditional proposition into an ordinary proposition whose truth value can be determined without selecting a specific value for the variable. This is accomplished by using the quantifier "for all x" (symbolically written $\forall x$) or "for some x" ($\exists x$).

For example, the conditional proposition

$$p(x): x^2 - 2x - 3 = 0$$

becomes an ordinary proposition when written

$$(\exists x)p(x): \text{For some } x, x^2 - 2x - 3 = 0 \quad (1)$$

which is true, or

$$(\forall x)p(x): \text{For all } x, x^2 - 2x - 3 = 0 \quad (2)$$

which is false.

The two quantifiers are closely related by negations. To see this, we need to find out how to negate quantified propositions.

If it is not true that, for all x, $x^2 - 2x - 3 = 0$, then what must be true? The negation cannot be

$$(\forall x) \sim p(x): \text{For all } x, x^2 - 2x - 3 \neq 0 \quad (3)$$

because statement (3) is false also, since $x = 3$ satisfies the equation. The negation of (2) is

$$(\exists x) \sim p(x): \text{For some } x, x^2 - 2x - 3 \neq 0$$

since (2) above is false because there exists at least one value of x such that $x^2 - 2x - 3 \neq 0$.

The negations of the two quantified conditional propositions are summarized in Theorems VII and VIII.

THEOREM VII.

$$\sim [(\forall x)p(x)] \Leftrightarrow (\exists x) \sim p(x)$$

THEOREM VIII.

$$\sim [(\exists x)p(x)] \Leftrightarrow (\forall x) \sim p(x)$$

The negation of a quantified statement requires two changes, a change in quantifiers and the negation of the following proposition.

EXERCISE 2.10

A. FUNDAMENTALS

Use the most appropriate quantifier, either $(\forall x)$ or $(\exists x)$, to change the following conditional propositions into true propositions.

1. $p(x)$: $3x - 4 < 2$ 2. $q(x)$: $(x-2)^2 \geq 0$
3. $p(t)$: $t + 1 < t + 2$ 4. $p(x)$: $\sqrt{x+2} > 0$
★5. (a) $p(x) \Rightarrow q(x)$: $x^2 = 16 \Rightarrow x = 4$
    (b) $p(x) \Rightarrow q(x)$: $x = 4 \Rightarrow x^2 = 16$

Find the truth sets for the following conditional propositions.

6. $p(x)$: $3x^2 - 2x - 1 = 0$
7. $q(x)$: $\sqrt{2x-1} = \sqrt{x+1}$
8. $p(t)$: $t - 2 = \sqrt{t-2}$
9. $q(t)$: $2 - t = \sqrt{t-2}$
10. $p(x)$: $\dfrac{1}{x} - \dfrac{2}{x} = 3$ 11. $q(x)$: $\dfrac{3}{x+1} = \dfrac{-1}{x-1}$
12. $p(t)$: $3t - 4 \geq 0$ 13. $q(t)$: $t^2 - 9 < 0$

B. ESSENTIALS

Decide whether or not $T_p \subseteq T_q$ to determine if $p(x) \Rightarrow q(x)$ is true for all $x \in$ reals.

14. (a) $p(x) \Rightarrow q(x)$: $x = 2 \Rightarrow x^2 - 4 = 0$
    (b) $p(x) \Rightarrow q(x)$: $x^2 - 4 = 0 \Rightarrow x = 2$
★15. (a) $p(x) \Rightarrow q(x)$: $x > 3 \Rightarrow x^2 - 9 > 0$
    (b) $p(x) \Rightarrow q(x)$: $x^2 - 9 > 0 \Rightarrow x < -3$
16. (a) $p(x) \Rightarrow q(x)$: $x = 3 \Rightarrow x^2 > 0$
    (b) $p(x) \Rightarrow q(x)$: $x = 0 \Rightarrow x^2 > 0$
★17. $p(x) \Rightarrow q(x)$: $\dfrac{x}{0} = 1 \Rightarrow x = 0$
18. $p(x) \Rightarrow q(x)$: $\sqrt{x} = -1 \Rightarrow x = 1$
★19. $p(x) \Rightarrow q(x)$: $|x| = -3 \Rightarrow x = 3$

Find $T_{p \wedge q}$ and $T_{p \vee q}$ using $T_{p \wedge q} = T_p \cap T_q$ and $T_{p \vee q} = T_p \cup T_q$.

20. $p(x)$: $2x - 1 > 0$, $q(x)$: $3x + 1 > 0$
21. $p(x)$: $x - 4 > 0$, $q(x)$: $2x - 1 < 0$
22. $p(x)$: $x - 2 > 0$, $q(x)$: $|x - 2| < 3$
23. $p(x)$: $2x - 3 \leq 0$, $q(x)$: $|x + 3| > 4$
24. $p(x)$: $x^2 + x - 2 < 0$, $q(x)$: $x^2 - 2x - 3 < 0$
25. $p(x)$: $\dfrac{x}{0} = 1$, $q(x)$: $x = 0$
26. $p(x)$: $|x| = -3$, $q(x)$: $x = 3$

2.11 LOGIC AND MATHEMATICAL PROOFS

There are several types of mathematical proofs that are constructed from the truth values discussed in the previous sections. Two of these are considered *direct proofs* of theorems stated as implications. First, let's discuss the most common argument used in proving implications directly.

Since an implication $p \Rightarrow q$ is false only if p is true and q is false, what we must accomplish to complete the proof is to show that the combination, p true and q false, cannot possibly occur. To do this, we simply assume p is true and show that the fact of q being true follows from this assumption. In other words, we must show that if p is true, q must be true. We can assume p is true for our proof, since if p is false, the implication is true regardless of the truth value of q.

EXAMPLE I. If x is an even integer, then x^2 is even.

Proof: First, assume x is an even integer (assume p is true). Then, since all even numbers are some multiple of 2, we have

$$x = 2n$$

where n is some integer, and

$$x^2 = 4n^2 \Rightarrow$$
$$x^2 = 2(2n^2)$$

Therefore, x^2 is an even integer (q is true) since it is written as a multiple of 2.

It is often easier, or sometimes even necessary, to restate an implication as its contrapositive in order to complete the proof. Since an implication and its contrapositive are logically equivalent, that is,

$$(p \Rightarrow q) \Leftrightarrow (\sim q \Rightarrow \sim p)$$

by proving the contrapositive true we also prove the original implication. We usually prove the contrapositive true by directly assuming its hypothesis is true ($\sim q$ true) and then showing that its conclusion cannot be false. Its conclusion is true when $\sim p$ is true.

EXAMPLE II. If x^2 is an even integer, then x is even.

Proof: First, we state the contrapositive,

if x is not even, then x^2 is not even $\Leftrightarrow$

if x is odd, then x^2 is odd

Now we assume that the hypothesis of the contrapositive is true, that is, that x is odd. Then

$$x = 2n + 1$$

where n is an integer, since x must follow an even number $2n$. Then we proceed to show that the conclusion of the contrapositive *must* be true. Squaring both sides yields

$$x^2 = (2n + 1)^2 \Rightarrow$$
$$x^2 = 4n^2 + 4n + 1 \Rightarrow$$
$$x^2 = 2(2n^2 + 2n) + 1$$

Therefore, x^2 is an odd number, since we were able to write it in the form $2m + 1$, where $m = 2n^2 + 2n$ is an integer. Now we know that the original theorem is true because we have proven that its contrapositive is true.

Now, let's look at a proof that a theorem, stated as an implication, is false. This is usually accomplished by finding a counterexample. Exhibiting one counterexample is a complete proof that the theorem is false. The counterexample must yield the true hypothesis–false conclusion combination, since that is the only combination that makes an implication false.

EXAMPLE III. If $x + y$ is divisible by m, then x or y is divisible by m, where x, y, and m are integers.

Proof: We need an example that makes p true and q false. Let $x = 1$, $y = 2$, and $m = 3$. Then $x + y$ is divisible by m, and p is true. Neither x nor y is divisible by m, so q is false. This exhibition of one counterexample completes the proof that the theorem is false.

The *indirect* method of proving a theorem is another way to proceed; it is often the best way. We assume that $p \Rightarrow q$ is false and then arrive at a contradiction. This means that our assumption that $p \Rightarrow q$ was false was incorrect and that $p \Rightarrow q$ must therefore be true. To assume an implication is false, we assume p is true and q is false, since the negation of $p \Rightarrow q$ is $p \wedge \sim q$.

EXAMPLE IV. Prove that there exist real numbers that are not rational.

Proof: First, let's restate this as an implication using a particular irrational number, say $\sqrt{2}$. Then,

if $x = \sqrt{2}$, then x is not a rational number

We then assume this is false, by assuming that $x = \sqrt{2}$ and that x is rational. Since x is assumed to be a rational number, we can write

$$x = \frac{p}{q}$$

where p and q are integers. We know that p and q do not have any common factors, for every rational number can be written uniquely in the form p/q, where the fraction is reduced to its lowest terms. Then

$$\sqrt{2} = \frac{p}{q} \Rightarrow$$
$$2 = \frac{p^2}{q^2} \Rightarrow$$
$$p^2 = 2q^2$$

Therefore, p^2 is even. Therefore, p is even. Then, p can be written as

$$p = 2n$$

where n is an integer. Therefore,

$$\sqrt{2} = \frac{2n}{q} \Rightarrow$$

$$2 = \frac{4n^2}{q^2} \Rightarrow$$

$$q^2 = 2n^2$$

Therefore, q^2 is even, and so is q. But this is a contradiction, since p and q cannot have any common factors, and the fact that both p is even and q is even means that both have a factor of 2. We must conclude that our assumption that the implication

$$x = \sqrt{2} \Rightarrow x \text{ is not rational}$$

was false was incorrect, since it led to a contradiction. Therefore, the statement must be true.

There is another common method of proof called *mathematical induction*, which is discussed in Section 2.12.

EXERCISE 2.11

A. FUNDAMENTALS

Use a direct proof of $p \Rightarrow q$ to prove the following by showing that if p is true, then q must be true.

1. If $a \cdot x = 0$ and $a \neq 0 \Rightarrow x = 0$ (a and x are real numbers). (HINT: If $a \neq 0$, then $1/a$ exists.)

2. If a is divisible by 3, then $a \cdot b$ is divisible by three. Assume $a, b \in$ integers.

★3. If a or b is divisible by 3, then $a \cdot b$ is divisible by three. Assume $a, b \in$ integers.

4. If a and b are divisible by 3, then $a + b$ is divisible by three. Assume $a, b \in$ integers.

Use a counterexample to prove the following implications are false. Your example must make p true and q false.

5. If $a \cdot b$ is divisible by 9, then a or b must be divisible by 9. Assume $a, b \in$ integers.

6. If $|x| < |y|$, then $x < y$.

7. If $|x| + |y| = 1$, then $x + y = 1$.

Use a direct proof of the contrapositive to prove the following by showing that if $\sim q$ is true (q false), then $\sim p$ must be true (p false).

8. If x^2 is odd, then x is odd.

★9. If $a \cdot b \neq 0$, then $a \neq 0$ and $b \neq 0$.

B. ESSENTIALS

Prove the following indirectly.

10. The additive identity for the real numbers is unique. (HINT: Assume there are two identities, x and y, such that $x \neq y$, $a + x = a$, and $a + y = a$. Then show that $x = y$.)

11. The multiplicative identity is unique.

12. The additive inverse of each real number is unique.

13. The multiplicative inverse of each real number is unique.

★14. The set of prime numbers is infinite.

2.12 MATHEMATICAL INDUCTION

Mathematical induction[1] is a method of proof that is used whenever we must verify a statement for all natural numbers, or for all natural numbers beyond some initial one. The use of the word *induction* is not related to its use in the scientific method, where inductive reasoning involves studying several examples in order to hypothesize the general rule. Math induction, like all proofs in mathematics, involves deductive reasoning. The induction part of the proof stems from the recursiveness of the problems themselves. Recursive definitions generate a statement from a preceding statement. For example, we can define the arithmetic sequence

$$2, 4, 6, 8, \ldots$$

as

$$a_1 = 2 \quad \text{and} \quad a_{i+1} = a_i + 2, \quad i = 1, 2, 3, \ldots$$

which says that the *successor* of a_i, namely, a_{i+1}, can be found by adding the common difference, 2, to a_i. A recursive (inductive) formula for the following geometric sequence is

$$3, 9, 27, 81, \ldots$$

$$a_1 = 3 \quad \text{and} \quad a_{i+1} = 3 \cdot a_i, \quad i = 1, 2, 3, \ldots$$

It is also possible to give an inductive definition for the set of natural numbers. Consider the following two statements describing a set.

1. The number 1 is in the set.
2. If any number k is in the set, then its successor, $k + 1$, must be in the set.

The only set completely defined by *both* of these statements is the set containing *all* natural numbers. Since $k = 1$ is in the set, according to statement 2, its successor, $k + 1 = 2$, must be there also. Now, since $k = 2$ is there, its successor, $k + 1 = 3$, must be in the set, etc.

Mathematical induction can also be used to describe the following situation: the falling of dominoes is analogous to proving a statement true for all natural numbers. Suppose we set up a string of dominoes in such a way that if any domino, say the kth one, falls, then its successor, the $(k + 1)$th one, will be struck, causing it to fall. *In addition*, we know that if any initial one is knocked over, by the inductive statement above we can determine that all the dominoes, from the initial one on, will be knocked over. We usually use the first domino as the initial one.

Proofs by mathematical induction always involve the same two separate steps. Suppose we have some conditional proposition $p(n)$, with n as the variable, which we wish to verify for all $n \in$ naturals. First we prove $p(n)$ is true for some initial value, usually $n = 1$. Then, we prove that if $p(n)$ is true for any particular natural number k between 1 and n, then $p(n)$ must be true for the next natural number, $n = k + 1$.

To prove $p(n)$ for all $n \in$ naturals, we must take the following steps:

1. Prove: $p(1)$.
2. Prove: $p(k) \Rightarrow p(k + 1)$, $1 \le k \le n$.

In the second step, $p(k)$ is called the induction hypothesis, and $p(k + 1)$ the induction conclusion.

EXAMPLE 1. Prove the following for all $n \in$ naturals:

$$p(n): 1 + 2 + \cdots + n = \sum_{i=1}^{n} i = \frac{n(n+1)}{2}$$

Proof:
1. Prove $p(1)$.

$$p(1): 1 = \frac{1(1+1)}{2}$$

is true since

$$\frac{1(1+1)}{2} = \frac{1 \cdot 2}{2} = 1$$

The following, which verify the theorem for $n = 2, 3,$ and 4, are not part of the proof.

$$p(2): 1 + 2 = \frac{2(2+1)}{2}$$

is true since

$$\frac{2(2+1)}{2} = \frac{2 \cdot 3}{2} = 3 \quad \text{and} \quad 1 + 2 = 3$$

$$p(3): 1 + 2 + 3 = \frac{3(3+1)}{2}$$

is true since

$$\frac{3(3+1)}{2} = \frac{3 \cdot 4}{2} = 6 \quad \text{and} \quad 1 + 2 + 3 = 6$$

$$p(4): 1 + 2 + 3 + 4 = \frac{4(4+1)}{2}$$

is true since
$$\frac{4(4+1)}{2} = 2(5) = 10$$
and
$$1 + 2 + 3 + 4 = 10$$

2. Prove $p(k) \Rightarrow p(k+1)$, or, in other words, if
$$1 + 2 + \cdots + k = \frac{k(k+1)}{2}$$
then
$$1 + 2 + \cdots + (k+1) = \frac{(k+1)[(k+1)+1]}{2}$$
or, in summation notation (see Appendix III),
$$\sum_{i=1}^{k} i = \frac{k(k+1)}{2} \Rightarrow \sum_{i=1}^{k+1} i = \frac{(k+1)(k+2)}{2}$$

Assume that $p(k)$ is true and prove that $p(k+1)$ must follow from this assumption. If
$$\sum_{i=1}^{k} i = \frac{k(k+1)}{2}$$
then
$$\sum_{i=1}^{k+1} i = \sum_{i=1}^{k} i + (k+1) \Rightarrow$$
$$\sum_{i=1}^{k+1} i = \frac{k(k+1)}{2} + (k+1) \Rightarrow$$
$$\sum_{i=1}^{k+1} i = \frac{k^2 + k}{2} + \frac{2k+2}{2} = \frac{k^2 + 3k + 2}{2} \Rightarrow$$
$$\sum_{i=1}^{k+1} i = \frac{(k+1)(k+2)}{2}$$
Therefore,
$$\sum_{i=1}^{k+1} i = \frac{(k+1)(k+2)}{2}$$
provided
$$\sum_{i=1}^{k} i = \frac{k(k+1)}{2}$$

We have shown that the formula, $n(n+1)/2$ for the sum of any n consecutive natural numbers, $1 + 2 + 3 + \cdots + n$, works for $n = 1$ and that if it works for any particular number, $n = k$, then it will work for its successor, $n = k + 1$. These *two* statements prove the formula works for all natural numbers. If it is true for $n = 1$, it must be true for $n = 2$, since if it works for any number, it must work for its successor. Now that it is true for $n = 2$, it must be true for $n = 3$, etc.

EXAMPLE II. Prove the following for all $n \in$ naturals.
$$p(n): (1 + x)^n \geq nx + 1 \quad \text{for } x \geq 0$$

Proof:
1. Prove $p(1): (1 + x)^1 \geq 1 \cdot x + 1$ is true since
$$1 + x \geq x + 1 \quad \text{for } x \geq 0$$
2. Prove $p(k) \Rightarrow p(k+1)$, that is,
$$(1+x)^k \geq kx + 1 \Rightarrow (1+x)^{k+1} \geq (k+1)x + 1$$
Assume $p(k)$, that is, that $(1+x)^k \geq kx + 1$. Then, prove that $p(k+1)$ follows from this assumption.
$$(1+x)^{k+1} = (1+x)^k (1+x) \Rightarrow$$
$$(1+x)^{k+1} \geq (kx+1) \cdot (1+x)$$
since, by hypothesis,
$$(1+x)^k \geq (kx + 1)$$
Therefore,
$$(1+x)^{k+1} \geq kx + kx^2 + 1 + x \Rightarrow$$
$$(1+x)^{k+1} \geq kx + x + kx^2 + 1 \Rightarrow$$
$$(1+x)^{k+1} \geq (k+1)x + kx^2 + 1 \geq (k+1)x + 1$$
since
$$kx^2 \geq 0$$
Hence
$$(1+x)^{k+1} \geq (k+1)x + 1$$
provided that
$$(1+x)^k \geq kx + 1$$
This, in addition to the fact that $p(1)$ is true, verifies $p(n)$.

EXAMPLE III. Prove the following formula for the nth partial sum of a geometric series.

$$p(n): \sum_{i=1}^{n} r^{i-1} = \frac{1-r^n}{1-r}$$

1. $p(1): r^{i-1} = r^0 = 1$ and $\dfrac{1-r^1}{1-r} = 1$
2. $p(k) \Rightarrow p(k+1)$

$$\sum_{i=1}^{k} r^{i-1} = \frac{1-r^k}{1-r} \Rightarrow \sum_{i=1}^{k+1} r^{i-1} = \frac{1-r^{k+1}}{1-r}$$

Proof:

$$\sum_{i=1}^{k+1} r^{i-1} = \sum_{i=1}^{k} r^{i-1} + r^{(k+1)-1} \Rightarrow$$

$$\sum_{i=1}^{k+1} r^{i-1} = \frac{1-r^k}{1-r} + r^k \Rightarrow$$

$$\sum_{i=1}^{k+1} r^{i-1} = \frac{1-r^k}{1-r} + \frac{r^k - r^{k+1}}{1-r} \Rightarrow$$

$$\sum_{i=1}^{k+1} r^{i-1} = \frac{1-r^{k+1}}{1-r}$$

Now, let's consider a statement that is not true and show how mathematical induction may break down.

EXAMPLE IV. Prove that $n^2 + n + 41$ generates all prime numbers.

Proof:
1. Prove $p(1)$.
$$1 + 1 + 41 = 43$$
which is prime.
2. $p(k) \Rightarrow p(k+1)$ cannot be proved since the theorem is false. To prove it false we only need to find one counterexample—one value of n for which $n^2 + n + 41$ is not a prime number. Can you find one?

EXERCISE 2.12

A. FUNDAMENTALS
Prove $p(1)$, $p(2)$, $p(3)$, and $p(4)$ for each of the following propositions.

⋆1. $p(n): 1 + 2 + \cdots + n = \dfrac{n(n+1)}{2} \Leftrightarrow$

$$p(n): \sum_{i=1}^{n} i = \frac{n(n+1)}{2}$$

2. $p(n): 1 + 3 + 5 + \cdots + (2n-1) = n^2 \Leftrightarrow$

$$p(n): \sum_{i=1}^{n} (2i-1) = n^2$$

⋆3. $p(n): 1^2 + 2^2 + \cdots + n^2 = \dfrac{n(n+1)(2n+1)}{6} \Leftrightarrow$

$$p(n): \sum_{i=1}^{n} i^2 = \frac{n(n+1)(2n+1)}{6}$$

4. $p(n): 1^3 + 2^3 + \cdots + n^3 = \dfrac{n^2(n+1)^2}{4} \Leftrightarrow$

$$p(n): \sum_{i=1}^{n} i^3 = \frac{n^2(n+1)^2}{4}$$

5. $p(n): 1^3 + 2^3 + \cdots + n^3$
$= (1 + 2 + 3 + \cdots + n)^2 \Leftrightarrow p(n): \sum_{i=1}^{n} i^3 = \left(\sum_{i=1}^{n} i \right)^2$

B. ESSENTIALS
Use mathematical induction to prove each of the following.

⋆6. $p(n): \sum_{i=1}^{n} i = \dfrac{n(n+1)}{2}$ (See Problem 1.)

7. $p(n): \sum_{i=1}^{n} (2i-1) = n^2$ (See Problem 2.)

⋆8. $p(n): \sum_{i=1}^{n} i^2 = \dfrac{n(n+1)(2n+1)}{6}$
(See Problem 3.)

9. ★(a) $p(n)$: $\sum_{i=1}^{n} \frac{1}{2^{i-1}} = \frac{2^n - 1}{2^{n-1}}$

(b) $p(n)$: $\sum_{i=1}^{n} \frac{2}{4^{i-1}} = \frac{2^{2i+3} - 2^5}{3(2^{2i})}$

10. (a) $p(n)$: $\sum_{i=1}^{n} \frac{1}{3^{i-1}} = \frac{3^n - 1}{2(3^{n-1})}$

(b) $p(n)$: $\sum_{i=1}^{n} \frac{1}{x^{i-1}} = \frac{x^n - 1}{(x-1)(x^{n-1})}$

11. Prove that $n^2 + n + 41$ generates prime numbers for all $n \in$ naturals.

12. Prove the following for all $m, n \in$ naturals by mathematical induction. (HINT: Induct on n only.)
(a) $x^m \cdot x^n = x^{m+n}$
(b) $x^m \cdot x^n = x^{m+n} \Rightarrow (x^m)^n = x^{m \cdot n}$

13. Prove:
(a) $n + 1 < 3n$ for all $n \in$ naturals
(b) $n < 1 + n^2$ for all $n \in$ naturals

14. ★(a) 3 divides $n^3 - n + 3$ evenly for all $n \in$ naturals
(b) 4 divides $7^n - 3^n$ evenly for all $n \in$ naturals

15. Prove
(a) $a^{2n} - b^{2n}$ is divisible by $a + b$ for all $n \in$ naturals
(b) $a^{2n-1} + b^{2n-1}$ is divisible by $a + b$ for all $n \in$ naturals

2.13 HISTORICAL COMMENTS ON SETS AND LOGIC

Set theory, as a body of mathematics, concerns infinite and finite sets. In 1851 Bolzano laid the foundation for the theory of sets by considering infinite sets and defining a "one-to-one correspondence" between the elements of two sets. Infinite sets were baffling to nineteenth century mathematicians because an infinite set can be put into a one-to-one correspondence with one of its subsets.

Georg Cantor, a Russian-born German, created set theory. A *set* is a collection of well-defined, distinguishable objects regarded as a single entity. His work with infinite sets was viewed variously as "a disease from which humanity will recover" and "fog on fog" to "a treatise on paradise" and "the greatest (work) of his age."

An example of Cantor's work is how he created a one-to-one correspondence between the integers and the rational numbers. If we list the rationals as illustrated below, and create the correspondence as indicated by the small numbers, each integer will be paired with exactly one rational number, and every rational number will be paired with an integer, even though all of the integers will appear in the top row of the array!

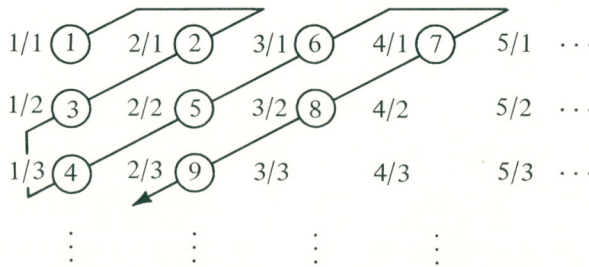

The seventeenth century, with men like Fermat, Descartes, and Euler, was a period of brilliant and somewhat haphazard discovery. The volume of new material created a need, indeed a demand, for increased precision and rigor. The eighteenth and nineteenth centuries saw mathematicians raising strong doubts about the validity of traditional mathematical reasoning. As early as 1666, Leibniz, in what he called his "school-boy essay," imagined a kind of "universal mathematics" that would give the structure and symbolism necessary to completely eliminate errors of reasoning. From then into the 1690s he spent time trying to create a symbolic logic. His work contained the concepts of disjunction, conjunction, negation, and equivalence in an algebraic context.

Before Leibniz introduced these concepts, mathematicians were analyzing each other's work according to the rules of logic of Aristotle.

Leibniz created the law of contradiction (a proposition cannot be both true and false) and the law of the excluded middle (a proposition must be either true or false). He developed the separate, systematic discipline of logic using mathematics as the basis, emphasizing that deductive proof is the only basis for establishing facts.

But mathematicians needed more. They sought to provide mathematics with a rigorous foundation upon which their ideas could be evaluated. A somewhat more effective step than Leibniz's was taken in 1847 by DeMorgan, who introduced quantified propositions. This was a major improvement over the narrow "to be" ("p is q" is true or false) relationships of Aristotle.

About the same time, George Boole, the father of symbolic logic, presented his work. Together, DeMorgan and Boole lead the reformation of Aristotelian logic toward a science of logic attached to mathematics.

Gottlob Frege (1848–1925) introduced many distinctions, the importance of which were first recognized by Bertrand Russell. For example, he stated that p implies q means either p is true and q is true or p is false.

The ideas outlined in Bertrand Russell's *Principles of Mathematics* of 1903, and detailed in Russell and Whitehead's *Principia Mathematica* (1910–1913), advanced symbolic logic by detailing a thorough treatment of logic in symbolic form, axiomatically. They introduced $\sim p$, $p \vee q$, $p \Rightarrow q$ (written then as $p > q$) with propositions and conditional propositions. They listed some elementary postulates from which the Aristotelian syllogisms and the principles of arithmetic and analysis are deduced as theorems. An example of one of their theorems is the principle of *reductio ad absurdum*, which states that if assuming p true results in p false, then p is false $[(p \Rightarrow \sim p) \Rightarrow \sim p]$.

They and their contemporaries thought that the goal of founding mathematics on formal rules of logic had been achieved, but in the early 1900s Kurt Gödel showed that it is impossible to construct a system embracing usual logic and *any* major branch of mathematics. He showed that no matter how one establishes a logical system there will be a true statement of the mathematics that cannot be proved.

Hilbert, von Neumann, and others developed Hilbert's proof theory from 1920 to 1930, yielding a method of establishing the consistency of any formal system where the consistency of number theory is central.

ENDNOTE

1. Mathematical induction was formally introduced by Maurolycus in 1575, but Euclid used it implicitly in his proof of the infinite number of primes. He showed that if there are n primes, then there must be $n + 1$ primes. Then, since there is one prime, there must be an infinite number. Maurolycus used math induction to prove

$$1 + 3 + \cdots + (2n - 1) = (n + 1)^2$$

for all natural numbers n.

PART II

Elementary Analysis of Functions

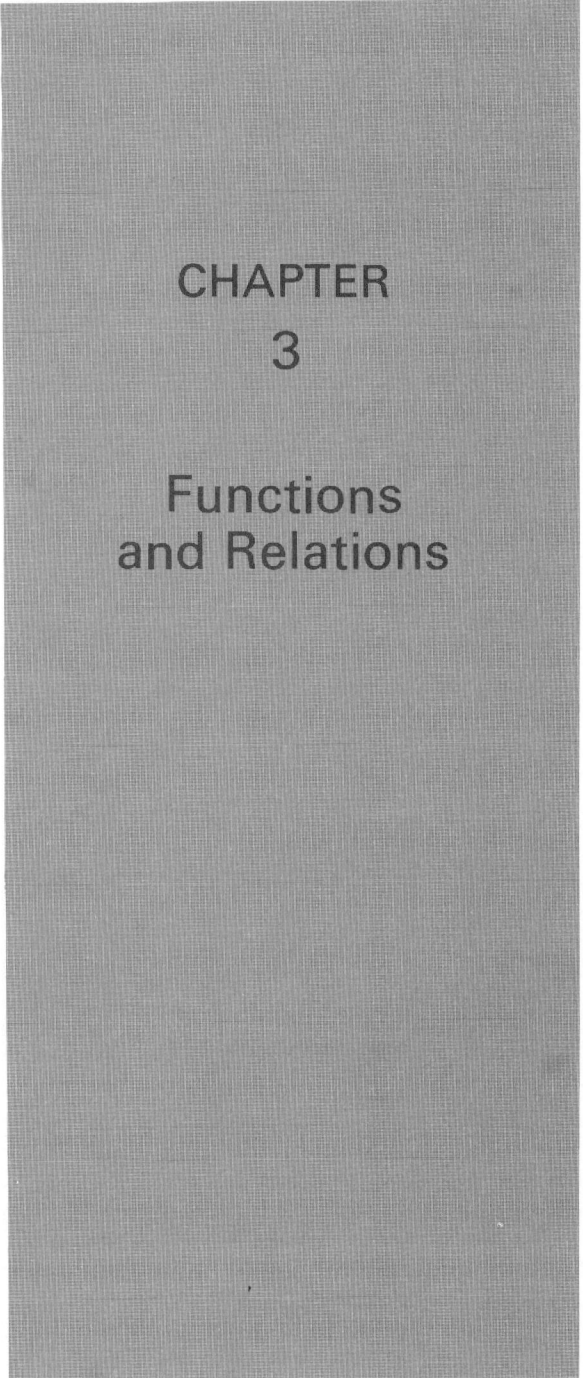

CHAPTER 3

Functions and Relations

3.1 INTRODUCTION

Studies of relationships between varying quantities are conducted in many disciplines in addition to mathematics. Psychologists are concerned with the relationship between behavioral characteristics and environment and heredity. Economists relate the quantity consumers are willing to buy to the per unit price. Physicists study numerous relationships, for example, the pressure of a gas as a function of temperature, the velocity of a falling object versus the time it has been falling, or how the force of the gravitational attraction between two objects is related to their masses and the distance between them. At first, our mathematical study of relationships will be restricted to two quantities (variables), one of which will usually have its values dependent upon the values chosen for the other variable.

We will therefore be able to classify the two variables as either *dependent* or *independent*. This is a causal relationship since changing values of the independent variable will *cause* a corresponding change in its dependent variable. In the relationships mentioned, notice that behavior is dependent on heredity and environment, the quantity purchased is dependent upon the price, etc. The role of independence and dependence is usually determined by the situation and may be completely reversed in two differing situations. For example, one day a physicist may be changing the temperature of an enclosed gas and measuring the resulting pressures. This means the pressure is dependent upon the temperature. The next day he may alter the pressure and measure the resulting temperatures. This means the temperature is dependent upon the pressure.

In studying such correspondences between two variables, it is usually necessary to establish whether or not the relationship is single-valued. Single-valued relationships are those that pair exactly one value of the dependent variable with each allowable value of the independent variable. Such relations are called **functions**. They may be defined by a formula, by a set of ordered pairs,

65

CHAPTER 3 Functions and Relations

or by a chart or graph of the relationship involved.

Since many definitions and theorems hold only for functions, we will now study functions in great detail.

3.2 FUNCTIONS

An easy way to remember the basic definitions, related notation, terminology, and properties of functional relationships between two variables is to keep the following three simple examples in mind. If you are not already completely familiar with the relations $y = x^2$, $y = x^3$, and $y^2 = x$, draw their graphs carefully—*before proceeding!* These functions are very common and we will be seeing them frequently in examples throughout the text. The curves are sketched in Figure 3.1.

These curves represent three markedly different relationships between the dependent variable (y) and the independent variable (x). The relationship $y = x^2$ is single-valued since each value of x generates a unique value of y. The relationship is not, however, a one-to-one correspondence between the two variables because two values of x generate the same value of y: $x = 2$ and $x = -2$ both yield $y = 4$. The relation $y = x^3$ is not only single-valued but it is also a one-to-one correspondence between the values of x and those related values of y. That is, not only for each value of x is there assigned exactly one value of y, but for each value of y there corresponds exactly one value of x ($x = 2 \rightarrow y = 8$ and $y = 8 \rightarrow x = 2$). The relation $y^2 = x$ is not single-valued, since one value of x usually generates two different values of y: $x = 4$ corresponds to $y = 2$ or $y = -2$. The only exception is $x = 0$, which yields only $y = 0$.

These differences are summarized in the following definitions for functions and relations involving only one independent and one dependent variable. The set of permissible values of the independent variable, which we will label X, is called the **domain**; the set of resulting values for the dependent variable, labeled Y, is called the **range**. Often both the range and domain are all real numbers.

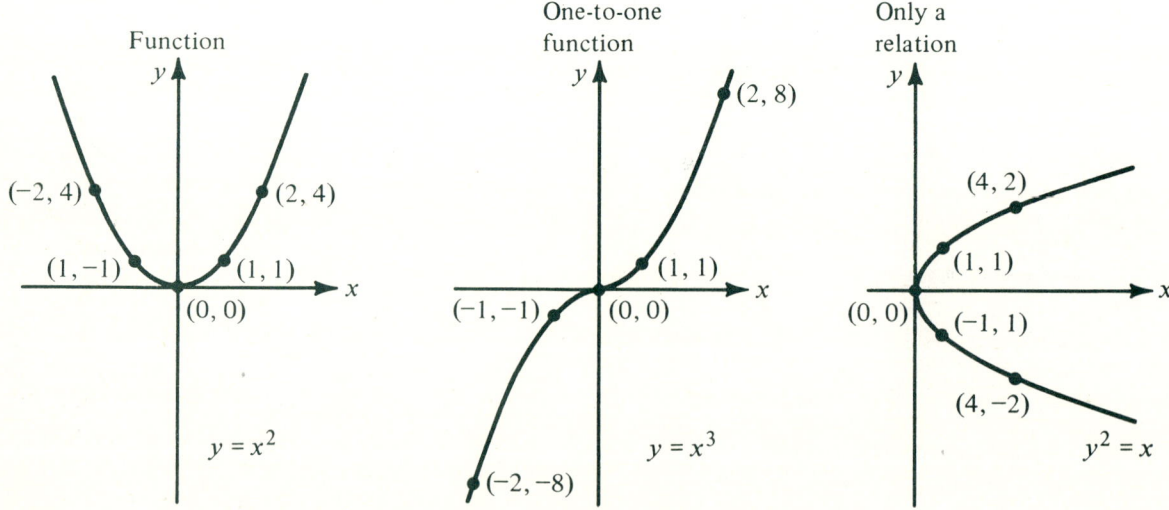

FIGURE 3.1. *Three Important Relations.*

SECTION 3.2   Functions

> **DEFINITION 3.1.**
> A *relation* is any well defined pairing of the elements of one set with the elements of a second set.

This pairing can be described by listing a set of ordered pairs whose first elements are from one set and whose second elements are from the other set.

EXAMPLE I.   For

$$X = \{a, b\} \quad \text{and} \quad Y = \{1, 2, 3\}$$

the set

$$R = \{(a, 1), (b, 2), (b, 3)\}$$

establishes a relationship between sets X and Y by pairing a from X with 1 from Y and b from X with both 2 and 3 from Y. Therefore, R is called a *relation* from X to Y.

Such a pairing can also be described by stating a rule of correspondence between the two sets, usually given as an equation involving two variables.

EXAMPLE II.   If

$$X = \{3, 6\} \quad \text{and} \quad Y = \{-1, 5\}$$

then the equation

$$y = 2x - 7$$

establishes a relationship between sets X and Y with $x = 3$ being paired with $y = -1$ and $x = 6$ with $y = 5$. Notice that this equation generates the set of ordered pairs $\{(3, -1), (6, 5)\}$, if the values of x are restricted to 3 or 6.

In each of the previous examples, set X, the first set, is called the *domain*, and the set of corresponding elements in the second set, Y, is called the *range*.

> **DEFINITION 3.2.**
> A *function* is a relation between two sets, X and Y, for which each $x \in X$ is paired with a *unique* $y \in Y$.

> **DEFINITION 3.3.**
> A *one-to-one function* is a function between two sets, X and Y, for which each $y \in Y$ is paired with a unique $x \in X$.

Figure 3.2 depicts Definitions 3.1, 3.2, and 3.3.

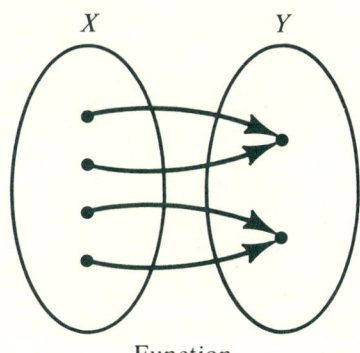
Function

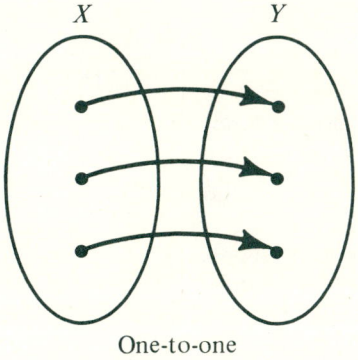
One-to-one function

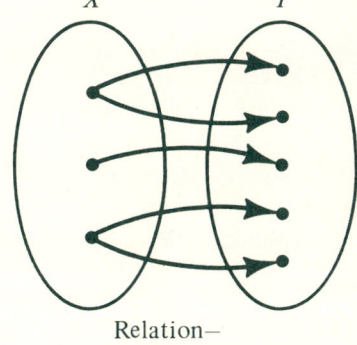
Relation— not a function

FIGURE 3.2.

CHAPTER 3 Functions and Relations

A function may pair several x's with the same y, but each x with only one y. In a one-to-one function, each x is paired with one y and each y with one x. A relation that is not a function[1] must pair some x with more than one y.

Examine the graphs of $y = x^2$, $y = x^3$, and $y^2 = x$ (Figure 3.1) to determine their domains and ranges. Then complete the following:

Domain of $y = x^2$ is $\{x | \underline{\hspace{2cm}}\}$

Range of $y = x^2$ is $\{y | y \geq 0\}$

Domain of $y = x^3$ is $\{x | x \in \text{reals}\}$

Range of $y = x^3$ is $\{y | \underline{\hspace{2cm}}\}$

Domain of $y^2 = x$ is $\{x | x \geq 0\}$

Range of $y^2 = x$ is $\{y | \underline{\hspace{2cm}}\}$

Study the domains of the following functions, which depict the most common restrictions on x.

EXAMPLE III. Find the domain of the functions in (b).

(a) $y = \dfrac{1}{x + 1}$, domain $\{x | x \neq -1\}$

(b) $y = \dfrac{3x + 2}{x^2 - 3x - 4}$,
domain $\{x | x \neq \underline{\;?\;} \text{ and } x \neq \underline{\;?\;}\}$

(c) $y = \sqrt{2x - 4}$, domain $\{x | x \geq 2\}$

The restrictions in (b) are $x \neq -1$ and $x \neq 4$.

An easy way to determine the range of a function is to examine its graph. In Example III(c), the range is $\{y | y \geq 0\}$, since $\sqrt{}$ means the principle or nonnegative square root.

Functional notation is extremely convenient, once we get accustomed to using it. We will often replace y with $f(x)$, read "f of x," to emphasize that the y-value is dependent upon the value of x. Then $f(x)$ is the value of the function f, at x. This is a shorthand notation since, in order to say, "the y-value associated with $x = 3$ in $y = 2x - 4$ is 2," we say,

$$f(x) = 2x - 4 \Rightarrow$$
$$f(3) = 2$$

Complete Example IV by simply replacing x with the value indicated ($f(5)$ means replace x with 5) to find the corresponding y-value of the function.

EXAMPLE IV.
(a) $f(x) = 3x - 2$
$f(1) = 3(1) - 2 = \underline{\;?\;}$
(b) $f(x) = 5x^2 - 2x - 1$
$f(2) = 5(2)^2 - 2(2) - 1 = \underline{\;?\;}$
(c) $f(x) = 3x^2 - 2x - 2$
$f(2) = 3(?)^2 - 2(?) - 2 = \underline{\;?\;}$
(d) $f(x) = x\sqrt{x + 1}$
$f(?) = 3\sqrt{3 + 1} = \underline{\;?\;}$

You may use any letters for the names of functions, but f, g, and h are commonly used in mathematics.

Our immediate objectives in the study of functions are (1) to become totally familiar with the notation and terminology of functions, (2) to understand the reason for making the definitions and developing the algorithms of functions, and (3) to make graphing functions a relatively easy, intuitive process with minimum reliance on the plotting of points and the arithmetic associated with point plotting.

EXERCISE 3.2

A. FUNDAMENTALS
Determine the y-value indicated for each of the following functions. Remember that $f(1)$ is the y-value for $x = 1$.
1. $f(x) = x^2 - x$
    (a) $f(1)$ (b) $f(0)$

2. $g(x) = \sqrt{1-x}$
   (a) $g(1)$ (b) $g(-3)$

3. $h(t) = 2t^3 - t + 4$
   (a) $h(s)$ (b) $h(-t)$

4. $f(x) = 3x$
   (a) $f(4x)$ (b) $f(5t)$

5. $h(x) = 3x^2$
   (a) $h(4x)$ (b) $h(5t)$

6. $g(t) = 2t^2 - t + 1$
   (a) $g(x+2)$ (b) $g(x-1)$

Decide which of the following relations are functions. The domains are all real numbers for which the equations are defined and real-valued.

7. (a) $y = x^2 + 1$ (b) $y = x^3 + 1$
   (c) $y = \sqrt{x} + 1$

8. (a) $y = \pm\sqrt{x}$ (b) $y^2 = x$
   (c) $y^2 = x - 2$

9. (a) $y \leq x$ (b) $y > 2x^2 - 1$
   (c) $y = 2x^2 - 1$

10. (a) $x^2 + y^2 = 4$
    (b) $9x^2 - 4y^2 = 36$

★11. (a) $y^2 = 1 - x$ (b) $|y| = \sqrt{1-x}$

Determine which of the following functions are one-to-one functions:

12. (a) $f(x) = x$ (b) $g(x) = x^3$
    (c) $h(x) = x^5$

13. (a) $y = x^2$ (b) $y = x^4$
    (c) $y = x^4 + 2x^2$

14. (a) $y = |x|$ (b) $f(x) = |x-1|$

B. ESSENTIALS

Find the domains of the following functions. Consider only the real numbers.

15. (a) $f(x) = \sqrt{x}$ (b) $g(x) = \sqrt{x+4}$
    (c) $h(x) = \sqrt{3-2x}$

16. (a) $f(x) = \sqrt{x^2+1}$
    (b) $g(x) = \sqrt{x^2-1}$
    (c) $y = \sqrt{x^2-x-2}$

17. (a) $y = \dfrac{1}{x-1}$ (b) $f(x) = \dfrac{3}{2-x}$

18. (a) $f(t) = \dfrac{1}{t^2 - 3t - 4}$
    (b) $g(z) = \dfrac{1}{z^2 - z - 12}$

19. (a) $y = \sqrt{x}$ (b) $y = \dfrac{1}{x}$
    (c) $y = \dfrac{1}{\sqrt{x}}$

★20. (a) $f(x) = \dfrac{2}{\sqrt{x^2+4}}$
    (b) $g(x) = \dfrac{3}{\sqrt{x^2-x-2}}$
    (c) $y = \dfrac{1}{\sqrt{x+1}-2}$

C. APPLICATIONS

21. Examine the graph of $y = f(x)$, where x_0, and its corresponding y-value, $f(x_0)$, are pictured. Locate $x_0 + 4$ and its corresponding y-value, $f(x_0 + 4)$. Draw the *secant line* from $(x_0, f(x_0))$ to $(x_0 + 4, f(x_0 + 4))$. Find its slope.

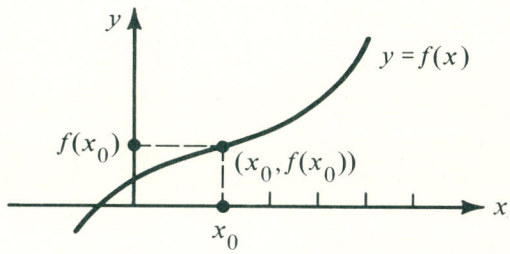

★22. For $f(x) = 2x + 1$, find
   (a) $f(x_0 + 4)$ (b) $f(x_0 + h)$
   (c) $\dfrac{f(x_0 + h) - f(x_0)}{h}$

Interpret (c) geometrically (see the illustration above).

CHAPTER 3 Functions and Relations

23. For $g(x) = 3 - 5x$, find

    (a) $g(x_0 + 4)$ (b) $\dfrac{g(x_0 + 4) - g(x_0)}{4}$

    (c) $\dfrac{g(x_0 + h) - g(x_0)}{h}$

24. For $f(x) = 2x^2 - x$, find
    (a) $f(2)$ (b) $f(3)$ (c) $f(5)$
    (d) $f(6)$

25. (a) For $f(x) = 2x$,
    (i) does $f(3 + 2) = f(3) + f(2)$?
    (ii) does $f(3 \cdot 2) = f(3) \cdot f(2)$?

    (b) Prove or disprove the following:
    (i) $f(a + b) = f(a) + f(b)$ for all functions f.
    (ii) $f(a \cdot b) = f(a) \cdot f(b)$ for all functions f.

26. (a) For $f(x) = -3x$, find $f(2x)$, $2f(x)$, $f(tx)$, and $tf(x)$.

    (b) Determine a class of functions for which $f(tx) = tf(x)$, where t is any real number.

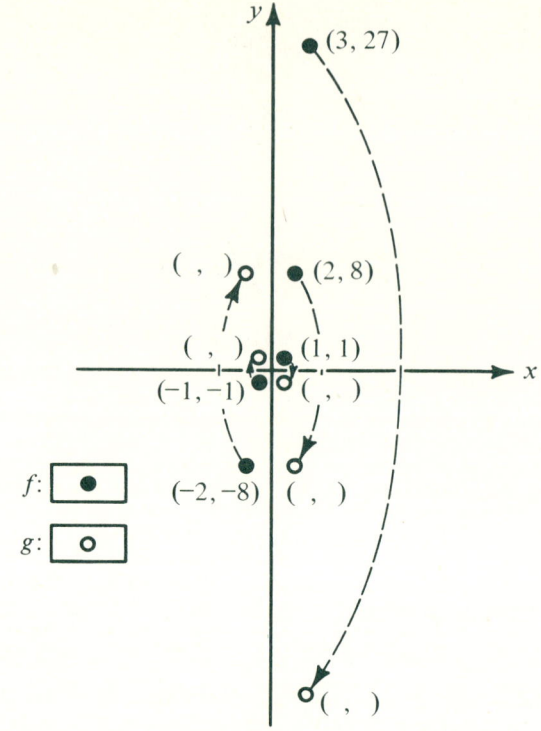

FIGURE 3.3.

3.3 CREATING NEW FUNCTIONS BY REFLECTION

We are now ready to study some general properties of functions and their graphs. We will use the following functions as examples:

$$y = x, \quad y = x^2, \quad y = x^3, \quad y = \sqrt{x}, \quad y = |x|$$

If you are not already completely familiar with each of these, refer to Section 1.11 before proceeding or you will not readily understand the following discussions.

First, we will learn how to create new functions from given ones in such a way that the *new* function will be predictably related to the original one.

Let's begin with the function f, defined by the following set of ordered pairs, and create a new function g, by negating the y-coordinates of the ordered pairs defining f. Then

$$f = \{(-2, -8), (-1, -1), (1, 1), (2, 8), (3, 27)\}$$

and the new function is

$$g = \{(-2, 8), (-1, 1), (1, -1), (2, -8), (3, -27)\}$$

By noticing that $f(3) = 27$ and $g(3) = -27$, $f(2) = 8$ and $g(2) = -8$, etc., we see that, in general,

$$g(x) = -f(x)$$

for all x in the domain of f. By plotting both sets of ordered pairs, we can see the geometric relationship between f and g. Label the points from g depicted in Figure 3.3.

Notice that each point of f was reflected about the x-axis when the sign of its y-coordinate was changed.

Let's repeat this method of creating a new function by changing the signs of the y-coordinates with a function defined by an equation, for example, $f(x) = x^2$. Then

$$g(x) = -f(x) \Rightarrow g(x) = -x^2$$

SECTION 3.3 Creating New Functions By Reflection

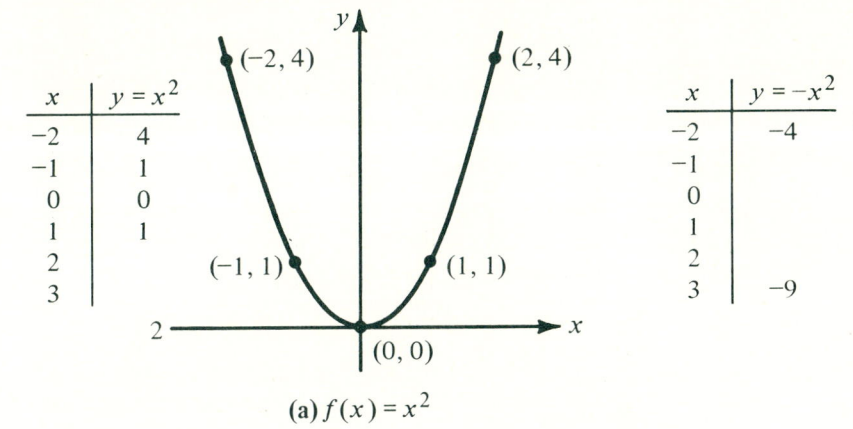

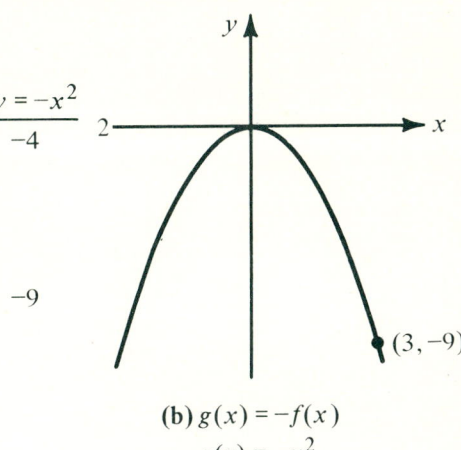

FIGURE 3.4.

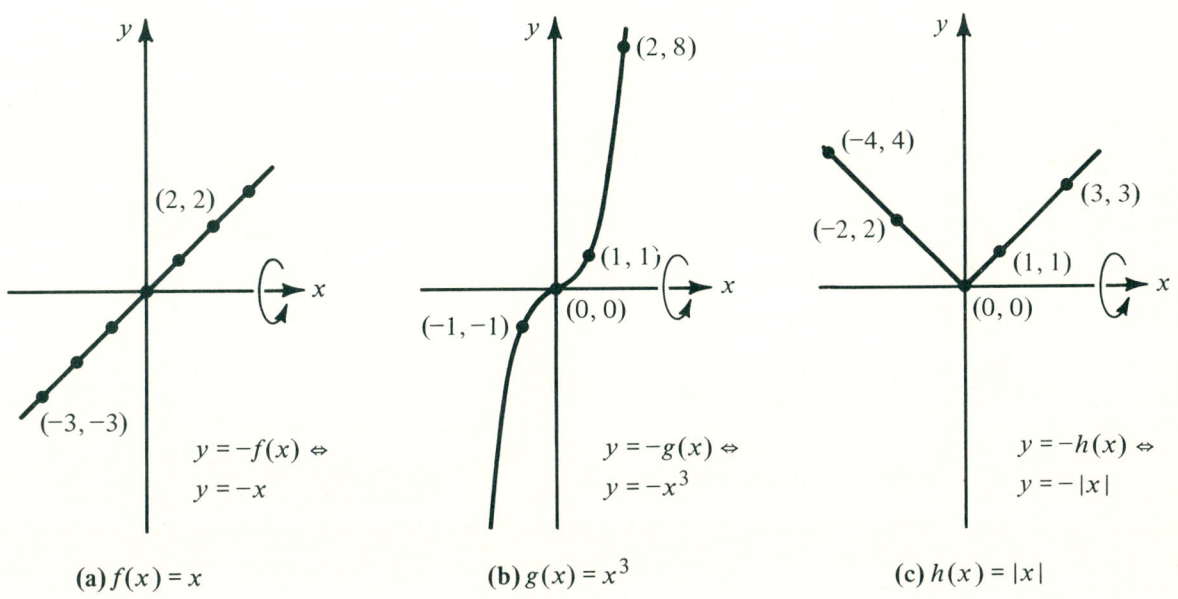

FIGURE 3.5.

The tables of some values (ordered pairs) and the resulting graphs are pictured in Figure 3.4.

Notice that $y = -x^2$ is simply $y = x^2$ reflected about the x-axis.

Now, draw the graph of $y = -f(x)$, $y = -g(x)$, and $y = -h(x)$ by reflecting the graph of each of the functions in Figure 3.5 about the x-axis. Your results for (a) and (b) should go up to the left side and down to the right side. For (c), the ∨ should become ∧.

Now, let's see what happens when we change the signs of the x-coordinates. Consider

$$f = \{(-2, 4), (-1, 1), (3, 9)\}$$

71

CHAPTER 3 Functions and Relations

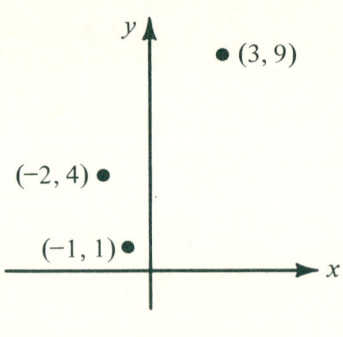

FIGURE 3.6.

and

$$g = \{(2, 4), (1, 1), (-3, 9)\}$$

By noticing that $f(-2) = 4$ and $g(2) = 4$, $f(-1) = 1$ and $g(1) = 1$, and $f(3) = 9$ and $g(-3) = 9$, we see that

$$g(x) = f(-x)$$

for each x in the domain of f. Geometrically, notice that changing the sign of the x-coordinate reflects the point about the y-axis. Plot the ordered pairs of function g in Figure 3.6. The ordered pairs for function f are already plotted.

Let's repeat this procedure of changing the sign of the x-coordinates for a function defined by an equation, for example, $f(x) = \sqrt{x}$, in order to create a new function g.

$$g(x) = f(-x) \Rightarrow$$
$$g(x) = \sqrt{-x}$$

The graphs are shown in Figure 3.7. Notice that $y = g(x)$ is simply $y = f(x)$ reflected about the y-axis. The domain of f is $x \geq 0$ and of g is $x \leq 0$.

Complete the graph of the new function $y = f(-x)$ by reflecting the graph of $y = f(x)$ about the y-axis for each of the functions in Figure 3.8.

The graphs of (a) and (b) go up on the left and down on the right. The graph of (c) does not result in a different formation, since $f(-x) = (-x)^2 = (-x)(-x) = x^2 = f(x)$.

Now, let's change the signs of both the x- and the y-coordinates. Can you guess what the geometric result will be? Let

$$f = \{(-2, -3), (-1, 2), (2, 4), (3, 5)\}$$

Then

$$g = \{(2, 3), (1, -2), (-2, -4), (-3, -5)\}$$

Since $f(-2) = -3$ and $g(2) = 3$, $f(-1) = 2$ and $g(1) = -2$, etc.,

$$g(x) = -f(-x)$$

for each x in the domain of f.

Describe the geometric result of changing the signs of both the x- and the y-coordinates of f by plotting the points of g on the system in Figure 3.9. The points for f are already plotted. Changing the signs of both the x- and y-coordinates of the ordered pairs composing the function f resulted in a new function whose plot could be obtained by reflecting the points from f about the origin.

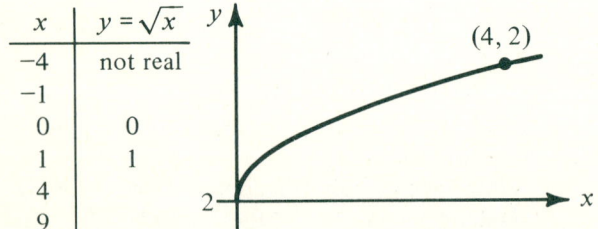
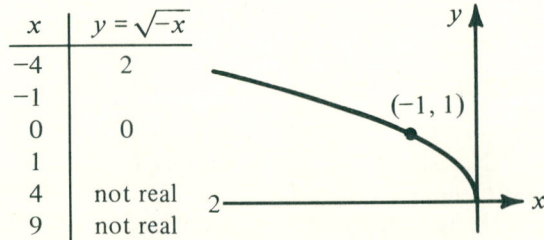

FIGURE 3.7.

SECTION 3.3 Creating New Functions By Reflection

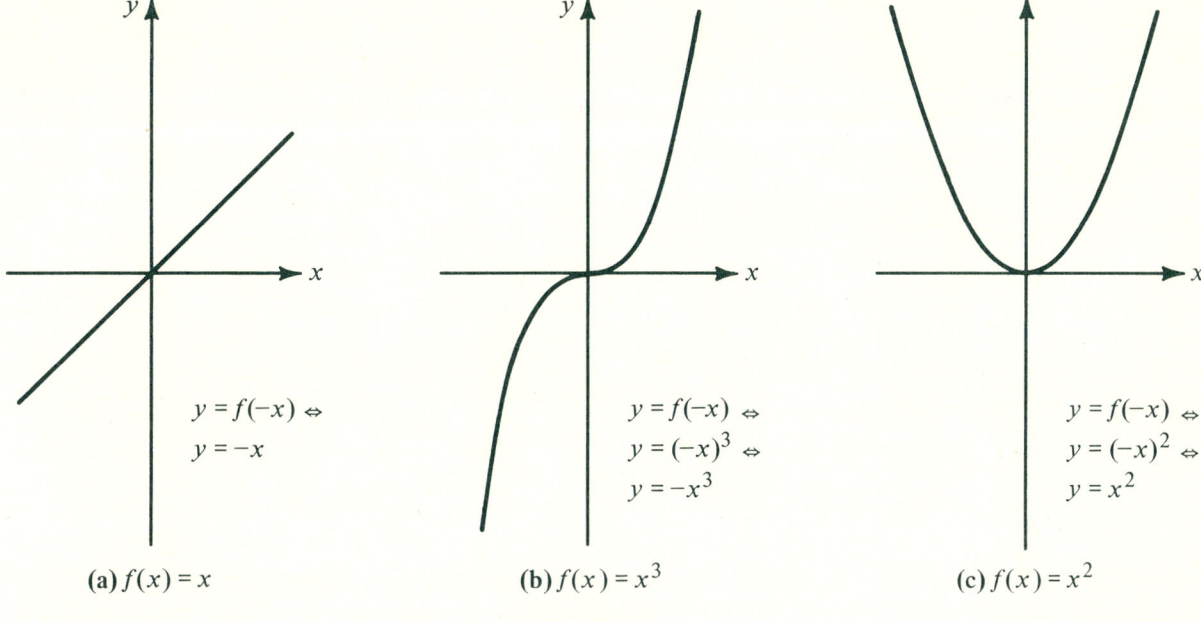

FIGURE 3.8.

Graph the new function $y = -f(-x)$ for the functions shown in Figure 3.10. Is $-f(-x)$ always different from $f(x)$? Your result for (b) should be the same as reflecting the curve about the x-axis,

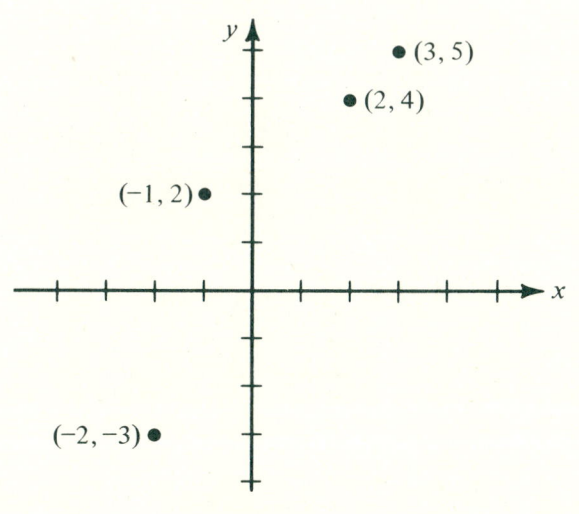

FIGURE 3.9.

since $-(-x)^2 = -x^2$. Your result for (c) should be the same as shown since $-(-x)^3 = x^3$.

A reflection about the origin can also be considered as two reflections, one about the x-axis and the other about the y-axis. This is consistent with the functional notation: since $y = f(-x)$ results in a reflection about the y-axis and $y = -f(x)$ a reflection about the x-axis, $y = -f(-x)$ will result in reflections about both axes. Consider $f(x) = \sqrt{x}$ in Figure 3.11(a). Then complete Figure 3.11(b) which should be a reflection of the graph of $y = f(x)$ about the y-axis, and Figure 3.11(c), which should be the reflection of Figure 3.11(b)) about the x-axis. The result should be the same as reflecting the first graph about the origin.

Repeat the two reflections in the reverse order: about the x-axis first, then about the y-axis. You should get the same final graph.

The results of our discussions on creating new functions as reflections of known ones are summarized below. The three proofs require using

CHAPTER 3 Functions and Relations

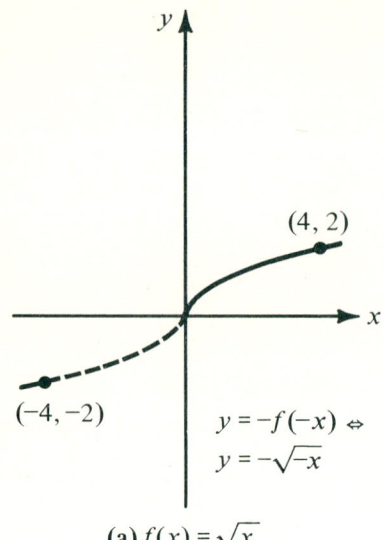

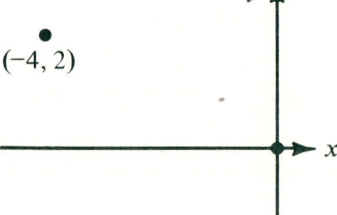

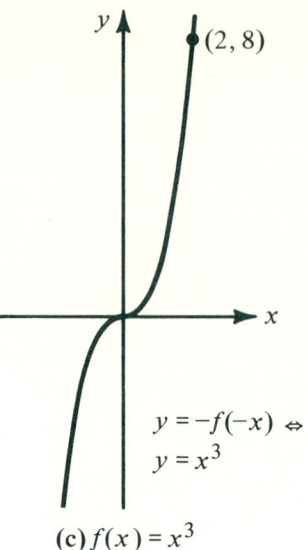

FIGURE 3.10.

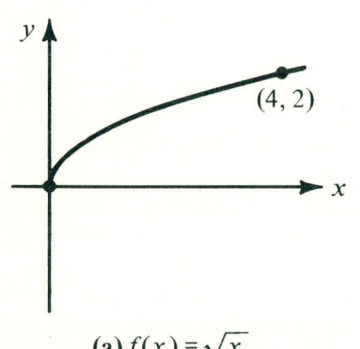

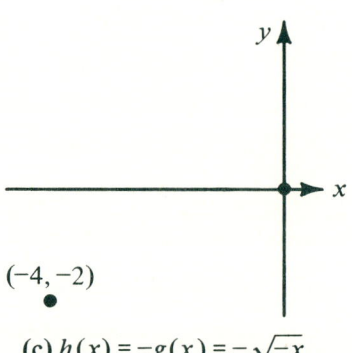

FIGURE 3.11.

plane geometry to establish symmetry about the x-axis for points (a, b) and $(a, -b)$, about the y-axis for points (a, b) and $(-a, b)$, and about the origin for points (a, b) and $(-a, -b)$.

> The above graphs of the following functions are obtained from the graph of $y = f(x)$ as follows:
> (i) For $y = -f(x)$, reflect f about the x-axis.
> (ii) For $y = f(-x)$, reflect f about the y-axis.
> (iii) For $y = -f(-x)$, reflect f about the origin.

EXERCISE 3.3

A. FUNDAMENTALS

Name the reflection necessary to obtain the graph of the following functions from the graph $f(x) = x$, $g(x) = x^2$, or $h(x) = \sqrt{x}$

1. (a) $y = f(-x) = -x$
   (b) $y = g(-x) = (-x)^2 = ?$
   (c) $y = h(-x) = ?$

74

SECTION 3.3 Creating New Functions By Reflection

2. (a) $y = -f(x) = -x$
   (b) $y = -g(x) = $?
   (c) $y = -h(x) = $?

3. (a) $y = -f(-x) = -(-x) = $ __?__
   (b) $y = -g(-x) = -(-x)^2 = $ __?__
   (c) $y = -h(-x) = $ __?__

Consider the function defined by the following sets of ordered pairs: $f = \{(-1, -4), (0, 7), (2, 5), (+1, 4), (-2, 5)\}$. Find the missing coordinate in each of the following:

4. (a) $f(2) = $ __?__   (b) $-f(2) = $ __?__
5. (a) $f(\_?\_) = 7$ (b) $f(\_?\_) = 4$
6. (a) $f(2) = $ __?__   (b) $f(-2) = $ __?__
7. (a) $-f(\_?\_) = -4$
   (b) $-f(-(\_?\_)) = -4$

B. ESSENTIALS
Find the missing value in each of the following:

8. (a) $f(x) = x^2$ and
       $g(x) = -f(x) \Rightarrow g(2) = $ __?__
   (b) $f(x) = x^3$ and
       $g(x) = f(-x) \Rightarrow g(-1) = $ __?__
   (c) $f(x) = x$ and
       $g(x) = -f(-x) \Rightarrow g(-4) = $ __?__
   (d) $f(x) = x$ and
       $g(x) = -f(x) \Rightarrow g(\_?\_) = 4$
   (e) $f(x) = x$ and
       $g(x) = -f(-x) \Rightarrow g(\_?\_) = -3$

★9. (a) $f(2) = 3$ and
       $g(x) = -f(x) \Rightarrow g(2) = $ __?__
   (b) $f(2) = 3$ and
       $g(x) = f(-x) \Rightarrow g(\_?\_) = 3$
   (c) $f(2) = 3$ and
       $g(x) = -f(-x) \Rightarrow g(?) = $ __?__

Find the set of ordered pairs that define the new function $y = g(x)$, where

10. $f = \{(1, 2), (2, 3), (-2, 4)\}$
    and $g(x) = -f(x)$

11. $f = \{(-1, 2), (1, 2), (2, 4), (-2, 4)\}$
    and $g(x) = f(-x)$

12. $f = \{(0, 3), (3, 2), (4, 5)\}$
    and $g(x) = -f(-x)$

Graph the new functions $y = f(-x)$, $y = -f(x)$, and $y = -f(-x)$, where f is defined by the following graphs:

13.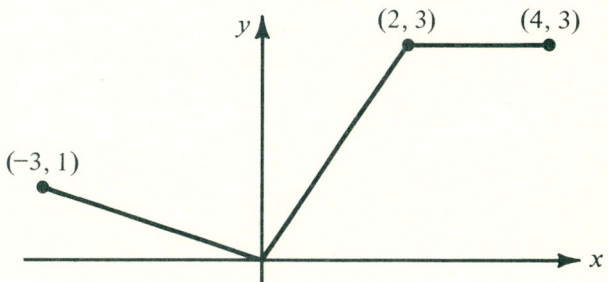

14.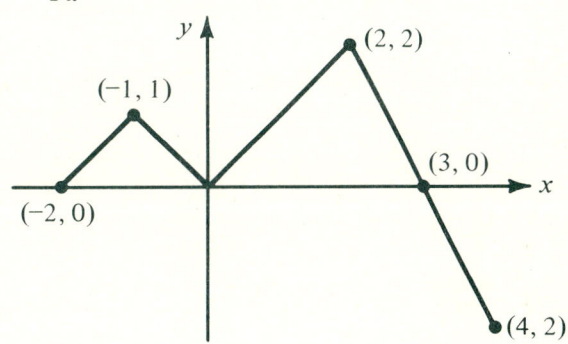

15.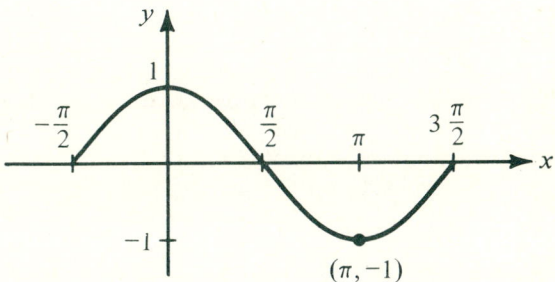

16.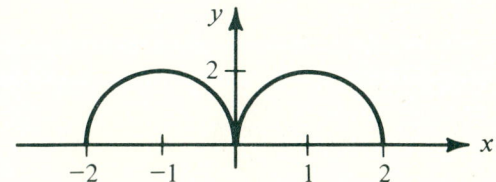

17.

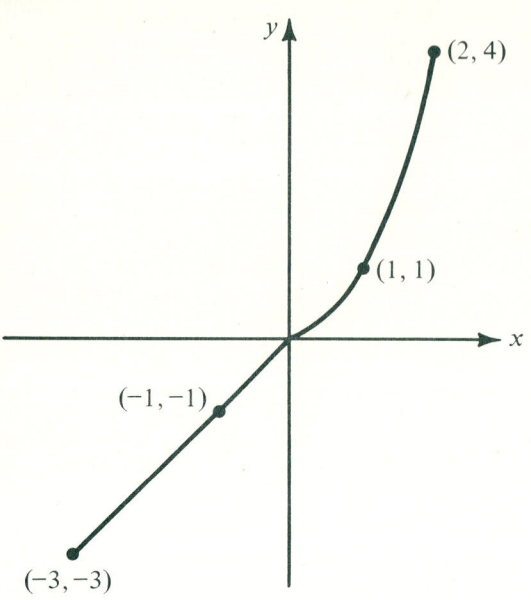

18.

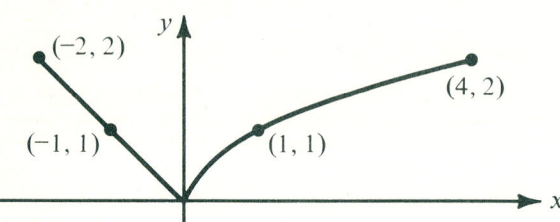

For each of the following problems, draw the three graphs on one coordinate-axis system.

19. (a) $f(x) = x^2$ (b) $y = -f(x)$
    (c) $y = f(-x)$

20. (a) $g(x) = x^3$ (b) $y = -x^3$
    (c) $y = -g(-x)$

21. (a) $h(x) = |x|$ (b) $y = -|x|$
    (c) $y = -h(-x)$

★22. (a) $f(x) = \sqrt{x}$ (b) $y = f(-x)$
    (c) $y = -f(-x)$

3.4 SYMMETRY—EVEN AND ODD FUNCTIONS (OPTIONAL)

For some functions, such as

$$f = \{(-2, 4), (-1, 1), (0, 0), (1, 1), (2, 4)\}$$

and

$$g = \{(-2, -8), (-1, -1), (1, 1), (2, 8)\}$$

changing the signs of the x-coordinates or of both coordinates may not yield a new function. Consider changing the signs of the x-coordinates of the ordered pairs of f:

$$f' = \{(2, 4), (1, 1), (0, 0), (-1, 1), (-2, 4)\}$$

and the signs of both coordinates of the ordered pairs of g:

$$g' = \{(2, 8), (1, 1), (-1, -1), (-2, -8)\}$$

This shows that

$$f(-x) = f(x)$$

and

$$-g(-x) = g(x)$$

Notice that there is a certain symmetry in each of the sets of ordered pairs. Recall that, in general, letting $y = f(-x)$ results in reflecting the graph of $y = f(x)$ about the y-axis and that letting $y = -g(-x)$ reflects the graph of $y = g(x)$ about the origin. Plotting the points comprising f and g will reveal why we did not get new functions by changing the signs. Complete Figure 3.12.

The set of points comprising f is symmetric about the y-axis. The set of points comprising g is symmetric about the origin.

Complete the graphs of the functions defined by the equations in Figure 3.13. Notice that $f(x) = x^2$ and $f(x) = |x|$ are both symmetric about the y-axis. Also,

$$f(x) = x^2 \Rightarrow f(-x) = (-x)^2 = x^2 \Rightarrow$$
$$f(x) = f(-x)$$

SECTION 3.4   Symmetry—Even and Odd Functions (Optional)

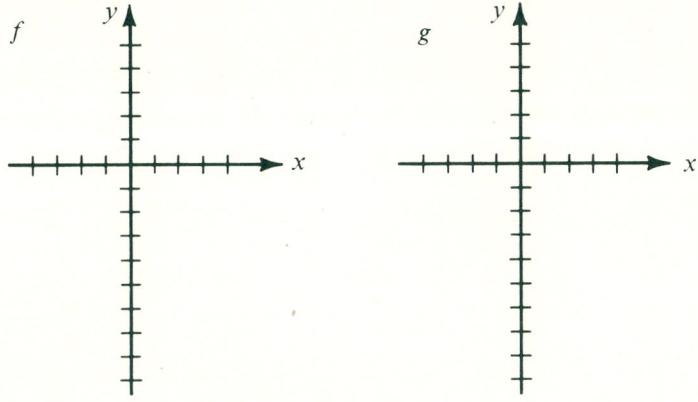

FIGURE 3.12.

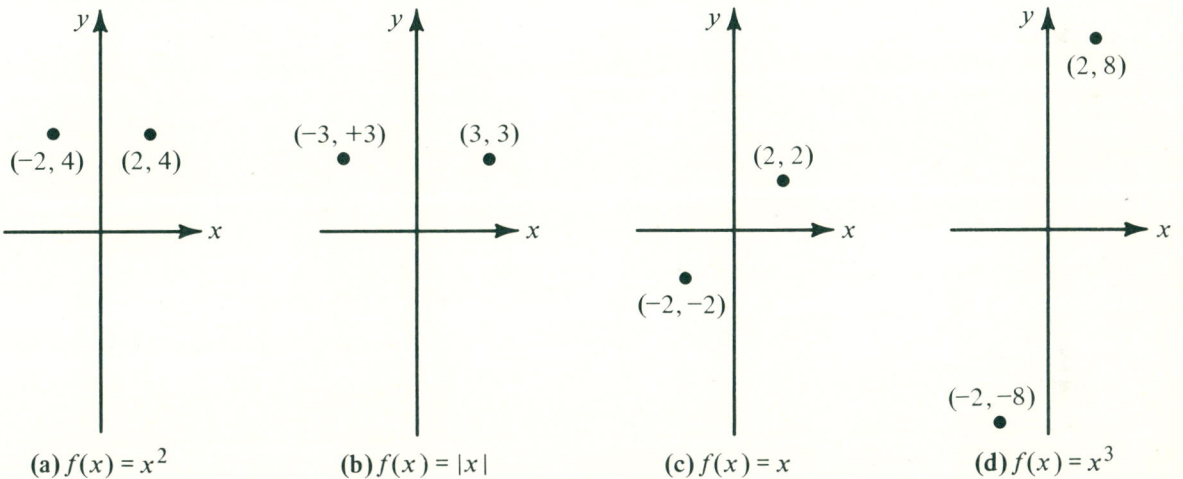

(a) $f(x) = x^2$ (b) $f(x) = |x|$ (c) $f(x) = x$ (d) $f(x) = x^3$

FIGURE 3.13.

and

$$f(x) = |x| \Rightarrow f(-x) = |-x| = |x| \Rightarrow$$
$$f(x) = f(-x)$$

The two functions $f(x) = x$ and $f(x) = x^3$ are symmetric about the origin, since

$$f(x) = x \Rightarrow -f(-x) = -(-x) = x$$
$$\Rightarrow f(x) = -f(-x)$$

and

$$f(x) = x^3 \Rightarrow -f(-x) = -(-x)^3 = x^3$$
$$\Rightarrow f(x) = -f(-x)$$

A function is *even* if its graph is symmetric about the y-axis. All functions containing only even powers of x are symmetric about the y-axis. A function is *odd* if its graph is symmetric about the origin. Notice that the odd functions $y = x$

77

and $y = x^3$ both have odd exponents. There are other types of odd functions, but we will not consider them here. These definitions are summarized below.

DEFINITION 3.4.
$y = f(x)$ is an even function if and only if $f(-x) = f(x)$. Its graph is symmetric about the x-axis.

DEFINITION 3.5.
$y = f(x)$ is an odd function if and only if $-f(-x) = f(x)$. Its graph is symmetric about the origin.

EXERCISE 3.4

A. FUNDAMENTALS

Determine whether the following functions are even, odd, or neither.

1. (a) $f(x) = x^2$ (b) $g(x) = x^2 + 1$
   (c) $h(x) = x^2 + x$

2. (a) $f(x) = x^3$ (b) $g(x) = x^3 + x$
   (c) $h(x) = x^3 + 1$

3. (a) $f(x) = \sqrt{x}$ (b) $g(x) = |x|$
   (c) $h(x) = \sqrt{x} + |x|$

Supply the missing number in each of the following.

4. (a) $y = f(x)$ is even, and
   $f(2) = 3 \Rightarrow f(-2) = \underline{\quad?\quad}$
   (b) $y = f(x)$ is even, and
   $f(-1) = -2 \Rightarrow f(?) = \underline{\quad?\quad}$

5. (a) $y = g(x)$ is odd, and
   $g(1) = 2 \Rightarrow g(-1) = \underline{\quad?\quad}$
   (b) $y = g(x)$ is odd, and
   $g(-1) = -2 \Rightarrow g(?) = \underline{\quad?\quad}$

B. ESSENTIALS

Replace x with $-x$ to determine if the following functions are even, odd, or neither. Explain carefully.

★6. $y = f(x) + g(x)$, where
   (a) $f(x) = x^2$ and $g(x) = |x|$
   (b) $f(x) = x^3$ and $g(x) = x$
   (c) $f(x) = x^2$ and $g(x) = x^3$
   (d) $f(x) = |x|$ and $g(x) = x$

7. $y = f(x) \cdot g(x)$, where
   (a) $f(x) = x^2$ and $g(x) = |x|$
   (b) $f(x) = x$ and $g(x) = x^3$
   (c) $f(x) = x^3$ and $g(x) = |x|$
   (d) $f(x) = x^2$ and $g(x) = x^3$

C. Prove the following theorems on even and odd functions.
[HINT: First rewrite each theorem as an implication involving $f(x)$ and $g(x)$.]

8. (a) The sum of two even functions is even.
   (b) The sum of two odd functions is odd.
   (c) The sum of an even and an odd function is neither even nor odd.

9. (a) The product of two even functions is even.
   ★(b) The product of two odd functions is even.
   (c) The product of an even and an odd function is odd.

★10. Every function can be expressed as the sum of an even and an odd function.

Discuss symmetry and then graph each of the following.

★11. $x^3y + 2xy = 0$

12. $xy^3 - 2xy = 0$

★13. $|x| + |x| = 1$

3.5 TRANSLATIONS, COMPRESSIONS AND EXPANSIONS

Consider the new function created by adding 1 to the y-coordinate of each of the following ordered pairs $(x, f(x))$.
 If
$$f = \{(-2, -3), (-1, -1), (1, 3), (2, 5)\}$$
then
$$g(x) = f(x) + 1 \Rightarrow$$
$$g = \{(-2, -2), (-1, 0), (1, 4), (2, 6)\}$$

SECTION 3.5 Translations, Compressions and Expansions

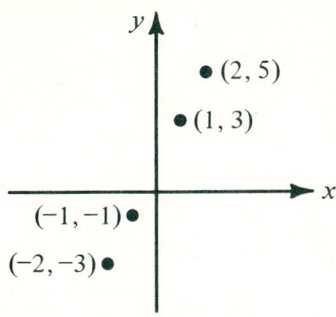

FIGURE 3.14.

By examining the graph of the two sets of ordered pairs, we can see why this is called a translation *up* by 1 unit. Plot the ordered pairs of g in Figure 3.14. The original function f is already plotted.

Complete the tables of values and the corresponding graphs for each of these examples of *vertical translations* (shifting up or down).

x or	$y = x^2$ $f(x) = x^2$	$y = x^2 + 2$ $y = f(x) + 2$	$y = x^2 - 1$ $y = f(x) - 1$
-2			
-1	1		
0			
1			
2	4		
3	9		

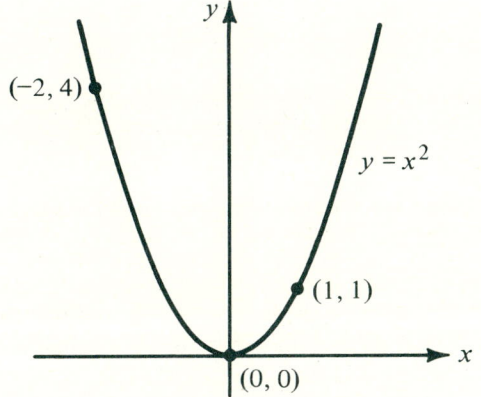

FIGURE 3.15.

x or	$y = \sqrt{x}$ $f(x) = \sqrt{x}$	$y = \sqrt{x} + 3$ $y = f(x) + 3$	$y = \sqrt{x} - 2$ $y = f(x) - 2$
-4	$\varnothing$		
-1	$\varnothing$		
0			
1			
2	$\sqrt{2} \simeq 1.4$		
3	$\sqrt{3} \simeq 1.7$		
4			

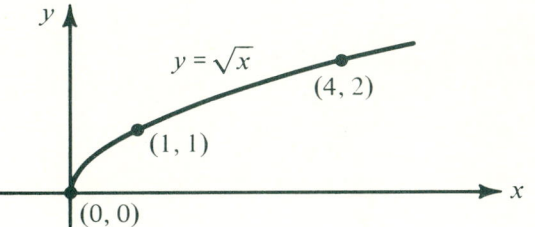

FIGURE 3.16.

VERTICAL TRANSLATIONS.
The graph of $y = f(x) + k$ is $y = f(x)$ translated vertically $|k|$ units, up for $k > 0$ or down for $k < 0$.

Consider the set of ordered pairs generated by adding 2 to the x-coordinates of each of the following ordered pairs.

If

$$f = \{(-4, -4), (-1, -2), (0, 0), (1, 2), (2, 4)\}$$

then

$$g(x) = f(x + 2) \Rightarrow$$
$$g = \{(-2, -4), (1, -2), (2, 0), (3, 2), (5, 4)\}$$

A plot of the two sets of ordered pairs reveals a horizontal translation to the right of 2 units. Plot the points depicting g. Those for f are already plotted.

79

CHAPTER 3   Functions and Relations

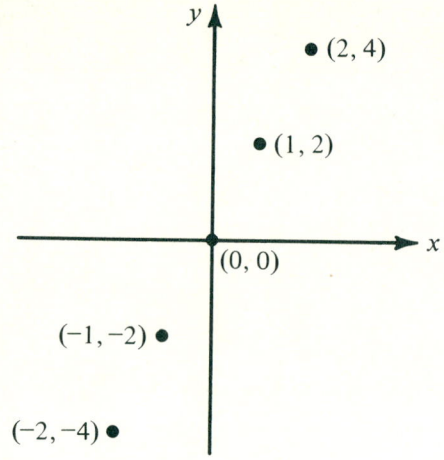

FIGURE 3.17.

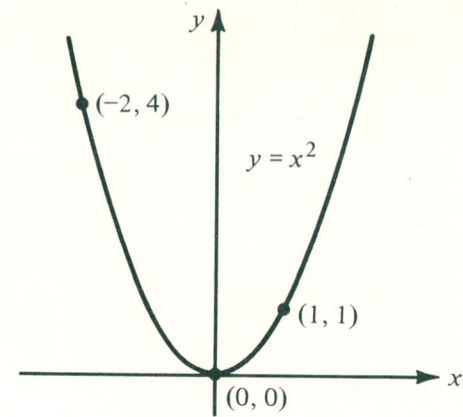

FIGURE 3.18.

By observing that
$$g(-2) = -4$$
and
$$f(-4) = -4 \Rightarrow g(-2) = f(-4)$$
and
$$g(2) = 0 \quad \text{and} \quad f(0) = 0 \Rightarrow g(2) = f(0), \text{ etc.}$$
we see that
$$g(x) = f(x - 2)$$

That is, if g is evaluated at x, f must be evaluated at a number 2 less than x to yield the same y-value. This seems reasonable, since g was created by increasing each x-coordinate of f by 2.

Now let's see what happens with functions defined by equations. Complete the following tables and their accompanying graphs.

x or	$y = x^2$ $f(x) = x^2$	$y = (x+2)^2$ $g(x) = f(x+2)$	$y = (x-1)^2$ $h(x) = f(x-1)$
-2			
-1	1		
0			
1			
2	4		
3	9		

x or	$y = \|x\|$ $f(x) = \|x\|$	$y = \|x+3\|$ $g(x) = f(x+3)$	$y = \|x-2\|$ $h(x) = f(x-2)$
-2	2		
-1			
0	0		
1	1		
2			
3	3		

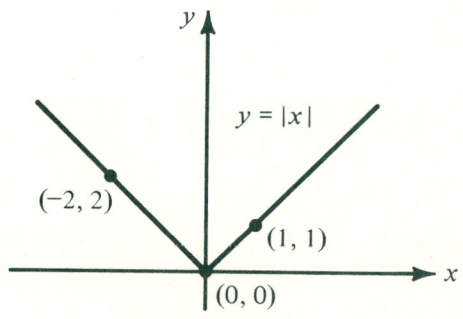

FIGURE 3.19.

Did you notice that $g(x) = f(x+2)$ and $g(x) = f(x+3)$ were translations to the *left* of the graph of $y = f(x)$ by 2 and 3 units, respectively? Similarly, $h(x) = f(x-1)$ and $h(x) = f(x-2)$ caused the graph of $y = f(x)$ to be translated to the *right* by 1 and 2 units, respectively.

SECTION 3.5   Translations, Compressions and Expansions

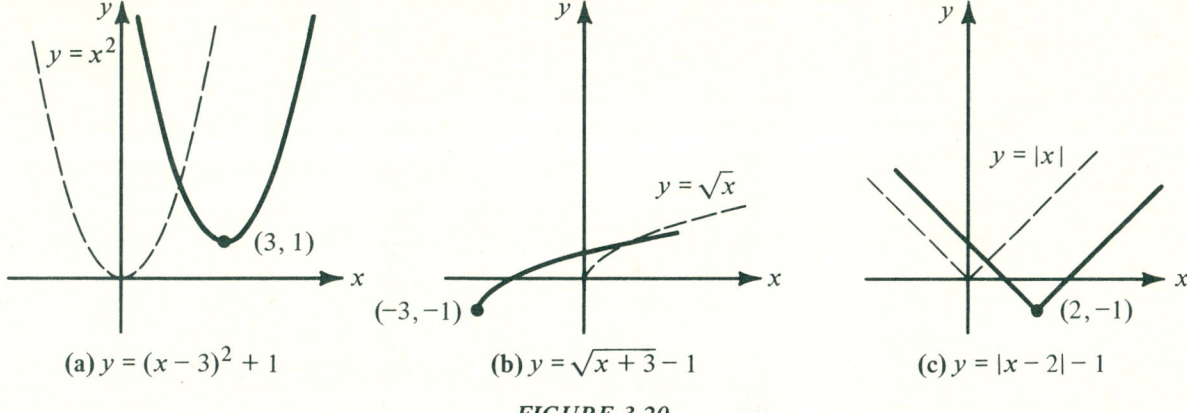

(a) $y = (x - 3)^2 + 1$ **(b)** $y = \sqrt{x + 3} - 1$ **(c)** $y = |x - 2| - 1$

FIGURE 3.20.

> HORIZONTAL TRANSLATIONS.
> The graph of $y = f(x + h)$ is $y = f(x)$ translated horizontally $|h|$ units, to the left for $h > 0$ or to the right for $h < 0$.

It is important to remember that horizontal translations are in the opposite direction as the sign of h, while the vertical translations are in the same direction as the sign of k.

Vertical and horizontal translations are combined in Figure 3.20.

Now let's consider the procedure involved in graphing functions such as

$$f(x) = -(x + 1)^2 + 3$$

and

$$g(x) = \sqrt{-(x - 2)} + 1$$

First, we see that $f(x)$ is related to $y = x^2$ and that $g(x)$ is related to $y = \sqrt{x}$, whose shapes we already know. First, look for the presence of negative signs, which indicate how to reflect the basic curve. We see that $f(x)$ is $y = x^2$ reflected about the x-axis (since $-y = -x^2$) and that $g(x)$ is $y = \sqrt{x}$ reflected about the y-axis (since x is replaced with $-x$ to get $\sqrt{-x}$). Finally, translations are noted: $f(x)$ is $y = -x^2$ translated to the left 1, because of $(x + 1)$, and up 3; $g(x)$ is translated to the right 2 and up 1. (See Figure 3.21, page 82.)

EXAMPLE II. The graphing procedure is shown in steps in Figure 3.21. You will soon be able to combine all of these into one step.

Notice that we indicated the reflections first and then the translations. This is the necessary order, as illustrated by Example III.

EXAMPLE III. Let $f(x) = \sqrt{x}$
(a) Translate up 2, then reflect about the x-axis.
(b) Reflect about the x-axis, then translate up 2.
(See Figure 3.22, page 83.)

The equations are different and the graphs are different. Remember: *reflect first and then translate!*

Now we will examine the last two ways the graph of a function can be changed. These two ideas are more difficult to employ directly when graphing a curve, but they are very powerful intuitive concepts that will help you metally picture the graph of a function. These changes are compressions and expansions, and they occur both toward and away from the x- and the y-axis.

First, recall that $f(2x) \neq 2f(x)$, in general. For example, if $f(x) = x^2$, $f(2x) = (2x)^2 = 4x^2$, while $2f(x) = 2x^2$. These are best understood relative to different axes. (See Figure 3.23, page 84.)

(a)

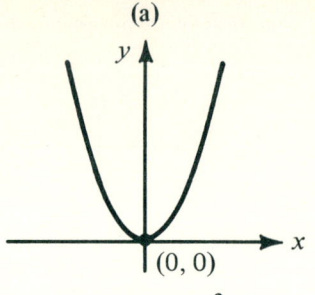

Step 1 $y = x^2$

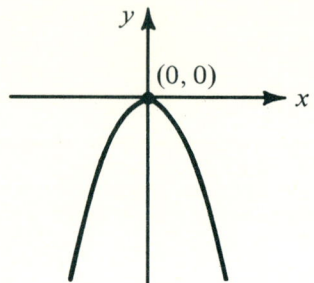

Step 2 $y = -x^2$

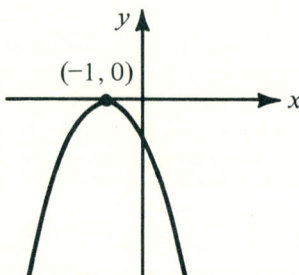

Step 3 $y = -(x + 1)^2$

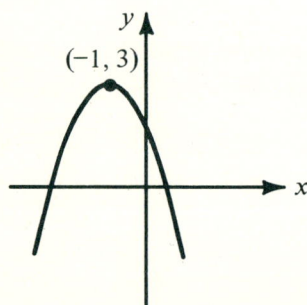

Step 4 $y = -(x + 1)^2 + 3$

(b)

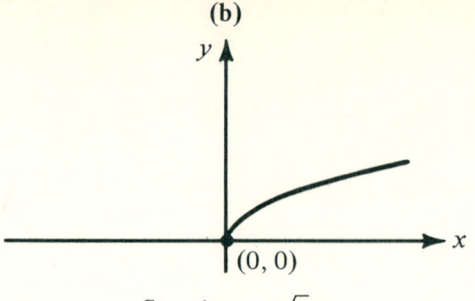

Step 1 $y = \sqrt{x}$

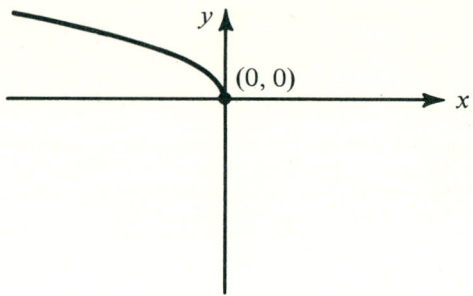

Step 2 $y = \sqrt{-x}$

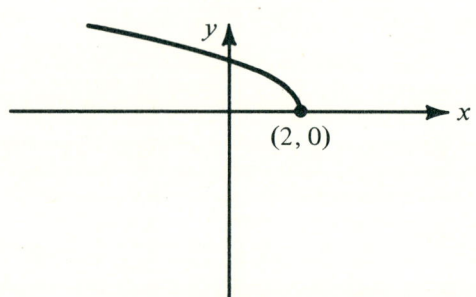

Step 3 $y = \sqrt{-(x - 2)}$

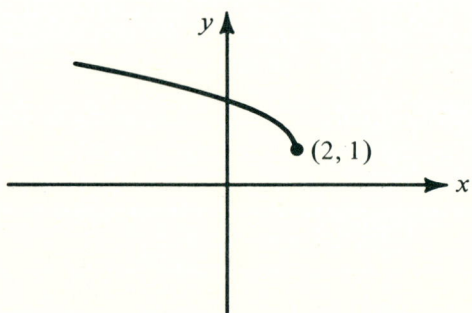

Step 4 $y = \sqrt{-(x - 2)} + 1$

FIGURE 3.21.

SECTION 3.5 Translations, Compressions and Expansions

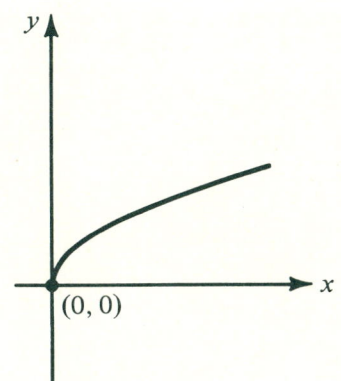

Step 1 $f(x) = \sqrt{x}$

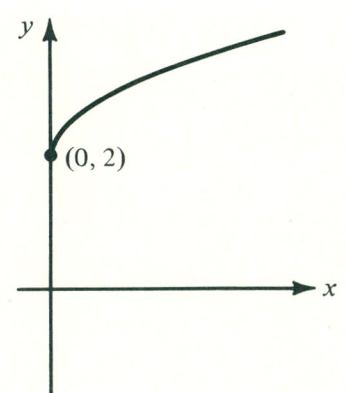

Step 2 $g(x) = f(x) + 2 = \sqrt{x} + 2$

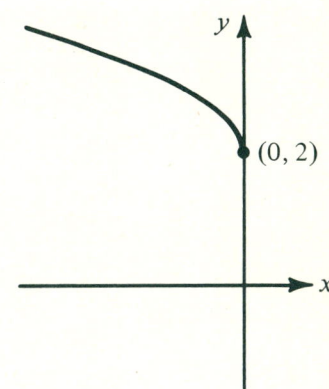

Step 3 $y = g(-x) = \sqrt{-x} + 2$

(a)

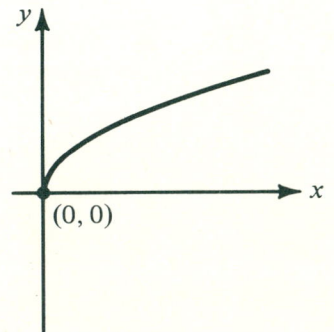

Step 1 $f(x) = \sqrt{x}$

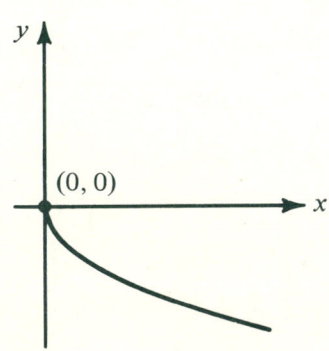

Step 2 $g(x) = -f(x) = -\sqrt{x}$

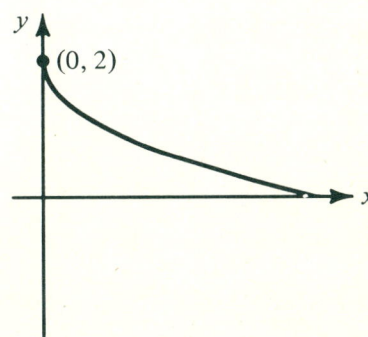

Step 3 $y = 9(x) + 2 = -\sqrt{x} + 2$

(b)

FIGURE 3.22.

83

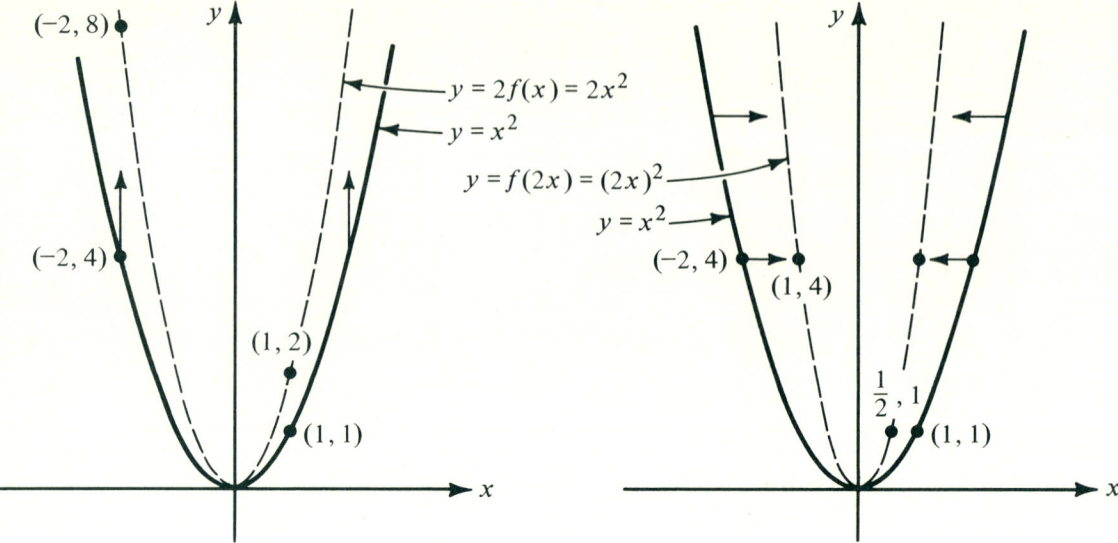

FIGURE 3.23.

We also have compression toward the x-axis, as in changing $f(x) = x^3$ into $y = \frac{1}{2}f(x) = \frac{1}{2}x^3$, and expansions away from the y-axis, as in $y = f(\frac{1}{2}x) = (\frac{1}{2}x)^3$. These are illustrated in Figure 3.24. The following compressions and expansions summarize these results for positive values of c. If c is negative, there is also a reflection involved.

> HORIZONTAL EXPANSIONS AND COMPRESSIONS.
> Consider the graph of $y = f(cx)$ for $c > 0$. For $c > 1$, it is a compression of $y = f(x)$ toward the y-axis by a factor of c. For $c < 1$, it is an expansion of $y = f(x)$ away from the y-axis by a factor of $1/c$.

> VERTICAL EXPANSIONS AND COMPRESSIONS.
> Consider the graph of $y = cf(x)$ for $c > 0$. For $c > 1$ it is an expansion of $y = f(x)$ away from the x-axis by a factor of c. For $c < 1$, it is a compression of $y = f(x)$ toward the x-axis by a factor of $1/c$.

EXERCISE 3.5

A. FUNDAMENTALS
Use one axis system to graph each of the following. Graph (a), (b), and (c) all on one system.

1. (a) $f(x) = x^2$ (b) $y = (x - 1)^2$
   (c) $y = x^2 + 2$

2. (a) $f(x) = x$ (b) $y = x + 2$
   (c) $y = f(x - 1)$

3. (a) $f(x) = \sqrt{x}$ (b) $y = f(x + 3)$
   (c) $y = f(x - 1) + 2$

B. ESSENTIALS

4. (a) $g(x) = |x|$ (b) $y = 1 - g(x)$
   (c) $y = 2 - g(x + 2)$

5. (a) $g(x) = x^2$ (b) $y = 2 - x^2$
   (c) $y = 3 - (x - 1)^2$

★6. (a) $f(x) = \sqrt{x}$ (b) $y = \sqrt{-x} + 1$
   (c) $y = \sqrt{-(x + 2)} + 2$

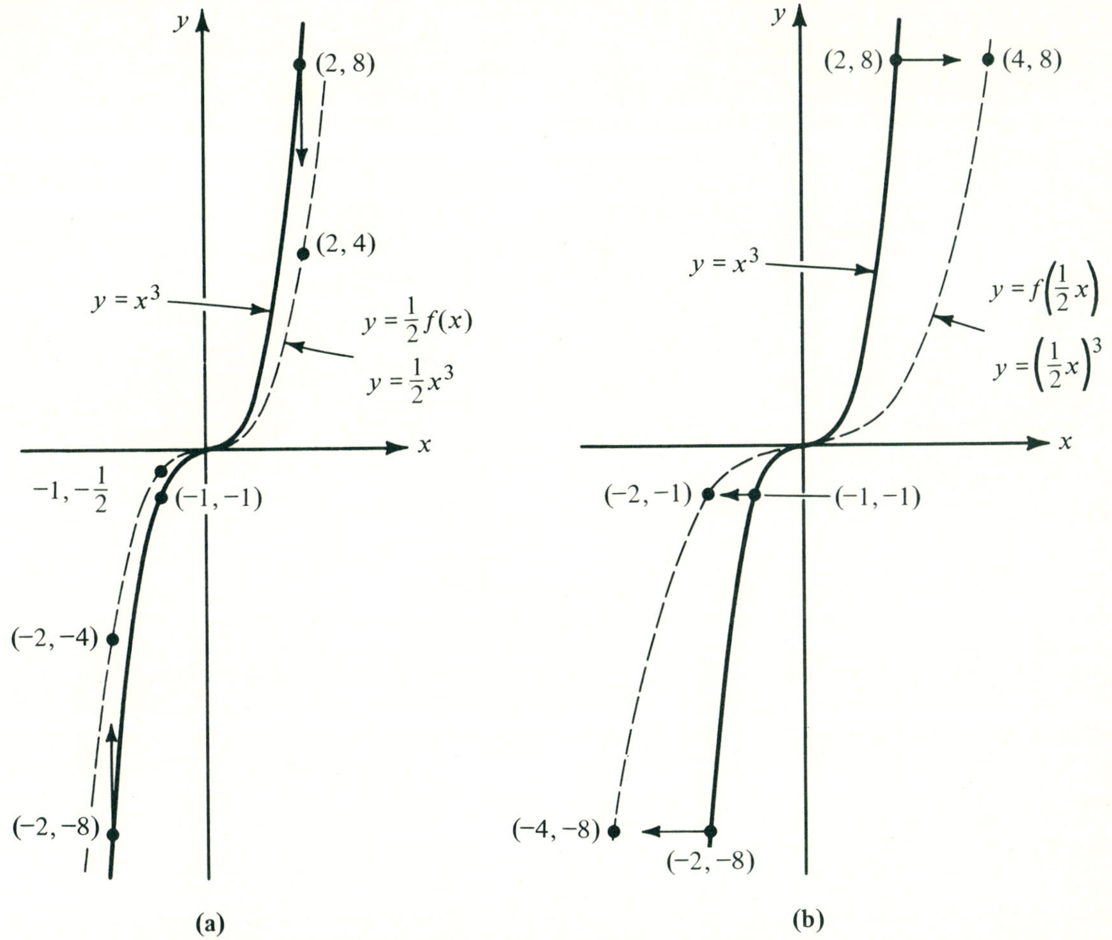

SECTION 3.5 Translations, Compressions and Expansions

FIGURE 3.24. *(a) Represents a Compression Toward the x-axis, Since All the y-values Become Half as Large. (b) Represents an Expansion Away from the y-axis by a Factor of 2, Since We Must Go Twice as Far on the x-axis to Any Given Value for y.*

Describe the translations, reflections, compressions, or expansions present in the following:

7. (a) $y = 2(x - 1)^2$
    (b) $y = -\frac{1}{2}(x + 2)^2 + 1$
8. (a) $y = 2\sqrt{-x + 1}$
    (b) $y = \sqrt{2(x - 1)} - 2$
9. (a) $y = |-3x| + 1$ (b) $y = -3|x + 1|$

Let $f(x)$ be defined by the following graph. Then graph the functions (10–16).

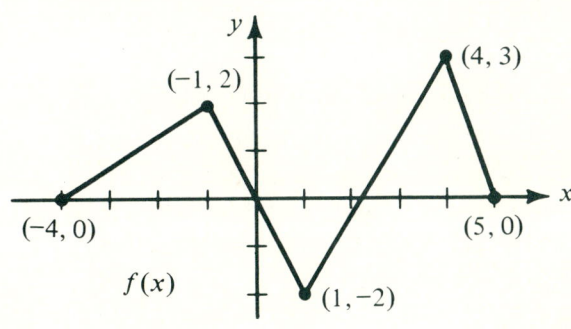

85

CHAPTER 3   Functions and Relations

10. (a) $y = f(-x) + 1$ (b) $y = -f(x + 1)$
11. $y = 2 - f(x - 1)$ 12. $y = 2f(x - 1)$
13. $y = 1 - \frac{1}{2}f(-x)$ 14. $y = f(x - 4)$
15. $y = f(4 - 2x)$ 16. $y = 2 - f(3 - x)$

Let $g(x)$ be defined by the following graph. Draw the graphs of the new functions indicated (17–19).

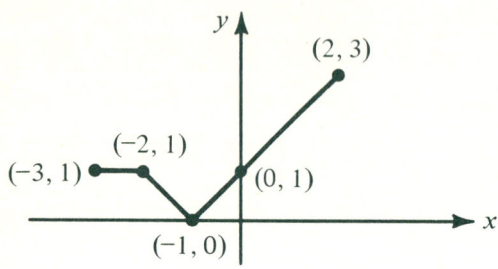

17. (a) $y = g(-x) + 2$ (b) $y = 2 - g(x)$
18. (a) $y = g(x + 1)$ (b) $y = g(2(x + 1))$
19. (a) $y = \frac{1}{2}g(x + 2) - 1$
    (b) $y = 2g(2 - x) + 3$

Graph the following functions:

★20. $f(x) = 2 - 3\sqrt{1 - x}$
21. $f(x) = 3 + |2 - x|$
22. $g(x) = (1 - x)^2 + 1$
23. $g(x) = 2 - 3(x - 1)^3$

3.6 INVERSES

Consider the following sets of ordered pairs:
$$f = \{(-1, 1), (0, 3), (1, 5), (2, 7)\}$$
and
$$f^{-1} = \{(1, -1), (3, 0), (5, 1), (7, 2)\}$$

What do you notice? These two sets represent functions that are said to be **inverses** of each other, since they are sets of inverted ordered pairs. The two sets can be generated with the functions:
$$f(x) = 2x + 3$$
and
$$f^{-1}(x) = \frac{x - 3}{2}$$

When we evaluate the function $f(x)$, first we multiply by 2, then we add 3. When we evaluate its inverse $f^{-1}(x)$ (read "f inverse of x"), first we subtract 3, then we divide by 2.

Notice that the function performs exactly the opposite (inverse) operations in precisely the reverse order.

Some common relationships that have the characteristics of inverse functions are given below.

f: John is Duke's son.

f^{-1}: Duke is John's father.

f: Putting on your socks and shoes.

f^{-1}: Taking off your shoes and socks.

f: Taking a plate out of the cupboard and dirtying it.

f^{-1}: Cleaning a plate and returning it to the cupboard.

There are three basic definitions of the inverse of a function, all of which are equivalent. Each has its advantage depending upon the context.

The first definition is related to the sets of ordered pairs comprising a function and its inverse.

> **DEFINITION 3.6.**
> If $(a, b) \in f$ (that is, $f(a) = b$), then $(b, a) \in f^{-1}$ (that is, $f^{-1}(b) = a$).

The second definition states that the inverse is the result of a process.

> **DEFINITION 3.7.**
> To find the inverse of a function $y = f(x)$,
>
> **1.** Solve the equation for x,
>
> then
>
> **2.** Rename the variables, calling y, "x" and x, "y".

The third definition of an inverse of a function is in Section 3.7. The algorithmic definition in Definition 3.7 is not acceptable mathematics but it yields a straightforward definition, as well a fast method for finding the inverse.

EXAMPLE I. Find the inverse of $f(x) = 2x + 3$.
1. Solve for x:

$$y = 2x + 3$$
$$2x = y - 3$$
$$x = \frac{y - 3}{2}$$

2. Rename the variables.

$$y = \frac{x - 3}{2}$$

This result agrees with our previous discussion. The graphs of the two functions appear in Figure 3.25.

> **INVERSES.**
> The graph of $y = f^{-1}(x)$ is the reflection about the line $y = x$ of the graph of $y = f(x)$.

By examining Figure 3.25, the first part of the proof is readily established. Since triangles ABP and ACP are congruent, $BP = PC$. For nonlinear functions, we use triangles OBP and OCP.

To completely establish the symmetry about the line $y = x$ for points $B(a, b)$ and $C(b, a)$, we must show that the line BPC is perpendicular to

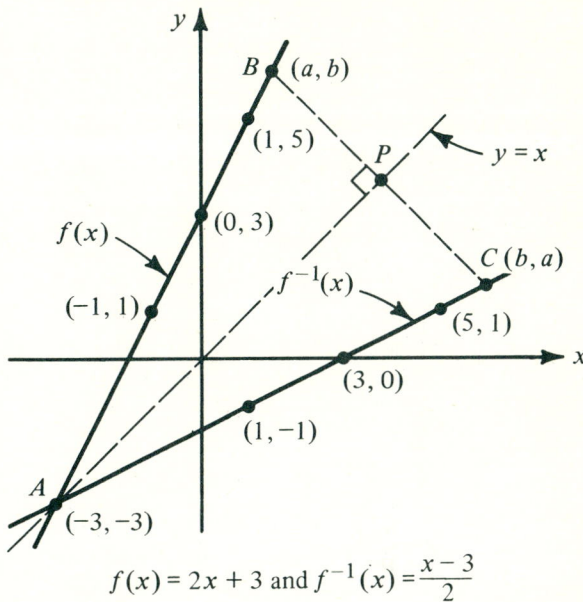

$f(x) = 2x + 3$ and $f^{-1}(x) = \dfrac{x-3}{2}$

FIGURE 3.25.

$y = x$. This follows directly from the slopes of the two lines.

Given the graph of the function, it is easy to draw the graph of the inverse of the function. We simply reflect the graph of the function about the line $y = x$. Do you see that this will cause the two graphs to intersect on the line $y = x$? Can you think of any graphs that are symmetric about the line $y = x$ and are therefore their own inverses?

The graphs of three functions and their inverses are shown in Figure 3.26.

Stop and derive the equations of the inverse for the three functions shown in Figure 3.26. Use the two-step procedure as in Example I.

Did you notice that the inverse of $y = x^2$ was not a function? Since ordered pairs, such as $(2, 4)$ and $(-2, 4)$, belong to the function, the ordered pairs $(4, 2)$ and $(4, -2)$ must belong to its inverse. This pairs one value from the domain ($x = 4$) with two different values from the range ($y = 2$ and $y = -2$). Therefore, the inverse is not a function. This will be true for every function that is not a

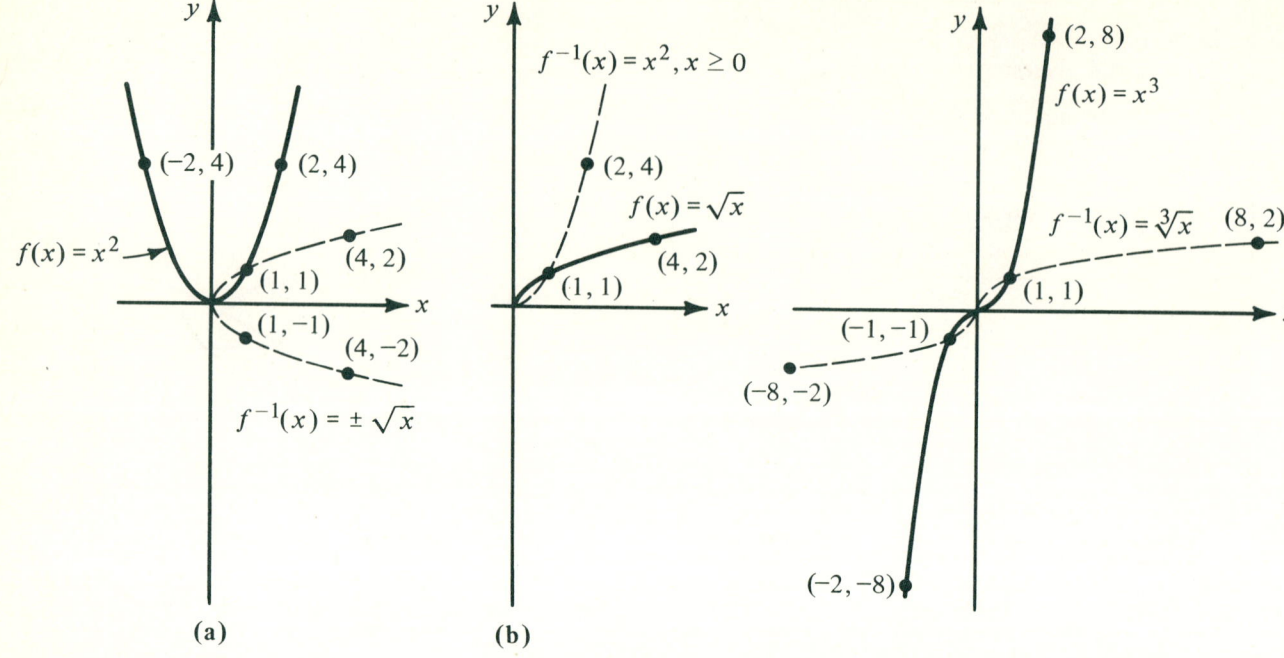

FIGURE 3.26.

one-to-one function. This is formally stated in Theorem I.

> **THEOREM I.**
> The inverse of a function $y = f(x)$ is a function if and only if f is a one-to-one function.

Proof:

CASE I. If f is a one-to-one function, then its inverse is a function.

Assume this is false. Then f is a one-to-one function and f^{-1} is not a function $[\sim(p \Rightarrow q) \Leftrightarrow p \wedge \sim q]$.

If f^{-1} is not a function, then for some x_0,

$(x_0, y_1) \in f^{-1}$ and $(x_0, y_2) \in f^{-1}$

where $y_1 \neq y_2$. But this means

$(y_1, x_0) \in f$ and $(y_2, x_0) \in f$

which contradicts our assumption that f was a one-to-one function. Therefore, our assumption that the original implication was false is incorrect.

CASE II. If f^{-1} is a function, then f is a one-to-one function.

Assume this is false. Then f^{-1} is ___?___ and f is ___?___.

If f is not a one-to-one function, then there exists a y_0 such that

$(a, y_0) \in f$ and $(b, y_0) \in f$

where $a \neq b$. But this means

$(?, ?) \in f^{-1}$ and $(?, ?) \in f^{-1}$

which is a contradiction of ___?___, completing the proof.

SECTION 3.6 Inverses

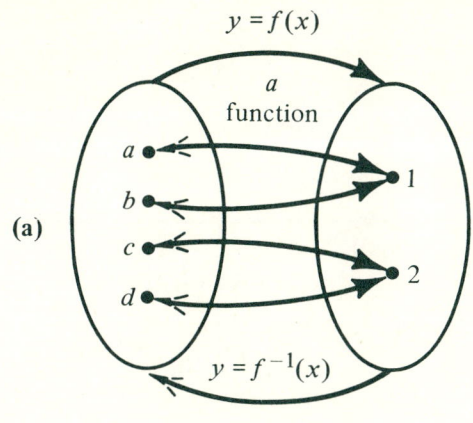
(a) Not a function

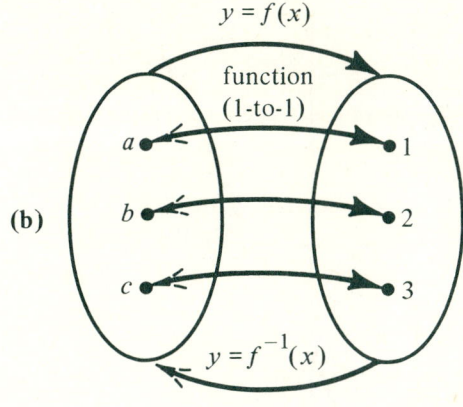
(b) A function

FIGURE 3.27.

The restriction of $x \geq 0$ on the inverse of $f(x) = \sqrt{x}$ is necessary so that the domain of $f(x)$ will equal the range of $f^{-1}(x)$, and the range of $f(x)$ will equal the domain of $f^{-1}(x)$. These restrictions follow easily from the first definition of an inverse (Definition 3.6).

These results are summarized in Figure 3.27. In part (a), $D_f = \{a, b, c, d\}$ and $R_f = \{1, 2\}$, but $D_{f^{-1}} = \{1, 2\}$ and $R_{f^{-1}} = \{a, b, c, d\}$, where D_f is the domain of f, $R_{f^{-1}}$ is the range of f^{-1}, etc. The same relationship applies to part (b), which depicts a one-to-one function from the left set to the right set. Its inverse (from right to left) is also a function.

Consider the following examples. In (a) below, the function $f(x)$ is first rotated about the y-axis, then the inverse of this result is graphed. In (b) the inverse of $f(x)$ is drawn first, then this result is spun about the y-axis.

EXAMPLE 1. For $f(x) = 2x - 4$, find (a) $[f(-x)]^{-1}$ and its graph and (b) $f^{-1}(-x)$ and its graph. The equations for the final curves in each example are found as follows:
(a) $f(x) = 2x - 4 \Rightarrow$
$f(-x) = 2(-x) - 4 = -2x - 4 \Rightarrow$
$y = -2x - 4$

1. Solve for x:
$$y = -2x - 4 \Rightarrow$$
$$2x = -y - 4 \Rightarrow$$
$$x = \frac{-y - 4}{2}$$

2. Rename:
$$y = \frac{-x - 4}{2} \Rightarrow$$
$$[f(-x)]^{-1} = -\frac{x + 4}{2}$$

(b) $f(x) = 2x - 4 \Rightarrow$
$y = 2x - 4$
1. Solve for x:
$$2x = y + 4 \Rightarrow$$
$$x = \frac{y + 4}{2}$$

2. Rename:
$$y = \frac{x + 4}{2} \Rightarrow$$
$$f^{-1}(x) = \frac{x + 4}{2} \Rightarrow$$
$$f^{-1}(-x) = \frac{-x + 4}{2}$$

CHAPTER 3   Functions and Relations

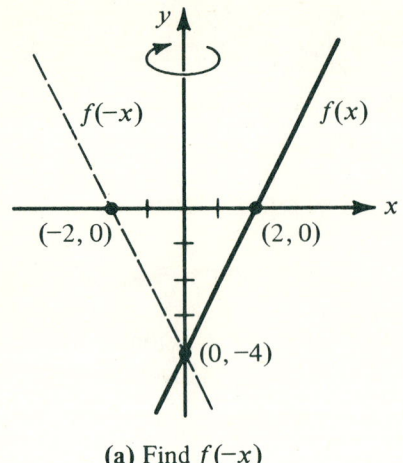

(a) Find $f(-x)$

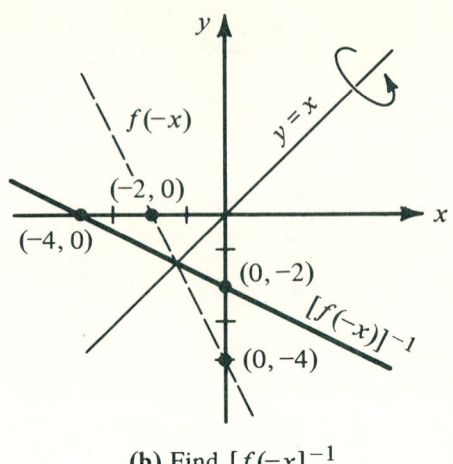

(b) Find $[f(-x)]^{-1}$

FIGURE 3.28.

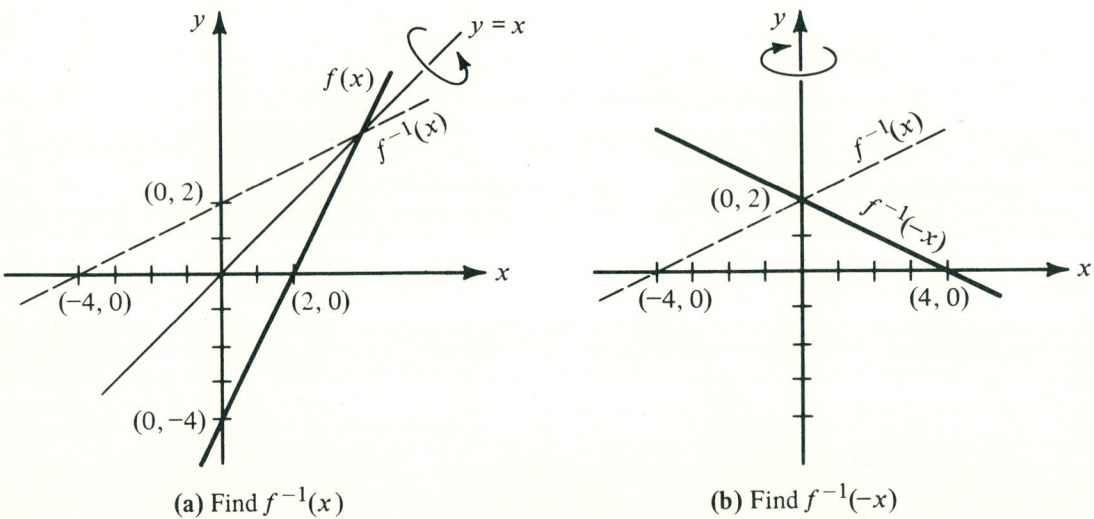

(a) Find $f^{-1}(x)$

(b) Find $f^{-1}(-x)$

FIGURE 3.29.

The fact that the two final equations are different is simply another illustration of the importance of correctly interpreting the role of the parentheses; $f^{-1}(-x)$ means the inverse evaluated at $-x$, but $[f(-x)]^{-1}$ means replace x with $-x$ in the original function $f(x)$, then find the inverse of this *new* function.

EXERCISE 3.6

A. FUNDAMENTALS
Find the equations for the inverses of the following functions and sketch both graphs.

1. (a) $f(x) = x + 2$ (b) $f(x) = x - 2$
2. (a) $g(x) = 3x$ (b) $g(x) = \frac{1}{3}x$
3. (a) $y = 2x + 3$ (b) $y = \frac{1}{5}x - 7$

Find the domains and ranges for each pair of relations:

4. (a) $y = \sqrt{x} + 3$ (b) $y = x^2 - 3$
5. (a) $y = x^3$ (b) $y = \sqrt[3]{x}$
6. (a) $y = \dfrac{1}{x - 3}$ (b) $y = \dfrac{3x + 1}{x}$
7. (a) $y = x$ (b) $y = \dfrac{1}{x}$

B. ESSENTIALS

(i) Graph each of the following functions, (ii) graph the inverse for each, (iii) find the equation for the inverse, and (iv) find the domain and range for each function and for its inverse.

8. $f(x) = \sqrt{-x}$ 9. $f(x) = 3x - 5$
10. $f(x) = -x^2$ 11. $f(x) = -x^3$

Let $f(x)$ be defined by the following graph:

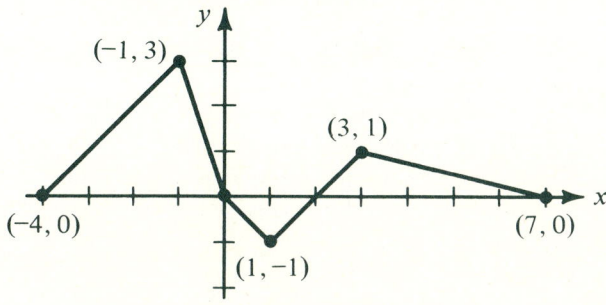

Then sketch the following:

12. $y = f^{-1}(x)$ 13. $y = f(-x)$
14. $y = -f(x)$ 15. $y = -f(-x)$

C. APPLICATIONS

16. (a) $y = -f^{-1}(x)$ (b) $y = [f(-x)]^{-1}$
17. (a) $y = f^{-1}(-x)$ (b) $y = [-f(x)]^{-1}$
18. $y = -f^{-1}(-x)$
★19. Find $y = -f^{-1}(x)$ and $y = [f(-x)]^{-1}$ for $f(x) = \sqrt{x}$.

20. Prove the relationships illustrated in problems 16, 17, and 19 for any function $y = f(x)$. [HINT: let $(a, b) \in f(x)$, then find the corresponding ordered pair for $f^{-1}(x)$, $-f^{-1}(x)$, etc.]

3.7 OPERATIONS ON FUNCTIONS

There are at least three basic ways of creating new functions from given ones: by addition, multiplication, or composition of functions.

The result of the addition of two functions is best described geometrically. Consider the functions $f(x) = |x + 4|$ and $g(x) = |x - 2|$, which are $y = |x|$ translated to the left 4 and to the right 2, respectively. Let their sum be

$$y = f(x) + g(x) \Rightarrow$$
$$y = |x + 4| + |x - 2|$$

This could be graphed by considering cases of $x + 4$ and $x - 2$ being both positive, etc., but it is much easier to graph each curve and then add them geometrically. This is accomplished by adding the y-coordinate (the height above the x-axis) of the lower curve to that of the upper one for several values of x, as indicated in Figure 3.30. The height

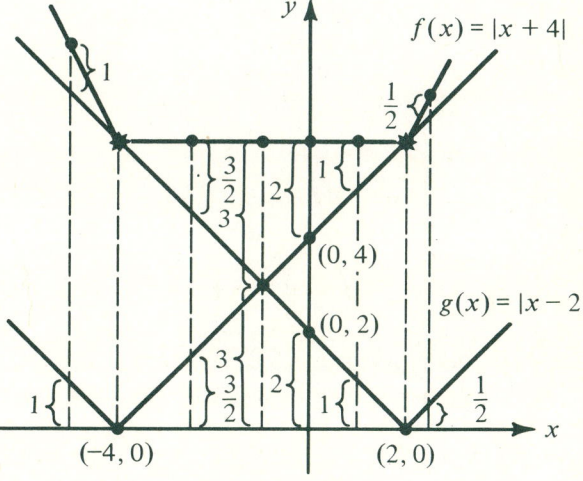

FIGURE 3.30.

CHAPTER 3   Functions and Relations

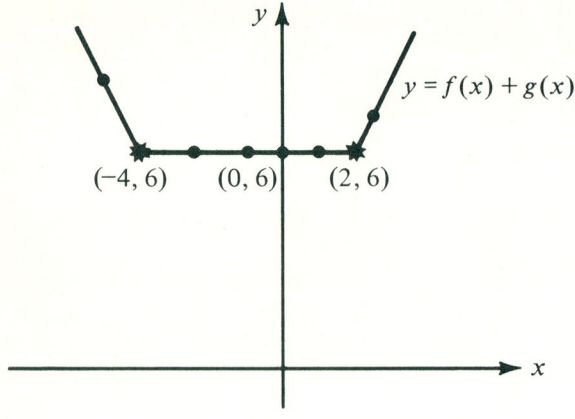

FIGURE 3.31.

of $f + g$ above the x-axis is the sum of the heights of f and g. This results in the graph in Figure 3.31. Notice the points on the graph of $f + g$ at $x = -4$ and $x = 2$, the x-intercepts of f and g. Notice that at each x, a vertical line whose length equals the sum of the distances to each curve will reach the new curve exactly.

The product of two functions is more difficult to examine, even with simple examples. Consider $f(x) = x^2$ and $g(x) = \sqrt{x + 1}$, and let their product be the new function

$$y = f(x) \cdot g(x) \Rightarrow$$
$$y = x^2\sqrt{x + 1}$$

The graphs of f, g, and $f \cdot g$ are in Figure 3.32.

Notice that between $x = -1$ and $x = 0$, x is extremely small (smaller toward $x = 0$). The result is a damping (making smaller) of $\sqrt{x + 1}$ in the product $x^2\sqrt{x + 1}$. Since neither of the factors x^2 or $\sqrt{x + 1}$ is ever negative, their product $y = x^2\sqrt{x + 1}$ is always positive or zero. As x gets bigger, both factors get bigger and their product grows larger at a faster rate than either factor.

The composition of functions is extremely important, since the vast majority of functions are compositions of other simpler functions. A function such as

$$f(x) = \sqrt{x^2 + 1}$$

for example, is composed of three functions in succession:

$$y = x^2, \qquad y = x + 1, \qquad y = \sqrt{x}$$

This can be seen by analyzing how we find $f(2)$. First we square the value, then add 1, then find the square root.

When dealing with the compositions of functions, the domains and ranges play essential roles, as we will discuss later.

A **composite function** always requires the successive use of the functions involved. The order in which we compose functions makes a big difference. Using the same functions as before, suppose

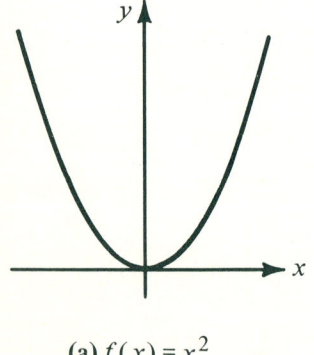

(a) $f(x) = x^2$

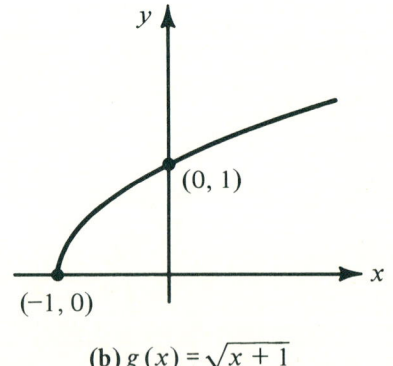

(b) $g(x) = \sqrt{x + 1}$

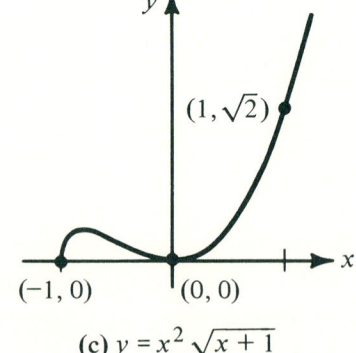

(c) $y = x^2\sqrt{x + 1}$

FIGURE 3.32.

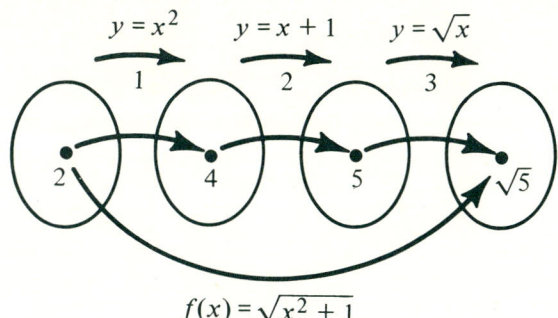

$f(x) = \sqrt{x^2 + 1}$

FIGURE 3.33.

we change the order to $y = x + 1$, then $y = x^2$, then $y = \sqrt{x}$. This yields:

$$g(x) = \sqrt{(x + 1)^2}$$

which is certainly not the same as $f(x)$ above.

Let's compose some functions. Evaluate each of the following as indicated for $f(x) = x + 1$, $g(x) = x^2$, and $h(x) = \sqrt{x}$.

1. $f(3) = \underline{}$, $h(4) = \underline{}$. Therefore, $x = 3 \Rightarrow \sqrt{x + 1} = \underline{}$. $\sqrt{x + 1}$ is the composition of f and h with f first.

2. $h(4) = \underline{}$, $f(2) = \underline{}$. Therefore, $x = 4 \Rightarrow \sqrt{x} + 1 = \underline{}$. This is the composition of the same functions as number 1, except in the reverse order. Notice that the results (2 and 3, respectively) are not equal.

3. $f(3) = \underline{} \Rightarrow g(f(3)) = g(4) = \underline{}$. $g(3) = \underline{} \Rightarrow f(g(3)) = f(9) = \underline{}$. Notice that $g(f(3)) = 16$ does not equal $f(g(3)) = 10$.

4. $g(f(2)) = g(\underline{}) = \underline{}$. $f(g(2)) = f(\underline{}) = \underline{}$.

5. $g(h(9)) = g(\underline{}) = \underline{}$. $h(g(9)) = h(\underline{}) = \underline{}$.

For Number 4 above, we often write $g(f(2))$ and $f(g(2))$ as $(g \circ f)(2) = 9$ and $(f \circ g)(2) = 5$; and for number 5 we write $(g \circ h)(9) = (h \circ g)(9) = 9$. Even though for certain values of x in the domain of h $(g \circ h)(x)$ may equal $(h \circ g)(x)$, composition of functions is not generally commutative. That is, usually $(f \circ g)(x) \neq (g \circ f)(x)$, as in 1–4 above.

Now let's examine some properties of composite functions. If you examine the following picture of $g(f(x))$, you will notice an important relationship between the range of f and the domain of g. This can cause important restrictions on the range and domain of the resulting composite function $(g \circ f)(x)$. The domain and range of $f(x) = x + 1$ are all real numbers, but the domain of $g(x) = \sqrt{x}$ is only those real numbers greater than or equal to zero. The resulting range of $g(x)$ is those real numbers greater than or equal to zero.

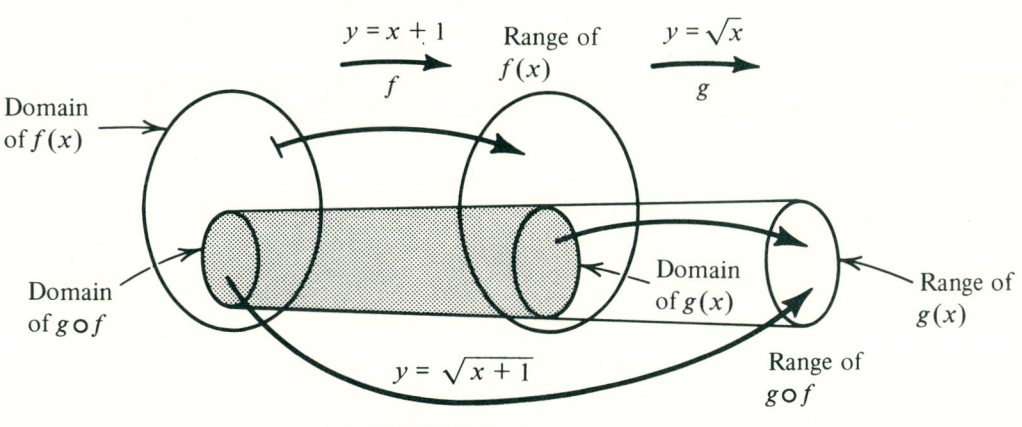

FIGURE 3.34.

The domain of the composite function $y = \sqrt{x+1}$ is:

$$x + 1 \geq 0 \quad \text{or} \quad x \geq -1$$

This is the subset of the domain of $f(x)$ that yields values of $y \geq 0$. These y-values are the elements of the domain of $g(x)$. The question of the domain of a function such as

$$y = \sqrt{x^2 - 4}$$

should have more meaning now. We recognize y as a composite of the two functions:

$$f(x) = \sqrt{x} \quad \text{and} \quad g(x) = x^2 - 4$$

Complete the following:

$$D_g = \{x \mid \underline{\quad ? \quad}\},$$
$$R_g = \{y \mid y \geq -4\}$$
$$D_f = \{x \mid x \geq 0\},$$
$$R_f = \{y \mid \underline{\quad ? \quad}\}$$
$$D_{f \cdot g} = \{x \mid x \leq -2 \text{ or } x > 2\},$$
$$R_{f \cdot g} = \{y \mid y \geq 0\}$$

Notice that the domain of $f(g(x))$ is a subset of the domain of $g(x)$. Such an analysis of the domains and ranges of composite functions is essential for graphing complicated functions. Most

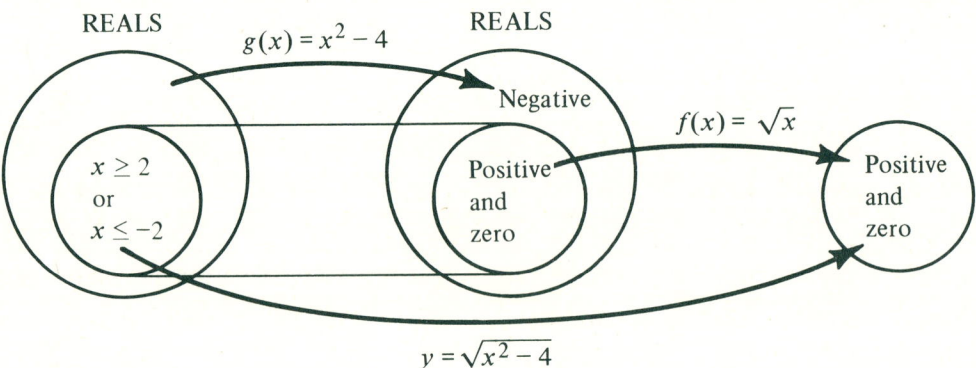

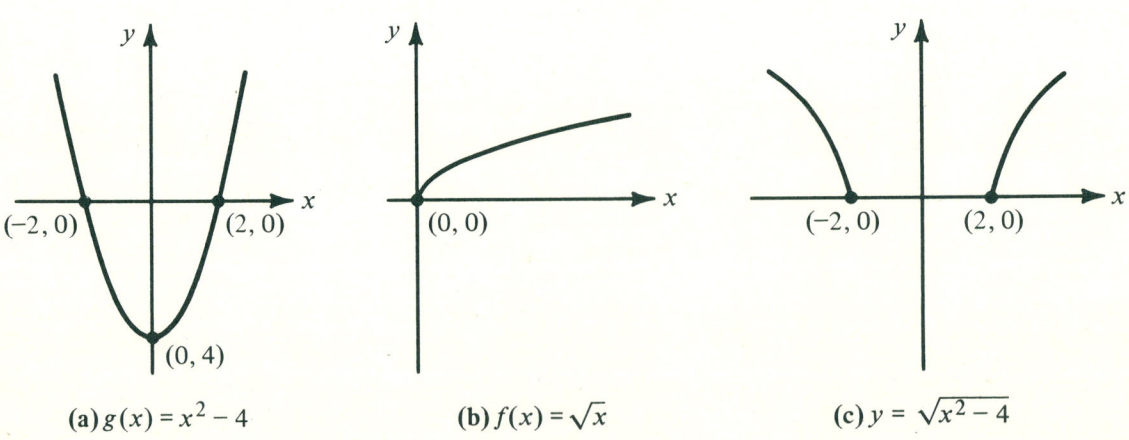

(a) $g(x) = x^2 - 4$ (b) $f(x) = \sqrt{x}$ (c) $y = \sqrt{x^2 - 4}$

FIGURE 3.35.

of the graphs that we will be drawing will not be this complicated, however.

Stop and find $f(g(x))$ and $g(f(x))$ for these two functions, which are inverses of each other.

$$f(x) = 3x - 4 \quad \text{and} \quad g(x) = \frac{x + 4}{3}$$

With the help of the composition of functions, we can now state the formal, precise definition of the inverse of a function.

DEFINITION 3.8.
The one-to-one functions $y = f(x)$ and $y = g(x)$ are inverses of each other if and only if

$$f(g(x)) = x \quad \text{and} \quad g(f(x)) = x$$

for all x in the domains of $f \circ g$ and $g \circ f$, respectively.

This can best be appreciated by studying the chart for $f(g(x))$ in Figure 3.36.

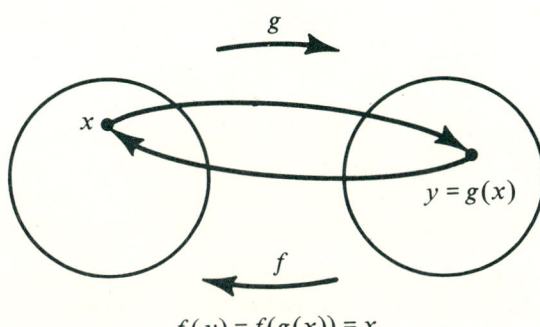

$$f(y) = f(g(x)) = x$$

FIGURE 3.36.

Notice that $y = f(x)$ must do exactly the opposite of what its inverse $y = g(x)$ does. Therefore, if g pairs x with some value y, f must pair y with x. Hence, composing a function and its inverse must go in circles, from x to y then back from y to x.

We can actually find equations of inverses using this definition.

EXAMPLE I. Find the inverse of $f(x) = 3x - 7$, using Definition 3.8.

If $g(x)$ is the inverse of $f(x) = 3x - 7$, then

$$f(g(x)) = x \quad \text{and} \quad g(f(x)) = x$$

Using the first of these with the function $f(x) = 3x - 7$, we get two results for $f(g(x))$ that can be equated:

$$f(g(x)) = x \quad \text{and} \quad f(g(x)) = 3g(x) - 7 \Rightarrow$$
$$3g(x) - 7 = x$$

This equation is simply solved for $g(x)$, the desired inverse.

$$3g(x) = x + 7$$
$$g(x) = \frac{x + 7}{3}$$

EXERCISE 3.7

A. FUNDAMENTALS

Graph parts (a), (b), and (c) of each of the following on the same set of axes:

1. (a) $f(x) = |x + 1|$ (b) $g(x) = |x - 1|$
   (c) $y = f(x) + g(x)$
2. (a) $f(x) = x^2$ (b) $g(x) = 2x - 3$
   (c) $y = f(x) + g(x)$
3. (a) $f(x) = x^3$ (b) $g(x) = x^2$
   (c) $y = f(x) + g(x)$

If $f(x) = x^2$, $g(x) = \sqrt{x}$, $h(x) = x^3$, and $k(x) = 2x - 1$, find the following:

4. (a) $f(g(9))$ (b) $g(f(9))$
5. (a) $g(f(-3))$ (b) $f(g(-3))$
6. (a) $h(k(2))$ (b) $k(h(2))$
7. (a) $(g \circ h)(-1)$ (b) $(f \circ h)(-1)$

B. ESSENTIALS

Graph the following exercises, each problem on a single graph:

8. (a) $f(x) = x^3$ (b) $g(x) = \sqrt{x - 2}$
   (c) $y = f(x) \cdot g(x)$

★9. (a) $f(x) = 1/x$ (b) $g(x) = |x + 3|$
    (c) $y = f(x) \cdot g(x)$

10. (a) $f(x) = -x^2$ (b) $g(x) = \sqrt{3 - x}$
    (c) $y = f(x) \cdot g(x)$

11. (a) $f(x) = \sqrt{x - 1}$ (b) $g(x) = x + 4$
    (c) $y = g(f(x))$

★12. (a) $f(x) = x^2 - 9$ (b) $g(x) = \sqrt{x}$
    (c) $y = g(f(x))$

13. (a) $f(x) = x^2 - 4x + 4$
    (b) $g(x) = \sqrt{x - 1}$ (c) $y = f(g(x))$

Use Definition 3.8 to find the inverses of the following:

14. $f(x) = 2x - 3$ ★15. $f(x) = \tfrac{1}{3}x + 5$

16. $f(x) = (x + 2)^2$

C. APPLICATIONS

17. Find an example of two functions $f(x)$ and $g(x)$, such that $f(g(x)) = x$ but $g(f(x)) \neq x$.

Decompose each of the following composite functions into two or more functions whose composition will yield the original function.

18. (a) $f(x) = \sqrt{x - 3}$ (b) $g(x) = \sqrt{x} - 3$
19. (a) $f(x) = x^2 + 1$ (b) $g(x) = (x + 1)^2$
20. (a) $f(x) = (2x)^3$ (b) $g(x) = (\tfrac{1}{2}x)^3$
21. $f(b) = x^2 + 6x + 9$
★22. $f(x) = \sqrt{x^2 - 8x + 1}$
23. $f(x) = \sqrt{x^2 + 6x + 9}$
24. $g(x) = \sqrt{2x^2 - x + 1}$

C. APPLICATIONS
Describe the geometrical effect composing the following pairs of functions as (i) $y = f(g(x))$ and (ii) $y = g(f(x))$ would have on the graph of $f(x)$.

25. $f(x) = x^2$
    $g(x) = x + 1$

26. $f(x) = \sqrt{x}$
    $g(x) = x - 3$

27. $f(x) = |x|$
    $g(x) = 2x$

28. $f(x) = x^3$
    $g(x) = \tfrac{1}{3}x$

29. Graph the functions $m(x) = 0.4$ gallons/mile, $s(x) = 50$ miles/hour, and their product $y = m(x) \cdot s(x)$. What are the dimensions (units) of the product?

30. The total revenue $R(x)$ from producing garbage disposals is given by the product $R(x) = p(x) \cdot x$. The demand curve, $p(x) = 16 - \sqrt{x}$, gives the relationship between the number of machines consumers are willing to buy, x, and their price, $p(x)$. Graph $y = x$, $p(x)$, and $R(x)$. What are the domains and ranges involved?

31. The total cost $C(x)$ of producing garbage disposals is $c(x) = F(x) + V(x)$. $F(x)$ represents the fixed costs — rent, salaries, etc., and $V(x)$ represents the variable costs — raw materials, hourly wages, etc. Graph $F(x)$ and $V(x)$, then sketch their sum $C(x)$. What are the domains and ranges of the functions $F(x) = 28$ and $V(x) = x^2 - x\sqrt{x}$?

32. The total profit $P(x)$ in the garbage disposal business is found from $P(x) = R(x) - C(x)$. Graph $P(x)$ from the graphs of $R(x)$ and $C(x)$ and estimate the quantity of garbage disposals that will yield maximum profit. What is this profit? What is the price? Compare $C(x_0)$ with $F(x_0)$ and $V(x_0)$, and $P(x_0)$ with $R(x_0)$ and $C(x_0)$ at the value x_0 that appears to yield the maximum profit.

33. The kinetic energy (KE) possessed by an object with a mass of m grams traveling with a velocity of v meters per second is given by $KE = \tfrac{1}{2}mv^2$ if the mass is constant. For a bullet fired straight up with a muzzle velocity of 300 meters per second, its velocity at any time t is given by $v = -9.8t + 300$. If the mass of the bullet is 10 grams, find the kinetic energy at $t = 1$ second, $t = 20$ seconds, and at maximum height. Notice that $KE = f(v)$ and $v = g(t) \Rightarrow KE = f(g(t))$.

• 34. The Chézy-Manning hydraulics equation $Q = (1.486/n)AR^{2/3}\sqrt{s}$ gives the flow rate Q in cubic feet per

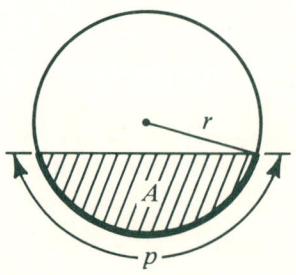

second for fluid flowing in a pipe. A is the cross-sectional area. R, the hydraulic radius, is equal to A/p, where p is the "wetted perimeter" shown above. S is the slope of the pipe. For a pipe for which $A = 9\pi$ square feet, $p = 2\pi$ feet, and $S = 1\% = 0.01$, the equation $n = 0.015 + 0.001t$ is an experimentally (empirically) calculated friction factor for time t in years. Find the flow rate Q for a new pipe and a ten-year-old pipe. Notice $Q = f(n)$ and $n = g(t) \Rightarrow Q = f(g(t))$.

• **35.** Using the following values, calculate the wetted perimeter p in the pipe for (a) a new pipe and (b) a ten-year-old pipe. Assume the flow rate Q is kept constant at $Q = 600$ cubic feet per second. Here, $p = f(n)$ and $n = g(t) \Rightarrow p = f(g(t))$. Use $A = 9\pi$ square feet, $S = 1\%$, and $n = 0.015 + 0.001t$, as in Problem 34.

36. In the illustration below, graph the three functions in (a), (b), and (c) for x, y, and u non-negative. Then consider (d), which depicts $u = f(x)$, $y = g(u)$, and the resulting composite function $y = g(f(x))$. The graph depicts the composite pairing that takes place in $y = \sqrt{25 - 2x}$. For $x = 8$, $u = 25 - 2x$ assigns $x = 8$ to $u = 16$. Then $y = \sqrt{u}$ pairs $u = 16$ with $y = 4$, as indicated by the arrows. This $y = 4$ is the final y-value paired with $x = 8$ on the composite $y = \sqrt{25 - x}$.

37. Draw the x, y, u composite picture for the following functions:

(a) $u = 25 - 8x$ $y = \sqrt{u}$ $y = \sqrt{25 - 8x}$

(b) $u = 16 - x^2$ $y = \sqrt{u}$ $y = \sqrt{16 - x^2}$

(c) $u = (x - 4)^3$ $y = |x|$ $y = |(x - 4)^3|$

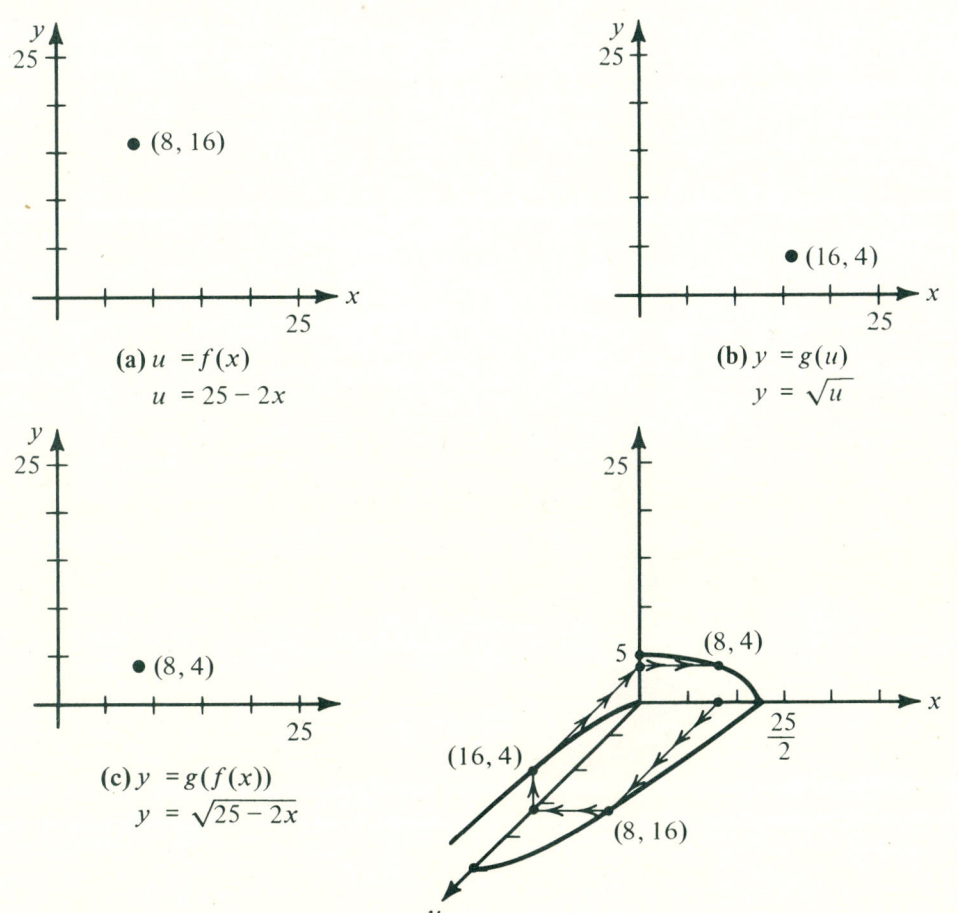

ENDNOTE

1. Functions have existed implicitly since ancient times as curves, as moving paths of projectiles, and as sets of points. One of the first explicit definitions of a function was given in 1667 by James Gregory. He defined a function as a quantity resulting from any combination of algebraic or other imaginable operations.

From 1665 on, Newton used the word "fluent" to describe any relationship between variables.

In 1673 Leibniz introduced the function concept to mean any quantity varying from point to point on a curve.

In 1697, Bernoulli adopted Leibniz's phrase, "function of x," which he symbolized as ϕx.

In 1734, Euler defined a function and introduced the notation $f(x)$. His notation for a function has persisted to this day.

CHAPTER 4

Linear Functions

4.1 THE SLOPE OF A STRAIGHT LINE

The linear functions and relations that we will be studying in this chapter have two things in common. First, their associated graphs are all straight lines. Second, they contain only terms involving x to the first power and y to the first power. (A term involving the product xy is considered a *second-degree* term.) In other words, they can be written in the form

$$Ax + By + C = 0$$

where A and B are not both zero and $A, B, C \in$ reals.

Let's begin by examining the two most basic examples of linear relations, a constant function $y = 3$ and a constant relation $x = 4$. See Figure 4.1.

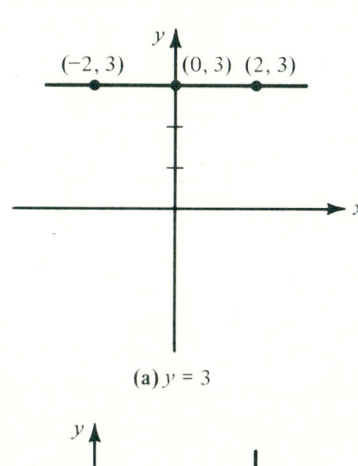

(a) $y = 3$

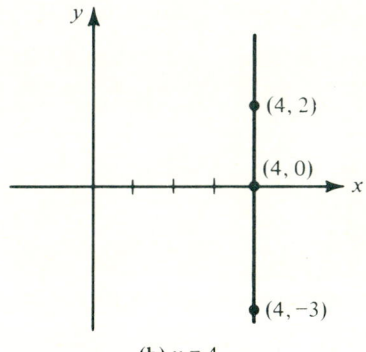

(b) $x = 4$

FIGURE 4.1.

These two constant relations have several properties not held by any other linear relation. The range of $y = 3$ contains only one value, 3; the domain of $x = 4$ contains only 4. Nonconstant linear relations usually have all real numbers as their domains and ranges. The only other exceptions are function that are explicitly restricted, such as the one represented by the graph in Figure 4.2, whose domain is $1 \leq x \leq 5$ and whose range is $2 \leq y \leq 3$.

Constant relations such as $x = 4$ are not functions and constant relations such as $y = 3$ are not one-to-one functions. All other linear equations are one-to-one functions over all real numbers (unless explicitly restricted).

Now, let's use the two linear functions $f(x) = x$ and $g(x) = -x$ to develop an important concept for straight lines called **slope**. The letter m usually stands for slope. The slope of the line $f(x) = x$ is considered positive since its graph increases to the right, and the slope of $g(x) = -x$ is considered negative since its graph decreases to the right. In general, **slope is the rate of change in y, relative to a specific change in x.** We use the symbols Δy (read "delta y") for a change in y and Δx for a change in x. Figure 4.3 depicts two such relative changes for each line.

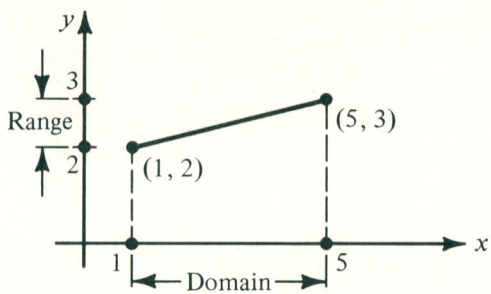

FIGURE 4.2.
A Linear Function wtih Domain and Range.

In part (a) of Figure 4.3, in moving from $x = 1$ to $x = 3$, we move along the line from point A to B. This causes a change in y from $y = 1$ to $y = 3$. The ratio of this change in y (Δy) to the specified change in x (Δx) is the slope (m) of the line passing through A and B.

$$m = \frac{\Delta y}{\Delta x} = \frac{3-1}{3-1} = \frac{2}{2} = 1$$

In moving from C to D, $\Delta y = -1 - (-3) = 2$ (we move up 2) and $\Delta x = -1 - (-3) = 2$ (we move

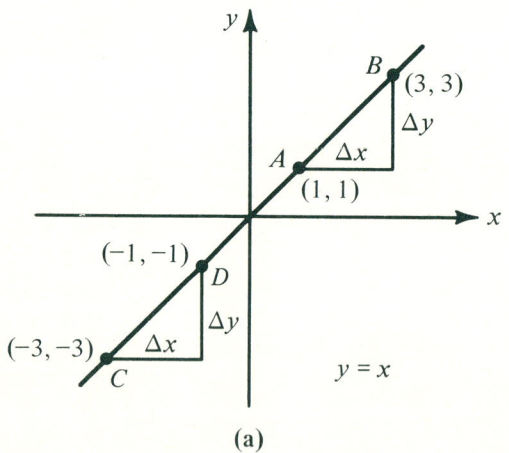

(a)

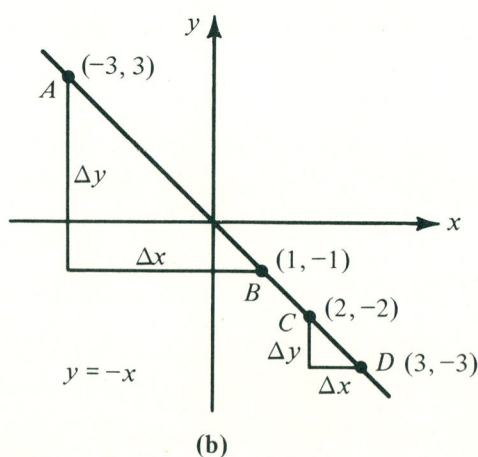

(b)

FIGURE 4.3.

to the right 2). Therefore, $m = 1$ for this relative rate of change also. The slopes between any 2 points on the same straight line will always be equal.

In part (b) of Figure 4.3, in moving from $x = -3$ to $x = 1$, we go from A to B, causing a decrease in y from $y = 3$ to $y = -1$. Therefore,

$$m = \frac{\Delta y}{\Delta x} = \frac{-1-3}{+1-(-3)} = \frac{-4}{4} = -1$$

In moving from C to D,

$$m = \frac{\Delta y}{\Delta x} = \frac{-3-(?)}{+3-(?)} = \frac{-5}{5} = -1$$

which says that if we move to the right 5, we must go down 5 to stay on the line $y = -x$.

We are now ready to extend our definition of slope to other lines. Remember that $c \cdot f(x)$ (where $c > 0$) causes an expansion away from the x-axis for $c > 1$ and a contraction toward the x-axis for $c < 1$. (If $c < 0$, there is also a reflection involved.) Study the graphs in Figure 4.4.

The graphs of $y = 2x$ and $y = -3x$ are expansions of $f(x) = x$ and $g(x) = -x$ away from the x-axis by factors of 2 and 3 respectively. This means that $y = 2x$ is rising twice as fast as $f(x) = x$ and $y = -3x$ is falling twice as fast as $g(x) = -x$. What does the fact that $y = \frac{1}{3}x$ is a contraction toward the x-axis of $f(x) = x$ by a factor of 3 mean about the rate at which $y = \frac{1}{3}x$ is rising? As you would expect, the slope of $y = 2x$ is twice that of $y = x$. For $y = -3x$, the slope downward $(m = -3)$ is three times that of $g(x) = -x$.

The general definition of the slope is given below.

DEFINITION 4.1.
The *slope* m of the line joining points (x_1, y_1) and (x_2, y_2) is given by

$$m = \frac{\Delta y}{\Delta x} = \frac{y_1 - y_2}{x_1 - x_2} = \frac{y_2 - y_1}{x_2 - x_1}, \quad x_1 \neq x_2$$

EXAMPLE 1. Find the slopes of the lines passing through the two given points A and B:

(a) $A(1, 3), B(4, 5) \Rightarrow m = \frac{\Delta y}{\Delta x} = \frac{y_2 - y_1}{x_2 - x_1}$

$$= \frac{5-3}{4-(?)} = \frac{?}{\underline{}}$$

(b) $A(-1, 4), B(3, -2) \Rightarrow m = \frac{\Delta y}{\Delta x} = \frac{y_2 - y_1}{x_2 - x_1}$

$$= \frac{-2-(?)}{3-(?)} = \frac{?}{\underline{}}$$

(c) $A(-2, -3), B(-1, -7) \Rightarrow m = \frac{\Delta y}{\Delta x} = \frac{y_2 - y_1}{x_2 - x_1}$

$$= \frac{(?)-(?)}{(?)-(?)}$$

$$= \frac{?}{\underline{}}$$

Your slopes should be $m = \frac{2}{3}$, $m = \frac{-3}{2}$, and $m = -4$.

The two constant relations $x = 4$ and $y = 3$ again deserve special consideration. The horizontal line, since it neither rises nor falls, is said to have zero slope; it is level. We see that

$$m = \frac{\Delta y}{\Delta x} = \frac{3-3}{-2-2} = 0$$

The vertical line $x = 4$ is as steep as a line can be.

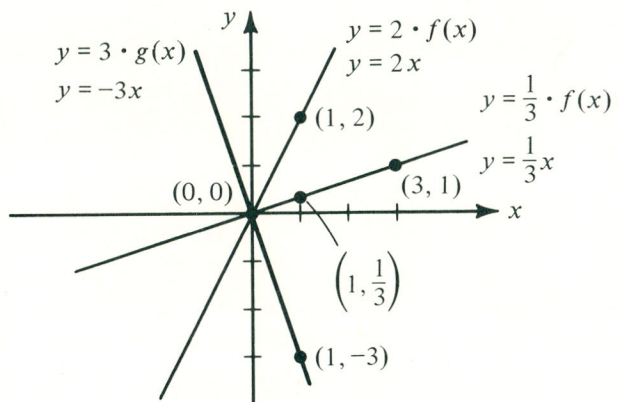

FIGURE 4.4.

CHAPTER 4 Linear Functions

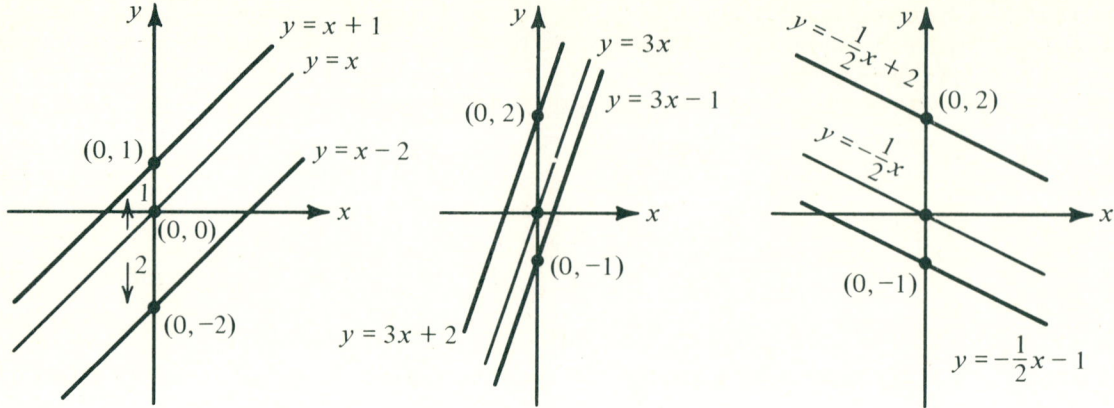

FIGURE 4.5.

We say its slope is undefined. From the slope formula we have

$$m = \frac{\Delta y}{\Delta x} = \frac{2-(-3)}{4-4} = \frac{5}{0}$$

The symbol for this is ∞, infinity. It is not a number, but simply a symbol standing, in this case, for an infinitely large, undefined slope.

We now have all the information needed to complete the discussion of the slope for any straight line.

Consider the pairs of graphs in Figure 4.5. Since $g(x) = f(x) + k$ is simply a translation up for $k > 0$ or down for $k < 0$ of the function $f(x)$, all straight lines are parallel to some multiple of $y = x$ or $y = -x$.

Since parallel lines have the same slope, the slope for any line has already been determined by our previous discussion. This is all summarized by the fact that every linear function can be expressed in the form

$$y = mx + b$$

where m is the slope and b is the translation, up or down, of the line $y = mx$. We call b the y-intercept, since equations of the form $y = mx + b$ have graphs that cross the y-axis at $(0, b)$.

EXAMPLE II. Find the slopes of the following lines:
(a) $\quad f(x) = \frac{1}{3}x - 5 \Rightarrow$
$$m = \tfrac{1}{3}$$
(b) $2y - 3x = 7 \Rightarrow$
$$2y = 3x + 7 \Rightarrow$$
$$y = \tfrac{3}{2}x + \tfrac{7}{2} \Rightarrow$$
$$m = \tfrac{3}{2}$$

EXAMPLE III. Use slope to determine whether or not the following points are collinear (lie on the same straight line):
(a) $A(1, 7)$, $B(-2, -2)$, $C(-1, 1)$

We will see if the slope of the line joining AB is the same as that joining BC. If it is, then, since the lines have point B in common, they are the same line and the points are collinear.

$$m_{AB} = \frac{\Delta y}{\Delta x} = \frac{-2-7}{-2-1} = \frac{-9}{-3} = 3$$

$$m_{BC} = \frac{\Delta y}{\Delta x} = \frac{1-(\ )}{-1-(\ )} = ?$$

(b) $A(1, 0)$, $B(-2, -1)$, $C(4, 1)$

$$m_{AB} = \frac{\Delta y}{\Delta x} = \frac{(\ )-(\ )}{(\ )-(\ )} = \frac{1}{3}$$

$$m_{BC} = \frac{\Delta y}{\Delta x} = \underline{\qquad} = ?$$

EXERCISE 4.1

A. FUNDAMENTALS

Find the slopes of the lines passing through the given points:

1. (a) $A(1, 2)$, $B(3, 2)$ (b) $A(2, 3)$, $B(11, 3)$
2. (a) $A(1, -1)$, $B(1, 3)$ (b) $A(2, 3)$, $B(2, 4)$
3. (a) $A(1, 3)$, $B(3, 5)$ (b) $A(3, 5)$, $B(1, 3)$
4. (a) $A(-1, 3)$, $B(0, 6)$ (b) $A(1, 7)$, $B(3, 13)$
5. (a) $A(-1, 3)$, $B(1, 11)$
   (b) $A(4, 1)$, $B(12, -1)$
6. (a) $A(\frac{1}{3}, \frac{4}{5})$, $B(1, \frac{14}{5})$
   (b) $A(-\frac{1}{3}, \frac{71}{72})$, $B(\frac{1}{4}, \frac{19}{24})$

Find the missing information in the following descriptions of linear functions:

7. $A(-1, 3)$, $B(0, ?)$, $m = 3$
8. $A(12, -1)$, $B(?, 0)$, $m = -\frac{1}{4}$
9. (a) $A(1, 3)$, $B(4, ?)$, $m = 1$
   (b) $A(1, 3)$, $B(?, 4)$, $m = -1$
10. (a) $A(2, 3)$, $B(-2, ?)$, $m = \frac{3}{4}$
    (b) $A(2, 3)$, $B(?, 7)$, $m = -\frac{4}{3}$

B. ESSENTIALS

Sketch the lines through the given pairs of points and compare their slopes.

11. (a) $A(0, 4)$, $B(-1, 3)$
    (b) $A(1, -9)$, $B(0, -8)$
12. (a) $A(2, 0)$, $B(0, -1)$
    (b) $A(1, 1)$, $B(3, -3)$
13. (a) $A(-6, 0)$, $B(3, 6)$
    (b) $A(-4, -3)$, $B(2, 1)$

C. APPLICATIONS

The slope of a line gives the rate of change of the dependent variable (usually y) relative to a change in the independent variable (usually x). Find the slopes of the following lines and explain the meaning of the relative rate of change each depicts.

14. (a) The number of gallons of gasoline in a gas tank after driving s kilometers is given by $g = 20 - 0.4s$.
    (b) The distance s traveled after any time t is $s = 50t$.
    (c) Interpret the product of the two slopes in Problems 14(a) and (b).

15. A free-falling object has a velocity given by $v = -19.6t + 20$, where t is in seconds and v is in meters per second. Interpret the slope of this line, its t-intercept, and its v-intercept.

16. The distance s, in meters, that a rocket is above the earth at any time t is given by $s = 1000t$. Interpret the slope of this line. What does the fact that $t = 0 \Rightarrow s = 0$ mean?

17. (a) In economics *marginal revenue* is the phrase used for the slope of a revenue function, such as $R(x) = 300x - 1000$, where x is quantity. What does marginal revenue mean? What is $R(0)$? What does $R(0)$ mean?
    (b) The marginal cost is the phase used to describe the slope of a cost function like $C(x) = 0.8x + 100$. Explain what the slope means if x is quantity. What does $C(0)$ mean? Can you explain what it means when marginal revenue equals marginal cost for some quantity level x?

18. In the following economic model, C is consumption, C_0 is autonomous consumption ($C_0 > 0$), and X is the national income:
$$C = bX + C_0, \quad 0 < b < 1$$
(a) Explain C_0.
(b) What happens to consumption as the national income increases?
(c) Explain b, the slope of the line.

Prove the following:

★19. If $y = m_1 x + b_1$ is parallel to $y = m_2 x + b_2$, then $m_1 = m_2$.

20. If $m_1 = m_2$, then $y = m_1 x + b_1$ is parallel to $y = m_2 x + b_2$.

CHAPTER 4 Linear Functions

★**21.** If $m_1 \cdot m_2 = -1$, then $y = m_1 x$ is perpendicular to $y = m_2 x$. (HINT: Prove that triangle *ABO* below is a right triangle.)

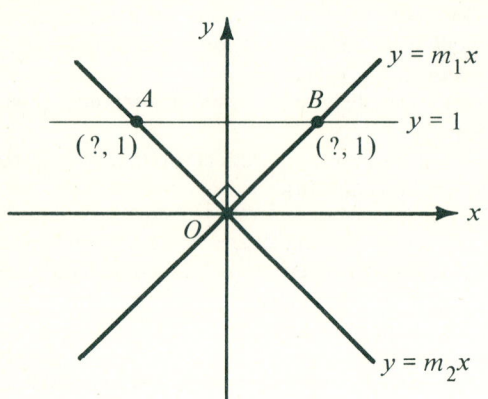

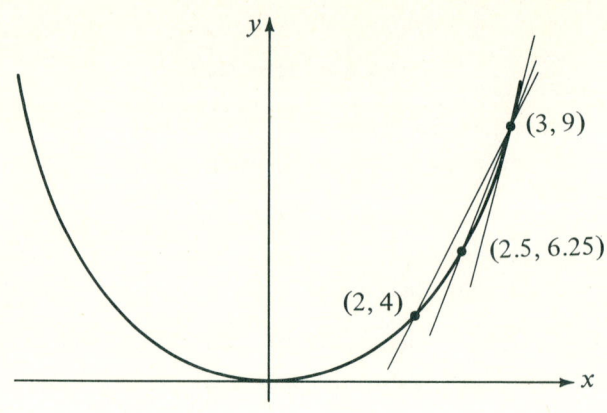

22. If $y = m_1 x$ is perpendicular to $y = m_2 x$, then $m_1 \cdot m_2 = -1$.

> **THEOREM I.**
> Parallel lines have equal slopes.

23. Prove Theorem I.

> **THEOREM II.**
> The slopes of perpendicular lines are negative reciprocals.

24. Prove Theorem II.

• **25.** Find the slopes of the following lines:
(a) l_1 through (2, 4) and (3, 9)
(b) l_2 through (2.5, 6.25) and (3, 9)
(c) l_3 through (2.7, 6.25) and (3, 9)
(d) l_4 through (2.9, 8.41) and (3, 9)
(e) l_5 through (2.99, 8.9401) and (3, 9)

This sequence of slopes (see illustration upper right column) is approaching the slope of the tangent line to the curve $y = x^2$ at $x = 3$. Can you guess what this line's slope really is?

4.2 THE DISTANCE FORMULA

The length of a section of a curve is an important mathematical concept called *arc length*. For linear curves, the length of a section of its arc is simply the distance *s* between the two points at the ends of that line segment. For example, to find the arc

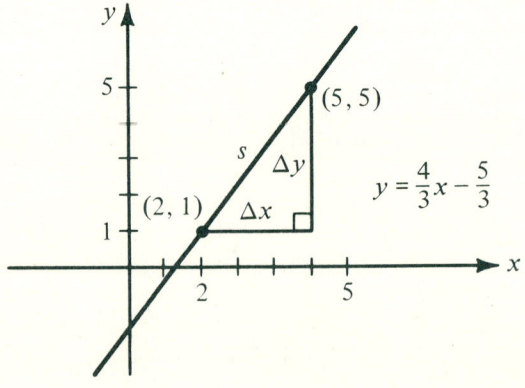

FIGURE 4.6.

length on $f(x) = \frac{4}{3}x - \frac{5}{3}$ between the points (2, 1) and (5, 5), simply apply the Pythagorean theorem,[1] as illustrated in Figure 4.6, to find the distance between them:

$$s^2 = (\Delta x)^2 + (\Delta y)^2 \Rightarrow$$
$$s^2 = (5 - 2)^2 + (5 - 1)^2 \Rightarrow$$
$$s^2 = 3^2 + 4^2 = 25 \Rightarrow$$
$$s = 5$$

We use the principle square root of 25, or $+5$, to keep distances positive.

The distance between two points is given formally in Theorem III. The following proof is a direct consequence of the Pythagorean theorem.

THEOREM III.
The distance s between two points

$$A(x_1, y_1) \quad \text{and} \quad B(x_2, y_2)$$

is given by

$$s = d(AB) = \sqrt{(x_2 - x_1)^2 + (y_2 - y_1)^2}$$

Proof:

$$s^2 = |\Delta x|^2 + |\Delta y|^2$$
$$s^2 = (x_2 - x_1)^2 + (y_2 - y_1)^2$$
$$s = \sqrt{(x_2 - x_1)^2 + (y_2 - y_1)^2}$$

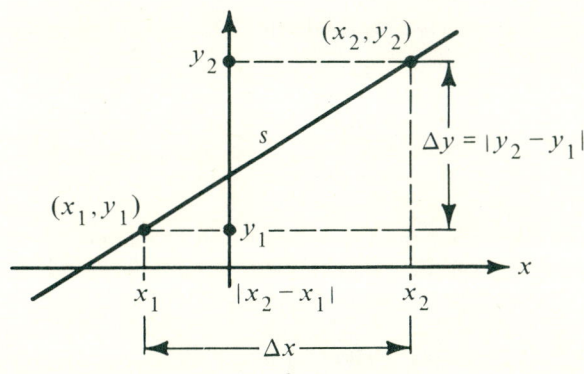

FIGURE 4.7.

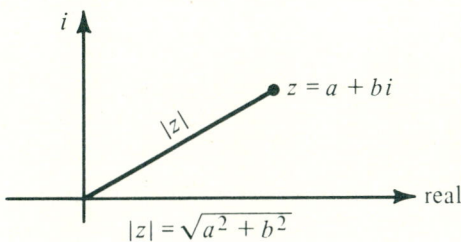

$$s = |x_2 - x_1|$$

FIGURE 4.8. *The Distance Between Two Points on the Real Number Line.*

$$|z| = \sqrt{a^2 + b^2}$$

FIGURE 4.9. *The Distance a Complex Number Is from the Origin.*

This **distance formula** is the third example of a definition of distance that we have seen thus far. The other two are recalled in Figures 4.8 and 4.9.

You will encounter two more definitions of distance, called *norms*. The one used in three-dimensional geometry is very similar to the two-dimensional distance formula above; the one used in the study of vectors is similar to that for complex numbers. There are many more!

EXAMPLE I. Find the distance between the points (1, 3) and $(-1, 7)$.

$$s = \sqrt{(x_2 - x_1)^2 + (y_2 - y_1)^2} \Rightarrow$$
$$s = \sqrt{(1 - (-1))^2 + (3 - 7)^2} \Rightarrow$$
$$s = \sqrt{2^2 + (-4)^2} = \sqrt{20} \Rightarrow$$
$$s = 2\sqrt{5} \approx 4.2$$

EXAMPLE II. Determine whether or not the following points are collinear: $A(-4, -2)$, $B(4, 4)$, $C(8, 7)$.

We will use the distance formula to see if

$d(AB)$ (the distance from A to B) plus $d(BC)$ equals $d(AC)$.

$d(AB) = \sqrt{(4-(-4))^2 + (4-(-2))^2} =$ _____

$d(BC) = \sqrt{(8-?)^2 + (7-?)^2} =$ _____

$d(AC) = \sqrt{(?-?)^2 + (?-?)^2} =$ _____

EXERCISE 4.2

A. FUNDAMENTALS

Find the distances between the following points [$d(AB)$]:

1. (a) $A(1, -1)$, $B(4, 3)$
   (b) $A(-1, -3)$, $B(2, 1)$
2. (a) $A(5, -2)$, $B(11, 6)$
   (b) $A(-14, -7)$, $B(-2, -2)$
3. (a) $A(3, -5)$, $B(7, -1)$
   (b) $A(3, -5)$, $B(5, -3)$

Determine whether or not the following points are collinear by using the distance formula:

4. $A(0, -7)$, $B(1, -2)$, $C(4, 13)$
5. $A(5, 4)$, $B(6, 5)$, $C(3, -1)$
6. $A(\frac{1}{2}, \frac{1}{2})$, $B(\frac{3}{2}, 0)$, $C(-1, -\frac{5}{4})$

Determine whether or not the following sets of points determine a right triangle.

7. $A(-1, -3)$, $B(4, y)$, $C(4, -3)$
8. $A(1, 0)$, $B(2, 5)$, $C(3, 2)$
9. $A(0, 1)$, $B(\frac{63}{5}, -\frac{11}{5})$, $C(3, 5)$

B. ESSENTIALS

10. Show that $d(AB) = d(BA)$ for any two points A and B.
11. Show that $d(AA) = 0$.

12. Use the following sets of points to illustrate the triangle inequality $d(AC) + d(CB) \geq d(AB)$:

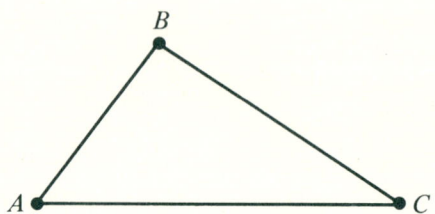

★(a) $A(1, -1)$, $B(7, 7)$, $C(7, -1)$
 (b) $A(1, 1)$, $B(4, 1)$, $C(3, 2)$

★13. Show that $(5, -3)$ is the midpoint of the line segment joining $A(2, -4)$ and $B(8, -2)$.

14. Sketch these points: $(0, 0)$, $(3, 1)$, $(6, 8)$, $(8, -4)$, $(6, -8)$, $(0, -8)$, $(-2, -4)$. Then put in $(3, -4)$. What do these points have in common? Compute the distance from $(8, -4)$ to $(3, -4)$, and from $(3, 1)$ to $(3, -4)$.

C. APPLICATIONS

15. (a) Find the set of points (x, y) located 4 units from $(0, 0)$. (HINT: your answer will be an equation that involves x and y.)
    (b) Find the set of points (x, y) for which the distance from the point to $(3, -4)$ is 5 units. (See Problem 14.)

★16. Find the set of points (x, y) for which the distance from the point to $(4, 0)$ plus the distance from the point to $(-4, 0)$ equals 10. Simplify the resulting equation.

17. Find the set of points (x, y) for which the distance from the point to $(5, 0)$ minus the distance from the point to $(-5, 0)$, in absolute value, equals 8.

18. Find the set of points (x, y) for which the distance from the point to the line $y = -3$ equals the distance from the point to $(0, 3)$.

19. Study the following areas and explain how they illustrate Pythagorean theorem.

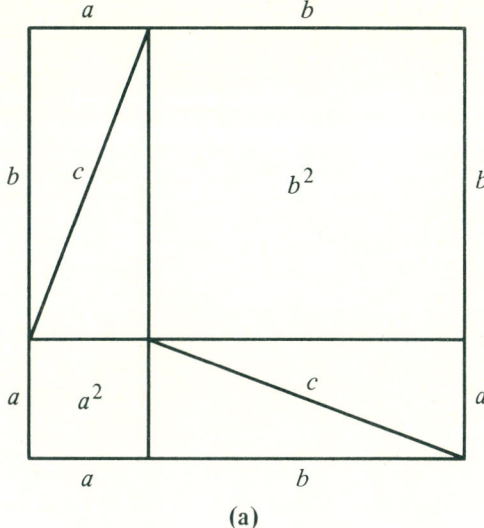

(a)

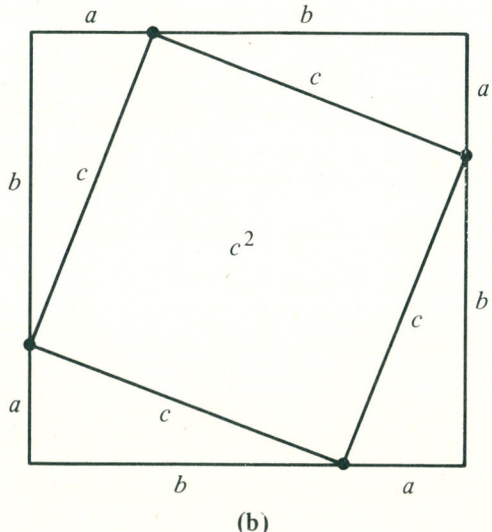

(b)

4.3 GRAPHING AND CURVE FITTING FOR LINEAR FUNCTIONS

At this point you should have no difficulty graphing a linear function such as $f(x) = 3x + 5$ or $g(x) = -2x - 1$. We can determine from the equations that the slopes and y-intercepts are $m = 3$ and $b = 5$ for $f(x)$, and $m = -2$ and $b = -1$ for $g(x)$. It is often necessary to also find the x-intercept when graphing a function. This is accomplished by simply setting $y = 0$ and solving for x. For $f(x)$ and $g(x)$ above, the x-intercepts are found as follows. For $f(x)$,

$$y = 3x + 5 \quad \text{and} \quad y = 0 \Rightarrow$$
$$3x + 5 = 0 \Rightarrow$$
$$x = -\tfrac{5}{3}$$

For $g(x)$,

$$y = -2x - 1 \quad \text{and} \quad y = 0 \Rightarrow$$
$$-2x - 1 = 0 \Rightarrow$$
$$x = -\tfrac{1}{2}$$

The graphs of these two functions, with the intercepts labeled, are shown in Figure 4.10.

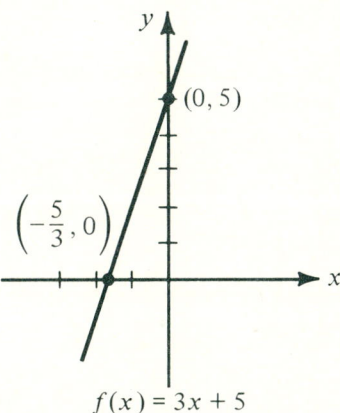

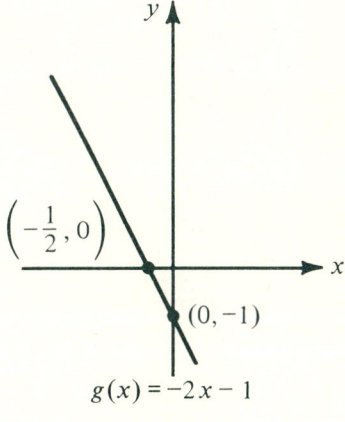

FIGURE 4.10.

Graphing linear functions, then, involves using the equation to find two points (usually the x- and y-intercepts), then drawing the line through them. The reverse problem of being given two points and then finding the equation of the linear function passing through them is called *curve fitting*.

Both of the methods we will use to find the equation of a linear function require calculating or knowing the slope m of the line, where

$$m = \frac{\Delta y}{\Delta x} = \frac{y_2 - y_1}{x_2 - x_1}, \quad x_1 \neq x_2$$

for two points (x_1, x_2), (y_1, y_2) on the line.

The first method is used whenever we know the slope of the line and one point on the line. To find the equation of the line through the point (x_0, y_0) with slope m, we need to translate the graph of $y = mx$ (which has the desired slope) horizontally "x_0" and vertically "y_0." The resulting equation is given below.

POINT-SLOPE FORM.
The equation of the line passing through the point (x_0, y_0) with a slope of m is

$$y = m(x - x_0) + y_0$$

This is called the *point-slope form* of the equation of a straight line.

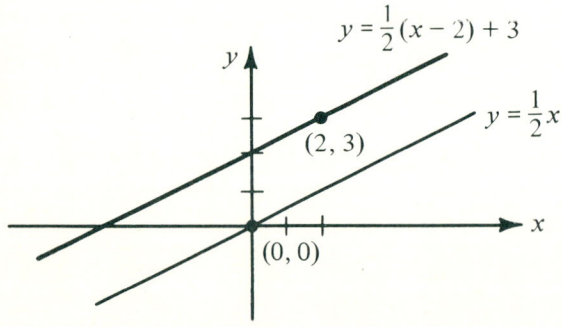

FIGURE 4.11.

EXAMPLE I. Find the equation of the line through the point (2, 3) with a slope of $\frac{1}{2}$.

All we must do is translate $y = \frac{1}{2}x$ to the right 2 and up 3. Therefore,

$$y = \tfrac{1}{2}(x - 2) + 3$$

is the equation of the line. It simplifies to

$$y = \tfrac{1}{2}x + 2$$

If we know two points on the desired line, we simply calculate the slope and proceed as before.

The second method for finding the equation of a line depends on the fact that all linear equations can be written in the *slope-intercept form*.

SLOPE-INTERCEPT FORM.
The equation of the line with slope m and y-intercept b is given by

$$y = mx + b$$

The line $y = mx + b$ crosses the y-axis at $(0, b)$.

EXAMPLE II. Find the equation of the line through (1, 3) and (4, 9).

We must evaluate m and b:

$$m = \frac{\Delta y}{\Delta x} = \frac{y_1 - y_2}{x_1 - x_2} = \frac{3 - 9}{1 - 4} = \frac{-6}{-3} = 2$$

Therefore, we now know the equation:

$$y = 2x + b$$

We realize that since both of the given points lie on the line, they must saisfy the equation of the line. To find b, therefore, we substitute $x = 4$ and $y = 9$ into the equation

$$y = 2x + b$$

yielding

$$9 = 2(4) + b$$

which can be solved for b, yielding
$$b = 1$$
Therefore, $y = mx + b$ becomes
$$y = 2x + 1$$
which is the desired equation.

The technique of solving a system of simultaneous linear equations may also be employed to find the equation of a line through two points. See Chapter 15.

EXERCISE 4.3

A. FUNDAMENTALS
Find the equations and sketch the following straight lines. Label the intercepts.

1. (a) Through $(1, 3)$ with $m = 3$
(b) Through $(3, -2)$ with $m = 3$

2. (a) Through $(3, -1)$ with $m = -\frac{1}{3}$
(b) Through $(-1, -3)$ with $m = -\frac{1}{3}$

3. (a) Through $(3, 4)$ and $(5, 7)$
(b) Through $(5, -1)$ and $(9, 5)$

4. (a) Through $(3, 3)$ and parallel to $y = 3x$
(b) Through $(3, 3)$ and parallel to $y = -\frac{1}{3}x$

5. (a) Through $(-2, -3)$ and perpendicular to $y = x$
(b) Through $(-2, -3)$ and perpendicular to $y = -x$

B. ESSENTIALS
Find the equations of the following straight lines:

★6. Through $(-1, -3)$ and parallel to $2x - 3y = 5$

★7. Through $(2, -1)$ and perpendicular to $3x + 4y - 7 = 0$

8. Through $(1, \frac{1}{2})$ and parallel to $3x - 5y + 7 = 0$

9. Through $(\frac{3}{4}, -\frac{1}{3})$ and perpendicular to $2x + 3y = 7$

10. The result of translating $2x - y + 3 = 0$ up 2 units $[y = f(x) + 2]$

11. (a) The result of revolving $y = 3x$ about the x-axis then translating it to the right 3 units.
(b) The result of translating $y = 3x$ to the right 3 units then revolving it about the x-axis.

12. Draw the curve $y = x^2$ and its tangent line at $(2, 4)$. The slope of this line, determined by methods of calculus, is $m = 4$. Find its equation.

C. APPLICATIONS
Consider the "average cost per unit" function:
$$C(x) = 3x + 5$$
Find the equations that depict each of the following translations.

★13. A translation up 2, due to an imposed tax of 2 per unit.

14. A translation down 3, due to a per unit subsidy of 3.

15. A translation to the right 4, due to a technological breakthrough which permits 4 more units to be built for any average cost figure.

16. (a) Find the equation of the line through $(0°, 32°)$ and $(100°, 212°)$.
(b) Find the equation of the line through $(32°, 0°)$ and $(212°, 100°)$.

•17. Find the equations of the following lines.

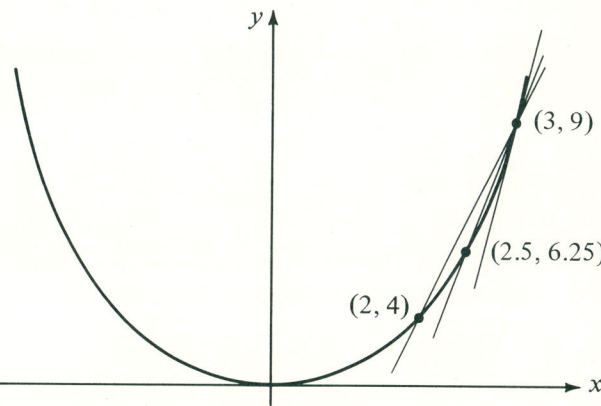

CHAPTER 4   Linear Functions

(a) l_1 through (2, 4) and (3, 9)
(b) l_2 through (2.5, 6.25) and (3, 9)
(c) l_3 through (2.7, 7.29) and (3, 9)
(d) l_4 through (2.9, 8.41) and (3, 9)

These are all called secant lines of the curve $y = x^2$. As the x-cordinate of the first point approaches $x = 3$, the secant line approaches the tangent to the curve $y = x^2$ at (3, 9). This is a central concept of calculus.

4.4 LINEAR FUNCTIONS AND THEIR INVERSES

First, let's find the equation of the inverse of a linear function. If we are given the equation of the function, it is not difficult to find 2 ordered pairs that satisfy the equation. Then, since if (a, b) satisfies $y = f(x)$, (b, a) must satisfy $y = f^{-1}(x)$, we can quickly find 2 points that must satisfy the equation of the inverse. For example, if $f(x) = 3x + 5$, then $f(0) = 5$ and $f(-2) = -1$. Therefore $(0, 5)$ and $(-2, -1)$ lie on the graph of $f(x)$, and $(5, 0)$ and $(-1, -2)$ must lie on the graph of $y = f^{-1}(x)$. By applying the curve fitting techniques of Section 4.3, we find the equation of the inverse through $(5, 0)$ and $(-1, -2)$ as follows:

$$m = \frac{\Delta y}{\Delta x} = \frac{-2 - 0}{-1 - 5} = \frac{-2}{-6} = \frac{1}{3}$$

and

$$y = m(x - x_0) + y_0 \Rightarrow$$
$$y = \tfrac{1}{3}[x - (-1)] - 2 \Rightarrow$$
$$y = \tfrac{1}{3}x + \tfrac{1}{3} - 2 \Rightarrow$$
$$y = \tfrac{1}{3}x - \tfrac{5}{3}$$

We can check this result by proving that $f(f^{-1}(x)) = x$ and $f^{-1}(f(x)) = x$.

$f(x) = 3x + 5$ and $f^{-1}(x) = \tfrac{1}{3}x - \tfrac{5}{3} \Rightarrow$
$f(f^{-1}(x)) = 3 \cdot f^{-1}(x) + 5 = 3(\tfrac{1}{3}x - \tfrac{5}{3}) + 5 \Rightarrow$
$f(f^{-1}(x)) = (x - 5) + 5 = x$

and

$f^{-1}(f(x)) = \tfrac{1}{3}f(x) - \tfrac{5}{3} = \tfrac{1}{3}(3x + 5) - \tfrac{5}{3} \Rightarrow$
$f^{-1}(f(x)) = x + \tfrac{5}{3} - \tfrac{5}{3} = x$

The quickest way to find the equation of the inverse of a function is the two-step procedure: (1) solve for x, and (2) rename the variables. The next example includes graphing a function, graphing its inverse, and finding the equation of the inverse. The domains and ranges for both the function and its inverse are all real numbers. Both are one-to-one functions.

EXAMPLE 1. Graph $f(x) = \tfrac{1}{2}x - 4$ and its inverse.

Solution:
1. First, let's find the intercepts.

$$f(0) = -4 \quad \text{and} \quad f(x) = 0 \Rightarrow$$
$$\tfrac{1}{2}x - 4 = 0$$
$$\tfrac{1}{2}x = 4$$
$$x = 8$$

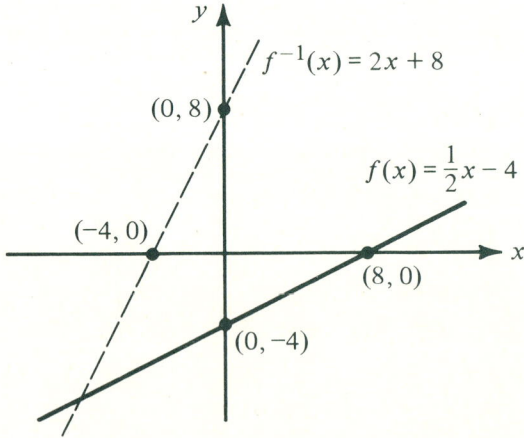

2. The graph of the inverse is plotted quickly by simply realizing that when $f(x)$ is reflected about the line $y = x$, its x-intercept moves to the y-axis (and is therefore the y-intercept of the inverse), and its y-intercept becomes the x-intercept of the inverse.

110

3. Now find the equation of the inverse function. First, solve for x:

$$f(x) = \tfrac{1}{2}x - 4 \Rightarrow$$
$$y = \tfrac{1}{2}x - 4 \Rightarrow$$
$$\tfrac{1}{2}x = y + 4 \Rightarrow$$
$$x = 2y + 8$$

Then, rename the variables:

$$y = 2x + 8 \Leftrightarrow$$
$$f^{-1}(x) = 2x + 8$$

EXERCISE 4.4

A. FUNDAMENTALS

Graph each of the following functions and their inverses:

1. (a) $f(x) = 3x$ (b) $y = -2x$
2. $y = 3x + 4$ 3. $y = \tfrac{3}{4}x - 3$
4. (a) $y = 3$ (b) $x = -2$
5. $y = x$

Find the equations of the inverses for the following:

6. (a) $y = 3$ (b) $x = -4$
7. (a) $y = 3x - 4$ (b) $y = \tfrac{1}{3}x - 5$

B. Find the equations of the inverses for the following functions, then prove that your result is the correct inverse by showing that $f(f^{-1}(x)) = x$ and $f^{-1}(f(x)) = x$.

8. $f(x) = 3x$ 9. $f(x) = 5x - 3$
10. $f(x) = \tfrac{1}{3}x - \tfrac{5}{3}$

11. Prove that the points $(1, 3)$ and $(3, 1)$ are equal distances away from the line $y = x$. This illustrates the symmetry about the line $y = x$ between a function, such as $y = 3x$, and its inverse, $y = \tfrac{1}{3}x$.

12. (a) Find the relationship between the slope of $y = mx$ and its inverse.

★(b) Find the inverse of $f(x) = mx + b$.

(c) Prove your result in (b). Show $f(f^{-1}(x)) = x$ and $f^{-1}(f(x)) = x$.

C. APPLICATIONS

Many uses of the inverse of a function require only that we solve for the other variable. The renaming of x and y is only necessary when we are graphing and want the horizontal axis (the axis of the independent variable) to be the x-axis in both cases. For the following, find the inverse by simply solving for the other variable. Do not rename variables.

★13. $F = \tfrac{9}{5}C + 32$, where F is the number of degrees Fahrenheit and C is the number of degrees Celsius.

14. (a) $C = 2.54I$, where I is the number of inches and C is the number of centimeters.

(b) $K = 2.2L$, where K is the number of kilograms and L is the number of pounds.

15. (a) $D = 0.20F$, where F is the number of francs and D is the number of dollars.

(b) $D = 0.36M$, where M is the number of marks and D is the number of dollars.

16. The "demand" function $p = 3x - 1.53$, where p is the price consumers are willing to pay to buy x amount (quality) of goods.

ENDNOTE

1. Pythagoras, "the father of mathematics," founded a religious brotherhood in southern Italy. The "Pythagoreans" studied number theory, music, and number mysticism. Though the attachment of Pythagoras's name to the theorem is relatively modern, the Pythagoreans probably knew of the theorem. The first written proof of the Pythagorean theorem (about 300 B.C.) is Proposition 47 of Book I of Euclid's *Elements*. The theorem may have been known by the Babylonians about 2000 B.C.

CHAPTER 5

Quadratic Functions

5.1 INTRODUCTION

In this chapter we will be working with functions of the form

$$f(x) = ax^2 + bx + c, \qquad a \neq 0, a, b, c \in \text{reals}$$

called **quadratic functions**. First, study the graphs of $f(x) = x^2$, its one possible reflection, and its translations in Figure 5.1. All quadratic functions can be graphed from the information available in these four examples (Figure 5.1). These curves are called *parabolas*. The two that open upward (hold water) are said to be *concave up*; the others are *concave down* (do not hold water). The lowest point on those that are concave up and the highest point on those that are concave down are called the *vertices* (plural of vertex) of the parabolas.

The fact that quadratic functions have a highest or lowest y-value, depending upon the concavity, makes them naturally suited for applications whose solutions demand either a maximization, such as in a profit function, or a minimization, such as in a cost function. These and other applications will be considered in this chapter.

We will also examine the curve-fitting problem that arises from having three given points instead of two as in the chapter on linear functions.

5.2 GRAPHING QUADRATIC FUNCTIONS

To obtain an accurate sketch of a quadratic function, we must find the x- and the y-intercepts and the vertex. Since three points uniquely determine a parabola (see Exercise 5.2, Problem 18), this is usually all we need to do. The intercepts can be found easily. To find the y-intercept, we simply find $f(0)$. To find the x-intercept(s) (if any), we set $f(x)$ equal to zero and then solve for x. To find the vertex, rewrite the quadratic in the form

$$f(x) = a(x - h)^2 + k$$

SECTION 5.2 Graphing Quadratic Functions

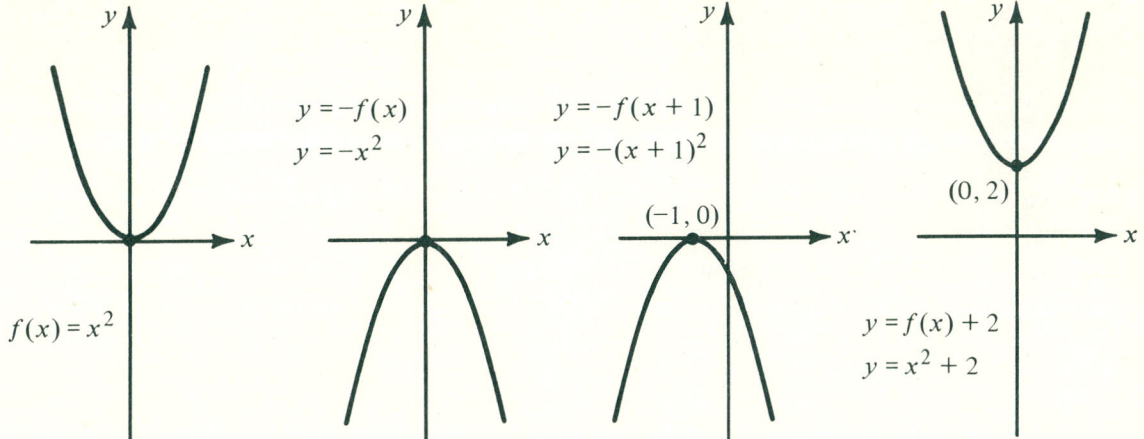

FIGURE 5.1.

by completing the square on the standard form, $f(x) = ax^2 + bx + c$. Then, the vertex is located at (h, k). This follows from our previous discussions of translations. The parabola will be concave up (hold water) for $a > 0$ and concave down (will not hold water) for $a < 0$. This follows from our discussions of reflections. Example I illustrates this process.

EXAMPLE I. Graph $f(x) = 2x^2 - 4x - 6$.
1. The y-intercept is $(0, -6)$, since
$$f(0) = -6$$

2. The x-intercepts are $(3, 0)$ and $(-1, 0)$, since
$$f(x) = 0 \Rightarrow$$
$$2x^2 - 4x - 6 = 0 \Rightarrow$$
$$x^2 - 2x - 3 = 0 \Rightarrow$$
$$\underline{} = 0 \Rightarrow$$

| $x - 3 = 0$ | or | $x + 1 = 0 \Rightarrow$ |
| $x = \underline{}$ | or | $x = \underline{}$ |

3. The vertex is $(1, -8)$, since
$$f(x) = 2x^2 - 4x - 6$$

by completing the square becomes
$$f(x) = 2(x^2 - 2x) - 6 \Rightarrow$$
$$f(x) = 2(x^2 - 2x + \mathbf{1}) - 6 - \mathbf{2} \Rightarrow$$
$$f(x) = 2(x - 1)^2 - 8$$

This is a translation of $y = 2x^2$ to the right 1 and down 8. The parabola is concave up, since $a = 2 > 0$. The graph is in Figure 5.2.

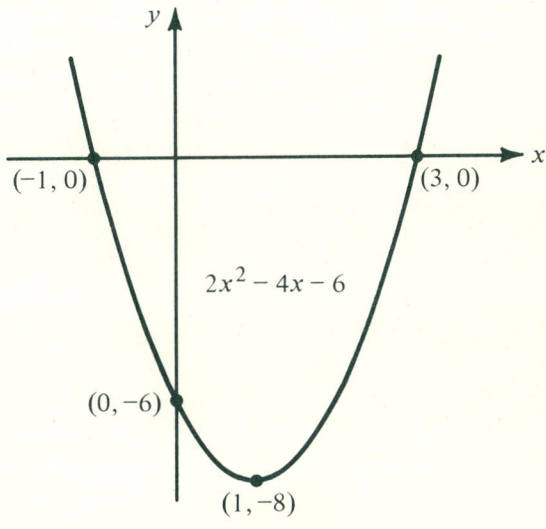

FIGURE 5.2.

Notice that the graph is symmetric about the line $x = 1$, which passes through its vertex.

Now let's plot the graph of
$$f(x) = -3x^2 - 6x + 4$$
which will be concave down, since $a = -3$ is negative. Since $f(x)$ does not factor, let's complete the square to find both the vertex of the parabola and its x-intercepts.

$$f(x) = -3x^2 - 6x + 4 \Rightarrow$$
$$f(x) = -3(x^2 + 2x) + 4 \Rightarrow$$
$$f(x) = -3(x^2 + 2x + 1) + 4 + 3$$

Notice that $(-3 \cdot 1) + 3$ equals 0. Now we have
$$f(x) = -3(x + 1)^2 + 7$$

Hence, the vertex is at $(-1, +7)$. Now, let's set $f(x)$ equal to zero and solve for x to find the x-intercepts.

$$f(x) = 0 \Rightarrow$$
$$-3(x + 1)^2 + 7 = 0 \Rightarrow$$
$$(x + 1)^2 = 7/3$$
$$x = -1 \pm \sqrt{\frac{7}{3}} = \frac{-3 \pm \sqrt{21}}{3}$$

Therefore, since $\sqrt{21} \approx 4.58$,
$$x = \frac{-3 + \sqrt{21}}{3} \approx \frac{-3 + 4.58}{3} = \frac{1.58}{3} \approx 0.53$$

or
$$x = \frac{-3 - \sqrt{21}}{3} \approx \frac{-3 - 4.58}{3} = \frac{-7.58}{3} \approx -2.53$$

Therefore, the x-intercepts are approximately $(0.53, 0)$ and $(-2.53, 0)$. Since $f(0) = 4$, the y-intercept is $(0, 4)$.

Now, we are ready to draw the graph. See Figure 5.3.

Notice that the domain of the function is all real numbers, but the range is bounded above by $y = 7$. $R = \{y \mid y \leq 7\}$.

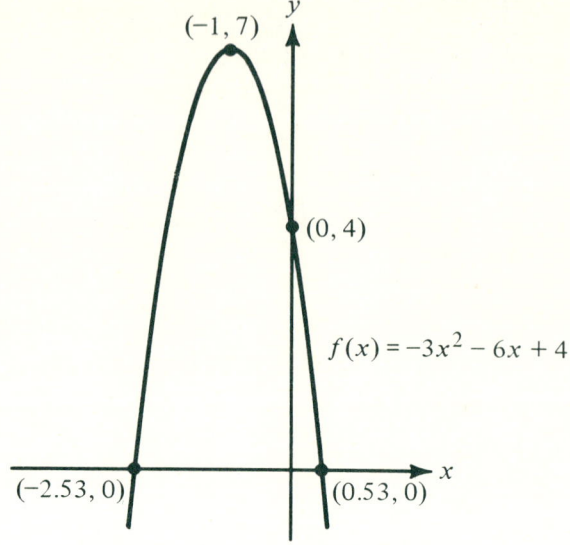

FIGURE 5.3.

The following theorem gives a useful formula for finding the vertex of a parabola.

THEOREM I. The vertex of the parabola given by the quadratic function
$$f(x) = ax^2 + bx + c, \quad a \neq 0$$
is
$$\left(-\frac{b}{2a}, -\frac{b^2 - 4ac}{4a}\right)$$

Proof: The proof of this theorem is based upon the fact that since
$$y = (x - h)^2 + k$$
is a translation of $y = x^2$, horizontally $|h|$ and vertically $|k|$, its vertex is located at (h, k). Now, simply complete the square for the general quadratic equation to write it in the above form.

$f(x) = ax^2 + bx + c, \quad a \neq 0 \Rightarrow$

$f(x) = a\left(x^2 + \dfrac{b}{a}x\right) + c \Rightarrow$

$f(x) = a\left[x^2 + \dfrac{b}{a}x + \left(\dfrac{b}{2a}\right)^2\right] + c - a \cdot \dfrac{b^2}{4a^2} \Rightarrow$

$f(x) = a\left(x + \dfrac{b}{2a}\right)^2 - \dfrac{b^2 - 4ac}{4a}$

Hence, the vertex is at $h = \dfrac{-b}{2a}$ and $k = -\dfrac{b^2 - 4ac}{4a}$.

The graph of $f(x) = ax^2 + bx + c \;\; a \neq 0$ has the following properties:

1. Its y-intercept is located at $(0, c)$.
2. Its vertex is located at

$$\left(-\dfrac{b}{2a}, \; -\dfrac{b^2 - 4ac}{4a}\right)$$

3. Its x-intercepts are located at

$$x = \dfrac{-b \pm \sqrt{b^2 - 4ac}}{2a}$$

where

1. There is one real value (two equal solutions) if $b^2 - 4ac = 0$.
2. There are two real values if $b^2 - 4ac > 0$.
3. There are two complex (no real) values if $b^2 - 4ac < 0$.

EXERCISE 5.2

A. FUNDAMENTALS

Find the intercepts for each of the following quadratic functions.

1. (a) $f(x) = x^2 + 4x + 3$
   (b) $g(x) = x^2 + 4x + 4$
   (c) $h(x) = x^2 + 4x + 5$

2. (a) $f(x) = -x^2 + 2x + 3$
   (b) $g(x) = -x^2 + 2x - 1$
   (c) $h(x) = -x^2 + 2x - 2$

3. (a) $h(x) = x^2 - x - 4$
   (b) $k(x) = -x^2 + x + 4$

Complete the square to find the vertices of the following parabolas.

4. (a) $y = (x - 2)^2 + 1$
   (b) $y = (x^2 + 6x + 9) - 3$
   (c) $y = x^2 + 8x + 5$

5. (a) $y = x^2 - 8x - 1$
   (b) $y = -x^2 + 8x + 1$

6. (a) $y = x^2 - 4x + 16$
   (b) $y = x^2 + 4x + 16$

7. (a) $y = x^2 + 6x + 1$
   (b) $y = 2x^2 + 12x + 2$

8. (a) $y = 2x^2 - 4x - 1$
   (b) $y = 8x^2 - 8x - 1$

B. ESSENTIALS

Graph the following quadratic functions. Label all intercepts and the vertex for each.

9. (a) $f(x) = x^2$ (b) $y = f(x - 1)$
   (c) $y = f(x + 3) - 2$

10. (a) $f(x) = x^2 + 4x + 4$
    (b) $y = f(1 - x)$ (c) $y = 3 - f(x + 1)$

11. (a) $f(x) = x^2 - x$ (b) $y = 1 + f(-x)$
    (c) $y = 2 - f(x + 1)$

12. (a) $f(x) = -x^2 - 2x + 8$
    (b) $y = x^2 + 2x - 8$

13. (a) $y = x^2 - 3x - 5$
    (b) $y = -x^2 + 3x + 5$

14. *(a) $y = 6x^2 + 5x - 6$
    (b) $y = 6x^2 - 7x - 6$

15. Find the range and domain of the functions in Problem 9.

16. Find the range of the functions in Problems 10 and 11.

CHAPTER 5   Quadratic Functions

C. APPLICATIONS

★17. Prove that the graph of $y = ax^2 + bx + c$ is symmetric about the line $x = -b/2a$. [HINT: Show that $f(-b/2a + x) = f(-b/2a - x)$ for all $x \in$ reals.]

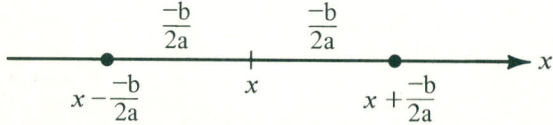

18. Show that 3 distinct points completely determine a parabola. [HINT: Assume there are 2 quadratic equations, $y = a_1 x^2 + b_1 x + c_1$ and $y = a_2 x^2 + b_2 x + c_2$, passing through $(0, y_1)$, $(1, y_2)$, and $(2, y_3)$.]

5.3 INVERSES OF QUADRATIC FUNCTIONS (OPTIONAL)

The graph of a function such as $y = \sqrt{x}$ can best be understood in the context of inverses. Since the inverse of the quadratic function $y = x^2$ is $y = \pm\sqrt{x}$, we know that the graph of $y = \pm\sqrt{x}$ can be found by reflecting the parabolic graph of $y = x^2$ about the line $y = x$. This means that $y = \sqrt{x}$ is the top half of this parabola, and $y = -\sqrt{x}$ is the bottom half.

One way to graph a function such as $y = 3 + \sqrt{4 - x}$ is to find the equation of its inverse, which will be a quadratic function, graph the resulting function, then reflect it about the line $y = x$ by switching the coordinates of the ordered pairs comprising the quadratic function.

EXAMPLE 1. Graph $y = 3 + \sqrt{4 - x}$.

Solution: The inverse of $y = 3 + \sqrt{4 - x}$ is found as follows.

1. Solve for x:

$$y = 3 + \sqrt{4 - x} \Rightarrow$$
$$\sqrt{4 - x} = y - 3 \Rightarrow$$
$$4 - x = (y - 3)^2 = y^2 - 6y + 9 \Rightarrow$$
$$x = -y^2 + 6y - 5$$

2. Rename the variables:

$$y = -x^2 + 6x - 5$$

The graph of $y = -x^2 + 6x - 5$ is shown with the graph of its inverse in Figure 5.5. The originally desired graph is the top half of the parabola opening to the left.

We are now ready to give a *complete* geometric analysis of the functional relationship

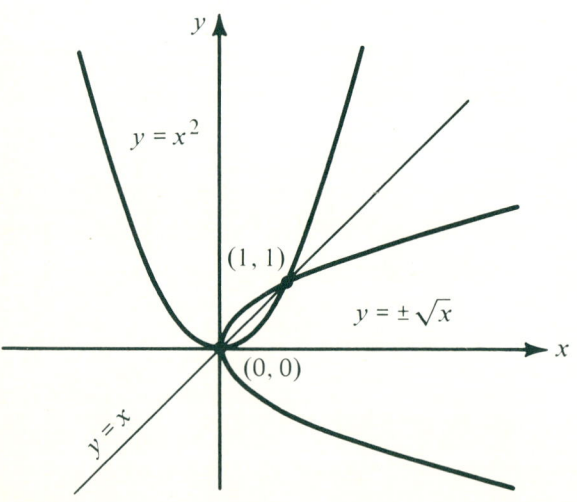

FIGURE 5.4. *A Function and Its Inverse Relation.*

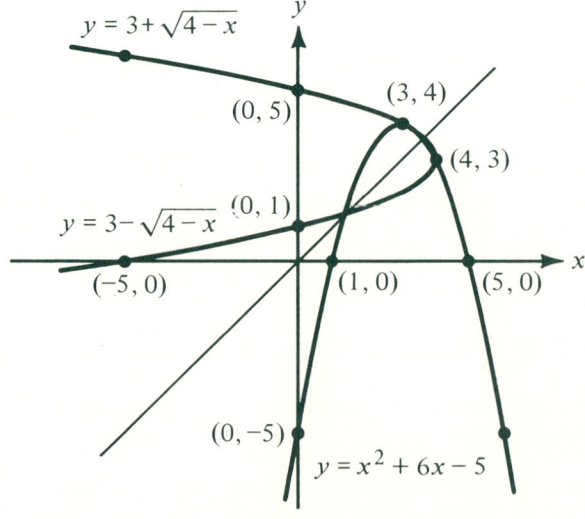

FIGURE 5.5.

established between two sets by a quadratic equation.

EXAMPLE II. Graph $f(x) = x^2 + 6x - 7$ and its inverse. Find the range and domain of each, and find the equation of the inverse relation.

Solution: Since $f(0) = -7$, the y-intercept is $(0, -7)$. Now let's calculate the x-intercepts.

$$f(x) = 0 \Rightarrow$$
$$x^2 + 6x - 7 = 0 \Rightarrow$$
$$\underline{}\ \underline{} = 0 \Rightarrow$$
$$\underline{} = 0 \quad \text{or} \quad \underline{} = 0 \Rightarrow$$
$$x = \underline{} \quad \text{or} \quad x = \underline{} \Rightarrow$$

The x-intercepts are $(-7, 0)$ and $(1, 0)$.
For the vertex,

$$x = \frac{-b}{2a} = \frac{-6}{2(1)} = -3 \Rightarrow$$

$$y = f(-3) = (-3)^2 + 6(-3) - 7 = -16$$

The vertex is $(-3, -16)$.
The graph is shown in Figure 5.6.

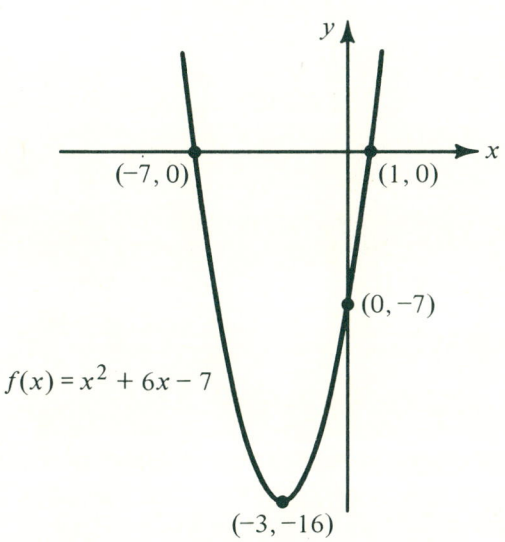

FIGURE 5.6.

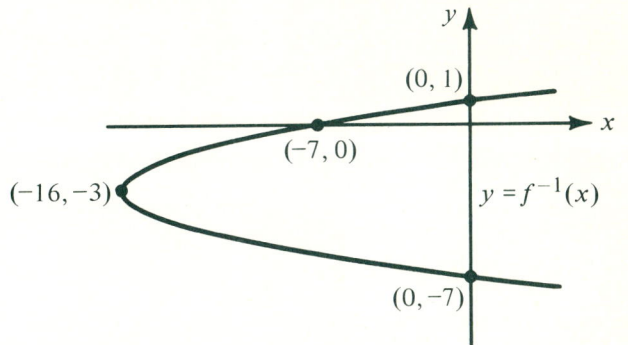

FIGURE 5.7.

The vertex of the inverse is $(-16, -3)$, its y-intercepts are $(0, -7)$ and $(0, 1)$, and its x-intercept is $(-7, 0)$. The graph of the inverse is shown in Figure 5.7.

The range and domain are

$$D_f = \{x \mid x \in \text{reals}\}$$
$$R_f = \{x \mid x \geq -16\}$$
$$D_{f^{-1}} = \{y \mid y \geq -16\}$$
$$R_{f^{-1}} = \{y \mid y \in \text{reals}\}$$

Notice that $D_f = R_{f^{-1}}$ and $R_f = D_{f^{-1}}$.

We must complete the square on $y = x^2 + 6x - 7$ to find the equation of its inverse.

1. Solving for x:

$$y = x^2 + 6x - 7 \Rightarrow$$
$$x^2 + 6x = y + 7 \Rightarrow$$
$$x^2 + 6x + 9 = y + 7 \Rightarrow$$
$$(x + 3)^2 = y + 16 \Rightarrow$$
$$x = -3 \pm \sqrt{y + 16}$$

2. Renaming the variables:

$$y = -3 \pm \sqrt{x + 16}$$

Therefore, the equation of the inverse of the quadratic equation $f(x) = x^2 + 6x - 7$ is the relation $f^{-1}(x) = -3 \pm \sqrt{x + 16}$.

EXERCISE 5.3

A. FUNDAMENTALS
Find the equations of the inverses of each of the following quadratic functions.

1. (a) $y = x^2$ (b) $y = (x+1)^2$
2. (a) $y = x^2 + 4$ (b) $y = 4 - x^2$
3. (a) $y = x^2 + 6x + 9$
   (b) $y = x^2 + 8x + 16$

4. Find the domains of the functions and their inverses from Problem 3.

B. ESSENTIALS
Graph the following functions and their inverses. Then, find the equations of the inverses, and the domains of each function and its inverse.

5. $f(x) = x^2 - x - 12$
6. $g(x) = 2x^2 - 6x - 3$
7. $h(x) = 2x^2 - 3x - 5$

Graph each of the following by finding the equation of its inverse, graphing the inverse, then reflecting this graph about the line $y = x$.

8. $y = -2 \pm \sqrt{x}$ 9. $y = \pm \sqrt{x + 9}$
10. $y = -1 \pm \sqrt{x - 2}$
11. $y = -1 \pm \sqrt{x + 16}$

5.4 MAXIMUMS AND MINIMUMS—AN APPLICATION

Many situations in life require consideration of the maximum outcome (profit, earnings, area, volume, etc.) or the minimum value (costs, materials, etc.). Some of these situations yield mathematical interpretations, called *models*, involving quadratic equations. In these models the maximum value for a concave-down parabola is at its vertex. For minimizations, the modeling quadratic will represent a concave-up parabola, whose minimum value is found at its vertex.

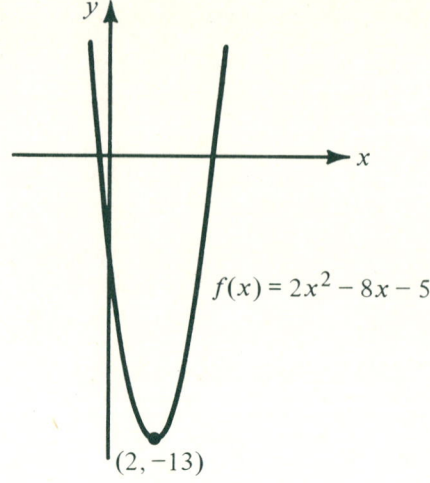

FIGURE 5.8.

EXAMPLE I. To find the minimum value of y in the following expression:

$$y = 2x^2 - 8x - 5$$

simply use the formula $x = -b/2a$, derived in Section 5.2, to find the x-coordinate of the vertex. Then substitute this value into the original quadratic equation to find the corresponding value of y. Hence,

$$x = \frac{-b}{2a} \Rightarrow$$

$$x = \frac{?}{?} = \frac{?}{?}$$

and

$$y = f(2) = 2(2)^2 - 8(2) - 5 = -13$$

Therefore, $y = -13$ is the minimum value for y, and it occurs at $x = 2$. The graph is shown in Figure 5.8.

EXAMPLE II. Find the maximum values of y in

$$y = -x^4 - 4x^2 + 3$$

Since this is quadratic in form, by substituting $u = x^2$ we have a quadratic equation in u:

$$y = -u^2 + 4u + 3$$

SECTION 5.4   Maximums and Minimums—An Application

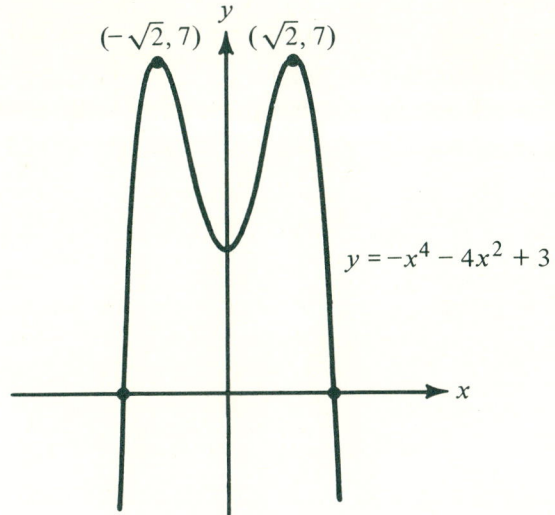

FIGURE 5.9.

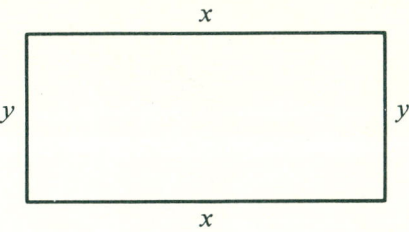

FIGURE 5.10.

where the maximum occurs at

$$u = \frac{-b}{2a} = \frac{-?}{2(?)} = \frac{?}{\underline{}}$$

Then, since $u = x^2$, the x-coordinates of the maximums are at

$$x = \pm\sqrt{u} = \pm\sqrt{2} \approx \pm 1.41$$

The maximum in either case is $y = 7$, since

$$f(+2) = f(-2) = -(\pm\sqrt{2})^4 - 4(\pm\sqrt{2})^2 + 3 = 7$$

The graph of this even function (symmetric about the y-axis) is shown in Figure 5.9.

Examples III, IV, and V present situations where the corresponding mathematical model involves a quadratic equation or an equation that is quadratic in form. In either case, the vertex is easily found, yielding the maximum or minimum depending upon the concavity.

EXAMPLE III. Find the dimensions of the rectangular field that will enclose the maximum pasture if 100 meters of fencing are available.

The area equals the length times the width, and the perimeter equals two times the length plus two times the width. Therefore, if x is the length and y the width, we have

$$A = xy \quad \text{and} \quad p = 2x + 2y = 100 \Rightarrow$$
$$A = xy \quad \text{and} \quad y = 50 - x \Rightarrow$$
$$A = x(50 - x) \Rightarrow$$
$$A = -x^2 + 50x$$

which is a concave-down parabola. Hence,

$$x = \frac{-b}{2a} = \frac{-50}{2(-1)} = 25 \text{ m}$$

is the value of x that will yield the maximum area. Then, since $y = 50 - x$,

$$y = 50 - 25 = 25 \text{ m}$$

And the maximum area is

$$A = x \cdot y = (25 \text{ m})(25 \text{ m}) = 625 \text{ m}^2$$

The procedure for solving these problems can be summarized as follows:

1. Draw a diagram and label all important parts.
2. Express the quantity to be maximized (or minimized) as a function of these variables.
3. Eliminate all but one independent variable. This will yield a quadratic equation.
4. Use $x = -b/2a$ to find the maximizing (or minimizing) value of the independent variable.
5. To find the actual maximum (or minimum) value, evaluate the original equation at the value from step 4.

EXAMPLE IV. A gutter is to be constructed from a rectangular piece of tin 10 centimeter by 100 centimeter by turning up equal segments on

CHAPTER 5   Quadratic Functions

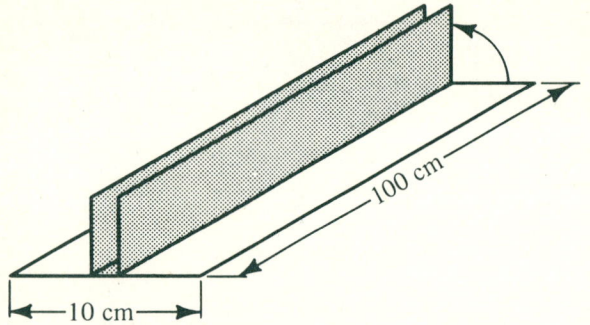

FIGURE 5.11.

each edge. Find the amount to turn up to maximize carrying capacity. Also determine the maximum capacity. See Figure 5.11.

$$V = bhl \Rightarrow$$
$$V = (10 - 2x) \cdot x \cdot (100) \Rightarrow$$
$$V = -200x^2 + 1000x$$

which is a concave-down parabola whose maximum is

$$x = \frac{-b}{2a} = \frac{-1000}{2(-200)} = 2.5 \text{ cm}$$

Therefore, the maximum capacity is

$$V = (10 - 2x) \cdot (x) \cdot (100)$$
$$V = [10 - 2(2.5)](2.5)(100)$$
$$V = 1250 \text{ cm}^3$$

EXAMPLE V. The height s that a free-falling object is above the ground at any time t is given by the formula

$$s = -\frac{1}{2}gt^2 + v_0 t + s_0$$

where g, which equals 9.8 meters per second per second (or 980 cm/sec/sec), is the acceleration due to gravity, v_0 is the initial velocity (the speed at which the object is traveling when we start timing), and s_0 is the initial height above the ground (the height at $t = 0$).

Find the maximum height a bullet reaches if it is fired straight up from the ground from a rifle whose muzzle velocity is 1200 meters per second.

Since

$$s = -\tfrac{1}{2}at^2 + v_0 t + s_0$$

and

$$a = 9.8 \text{ m/sec}^2$$
$$v_0 = 1200 \text{ m/sec}$$
$$s_0 = 0$$

we have

$$s = -4.9t^2 + 1200t$$

Hence, s is maximized at

$$t = \frac{-b}{2a} = \frac{-1200}{2(-4.9)} \approx 125 \text{ sec}$$

or

$$t = 2.08 \text{ min}$$

The maximum height is

$$s = -4.9(125)^2 + 1200(125)$$
$$= 63{,}537.5 \text{ m}$$
$$\approx 63.5 \text{ km}$$

The graph of this distance function is in Figure 5.12.

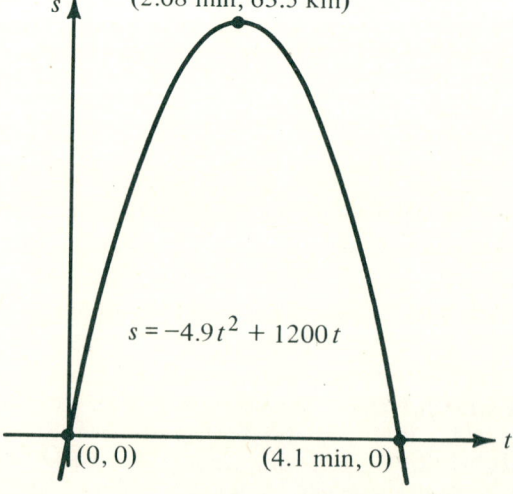

FIGURE 5.12.

SECTION 5.4 Maximums and Minimums—An Application

Answer the following questions about a bullet whose height s above the ground at any time t is given by $s = -4.9t^2 + 1200t$.

1. What is the height s initially? _____
2. What is the height 1 second after the shot? _____
3. When does the bullet hit the ground? (HINT: What is the value of s when it hits the ground?) _____
4. At what times is the bullet 1195.1 meters above the ground?

   $s = 1195.1 \Rightarrow -4.9t^2 + 1200t = 1195.1 \Rightarrow$

   $\underline{\quad ? \quad} = 0 \Rightarrow$

   $(\underline{\quad ? \quad})(t - 1) = 0 \Rightarrow$

   $\underline{\quad ? \quad} = 0 \text{ or } \underline{\quad ? \quad} = 0 \Rightarrow$

   $t = \underline{\quad ? \quad} \text{ or } t = \underline{\quad ? \quad}$

5. If the velocity at any time t is given by $v = -9.8t + 1200$, what is the velocity at impact? (Hint: See number 3 above.) _____

The answers are:

1. 0 meters, since initial height is at $t = 0$
2. 1195.1 meters
3. $t = 1200/4.9 \approx 245$ seconds, when $s = 0$
4. 1 second and $1195.1/4.9 \approx 243$ seconds
5. -1201 meters per second, since it hits at $t = 245$ seconds

EXERCISE 5.4

A. FUNDAMENTALS
Find the minimum values of y for each of the following:

1. (a) $y = x^2 + 6x + 9$
   (b) $y = x^2 + 8x + 16$
2. (a) $y = x^2 - 9$ (b) $y = x^2 - 16$
3. (a) $f(x) = x^2 - 3x - 4$
   (b) $y = f(x - 2)$

Find the maximum values of y for each of the following:

4. (a) $y = 9 - x^2$ (b) $y = 16 - 4x^2$
5. (a) $y = -x^2 + 10x - 25$
   (b) $y = -x^2 + 12x - 36$
6. (a) $g(x) = -x^2 + 2x + 8$
   (b) $y = g(x) + 1$

B. ESSENTIALS
Find the maximum height a projectile attains if its height above the ground at time t is given by the following equations.

7. $s = -4.9t^2 + 19.6t + 20$
8. $s = -4.9t^2 + 58.8t + 3$
9. $s = -4.9t^2 + 98t + 100$

Find the maximum or minimum value of y for each of the following. Indicate whether you have found a maximum or a minimum.

10. (a) $y = 2x^2 + x - 15$
    (b) $y = 15 - 2x^2 - x$
11. (a) $y = x^2 - 8x - 9$
    (b) $y = x^4 - 8x^2 - 9$
12. (a) $y = -x^4 + 18x^2 + 48$
    (b) $y = -x + 18\sqrt{x} + 48$

13. A rectangular pasture is to be fenced along a river with the river acting as the fence on one side. Find the dimensions of field enclosing the maximum area if 100 meters of fencing is used.

14. Find the two numbers whose sum is 120 and whose product is as large as possible.

15. The total cost of producing x radios is given by the following formula:

$$C(x) = 3.25x^3 - 6.50x^2 + 15.75x$$

Find the quantity that should be produced to minimize average cost per radio. Find this minimum average cost. (HINT: Average cost $= C(x)/x$.)

16. The total revenue obtained by selling x radios is given by the following formula:

$$R(x) = -7.50x^2 + 45.00x$$

Find the quantity that should be produced to maximize total revenue.

⋆**17.** Total profit is found by subtracting total cost from total revenue: $P(x) = R(x) - C(x)$. Find the maximum profit if total cost and total revenue are given by the following formulas:

$$R(x) = -2.50x^2 + 40.00x$$
$$C(x) = 2.50x^2 - 35.00x$$

18. (a) Find the dimensions of the rectangle with the largest area that can be inscribed in the circle $x^2 + y^2 = 16$. (HINT: Maximize the square of the area.)

(b) Find the dimensions of the largest rectangle that can be inscribed in the semicircle $y = \sqrt{16 - x^2}$. (HINT: Maximize the square of the area.)

⋆**19.** The drawing below depicts two ships A and B. Ship A is initially 200 kilometers due south of B. B is traveling due east at 15 kilometers per hour, and A is traveling due north at 20 kilometers per hour. Find the time at which the two ships are closest together. (HINT: Use the formula: distance = rate × time to find the distances each ship has traveled during some fixed time t.)

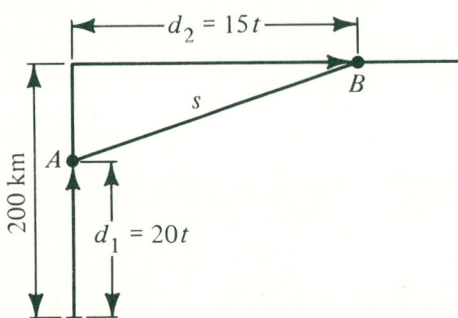

20. Repeat Problem 19 if ship A travels 4 kilometers per hour and ship B travels 3 kilometers per hour.

21. Repeat Problem 19 if ship A travels 8 kilometers per hour and ship B travels 6 kilometers per hour.

22. There is one method of finding an equation that approximately describes a set of data (points) that requires finding the equation of the straight line that best fits the data. By *best fits*, we mean that the sum of the squares of the vertical distances each point is located away from the line must be made as small as possible. For the collection of points pictured below, $y = b$ will be the best-fitting horizontal line for the value of b that minimizes s, where

$$s = d_1{}^2 + d_2{}^2 + d_3{}^2 + d_4{}^2 + d_5{}^2$$

Find b that minimizes s if $d_1 = 4 - b$, $d_2 = 3 - b$, $d_3 = b - 2$, ...

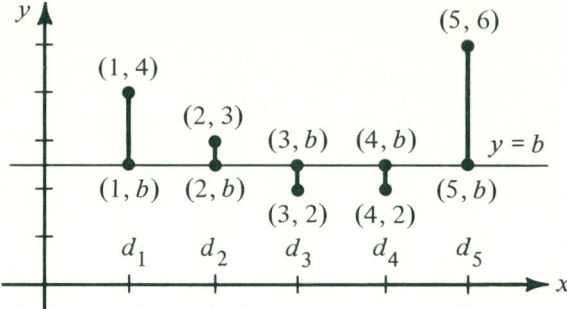

23. Find the horizontal line that best fits the following set of measurements of boiling water, where temperature is given in degrees centigrade.

measurement	1	2	3	4	5
temperature	98	99	102	105	96

24. Find the equation of the form $y = mx$ that best fits this set of measurements of the velocity of a rock dropped into a well.

time (in seconds)	0	1	2	3	4
velocity (in meters per second)	0	−11	−21	−28	−39

5.5 CURVE FITTING (OPTIONAL)

In this section, we are given data that will allow us to find the equation of the parabola. This requires a minimum of three nonlinear data points. If the points include the vertex or any intercepts, the problem is much easier. Consider these examples.

EXAMPLE I. Find the equation of the parabola through the points (2, 0), (4, 0), and (0, 8). We know that $f(x)$ must be of the form

$$f(x) = a(x - 2)(x - 4)$$

and that $f(0) = 8$. Therefore,

$$f(0) = a(-2)(-4) = 8 \Rightarrow$$
$$a = \underline{\quad ? \quad}$$

and the equation must be

$$f(x) = 1 \cdot (x - 2)(x - 4) \Rightarrow$$
$$f(x) = x^2 - 6x + 8$$

Check to make sure this equation satisfies the above conditions; that is, does $f(2) = 0$, $f(4) = 0$, and $f(0) = 8$?

EXAMPLE II. It is experimentally determined that a toy cannon sends a projectile to its maximum height of 5 meters in 2.5 seconds. Find the equation modeling its flight.

We know the vertex of this parabola is at (2.5, 5). Hence the equation, written in the form

$$s = a(t - h)^2 + k$$

becomes

$$s = a(t - \underline{\quad ? \quad})^2 + \underline{\quad ? \quad}$$

Then, assuming that the cannon was fired from the ground, we have

$$t = 0 \Rightarrow s = 0$$

and

$$0 = a(0 - 2.5)^2 + 5 \Rightarrow$$
$$a = \underline{\quad ? \quad}$$

Hence,

$$s = -0.8(t - 2.5)^2 + 5 \Rightarrow$$
$$s = -0.8t^2 + 4t$$

In general, there is a unique quadratic function that is completely determined by 3 given points. (See Problem 20, Exercise 5.4.) If these 3 points are called (x_0, y_0), (x_1, y_1), and (x_2, y_2), then Theorem II tells us how to find the corresponding quadratic function. This method is especially well-suited for hand-held calculators. *Interpolating* simply means passing between (through) the given points. The quadratic equation is written in this unusual form to facilitate the proof.

THEOREM II.
The "interpolating" quadratic function satisfied by the 3 distinct points (x_0, y_0), (x_1, y_1), and (x_2, y_2) is

$$f(x) = a_0 + (x - x_0)a_1 + (x - x_1)(x - x_0)a_2$$

where

$$a_0 = y_0$$

$$a_1 = \frac{y_1 - y_0}{x_1 - x_0}$$

and

$$a_2 = \frac{y_2 - y_0 - (x_2 - x_0)a_1}{(x_2 - x_0)(x_2 - x_1)}$$

Proof: The proof simply amounts to deriving the formulas for a_0, a_1, and a_2. To derive the formulas, use the fact that, since the graph of $f(x)$ must pass through the given points (x_0, y_0), (x_1, y_1), and (x_2, y_2), the function must be satisfied by each of the 3 points. That is, $f(x_0)$ must equal y_0, $f(x_1)$ must equal y_1, and $f(x_2)$ must equal y_2. We will derive the formulas for a_0 and a_1, but the derivation of a_2 is left as an exercise.

The derivation of a_0 is as follows:

$$f(x_0) = y_0 \Rightarrow$$
$$f(x_0) = a_0 + (x_0 - x_0)a_1$$
$$+ (x_0 - x_1)(x_0 - x_0)a_2 \Rightarrow$$
$$y_0 = a_0 + 0 + 0 \Rightarrow$$
$$y_0 = a_0$$

CHAPTER 5 Quadratic Functions

The derivation of a_1 is as follows:

$$f(x_1) = y_1 \Rightarrow$$

$$f(x_1) = a_0 + (x_1 - x_0)a_1 + (x_1 - x_1)(x_1 - x_0)a_2 \Rightarrow$$

$$y_1 = a_0 + (x_1 - x_0)a_1 \Rightarrow$$

$$(x_1 - x_0)a_1 = y_1 - a_0 = y_1 - y_0 \Rightarrow$$

$$a_1 = \frac{y_1 - y_0}{x_1 - x_0}$$

EXAMPLE III. Find the quadratic function whose graph passes through the points $(1, 3)$, $(2, -1)$, and $(4, 3)$.

Let $(x_0, y_0) = (1, 3)$, $(x_1, y_1) = (2, -1)$, and $(x_2, y_2) = (4, 3)$.

Then

$$a_0 = 3$$

$$a_1 = \frac{y_1 - y_0}{x_1 - x_0} = \frac{-1 - 3}{2 - 1} = -4$$

and

$$a_2 = \frac{y_2 - y_0(x_2 - x_0)a_1}{(x_2 - x_0)(x_2 - x_1)} \Rightarrow$$

$$a_2 = \frac{3 - 3 - (4 - 1)(-4)}{(4 - 1)(4 - 2)} \Rightarrow$$

$$a_2 = \frac{0 - 3(-4)}{3 \cdot 2} = 2$$

Therefore,

$$f(x) = a_0 + (x - x_0)a_1 + (x - x_1)(x - x_0)a_2 \Rightarrow$$

$$f(x) = \underline{\ ?\ } + (x - \underline{\ ?\ })(\underline{\ ?\ }) + (x - \underline{\ ?\ })(x - \underline{\ ?\ })(\underline{\ ?\ }) \Rightarrow$$

$$f(x) = 3 - 4x + 4 + 2x^2 - 6x + 4 \Rightarrow$$

$$f(x) = \underline{\ ?\ } x^2 - \underline{\ \ \ \ \ } x + \underline{\ ?\ }$$

Check $f(x_0) = f(1), f(x_1) = f(2)$, and $f(x_2) = f(4)$ to see if the resulting y-values are 3, −1, and 3, respectively.

In Chapter 15 we will see another method for finding the equation of the parabola passing through 3 points using simultaneous equations.

EXERCISE 5.5

A. FUNDAMENTALS

Find the equation of the parabola whose graph passes through the following points:

1. (a) $(3, 0), (-3, 0), (0, -9)$
   (b) $(5, 0), (-5, 0), (0, -25)$

2. (a) $(4, 0), (0, 16), (8, 16)$
   (b) $(-1, 0), (0, 1), (-2, 1)$

3. (a) $(1, 0), (3, 0), (0, 3)$
   (b) $(1, 0), (3, 0), (0, -3)$

4. (a) vertex at $(3, 3)$ and through $(0, 1)$
   (b) vertex at $(-3, 3)$ and through $(0, 1)$

5. (a) vertex at $(2, -2)$ and through $(5, 0)$
   (b) vertex at $(2, 2)$ and through $(5, 0)$

B. ESSENTIALS

Find the interpolating quadratic for the following points:

6. (a) $(0, 0)(1, 1)(2, 4)$
   (b) $(0, 0)(1, -1)(2, -4)$

7. ★(a) $(1, 5), (-2, 8), (3, 13)$
   (b) $(1, -5), (2, -8), (-3, 13)$

8. (a) $(1, 0), (2, -2), (3, 0)$
   (b) $(1, 16), (2, 30), (3, 48)$

Find the quadratic equation whose graph passes through the following points. Use the method you think is most efficient.

9. (a) $(-2, 0), (4, 0), (1, -9)$
   (b) $(5, 0), (-3, 0), (1, 32)$

10. (a) $(4, 0), (\frac{3}{2}, 0), (0, 12)$
    (b) $(-4, 0), (-\frac{3}{2}, 0), (0, 12)$

11. (a) $(1, 3), (-1, 1), (2, 7)$
    (b) $(1, 6), (0, 2), (-2, 6)$

C. APPLICATIONS

12. Derive the formula for a_2 in the theorem for the interpolating quadratic function. (Theorem II).

Find the quadratic function that describes the motion of the free-falling object, given the following information.

Remember that $s = -4.9t^2 + v_0 t + s_0$ is the general equation.

13. A ball is thrown straight up with an initial velocity v_0 of 20 meters per second from a building 200 meters high.

14. A projectile is launched straight up from the ground with an initial velocity of 1200 meters per second.

★15. An object is dropped from a tower 50 meters high and the height of the object above the ground at various times is measured as follows: At $t = 2$, $s = 30.4$, (time t in seconds, distance s in meters).

16. Assume the projectile in Problem 15 is thrown with some unknown initial velocity and at $t = 1$ second, $s = 35.3$ m.

17. Find the minimum cost per radio if we know that an \$8.00 expense is incurred if none are produced, it costs \$13.00 per radio if we produce 5, and \$5.00 if we produce 10. (The numbers of radios are in thousands.) Assume the relationship is quadratic.

18. Find the maximum profit from producing radios if we know that the profit from producing 2 radios is \$3.00, the profit from 3 radios is \$4.00, and we break even if we produce 1 radio. (The numbers of radios and amounts of dollars are in thousands.)

We can use a method called "finite differences" to find the equations of parabolas. For example, to find the equation of the parabola passing through (1, 5), (2, 11), (3, 19), and (4, 29) we construct the following table. To see how this table is constructed, fill in the bottom row as follows. When $x = 4$, $y = 29$; put a 29 in the y column. The first change in y, Δy, is found by subtracting 19 from 29. The second change in y, $\Delta^2 y$, is found by subtracting the two previous, consecutive changes in y (subtract 8 from 10). The fifth column is simply $ax^2 + bx + c$ evaluated at $x = 4$. The second Δy column is the difference of consecutive values of $y = ax^2 + bx + c$, and the last column reflects the differences from consecutive entries in the previous column.

x	y	Δy	$\Delta^2 y$	$y = ax^2 + bx + c$	Δy	$\Delta^2 y$
1	5	6		$a + b + c$	$3a + b$	
2	11	8	2	$4a + 2b + c$	$5a + b$	$2a$
3	19		?	$9a + 3b + c$		?
4	?	?		?	?	

Then, since the corresponding second changes in y (Δy^2) must be equal,
$$2a = 2 \Rightarrow$$
$$a = 1$$

Then, since corresponding first changes in y (Δy) must be equal,
$$3a + b = 6$$

Thus $a = 1$ from above means
$$3 \cdot 1 + b = 6$$
$$b = 3$$

Then, since corresponding y-values must be equal,
$$a + b + c = 5$$

Thus $a = 1$, $b = 3$ means
$$1 + 3 + c = 5$$
$$c = 1$$

The equation is
$$y = ax^2 + bx + c$$
or
$$y = x^2 + 3x + 1$$

Use this technique to solve Problems 19–21.

19. (1, 6), (2, 15), (3, 28)

20. (1, 4), (2, 13), (3, 26)

21. (1, 6), (2, 13), (3, 26)

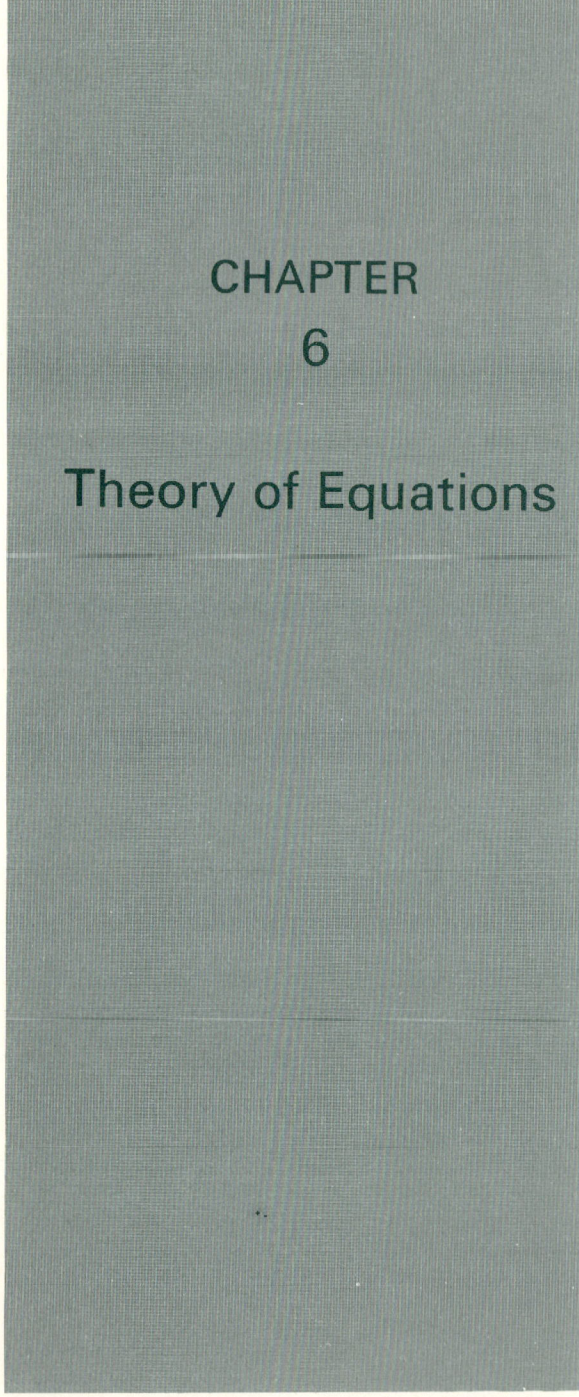

CHAPTER 6

Theory of Equations

6.1 INTRODUCTION TO POLYNOMIALS

Now we are ready to study the graphing of a wider class of functions called *polynomial functions*. First, we must decide exactly which functions are polynomial functions (usually called simply *polynomials*) and which are not. The following functions are polynomials: $y = 2x^3 - 3x^2 + 2x + 1$, $y = 3x + 4$, $y = 2x^2 - 3x + 1$, and $y = \sqrt{2}x$, $y = 2$. The following functions are not polynomials: $y = 1/x$, $y = \sqrt{x}$, $y = (x+1)/(x-3)$, $y = 2^x$, and $y = |x+1|$.

The first difference between some of the functions in the two lists is found in the exponents involved. Polynomials contain only positive integers or zero as exponents. The functions $y = 1/x$ and $y = \sqrt{x}$ do not meet this restriction because $1/x = x^{-1}$ and $\sqrt{x} = x^{1/2}$. The other restriction is that only sums of real numbers times natural-number powers of x or x^0 are polynomials. This means that quotients, exponentials, and absolute value expressions are not polynomials.

> POLYNOMIALS.
> A polynomial can be written in the form
> $$f(x) = a_n x^n + a_{n-1} x^{n-1} + \cdots + a_1 x^1 + a_0 x^0$$
> where the coefficients a_i are real numbers and the exponents are nonnegative integers.

The highest power of x present in the equation is called the *degree* of the polynomial, and its coefficient is called the *leading coefficient*. In the box above if $a_n \neq 0$, then a_n is the leading coefficient and n is the degree of the polynomial.

6.2 SYNTHETIC DIVISION

One of the more important tasks in the graphing of polynomials is that of determining the x-intercepts, called the *zeros* or *roots* of a polynomial.

Since polynomials involve numerous and often high powers of x, it could be rather tedious to determine whether or not $x = 4$, for example, is a zero for a polynomial such as

$$f(x) = 2x^3 + 3x^2 - 4x - 3$$

Luckily, there is a very easy way, called *synthetic division*. It is based upon the fact that if $x = 4$ is a zero of a polynomial, then $(x - 4)$ must be a factor. Therefore, by dividing $2x^3 + 3x^2 - 4x - 3$ by $x - 4$, we will be able to see whether or not it is a factor. If it is, the remainder in the division process will equal zero.

Synthetic division is simply a short, quick way of performing long division. To see how it works, closely compare the following two arrays:

Long Division

$$\begin{array}{r} 2x^2 - 5x + 16 \\ x + 4 \overline{\smash{\big)}\,2x^3 + 3x^2 - 4x - 3} \\ \underline{2x^3 + 8x^2 } \\ -5x^2 - 4x \\ \underline{-5x^2 - 20x } \\ 16x - 3 \\ \underline{16x + 64} \\ -67 \end{array}$$

Synthetic Division

$$\begin{array}{r|rrrr} -4 & 2 & 3 & -4 & -3 \\ & & -8 & 20 & -64 \\ \hline & 2 & -5 & 16 & -67 \end{array}$$

The x's, which are simply position markers, are eliminated in synthetic division. We only use synthetic division for divisors of the form $(x \pm c)$. Notice that one difference between the two is the 4 in the divisor; $x + 4$ is changed to -4 in synthetic division. This allows us to add instead of subtract by automatically changing the signs of the products generated. Both methods are started by simply writing the leading coefficient (in this case 2) in the first place reserved for the answer (the quotient). Next, in long division we multiply 2 times 4 and subtract, and in synthetic division we multiply 2 times -4 and add. This is continued until we have exhausted the dividend, $2x^3 + 3x^2 - 4x - 3$.

Obviously, $(x + 4)$ is not a factor, since the remainder was -67, not 0. We have not wasted our time completely, however, because the remainder happens to be the value of y corresponding to $x = -4$.

We now have

$$\frac{2x^3 + 3x^2 - 4x - 3}{x + 4} = 2x^2 - 5x + 16 + \frac{-67}{x + 4}$$

where $-67/(x + 4)$ is the remainder. Therefore, by multiplying both sides by $x + 4$, we have

$$f(x) = 2x^3 + 3x^2 - 4x - 3 \Rightarrow$$
$$f(x) = (2x^2 - 5x + 16)(x + 4) + (-67) \Rightarrow$$
$$f(-4) = [2(-4) = 5(-4) + 16](-4 + 4) - 67$$
$$= 0 - 67 \Rightarrow$$
$$f(-4) = -67$$

This result is summarized in the Remainder Theorem.

THE REMAINDER THEOREM.
If $p(x)$ is any polynomial function, then it can be written in the form

$$p(x) = q(x)(x - c) + r$$

where $q(x)$ is the quotient in the division of $p(x)$ by $x - c$, r is the remainder, and

$$p(c) = r$$

EXAMPLE I. Write $p(x) = 2x^3 - 3x - 4$ in the form

$$p(x) = q(x) \cdot (x - 3) + r$$

We will use synthetic division to divide

$$\frac{2x^3 - 3x - 4}{x - 3}$$

First, we change the sign of the divisor. Then we write down all of the coefficients without the x's. Since the only way to keep track of the position of the numbers is by the column in which they are written, we must put a zero in for the missing x^2 term.

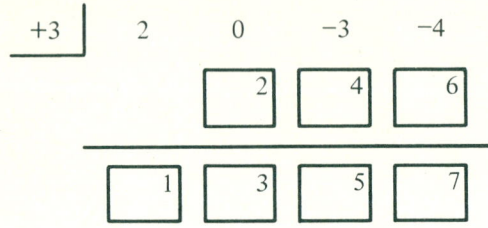

FIGURE 6.1.

Now, fill in the squares as we go. (If you think you see how to proceed, stop and fill in the squares and then check your results with those that follow.) First, drop the leading coefficient straight down to fill box one. Then, multiply our divisor ($+3$) by the contents of box one ($+2$), and put your answer in box two. Now, add 0 and box two to get box three. The product of the divisor (3) times box three goes in box four. The sum of -3 and box four goes in box five. The result of multiplying 3 times box five goes in box six, and -4 plus box six yields the remainder, box seven. The numbers in the boxes below the line are the coefficients of the quotient polynomial $q(x)$.

Now we have

$$\frac{2x^3 - 3x - 4}{x - 3} = 2x^2 + 6x + 15 + \frac{41}{x - 3}, \quad x \neq 3$$

Multiplying both sides by $x - 3$ yields

$$2x^3 - 3x - 4 = (2x^2 + 6x + 15)(x - 3) + 41$$

for all $x \in$ reals. Therefore,

$$p(3) = 41$$

Notice that since we are using only linear divisors, the degree of the quotient will always be one less than that of the dividend. That is, the degree of $p(x)$ is one more than the degree of $q(x)$, where $p(x) = q(x)(x - c) + r$.

EXERCISE 6.2

A. FUNDAMENTALS

Use synthetic division to do the following divisions:

1. (a) $\dfrac{2x^3 - x^2 - 2x + 1}{x + 1}$

   (b) $\dfrac{2x^3 - x^2 - 2x + 1}{x - 1}$

2. (a) $\dfrac{3x^3 + x^2 - 12x - 4}{x + 4}$

   (b) $\dfrac{3x^3 + x^2 - 12x - 4}{x + 5}$

3. (a) $\dfrac{x^4 + 2x^3 - x^2 - 4x - 2}{x - 1}$

   (b) $\dfrac{x^4 + 2x^3 - x^2 - 4x - 2}{x - 2}$

Determine which of the following are polynomials. For those that are polynomials, identify the leading coefficient (a_n), and give the degree of the polynomial.

4. (a) $3x^4 - 2x^3 + x - 1$
   (b) $-2x^3 + 3x^4 + x - 1$

5. (a) $(x + 1)(x - 1)$ (b) $\dfrac{x + 1}{x - 1}$

6. (a) $(2x + 1)^2(x - 1)^3$
   (b) $(2x + 1)^{1/2}(x - 1)^{1/3}$

7. (a) $f(x) = 2x^3 - 7x$
   (b) $g(x) = 4x^3 - 8x^2 - 1$
   (c) $y = f(x) + g(x)$
   (d) $y = f(x) \cdot g(x)$

8. (a) $f(x) = 2x^4 - x^2 - 1$
   (b) $g(x) = 4x^3 - x + 4$
   (c) $y = f(x) + g(x)$
   (d) $y = f(x) \cdot g(x)$

9. (a) $f(x) = 3x^4 - x^2 - 1$
   (b) $g(x) - 3x^4 + x^3 + x$
   (c) $y = f(x) + g(x)$
   (d) $y = f(x) \cdot g(x)$

B. ESSENTIALS

Use synthetic division to write each of the following quotients in the form $f(x) = (x - c) \cdot q(x) + r$.

★10. (a) $\dfrac{x^3 + 4x^2 + x - 6}{x - 1}$

   (b) $\dfrac{x^3 + 4x^2 + x - 6}{x + 2}$

   (c) $\dfrac{x^3 + 4x^2 + x - 6}{x + 3}$

11. (a) $\dfrac{x^4 + 10x^3 + 36x^2 + 54x + 27}{x + 3}$

    (b) $\dfrac{x^4 + 10x^3 + 36x^2 + 54x + 27}{x + 1}$

12. (a) $\dfrac{2x^3 + 7x^2 + 2x - 3}{x + 4}$

    (b) $\dfrac{2x^3 + 7x^2 + 2x - 3}{x + 5}$

    (c) $\dfrac{2x^3 + 7x^2 + 2x - 3}{x + 6}$

13. (a) $\dfrac{2x^4 + 3x^3 - 17x^2 - 27x - 9}{x + 3}$

    (b) $\dfrac{2x^4 + 3x^3 - 17x^2 - 27x - 9}{x + 4}$

14. (a) $\dfrac{2x^3 + 7x^2 - 3}{x - 4}$

    (b) $\dfrac{2x^3 + 7x^2 - 3}{x - 5}$

15. (a) $\dfrac{x^4 - 5x^2 + 4}{x - 3}$

    (b) $\dfrac{x^4 - 5x^2 + 4}{x - 4}$

16. Find $f(0)$, $f(1)$, $f(3)$, and $f(-3)$ for $f(x) = 2x^4 + x^3 + 3x^2 - x - 5$.

C. APPLICATIONS

17. Prove that the degree of the sum of two polynomials is less than or equal to the higher degree of the polynomials.

18. State and prove a theorem analogous to the one in Problem 17 for products of polynomials.

6.3 RATIONAL ZEROS OF POLYNOMIAL FUNCTIONS

Since one of the essentials in graphing is to find the x-intercepts, before we begin graphing polynomial functions, we must learn how to find the values of x that correspond to $y = 0$, the roots or zeros of the polynomial.

If we can determine the complete factorization of a polynomial, we can find its zeros. For example, if we could determine that

$$f(x) = 2x^4 - x^3 - 14x^2 - 5x + 6$$

is factorable into

$$f(x) = (2x - 1)(x + 2)(x - 3)(x + 1)$$

then we would know that its zeros are $x = \frac{1}{2}$, $x = -2$, $x = 3$, and $x = -1$. Unfortunately, it is difficult to factor polynomials with degrees greater than 2. We must therefore resort to a "trial and error" method, using some theorems to reduce the number of possible zeros and using synthetic division to test each candidate.

Let's use synthetic division on the above (unfactored) polynomial. We will first test $x = -1$ as a possible zero [$(x + 1)$ as a possible factor].

If $(x + 1)$ is a factor, then dividing $2x^4 - x^3 - 14x^2 - 5x + 6$ by $x + 1$ should result in a zero remainder:

$$\begin{array}{r|rrrrr} -1 & 2 & -1 & -14 & -5 & 6 \\ & & -2 & 3 & 11 & -6 \\ \hline & 2 & -3 & -11 & 6 & 0 \end{array}$$

Now, by The Remainder Theorem, Section 6.2, we know that $f(x) = (x + 1)(2x^3 - 3x^2 - 11x + 6)$.

Now we are faced with factoring $2x^3 - 3x^2 - 11x + 6$, a polynomial of 1 less degree than the original one. Let's try $x = -2$:

$$\begin{array}{r|rrrr} -2 & 2 & -3 & -11 & 6 \\ & & ? & & \\ \hline & & ? & & 0 \end{array}$$

Now, since we have no remainder,

$$2x^3 - 3x^2 - 11x + 6 = (x+2)(2x^2 - 7x + 3)$$

and

$$f(x) = (x+1)(x+2)(2x^2 - 7x + 3) \Rightarrow$$
$$f(x) = (x+1)(x+2)(2x-1)(x-3)$$

The real problem is to decide which numbers to try as possible zeros.

Suppose p/q, where p and q are integers, is a rational zero for $y = f(x)$ above. Then $(qx - p)$ would have to be a factor of $f(x)$, since

$$qx - p = 0 \Rightarrow$$
$$qx = p \Rightarrow$$
$$x = \frac{p}{q}$$

By carefully examining the multiplication of

$$(x+1)(x+2)(2x-1)(x-3)$$

which yields

$$2x^4 - x^3 - 14x^2 - 5x + 6$$

we can determine something about p/q, relative to the leading coefficient 2 and the constant term 6. Notice that $2x^4$ results from the product of the first term in each factor, $x \cdot x \cdot 2x \cdot x$. Therefore, if $(qx - p)$ is a factor, (qx) times the first terms in the remaining factors must yield $2x^4$. Then, *q must evenly divide the leading coefficient*, since leading coefficient has q as a factor. Similarly, the constant term 6 is the product of the last terms in each of the factors, $(1)(2)(-1)(-3)$. Hence, if $(qx - p)$ is a factor, then p times the second terms in the remaining factors must equal 6. Therefore, *p must evenly divide the constant term*. These results are summarized in Theorem I, which is followed by part of the proof.

THEOREM I.
If p/q is a rational zero of a polynomial function

$$f(x) = a_n x^n + a_{n-1} x^{n-1} + \cdots + a_0$$

then p must divide the constant term a_0, and q must divide the leading coefficient a_n.

Proof: If p/q is zero, then $f(p/q) = 0$ or

$$a_n \left(\frac{p}{q}\right)^n + a_{n-1} \left(\frac{p}{q}\right)^{n-1} + \cdots$$
$$+ a_1 \left(\frac{p}{q}\right)^1 + a_0 \left(\frac{p}{q}\right)^0 = 0 \Rightarrow$$
$$a_n \frac{p^n}{q^n} + a_{n-1} \frac{p^{n-1}}{q^{n-1}} + \cdots + a_1 \frac{p}{q} + a_0 = 0$$

Then, by multiplying both sides by q^n, we have

$$a_n p^n + a_{n-1} p^{n-1} q + \cdots$$
$$+ a_1 p q^{n-1} + a_0 q^n = 0 \Rightarrow$$
$$a_n p^n = -a_{n-1} p^{n-1} q - \cdots$$
$$- a_1 p q^{n-1} - a_0 q^n \Rightarrow$$
$$a_n p^n = q(-a_{n-1} p^{n-1} - \cdots$$
$$- a_1 p^{n-2} - a_0 q^{n-1})$$

Now, since q divides the right side of the equation, it must also divide the left. But since p and q have no common factors (since we always assume p/q is a rational number that has been reduced to lowest terms), the only way q can divide $a_n p^n$ is for q to divide a_n, the leading coefficient. The other part of the proof, that p must divide a_0, is included as an exercise.

We will use the Theorem I to compile a list of all possible rational zeros for a given polynomial.

Then we will use synthetic division to check each candidate. The only possible zeros not included in this list are irrational ones, such as $x = \sqrt{2}$, which we will examine in Section 6.5.

Two illustrations of the application of this theorem are included in Example I.

EXAMPLE I. Find the list of rational candidates for the zeros of (a) $p(x) = x^3 - 2x^2 - 6x + 8$ and (b) $p(x) = 3x^4 - 2x^2 + 7x - 10$.

Solution:
(a) Since p must divide the constant term 8 evenly,
$$p = \pm 1, \pm 2, \pm 4, \pm 8$$
and since q must divide the leading coefficient 1 evenly,
$$q = \pm 1$$
Therefore,
$$\frac{p}{q} = \frac{p}{\pm 1} = \pm 1, \pm 2, \pm 4, \pm 8$$

(b) Since p must divide the constant term __?__,
$$p = \pm 1, \pm \underline{}, \pm \underline{}, \pm 10$$
and since q must divide the leading coefficient __?__,
$$q = \pm 1, \pm \underline{}$$
Therefore, p/q equals the divisors of p divided by ± 1 or ± 3, or
$$x = \frac{p}{q} = \pm 1, \pm 2, \pm 5, \pm 10, \pm \frac{1}{3}, \pm \frac{2}{3}, \pm \frac{5}{3}, \pm \frac{10}{3}$$

Theorems II and III help us eliminate some of the candidates from these lists as we proceed with the synthetic division.

THEOREM II.
If $p(x) = (x - c) \cdot q(x) + r$, and if the coefficients of $q(x)$ and the remainder r are all positive for $c > 0$, then there are no zeros of $p(x)$ greater than $x = c$.

THEOREM III.
If $p(x) = (x - c) \cdot q(x) + r$, and if the coefficients of $q(x)$ and r alternate in sign (positive, negative, positive, ...), then there are no zeros of $p(x)$ less than $x = c$, provided $c < 0$.

Instead of proving these theorems in general, let's work through the proof with a particular polynomial, for example,
$$p(x) = x^4 + 2x^3 + 4x + 8$$
The following does not constitute a proof. It is simply a plausability argument. The formal proof is a natural generalization of the following argument.

Let's illustrate this argument for Theorem II. For some positive value of c, for example, $c = 2$, synthetic division gives

$$\underline{2 \rfloor \quad 1 \quad 2 \quad 4 \quad 8}$$
$$?$$

Therefore,
$$p(x) = (x - c)q(x) + r \Leftrightarrow$$
$$p(x) = (x - 2)(x^3 + 4x^2 + 8x + 20) + 8$$
where the coefficients of
$$q(x) = x^3 + 4x^2 + 8x + 20$$
and the remainder, $r = 8$, are all positive. We will now show that no value larger than $x = 2$ can possibly be a zero for $p(x)$. If $x_0 > 2$, then $x_0 - 2$ is positive, $r = 8$ is positive, and $q(x_0) = x_0^3 + 4x_0^2 + 8x_0 + 20$ is positive. Therefore,
$$p(x_0) = (x_0 - 2)(x_0^3 + 4x_0^2 + 8x_0 + 20) + 8 > 0$$
This means that
$$p(x_0) \neq 0 \quad \text{for any } x_0 > 2$$
as expected.

Now let's turn to Theorem III. Consider
$$p(x) = x^4 + x^3 - x^2 + x + 2$$

131

and a negative value of c, such as $c = -2$. The synthetic division yields

$$\begin{array}{r|rrrrr} -2 & 1 & 1 & -1 & 1 & 2 \\ & & & & & \\ \hline & & & ? & & \end{array}$$

We then have

$$p(x) = (x + 2)(x^3 - x^2 + x - 1) + 4$$

where the coefficients of

$$q(x) = x^3 - x^2 + x - 1$$

and $r = 4$ alternate in sign. We will now show that no value of x smaller than $x = -2$ can possibly be a zero for $p(x)$.

If $x_0 < -2$, then $x_0 + 2 < 0$ and $q(x_0) = x_0^3 - x_0^2 + x_0 - 1 < 0$, since x_0 would be negative and *every* term in $q(x_0)$ would be negative. If x_0 were -3, for example, $(-3)^3 < 0$, $-(-3)^2 = -9 < 0$, etc. Then, since $r > 0$, we have

$$p(x_0) = (x_0 + 2)(x_0^3 - x_0^2 + x_0 - 1) + r > 0$$

since the product of the negative factors $x_0 + 2$ and $x_0^3 - x_0^2 + x_0 - 1$ must be positive.

Therefore,

$$p(x_0) \neq 0$$

for any value of $x < -2$.

The statements and discussions of the preceding theorems may make them seem more complicated than they really are. Below is an example that illustrates their use.

EXAMPLE II. Find all rational zeros for

$$p(x) = 2x^4 + x^3 - 19x^2 - 9x + 9$$

The possible rational zeros are found by knowing that

$p = \pm 1, \pm?, \pm?$ and $q = \pm 1, \pm?, \pm?$

Hence,

$$\frac{p}{q} = \pm 1, \pm 3, \pm 9, \pm \frac{1}{2}, \pm \frac{3}{2}, \pm \frac{9}{2}$$

Notice what happens if we attempt to divide by $x - 4$. Using synthetic division, we have

$$\begin{array}{r|rrrrr} 4 & 2 & 1 & -19 & -9 & 9 \\ & & 8 & 36 & 68 & 236 \\ \hline & 2 & ? & ? & ? & ? \end{array}$$

This illustrates Theorem II. Since we divided by a positive number and the signs of the bottom line [the coefficients of $q(x)$ and $r = 245$] are all positive, we know that there will be no zeros of $p(x)$ greater than $x = 4$. Hence, we do not have to try $x = 9$, even though it is on our list.

Now let's try $x = -4$:

$$\begin{array}{r|rrrrr} -4 & 2 & 1 & -19 & -9 & 9 \\ & & -8 & 28 & -36 & 180 \\ \hline & 2 & ? & ? & ? & ? \end{array}$$

This illustrates Theorem III, because we divided by a negative and the signs of the third line alternate. Hence, there are no zeros for $p(x)$ less than $x = -4$. Now we do not have to try $x = -9$.

To proceed, try $x = 3$:

$$\begin{array}{r|rrrrr} 3 & 2 & 1 & -19 & -9 & 9 \\ & & ? & ? & ? & ? \\ \hline & ? & ? & ? & ? & 0 \end{array}$$

Hence,

$$p(x) = 2x^4 + x^3 - 19x^2 - 9x + 9 \Rightarrow$$
$$p(x) = (x - 3)(2x^3 + 7x^2 + 2x - 3)$$

Now try $x = -1$ by dividing $x + 1$ into the unfactored third degree polynomial $2x^3 + 7x^2 + 2x - 3$:

$$\begin{array}{r|rrrr} -1 & 2 & 7 & 2 & -3 \\ & & ? & ? & 3 \\ \hline & ? & ? & ? & 0 \end{array}$$

Therefore,

$$2x^3 + 7x^2 + 2x - 3 = (x + 1)(2x^2 + 5x - 3)$$
$$= (x + 1)(2x - 1)(x + 3)$$

and

$$p(x) = (x - 3)(x + 1)(2x - 1)(x + 3)$$

The zeros for $p(x)$ are $x = 3$, $x = -1$, $x = +\frac{1}{2}$, and $x = -3$.

SECTION 6.3   Rational Zeros of Polynomial Functions

Theorems II and III only allow us to find *positive upper bounds* (numbers for which any larger values will not work) and *negative lower bounds* (numbers for which any smaller values will not work) on the set of zeros for a polynomial. To divide by a positive and have the signs alternate or to divide by a negative and have all the signs turn out positive (or negative) means nothing.

The following, called Descartes' Rules of Signs, are helpful in determining when we have found all the negative or positive zeros. Here we count the zero $x = -2$ in $(x + 2)^2$ as 2 negative roots and the zero $x = 1$ in $(x - 1)^3$ as 3 roots. The rules follow:

1. The number of positive real zeros of a polynomial $p(x)$ is equal to or less by some even integer than the number of changes in the signs of the coefficients present in $p(x)$, where $p(x)$ is written in descending powers of x.
2. The number of negative real zeros of $p(x)$ is equal to or less by some even number than the changes in sign in $p(-x)$, where $p(x)$ is written in descending powers of x.

EXAMPLE III. In the equation

$$p(x) = 2x^4 + x^3 - 19x^2 - 9x + 9$$
$$\qquad\qquad\quad (1) \qquad\qquad (2)$$

There are two sign changes marked. Therefore, there are either 2 or no positive zeros. And since

$$p(-x) = 2x^4 - x^3 - 19x^2 + 9x + 9$$
$$\qquad\qquad\quad (1) \qquad\qquad (2)$$

has 2 sign changes, there are either 2 or no negative zeros. This agrees with our previous factorization of $p(x)$ into

$$p(x) = (x - 3)(x + 1)(2x - 1)(x + 3)$$

which shows 2 positive and 2 negative zeros. The graph of this function is given in Figure 6.2.

EXERCISE 6.3

A. FUNDAMENTALS
Use synthetic division and the theorems introduced in this section to completely factor the following polynomials.

1. (a) $f(x) = 2x^3 - x^2 - 2x + 1$
   (b) $g(x) = -2x^3 + x^2 + 2x - 1$
2. (a) $f(x) = 3x^3 + x^2 - 12x - 4$
   (b) $g(x) = -3x^3 + x^2 + 12x - 4$
3. (a) $y = 2x^4 + x^3 + 3x^2 - x - 5$
   (b) $y = 2x^4 - x^3 + 3x^2 + x - 5$
4. (a) $y = x^4 + 10x^3 + 36x^2 + 54x + 27$
   (b) $y = x^4 - 10x^3 + 36x^2 - 54x + 27$
★5. $y = 9x^4 + 36x^3 + 35x^2 - 4x - 4$
6. $y = 6x^4 + x^3 - 46x^2 - 39x + 18$

7. Use Descartes' Rules of Signs on Problems 2(a), 3(a), and 4(a) to determine the number of possible positive zeros and the number of possible negative zeros.

B. ESSENTIALS
Use synthetic division to find the smallest natural number, c, that yields all positive coefficients for $q(x)$ and yields $r > 0$ for $f(x) = (x - c)q(x) + r$. Since there are no zeros for $f(x)$ greater than $x = c$, c is an upper bound for this set. If c is the smallest number that is an

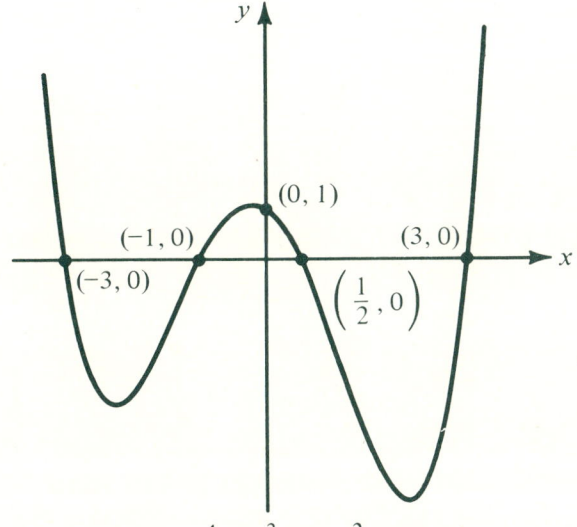

$p(x) = 2x^4 + x^3 - 19x^2 - 9x + 9$

FIGURE 6.2.

upper bound for the set of zeros, c is the *least upper bound*, LUB.

8. (a) $f(x) = 2x^2 + 3x + 1$
   (b) $y = f(-x)$

9. ★(a) $f(x) = x^4 - 2x^3 - x^2 - 1$
   (b) $y = f(-x)$

10. (a) $f(x) = 6x^4 - 7x^3 - 12x^2 + 3x + 2$
    (b) $y = f(-x)$

11. Find the largest negative integer that is a lower bound for the sets of zeroes in Problems 8, 9, and 10. This is the GLB (greatest lower bound) for the set.

C. APPLICATIONS

12. Prove that if $x = p/q$ is a zero for some polynomial $y = p(x)$, then p must divide the constant term.

13. Prove Theorem II.

6.4 GRAPHING POLYNOMIAL FUNCTIONS

When we are graphing polynomial functions, there are several things we should get into the habit of checking before we begin plotting points. First we should check the degree of the polynomial and then the sign of the leading coefficient. These tell us the behavior of the graph of the polynomial to the far left and right. The best way to remember this information is to keep the graphs of $y = x$ and $y = x^2$ in mind as examples of an odd and an even-degree polynomial with a positive leading coefficient a_n (in these cases $+1$).

For polynomials whose leading coefficients are negative, simply reflect the 2 curves about the x-axis and note their extreme values.

The degree of polynomial also yields information about the maximum number of times it may turn and the maximum number of zeros.

> THEOREM IV.
> An nth-degree polynomial may have, at most, $n - 1$ turning points.

> THEOREM V.
> An nth-degree polynomial may have, at most, n zeros.

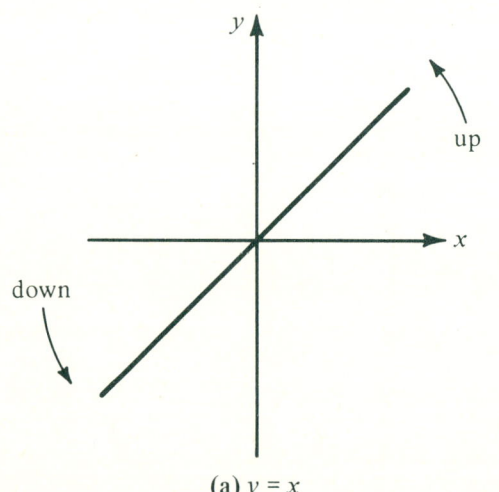
(a) $y = x$

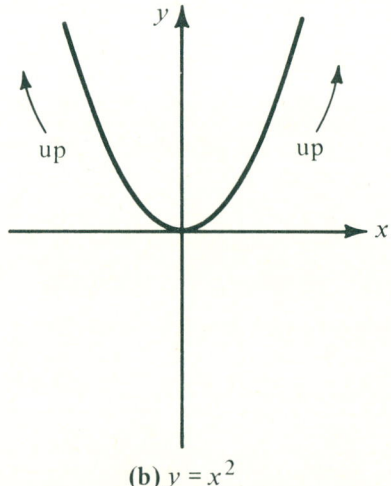
(b) $y = x^2$

FIGURE 6.3. *(a) Odd-degree Polynomials with a Positive Leading Coefficient, Eventually Go Down to the Left and Up to the Right. (b) Even-degree Polynomials with $a_n > 0$, Eventually Go Up to the Left and Up to the Right.*

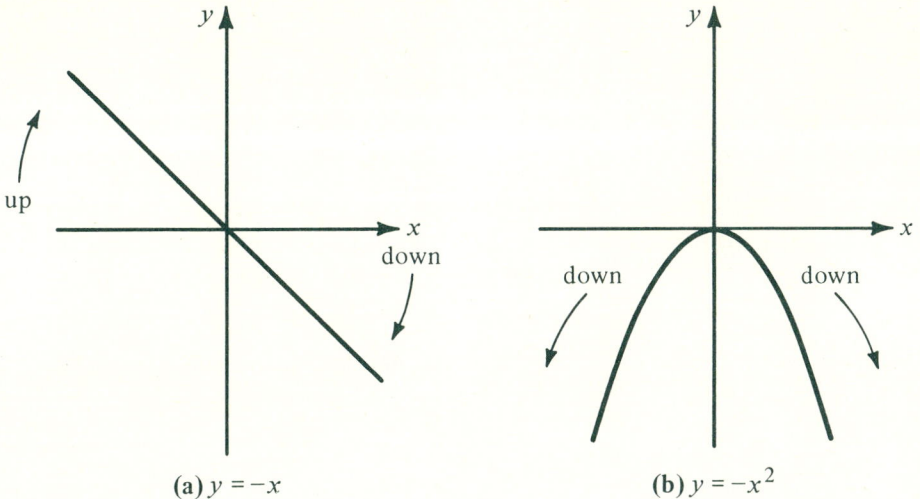

FIGURE 6.4. (a) Odd-degree Polynomials with $a_n < 0$ Go Up to the Left and Down to the Right. (b) Even-degree Polynomials with $a_n < 0$ Go Down to the Left and Right.

The above graphs illustrate the four possible graphs of third-degree polynomials with positive leading coefficients. All we know, in general, is that a third-degree polynomial may have *at most* 3 zeros (x-intercepts) and 2 turning points. An asterisk (∗) marks the zeros and an open dot (◦) marks the turning points in Figure 6.5. Notice that all the graphs go up to the right and down to the left.

We will, of course, use synthetic division to find the x-intercepts and the complete factorization of a given polynomial. Once we have a polynomial expressed in factored form, each factor is raised to some power. This power is called the *multiplicity* of the zero and is helpful in graphing the polynomial.

In Figure 6.6 (page 136), $x = 1$ is a zero of multiplicity 1, $x = -2$ is a zero of multiplicity 2, and $x = 4$ has a multiplicity of 3.

Figure 6.6 shows how to interpret the information supplied by the multiplicity of a zero. At the zeros with odd multiplicity, $x = 1$ and $x = 4$, the graph cuts through the x-axis. At the zeros with even multiplicity, the graph only touches the x-axis

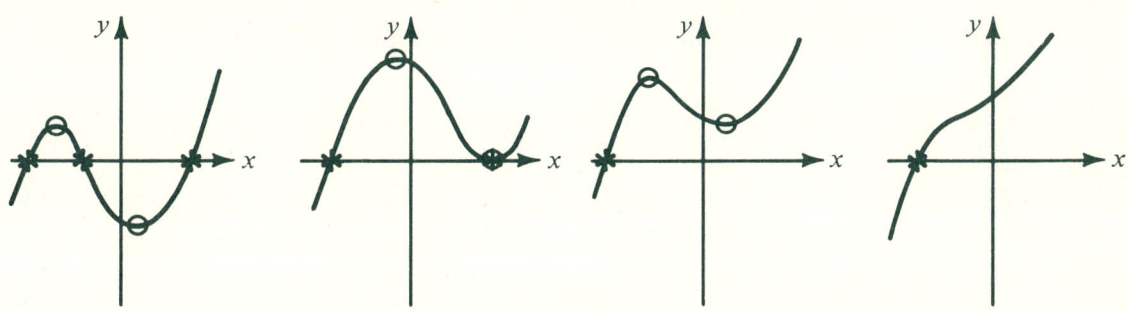

FIGURE 6.5.

CHAPTER 6   Theory of Equations

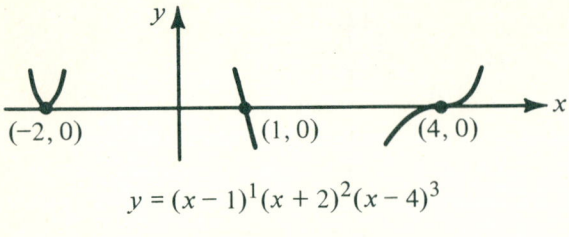

$$y = (x-1)^1(x+2)^2(x-4)^3$$

FIGURE 6.6.

rather than passing through it. To understand this, we need to investigate the behavior of the polynomial near each of its zeros. First, let's consider values close to $x = 1$.

$$p(x) = (x-1)(x+2)^2(x-4)^3$$
$$p(0.9) \approx (-0.1)(8.41)(-29.8) = 25.05$$
$$p(0.999) \approx (-0.001)(8.99)(-27.03) = 0.24$$
$$p(1.0001) \approx (0.0001)(9.001)(-26.99) = -0.024$$
$$p(1.1) \approx (0.1)(9.61)(-24.39) = -23.44$$

Notice that the values of the factors $(x+2)^2$ and $(x-4)^3$ varied relatively little compared to the variations in the values of $p(x)$. These wide variations in $p(x)$ are caused by the wide variations in the value of one factor, $(x-1)$. That is why the curve is close to linear near $x = 1$. Notice that the sign change is a result of this factor. At $x = 1$,

$$(x+2)^2(x-4)^3 = -243$$

means the curve behaves like

$$y = -243(x-1)$$

near $x = 1$.

Close to $x = -2$, the curve is dominated by the factor $(x+2)^2$. Since

$$x = -2 \Rightarrow (x-1)(x-4)^3 = +864$$

$p(x)$ looks like

$$y = 864(x+2)^2$$

close to $x = -2$. For now, the sign of the other factors near $x = -2$ is their most important contribution. If we had a negative value for the product of the other factors at $x = -2$, the curve would

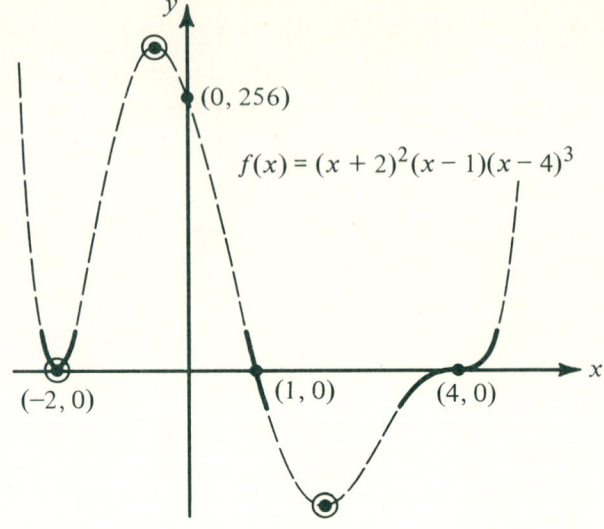

FIGURE 6.7.

look like that of $y = -(x+2)^2$, a concave-down parabola *touching* the axis at $x = -2$.

See Figure 6.7 for the behavior of $p(x)$ near $x = 4$. Notice that the degree of the polynomial is even, 6, and the leading coefficient is positive, $+1$. Hence the curve goes up to the right and left, has no more than 6 zeros (3 distinct ones actually), and has no more than 5 turning points. The 3 points circled are turning points.

Following is a complete example of everything we have been discussing.

EXAMPLE I. Graph and label all intercepts for

$$f(x) = -x^5 - 4x^4 + 4x^3 + 34x^2 + 45x + 18$$

(1) y-intercept:
$$f(0) = 18$$

(2) x-intercept:
$$f(x) = 0 \Rightarrow$$
$$x^5 + 4x^4 - 4x^3 - 34x^2 - 45x - 18 = 0$$

By Descartes' Rules of Signs, since $f(x)$ has only 1 sign change (from $+4x^4$ to $-4x^3$) there is, at most, 1 positive real zero. Also, since,

$$f(-x) = x^5 - 4x^4 - 4x^3 + 34x^2 - 45x + 18$$

has 4 sign changes, there are either 4, 2, or no negative real zeros. If $x = p/q$ is a rational zero, then q divides 1 and p divides 18. Hence, $p/q = \pm 1, \pm 2, \pm 3, \pm 6, \pm 9, \pm 18$. Let's use synthetic division to try these candidates.

For $x = 6$,

$$\begin{array}{r|rrrrrr} 6 & 1 & 4 & -4 & -34 & -45 & -18 \\ & & 6 & 60 & 336 & 1812 & 10{,}602 \\ \hline & 1 & 10 & 56 & 302 & 1767 & 10{,}594 \end{array}$$

Since 6 is positive and all coefficients and the remainder are positive, we need try no larger number. Hence, 6, 9, and 18 are not zeros.

For $x = -6$,

$$\begin{array}{r|rrrrrr} -6 & 1 & 4 & -4 & -34 & -45 & -18 \\ & & -6 & 12 & -48 & 492 & -2682 \\ \hline & 1 & -2 & 8 & -82 & 447 & -2700 \end{array}$$

Since we divided by a negative and the signs on the third line alternate, no smaller number will work. Hence, $x = -6$, $x = -9$, and $x = -18$ are not zeros.

Thus, all zeros must be between -6 and 6. We call -6 a lower bound for the possible zeros and $+6$ an upper bound. Trying candidates between -6 and $+6$ yields the following.

For $x = 3$,

$$\begin{array}{r|rrrrrr} 3 & 1 & 4 & -4 & -34 & -45 & -18 \\ & & 3 & 21 & 51 & 51 & 18 \\ \hline & 1 & 7 & 17 & 17 & 6 & 0 \end{array}$$

Hence we have $(x - 3)(x^4 + 7x^3 + 17x^2 + 17x + 6) = 0$. This is our only positive zero. Now, to factor the fourth-degree polynomial, whose candidate list of rational zeros is $\pm 1, \pm 2, \pm 3, \pm 6$, let's try any negative candidate, for example, -1.

For $x = -1$,

$$\begin{array}{r|rrrrr} -1 & 1 & 7 & 17 & 17 & 6 \\ & & ? & & & \\ \hline & & ? & & & 0 \end{array}$$

We now have
$$(x - 3)(x + 1)(x^3 + 6x^2 + 11x + 6) = 0$$

Trying $x = -2$ by dividing $x + 2$ into the cubic polynomial yields the following:

$$\begin{array}{r|rrrr} -2 & 1 & 6 & 11 & 6 \\ & & ? & & \\ \hline & & ? & & \end{array}$$

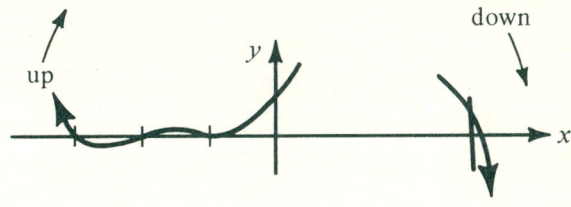

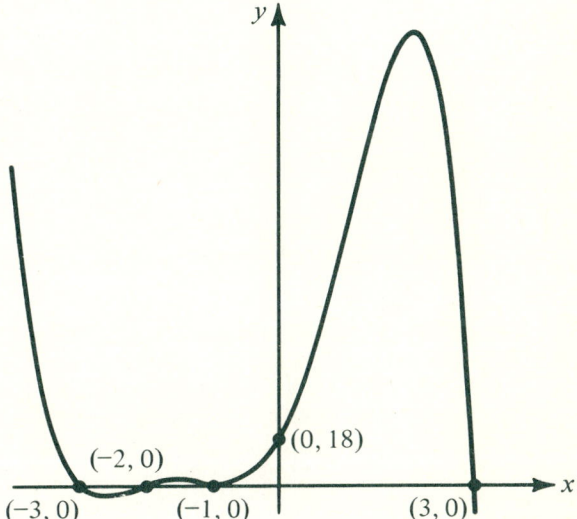

$f(x) = -x^5 - 4x^4 + 4x^3 + 34x^2 + 45x + 18$

FIGURE 6.8.

Hence,
$$(x - 3)(x + 1)(x + 2)(x^2 + 4x + 3) = 0 \Rightarrow$$
$$(x - 3)(x + 1)(x + 2)(x + 1)(x + 3) = 0 \Rightarrow$$
$$(x + 1)^2(x + 2)(x + 3)(x - 3) = 0$$

We are finally ready to graph this fifth-degree polynomial, which has at most 5 zeros and 4 turning points. Since it is an odd-degree polynomial with a negative leading coefficient, it must behave like $y = -x$, in that it goes up to the left and down to the right. $x = -1$ is a zero of multiplicity 2. See Figure 6.8. We cannot draw a perfect graph without finding the turning points. Unfortunately, the only way—short of trying an extremely large number of values of x in the polynomial one at a time—involves calculus and is therefore beyond the scope of this text.

We should note in passing that when we divided by $x = 3$ and all the signs were positive, we knew that no larger value of x would be a zero (even though $r = 0$).

EXERCISE 6.4

A. FUNDAMENTALS

Graph the following polynomial functions, labeling all rational intercepts.

1. (a) $p(x) = (x - 1)^2(x + 2)$
   (b) $p(x) = (x - 1)(x + 2)^2$
2. (a) $p(x) = (x + 1)^2(x + 3)$
   (b) $p(x) = (x + 1)^3(x + 3)$
3. (a) $p(x) = (x - 1)^2(x - 3)$
   (b) $p(x) = (x - 1)^2(3 - x)$
4. (a) $p(x) = (x - 4)(x - 2)(x + 2)$
   (b) $p(x) = (x - 4)(x - 2)(x + 2)(x + 4)$
5. (a) $p(x) = (x - 2)^2(x + 1)(x + 3)^3$
   (b) $p(x) = (x - 2)^3(x + 1)^2(x + 3)$

The following can be factored by inspection or by using synthetic division. Use either method to find the x-intercepts, then graph each of the following polynomial functions.

6. (a) $p(x) = x^2 - 3x - 40$
   (b) $p(x) = x^4 - 3x^2 - 40$
7. (a) $p(x) = x^2(x - 3) - 4(x - 3)$
   (b) $p(x) = x^2(x + 1) - 9(x + 1)$
8. ★(a) $p(x) = 2x^3 + x^2 - 6x - 3$
   (b) $p(x) = -2x^3 - x^2 + 2x + 1$
9. (a) $p(x) = 3x^3 - 2x^2 - 75x + 50$
   (b) $y = p(-x)$ (c) $y = -p(x)$

B. ESSENTIALS

Graph and label all rational intercepts for the following:

10. $p(x) = x^3 + 2x^2 - x - 2$
11. $p(x) = x^3 + 5x^2 - 2x - 24$
★12. $y = x^4 - 4x^3 + 3x^2 + 4x - 4$
13. $y = -x^4 - 6x^3 - 8x^2 + 6x + 9$
14. $y = x^5 + 5x^4 - 10x^3 + 10x^2 - 5x + 1$
15. $y = 4x^3 - 12x^2 - x + 3$
★16. $y = -3x^3 - x^2 + 20x - 12$
17. $y = -x^5 + 2x^4 + 6x^3 - 5x^2 - 8x - 12$
18. $p(x) = 6x^4 + 5x^3 - 38x^2 + 5x + 6$
19. $y = -x^8 - 4x^7 + 12x^5 + 6x^4 - 12x^3 - 8x^2 + 4x + 3$
20. $p(x) = x^7 + 4x^6 - 3x^5 - 30x^4 - 24x^3 + 48x^2 + 80x + 32$
21. $y = x^4 + 2x^2 + x + 1$
22. $y = x^7 + x^6 - 14x^5 - 10x^4 + 53x^3 + 5x^2 - 72x + 36$

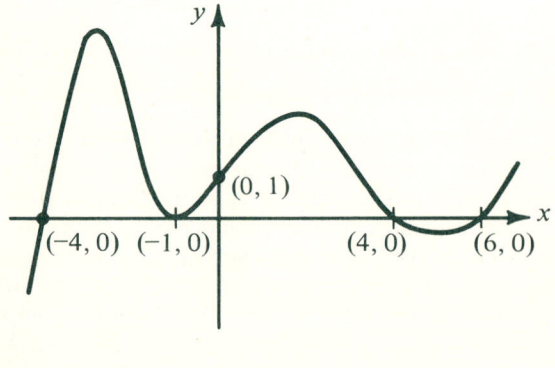

(a)

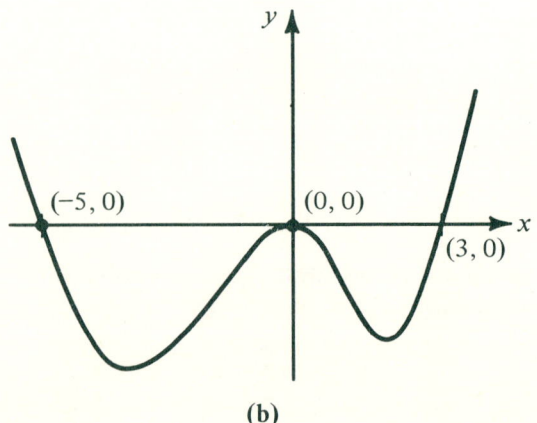

(b)

23. Find the lowest-degree polynomial that describes each graph on page 138.

24. Graph the following. Use the graphs from (a) to draw those for (b) and (c).
(a) $f(x) = x^4 - 4x^2$ and $y = -f(x)$
(b) $y = |x^4 - 4x^2|$
(c) $|y| = x^4 - 4x^2$

25. Graph the following.
(a) $f(x) = 4x^3 + 8x^2 - x - 2$ and $y = -f(x)$
(b) $y = |f(x)|$
(c) $|y| = f(x)$

C. APPLICATIONS

• 26. The interpolating polynomial for third-degree polynomials is the natural generalization of that for quadratic functions in Section 5.5. The third-degree polynomial that passes through 4 given points (x_0, y_0), (x_1, y_1), (x_2, y_2), and (x_3, y_3) is

$$f(x) = a_0 + (x - x_0)a_1 + (x - x_1)(x - x_0)a_2 + (x - x_2)(x - x_1)(x - x_0)a_3$$

where

$$a_0 = y_0$$

$$a_1 = \frac{y_1 - y_0}{x_1 - x_0}$$

$$a_2 = \frac{y_2 - y_0 - (x_2 - x_0)a_1}{(x_2 - x_0)(x_2 - x_1)}$$

and

$$a_3 = \frac{y_3 - y_0 - (x_3 - x_0)a_1 - (x_3 - x_1)(x_3 - x_0)a_2}{(x_3 - x_2)(x_3 - x_1)(x_3 - x_0)}$$

The new coefficient, a_3, is derived by using the formula $f(x_3) = y_3$ and solving for a_3. Use these formulas to find the third-degree polynomial whose graph passes through the following points.
(a) $(-3, 5)$, $(-2, 8)$, $(-1, 5)$, $(1, 5)$
(b) $(-2, -2)$, $(-1, 2)$, $(0, 2)$, $(2, 8)$

• 27. Find the fourth-degree polynomial whose graph passes through the following 5 points, by generalizing the interpolating polynomial for the cubics above to fourth-degree polynomials and finding a_4.

$(-2, 2)$, $(-1, 4)$, $(0, 3)$, $(1, 0)$, $(2, 1)$

28. The method of finite differences in Section 5.5 is easily extended to a polynomial of any degree. We simply continue to find the consecutive differences until the differences are all equal or we run out of points. In either case, the number of the final difference (second difference, third difference, etc.) is the degree of the polynomial needed. Use finite differences to find the polynomials whose graphs pass through the following given points.
(a) $(1, 1)$, $(2, 2)$, $(3, 4)$, $(4, 5)$
(b) $(1, 2)$, $(2, 5)$, $(3, 24)$, $(4, 71)$, $(5, 158)$

6.5 IRRATIONAL ZEROS OF POLYNOMIAL FUNCTIONS

Since there is no direct method available for determining the irrational zeros of polynomials with degrees greater than 4, we must resort to approximating them. The following two examples illustrate two somewhat related techniques for approximating the irrational zeros of polynomials. Both of these, and several other methods, are easily programmed on a computer.

EXAMPLE I. For

$$f(x) = x^3 + 2x^2 - x - 4$$

the possible rational zeros are ± 1, ± 2, ± 4. Trying $x = 1$ and $x = 2$ yields

$$\underline{1|}\ \ 1\ \ \ 2\ \ \ -1\ \ \ -4$$
$$\ \ \ \ \ \ \ \ \ \ \ \ \ \ ?$$
$$\overline{\ \ \ \ \ \ \ \ \ \ \ \ \ \ ?\ \ \ \ \ \ \ \ \ \ \ \ \ }$$

$$\underline{2|}\ \ 1\ \ \ 2\ \ \ -1\ \ \ -4$$
$$\ \ \ \ \ \ \ \ \ \ \ \ \ \ ?$$
$$\overline{\ \ \ \ \ \ \ \ \ \ \ \ \ \ ?\ \ \ \ \ \ \ \ \ \ \ \ \ }$$

We know 2 points on the curve, $(1, -2)$ and $(2, 10)$. In order for the curve to pass through both of these points, it must cross the x-axis between them, because one has a positive y-value and the other has a negative y-value. Since there are no rational candidates for zeros between $x = 1$ and $x = 2$, we know that this zero is irrational. Let's try $x = \frac{3}{2}$, the midpoint of the interval $[1, 2]$:

$$\underline{\tfrac{3}{2}|}\ \ 1\ \ \ 2\ \ \ \ -1\ \ \ \ -4$$
$$\ \ \ \ \ \ \ \ \ \ \ \ \ \ \tfrac{3}{2}\ \ \ \ \tfrac{21}{4}\ \ \ \ \tfrac{51}{8}$$
$$\overline{\ \ \ \ \ \ \ \ 1\ \ \ \ \tfrac{7}{2}\ \ \ \ \tfrac{17}{4}\ \ \ \ \tfrac{19}{8}}$$

At $x = \frac{3}{2}$, the value $\frac{19}{8}$ is positive. We now have the irrational zero trapped between $x = 1$ and $x = \frac{3}{2}$. Trying the midpoint of this interval, $x = \frac{5}{4}$, yields

$$\begin{array}{r|rrrr} \frac{5}{4} & 1 & 2 & -1 & -4 \\ & & \frac{5}{4} & \frac{65}{16} & \frac{245}{64} \\ \hline & 1 & \frac{13}{4} & \frac{49}{16} & -\frac{11}{64} \end{array}$$

Now, since the y-value changes sign from negative at $x = \frac{5}{4}$ to positive at $x = \frac{3}{2}$, we know the zero is between $x = \frac{5}{4} = 1.25$ and $x = 1.5$. We simply continue checking the sign of the y-value corresponding to the midpoint until we have the zero trapped between two numbers that are close enough together to yield the desired accuracy. The next midpoint, $x = \frac{11}{8}$, yields a positive remainder; hence the zero is between $\frac{5}{4}$ and $\frac{11}{8}$. This midpoint, $\frac{21}{16}$, yields a positive remainder, trapping the zero between $x = \frac{5}{4}$ and $x = \frac{21}{16}$; thus $1.25 < x < 1.3125$. The next midpoint, $x = \frac{41}{32} = 1.28125$, yields a positive remainder; hence $1.25 < x < 1.28$. If we let $x = 1.266$, the midpoint of the last interval, we know we are within 0.015 of the irrational zero. This is certainly close enough for any graph we may want to draw, so let's stop!

There is another method that is not iterative (continued repetition of the same process), so its accuracy is not as apparent. Example II illustrates the technique.

EXAMPLE II. For

$$f(x) = x^4 - 2x^3 + x - 6$$

the possible rational zeros are $x = \pm 1, \pm 2, \pm 3, \pm 6$. Trying $x = -2$ and $x = -1$ yields two points on the curve, $(-2, 24)$ and $(-1, -4)$. Since there are no candidates for rational zeros between -2 and -1, this zero must be irrational. The solid line in Figure 6.9 is a possible sketch of this section of the polynomial. If we approximate the polynomial between these 2 points with a straight line, we can approximate the irrational zero with the x-intercept of the linear approximation. The equation of our approximating line (the dotted

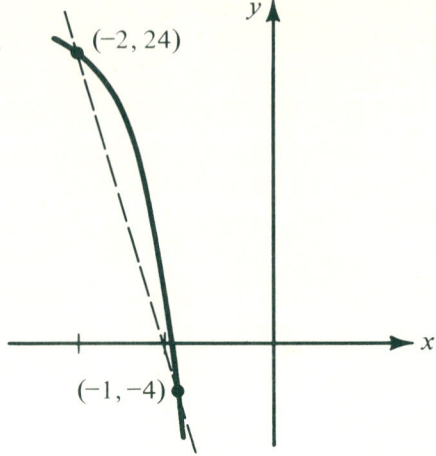

FIGURE 6.9.

line above) is found using its slope, $m = -\frac{28}{3}$, and one of the 2 points:

$$y - y_0 = m(x - x_0) \Rightarrow$$
$$28x + 3y + 40 = 0$$

The x-intercept for this line is $(-\frac{10}{7}, 0)$. Hence, we can use $x = -\frac{10}{7} \approx -1.43$ as an approximation of the irrational zero between $x = -2$ and $x = -1$. This method is called *linear interpolation*.

A more common format for linear interpolation involves some plane geometry, as illustrated in Figure 6.10. We can use the similar triangles

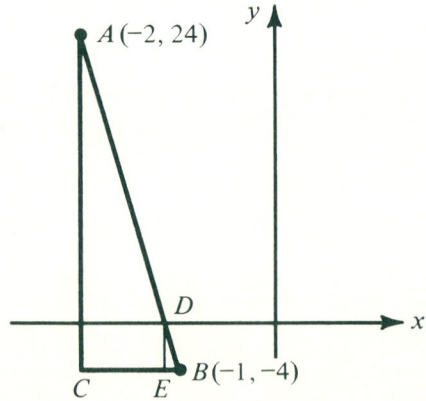

FIGURE 6.10.

ABC and *BDE* to find the *x*-intercept for the linear approximation. $AC = 28$, $BC = 3$, $DE = 4$, and we need to find *BE*. Then, since corresponding parts of similar triangles are proportional, we have

$$\frac{BC}{AC} = \frac{BE}{DE} \Rightarrow$$

$$\frac{3}{28} = \frac{BE}{4} \Rightarrow$$

$$BE = \frac{3}{7}$$

Hence, point *D*, our *x*-intercept, is located at $-1 + (-\frac{3}{7}) = -\frac{10}{7}$.

EXERCISE 6.5

A. FUNDAMENTALS
Find the indicated irrational zero using the linear interpolation method.

1. $f(x) = x^3 - x - 3$; between $x = 1$ and $x = 2$.

2. $g(x) = -x^3 + 2x^2 + x - 1$; between $x = 1$ and $x = \frac{3}{2}$.

3. $p(x) = x^4 - 10x^2 + 10x - 5$; between $x = -3$ and $x = -4$.

[HINT: To avoid working with negative numbers, let $y = p(-x)$.]

4. $y = x^4 + 2x^3 - x^2 + 2x - 3$; between $x = 0$ and $x = 1$.

5. $y = 2x^4 + x^3 - 3x^2 + 4x - 5$; between $x = 1$ and $x = 2$.

Find the indicated irrational zero to 3 decimal places using the first method discussed.

6. $f(x) = -x^3 + 2x^2 + x - 1$; between $x = -1$ and $x = 0$.

[HINT: Find the zero between $x = 0$ and $x = +1$ for $y = f(-x)$.]

7. $y = x^3 - 3x^2 + x - 1$; between $x = 2$ and $x = 3$.

8. $f(x) = -2x^3 + x^2 + 3x + 10$; between $x = -2$ and $x = -1$.

9. $p(x) = 2x^4 - 3x^3 - 4x^2 - 12x + 1$; between $x = 2$ and $x = 3$.

B. ESSENTIALS
Graph the following polynomials. Label all zeros, rational and irrational.

10. $f(x) = x^3 - 2x^2 - 8x - 5$
11. $p(x) = -2x^4 + x^3 + 13x^2 + 9x - 18$
12. $y = x^4 - 2x^3 + 2x^2 - 3x + 2$
13. $y = -x^5 + 2x^4 + 7x^3 - 7x^2 - 12x - 4$

C. APPLICATIONS
Find the following to 3 decimal places.

14. $\sqrt{7}$ (HINT: Find the irrational zero of $x^2 - 7$ between $x = 2$ and $x = 3$.)

15. $\sqrt[3]{21}$ 16. $\sqrt[5]{13}$ 17. $\sqrt[4]{7}$

18. $\sqrt{125}$

6.6 COMPLEX ZEROS OF POLYNOMIAL FUNCTIONS (optional)

We shall now complete our study of polynomial functions by considering *complex polynomials*—polynomials with complex numbers as coefficients, or polynomials that have complex as well as real zeros. Let's begin with the Fundamental Theorem of Algebra, first established by the German mathematician Gauss in 1799. Unfortunately, its proof is beyond the scope of this book.

THEOREM VI.

Fundamental Theorem of Algebra: Every complex polynomial of degree $n > 0$ has at least one root over the complex numbers.

The Fundamental Theorem of Algebra leads directly to a very powerful corollary, shown in Theorem VII.

CHAPTER 6   Theory of Equations

> **THEOREM VII.**
> Every complex polynomial of degree n has exactly n roots.

Proof: Let
$$f(z) = c_n z^n + c_{n-1} z^{n-1} + \cdots + c_1 z^1 + c_0$$
be a complex polynomial with complex zeros. Then, by the Fundamental Theorem, it must have at least 1 zero, say $z = r_1$. Then we can write $f(z)$ as
$$f(z) = (z - r_1) f_2(z)$$
where $f_2(z)$ is a complex polynomial of degree $n - 1$. Now $f_2(z)$ has at least 1 zero, say $z = r_2$. Hence,
$$f_2(z) = (z - r_2) f_3(z)$$
and
$$f(z) = (z - r_1)(z - r_2) f_3(z)$$
We can apply the Fundamental Theorem n times, until we have
$$f(z) = (z - r_1)(z - r_2) \cdots (z - r_n)$$
where each of the r's need not be distinct. Then we have exactly n zeros for $f(z)$.

These zeros are all there are, for if $z = r^*$ were another zero, then $f(r^*)$ would equal zero. We would then have
$$(r^* - r_1)(r^* - r_2) \cdots (r^* - r_n) = 0$$
which would mean one of the factors was equal to zero. Then, r^* would have to equal one of the zeros $r_1, r_2, \ldots, r_n$; it could not be a different zero. Hence, there are exactly n zeros for the nth-degree polynomial $f(z)$.

Now we will restrict the discussion to real polynomials with complex zeros—that is, polynomials with only real numbers for coefficients.
This theorem describes the nature of the complex zeros of real polynomials.

> **THEOREM VIII.**
> For any real polynomial
> $$f(z) = a_n z^n + a_{n-1} z^{n-1} + \cdots + a_1 z^1 + a_0 z^0$$
> if $z = a + bi$ is a complex zero ($b \neq 0$), then $\bar{z} = a - bi$ is also a complex zero.

Proof: Let
$$f(z) = a_n z^n + \cdots + a_1 z^1 + a_0$$
Then $f(z)$ can be written as
$$f(z) = (z - r)(z - \bar{r}) q(z) + (cz + d)$$
where $r = a + bi$ and $\bar{r} = a - bi$. This simply says that if we divide $f(z)$ by a quadratic $(z - r)(z - r)$, then the remainder will at most be linear in z. Now,
$$(z - r)(z - \bar{r}) = [z - (a + bi)][z - (a - bi)]$$
$$= z^2 - 2az + a^2 + b^2$$
is a real polynomial, and the fact that $r = a + bi$ is a zero of $f(z)$ means that
$$f(r) = 0 \Rightarrow$$
$$f(r) = (r - r)(r - \bar{r}) q(z) + (cr + d) = 0 \Rightarrow$$
$$cr + d = 0 \Rightarrow$$
$$c(a + bi) + d = 0 \Rightarrow$$
$$(ca + d) + bci = 0 + 0i \Rightarrow$$
$$ca + d = 0 \quad \text{and} \quad bc = 0$$
Where,
$$b \neq 0 \Rightarrow c = 0$$
and
$$ca + d = 0 \Rightarrow d = 0$$
Therefore,
$$f(z) = (z - r)(z - \bar{r}) \cdot q(z) + 0$$
and $\bar{r} = a - bi$ is a zero whenever $\bar{r} = a + bi$ is a zero, since $f(\bar{r}) = 0$.

The next theorem tells us that the quadratic formula will yield all complex zeros for real polynomials.

Theorem IX.
All real polynomial functions can be expressed as a product of linear and quadratic factors.

Proof: Since the set of real polynomials is a subset of the set of complex polynomials, Theorem II states we can write any nth-degree polynomial $f(z)$ as

$$f(z) = (z - r)(z - r) \cdots (z - r)$$

If any zero, $r_1, r_2, \ldots, r_n$, is a real number, then we are finished with that particular factor, since it is linear and therefore satisfies the conclusion of the theorem.

If any zero, for example, r_i, is complex, then by Theorem IV its complex conjugate $\bar{r}_i$ must also be a zero. But if

$$r_i = a + bi$$

then

$$\bar{r}_i = a - bi$$

and

$$(z - r_i)(z - \bar{r}_i) = z^2 - 2az + (a^2 + b^2)$$

which is a real, quadratic factor. This completes the proof, since every zero is either real or complex.

EXERCISE 6.6

Find the zeros for the following real polynomials. Include complex zeros.

1. (a) $p(x) = x^3 - 1$ (b) $p(x) = x^4 - 1$
2. (a) $y = x^2 + 8$ (b) $y = x^3 + 8$
3. (a) $f(x) = x^4 + x^3 - x - 1$
   (b) $f(x) = x^4 + 2x^3 + x^2 - 8x - 20$
4. $y = z^4 + z^3 + 10z^2 + 9z + 9$ (HINT: Try $z = 3i$ and its conjugate.)
5. $y = z^4 - 2z^3 + 6z^2 - 8z + 8$ (HINT: Try $z = 1 - i$ by synthetic division.)
6. $y = z^4 - 2z^3 - 2z^2 + 6z + 5$ (HINT: Try $z = 2 + i$.)

6.7 HISTORICAL COMMENTS ON POLYNOMIALS

Second-degree polynomials and their positive roots were known to the ancient Babylonians, Egyptians, and Greeks.

In 1637, Descartes published *La Géométrie* in which he presented his original ideas and the prevailing ideas on the theory of equations. Much of our study is based on his work. Descartes stated that an nth-degree equation could have n roots, and that $f(x)$ is divisible by $x - a$, where $a > 0$, if and only if $f(a) = 0$. He established the modern method of finding rational roots and he introduced the Rule of Signs and the method of undetermined coefficients.

He used this method to factor polynomials as follows. Suppose $x^2 + 5x + 4$ factors into $(x + a) \cdot (x + b)$. Then,

$$x^2 + 5x + 4 = x^2 + (a + b)x + a \cdot b$$

where $a + b$ and $a \cdot b$ are our undetermined coefficients. Since two polynomials cannot be equal unless their corresponding coefficients are identical, we have

$$a + b = 5 \quad \text{and} \quad a \cdot b = 4$$

These equations are then solved for a and b, which yields the factors of $x^2 + 5x + 4$.

In 1770, Lagrange presented a method for approximating real roots of polynomials of degrees higher than 4. Cardan recognized that an nth-degree polynomial has n zeros and that complex roots occur in pairs, which Newton later proved. Gauss proved the Fundamental Theorem of Algebra, in 1797, at the age of 20.

CHAPTER 6 Theory of Equations

Two young mathematicians, Abel and Galois, contributed greatly to the theory of equations. At the age of 22, Abel presented a rather long, involved proof that a formula similar to the quadratic formula does not exist for polynomials of a degree higher than four. But since Abel's proof contained a slight error, Galois is credited with the first definitive proof of this fact. The night before the twenty-year-old Galois was to die in a duel, he outlined a comprehensive theory of roots of equations in a letter to a friend. His ideas, introduced in this hastily written outline (his formal papers sent to Cauchy and Fourier were lost), were to keep mathematicians busy for years.

CHAPTER 7

Quotients of Polynomials: Rational Functions

7.1 INTRODUCTION

In this chapter, we will study *rational functions*. Rational functions can be expressed as

$$y = \frac{p(x)}{q(x)}, \qquad q(x) \neq 0 \text{ for all } x$$

where $p(x)$ and $q(x)$ are polynomials. (Recall that a rational number is a real number that can be expressed as a quotient of two integers p/q, where $q \neq 0$.)

First, since zero plays an unusual and important role in the process of division, let's make sure we really understand the three very different ways it can appear.

First, if we divide any nonzero number into zero, the quotient is zero. Thus

$$\frac{0}{3} = 0$$

since $3 \cdot 0 = 0$.

Second, division of any nonzero number by zero is undefined. That is,

$$\frac{4}{0} \text{ is not defined}$$

since there is no number c such that $0 \cdot c = 4$.

Third, division of zero by zero is meaningless. That is,

$$\frac{0}{0} \text{ is not defined}$$

since there are many numbers c such that $0 \cdot c = 0$. This third case is called *indeterminate*, because, according to the "theory of limits" in calculus, practically anything can happen. The second case is not very complicated, because when it occurs, we can usually determine exactly what is happening. Consider the following division problems where the denominator is approaching zero.

$$4 \div 2 = 2 \qquad -4 \div 2 = -2$$
$$4 \div 1 = 4 \qquad -4 \div 1 = -4$$

CHAPTER 7 Quotients of Polynomials: Rational Functions

$4 \div \dfrac{1}{2} = 4 \cdot 2 = 8$ $\qquad -4 \div \dfrac{1}{2} = -8$

$4 \div \dfrac{1}{10} = 4 \cdot 10 = 40$ $\qquad -4 \div \dfrac{1}{10} = -40$

$4 \div \dfrac{1}{100} = 400$ $\qquad -4 \div \dfrac{1}{100} = -400$

$4 \div \dfrac{1}{10,000} = 40,000$ $\qquad -4 \div \dfrac{1}{10,000} = -40,000$

$4 \div \dfrac{1}{10^6} = 4 \times 10^6$ $\qquad -4 \div \dfrac{1}{10^6} = -4 \times 10^6$

(4 million) $\qquad\qquad$ (negative 4 million)

We write these trends as follows: as x approaches zero ($x \to 0$),

$$\dfrac{4}{x} \to +\infty \quad \text{and} \quad \dfrac{-4}{x} \to -\infty$$

meaning that $4/x$ gets larger and larger as x gets closer to zero, and $-4/x$ gets smaller and smaller. The symbol ∞, for infinity, does not represent a number. It is used to represent the concept that $4/x$ becomes indefinitely large as x approaches zero.

In summary, for $x > 0$, $x \to 0 \Rightarrow$

$$\dfrac{c}{x} \to \begin{cases} \infty & \text{for } c > 0 \\ -\infty & \text{for } c < 0 \\ \text{any indeterminant real number} & \text{for } c = 0 \end{cases}$$

and

$$\dfrac{0}{k} = 0$$

whenever k is any nonzero real number ($k \neq 0$).

7.2 LIMITS, CONTINUITY, AND ASYMPTOTES

Consider the graph of $y = 1/x$ in Figure 7.1. Notice that as x gets closer to zero on the *right* side of the y-axis, y increases indefinitely, but as x gets closer to zero on the *left* side of the y-axis, y decreases indefinitely.

For x approaching zero from the right side we write

$$x \to 0^+ \Rightarrow y \to +\infty$$

For x approaching zero from the left side we write

$$x \to 0^- \Rightarrow y \to -\infty$$

We call the y-axis ($x = 0$) a *vertical asymptote*, since the curve approaches this line as x approaches zero. Since x cannot equal zero, the curve never actually touches this vertical asymptote.

This graph approaches the x-axis ($y = 0$) as x varies to $\pm \infty$. We call the x-axis a *horizontal asymptote* and write

$$x \to \infty \Rightarrow y \to 0$$

and

$$x \to -\infty \Rightarrow y \to 0$$

We will explain why the x-axis is a horizontal asymptote and what causes a curve to be asymptotic to the x-axis following Example I.

EXAMPLE I. Use the graph of $f(x) = 1/x$ in Figure 7.1 to complete the graphs of the following functions.

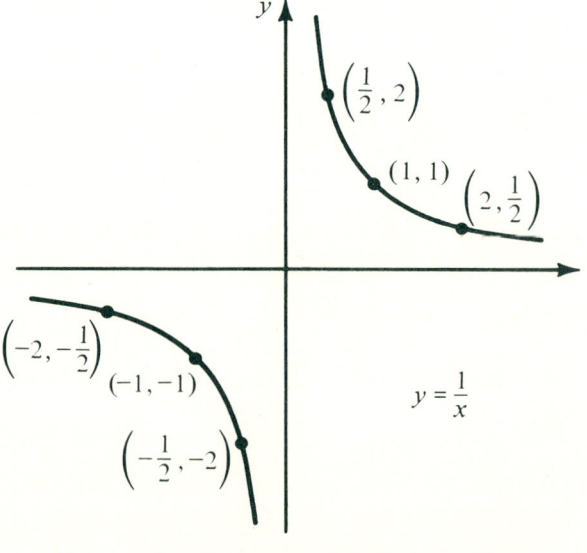

FIGURE 7.1.

SECTION 7.2 Limits, Continuity, and Asymptotes

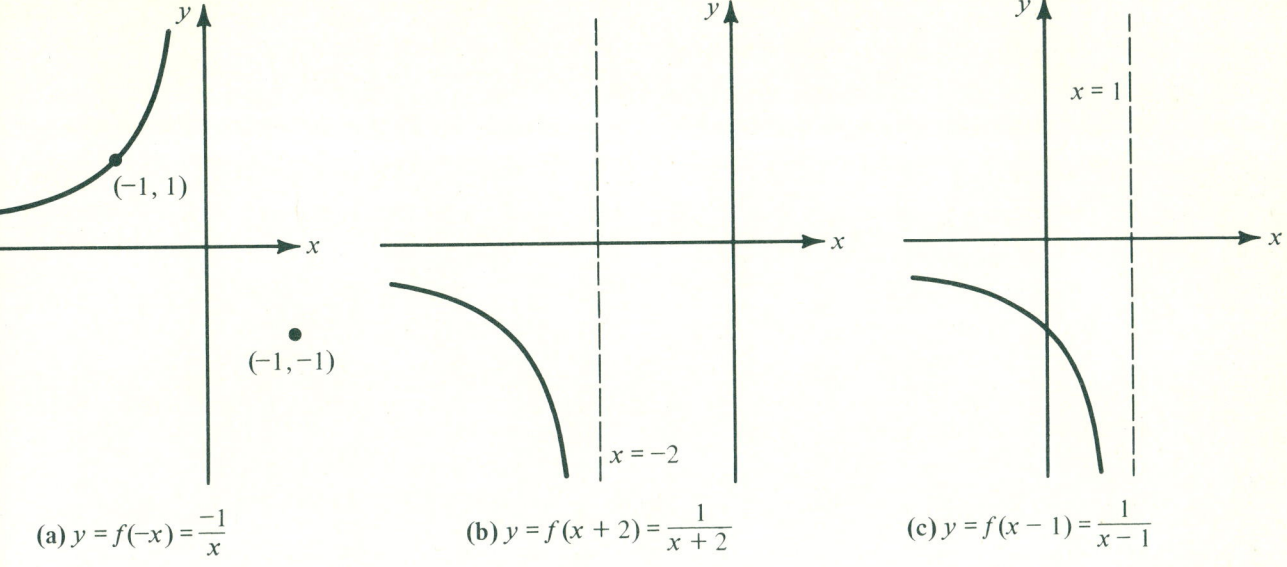

FIGURE 7.2.

(a) $y = f(-x) = \dfrac{-1}{x}$

(b) $y = f(x+2) = \dfrac{1}{x+2}$

(c) $y = f(x-1) = \dfrac{1}{x-1}$

Notice that the vertical asymptotes are $x = 0$, $x = -2$, and $x = 1$, since

$$x \to 0 \Rightarrow y = -\frac{1}{x} \to \infty$$

$$x \to -2 \Rightarrow y = \frac{1}{x+2} \to \infty$$

and

$$x \to 1 \Rightarrow y = \frac{1}{x-1} \to \infty$$

Now, let's investigate the horizontal asymptote $y = 0$. By comparing the rate at which expressions involving different powers of x increase, we can determine the conditions that will cause the x-axis to be an asymptote.

x	x^2	x^3	x^4
1	1	1	1
10	100	1000	10,000
25	625	15,625	390,625
50	2500	125,000	6,250,000
125	15,625	1,953,125	244,140,625

We can see that even for x, increasing only to the relatively small value of 125, the higher the power of x, the faster its rate of growth. Even a difference of only 1 in the exponent causes a tremendous change in the growth rate. For example, from $x = 25$ to $x = 50$, x^2 grows by 1875, x^3 grows by 109,375, and x^4 grows by 5,259,375.

Functions such as

$$f(x) = \frac{1}{x}, \qquad g(x) = \frac{x-1}{x^2+1}, \qquad h(x) = \frac{2x^2 + 700}{x^3 - 1}$$

which are ratios of two polynomials of varying degree, all have higher-degree polynomials in the denominators. This means that as $x \to \infty$, the denominators will grow significantly faster than the numerators and we will be dividing by increasingly

147

CHAPTER 7   Quotients of Polynomials: Rational Functions

larger and larger numbers relative to the slower-growing numerators. Then, as x approaches ∞, the corresponding y-values must get smaller—approaching zero in each case (see Appendix V).

Therefore, we have the following theorem.

THEOREM I.
The rational function

$$y = \frac{p(x)}{q(x)}$$

will have the x-axis, $y = 0$, as a *horizontal asymptote* whenever the degree of $q(x)$ is larger than that of $p(x)$.

EXAMPLE II.   Graph

$$f(x) = \frac{x}{x^2 - 4}$$

Since the degree of the denominator is greater than that of the numerator, the x-axis is an asymptote.

For the intercepts, we have

$$f(0) = \underline{\quad ? \quad}$$

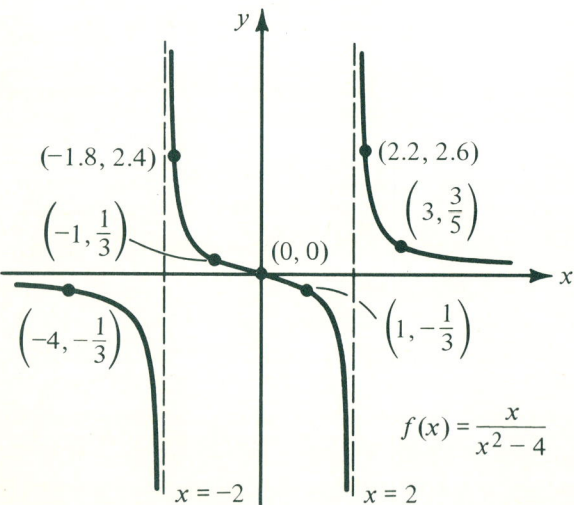

FIGURE 7.3.

and

$$f(x) = 0 \Rightarrow \frac{x}{x^2 - 4} = 0 \Rightarrow x = \underline{\quad ? \quad}$$

The vertical asymptotes occur when

$$x^2 - 4 = 0$$

That is,

$$x = \underline{\quad ? \quad} \quad \text{or} \quad x = \underline{\quad ? \quad}$$

since

$$x \to \pm 2 \Rightarrow f(x) \to \infty$$

The graph, with the asymptotes and intercepts labeled, is shown in Figure 7.3.

Example III illustrates the lack of information supplied by the indeterminate form $\frac{0}{0}$.

EXAMPLE III.   Graph each of the following rational functions.

(a) $f(x) = \dfrac{(x - 1)(x + 2)}{(x + 2)}$ (b) $g(x) = \dfrac{(x + 2)^2}{x + 2}$

(c) $h(x) = \dfrac{x + 2}{(x + 2)^2}$

For all values of x, except $x = -2$, we have $f(x) = x - 1$, $g(x) = x + 2$, and $h(x) = 1/(x + 2)$. These are very easily graphed. Once this is completed, we have to investigate the behavior of each curve near $x = -2$. We still use the above simplified equations, since *near* $x = -2$ is not quite the same as $x = -2$. As $x \to -2$, $f(x) \to -3$, $g(x) \to 0$, and $h(x) \to \infty$. Now we know that for $f(x)$ there is a hole at $(-2, -3)$, since x cannot equal -2. For $g(x)$ there is a hole at $(-2, 0)$, the x-intercept. For $h(x)$ there is an asymptote at $x = -2$, exactly as exists for $y = 1/(x + 2)$. The graphs of these three functions are shown in Figure 7.4.

Points such as $x = -2$ in the above examples are called *points of discontinuity*. Since the functions are not defined at $x = -2$, they are not continuous functions. You can draw the graphs of *continuous functions* without lifting your pencil from the paper (see Appendix V). Vertical asymptotes and "holes" are the most common types of discontinuities.

148

SECTION 7.2 Limits, Continuity, and Asymptotes

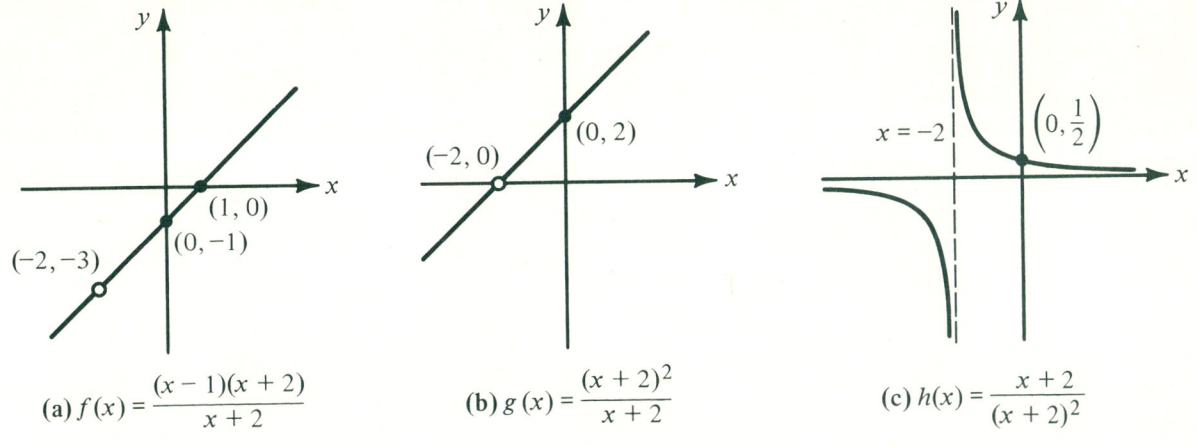

(a) $f(x) = \dfrac{(x-1)(x+2)}{x+2}$

(b) $g(x) = \dfrac{(x+2)^2}{x+2}$

(c) $h(x) = \dfrac{x+2}{(x+2)^2}$

FIGURE 7.4.

The step function in Section 1.11 is an example of a discontinuous function that is defined for all real numbers. It has "jump" discontinuities.

Mathematicians commonly use abbreviations when discussing asymptotes and continuity. For example,

$$x \to 2 \Rightarrow \frac{1}{x-2} \to \infty$$

is written as

$$\underset{x \to 2}{\text{limit}}\; \frac{1}{x-2} = \infty$$

This is read as "the limit as x approaches 2, of 1 divided by $x - 2$ is infinite." The limit

$$x \to 2 \Rightarrow \frac{(x+1)(x-2)}{x-2} \to 3$$

is written

$$\underset{x \to 2}{\text{limit}}\; \frac{(x+1)(x-2)}{x-2} = 3$$

This notation is very useful in calculus and is therefore worth learning. We call 3 the limit of $(x + 1)(x - 2)/(x - 2)$, as x approaches 2. The notation has a conventional restriction attached. That is, *x approaches but never equals* 2 in the above examples.

Let's summarize what we know about rational functions using the "limit notation."

For rational function

$$f(x) = \frac{p(x)}{q(x)}$$

(a) The function is asymptotic to the x-axis, if

$$\underset{x \to \pm\infty}{\text{limit}}\; f(x) = 0$$

(b) The function is asymptotic to $x = x_0$, if

$$\underset{x \to x_0}{\text{limit}}\; f(x) = \pm\infty$$

(c) The function has a hole at (x_0, y_0), if $f(x_0) = \frac{0}{0}$ and

$$\underset{x \to x_0}{\text{limit}}\; f(x) = y_0$$

For a more complete explanation of the limiting process, see Appendix V.

EXERCISE 7.2

A. FUNDAMENTALS
Find any points of discontinuity in each of the following rational functions. Classify the points according to

149

CHAPTER 7 Quotients of Polynomials: Rational Functions

whether the graph will have an asymptote there or simply a hole caused by 1 missing point.

1. $y = \dfrac{x^2 + 2x + 1}{x^2 + 3x + 2}$
2. $y = \dfrac{2x^3 - 4x + 9}{2x^2 + 3x + 1}$
3. $y = \dfrac{x^3 + x^2 - 9x - 9}{x^3 + x^2 - 4x - 4}$
4. $y = \dfrac{x^4 - x^2 + 1}{x^4 - x^2 - 12}$
5. $y = \dfrac{2x^2 - x - 3}{x^3 - 3x^2 - 4x + 12}$
6. $y = \dfrac{2x^3 - 14x - 12}{x^4 + x^3 - 7x^2 - x + 6}$

7. Which of the rational functions in Problems 1–6 have graphs that are asymptotic to the x-axis?

Recalling that the graph of the odd function $f(x) = 1/x$ is asymptotic to the y-axis because $1/x$ is undefined at $x = 0$, and it is asymptotic to the x-axis because $\lim\limits_{x \to \pm\infty} 1/x = 0$, graph the following:

8. (a) $y = \dfrac{1}{x + 1}$ (b) $y = \dfrac{2}{x + 1}$
   (c) $y = \dfrac{1}{2x + 1}$ (d) $y = \dfrac{2}{2x + 1}$

9. (a) $y = \dfrac{1}{x - 3}$ (b) $y = \dfrac{4}{x - 3}$
   (c) $y = \dfrac{4}{5x - 3}$

10. (a) $y = -\dfrac{1}{x}$ (b) $y = \dfrac{-1}{x - 1}$
    (c) $y = \dfrac{-3}{2x - 1}$

B. ESSENTIALS

Complete the graph of the even function $f(x) = 1/x^2$ below as a basis for the following exercises:

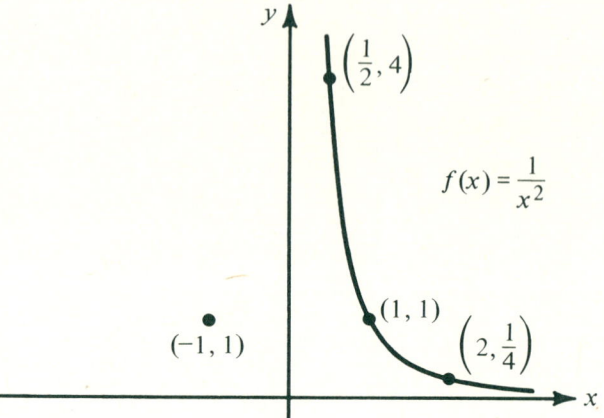

11. (a) $y = \dfrac{1}{(x + 1)^2}$ ★(b) $y = \dfrac{2}{(x + 1)^2}$
    (c) $y = \dfrac{1}{(2x + 1)^2}$ (d) $y = \dfrac{3}{(4x + 1)^2}$

12. (a) $y = \dfrac{1}{(x - 2)^2}$ (b) $y = \dfrac{1}{(3x - 2)^2}$
    (c) $y = \dfrac{4}{(2x - 2)^2}$

13. (a) $y = \dfrac{-1}{x^2}$ (b) $y = \dfrac{-1}{(x - 3)^2}$
    (c) $y = \dfrac{-2}{(4 - 3x)^2}$

Complete the graph of the even function $f(x) = 1/(x^2 + 1)$ below as a basis for the following exercises:

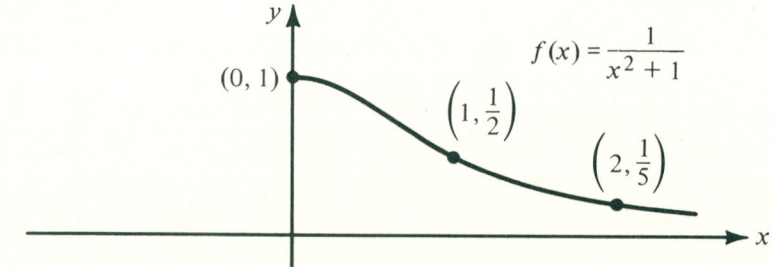

14. (a) $f(x) = \dfrac{1}{(x-1)^2 + 1}$

 (b) $y = \dfrac{1}{(x+2)^2 + 1}$

 ★(c) $y = \dfrac{1}{x^2 + 2x + 3}$

15. (a) $y = \dfrac{-1}{x^2 + 1}$ (b) $y = \dfrac{-2}{x^2 + 1}$

 (c) $y = \dfrac{-1}{x^2 + 4}$

16. (a) $y = \dfrac{-2}{x^2 + 6x + 14}$

 (b) $y = \dfrac{3}{x^2 - 3x + 5}$

7.3 INTRODUCTION TO GRAPHING RATIONAL FUNCTIONS

Fortunately, it requires very little arithmetic and, usually, only a little algebra to graph a rational function $f(x) = p(x)/q(x)$.

1. To determine the y-intercept find $y = f(0)$.
2. To determine the x-intercept(s), let $f(x) = 0$ and solve for x. This will result in simply finding the zeros for $p(x)$.
3. To determine the vertical asymptotes, find the zeros for $q(x)$. These values must be different from those in number two, or we may only have a hole in the graph instead of a vertical asymptote.
4. To determine if the x-axis is an asymptote, determine whether the degree of $q(x)$ is greater than that of $p(x)$. If so,

$$\lim_{x \to \pm \infty} \frac{p(x)}{q(x)} = 0$$

and the x-axis is the horizontal asymptote.

With a little practice, you will find that these steps will provide enough information to allow you to complete the sketch of the curve. Remember that you can always substitute any value of x (which does not result in division by zero) into the function, find its corresponding y-value, and plot that point. At first, you may have to resort to this for several values of x, but as your intuition for graphing rational functions improves with practice, you will seldom have to plot specific points. The following examples illustrate a variety of types of rational functions.

Compare the graph of Figure 7.6, $y = \dfrac{x^2 - 4x - 5}{x + 1}$, with that of $y = x - 5$ shown in Figure 7.5 below.

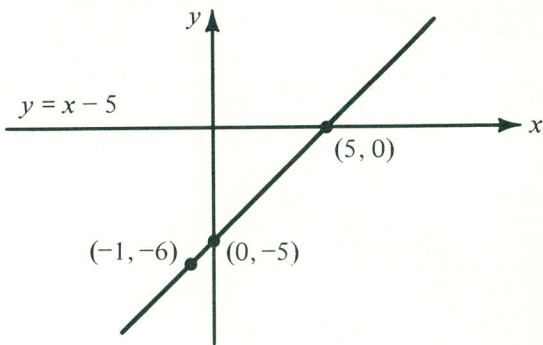

FIGURE 7.5.

EXAMPLE I. Determine why the function

$$f(x) = \frac{x^2 - 4x - 5}{x + 1}$$

is equivalent to the following:

$$g(x) = x - 5 \quad \text{and} \quad x \neq -1$$

(1) $f(0) = \underline{\quad ? \quad}$
(2) $f(x) = 0 \Rightarrow x - 5 = 0 \Rightarrow x = \underline{\quad ? \quad}$
(3) There are no vertical asymptotes, since

$$f(-1) = \tfrac{0}{0} \quad \text{and} \quad \lim_{x \to -1} f(x) = -6$$

means there is a hole at $(-1, -6)$.
(4) The x-axis is not an asymptote, since the degree of the bottom is not larger than that of the top.

Now graph $y = g(x)$, $x \neq -1$. It is important to label the hole in the graph.

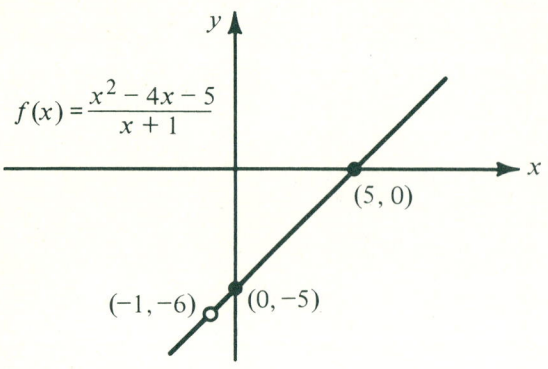

FIGURE 7.6.

EXAMPLE II. For the function

$$f(x) = \frac{x + 3}{2x^2 - 7x - 4}$$

complete the following:

(1) $f(0) = $ ___?___

(2) $f(x) = 0 \Rightarrow \dfrac{x + 3}{2x^2 - 7x - 4} = 0 \Rightarrow x = $ ___?___

(3) Vertical asymptotes:

$$2x^2 - 7x - 4 = 0 \Rightarrow$$

$$(\underline{})(\underline{}) = 0 \Rightarrow$$

$$\underline{} = 0 \quad \text{or} \quad \underline{} = 0 \Rightarrow$$

$$x = \underline{} \quad \text{or} \quad x = \underline{}$$

Note that neither of these agrees with $x = -3$ from item 2. Therefore, these values do not yield $\frac{0}{0}$. They are vertical asymptotes.

(4) The x-axis is an asymptote, since

$$\lim_{x \to \pm \infty} \frac{x + 3}{2x^2 - 7x - 4} = 0$$

The degree of the denominator being larger than that of the numerator means that the bottom grows at a *much* faster rate than the top, causing the quotient to approach zero as x approaches ∞. See Figure 7.7. Notice that the graph crosses

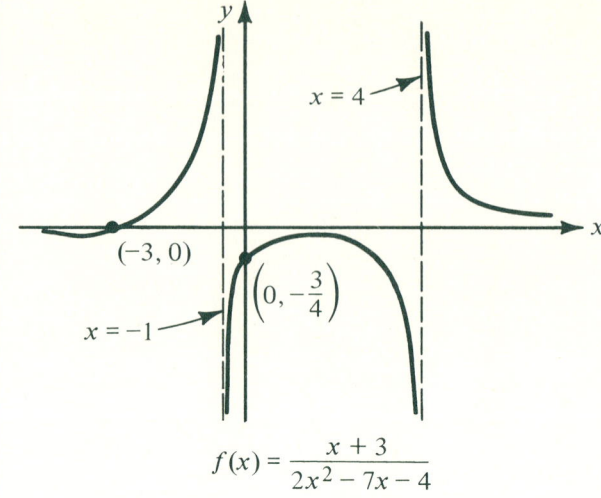

FIGURE 7.7.

its horizontal asymptote, $y = 0$, at $x = -3$, but it is still asymptotic to the left and right.

The limit $f(x) = 0$ means *only* that as x $x \to \pm \infty$ grows arbitrarily large (or small), the value of $f(x)$ becomes arbitrarily close to, but never equal to zero. For any value of x not arbitrarily large (or small), the corresponding y-value may not be close to zero at all. It may even equal zero.

Remember the following as you graph rational functions:

(a) The curve cannot cross a vertical asymptote.
(b) The curve can only cross the x-axis at the intercepts found by setting $f(x) = 0$.
(c) The curve can only cross the y-axis at $y = f(0)$.

The following two examples illustrate the effect of a factor of multiplicity 2 in the numerator and denominator. Both are asymptotic to the x-axis, since the polynomial with the larger degree is in the denominator.

EXAMPLE III. Figure 7.8 is a graph of the following curve.

$$f(x) = \frac{(x + 1)^2}{(x + 2)(x - 2)(x^2 + x + 1)}$$

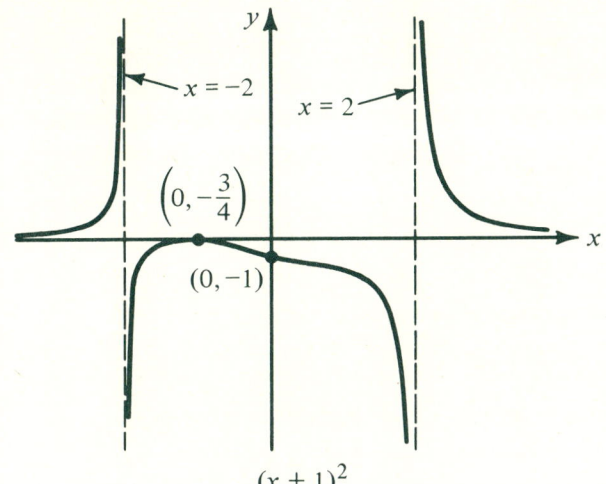

$$f(x) = \frac{(x+1)^2}{(x+2)(x-2)(x^2+x+1)}$$

FIGURE 7.8.

EXAMPLE IV. Figure 7.9 is a graph of the following curve.

$$f(x) = \frac{4}{(x+2)^2}$$

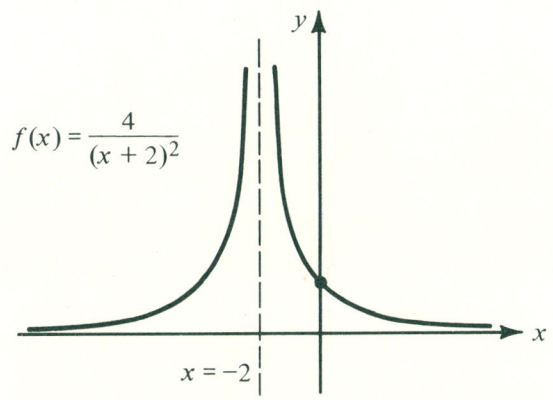

FIGURE 7.9.

Notice that as $x \to -2$ from the right,

$$\lim_{x \to -2^+} f(x) = +\infty$$

and as $x \to -2$ from the left,

$$\lim_{x \to -2^-} f(x) = +\infty$$

since $(x+2)^2$ is nonnegative for all values of x.

EXERCISE 7.3

A. FUNDAMENTALS
Graph the following rational functions, labeling all intercepts and asymptotes. Notice how the signs of the y-values of the first graph affect the second graph.

1. (a) $y = x^2 - 1$ (b) $y = \dfrac{1}{x^2 - 1}$

2. (a) $y = x^2 - 4$ (b) $y = \dfrac{1}{x^2 - 4}$

3. (a) $y = 4x^2 - 25$ (b) $y = \dfrac{4}{4x^2 - 25}$

4. (a) $y = x^2 - x - 12$
   (b) $y = \dfrac{-4}{x^2 - x - 12}$

5. (a) $y = x^2 + x - 6$ (b) $y = \dfrac{x+1}{x^2 + x - 6}$

6. (a) $y = x^3 - 8$ (b) $y = \dfrac{x^2 - 4}{x^3 - 8}$

B. ESSENTIALS
Graph the following rational functions. Write each expression as a single fraction first. Label all intercepts and asymptotes.

★7. $y = \dfrac{2}{x^2 - 1} + \dfrac{2}{x + 1}$

8. $y = \dfrac{1}{x^2 - 4} + \dfrac{1}{x + 2}$

9. $y = \dfrac{1}{x + 1} - \dfrac{1}{x - 3}$

10. $f(x) = \dfrac{x}{x^2 + 2x + 1} - \dfrac{2}{x + 1}$

11. $y = \dfrac{2x}{2x - 1} - \dfrac{x}{x - 4}$

CHAPTER 7   Quotients of Polynomials: Rational Functions

12. $f(x) = \dfrac{9}{8x^2 - 8} + \dfrac{15}{8x^2 - 72}$

13. $g(x) = \dfrac{1}{x^2 - 4x + 3} - \dfrac{2}{x^2 + x - 12}$

14. $y = \dfrac{1}{x^3 + 2x^2 - x - 2} + \dfrac{1}{x^2 - 3x - 10}$

15. $h(x) = \dfrac{2x}{x^2 + 4x + 3} + \dfrac{1}{x + 3} - \dfrac{1}{x + 1}$

16. $y = \dfrac{15}{x^2 - x - 2} + \dfrac{4}{x + 1} - \dfrac{2}{x - 1}$

Solve the following inequalities geometrically.

★17. $\dfrac{x^2 + x - 2}{x^4 - 3x^2 - 4} \geq 0$

18. $\dfrac{x^2 - 4x + 4}{x^3 - 2x^2 - x + 2} < 0$

19. $\dfrac{2x^3 - 5x^2 - 4x + 3}{x^4 + 2x^3 - 7x^2 - 8x + 12} \geq 0$

20. Vilfredo Pareto (1848–1923), the Italian sociologist and economist, formulated a "natural" law of distribution of income. The law states that the distribution of income will be the same regardless of taxation or the political and social structure. Recent studies have refuted the universality of this law to some extent. However, if the "natural" law of distribution of income is restricted to a level *above* subsistence, data indicate that the mathematical model for distribution of income is

$$N = \dfrac{p}{x^b}$$

where b varies with the population and N equals the number of people whose income is greater than or equal to x dollars in a population of size p.

Pareto's law[1] for a population of $2\tfrac{1}{2}$ million, using $b = \tfrac{1}{2}$, is written as follows:

$$N = \dfrac{2.5 \times 10^6}{\sqrt{x}}$$

(a) Graph this function for $N \leq p$.
(b) How many millionaires are there?
(c) What is the lowest income of the top 100 incomes?
(d) How many incomes are less than $8100.00?

Repeat (a) through (d) for a population of 3125 million, using $b = \tfrac{3}{2}$. For (d), find how many incomes are between $27,000.00 and $64,000.00.

21. Product-transformation equations express the relationship between the amounts of two commodities produced by the same production process, such as milk and yogurt in a creamery, gas and stove oil in a refinery, and steaks and hamburgers in a butcher shop. The maximum production levels of each are the intercepts of the product-transformation equations. Find the largest amounts indicated by the following equations, and draw the graphs of the equations.

(a) For milk, x, and yogurt, y, in liters per day, the product-transformation equation is

$$y = \dfrac{5994 - 2x}{x + 3}$$

(b) For gas, g, and stove oil, s, in barrels per day, the product-transformation equation is

$$s = \dfrac{15{,}996 - g}{g + 4}$$

●(c) If the demand for gasoline is one hundred times that for stove oil, how much of each should be produced?

22. Graph the following, using the curves of (a) to determine (b) and (c).

(a) $f(x) = \dfrac{3x - 6}{2x^2 + 5x - 3}$ and $y = -f(x)$

(b) $y = |f(x)|$
(c) $|y| = f(x)$

23. Graph the following.

(a) $y = \dfrac{2x}{3x^2 - 27}$ (b) $y = \dfrac{|2x|}{3x^2 - 27}$

(c) $y = \dfrac{2x}{|3x^2 - 27|}$

7.4 FURTHER GRAPHING OF RATIONAL FUNCTIONS

There are two other types of asymptotes that commonly appear in studies of rational functions. One is a *horizontal asymptote* other than $y = 0$, the x-axis. The other is an asymptote that is neither horizontal nor vertical, called an *oblique asymptote*.

The following two examples depict these two types of asymptotes.

SECTION 7.4   Graphing of Rational Functions

EXAMPLE I. Figure 7.10 is a graph of the function

$$f(x) = \frac{2x^2 - 7x - 6}{x^2 - 2x - 3}$$

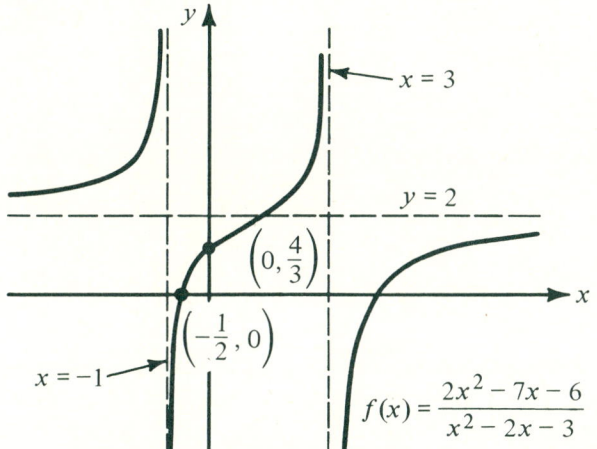

FIGURE 7.10.

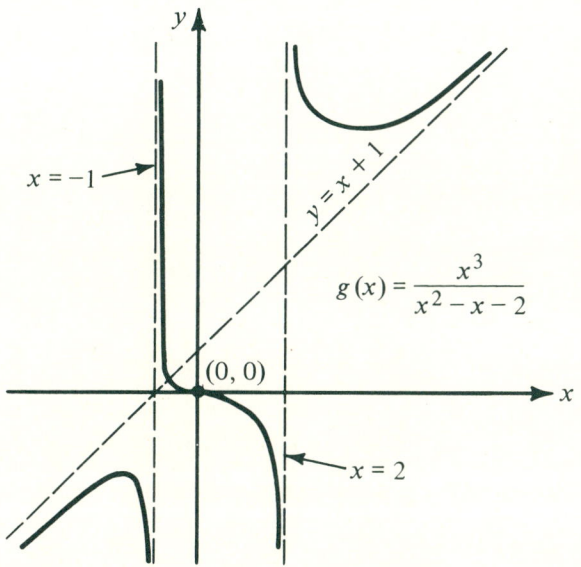

FIGURE 7.11.

EXAMPLE II. Figure 7.11 is a graph of the function

$$g(x) = \frac{x^3}{x^2 - x - 2}$$

These functions are studied together because they are both "improper" rational expressions (as, $\frac{9}{4}$ is an improper rational number), for the denominator can be divided into the numerator to yield a "proper" rational expression (for example, 4 can be divided into 9 in $\frac{9}{4}$ to yield $2\frac{1}{4}$ or $2 + \frac{1}{4}$).

For the function in Example I, long division gives

$$f(x) = 2 + \frac{-3x + 2}{x^2 - 2x - 3}$$

For the function in Example II, it gives

$$g(x) = x + 1 + \frac{3x + 2}{x^2 - x - 2}$$

Notice that as x grows very large, the remainders approach zero, since the degrees of the denominators are larger than the degrees of the numerators. This means that $y = f(x)$ approaches $y = 2$, as x becomes large (or small) without bound. Similarly, $y = g(x)$ approaches $y = x + 1$, since the remainder, $(3x + 2)/(x^2 - x - 2)$, gets closer and closer to zero, as x approaches $\pm\infty$. This is why we have a horizontal asymptote of $y = 2$ for $y = f(x)$ and an oblique asymptote of $y = x + 1$ for $y = g(x)$. Both types of asymptotes can readily be determined by rewriting such improper rational functions as proper rational functions. Carefully examine the long division involved in the two preceding examples. We can see that horizontal asymptotes (other than $y = 0$) will occur only when the degrees of the numerator and denominator are equal; oblique asymptotes will occur only when the degree of the denominator is exactly 1 less than the degree of the numerator.

155

Otherwise, the quotient would not be linear (have x raised only to the first power). If the degree of the denominator is more than 1 less than that of the numerator, there are no near horizontal or oblique asymptotes.

Also notice that for

$$f(x) = 2 + \frac{-3x + 2}{x^2 - 2x - 3}$$

the remainder is negative for $x > 3$ (try $x = 4$ in the remainder) and the graph is below the asymptote $y = 2$ the right of the asymptote $x = 3$. Since $(-3x + 2)/(x^2 - 2x - 3)$ is positive for $x < -1$ (try $x = -2$ in the remainder), the graph is above the asymptote $y = 2$, to the left of the asymptote $x = -1$.

The same argument applies to $g(x)$. Since $r(x) = (3x + 2)/(x^2 - x - 2)$ is positive for $x > 2$, the curve is above the asymptote on the right. Since the remainder is negative for $x < -1$, the graph is below $y = x + 1$ to the left.

We complete our discussion of rational functions with Theorem II.

THEOREM II.
For
$$f(x) = \frac{a_m x^m + a_{m-1} x^{m-1} + \cdots + a_1 x^1 + a_0}{a_n x^n + a_{n-1} x^{n-1} + \cdots + a_0}$$

(a) If $n > m$, $y = 0$ is the *horizontal asymptote*.
(b) If $n < m$, when m is 1 greater than n there is an *oblique asymptote*; when m is more than 1 greater than n there are no horizontal or oblique asymptotes.
(c) If $n = m$, $y = a_m/a_n$ is the *horizontal asymptote*.

Since for these rational expressions *vertical asymptotes* only arise from division by zero, they are found by setting the denominator equal to zero and solving for x. These values are vertical asymptotes provided they are not also zeros of the numerator. If they are, we may have a hole in the graph or a vertical asymptote, depending upon the limit. If the limit exists, there is a hole.

EXERCISE 7.4

A. FUNDAMENTALS
Name the horizontal or oblique asymptote of each of the following rational functions:

1. (a) $f(x) = \dfrac{2x - 3}{x + 4}$ (b) $y = f(x) + 2$

2. (a) $y = \dfrac{1}{x} - 2$ (b) $y = \dfrac{1 - 3x}{x}$

3. (a) $y = \dfrac{x^2 - 1}{x}$ (b) $y = \dfrac{x^3 - 1}{x^2}$

4. $f(x) = \dfrac{3x^2 - 4x + 1000}{x^2 - 1}$

5. $f(x) = \dfrac{3x^3 - 2x^2 - 1}{x^2 - 2x - 1}$

6. $y = \dfrac{12x^4 - 3x^2 - 1}{3x^4 - x - 4}$

Graph the following rational functions, labeling all intercepts and asymptotes.

7. $y = \dfrac{1}{x} + 2 = \dfrac{2x + 1}{x}$

8. $y = \dfrac{1}{x^2} - 3 = \dfrac{1 - 3x^2}{x^2}$

9. $y = \dfrac{1}{x - 1} - 3 = \dfrac{4 - 3x}{x - 1}$

10. $y = \dfrac{1}{(x + 2)^2} + 4 = \dfrac{4x^2 + 16x + 17}{x^2 + 4x + 4}$

11. $y = x + \dfrac{x + 1}{x^2} = \dfrac{x^3 + x + 1}{x^2}$

★12. $y = 2x - 1 + \dfrac{x - 3}{x^2 - 4} = \dfrac{2x^3 - x^2 - 7x + 1}{x^2 - 4}$

B. ESSENTIALS

13. $f(x) = \dfrac{2x^2 - x - 3}{3x^2 + 7x + 4}$

14. $f(x) = \dfrac{2x^3 - 2x^2 - 2x + 2}{2x^2 - 8}$

15. $y = \dfrac{x^2 + x - 6}{2x^2 - 2x - 12}$

16. $y = \dfrac{x - 1}{x^3 - 4x}$

17. $f(x) = \dfrac{x^3 - 8}{x^2 - 4}$

18. $y = \dfrac{2x^3 + 2x^2 - 8x - 8}{x^2 + x - 2}$

★19. $f(x) = \dfrac{2x^4 - 10x^2 + 8}{x^4 - 2x^3 - 15x^2}$

20. $f(x) = \dfrac{x^2 + 2x + 1}{x^3 - 7x + 6}$

21. (a) $y = \dfrac{x^3 - 3x + 2}{x^2 + 4x + 3}$

    (b) $y = \dfrac{x^2 + x - 2}{x^3 + 5x^2 + 7x + 3}$

22. (a) $f(x) = \dfrac{2x^3 + 7x^2 + 4x - 4}{x^3 - 4x^2 + x + 6}$

    (b) $f(x) = \dfrac{2x^2 + 3x - 2}{x^4 - 6x^3 + 9x^2 + 4x - 12}$

23. The following function yields the cost, c, of completing a project in t number of months. Graph the function, determine the suitable domain and range, and justify the asymptotes.

$$c = \dfrac{t^2 - 25t + 160}{2t - 2}$$

24. The following represents the cost, c, of a certain electrical component delivered in t days. Graph the function, determine the domain and range, and explain the asymptotes. Such a curve is used in cost expediting.

$$c = \dfrac{0.03t^3 - 0.3t^2 + 0.73t - 12.3}{t^2 - 9}$$

25. The following function represents the velocity, v, in kilometers per hour, of a rocket t hours after launching. The power, supplied by nuclear fusion, is inexhaustible. Graph and justify the asymptotes. What is c? What are the domain and range?

$$v = \dfrac{100ct^2}{100t^2 + 1}$$

• 26. Graph and explain the following modification of Problem 25. The new horizontal asymptote depicts terminal velocity.

$$v = \begin{cases} \dfrac{100ct^2}{100t^2 + 1}, & 0 \le t \le 1 \\ \dfrac{-1.01ct + 201.01c}{101t + 101}, & t \ge 1 \end{cases}$$

Graph and label the following functions and their inverses.

27. $f(x) = \dfrac{-1}{x - 4}, \quad x < 4$

28. $g(x) = \dfrac{8}{x^2 - 4x + 4}, \quad x \le 2$

29. $y = \dfrac{2x^2 - 4x + 6}{x^2 - 2x + 1}, \quad x < 1$

30. $f(x) = \dfrac{x^4 - 1}{x^2}, \quad x > 0$

• 31. Graph and label $f(x) = 1/(2 + 4^{1/x})$.

7.5 PARTIAL FRACTIONS (OPTIONAL)

It is a common procedure to combine two fractions by addition in order to yield one fraction as their sum, but sometimes (in calculus, for example) it is necessary to do the reverse. That is, given a complicated fraction, one might want to break it down into less complicated parts, the sum of which will equal the original fraction.

For example, since

$$\dfrac{2}{x + 3} + \dfrac{4}{x - 2} = \dfrac{6x + 8}{x^2 + x - 6}$$

we should be able to go backwards from the sum $(6x + 8)/(x^2 + x - 6)$ to the fractions $2/(x + 3)$ and $4/(x - 2)$ that add up to $(6x + 8)/(x^2 + x - 6)$.

CHAPTER 7   Quotients of Polynomials: Rational Functions

This is accomplished as follows. First, write
$$\frac{6x + 8}{x^2 + x - 6} = \frac{6x + 8}{(x + 3)(x - 2)} = \frac{A}{x + 3} + \frac{B}{x - 2}$$

Then simply find A and B:
$$\frac{6x + 8}{(x + 3)(x - 2)} = \frac{A}{x + 3} + \frac{B}{x - 2} \Rightarrow$$
$$\frac{6x + 8}{(x + 3)(x - 2)} = \frac{A(x - 2) + B(x + 3)}{(x + 3)(x - 2)} \Rightarrow$$
$$6x + 8 = Ax - 2A + Bx + 3B \Rightarrow$$
$$6x + 8 = (A + B)x + (-2A + 3B)$$

But these two polynomials can be equal for all values of x only if they are identical. Hence, the coefficient of the x term on the left, 6, must equal the coefficient of the x term on the right, $A + B$. Also, the two constant terms, 8 and $-2A + 3B$, must be identical. This yields the two equations

$$A + B = 6$$

and

$$-2A + 3B = 8$$

which when solved for A and B yield

$$A = 2 \quad \text{and} \quad B = 4$$

This technique of setting the corresponding coefficients of two equal polynomials equal to each other is called *equating coefficients*. The process of rewriting a fraction as a sum of other simpler fractions is called *the decomposition of the expression into partial fractions*.

Since every real polynomial can be written as a product of linear and quadratic factors, we know that we can decompose every rational expression into the sum of proper fractions of the form

$$\frac{A}{(x - a)^n} \quad \text{and} \quad \frac{Ax + B}{(ax^2 + bx + c)^n}$$

This decomposition by partial fractions is unique.

Below are some examples of the decomposition of rational expressions into partial fractions.

Instead of equating coefficients, as in Example I, to find the numerators, we will use the fact that two polynomials cannot be equal unless they yield equal values of y for every value of x. We will choose particular values of x that will eliminate some of the unknowns (hopefully, all but one). This procedure will usually have to be repeated using several different values of x, one for each unknown letter in the partial fractions.

EXAMPLE II. We will decompose the following rational expression:

$$\frac{5x^2 - 2x + 3}{x^3 - x^2 + 2} = \frac{5x^2 - 2x + 3}{(x + 1)(x^2 - 2x + 2)}$$
$$= \frac{A}{x + 1} + \frac{Bx + C}{x^2 - 2x + 2} \Rightarrow$$
$$A(x^2 - 2x + 2) + (Bx + C)(x + 1)$$
$$= 5x^2 - 2x + 3$$
$$x = -1 \Rightarrow A(5) + 0 = 5 + 2 + 3 \Rightarrow A = 2$$
$$x = 0 \Rightarrow A(2) + C = 3 \Rightarrow$$
$$4 + C = 3 \Rightarrow$$
$$C = -1$$

and
$$x = 1 \Rightarrow A(1) + (B + C)(2) = 6 \Rightarrow$$
$$2 + 2B - 2 = 6 \Rightarrow$$
$$B = 3$$

Therefore,
$$\frac{5x^2 - 2x + 3}{x^3 - x^2 + 2} = \frac{2}{x + 1} + \frac{3x - 1}{x^2 + 2x + 2}$$

By using $x = -1$, the B and C were eliminated, allowing us to solve for A. Letting $x = 0$ left only A and C, and we knew A. Letting $x = 1$ left A, B, and C, but since we know A and C at that stage, we were able to solve for B.

EXAMPLE III. Next, let's decompose the rational expression

$$\frac{x^2 + 7x + 9}{x^3 + 4x^2 + 5x + 2} = \frac{A}{x + 1} + \frac{B}{(x + 1)^2} + \frac{C}{x + 2}$$

158

CHAPTER 7 Quotients of Polynomials: Rational Functions

Notice that since $x + 1$ is a factor of multiplicity 2, we must allow for its appearance to the first or second power to cover all possible combinations. Then

$A(x + 1)(x + 2) + B(x + 2)$
$\qquad + C(x + 1) = x^2 + 7x + 9$

$x = -1 \Rightarrow B(1) = 3 \Rightarrow$
$\qquad B = 3$

$x = -2 \Rightarrow C(\underline{\ ?\ }) = -1 \Rightarrow$
$\qquad C = \underline{\ ?\ }$

and

$x = 0 \Rightarrow A(1)(2) + B(2) + C(1) = 9 \Rightarrow$
$\underline{\qquad ?\qquad} = 9 \Rightarrow$
$(\ )A = \underline{\ ?\ } \Rightarrow$
$A = 2$

Therefore

$$\frac{x^2 + 7x + 9}{x^3 + 4x^2 + 5x + 2} = \frac{2}{x+1} + \frac{3}{(x+1)^2} + \frac{-1}{x+2}$$

EXAMPLE IV. Decompose the rational expression

$$\frac{x^3 + 4x^2 + 6x + 1}{x^2 + 2x + 1} = x + 2 + \frac{x - 1}{x^2 + 2x + 1}$$

Since this is an improper rational expression, the first step was long division. Then

$$\frac{x-1}{x^2+2x+1} = \frac{x-1}{(x+1)^2} = \frac{A}{x+1} + \frac{B}{(x+1)^2}$$

Therefore,

$A(x + 1) + B = x - 1$

$x = -1 \Rightarrow B = \underline{\ ?\ }$

and

$x = 0 \Rightarrow \underline{\qquad ?\qquad} = -1 \Rightarrow$
$\underline{\ ?\ } + (-2) = -1 \Rightarrow$
$A = \underline{\ ?\ }$

Then

$$\frac{x^3 + 4x^2 + 6x + 1}{x^2 + 2x + 1} = x + 2 + \frac{1}{x+1} + \frac{-2}{(x+1)^2}$$

EXERCISE 7.5

A. FUNDAMENTALS

Decompose the following into partial fractions:

1. $\dfrac{3}{x^2 - x - 12}$
2. $\dfrac{3x + 1}{x^2 - x - 2}$
3. $\dfrac{3x^3 + x - 6}{x^2 - 1}$
4. $\dfrac{3x^2 + 3x + 1}{x^3 + x}$

B. ESSENTIALS

5. $\dfrac{3x}{x^2 + 2x + 1}$
6. $\dfrac{5x^2 + x + 6}{x^3 + x^2 - x - 1}$
7. $\dfrac{6x^2 - 5x + 6}{x^3 - x^2 - 8x + 12}$
8. $\dfrac{2x^4 + 3x^3 + 5x^2 + 4x + 1}{x^3 + x^2 + x}$
9. $\dfrac{x^5 + x^4 + 3x^3 + x^2 + 6x}{x^4 + 2x^2 + 1}$

Graph each of the following rational functions by first decomposing them into partial fractions, then graphing each partial fraction separately, and then adding the parts geometrically, as in Section 3.7.

10. $f(x) = \dfrac{2x}{x^2 - 1}$
★11. $y = \dfrac{2x^2 + 2}{x^4 - 2x^2 + 1}$
12. $f(x) = \dfrac{2x^2}{x^3 + x^2 + x + 1}$

ENDNOTE

1. Draper, Jean E., and Klingman, Jane S. *Mathematical Analysis: Business and Economic Applications.* 2d. ed. New York: Harper & Row, 1972.

CHAPTER 8

Exponential and Logarithmic Functions

8.1 INTRODUCTION

Although polynomials and quotients of polynomials are usually the easiest of all functions to use, they cannot be used to accurately describe many situations. We need another class of functions that are not sums, differences, products, or quotients of polynomials. Such functions are called **transcendental functions**. In this chapter, we will study two transcendental functions and some situations they model. First, we will examine **exponential functions**, functions whose independent variables are in the exponent, for example,

$$y = 2^x, \qquad y = 10^{x+1}, \qquad \text{and} \qquad y = (\tfrac{1}{2})^x$$

Next we will study the inverses of these exponential functions, **logarithmic functions**.

8.2 EXPONENTIAL FUNCTIONS

Since we are going to be working extensively with exponents,[1] it is essential that you know the four laws of exponents below. Remember that $b > 0$.

(1) $b^m/b^n = b^{m-n}, \qquad m, n \in$ rationals
(2) $b^m \cdot b^n = b^{m+n}, \qquad m, n \in$ rationals
(3) $(b^m)^n = b^{m \cdot n}, \qquad m, n \in$ rationals
(4) $b^{1/m} = \sqrt[m]{b}, \qquad m \in$ naturals

The following are natural consequences of the laws of exponents.

(1) $b^0 = 1$
(2) $b^{-1} = 1/b$
(3) $b^{m/n} = \sqrt[n]{b^m}, \qquad m, n \in$ integers

Part of our task in this chapter will be to extend the values of the exponents m and n to include all real numbers. However, we will still restrict the base b to positive real numbers. The following examples of expressions involving negative bases illustrate the problems that arise if we do not do so: The expression $(-4)^{1/2} = 2i$ is not a real number, $[(-2)^2]^{1/2} \neq [(-2)^{1/2}]^2$, and $(-16)^{-1/4} = 1/2i$ is not a real number.

An effective way to initially understand exponential functions is to study a particular example carefully, then make changes and notice what happens. Consider $f(x) = 2^x$, whose small base, 2, will allow us to plot some points easily.

Examine the following two tables of values for trends. In the first table, x is growing larger, and in the second, x is getting smaller.

x	$y = 2^x$	x	$y = 2^x$
1	2	0	$2^0 = 1$
3	8	-1	$\frac{1}{2} = 0.5$
5	32	-3	$\frac{1}{8} = 0.125$
7	168	-5	$\frac{1}{32} = 0.031$
11	2048	-7	$\frac{1}{128} = 0.0078$

From the first table we see that a small change in x results in a much larger change in y. For example, a change of 2 in x, from $x = 5$ to $x = 7$, causes a change of 94 in y, from $y = 32$ to $y = 128$. This means that the graph rises sharply to the right. From the second table, what is y approaching as x gets smaller? In other words, what is

$$\lim_{x \to -\infty} 2^x = \underline{\quad ? \quad}$$

Since the limit is zero, $y = 0$ (the x-axis) is an asymptote to the curve. Do you see any negative y-values? Is $2^x = 0$ for any x?

This information is summarized in Figure 8.1.

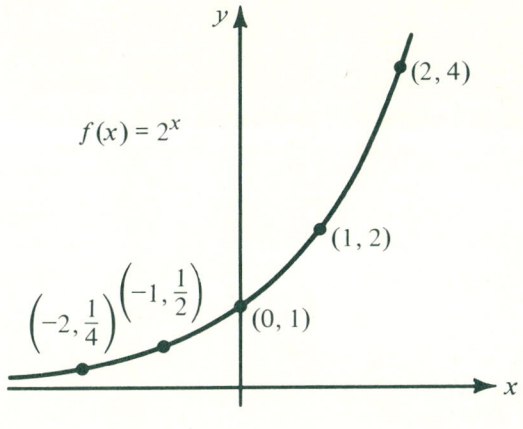

FIGURE 8.1.

The domain is all real numbers, but the range is only positive values of y. The graph depicts a one-to-one function.

Complete the graphs in Figure 8.2 by translating or reflecting the curve in Figure 8.1 about the appropriate axis.

The first graph is $f(x) = 2^x$, reflected about the y-axis. The second is $y = 2^x$, reflected about the x-axis (notice that $y = -2^x$ is always negative). The third is $y = 2^x$, translated up 1, and the last is $f(x) = 2^x$, shifted to the left 1.

All but one of the five graphs in Figure 8.2 have the x-axis as an asymptote The last one shows that only an up or down translation $(y = f(x) + k)$ results in a new horizontal asymptote.

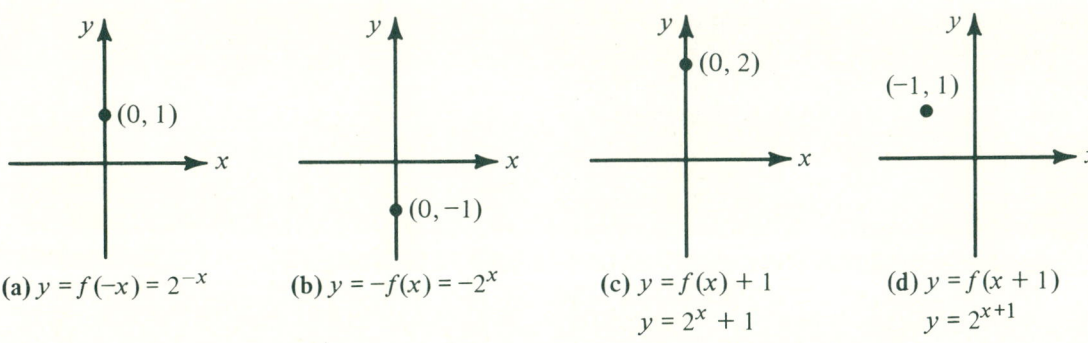

(a) $y = f(-x) = 2^{-x}$
(b) $y = -f(x) = -2^x$
(c) $y = f(x) + 1$
 $y = 2^x + 1$
(d) $y = f(x + 1)$
 $y = 2^{x+1}$

FIGURE 8.2.

CHAPTER 8 Exponential and Logarithmic Functions

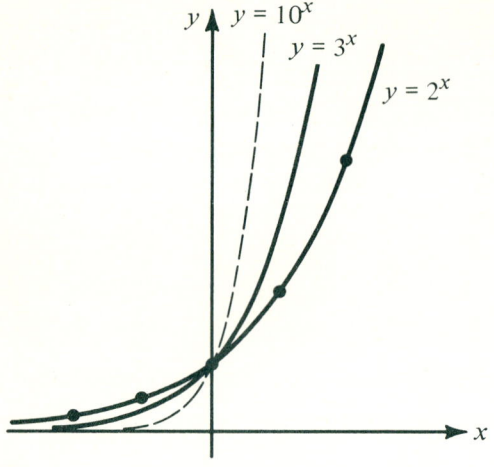

FIGURE 8.3.

Now let's change the base of the exponential function. The shapes of all exponential curves are very similar, with higher bases causing both a faster growth to the right and a faster approach to the asymptote to the left. (See Figure 8.3.)

For fractional bases, we often simply rewrite the function in an equivalent form with a whole-number base. For example,

$$y = (\tfrac{1}{2})^x = (2^{-1})^x = 2^{-x}$$

We will not encounter negative numbers as bases, since such exponential functions are not continuous. For $y = (-2)^x$, for example, $x = 1 \Rightarrow y = -2$, but $x = 2 \Rightarrow y = +4$. The curve is not defined at values of x like $\tfrac{3}{2}$ and $\tfrac{9}{4}$ that involve even roots of negative numbers and therefore are not real numbers.

EXERCISE 8.2

A. FUNDAMENTALS
Evaluate each of the following exponential functions at the indicated values of the independent variable.

1. (a) $y = 2^x$ at $x = -2$, $x = 0$, $x = 2$
   (b) $y = 3^x$ at $x = -2$, $x = 0$, $x = 2$
   (c) $y = 5^x$ at $x = -2$, $x = 0$, $x = 2$

2. (a) $y = 6^x$ at $x = -1, 1, 2$
   (b) $y = 2 \cdot 3^x$ at $x = -2, -1, 0$
   (c) $y = 3 \cdot 2^x$ at $x = 1, 2, 3$

3. (a) $y = -2^x$ at $x = 1, 2, 3$
   (b) $y = (-2)^x$ at $x = 1, 2, 3$
   (c) $y = -(-2)^x$ at $x = 1, 2, 3$

4. (a) $y = 2^x$ at $x = -2, -1, 0$
   (b) $y = 2^{-x}$ at $x = 0, 1, 2$
   (c) $y = (\tfrac{1}{2})^x$ at $x = 0, 1, 2$

5. (a) $y = 10^x$ at $x = -1, -2, -3$
   (b) $y = 10 \cdot 10^x$ at $x = -1, -2, -3$
   (c) $y = 10^{x+1}$ at $x = -1, -2, -3$

6. (a) $y = \tfrac{1}{3} \cdot 3^x$ at $x = 0, 1, 2$
   (b) $y = 3^{x-1}$ at $x = 0, 1, 2$
   (c) $y = \tfrac{1}{4} \cdot 2^x$ at $x = 0, 1, 2$

Solve each of the following equations for x.

7. (a) $2^x = 2^3$ (b) $2^x = 16$

8. (a) $10^x = 10^{-1}$ (b) $10^x = \dfrac{1}{100}$

9. (a) $3^{x+1} = 3^{-2}$ (b) $3^{x+1} = 1$

10. (a) $4^x \cdot 4^x = 16$ (b) $\dfrac{4^{x+1}}{4^2} = \dfrac{1}{16}$

Graph each pair of functions on the same axes system.

11. (a) $y = 2^x$ (b) $y = 3^x$
12. (a) $y = 3^x$ (b) $y = -3^x$
13. (a) $y = 5^x$ (b) $y = 5^{-x}$
★14. (a) $y = 7^{-x}$ (b) $y = (\tfrac{1}{3})^x$

B. ESSENTIALS

Graph each of the following pairs of exponential functions on the same axes system. Find the horizontal asymptotes.

15. (a) $y = 3^{x+1}$ (b) $y = 3^{x-1}$
16. (a) $y = 2^{-x} + 1$ (b) $y = 2^x - 1$
17. (a) $y = 3^{x-1}$ (b) $y = 3^{1-x}$
18. (a) $y = 2^{-x} + 3$ (b) $y = 2^{1-x} + 3$
19. (a) $f(x) = -2^x$ (b) $y = f(x+2)$
20. (a) $f(x) = 3^{-x}$ ★(b) $y = 1 - f(x+2)$
21. (a) $f(x) = 2^x$ (b) $y = 2 \cdot f(x)$
    (c) $y = f(2x)$

C. APPLICATIONS

The graphs of the following products of functions depict the damping effect of exponential functions. Sketch them carefully.

★22. $y = 2x \cdot 3^{-x}$ 23. $y = x^2 \cdot 2^{-x}$
24. $y = x^3 \cdot 5^{-x}$

Carefully consider the effect of the absolute-value symbol or the squaring of part of the following functions, then graph.

25. (a) $y = 2^x$ (b) $y = |2^x|$
26. (a) $y = 3^x$ (b) $y = 3^{|x|}$
27. (a) $y = -2^x$ (b) $y = -2^{x^2}$

We can use linear interpolation to approximate the values of numbers with irrational exponents. For example to approximate the value $4^{\sqrt{2}}$, we find $y_0 \approx 4^{\sqrt{2}}$ by finding DE, since $y_0 = 4 + DE$. Since triangles ABC and ADE are similar,

$$\frac{DE}{BC} = \frac{AD}{AC} \Rightarrow$$

$$\frac{DE}{8-4} = \frac{\sqrt{2}-1}{3/2 - 1} \Rightarrow$$

$$DE = 8(\sqrt{2} - 1) \approx 8(1.41 - 1) = 3.3$$

Therefore,

$$y_0 = 4 + DE = 7.3$$

and

$$4^{\sqrt{2}} \approx 7.3$$

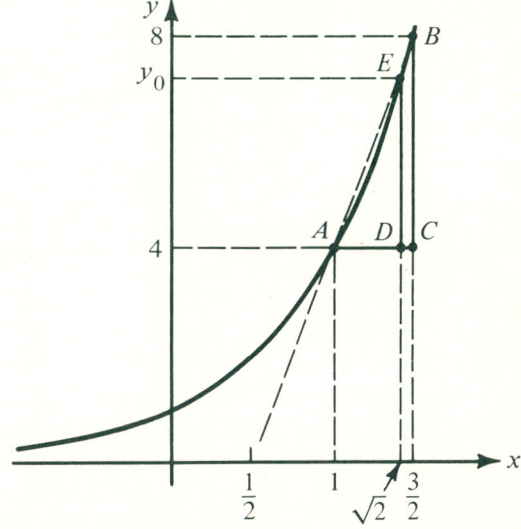

Use this approach to solve Problems 28–31.

28. $4^{\sqrt{3}}$ 29. $9^{\sqrt{5}}$ 30. $10^{\sqrt{2}}$ 31. 10^{π}

8.3 A NEW BASE AND SOME APPLICATIONS FOR EXPONENTIAL FUNCTIONS

Intuition is a powerful tool in mathematics, but unless it is accompanied by sound understanding of the fundamental underlying concepts, it can lead to erroneous conclusions. For example, consider

$$\lim_{x \to 0} (1 + x)^{1/x}$$

which appears to be of the form 1^∞. It seems, intuitively, that this limit should equal 1, but it doesn't. The form 1^∞ is another indeterminate form, similar to 0/0, and a limit of this form may equal values other than 1, depending upon the context. The graph of $y = (1 + x)^{1/x}$ in Figure 8.4

CHAPTER 8   Exponential and Logarithmic Functions

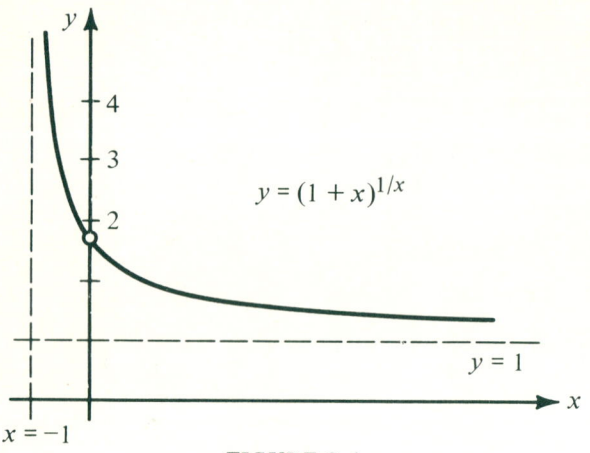

FIGURE 8.4.

indicates that this limit, as x approaches zero, is closer to 3 than to 1.

If we use a hand calculator to evaluate $(1 + x)^{1/x}$ at values of x approaching zero, we get a table like the one below.

x	$y = (1 + x)^{1/x}$
1	2
$\frac{1}{2}$	$\frac{9}{4} \approx 2.25$
$\frac{1}{4}$	2.4414
$\frac{1}{8}$	2.5658
$\frac{1}{10}$	2.5937
$\frac{1}{100}$	2.7048
$\frac{1}{10,000}$	2.7181
$\frac{1}{100,000}$	2.7183
$\frac{1}{1,000,000}$	2.71828

This indicates that the above limit does exist and that its value may be about 2.718. This is only an approximate value of the limit because the limiting value is an irrational number. Even if we continued to calculate, the result would never be a terminating or repeating decimal. We call this limiting, irrational number e and define it as

$$e = \lim_{x \to 0}(1 + x)^{1/x}$$

This number, like π, is of major importance in mathematics and physics, and the exponential function

$$y = e^x$$

is an essential, fundamental function.

In this section, we will study two closely related applications involving $y = e^x$ as well as other exponential functions.

First, let's examine exponential growth. These are situations where the rate at which y increases (grows) is directly proportional to the value of y. That is, the larger y is, the faster it is growing. Population growth involving people, hares, bacteria, etc., is usually an exponential growth situation, since the presence of more organisms results in a greater *rate* of population growth. The

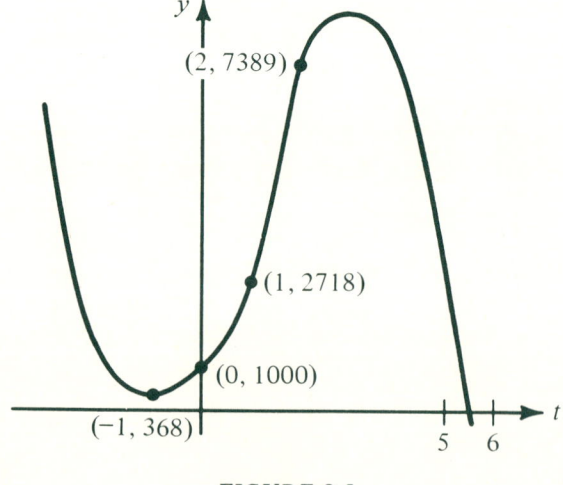

FIGURE 8.5.

SECTION 8.3   A New Base and Some Applications for Exponential Functions

problem, then, is to find a suitable mathematical model for such situations.

At the beginning of a study of the rate of population growth of a husk of hares, there were 1000 hares ($t = 0$, $y = 1000$). One year later, there were 2718 ($t = 1$, $y = 2718$). Two years later there were 7389. A previous study shows that one year before the study began, there were 368 hares.

We now have 4 points that could be described by a third-degree polynomial, as shown in Figure 8.5. This model is unsatisfactory because it does not describe the situation beyond our given values. It indicates more hares the further back in time we go and less hares (even negative numbers) if we look ahead.

A curve that more reasonably describes the situation is drawn in Figure 8.6. It is the graph of an exponential function. It shows the population increasing fast; the larger the number, the faster the increase. As we go back in time (t decreases), the number of hares approaches zero.

Consider the growth in the population of the amoeba, a one-celled animal that multiplies by dividing. That is, it multiplies through mitosis or cell division. If we begin our study with a population of one amoeba, and assume that an amoeba divides every day, then we have the following number of amoebae, N, at each indicated time, t.

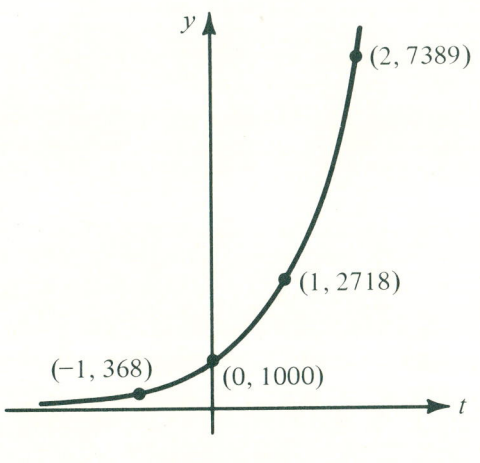

FIGURE 8.6.

Complete the count. At the beginning, $t = 0$, $N = 1$. One day later, $t = 1$, $N = 2$. In one more day, $t = 2$, $N = 4$.

$$t = 3, \quad N = ?$$
$$t = 4, \quad N = ?$$
$$t = 5, \quad N = ? = 2^?$$

In general, at any time t, there are $N = 2^t$ amoebae. Check your entries above using this formula.

Unfortunately, since we do not have a table for the values of 2^t, N is difficult to find for large values of t. (If you have a calculator with the y^x function, this is not a problem. If so, predict how many amoebae there will be in one year. You should get about 7.5×10^{109}.) It is common to model such exponential growth situations using base e.

With calculus, it can be established that **all such population growth problems can be modeled by an equation of the form**

$$y = y_0 e^{kt}$$

where y_0 is the initial population, since $t = 0 \Rightarrow y = y_0$, and k is a proportionality constant that has a different fixed value for each situation. The independent variable, t, usually represents time.

Then, for our husk of hares above,

$$t = 0 \Rightarrow y = 1000$$

means our equation

$$y = y_0 e^{kt}$$

becomes

$$y = 1000 e^{kt}$$

Then, $t = 1 \Rightarrow y = 2718$ means that

$$2718 = 1000 e^{k \cdot 1} \Rightarrow$$

$$e^k = \frac{2718}{1000} = 2.718$$

By consulting a table of values for the exponential function $y = e^x$ (Appendix VI), we see that

$$e^1 \approx 2.718 \Rightarrow k = 1$$

Therefore,
$$y = 1000e^t$$
describes our population of hares. We can now use this function to predict the future population. We predict that the population of hares four years after the beginning of the study will be
$$y = 1000e^4$$
From Table I
$$e^4 \approx 54.598 \Rightarrow$$
$$y \approx 54{,}598 \text{ hares}$$

Exponential decay is very similar to exponential growth except that the rate at which y *decreases* is proportional to the size of y. That is, the greater the quantity y present, the faster the quantity decreases or decays. Radioactive elements decay this way. Thus, if we start with two chunks of radium, one large and the other very small, eventually the two chunks will contain nearly the same amount of radium. The larger chunk will decay at a greater rate (in grams per year) than the smaller one. Both have the same percentage rate of decay, which is determined by the half-life of radium. This property is related to the half-life of radioactive elements. *Half-life* is defined as the amount of time required for a radioactive element to decay to half its present mass. Such **exponential decay situations are modeled by the function**
$$y = y_0 e^{-kt}$$

To date their archeological findings, archeologists use the fact that the half-life of carbon-14, C_{14}, is 5730 years. Radioactive C_{14} is continually produced by the reaction of neutrons created from cosmic-ray bombardment of our atmosphere with available nitrogen. This is absorbed by plants, and hence by all living things. The amount of C_{14} in any living organism is maintained at a constant level. In 1946 Willard Libby, the Nobel prize-winning chemist, reasoned that since absorption of C_{14} ceases at death, the amount present in a bone or a stick from a campfire will reduce by radioactive decay at a rate of 50% every 5730 years.

If there are 100 kilograms of C_{14} present in an object now, we can predict how much will be left in 1000 years. Since we begin with 100 kilograms,
$$y_0 = 100$$
and
$$y = y_0 e^{-kt} \Rightarrow$$
$$y = 100 e^{-kt}$$
Since the half-life of C_{14} is 5730 years, $t = 5730 \Rightarrow$ $y = 50$ kilograms, half of the original 100 kilograms. Therefore,
$$50 = 100 e^{-k(5730)} \Rightarrow$$
$$e^{-5730k} = 0.5$$

Then, from the exponential table in Appendix VI we see that
$$e^{-0.69} \approx 0.5$$
Therefore,
$$e^{-0.69} \approx e^{-5730k} \Rightarrow$$
$$-0.69 \approx -5730k \Rightarrow$$
$$k \approx 0.00012$$
Then, the equation is
$$y = 100 e^{-0.00012x}$$
In 1000 years, we will have approximately
$$y = 100 e^{-0.00012(1000)} \Rightarrow$$
$$y = 100 e^{-0.12} \Rightarrow$$
$$y \approx 88.7 \text{ kg of carbon-14}$$

There is another application of exponential functions that is somewhat different from the two preceding ones. It concerns loans involving compound interest rates.

The amount of money generated by investing an initial amount called the principal, P, at an annual interest rate, i, is modeled by an exponential function as follows.

SECTION 8.3 A New Base and Some Applications for Exponential Functions

At the end of one year, $2000.00 invested at 6% interest earns $120.00, which is 6% of $2000.00. The new amount in the account is $2000.00 + $120.00. In general, after one year, a principal, P, invested at an annual interest rate, i, becomes

$$A_1 = P + i \cdot P = P(1 + i)$$

This new amount becomes the principal for the next year. At the end of two years,

$$A_2 = A_1 + i \cdot A_1 = A_1(1 + i) = [P(1 + i)](1 + i) \Rightarrow$$
$$A_2 = P(1 + i)^2$$

Repeating this compounding over n years yields an amount, A, from a principal, P, invested at an annual interest rate of i

$$A = P(1 + i)^n$$

If interest is compounded semiannually, an interest rate of $i/2$ is calculated twice a year, or $2n$ times. Then after six months, a principal, P, becomes

$$A_{1/2} = P + i/2 \cdot P = P(1 + i/2)$$

After one year, we have

$$A_1 = A_{1/2} + i/2 \cdot A_{1/2} = A_{1/2}(1 + i/2) \Rightarrow$$
$$A_1 = [P(1 + i/2)](1 + i/2) \Rightarrow$$
$$A_1 = P(1 + i/2)^2$$

After n years, a principal, P, invested at an interest rate of i compounded k times a year becomes

$$A = P(1 + i/k)^{nk}$$

EXAMPLE I. Compare the principal, after five years, on an initial amount of $2000.00 invested at 6% per year (a) compounded annually and (b) compounded quarterly.

(a) For the annually compounded interest,

$$A = P(1 + i)^n$$

where $P = \$2000.00$, $i = 0.06$, and $n = 5$, yields

$$A = \underline{\quad ? \quad}(1 + \underline{\quad\quad})^? \Rightarrow$$
$$A = (2000)(1.06)^5 \Rightarrow$$
$$A = (2000)(1.34) = \$2676.45$$

(b) For the quarterly compounded interest,

$$A = P(1 + i/k)^{nk}$$

where $P = \underline{\quad ? \quad}$, $i/k = ?/? = 0.015$, and $nk = 20$ (since $n = $ five years), yields

$$A = 2000(1 + \underline{\quad ? \quad})^? \Rightarrow$$
$$A = 2000(1.015)^{20} \Rightarrow$$
$$A = 2000(1.347) = \$2693.71$$

Quarterly compounding resulted in $17.26 more in five years.

Now let's investigate what happens as the number of compounding periods increases toward "continuously compounded" interest.

Since we will assume the principal remains fixed for each case, we are interested in the value of

$$(1 + i/k)^{nk}$$

as the number of compounding periods, k, increases. For ease of calculation let's use an annual (nominal) interest rate of 10% on an amount invested for ten years, $i = 0.1$ and $n = 10$. Then

$$(1 + i/k)^{nk} = (1 + 0.1/k)^{10k}$$

Using a hand calculator, we see that for quarterly compounding ($k = 4$),

$$(1 + i/k)^{nk} = (1 + 0.1/4)^{40} \approx 2.685$$

For monthly compounding ($k = 12$),

$$(1 + i/k)^{nk} = (1 + 0.1/12)^{120} \approx 2.707$$

For daily compounding ($k = 365$),

$$(1 + i/k)^{nk} = (1 + 0.1/365)^{3650} \approx 2.718 \approx e$$

This result suggests that

$$A = P(1 + 0.1/k)^{10k}$$

might become

$$A = Pe$$

as the number of compounding periods, k, increases.[1] In fact, since

$$\lim_{k \to \infty}(1 + 1/k)^k = \lim_{k \to 0}(1 + k)^{1/k} = e$$

167

CHAPTER 8   Exponential and Logarithmic Functions

we have, for an investment of P dollars continuously compounded at an annual interest rate i for n years,

$$A = Pe^{in}$$

EXAMPLE II. Assume that interest on the amount in Example I was continuously compounded at 6% per year for five years. Find the final amount.

$$A = Pe^{in}$$

where $P = \$2000.00$, $i = 0.06$, and $n = 5$, yields

$$A = (\quad)e^{?} \Rightarrow$$
$$A = (2000)e^{0.3} \Rightarrow$$
$$A = (2000)(?) = \$2699.72$$

EXAMPLE III. Find the amount in a continuously compounded savings program after five years, if the nominal interest rate is $6\frac{3}{4}\%$ and the principal is $1000.00.

$$S = Pe^{rn} \Rightarrow$$
$$S = 1000e^{(0.0675)(5)}$$
$$= 1000e^{0.3375} = \$1401.44$$

EXERCISE 8.3

A. FUNDAMENTALS

1. If the population of a town at any time, t, is given by

$$y = 40{,}000 e^{0.17t}$$

(a) What is the population at $t = 0$ (the initial population)?
(b) What was the population one year earlier?
(c) What will the population be in three years?

2. The equation modeling the radioactive decay of 20 grams of carbon-14 is given by

$$y = 20e^{-0.00012t}$$

(a) How many grams will be left in 1000 years?
(b) How many grams will be left in 5000 years? (Recall that the half-life of carbon-14 is 5730 years.)

3. Barometric pressure, measured in centimeters of mercury, is given for any altitude of h kilometers by

$$P = 76e^{-h/8}$$

(a) What is the barometric pressure at sea level?
(b) What is the barometric pressure at the top of Mount Rainier in the state of Washington, if it is 4250 meters high?
(c) What is the barometric pressure in Denver, the mile-high city?

B. ESSENTIALS

4. When was a 10-liter tank half full, if the volume of water entering it doubled every minute and it took 20 minutes for the tank to fill? (HINT: No equation is necessary.)

5. If paper is 0.006 centimeters thick, how high is a stack of paper formed by cutting a 20 by 30 centimeter sheet of paper in half 24 times? (First you have 1, then 2, then 4 pieces, etc.)

6. If a paper molecule is 10^{-9} centimeters in diameter, how many times can you cut the paper in Problem 5 without destroying the paper molecule?

★7. The light intensity of a star is related to its magnitude (brightness) by the equation

$$I = I_0 \, 10^{-0.4M}$$

where M is the magnitude of a star whose light intensity is I, and I_0 is the light intensity of a star of zero magnitude. Compare the light intensities for the sun, a -27 magnitude star, and Alpha Centauri, a star of -0.3 magnitude (the second-closest star). A star of magnitude 6 is just visible from earth.

★8. If the number of bacteria in a culture is 10,000 at the beginning of a study and 15,000 two days later, how many will there be one week after the beginning of the study?

9. Compare the cost of a $20,000.00 mortgage at an 8% annual interest rate for (a) a twenty-year mortgage and (b) a twenty-nine-year mortgage.

10. Compare the yield on $5000.00 that (a) earns an annual interest of $7\frac{1}{4}\%$ over three years and (b) is continually compounded at the same annual rate over three years.

• **11.** Find the time (to the nearest year) that it takes to double an investment that is compounded annually at 6% interest.

12. Extend the investment equation to find the value of a piece of machinery after ten years, if it depreciates at a rate of 10% per year and costs $50,000.00 initially.

• **13.** Gompertz curves, of the form $N = ca^{R^t}$, are useful models for many situations, including the growth of certain groups of people.

Treating N as a function of t, graph the Gompertz curve for a certain company where the size at maturity is $c = 1000$, the initial size is 100, and the rate of growth is $R = 20\%$ per year. How many employees will there be after two years? How long will it take the company to double in size?

14. Curves of the type $y = a(1 - e^{-kt})$ are used by psychologists to model the learning of particular tasks such as typing and riding a bicycle—learning that has been reinforced t times.
(a) Graph the curve for learning how to ride a bicycle, where y is the number of meters ridden without a fall and t is the number of weeks you have been riding. Use $a = 100$ and $k = 0.01$.
(b) Estimate the number of weeks of riding that will be necessary before you will be able to ride 20 kilometers without a fall.

15. Graph the "hyperbolic trig" functions

$$y = \sinh x = \frac{e^x - e^{-x}}{2}$$

and

$$y = \cosh x = \frac{e^x + e^{-x}}{2}$$

from the graphs of $y = \tfrac{1}{2}e^x$ and $y = \tfrac{1}{2}e^{-x}$.

8.4 LOGARITHMIC FUNCTIONS

The way we define and use logarithms is very similar to the way we handle square roots. Fundamentally, there is no way to solve the equation $y = x^2$ for x. This difficulty is overcome by defining the square root symbol ($\sqrt{}$) and the process of taking square roots in terms of the **inverse** of squaring. That is, $\sqrt{9} = 3$ because $3^2 = 9$. To understand square rooting, we think of its inverse—squaring. Logarithms and logarithmic functions are defined analogously. They are interpreted as the inverse of an exponential expression. There is no algebraic way of solving the exponential equation $y = b^x$ for x. The symbol we invented to solve for x, "$\log_b$," is read, "log to the base b." As with $\sqrt{}$, the only way to understand a logarithm is by examining its inverse. We will use the following definition of a logarithm in terms of its inverse exponential expression whenever we work with logs. We read

$$y = \log_b x$$

as, "y equals the log to the base b of x," where "log" is short for "logarithm."

DEFINITION 8.1.

$$y = \log_b x \Leftrightarrow$$
$$x = b^y, \qquad b > 0$$

This definition makes the exponential function $g(x) = b^x$ and the logarithmic function $f(x) = \log_b x$ inverses of each other.

Now we can find the equation of the inverse of the exponential function

$$y = b^x$$

by first using Definition 8.1 to solve for x,

$$x = \log_b y$$

and then renaming the variables,

$$y = \log_b x$$

We often write $f^{-1}(x)$ for y when we are dealing with inverses.

Let's use this two-step process involving the definition of a logarithm to find the equation of the inverse for each of the following exponential functions. The inverse of each of these exponential functions is a logarithmic function with the same base.

EXAMPLE I. Find the equation of the inverse for each of the following:

(a) $y = 3^x \Leftrightarrow x = \log_3 y$

$y = f^{-1}(x) = \log_3 x$

(b) $y = 10^x \Leftrightarrow x = \log_{10} y$

$f^{-1}(x) = \log_{10} x$

(c) $y = e^x \Leftrightarrow x = \log_e y$

$y = \log_e x$

(d) $y = 10^{x+1} \Leftrightarrow x + 1 = \log_{10} y$

$y = -1 + \log_{10} x$

(e) $y = e^x - 1 \Rightarrow y + 1 = e^x \Leftrightarrow x = \log_e(y + 1)$

$f^{-1}(x) = \log_e(x + 1)$

For $y = \log_{10} x$, we will usually write $y = \log x$; for $y = \log_e x$, we write $y = \ln x$. Any other bases must be written explicitly, since these are the most common bases.

To emphasize the importance of automatically translating logarithmic statements into the equivalent exponential form, let's find the logs of some specific numbers using various positive bases.

EXAMPLE II. Find the logs for each of the following:

$y = \log_2 8 \Leftrightarrow 2^y = 8 \Rightarrow y = 3 \Rightarrow \log_2 8 = 3$

$y = \log_4 16 \Leftrightarrow 4^y = ? \Rightarrow y = 2 \Rightarrow \log_4 16 = 2$

$y = \log_{10} 1 \Leftrightarrow 10^y = 1 \Rightarrow y = ? \Rightarrow \log_{10} 1 = 0$

$y = \log_e 1 \Leftrightarrow e^y = ? \Rightarrow y = ? \Rightarrow \log_e 1 = ?$

$y = \log_{10}(\frac{1}{100}) \Leftrightarrow ? \Rightarrow y = ? \Rightarrow \log_{10}(\frac{1}{100}) = ?$

The last three values of y are 0, 0, and -2. Logarithms are similar to square roots in that logs also often result in irrational values. For example,

$$y = \log_{10} 2 \Leftrightarrow$$

$$10^y = 2$$

creates a problem because no rational value of y will work. We know that

$$10^1 = 10$$

$$10^{1/2} = \sqrt{10} \approx 3.162$$

$$10^{1/3} = \sqrt[3]{10} \approx 2.153$$

$$10^{3/10} = \sqrt[10]{10^3} \approx 1.9953$$

$$10^{0.301} \approx 1.99986 \approx 2$$

which indicates that $\log_{10} 2 \approx 0.301$.

Now we are ready to graph logarithmic functions. Since the logarithmic functions are simply the inverses of exponential functions, let's begin the study of the graphs of logarithmic functions by examining the exponential function $y = 2^x$ and its inverse. Their graph is shown in Figure 8.7. From the graph, we see that the domain and range for $y = \log_2 x$ are

$$D = \{x \mid x > 0\}$$

and

$$R = \{y \mid y \in \text{reals}\}$$

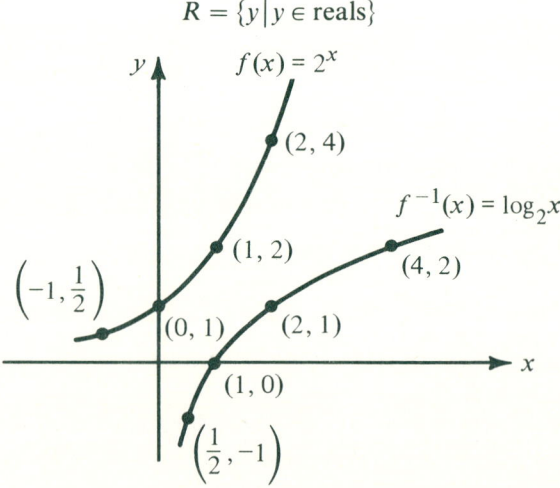

FIGURE 8.7.

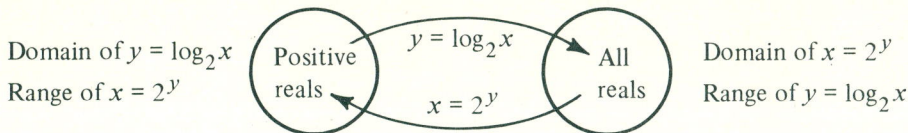

Domain of $y = \log_2 x$
Range of $x = 2^y$

Positive reals

$y = \log_2 x$
$x = 2^y$

All reals

Domain of $x = 2^y$
Range of $y = \log_2 x$

FIGURE 8.8.

It is easy to remember that the domain of

$$y = \log_2 x$$

is restricted to positive values, because

$$y = \log_2 x \Leftrightarrow x = 2^y$$

requires that $x > 0$, since $2^y > 0$ for all real values of y. Notice that "$\log_2 x$" is the name of the *exponent* in

$$x = 2^y$$

The domains and ranges are depicted in Figure 8.8. Remember that we rename the variables for the inverse to keep the roles of independent and dependent variables (x and y, respectively) consistent when graphing. They are not renamed in Figure 8.8, because we are not graphing.

EXAMPLE III. Now that we know the graph of $y = \log_2 x$, let's reflect and translate it to get the graphs of the following functions. Draw the graph of $y = f(-x)$ (a reflection about the y-axis) in Figure 8.9(a), and the reflection about the origin, $y = -f(-x)$, in Figure 8.9(b). From part (c) in Figure 8.9, we see that only horizontal translations change the asymptote. In (c), $y = f(x + 2)$ changed the asymptote from the y-axis, $x = 0$, to $x = -2$. Thus, the function $y = f(x + 2)$ is a translation of the curve $f(x) = \log_2 x$ and its asymptote 2 units to the left.

(a)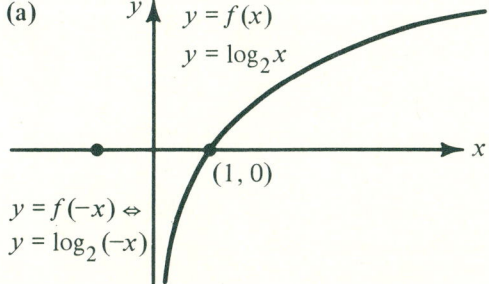

(b)

(c)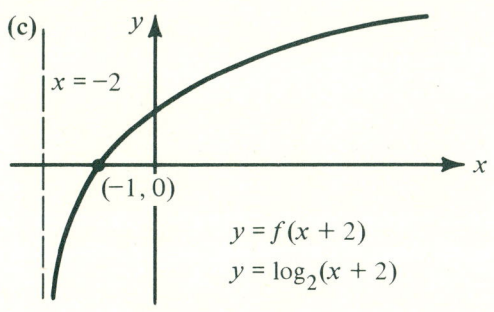

(d)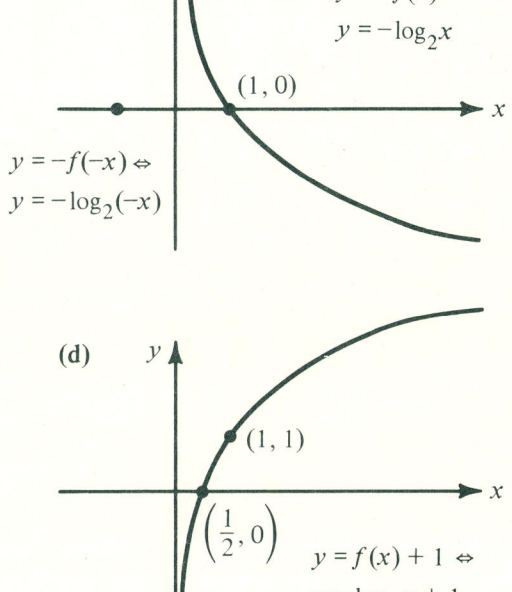

FIGURE 8.9.

CHAPTER 8 Exponential and Logarithmic Functions

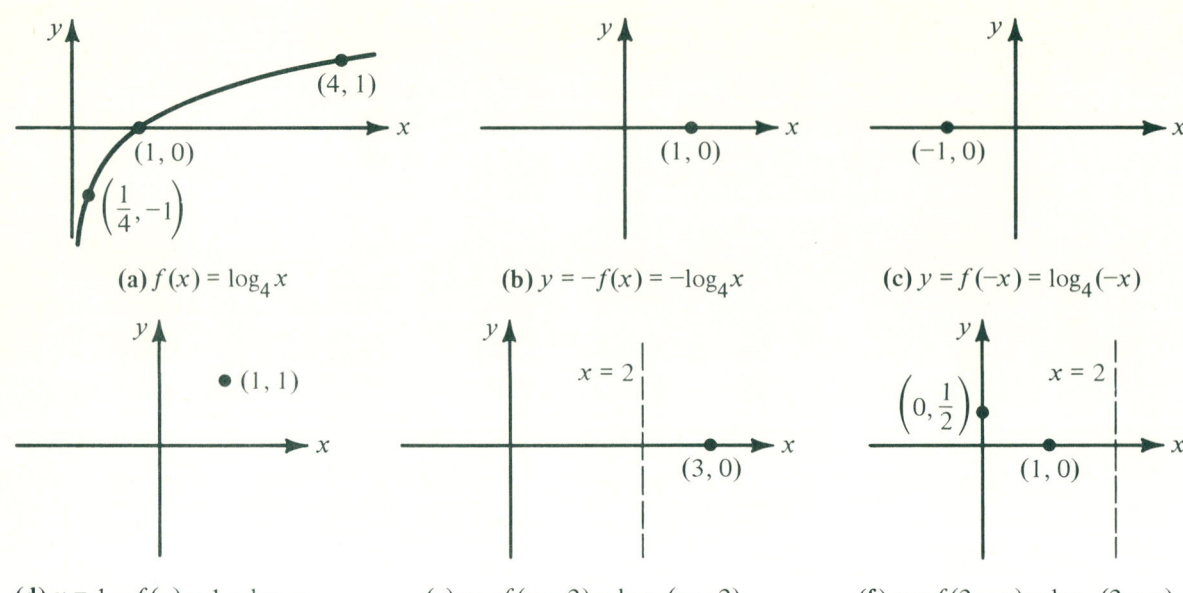

FIGURE 8.10.

EXAMPLE IV. Now, complete the graph of the translations or reflection of $f(x) = \log_4 x$, as shown in Figure 8.10.

The domains of the functions in Figure 8.10 are all real numbers on one side of their vertical asymptotes. $x = 0$ for (b), (c), and (d); $x = 2$ for (e) and (f). The domains of the functions are the following.

For (b),
$$D = \{x \mid x > 0\}$$

For (c),
$$D = \{x \mid x < 0\}$$

For (d),
$$D = \{x \mid x > 0\}$$

For (e),
$$D = \{x \mid x > 2\}$$

And for (f),
$$D = \{x \mid x < 2\}$$

Now we are ready to graph a logarithmic function that is both a reflection and a translation of a simple function and its inverse.

EXAMPLE V. Graph $f(x) = 1 - \log_3 (x - 3)$ and its inverse. Find the exponential equation of its inverse.

The graph of f is the graph of $y = \log_3 x$, reflected about the x-axis and translated to the right 3 and up 1. The domain of f is all values of x such that
$$x - 3 > 0 \Rightarrow$$
$$x > 3$$

Therefore, there is no y-intercept.

For the x-intercept, set f equal to zero and solve for x.
$$f(x) = 0 \Rightarrow 1 - \log_3 (x - 3) = 0 \Rightarrow$$
$$\log_3 (x - 3) = 1 \Leftrightarrow$$

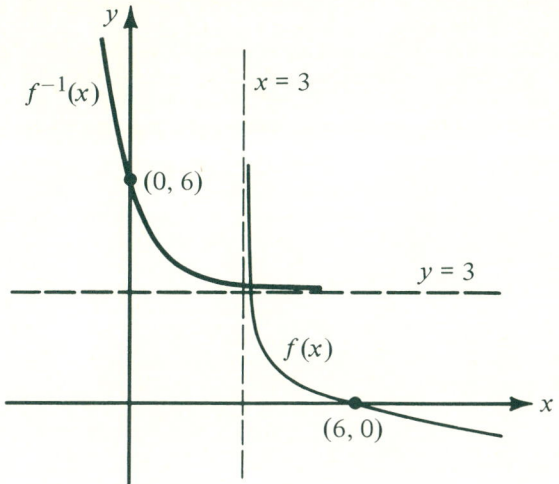

$f(x) = 1 - \log_3(x - 3)$ and $f^{-1}(x) = 3 + 3^{1-x}$

FIGURE 8.11.

By the definition of logs,
$$\log_3(x - 3) = 1 \Leftrightarrow x - 3 = 3^1$$

Therefore,
$$x = \underline{}?$$

To find the equation of the inverse, first solve for x:
$$y = 1 - \log_3(x - 3) \Rightarrow$$
$$\log_3(x - 3) = 1 - y \Leftrightarrow$$
$$x - 3 = 3^{1-y} \Rightarrow$$
$$x = 3 + 3^{1-y}$$

Then, rename the variables:
$$y = 3 + 3^{1-x}$$

The graphs are given in Figure 8.11. Notice the symmetry about the line $y = x$. What are the asymptotes of each graph?

EXERCISE 8.4

A. FUNDAMENTALS
Use the definition of logarithms to write each of the following exponential expressions as an equivalent logarithmic expression.

1. (a) $3^0 = 1$ (b) $10^0 = 1$ (c) $e^0 = 1$
2. (a) $5^1 = 5$ (b) $10^1 = 10$ (c) $e^1 = e$
3. (a) $2^{-1} = \frac{1}{2}$ (b) $10^{-2} = \frac{1}{100}$ (c) $10^{-3} = 0.001$
4. (a) $2^3 = 8$ (b) $3^4 = 81$ (c) $4^3 = 64$

Graph each of the following pairs of functions on the same coordinate system.

5. (a) $y = 2^x$ (b) $y = \log_2 x$
6. (a) $y = 10^x$ (b) $y = \log x$
7. (a) $y = e^x$ (b) $y = \ln x$
8. (a) $y = 2^{-x}$ (b) $y = -\log_2 x$
9. (a) $y = -3^x$ (b) $y = \log_3(-x)$
10. (a) $y = -4^{-x}$ (b) $y = -\log_4(-x)$
11. (a) $y = \log_2 x + 1$ (b) $y = \log_2(x + 1)$

B. ESSENTIALS
Graph each of the following functions and its inverse on the same set of axes. Find the domain of each function and of its inverse. Find the equation of the inverse function.

12. $y = 3^x$ 13. $y = 2^{x+1}$ 14. $y = 3^x - 1$
15. $y = 1 - 2^x$ 16. $y = 4 - 2^{x+1}$
17. $y = 3^{1-x} + 2$ ★18. $y = e^{x+1} - 1$
19. $y = \log_3 x$ 20. $y = \log_3(x + 2)$
21. $y = 3 - \log_2 x$ 22. $y = 2 - \log_4(1 - x)$
★23. $y = 4 - \log_5(x + 3)$
24. $y = 3 - \log_3(3x - 1)$
25. $y = 4 - \log_2(2x + 3)$

C. APPLICATIONS
Carefully examine the effect of the absolute value sign or the squaring, then graph each of the following.

26. $y = \log |x|$ 27. $y = |\log x|$
28. $y = |\log |x||$
29. (a) $y = 2 \log x$ (b) $y = \log x^2$
30. (a) $y = 3 \log_2 x$ (b) $y = \log_2 x^3$

CHAPTER 8   Exponential and Logarithmic Functions

31. (a) $y = \log_3 x + 1$ (b) $y = \log_3 3x$
32. (a) $y = \log_4 \frac{1}{16}$ (b) $y = \log_4 x - 2$
33. Explain why 1 cannot be used as a logarithmic base.

8.5 EQUATIONS INVOLVING LOGARITHMS

We shall begin this section by learning the three basic properties of logarithms which we will apply to solving logarithmic equations.

These three properties of logarithms are very similar to the laws of exponents, since, after all, *a logarithm is itself an exponent*. That is,

$$x = 10^y \Leftrightarrow$$
$$y = \log_{10} x$$

means that $\log_{10} x$ is the exponent of 10 that will yield x. Therefore, each proof relies on the corresponding property of exponents. We shall prove the second theorem here and consider the proofs of the other two (which are quite similar) in the exercises. The base b is any positive real number. Both M and N must be positive.

THEOREM I.
$$\log_b M \cdot N = \log_b M + \log_b N$$

THEOREM II.
$$\log_b \frac{M}{N} = \log_b M - \log_b N$$

THEOREM III.
$$\log_b M^p = p \log_b M$$

Proof: Now, let's prove Theorem II. First, let $s = \log_b M/N$, $t = \log_b M$, and $v = \log_b N$.

This will allow us to convert each logarithmic expression into an equivalent exponential expression so we can use the laws of exponents. We must show that

$$s = t - v$$

where

$$s = \log_b \frac{M}{N} \Leftrightarrow \frac{M}{N} = b^s$$
$$t = \log_b M \Leftrightarrow M = b^t$$

and

$$v = \log_b N \Leftrightarrow N = b^v$$

Therefore,

$$\frac{M}{N} = b^s \quad \text{and} \quad \frac{M}{N} = \frac{b^t}{b^v} = b^{t-v} \Rightarrow$$
$$b^s = b^{t-v} \Rightarrow$$
$$s = t - v$$

Therefore,

$$\log_b \frac{M}{N} = \log_b M - \log_b N$$

Use the definition of a logarithm,

$$y = \log_b x \Leftrightarrow x = b^y$$

to solve the following equations for x:

(1) $\log_3 x = 2 \Leftrightarrow x = 3^2 \Rightarrow x = ?$
(2) $\log_2 8 = x \Rightarrow x = ?$
(3) $\log_4 (x + 1) = 2 \Leftrightarrow x + 1 = 4^2 \Rightarrow x = ?$
(4) $\log_3 (2x - 1) = -1 \Leftrightarrow ? = 3^{-1} \Rightarrow x = ?$
(5) $\log_2 2x(x - 1) = 2 \Leftrightarrow 2x(x - 1) = \underline{} \Rightarrow$
$\underline{} = 0 \Rightarrow ()() = 0 \Rightarrow$
$? = 0 \quad \text{or} \quad ? = 0 \Rightarrow$
$x = ? \quad \text{or} \quad x = ?$

Since for $x = -1$, $2x(x - 1)$ is negative, $x = -1$ is not in the domain of $\log_2 2x(x - 1)$. The answers are (1) 9, (2) 3, (3) 15, (4) 2/3, and (5) 2.

174

Now, we will use the three basic properties of logs and the definition of a log to solve the following logarithmic equations.

EXAMPLE I. Solve the following equation for x:

$$\log_2 x + \log_2 (x - 1) = 1$$

Solution: By Theorem I,

$$\log_2 x + \log_2 (x - 1) = 1 \Rightarrow$$
$$\log_2 x(x - 1) = 1$$

By the definition of logs, this logarithmic equation is equivalent to the following exponential equation:

$$x(x - 1) = 2^1 \Rightarrow$$
$$x^2 - x = 2 \Rightarrow$$
$$\underline{\quad ? \quad} = 0 \Rightarrow$$
$$\underline{\quad ? \quad} = 0 \Rightarrow$$
$$x = 2 \quad \text{or} \quad x = -1 \Rightarrow$$
$$x = 2$$

We must reject $x = -1$ because the domain of the logarithm function $y = \log_2 x$ is $x > 0$.

EXAMPLE II. Solve the following for x:

$$\log_3 (x + 2) + \log_3 (x + 1) - \log_3 (x + 6) = 1$$

Solution: By Theorems I and II, we have

$$\log_3 \frac{(x + 2)(x + 1)}{x + 6} = 1$$

Then, by the definition of logs, this is equivalent to

$$\frac{x^2 + 3x + 2}{x + 6} = 3^1 \Rightarrow$$
$$x^2 + 3x + 2 = 3 \cdot \; ? \Rightarrow$$
$$x^2 = \underline{\quad ? \quad} \Rightarrow$$
$$x = \underline{\quad ? \quad} \Rightarrow$$
$$x = 4$$

Again, $x = -4$ must be rejected because it is not in the domains of all the functions involved.

EXAMPLE III. Solve the following equation for x:

$$\log_4 \sqrt{2x - 3} = -1$$

Solution:

$$\log_4 \sqrt{2x - 3} = -1 \Rightarrow$$
$$\log_4 (2x - 3)^{1/2} = -1 \Rightarrow$$
$$\tfrac{1}{2} \log_4 (2x - 3) = -1 \Rightarrow$$
$$\log_4 (2x - 3) = -2 \Leftrightarrow$$
$$2x - 3 = 4^{-2} \Rightarrow$$
$$2x - 3 = \tfrac{1}{16} \Rightarrow$$
$$2x = \tfrac{49}{16} \Rightarrow$$
$$x = \tfrac{49}{32}$$

The fact that we only have tables of logarithms for base 10 and base e, and that it is more convenient to study logarithms using smaller bases, sometimes causes a problem, as illustrated by Example IV.

EXAMPLE IV. Find the y-intercept of

$$f(x) = \log_3 (x + 2)$$

Solution $f(0) = \log_3 2$ is not in our log tables. If we let

$$y = \log_3 2$$

then, by the definition of logs,

$$2 = 3^y$$

Now, taking the log to the base 10 of both sides yields the y-intercept.

$$\log_{10} 2 = \log_{10} 3^y \Rightarrow$$
$$\log 2 = y \log 3 \Rightarrow$$
$$y = \frac{\log 2}{\log 3} \approx \frac{0.3010}{0.4771} \approx 0.631$$

In finding the y-intercept above, we saw that

$$\log_3 2 = \frac{\log_{10} 2}{\log_{10} 3}$$

CHAPTER 8   Exponential and Logarithmic Functions

This generalizes to Theorem IV, which tells us how to switch bases of logarithms.

THEOREM IV.

$$\log_a x = \frac{\log_b x}{\log_b a}, \quad a > 0, b > 0$$

There are two other theorems on logs that are important consequences of the fact that $y = b^x$ and $y = \log_b x$ are inverses of each other and that for inverses $f(f^{-1}(x)) = x$ and $f^{-1}(f(x)) = x$.

THEOREM V.

$$\log_b b^x = x, \quad b > 0$$

THEOREM VI.

$$b^{\log_b x} = x, \quad b > 0$$

EXERCISE 8.5

A. FUNDAMENTALS
Solve each of the following logarithmic equations for x by first rewriting each in its equivalent exponential form.

1. (a) $\log_2 x = 1$ (b) $\log x = 1$
   (c) $\ln x = 1$
2. (a) $\log x = 2$ (b) $\log x = -2$
   (c) $\log x = 0$
3. (a) $\log_2 x = 3$ (b) $\log_2 (-x) = 3$
   (c) $\log_2 (x + 1) = 3$
4. (a) $\log_3 x = 2$ (b) $\log_3 (1 - x) = 2$
5. (a) $\log x = -1$ (b) $\log (x + 2) = -1$
6. (a) $\log_2 (x^2 - x) = 1$
   (b) $\log_3 (x^2 + 8x) = 2$

B. ESSENTIALS
Solve each of the following equations for x. Remember that since the domain of the logarithmic function $y = \log_b x$ is $x > 0$, there may be extraneous solutions.

7. (a) $\log_3 x + \log_3 (x - 2) = 1$
   (b) $\log_2 x + \log_2 (x + 3) = 2$
   ★(c) $\log (x + 2) = -1 - \log x$
8. (a) $\log x + \log (x - 3) = 1$
   (b) $\log_3 3 + \log_3 (x^2 - 3x - 1) = 2$
   (c) $\log_2 x = 4 - \log_2 (2x - 31)$
9. (a) $\log_2 x - \log_2 (x - 1) = 1$
   (b) $\log_2 2x^2 - \log_2 (2x + 3) = 1$
   (c) $\log_3 9 - \log_3 (x - 2) = 1$
10. $\log_3 \left(\dfrac{1}{x+1}\right) = 1 - \log_3 (x + 4)$
★11. $\ln x = \ln (x - 6) - \ln (x - 4)$
12. $\log_2 (x - 4) = 3 - \log_2 (x + 3)$
13. $1 + 2 \log_2 x = \log_2 (5x - 3)$
14. $\frac{1}{2} \ln (x + 5) = \ln (x - 1)$
15. $\log_2 3x - \log_2 (x + 3) = \log_2 (x - 1) + 1$
16. $\log_3 x - \log_3 (x + 1) = \log_3 (x + 4) - 2$
★17. $\log_2 \sqrt{x + 6} - \log_2 (x - 1) = \frac{3}{2}$

Use the fact that $\log 2 = 0.30$ and $\log 3 = 0.47$, and the three properties of logs, to find each of the following:

18. $\log 6$ 19. $\log 9$ 20. $\log 36$
21. $\log 1.5$ 22. $\log 81$ 23. $\log \sqrt{2}$
24. $\log \frac{4}{27}$ 25. $\log 30$

Solve the following equations for x.

26. (a) $\log_x 16 = 4$ (b) $\log_x \frac{1}{16} = 2$
27. (a) $\log_x x^{4x} = 8$ (b) $x^{\log_x 10x} = \frac{1}{10}$
    (c) $\log [\log (\log x)] = 1$
28. (a) $\log |x + 1| = -3$
    (b) $|\log (x - 1)| = 1$

29. Find the y-intercepts for the graphs of the following logarithmic functions.
(a) $f(x) = \log_3 (4x + 5)$
(b) $g(x) = 1 - \log_4 (x + 3)$

30. Use logarithms to find the x-intercepts of the graphs of these exponential functions.

(a) $y = 3 - 4^{x+2}$
(b) $f(x) = 2 - e^{3-2x}$

31. Prove Theorem I.

32. Prove Theorem III.

33. Prove Theorem IV.

8.6 COMPUTATIONS WITH LOGARITHMS (OPTIONAL)

As sophisticated calculators become more readily available, the importance of logarithms in doing complicated arithmetic will decrease. But for now, logarithms are still helpful in complicated arithmetic problems. They probably will continue to be essential for extracting roots, especially cube roots and higher, for many years to come.

The usefulness of logarithms in accomplishing arithmetic stems from the three basic properties of logs. Fill them in and check your answers with those in Section 8.5.

(a) $\log M \cdot N = $ _____
(b) $\log M/N = $ _____
(c) $\log M^p = $ _____

These properties reduce multiplication and division problems to simple addition and subtraction operations, respectively. Raising numbers to powers requires only one multiplication.

EXAMPLE 1. By examining the logarithms of the following numbers you will see why we only need tables for the logarithms of numbers between 1 and 10.

$$\log 2.54 \approx 0.4048$$
$$\log 254 = \log \left(\tfrac{254}{100} \cdot 100\right)$$
$$= \log (2.54 \times 100)$$
$$= \log 2.54 + \log 100$$
$$\approx 0.4048 + 2 = 2.4048$$

$$\log .254 = \log \left(\tfrac{2.54}{10}\right) = \log (2.54 \times 10^{-1})$$
$$= \log 2.54 - \log 10$$
$$= 0.4048 - 1 = -0.5951$$

Since 2540 is $2.54 \times 1000 = 2.54 \times 10^3$, its log is $0.4048 + 3$. The number 0.4048, called the **mantissa**, is the same for all these numbers above. The 3 in log 2540 is called the **characteristic** and is easily found by counting the number of places the decimal is positioned away from standard position (one place to the right).

TABLE 8.6. *Logarithms*

NUMBER	MANTISSA	CHARACTERISTIC
254.0	$\log 2.54 = 0.4048$	$+2$
.00081	$\log 8.1 = 0.9085$	-4
27.8	$\log 2.78 = 0.4440$	$+1$

In this section we will also have to find antilogs of numbers. For example,

NUMBER	ANTILOG
0.5740	3.75
2.5740	375.0
-1.4260	.0375

The antilog of 0.5740 is 3.75, since $\log 3.75 = 0.5740$. To find the antilog of -1.4260 we must first write it with a positive mantissa, since the mantissa must be a number between 0 and 1 in order to use our tables. We accomplish this by adding and subtracting 2.

$$-1.4260 = (2 - 1.4260) - 2$$
$$= 0.5740 - 2$$

whose antilog is found by locating the antilog for 0.5740 and then moving the decimal two places to the left of standard position.

There is one more technique we must examine before we begin doing arithmetic problems. Many of the numbers that we will encounter will not be

CHAPTER 8   Exponential and Logarithmic Functions

found in the log table in this book. To solve this problem we will approximate such values using linear interpolation.

For example, suppose we want to find the log of 2.753. Since this number is between 2.75 and 2.76, two consecutive entries in our table, we can generate the picture in Figure 8.12. The value we wish to find is indicated by the question mark on the y-axis. We shall assume that the line joining A and B is a straight line, and then use similar triangles to approximate the desired value. Notice that the line is almost straight. The closer together our values on the x-axis are located, the closer the curve between them is to a straight line. Also, the larger our values are, the more linear the logarithmic curve becomes.

Figure 8.13 depicts a section of the curve $y = \log_{10} x$, the linear approximation through A and B, the desired value y_1, and the desired approximation y_2. We must calculate Δy. Since triangle ABC is similar to triangle ADE, their corresponding parts are proportional. Hence,

$$\frac{DE}{BC} = \frac{AE}{AC} \Rightarrow$$

$$\frac{\Delta y}{0.4409 - 0.4393} = \frac{2.753 - 2.75}{2.76 - 2.75} = \frac{0.003}{0.01} \Rightarrow$$

$$\Delta y = (0.0016)(0.3) \Rightarrow$$

$$\Delta y = 0.00045 \approx 0.0005 \Rightarrow$$

$$y_1 \approx y_2 = 0.4393 + \Delta y \Rightarrow$$

$$y_1 \approx 0.4398$$

The process can be simplified greatly using the following scheme.

$$10\left[3\left[\begin{array}{l}\log 2.75 = 0.4393 \\ \log 2.753 = y \\ \log 2.76 = 0.4409\end{array}\right]\Delta y\right]0.0016$$

$$\frac{3}{10} = \frac{\Delta y}{0.0016} \Rightarrow$$

$$\Delta y = (0.3)(0.0016)$$

$$\Delta y \approx 0.0005$$

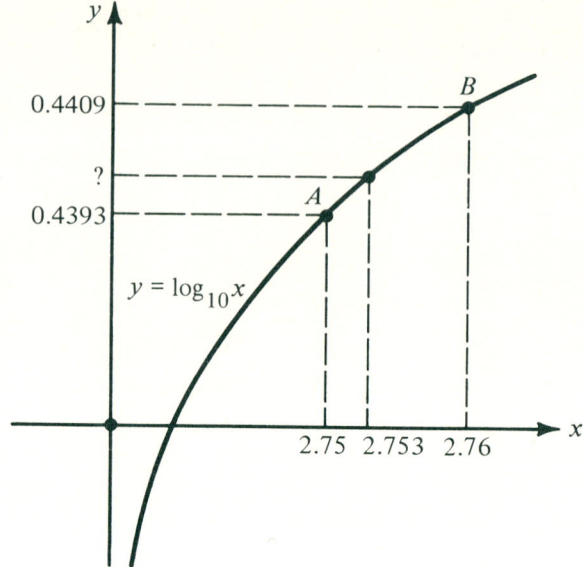

FIGURE 8.12.

Then

$$y = 0.4393 + \Delta y \Rightarrow$$

$$y = 0.4393 + .0005 \Rightarrow$$

$$y = 0.4398$$

With a little practice, linear interpolation can be carried out very quickly, almost mentally. We

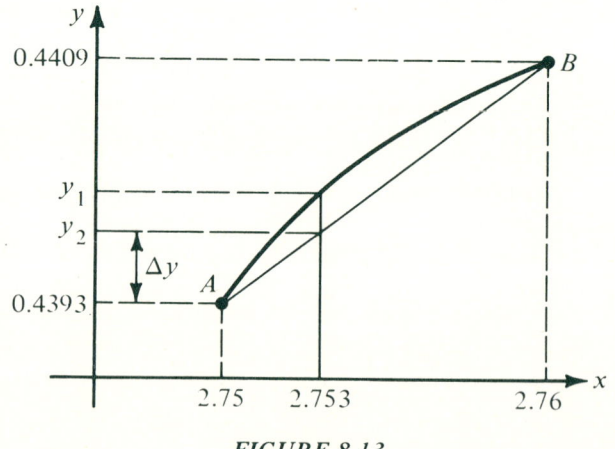

FIGURE 8.13.

also use linear interpolation to approximate antilogs of numbers that are not in our table.

EXAMPLE II. Find the antilog of 0.0112.
By examining the tables, we find

$$0.01\left[\Delta x\begin{bmatrix}\log 1.02 = 0.0086\\ \log x\ \ \ = 0.0112\\ \log 1.03 = 0.0128\end{bmatrix}0.0026\right]0.0042$$

$$\frac{\Delta x}{0.01} = \frac{0.0026}{0.0042} \Rightarrow$$

$$\Delta x \approx 0.006$$

Hence,
$$\log x = \log(1.02 + \Delta x) \Rightarrow$$
$$\log x \approx \log 1.026 \approx 0.0112$$

Now, let's use these tools to carry out a complicated arithmetic problem.

EXAMPLE III. Solve the equation
$$x = \frac{\sqrt[3]{1.81}}{(527)^3 \sqrt{0.0118}}$$

Applying the three properties of logs, we have

$$\log x = \log \frac{(1.81)^{1/3}}{(527)^3(0.0118)^{1/2}} \Rightarrow$$

$$\log x = \log(1.81)^{1/3} - \log[(527)^3 \cdot (0.0118)^{1/2}]$$
$$= \tfrac{1}{3}\log(1.81) - 3\log(527) - \tfrac{1}{2}\log(0.0118)$$
$$= \tfrac{1}{3}(0.2577) - 3(2.7218) - \tfrac{1}{2}(0.0719 - 2)$$
$$= 0.859 - 8.1654 + 0.8640$$
$$= -6.4424$$
$$= -6.4424 + (7 - 7)$$

Hence,
$$\log x = 0.5576 - 7$$

We will ignore the characteristic, −7, until the last step.

SECTION 8.6 Computations with Logarithms (Optional)

$$0.01\left[\Delta x\begin{bmatrix}\log 3.61 = 0.5575\\ \log x\ \ \ = 0.5576\\ \log 3.62 = 0.5587\end{bmatrix}0.0001\right]0.0012$$

$$\frac{\Delta x}{0.01} = \frac{0.0001}{0.0012} \Rightarrow$$

$$\Delta x = 0.0008 \Rightarrow$$

$$x \approx (3.61 + \Delta x) \times 10^{-7} \Rightarrow$$

$$x \approx 3.6108 \times 10^{-7} \Rightarrow$$

$$x \approx 0.00000036108$$

EXERCISE 8.6

A. FUNDAMENTALS
Find the following common logarithms (base-10 logarithms).

1. (a) log 2.51 (b) log 251
2. (a) log 3.71 (b) log 0.0371
3. (a) log (2.71 × 10³) (b) log (2.71 × 10⁻⁴)

Find the antilogs of the following numbers.

4. (a) 0.5276 (b) 2.5276
5. (a) 0.7396 (b) 0.7396 − 3
6. (a) 0.9731 − 4 (b) −2.0269

Use logarithms to perform the following arithmetic calculations.

7. (a) (31)(7) (b) $\dfrac{31}{7}$
8. (a) $\dfrac{15 \cdot 10}{4}$ (b) $\dfrac{2 \cdot 73}{5}$
★9. (a) $\sqrt{16}$ (b) $\sqrt[3]{64}$
10. (a) $\sqrt{225}$ (b) $\sqrt[3]{125}$

B. ESSENTIALS
Use linear interpolation to find the following logarithms.

11. (a) log 2.785 (b) log 371.5
12. (a) log 0.01389 (b) log 0.1381

Use linear interpolation to find the antilogs of the following.

13. (a) 0.5533 (b) 3.6707
14. (a) 0.6305 − 2 (b) − 3.3666

Use logarithms to perform the following arithmetic calculations.

15. (a) $\sqrt{2.75}$ (b) $\sqrt{\dfrac{2.75}{0.013}}$
16. (a) $\sqrt[3]{0.00157}$ ★(b) $\sqrt[3]{(0.00157)(273)}$
17. (a) $(374.1)^4$ (b) $\left[\dfrac{(283.2)(17.3)}{821}\right]^4$

C. APPLICATIONS

18. The general gas law in physics states that PV/T is always constant for a given gas, where P is pressure, V is volume, and T is temperature. Use logarithms to find the new pressure if 400 milliliters of a gas with a pressure of 763.7 millimeters at 27°C is compressed to 273.5 milliliters at 27°C.

19. It is possible to fit a meter stick into a matchbox if the meter stick is moving fast enough. Find the contracted length, L, of a meter stick moving at a velocity of 272,300 kilometers per second, if

$$L = L_0 \sqrt{1 - \dfrac{v^2}{c^2}}$$

where c = 299,800 kilometers per second is the speed of light.

8.7 APPLICATIONS OF LOGARITHMS

If you were to examine the scale on a slide rule you would see something similar to Figure 8.14. Figure 8.14 is called a *logarithmic scale*. Notice that the distance from 1 to 8 is three times the distance from 1 to 2. Compare the distance from 1 to 9 to that from 1 to 3. The scale depicts exponential growth. The distances from 2 to 8 and from 3 to 9 correspond to tripling and doubling, respectively, on a linear scale.

A logarithmic scale is created on the right in Figure 8.15 by marking off the logs of the numbers but labeling them with the numbers themselves.

This logarithmic scale is the basis of the slide rule. Suppose we have a rule that is 1 meter in length with which we wish to multiply 2 × 3. The distance from 1 to 2 on the D scale equals the log of 2, or 0.301 meters. The distance from 1 to 3 on the C scale equals the log of 3, or 0.471 meters. The number 6 on the D scale is located at the sum of the distances

$$0.301 \text{ m} + 0.471 \text{ m} = 0.771 \text{ m}$$

away from 1. This distance, which is the log of 6, is labeled 6 on the D scale. Then, multiplication on the slide rule is based upon

$$2 \cdot 3 = 6 \Leftrightarrow$$
$$\log 2 + \log 3 = \log 6$$

The numbers on the C and D scales are actually located at a distance from 1 equal to their logs.

There are several other applications involving logarithmic scales. We will investigate two of the more familiar ones, the Richter scale for earthquakes and the decibel scale for sound.

A familiar use of a logarithmic scale is the Richter scale for earthquakes. This scale is logarithmic, because the large range of the amplitudes of quakes makes straightforward comparisons difficult. On semilog plots (one axis is a normal scale, the other axis is logarithmic) curves representing different earthquakes are generally parallel. These plots are the amplitude of the ground motion at varying distances from the epicenter of the quake, as recorded on seismographs. The

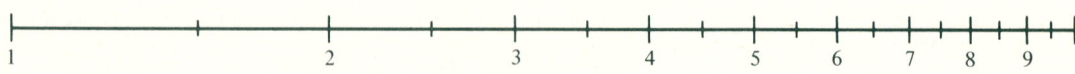

FIGURE 8.14.

SECTION 8.7   Applications of Logarithms

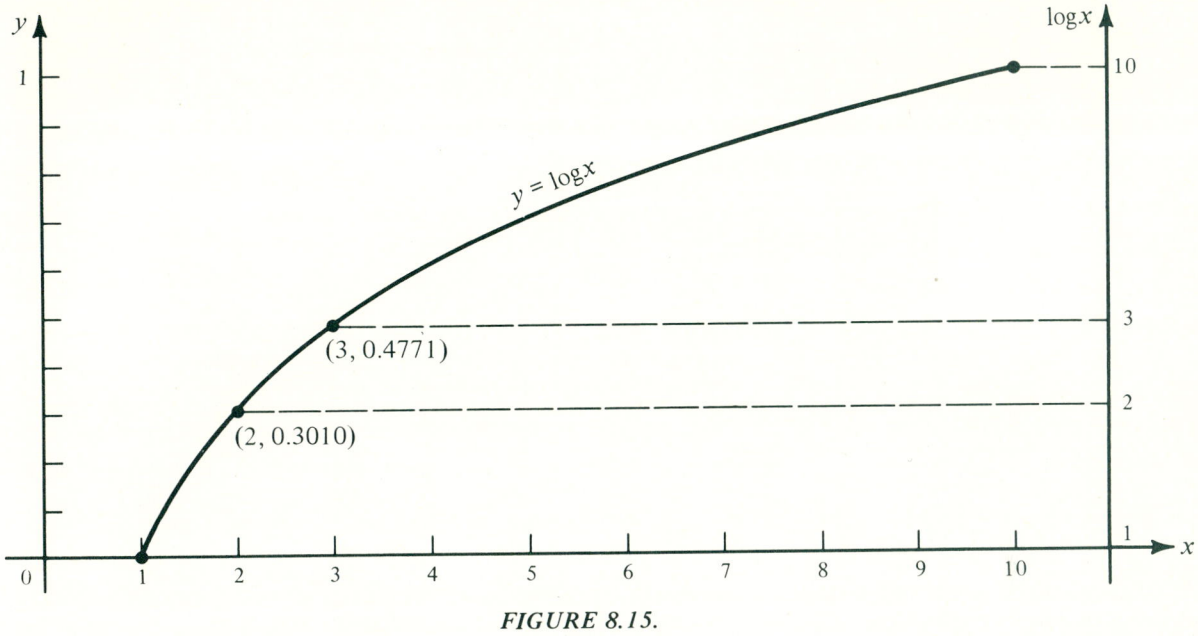

FIGURE 8.15.

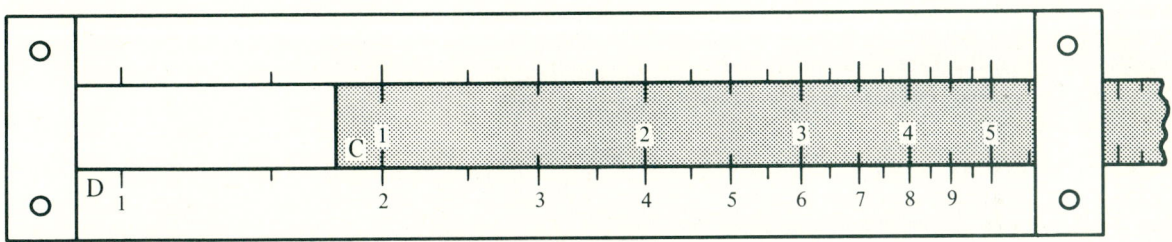

FIGURE 8.16.

parallel nature of plots of different quakes shows that the logs of their amplitudes (the vertical distance between the curves above) is roughly a constant. Therefore,

$$\log A_1 - \log A_2 = C \Rightarrow$$

$$\log \frac{A_1}{A_2} = C \Rightarrow$$

$$\frac{A_1}{A_2} = 10^C$$

This means the amplitudes of two quakes have a constant ratio at each distance from their epicenters. All that remained was to establish a "zero" earthquake to which all others could be compared. This zero quake was arbitrarily defined to be the smallest earthquake that had then been recorded. As a result, it is possible to have earthquakes with negative magnitudes. The largest magnitude assigned has been 8.9, although, mathematically there is no maximum value.

The plot in Figure 8.17 also shows that the magnitude of an earthquake decreases rapidly away from its epicenter.

CHAPTER 8   Exponential and Logarithmic Functions

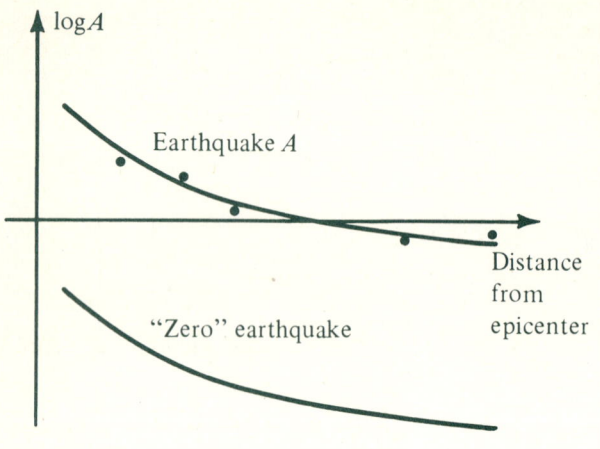

FIGURE 8.17.

The magnitude, M, of an earthquake, A, is defined as the vertical distance between its graph of seismic readings and the graph of the "zero" earthquake. Therefore,

$$M = \log A - \log A_0$$

Then, if earthquake A registers a magnitude of 6 ($M_A = 6$) on the Richter scale, and B registers 4, the difference between the two quakes is a Richter reading of 2, but

$$M_A - M_B =$$
$$(\log A - \log A_0) - (\log B - \log A_0) \Rightarrow$$
$$6 - 4 = \log A - \log B \Rightarrow$$
$$2 = \log \frac{A}{B} \Rightarrow$$
$$10^2 = \frac{A}{B} \Rightarrow$$
$$A = 100B$$

Thus A was actually 100 times stronger than B.

There is also an equation that approximates the energy released by an earthquake of magnitude M on the Richter scale. It is

$$\log E = 11.4 + 1.5M$$

where E is the energy in ergs.

The *decibel scale* for measuring sound is also a logarithmic scale. Here, the least audible sound (the threshold of hearing for youths, or "zero" sound) is used as the base. The decibel, db, is a measure of the intensity of a sound; it is defined as 10 times the difference of the logs of the sound and the "zero" sound. Then,

$$A = 10[\log \text{ (the sound)} - \log (\text{"zero" sound})] \Rightarrow$$
$$A = 10(\log S - \log Z) \Rightarrow$$
$$A = 10 \log \frac{S}{Z}$$

Therefore, an increase of 10 decibels means a tenfold increase in sound, since if $A_2 = A_1 + 10$, we have

$$A_2 = 10(\log S_2 - \log Z)$$

and

$$A_1 = 10(\log S_1 - \log Z)$$

Therefore,

$$A_2 - A_1 =$$
$$10(\log S_2 - \log Z) - 10(\log S_1 - \log Z) \Rightarrow$$
$$A_2 - A_1 = 10(\log S_2 - \log S_1) \Rightarrow$$
$$10 \text{ db} = 10 \log \frac{S_2}{S_1} \Rightarrow$$
$$1 = \log \frac{S_2}{S_1} \Rightarrow$$
$$\frac{S_2}{S_1} = 10^1 \Rightarrow$$
$$S_2 = 10 S_1$$

Therefore, the second sound, while only 10 decibels higher than the first, is 10 *times* as loud as the first. A 20-decibel increase results from a hundredfold increase in sound. Some approximate decibel readings are shown on page 183.

SECTION 8.7 Applications of Logarithms

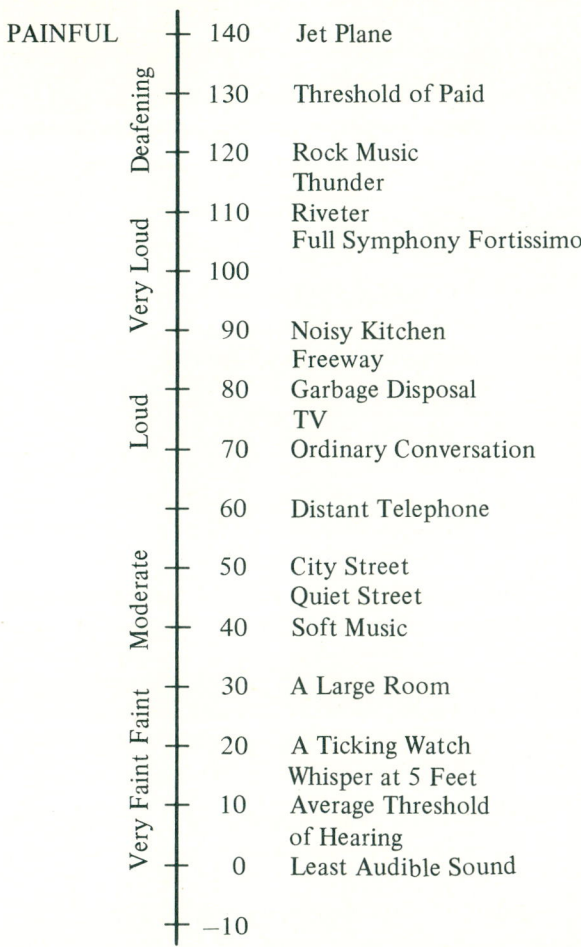

EXERCISE 8.7

1. How much stronger is an earthquake that measures 8 on the Richter scale than one that measures 4?

2. Show that San Francisco's 1906 earthquake that measured 8.4 on the Richter scale is 250 times stronger than one measuring 6.

★**3.** Compare Seattle's 1964 quake measured at 6.2 with Alaska's 1964 quake measured at 7.6. Then, compare Alaska's 1964 quake with Managua's 1973 quake that measured 7 on the Richter scale.

4. Leaves rustling in a breeze have a sound intensity of 20 decibels. How many times louder is a soft whisper of 10 decibels?

5. The sound intensity of Niagara Falls is a billion times greater than the least audible sound. How many decibels is this?

6. The magnitude of a star (its brightness) is measured on a logarithmic scale with $M = 2.5 \log(I_0/I)$. (See Problem 7, Exercise 8.3.) Compare the brightness of the sun with that of Regulus, a star of magnitude 1.3 found in the constellation Leo. How many times brighter is the sun?

★**7.** In determining the polluting strength of a factory or city's effluent, the measurement of the biochemical oxygen demand, BOD, is among the most important tests performed. Since the oxygenating organisms present increase at an exponential rate, the rate of deoxidation (reduction in BOD) is proportional to the amount of organic matter present. Then the amount of oxidizable organic matter, L, present at any time, t, is

$$L = L_0 \, 10^{-kt}$$

If a lake has a BOD of 200 milligrams/liter initially and 150 milligrams/liter 2 days later, when will it be drinkable? To be drinkable, it must have a BOD of 0.5 milligrams/liter?

8. In pure water, a very small number of H_2O molecules break up (ionize) to form H^+ and ^-OH parts (ions) according to the equation

$$H_2O \rightleftarrows H^+ + {}^-OH$$

where the hydrogen ions often associate themselves with one or more water molecules. The following relationships are known: First, one unit of water (called a mole) weighs 18 grams and has a volume of 18 milliliters. Second, in pure water the H^+ concentration (abbreviated $[H^+]$) is 1×10^{-7} moles per liter. The $[^-OH]$ is also 1×10^{-7} moles per liter. Third, the product of the H^+ and ^-OH concentrations is called the equilibrium constant K. For pure water,

$$K = [H^+][^-OH] = (1 \times 10^{-7})(1 \times 10^{-7}) \Rightarrow$$
$$K = 1 \times 10^{-14}$$

The $[H^+]$ and the $[^-OH]$ may vary according to the

acidity (more H^+ ions) or the basicity (more ^-OH ions), but the product of the concentrations of the ions,

$$[H^+] \cdot [^-OH]$$

always equals K.

Determine the $[^-OH]$ if $[H^+] = 0.001$ moles per liter.

Since these concentrations involve such small numbers, logarithms are used to describe the acidity or basicity of a solution. The resulting expressions are called pH and pOH, where pH is the negative of the log of the hydrogen ion concentration, $[H^+]$

$$pH = -\log [H^+]$$

and

$$pOH = -\log [^-OH]$$

The pH and pOH for pure water are both 7, since

$$[H^+] = [OH^-] = 1 \times 10^{-7}$$

and

$$-\log (1 \times 10^{-7}) = 7$$

In the basic solution in this problem,

$$pH = -\log (1 \times 10^{-3}) = 3$$

and

$$pOH = -\log [^-OH] = -\log \left(\frac{1 \times 10^{-14}}{1 \times 10^{-3}}\right) = 11$$

9. What is the concentration of hydrogen ions $[H^+]$ in a cup of coffee whose pH is 5.7? What is the $[OH^-]$?

10. How many times more acidic is garden soil with a pH of 4.5 than garden soil with a pH of 6?

11. In a 0.02 molar solution of a strong acid such as hydrochloric acid, $[H^+] = 0.02M$, since the acid completely ionizes. Calculate this pH and pOH.

12. Consider a weak acid, such as acetic acid, whose ionization equation is

$$CH_3COOH \rightleftarrows CH_3COO^- + H^+$$

We find $[H^+]$ from the equation

$$\frac{[CH_3COO^-][H^+]}{[CH_3COOH]} = 1.8 \times 10^{-5}$$

for 1.8×10^{-5} is the equilibrium constant for this acid. For a $2M$ solution, $[H^+] = [CH_3COO^-]$ and $[CH_3COOH] = 2 - [H^+]$. Calculate this pH.

8.8 MORE APPLICATIONS INVOLVING LOGARITHMS

Another important application of logarithms involves one or two logarithmic scales like the one shown below. Experimental data are plotted on

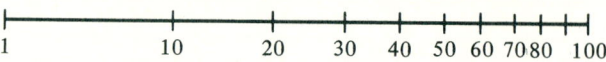

coordinate systems with logarithmic scales on both axes (log-log systems) or on only the vertical axis (semilog systems). This allows experimentalists to find the equation that models the situation when the equation is a power function of the form

$$y = n \cdot x^m$$

or an exponential function of the form

$$y = n \cdot 10^{mx}$$

This application will become clear when we see that **(1) a power function graphed on log-log graph paper will be graphed as a straight line, and (2) a linear plot will result when we graph an exponential function on semilog graph paper.** The slope and y-intercept of these straight lines yield all the information needed.

To see how linear plots result from power functions, take the log of both sides of

$$y = n \cdot x^m$$

to yield

$$\underline{\hspace{2cm}} = \underline{\hspace{2cm}}$$

Then, since $\log M \cdot N = \log M + \log N$, and $\log M^p = p \log M$, we can rewrite the right side as

$$\log y = \log \underline{\hspace{1cm}} + \underline{\hspace{1cm}} \log \underline{\hspace{1cm}}$$

If we change the order, we have

$$\log y = m \log x + \log n$$

SECTION 8.8 More Applications Involving Logarithms

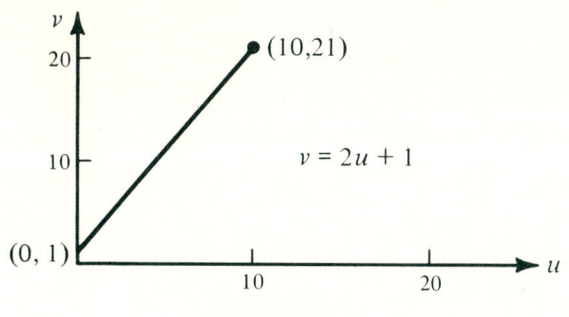

FIGURE 8.18.

Graph the linear function $v = 2u + 1$ on the axis system provided in Figure 8.18.

Did you draw a straight line whose slope is $+2$ and whose v-intercept is $(0, 1)$?

Since the domain of a logarithmic function contains only positive reals, we will restrict our graph to the first quadrant. The graph of

$$v = 2u + 1$$

is redrawn with the logarithmic scales on the horizontal and vertical axes in Figure 8.19.

Now let's see which power function we have graphed on the "log-log" paper. That is, the linear graph in u and v above represents a power function in x and y where

$$y = nx^m \Leftrightarrow v = 2u + 1$$

Now, let's make a change of variables by letting $u = \log x$, $v = \log y$, and the constant $\log n = b$. These substitutions yield a linear equation in u and v:

$$v = mu + b$$

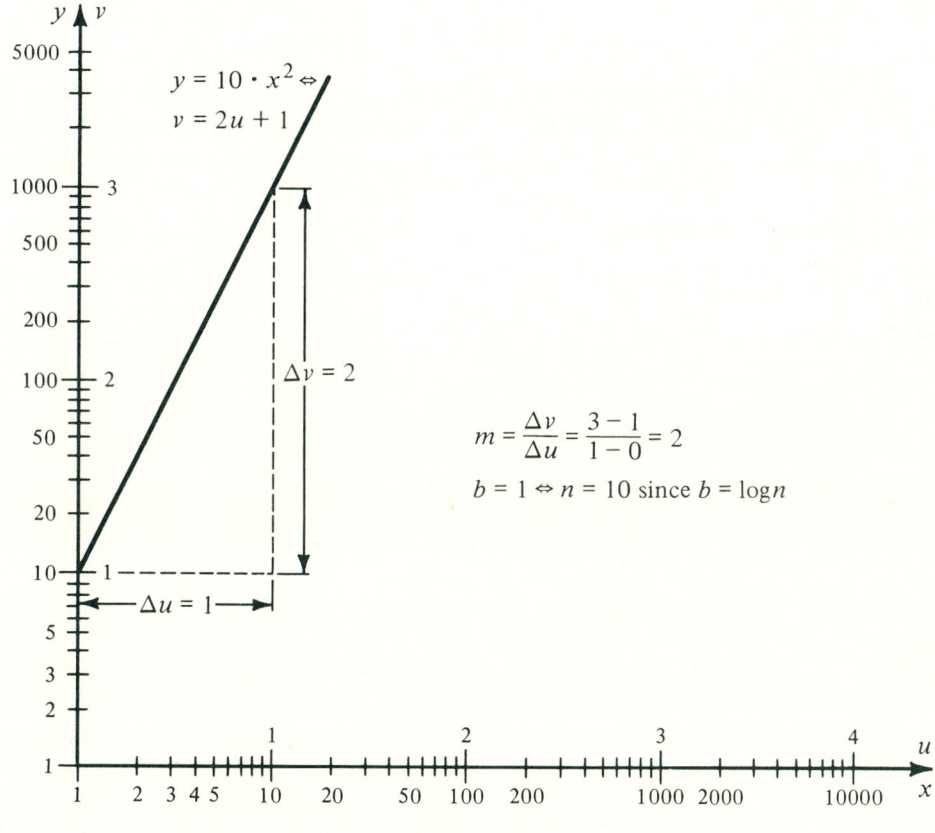

FIGURE 8.19.

185

CHAPTER 8 Exponential and Logarithmic Functions

with
$$u = \log x, \quad v = \log y, \quad 1 = \log n$$

The m in $y = nx^m$ is the slope of the line, as shown above. In $v = 2u + 1$,

$$m = \frac{\Delta u}{\Delta v} = 2$$

Since the v-intercept, $v = 1$, is $\log n$, we have

$$\log n = 1 \Leftrightarrow$$
$$n = 10$$

Therefore, the power function whose log-log plot is shown above is

$$y = 10 \cdot x^2$$

Since both axes are logarithmic scales in log-log plots, the slope of the line (the exponent in the power function) can be determined by simply measuring some run, Δu, and its corresponding rise, Δv, using any convenient scale. We often use a centimeter scale for such measurements. The multiplier, n, appears on the graph also. It is the y-intercept.

Such a log-log plot is used to find the mathematical model for empirical (experimental) data.

If data plotted on semilog graph paper yield a straight line, we know the model will be an exponential equation. The graph will yield the slope and y-intercept, which is all we need to know to determine the exponential equation.

EXAMPLE I. Find the equation that describes the following data.

x	1	10	15.5
y	100	1.99	1

Plot the data on a log-log coordinate system. (See Figure 8.20.) Since the data yield a linear graph on a log-log coordinate system, the modeling equation is a power function of the form

$$y = nx^m$$

where the v-intercept (0, 2) yields $b = 2$ and

$$b = \log n \Rightarrow n = 10^2 = 100$$

The slope of the line, m, can be found by measuring Δu and its corresponding Δv or by calculating the (u, v) coordinates for 2 data points as follows. The point (15.5, 1) yields

$$u = \log x = \log 15.5 = 1.18$$

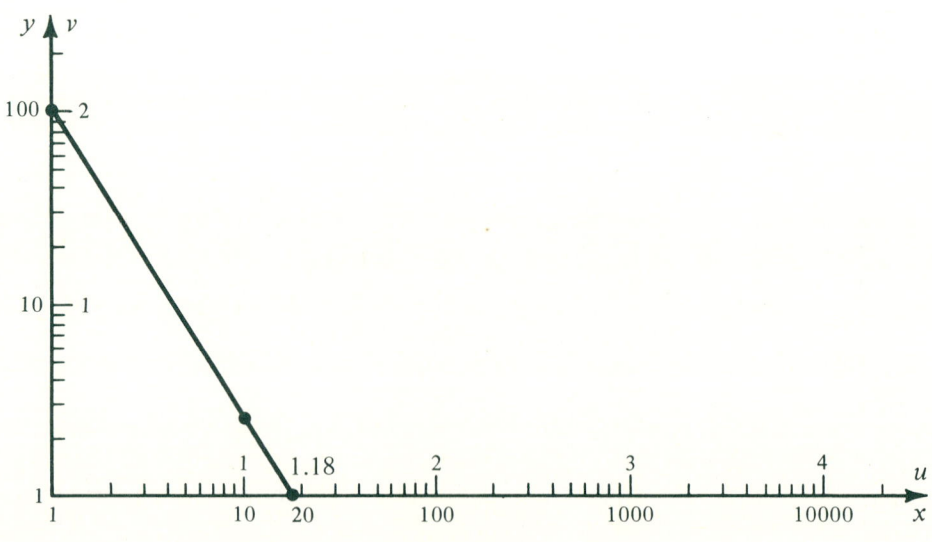

FIGURE 8.20.

and
$$v = \log y = \log 1 = 0$$
The point (1, 100) yields
$$u = \log x = \log \underline{} = \underline{}$$
and
$$v = \log y = \log \underline{} = \underline{}$$
Therefore, the slope of the line in the *u, v*-coordinate system is
$$m = \frac{\Delta u}{\Delta v} = \frac{0 - 2}{1.18 - 0} = \frac{-2}{1.18} = -1.7$$

Therefore, the equation that fits the given data is
$$y = 100 \cdot x^{-1.7}$$

Take the log of both sides of the following exponential function to see why its semilog plot is a straight line.
$$y = n \cdot 10^{mx} \Rightarrow$$
$$\log y = \underline{} + \underline{} \Rightarrow$$
$$\log y = mx + \log n$$

where a substitution of $v = \log y$ and b for the constant $\log n$ yields
$$v = mx + b$$

EXAMPLE II. The following data depict Newton's law of cooling, where x is time and T is the temperature difference between the inside and outside of a house. The heat is turned off at $x = 0$.

x	0	1	4 hours
T	23.2°	10.98°	1.16° centigrade

Since the data yield a linear graph on a semilog coordinate system, an equation that fits the data is an exponential function of the form
$$T = n \cdot 10^{mx}$$

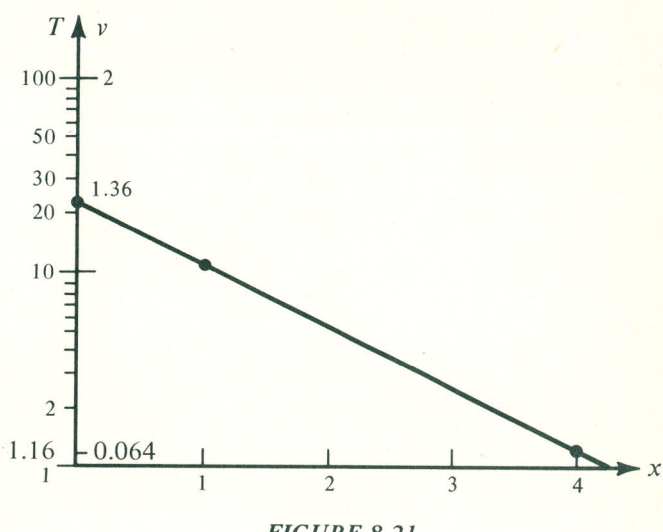

FIGURE 8.21.

where
$$v = \log T \quad \text{and} \quad b = \log n$$
Since the *v*-intercept is $b = 1.36$,
$$\log n = 1.36 \Rightarrow n = 23.2$$
The slope of the line, m, is
$$m = \frac{\Delta v}{\Delta x}$$

where the points (0, 23.2) and (4, 1.16) in (x, T) correspond to the points (4, 0.064) and (0, 1.36) in (x, v) since $v = \log T$. Therefore,
$$m = \frac{\Delta v}{\Delta x} = \frac{0.064 - 1.36}{4 - 0} = -0.324$$

and the equation of this exponential function,
$$T = n \cdot 10^{mx}$$
is
$$T = (23.2)10^{-0.324x}$$

Since the scales on the vertical and horizontal axes are different on semilog plots, the slope *cannot* be determined by measuring Δx and Δv on the graph as it can for log-log plots.

CHAPTER 8   Exponential and Logarithmic Functions

EXERCISE 8.8

A. FUNDAMENTALS

Graph the following curves on a log-log coordinate system by graphing data and then finding the equation.

1. (a) $y = 10x^3$ (b) $y = 100x^3$
2. (a) $y = 100x^{-1}$ (b) $y = 100x^{-4}$
3. (a) $y = x^2$ (b) $y = 100x^2$
4. (a) $y = 80x^{-2.3}$ (b) $y = 8x^{2.3}$

Graph the following curves on a semilog coordinate system.

5. (a) $y = 25(10^{-x})$ (b) $y = 100(10^{-x})$
6. (a) $y = 40(10^{-2x})$ (b) $y = 40(10^{+2x})$
7. (a) $y = 90(10^{-3.1x})$ (b) $y = 9(10^{3.1x})$

B. ESSENTIALS

Graph the following data on log-log and semilog graph paper to determine if the relationship is a power function of the form

$$y = nx^m$$

or an exponential function of the form

$$y = n10^{mx}$$

Then find the values of m and n from your linear graph.

8. From Newton's law of cooling,

x	1	3	5	hours
T	7.96	1.26	0.2	°centigrade

9. A fixed amount of money, P, invested at a varying rate of interest, $x - 1$, for a fixed number of years, m, yields the following returns, s.

x	1.01	1.03	1.06
s	4204	4637	5353

10. The intensities, I, of the apparent brightnesses of stars of various magnitudes, M, are given as multiples of the intensity of a star of zero magnitude, I_0. (A star of magnitude 6 can barely be seen with the naked eye.)

M	1	2	6
I	$0.4I_0$	$0.06I_0$	$0.004I_0$

11. The magnitude of an earthquake, M, and its value on the Richter scale relative to the amplitude of the ground motion, A, are given in terms of the zero quake, A_0.

	Seattle, 1964	Alaska, 1964	San Francisco, 1906
M	6.2	7.6	8.4
A	$1584893 A_0$	$39810717 A_0$	$251188643 A_0$

12. Continuously compounded interest generates the formula

$$A = Pe^{rn}$$

where r is the annual interest rate, P is the principal, and A is the amount after n years. Make a semilog graph using a log scale of base e to find the equation for the following data.

(a)

r	$5\frac{3}{4}\%$	6%	$6\frac{1}{4}\%$
A	17,771	18,221	18,682

(b)

n	1	2	10
A	10,618	11,275	18,221

What was the principal in each case? How many years was the investment in (a)? What was the rate in (b)?

13. The vapor pressure of a substance is the pressure exerted by the molecules of that substance in a gaseous state. ($\exp_{10} x \Leftrightarrow 10^x$)

$$\text{VP} = C \cdot \exp_{10}\left(-\frac{H}{4.58} \cdot \frac{1}{T}\right)$$

or

$$\log \text{VP} = \frac{-H}{4.58} \cdot \frac{1}{T} + \log C$$

shows how the vapor pressure, VP, rises rapidly as the temperature, T, increases. H is the amount of heat, in calories, required to vaporize one mole of the liquid against a constant pressure. Since this equation is of the form

$$y = mx + b$$

where

$$y = \log VP, \quad m = \frac{-H}{4.58}, \quad x = \frac{1}{T}, \quad b = \log C$$

a plot on a semilog coordinate system (log VP versus $1/T$) will allow us to find H from the following measured data for ethanol, C_2H_5OH. Find H.

$1/T$	0.00366	0.00330	0.00300	°Kelvin
VP	13	80	350	mm of Hg

8.9 HISTORICAL COMMENTS ON LOGARITHMS (OPTIONAL)

Surprisingly, exponential notation was developed long after logarithms, which were invented by John Napier in 1594 as an aid to arithmetic. Napier's logarithms used $(1 - 10^{-7})^{10^7}$, a number close to $1/e$, as their base. Napier's logarithms were not readily or universally accepted. The German astronomer, Kepler, for example, refused to use logarithms until they were rigorously explained.

In 1600 Joost Bürgi, an assistant to Kepler, independently compiled a table of logarithms. In 1620 he finally published his work, which contained the first table of antilogs.

Henry Briggs computed the first table of base-10 logarithms in 1615.

Napier, Bürgi, and Briggs's tables were calculated by tedious repeated multiplications and extractions of square roots. Base-10 logarithms, for example, were calculated by first taking successive roots of 10,

$$\sqrt{10}, \sqrt{\sqrt{10}}, \sqrt{\sqrt{\sqrt{10}}}, \ldots$$

until Briggs found a number *slightly* larger than 1. It took 54 successive square roots. Calling this number x, he had $10^{(1/2)^{54}} = x$ and $\log_{10} x = \left(\frac{1}{2}\right)^{54}$ as a starting point for the computations. He then used the theorems on logs to build his table.

But Napier and Bürgi founded their inventions on comparing arithmetic and geometric sequences, since the terms of an arithmetic sequence are the logarithms of the corresponding terms of the geometric sequence:

$$\ldots, -2, -1, 0, 1, 2, 3, \ldots$$

and

$$\ldots, a^{-2}, a^{-1}, a^0, a^1, a^2, a^3, \ldots$$

The intriguing idea was that multiplication and division in the geometric sequence correspond to simple addition and subtraction in the arithmetic one.

Napier chose the name logarithm, meaning the "number of the ratio," to refer to a common ratio in a sequence of numbers used in calculating his tables.

The modern definition of a logarithmic function as the inverse of an exponential function is attributed to Euler. He also introduced the natural base, e, and extended logarithms to negative and complex numbers in 1747.

ENDNOTE

1. Oresme used exponents on numbers in the thirteenth century, but positive exponents and the square root symbol were first used systematically in algebraic expressions by Descartes in "La Géométrie" in 1637. Newton used positive and negative rational exponents in his writings. Gauss used exponents, but at first he preferred to write xx for x^2, since it was equally easy to write. His change in 1801 made it official.

CHAPTER 9

The Circular Functions

9.1 INTRODUCTION

Chapters 9–12 constitute our study of trigonometry and the circular functions. In this chapter we will define and study the circular functions whose domains are real numbers and whose geometric interpretations are based on a circle of radius one, called the unit circle.

In Chapter 10 we will define and study the trigonometric functions whose domains are elements expressed in degrees and whose geometric interpretations are based on right triangles.

Whether you first study Chapter 11, which is functionally oriented, or Chapter 10, the analytic approach, you will soon see the natural parallel that exists between the two chapters. By the end of Chapter 12, which incorporates both definitions, you will be quite comfortable with both approaches.

9.2 THE RADIAN

The definitions of the circular functions are based upon a section of arc of the unit circle pictured in Figure 9.1. Notice that each section of arc, s, subtends a central angle, θ, whose initial side is on the x-axis and whose terminal side is the radial line labeled r.

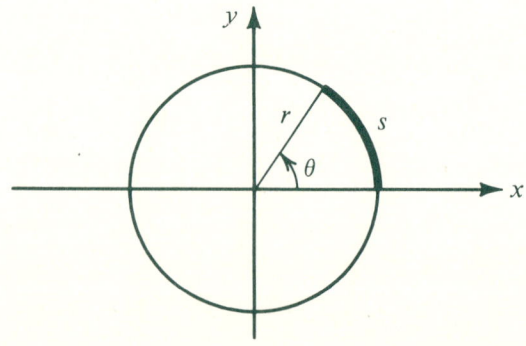

FIGURE 9.1. *The Unit Circle,* $x^2 + y^2 = 1$.

Now, let's establish the relationship between s, θ, and r, where $r = 1$. If we were to measure the angle θ ($\angle\theta$) in degrees,[1] the relationship between s and θ would be unduly complicated. We are at liberty to define the units of measurement for θ in such a way that θ and r are related by the simple equation

$$s = r\theta$$

for any radius r. Then, for $r = 1$,

$$s = r\theta \Rightarrow$$
$$s = \theta$$

Now we need to establish the relationship between measuring θ in terms of s and measuring θ in degrees. The units of measure for measuring θ in terms of s are called *radians*.[2] We know that the circumference of a circle is given by

$$C = 2\pi r$$

Now, if r and θ are to be related by the equation

$$s = r\theta$$

then, when θ measures a complete circle, the arc length, s, must equal the circumference, C. Therefore,

$$s = C \Rightarrow$$
$$r\theta = 2\pi r \Rightarrow$$
$$\theta = 2\pi$$

This means that 2π radians is a complete circle. Since 360° is also a complete circle,

$$2\pi = 360°$$

This explains the following definition.

DEFINITION 9.1.
A radian is a real number measure and π radians = 180 degrees

---

We will use radian measure for measuring central angles, θ, and lengths of arcs, s. Then, for the unit circle (where $r = 1$), $s = r\theta$ gives us

$$s = \theta$$

for s and θ measured in radians. *This results in radian measures being dimensionless.* That is, if r and s are measured in meters, then we have

$$s = r\theta \Leftrightarrow$$
$$s \text{ meters} = (r \text{ meters})(\theta \text{ radians})$$

We are forced to ignore the word "radians" in calculations so that the equation will be balanced dimensionally:

$$\text{meters} = \text{meters}$$

The next examples will help you become familiar with radians before we proceed to the circular functions.

EXAMPLE 1. Find the radian measure of the following angles. Remember that π radians = 180°.

$\theta = 90° = \underline{\quad?\quad}$ radians (a right angle)

$\alpha = \underline{\quad?°\quad} = \pi$ radians (a straight angle)

$\beta = 270° = \underline{\quad?\quad}$ radians (3 right angles, $3 \cdot 90°$)

$\phi = \underline{\quad?°\quad} = -\pi/2$ radians (a negative angle)

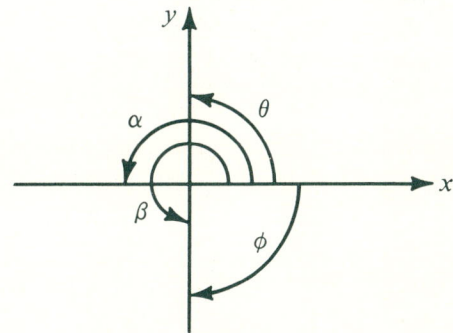

FIGURE 9.2.

We say the measure of θ is $\pi/2$ radians, or simply $\pi/2$. Then $\alpha = 180° = \pi$, $\beta = 270° = 3\pi/2$, and $\phi = -90° = -\pi/2$.

CHAPTER 9 The Circular Functions

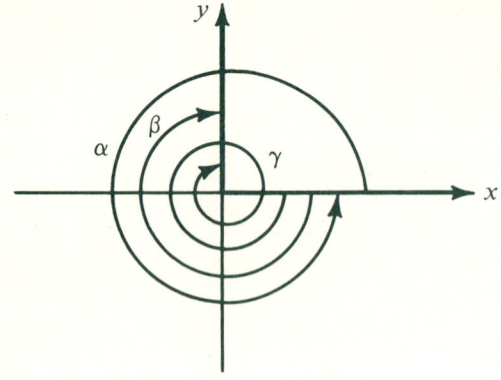

FIGURE 9.3.

EXAMPLE II. Find the radian measure of the angles shown in Figure 9.3. ($\pi \approx 3.14$)

$\alpha = +360° = 2\pi \approx 6.283$ (radians)

$\beta = -$ __?°__ $= -3\pi/2 \approx$ __?__

$\gamma = \underline{-540°} =$ __?__ $\approx$ __?__

Notice that the positive x-axis was used as the terminal side of all the angles. The angle $\beta = -270° \approx -4.712$, and the angle $\gamma = -7\pi/2 \approx -10.996$. We can use any *ray* (half line) for the initial side of an angle, but usually the positive x-axis is the most convenient.

There are only a few important angles whose radian measures we must recognize. Convert each of the following to degrees.

$0 = 0°$ $3\pi/4 = 3(\pi/4) =$ __?°__

$\pi/6 = 180°/6 =$ __?°__ $\pi = 180°$

$\pi/4 =$ __?°__ $3\pi/2 = 3(\pi/2) =$ __?°__

$\pi/3 =$ __?°__ $5\pi/4 =$ __?°__

$\pi/2 =$ __?°__ $2\pi =$ __?°__

The above are all positive angles and are multiples of $\pi/2 = 90°$, $\pi/3 = 60°$, $\pi/4 = 45°$, or $\pi/6 = 30°$, the four basic angles.

Draw the following angles on the axis system provided, making sure your arrow indicates the sign of the angle (see Figure 9.2).

$\alpha = -\pi/2$ $\theta = -11\pi/6$

$\beta = -3\pi/4$ $\phi = 15\pi/6$

$\gamma = 7\pi/6$ $\omega = -11\pi/2$

The terminal sides of your angles should be the negative y-axis for α, in quadrant III for β and γ, in quadrant I for θ, the positive y-axis for ϕ, and the positive y-axis for ω.

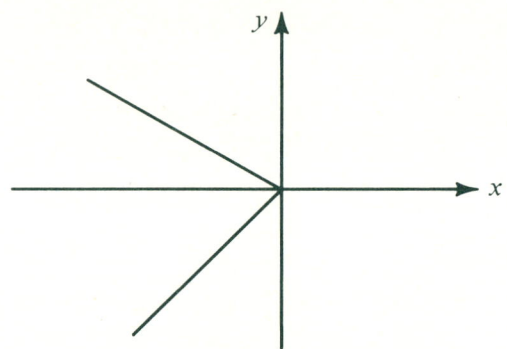

FIGURE 9.4.

EXERCISE 9.2

A. FUNDAMENTALS

Convert the following measurements to radians.

1. (a) 0° (b) 360° (c) 720°
2. (a) 180° (b) −180° (c) 540°
3. (a) 45° (b) 135° (c) 225°
4. (a) 30° (b) 150° (c) 210°
5. (a) 60° (b) 120° (c) 240°

Convert the following measurements to degrees.

6. (a) π (b) 2π (c) 3π
7. (a) $\pi/2$ (b) $3\pi/2$ (c) $5\pi/2$
8. (a) $\pi/6$ (b) $5\pi/6$ (c) $-\pi/6$
9. (a) $\pi/4$ (b) $7\pi/4$ (c) $-11\pi/4$
10. (a) $2\pi/3$ (b) $-5\pi/3$ (c) $-7\pi/3$

Draw a circle of the given radius that depicts the given angle and its corresponding arc of the circle. Notice that the arcs may be positive, zero, or negative.

11. (a) $r = 2, \theta = 5\pi/6$
    (b) $r = 2, \theta = -5\pi/6$
12. (a) $r = 1, \theta = 5\pi/4$
    (b) $r = 1, \theta = -5\pi/4$
13. (a) $r = 1, \theta = 2\pi/3$
    (b) $r = 1, \theta = -4\pi/3$

B. ESSENTIALS

Draw a circle of radius one. Then find the coordinates of the point on the circle that is on the terminal side of each of the following arcs.

14. (a) $\pi/2$ (b) $-3\pi/2$
15. (a) π (b) $-\pi$
16. (a) $3\pi/2$ (b) $-\pi/2$
17. (a) $7\pi/2$ (b) $11\pi/2$

Convert the following to radian measure. (HINT: Use $60' = 1°$ and $60'' = 1'$ to convert to decimal degrees first.)

★18. 25°6'   19. 118°30'   20. 19°30'30''
21. 247°20'15''
22. (a) 1°   (b) 1'   (c) 1''
23. −280°13'18''
24. Convert 1 radian to degrees, minutes, and seconds.

9.3 THE COSINE AND SINE FUNCTIONS

The domains of the circular functions are all real numbers. We shall begin by showing how every real number can be made to correspond to some length of arc of the unit circle. Each real number will then have a corresponding central angle (measured in radians) and a point on the unit circle itself corresponding to the endpoint of its arc. This can be accomplished by *wrapping* the real number line around the unit circle (Figure 9.5).

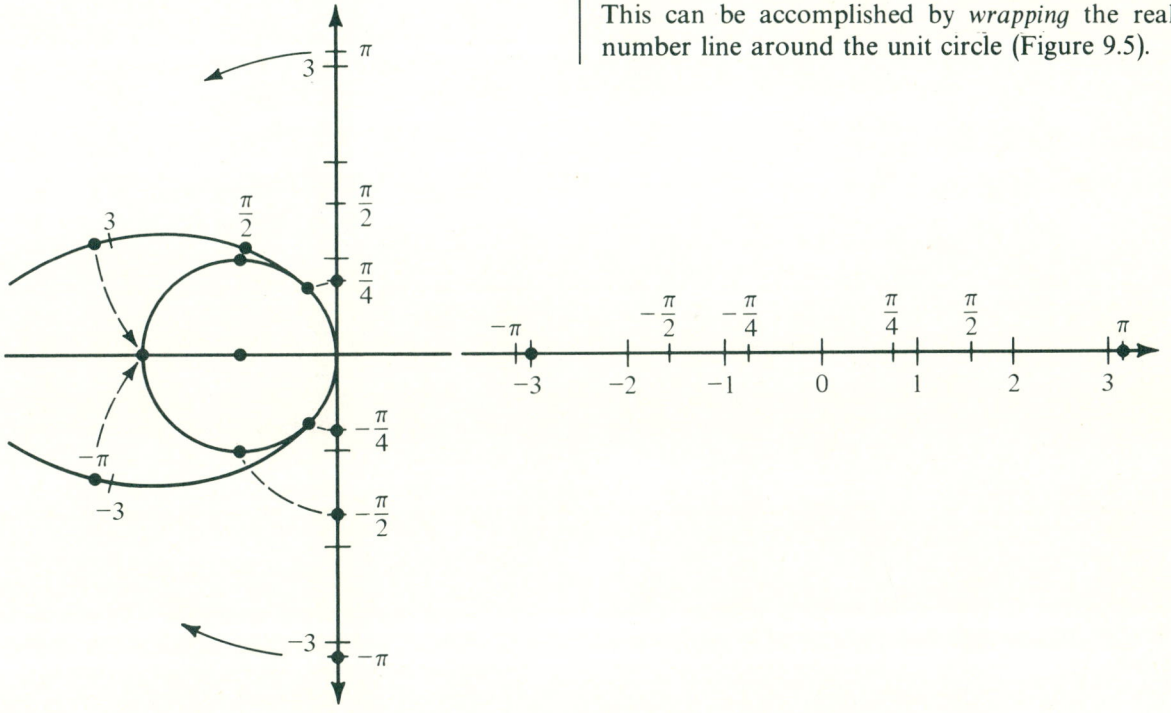

FIGURE 9.5. *Wrapping the Real Number Line Around the Unit Circle.*

CHAPTER 9 The Circular Functions

Give the (x, y) coordinates of the location of the following real numbers as the real number line is wrapped around the unit circle. These ordered pairs are called the terminal points or reference points for the real numbers. First we will look at some positive reals (see Figure 9.6):

$$0 \to (1, 0)$$
$$\pi/2 \to (0, 1)$$
$$\pi \to (-1, 0)$$
$$3\pi/2 \to (?, ?)$$
$$2\pi \approx 6.28 \to (?, ?)$$
$$7\pi/2 \to (?, ?)$$
$$57\pi/2 \approx 89.5 \to (?, ?)$$

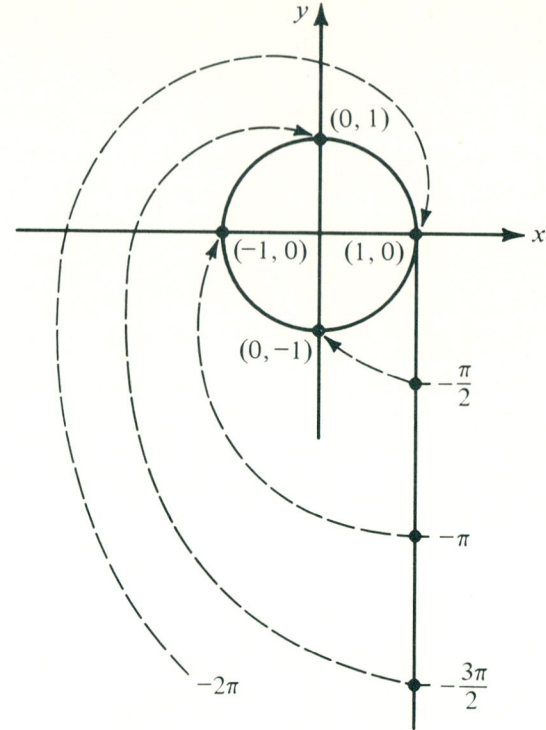

FIGURE 9.7. The Negative Reals

Then the negative reals (see Figure 9.7):

$$-\pi/2 \to (?, ?)$$
$$-\pi \to (?, ?)$$
$$-3\pi/2 \to (?, ?)$$
$$-2\pi \to (1, ?)$$

Finding the coordinates corresponding to a real number like 2 is much more complicated. The angle and arc length associated with 2 are pictured in Figure 9.8.

$$r = 1 \quad \text{and} \quad s = r\theta \Rightarrow$$
$$\theta = s = 2 = 2 \cdot 180°/\pi \approx 115°$$

We will learn how to find the coordinates (x_0, y_0) later. They happen to be

$$x_0 \approx -0.416 \quad \text{and} \quad y_0 \approx 0.909$$

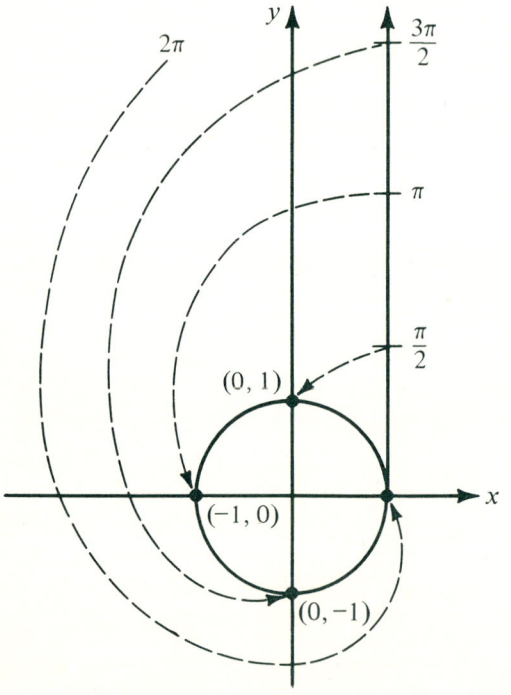

FIGURE 9.6. The Positive Reals

SECTION 9.3 The Cosine and Sine Functions

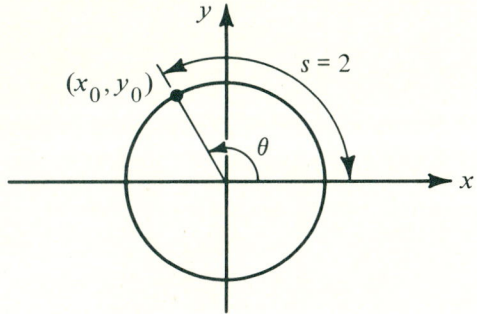

FIGURE 9.8.

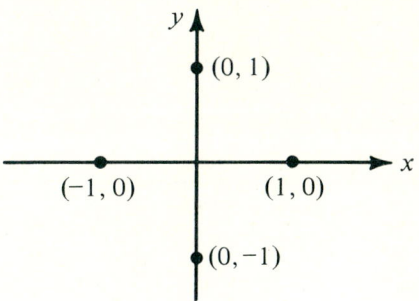

FIGURE 9.10.

In summary, we have demonstrated that every real number can be made to correspond to some point of the unit circle by wrapping the real number line around the unit circle. In this way, every real number is paired with a length of arc, s, and (for the unit circle $r = 1$) with a central angle θ whose measure in radians has the same value as s. Then every real number is also associated with an ordered pair (x, y), called its **reference point**, located on the unit circle at the end of the corresponding section of arc and on the terminal side of angle θ.

We are now ready to define two of the six circular functions. See Figure 9.9.

DEFINITION 9.2.
The *cosine* of the real number θ is the value of the x-coordinate of its reference point on the unit circle.

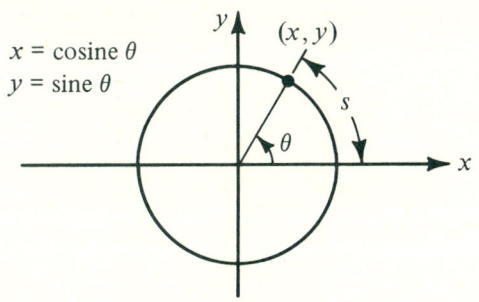

FIGURE 9.9.

DEFINITION 9.3.
The *sine* of the real number θ is the value of the y-coordinate of its reference point on the unit circle.

Find the corresponding values of cosine θ and sine θ, which we abbreviate as cos θ and sin θ, respectively, for the following values of θ. Draw a unit circle in Figure 9.10 for reference.

(1) $\theta = \pi/2 \Rightarrow x = 0$ and $y = 1 \Rightarrow$
  $\cos \pi/2 = 0$ and $\sin \pi/2 = 1$

(2) $\theta = \pi \Rightarrow x = -1$ and $y = 0 \Rightarrow$
  $\cos \pi = -1$ and $\sin \pi = ?$

(3) $\theta = 3\pi/2 \Rightarrow x = ?$ and $y = -1 \Rightarrow$
  $\cos 3\pi/2 = 0$ and $\sin 3\pi/2 = ?$

(4) $\theta = -\pi/2 \Rightarrow x = ?$ and $y = ? \Rightarrow$
  $\cos(-\pi/2) = ?$ and $\sin(-\pi/2) = ?$

You should have found that the sine of π equals the value of the y-coordinate of the point $(-1, 0)$, since the arc from $(0, 0)$ to $(-1, 0)$ on the unit circle has a measure of π. For $\theta = 3\pi/2$, $x = 0$ and $\sin 3\pi/2 = -1$. For $\theta = -\pi/2$, $x = 0$ and $y = -1 \Rightarrow \cos(-\pi/2) = 0$ and $\sin(-\pi/2) = -1$.

Now let's find the values for the cosine and sine of the other fundamental angles, $\pi/6$, $\pi/4$, and $\pi/3$, by finding the x- and y-coordinates of their reference points.

CHAPTER 9   The Circular Functions

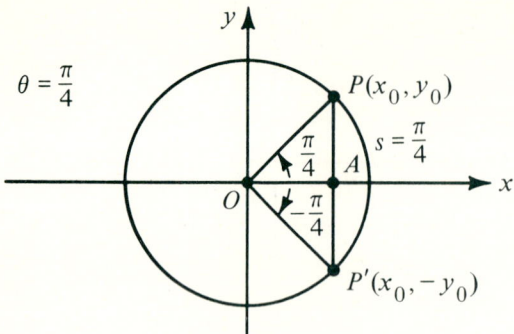

FIGURE 9.11.

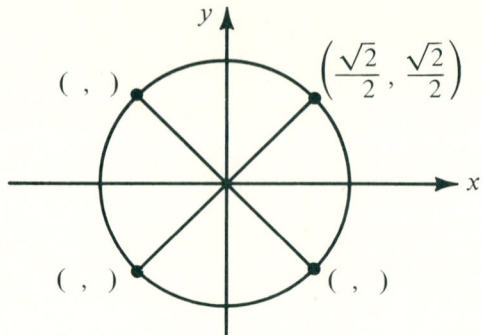

FIGURE 9.13.

For $\theta = \pi/4$, we have Figure 9.11. Triangle OPP' is a right triangle. The legs each equal 1, since they are radii of the unit circle. Then, by the Pythagorean theorem, the hypotenuse PP' is $\sqrt{1^2 + 1^2} = \sqrt{2}$. This is twice the y-coordinate. Therefore,

$$y_0 = \sqrt{2}/2$$

Triangle OAP is also a right triangle. See Figure 9.12. Therefore,

$$x_0 = \sqrt{1^2 - (\sqrt{2}/2)^2} = \underline{\qquad ?\qquad}$$

Since the cosine equals the x-coordinate of the reference point, and the sine is the y-coordinate, we have

$$\cos \pi/4 = \sqrt{2}/2$$

and

$$\sin \pi/4 = \sqrt{2}/2$$

Fill in the missing coordinates in Figure 9.13. Now we see that $\cos 3\pi/4 = -\sqrt{2}/2$, $\sin 3\pi/4 = \sqrt{2}/2$, $\cos 5\pi/4 = -\sqrt{2}/2$, $\sin 5\pi/4 = -\sqrt{2}/2$, $\cos 7\pi/4 = \sqrt{2}/2$, and $\sin 7\pi/4 = -\sqrt{2}/2$.

Now, let's find the coordinates corresponding to $\theta = \pi/6$. See Figure 9.14.

$$r = 1 \quad \text{and} \quad \theta = \pi/6 \Rightarrow$$
$$s = \pi/6$$

First, let's establish that triangle OPP' is an equilateral triangle. The measure of angle $QP'P$ equals one half of its intercepted arc $\widehat{PQ}$, since it is an inscribed angle.

$$\widehat{PQ} = \pi - 2(\pi/6) = 2\pi/3 \Rightarrow$$
$$\angle OP'P = \widehat{PQ}/2 = \pi/3$$

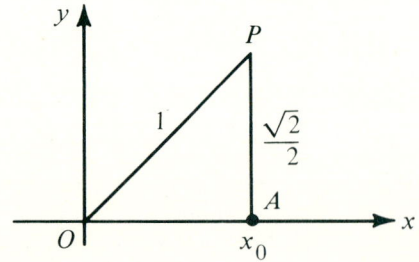

FIGURE 9.12.

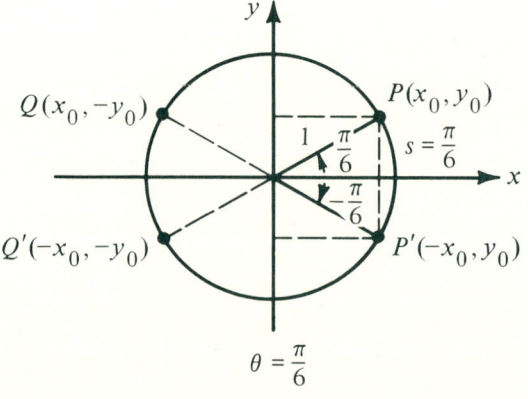

FIGURE 9.14.

196

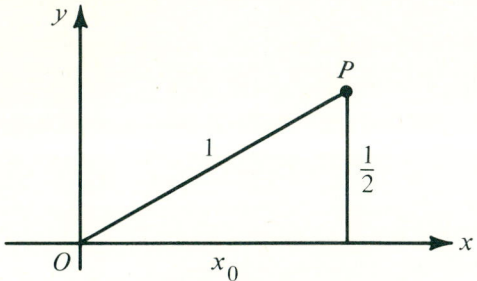

FIGURE 9.15.

Then, since $OP = OP' = 1$,
$$\angle OP'P = \angle OPP' = \pi/3$$
and
$$\angle POP' = \pi - 2(\pi/3) = \pi/3$$

We have now established that $\triangle OPP'$ is equiangular; hence it is equilateral. Therefore,
$$PP' = 1$$
and
$$y_0 = PP'/2 = \tfrac{1}{2}$$

Then, from the right triangle in Figure 9.15, we have
$$x_0 = \sqrt{1^2 - (\tfrac{1}{2})^2} = \sqrt{3/4} = \sqrt{3}/2$$

Therefore,
$$\cos \pi/6 = \sqrt{3}/2$$
and
$$\sin \pi/6 = \tfrac{1}{2}$$

Now, find the cosine and sine of the following real numbers. You may need to refer to Figure 9.16.

$\cos 5\pi/6 =$ __?__ , $\sin 5\pi/6 =$ __?__

$\cos 7\pi/6 =$ __?__ , $\sin 7\pi/6 =$ __?__

$\cos 11\pi/6 =$ __?__ , $\sin 11\pi/6 =$ __?__

$\cos(-5\pi/6) =$ __?__ , $\sin(-5\pi/6) =$ __?__

Your answers should be either $\sqrt{3}/2$ or $-\sqrt{3}/2$ for the cosines (positive if the reference point has

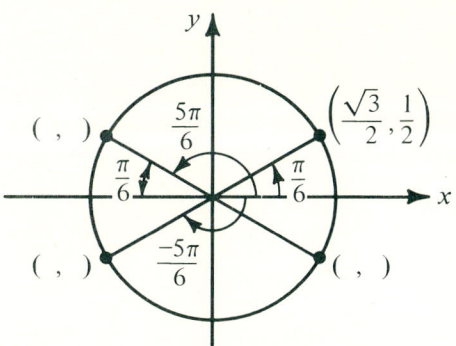

FIGURE 9.16.

a positive x-coordinate, since $x = \cos \theta$) and either $\tfrac{1}{2}$ or $-\tfrac{1}{2}$ for the sines (positive if the reference point has a positive y-coordinate, since $y = \sin \theta$).

Figure 9.17 depicts the values of the cosine and the sine of $\pi/3$. The proof is left as an exercise.

The important values of the cosine and sine are summarized in the table below. They will occur so often that it is necessary to learn them.

	0	$\pi/6$	$\pi/4$	$\pi/3$	$\pi/2$
$x = \cos \theta$	1	$\sqrt{3}/2$	$\sqrt{2}/2$	$\tfrac{1}{2}$	0
$y = \sin \theta$	0	$\tfrac{1}{2}$	$\sqrt{2}/2$	$\sqrt{3}/2$	1

The signs of the cosine and sine are easy to remember because the cosine is positive in the quadrants where x is positive (since $x = \cos \theta$) and the sine is positive in the quadrants where y is

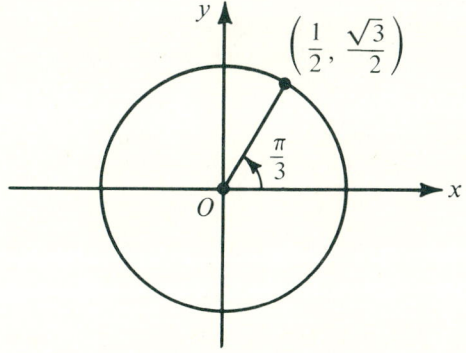

FIGURE 9.17.

CHAPTER 9 The Circular Functions

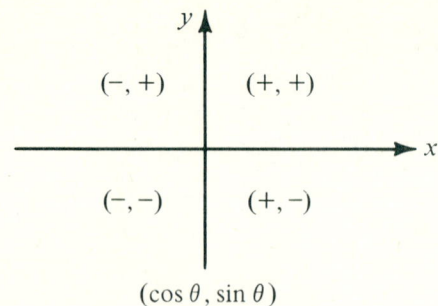

FIGURE 9.18.

positive (since $y = \sin\theta$). This means that the cosine is positive only when the reference point lies in the first or fourth quadrants, and the sine is positive only when the reference point lies in the first or second quadrants. See Figure 9.18.

EXERCISE 9.3

A. FUNDAMENTALS

Find the values of the cosine and sine for the indicated real numbers.

1. (a) $\cos \pi/3$ (b) $\cos 7\pi/3$
   (c) $\cos 13\pi/3$
2. (a) $\sin(-\pi/6)$ (b) $\sin(-13\pi/6)$
   (c) $\sin(-25\pi/6)$
3. (a) $\cos \pi/4$ (b) $\cos 3\pi/4$
4. (a) $\cos(-\pi/4)$ (b) $\cos(-3\pi/4)$
5. (a) $\cos \pi/3$ (b) $\cos(-\pi/3)$
   (c) $\cos \pi/6$ (d) $\cos(-\pi/6)$
   (e) $\cos 2\pi/3$ (f) $\cos(-2\pi/3)$
   (g) $\cos 11\pi/6$ (h) $\cos(-11\pi/6)$
6. (a) $\sin \pi/3$ (b) $\sin(-\pi/3)$
   (c) $\sin \pi/6$ (d) $\sin(-\pi/6)$
   (e) $\sin 5\pi/6$ (f) $\sin(-5\pi/6)$
7. (a) $\sin \pi/4$ (b) $\sin 3\pi/4$
8. (a) $\sin 5\pi/4$ (b) $\sin 11\pi/4$
9. (a) $\cos 22\pi/3$ (b) $\cos 20\pi/3$
10. (a) $\sin 41\pi/6$ (b) $\sin(-31\pi/6)$

B. ESSENTIALS

Evaluate the following expressions for the indicated value of the variable. Look for relationships—they are not coincidences.

11. $\sin 2\theta$ and $2 \sin \theta$ for
    (a) $\theta = \pi/4$ (b) $\theta = \pi/6$
12. $1/2 \cos \theta$ and $\cos \theta/2$ for
    (a) $\theta = \pi/3$ (b) $\theta = \pi/2$
13. $\sin^2 \theta + \cos^2 \theta$ for
    (a) $\theta = \pi/3$ (b) $\theta = \pi/4$ (c) $\theta = \pi/6$
14. $\sin 2\theta$ and $\sin \theta \cos \theta$ for
    (a) $\theta = \pi/3$ (b) $\theta = \pi/4$ (c) $\theta = \pi/6$
15. $\cos 2\theta$ and $\cos^2 \theta - \sin^2 \theta$
    (a) $\theta = \pi/3$ (b) $\theta = \pi/4$ (c) $\theta = \pi/6$
★16. $\cos^2 \theta/2$ and $1 + \cos \theta$ for
    (a) $\theta = \pi/3$ (b) $\theta = \pi/2$ •(c) $\theta = \pi/6$

Find all of the values of θ between 0 and 2π for which the following are true.

17. (a) $\sin \theta = \frac{1}{2}$ (b) $\sin \theta = -\frac{1}{2}$
18. (a) $\cos \theta = \sqrt{3}/2$ (b) $\cos \theta = -\sqrt{3}/2$
19. (a) $\sin \theta = \sqrt{2}/2$ (b) $\sin \theta = -\sqrt{2}/2$
20. (a) $\sin \theta = \sqrt{3}/2$ (b) $\sin 2\theta = \sqrt{3}/2$
★21. (a) $\cos \alpha = \frac{1}{2}$ (b) $\cos(\beta/2) = \frac{1}{2}$
22. (a) $\sin \alpha = -\sqrt{3}/2$
    (b) $\sin(-\beta) = -\sqrt{3}/2$
23. (a) $\cos \alpha = -\sqrt{2}/2$
    (b) $\cos(-\beta/2) = -\sqrt{2}/2$
24. $\cos \theta > \sin \theta$
25. (a) $\cos \theta = \sin \theta$
    (b) $\cos \theta < \sin \theta$
26. (a) $\cos \theta = -\sin \theta$
    (b) $\cos \theta = \sin(-\theta)$

C. APPLICATIONS

27. Prove that $\cos \pi/3 = \frac{1}{2}$ and $\sin \pi/3 = \sqrt{3}/2$. (HINT: Show that $\triangle OPP'$ is equilateral with $PP' = 1$, as we did for $\triangle OPP'$ on page 197.)

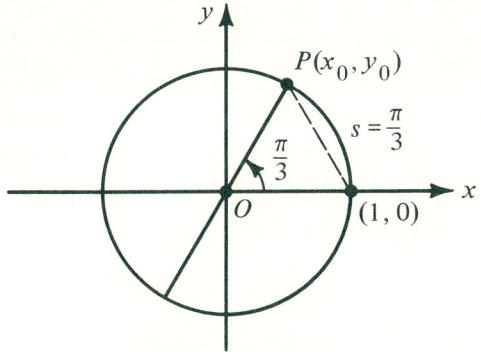

28. Use the distance formula and the fact that equal arcs of a circle subtend equal chords to prove that $\cos \pi/6 = \sqrt{3}/2$ and $\sin \pi/6 = \frac{1}{2}$. [HINT: The distance from P to $(0, 1)$ must equal the distance from P to P'. Remember that $x_0^2 + y_0^2 = 1$, since (x_0, y_0) lies on the unit circle.]

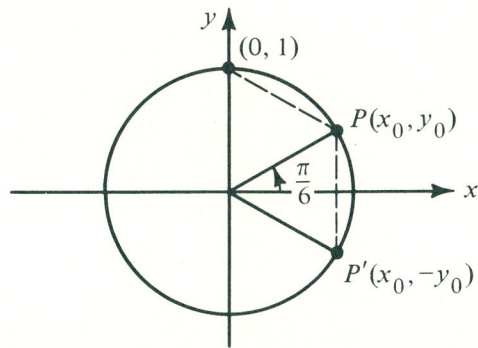

29. Plot the indicated ordered pairs $(\theta, f(\theta))$ for $f(\theta) = \cos \theta$ on the axis system provided, then connect the points with a smooth curve.

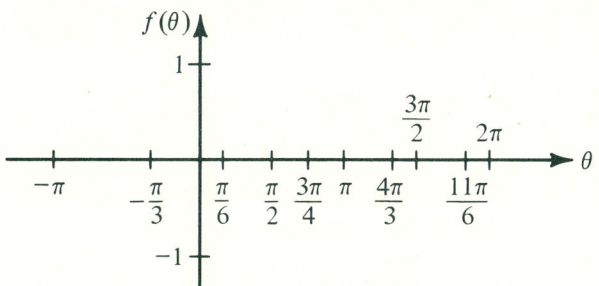

30. Repeat Problem 29 for $f(\theta) = \sin \theta$.

9.4 GRAPHING THE COSINE AND SINE FUNCTIONS

Before we graph these two circular functions, there is one property of the cosine and sine that you may have noticed already that needs to be stated formally. Their values repeat every 2π. This is explained by the wrapping nature of the definitions of sine and cosine, as illustrated in Figure 9.19. This figure shows that θ, $\theta + 2\pi$, and $\theta + 4\pi$ all have the same terminal side. This means that the reference point (x_0, y_0) on the unit circle will be the same for each of them. Therefore,

$$\cos \theta = \cos (\theta + 2\pi) = \cos (\theta + 4\pi) = \cdots = x_0$$

and

$$\sin \theta = \sin (\theta + 2\pi) = \sin (\theta + 4\pi) = \cdots = y_0$$

We call this repeating property "periodicity" and say that the **cosine and sine have periods of 2π**. This is summarized in the two following theorems on periodicity.

THEOREM I.
For all real numbers θ,

$$\cos [\theta + K(2\pi)] = \cos \theta$$

where $K \in$ integers

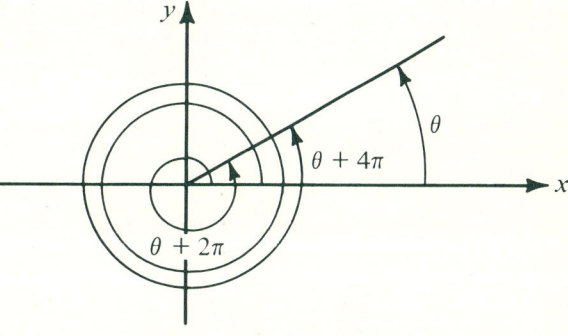

FIGURE 9.19.

CHAPTER 9   The Circular Functions

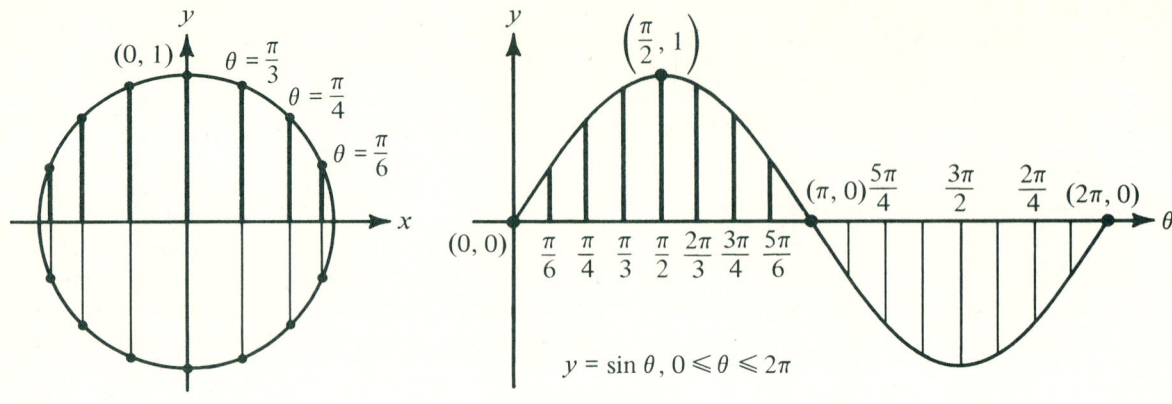

FIGURE 9.20.

THEOREM II.
For all real numbers θ,
$$\sin(\theta + 2\pi K) = \sin\theta$$
where $K \in$ integers

We will use the concept of the periodicity of the cosine and sine functions in our graphing by drawing the graph for only one period since we know that the trig functions simply repeat themselves beyond this. It is essential to remember that these graphs repeat themselves periodically, even though they are not drawn that way.

Now let's graph the sine function. Consider the vertical components of various points on the unit circle. Since these are the values of $\sin\theta$, they measure $f(\theta)$ for $f(\theta) = \sin\theta$ for each value of θ. See Figure 9.20.

If we draw the graph over more than one period, the periodicity of the sine function is evident. See Figure 9.21.

The maximum value of $\sin\theta$ is 1 and the minimum value is -1. We say that the **amplitude** of $y = \sin\theta$ is 1 and that the range of the sine function is

$$R = \{y \in \text{reals} \mid -1 \leq y \leq 1\}$$

Now we are ready to graph the cosine function. This graph can be obtained by plotting values from the unit circle much as we did for the sine curve. This time, the horizontal component of each reference point is the value of $f(\theta) = \cos\theta$ for each value of θ. See Figure 9.22.

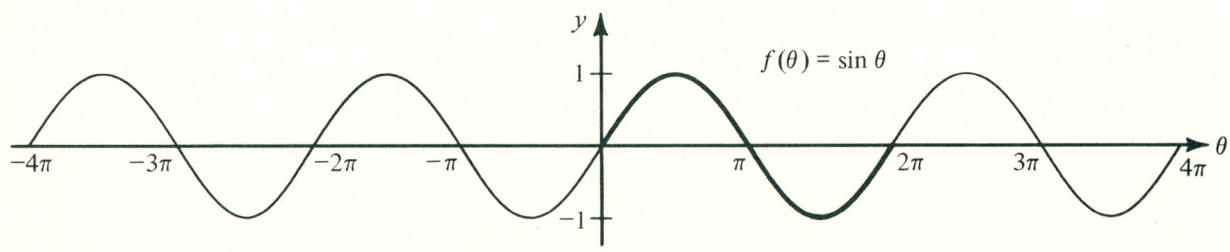

FIGURE 9.21.

200

SECTION 9.4   Graphing the Cosine and Sine Functions

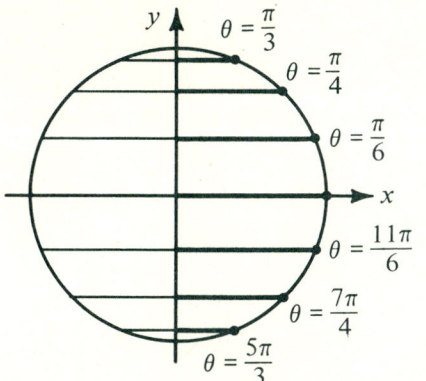
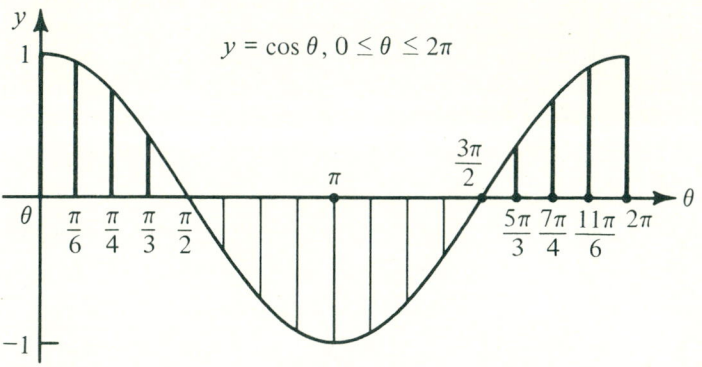

FIGURE 9.22.

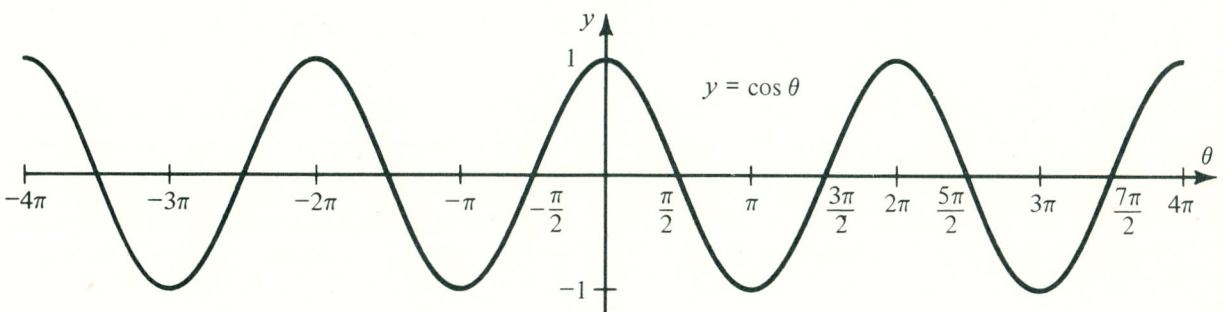

FIGURE 9.23.

Extending the graph of the cosine function over more than one period shows the repetition of the graph in Figure 9.22 every 2π. See Figure 9.23.

The graphs of the cosine and the sine functions are basically similar. In fact, we will see that they are simply translations by $\pi/2$ of each other.

By recalling that $y = 3f(x)$ is an expansion away from the x-axis by a factor of 3, we can easily graph the following functions. Notice that only one period is depicted, since the remainder of the graph is simply a repetition of this picture.

EXAMPLE I.   Graph $f(\theta) = 3 \cos \theta$.
See Figure 9.24.

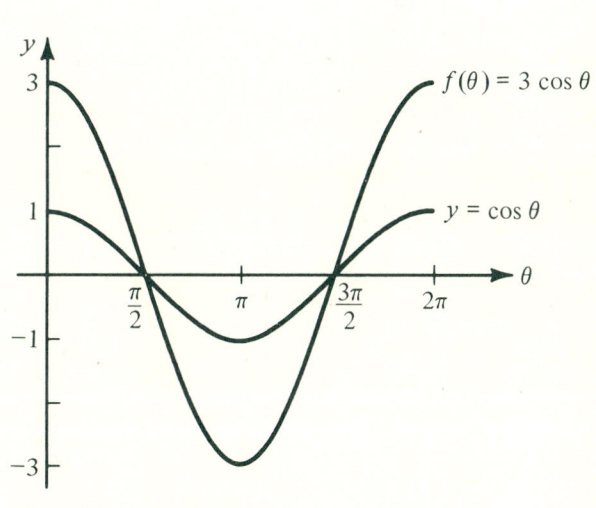

FIGURE 9.24.

201

EXAMPLE II. Graph $f(\theta) = \frac{1}{2} \sin \theta$.
See Figure 9.25.

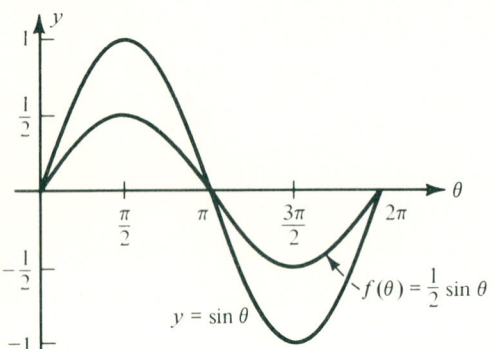

FIGURE 9.25.

Notice that in Example I the amplitude is 3 and in Example II the amplitude is $\frac{1}{2}$.

By recalling that $y = f(2x)$ is a compression toward the y-axis by a factor of 2, we can readily graph the following functions.

EXAMPLE III. Graph $f(\theta) = \cos 2\theta$.
See Figure 9.26.

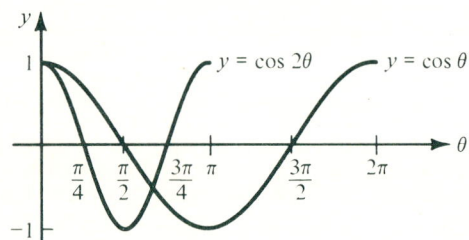

FIGURE 9.26.

EXAMPLE IV. Graph $f(\theta) = \sin \theta/2$.
See Figure 9.27.

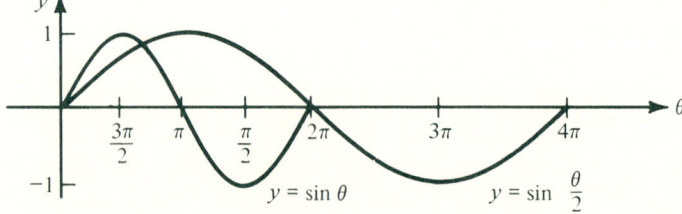

FIGURE 9.27.

The graph in Example III illustrates a function whose period is π, and the graph in Example IV illustrates one whose period is 4π. In general,

$$f(\theta) = \cos B\theta$$

and

$$f(\theta) = \sin B\theta$$

will have periods of

$$p = (2\pi)/B$$

When working with trigonometric functions, a horizontal translation is called a **phase shift**. The function $f(x) = \cos (x - \pi/4)$, a translation of $y = \cos x$ to the right by $\pi/4$, is called a positive phase shift of $\pi/4$. What are the phase shifts for the following functions?

$$f(\theta) = \cos (x - \pi/6) \underline{\qquad}$$

and

$$g(\theta) = \sin (x + 3\pi/4) \underline{\qquad}$$

They are positive $\pi/6$ and negative $3\pi/4$, respectively, as illustrated in Figures 9.28 and 9.29.

EXAMPLE V. Graph $f(x) = \cos (x - \pi/6)$ and $y = \cos x$ on one axis system.
See Figure 9.28.

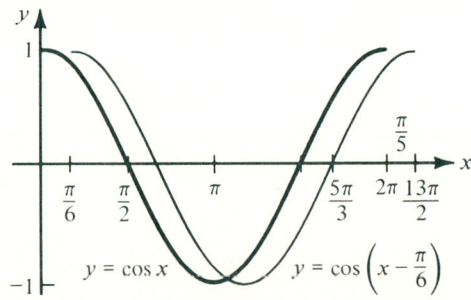

FIGURE 9.28.

SECTION 9.4   Graphing the Cosine and Sine Functions

EXAMPLE VI. Graph $f(x) = \sin(x + 3\pi/4)$ and $y = \sin x$ on one axis system.
See Figure 9.29.

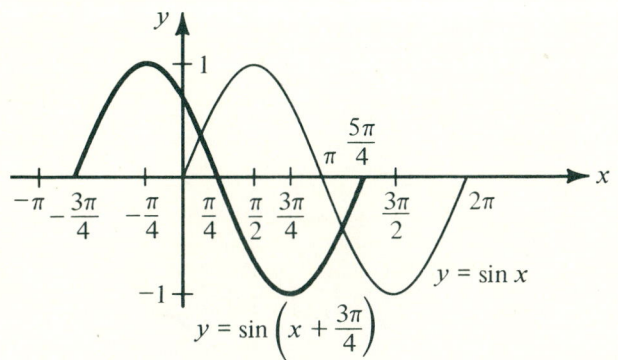

FIGURE 9.29.

Notice that the amplitudes of the graphs in Examples V and VI are both 1 and the periods are both 2π.

Vertical translations, called "bias," are common in electronics applications, but they are seldom emphasized at this level.

EXERCISE 9.4

A. FUNDAMENTALS
Match the function with its graph.
  1. (i) $y = \cos \theta$ _____ .
     (ii) $y = \sin \theta$ _____ .

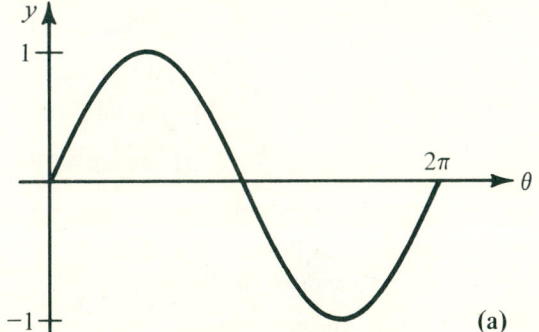

(a)

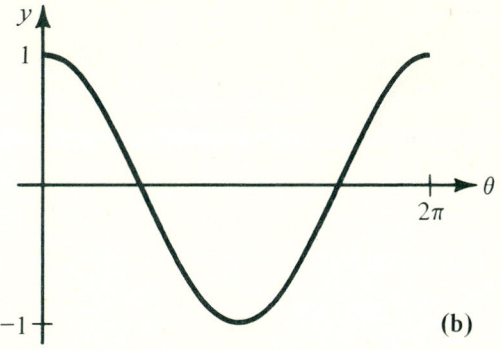

(b)

  2. (i) $y = \cos 2\theta$ _____ .
     (ii) $y = \sin 2\theta$ _____ .

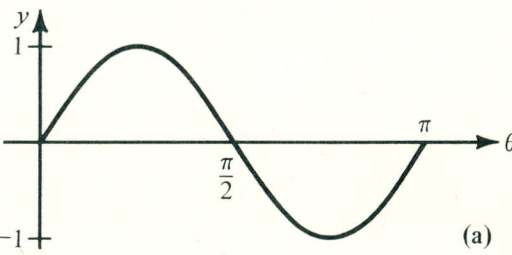

(a)

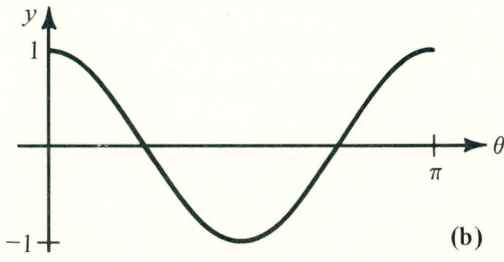

(b)

  3. (i) $y = \cos \dfrac{\theta}{2}$ _____ .
     (ii) $y = \cos 2\theta$ _____ .

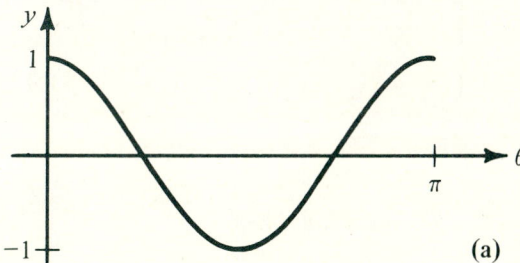

(a)

203

CHAPTER 9   The Circular Functions

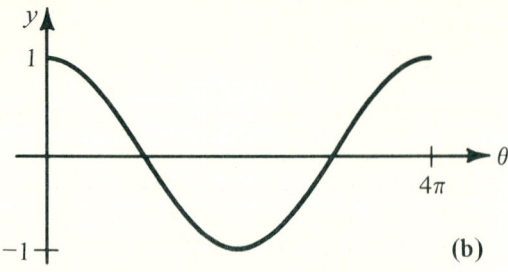
(b)

4. (i) $y = \frac{1}{2} \sin \theta$ _____ .
   (ii) $y = 2 \cos \theta$ _____ .

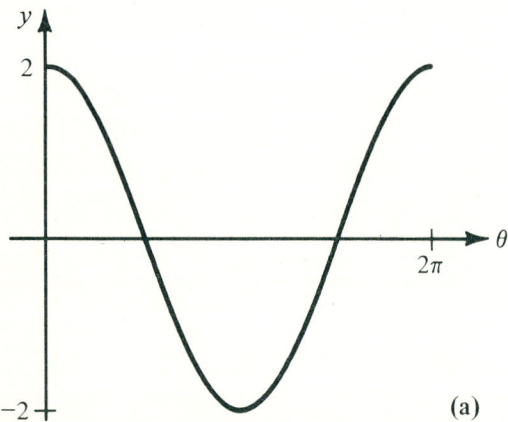
(a)

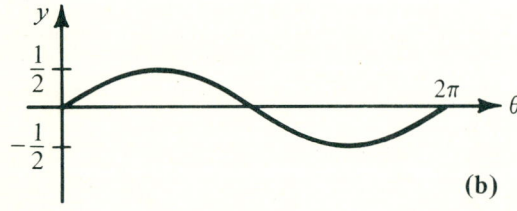
(b)

5. (i) $y = \sin (\theta - \pi/2)$ _____ .
   (ii) $y = \cos (\theta + \pi/4)$ _____ .

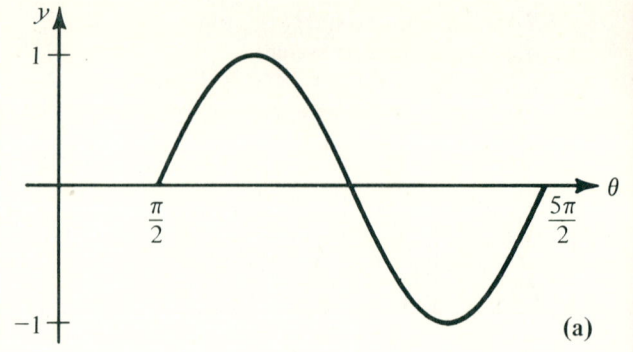
(a)

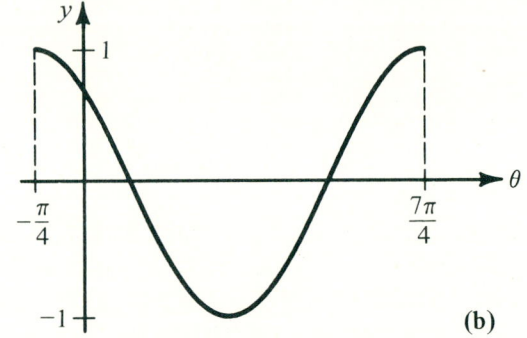
(b)

B. ESSENTIALS
Graph each of the following pairs of functions on the same coordinate system.

6. (a) $f(\theta) = 2 \sin \theta$ (b) $g(\theta) = \sin 2\theta$
7. (a) $y = 2 \cos x$ (b) $y = \cos 2x$
8. (a) $y = \frac{1}{2} \sin \theta$ (b) $y = \sin (\theta/2)$
9. (a) $y = \sin (\theta + \pi/4)$
   (b) $y = \sin (\theta - \pi/4)$
10. (a) $s = \cos t$ (b) $s = \sin (t - \pi/2)$
11. (a) $y = \cos (\theta + \pi/2)$ (b) $y = \sin \theta$
12. (a) $y = -\sin x$ (b) $y = \sin (-x)$
13. (a) $y = \cos (-\theta)$ (b) $y = -\cos \theta$
14. (a) $y = \sin (\pi - \alpha)$ (b) $y = \cos (\pi - \alpha)$
15. (a) $y = \cos (\theta - \pi/4)$
    (b) $y = \sin (\theta - 3\pi/4)$
★16. $f(\theta) = 2 \sin 3(\theta - \pi/6)$

17. $y = 2 \cos 3(\theta - \pi/6)$

18. $y = 3 \sin 2(x + \pi/3)$

19. $y = \frac{1}{2} \cos 4(\theta + \pi/6)$

20. $y = 2 \cos \frac{1}{4}(\theta - \pi/2)$

21. (a) $y = \cos \theta$ (b) $y = \sin \theta$
    (c) $y = \cos \theta + \sin \theta$

22. (a) $y = \cos \theta$ (b) $y = \theta$
    (c) $y = \theta + \cos \theta$

23. (a) $y = \theta \cos \theta$ •(b) $y = \theta^2 \cos \theta$
    •(c) $y = \theta^3 \cos \theta$

•24. $y = (\cos \theta)/\theta$ •25. $y = (\sin \theta)/\theta$

•26. $y = 2^x \cos x$ •27. $y = 3^\theta \sin \theta$

28. Prove that the graph of $f(\theta) = \cos \theta$ is simply a translation to the left by $\pi/2$ of the graph of $g(\theta) = \sin \theta$.

Graph the following on the same coordinate system.

29. (a) $f(x) = 2 \sin(x - \pi/4)$
    (b) $g(x) = 2 \cos(x - \pi/4)$
    (c) $y = f(x) + g(x)$

30. (a) $f(x) = 3 \sin x$ (b) $g(x) = 5 \cos x$
    (c) $y = f(x) + g(x)$

31. (a) $f(x) = 2 \sin(x - \pi/6)$
    (b) $g(x) = 3 \cos(x + \pi/6)$
    (c) $y = f(x) + g(x)$

•32. (a) $f(x) = \sin x + \sin 2x$
    (b) $g(x) = f(x) + \sin 4x$
    (c) $h(x) = g(x) + \sin 6x$

9.5 MORE CIRCULAR FUNCTIONS

Although the unit circle is the most convenient circle for defining the circular functions, any circle will work. To complete our introduction of the circular functions, we will consider the general circle $x^2 + y^2 = r^2$. First, we want to extend the definitions of the cosine and sine functions to a circle of radius r.

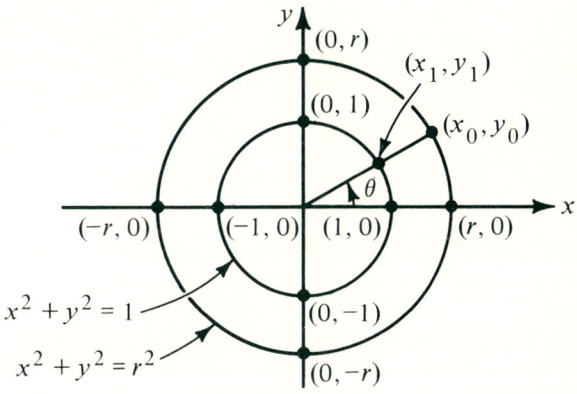

FIGURE 9.30.

Consider Figure 9.30. We know that on the unit circle,

$$\cos \theta = x_1 \quad \text{and} \quad \sin \theta = y_1$$

We wish to consistently redefine them in terms of x_0, y_0, and r. That is, we want our first definitions, where $r = 1$, to be special cases of these new, more general ones. See Figure 9.31. The two right

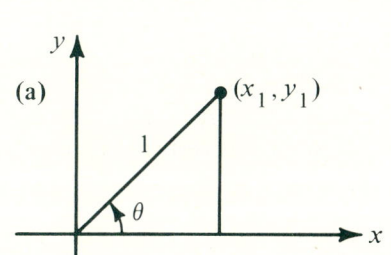
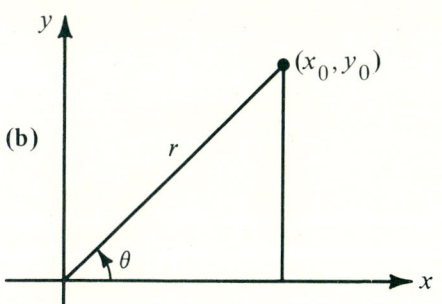

FIGURE 9.31.

triangles are similar triangles because the angles of one triangle equal the corresponding angles of the other. By recalling that corresponding sides of similar triangles are proportional, we have

$$x_1/1 = x_0/r \quad \text{and} \quad y_1/1 = y_0/r$$

But, $x_1 = \cos \theta$ and $y_1 = \sin \theta$ means that we have

$$\cos \theta = x_0/r \quad \text{and} \quad \sin \theta = y_0/r$$

for any circle of radius r, where $r \neq 0$.

There are four other ratios involving x, y, and r that complete the list of the circular functions. They are named tangent (tan), secant (sec), cotangent (cot), and cosecant (csc). See Definition 9.4 and Figure 9.32.

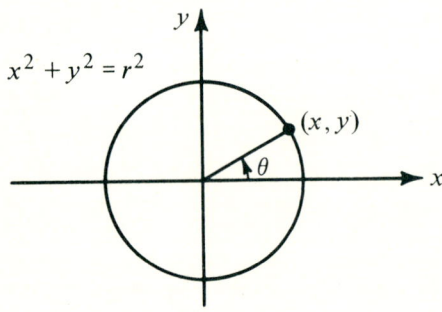

FIGURE 9.32.

DEFINITION 9.4.
The six circular functions defined in terms of the circle of radius r are

$$\sin \theta = y/r \qquad \csc \theta = r/y$$
$$\cos \theta = x/r \qquad \sec \theta = r/x$$
$$\tan \theta = y/x \qquad \cot \theta = x/y$$

We will draw the graphs of these four new circular functions by carefully examining the relationship between each function and the sine or cosine functions. Then, we will use our knowledge of these two graphs to draw the new ones.

First, let's use the above definitions to show that each of the four new functions can be expressed in terms of sine or cosine as follows.

$$\csc \theta = \frac{r}{y} = \frac{1}{y/r} = \frac{1}{\sin \theta}$$

$$\sec \theta = \frac{r}{x} = \frac{1}{?} = \frac{1}{?}$$

$$\tan \theta = \frac{y}{x} = \frac{y/r}{x/r} = \frac{?}{?}$$

$$\cot \theta = \frac{x}{y} = \frac{?/?}{?/?} = \frac{?}{?}$$

This is summarized by the four equations in Theorem III, called *identities* because the equations hold for *all* values of the variable in the domains of the functions involved. That is, the denominators on the right cannot equal zero.

THEOREM III.

$$\csc \theta = 1/(\sin \theta)$$
$$\sec \theta = 1/(\cos \theta)$$
$$\tan \theta = (\sin \theta)/(\cos \theta)$$
$$\cot \theta = (\cos \theta)/(\sin \theta)$$

Now, let's draw the graph of the cosecant function. Since

$$\csc \theta = 1/(\sin \theta)$$

the cosecant function will repeat its values as often as the sine function does. This means that the period of the cosecant function is also 2π.

The above identity also tells us that the cosecant function has vertical asymptotes whenever

$$\sin \theta = 0$$

which is for

$$\theta = K\pi, \qquad K \in \text{integers}$$

This identity also indicates that the sign of csc θ, where $\theta \neq K\pi$ and $K \in$ integers, will be the same as that of sin θ. That is, the cosecant will be positive in the first and second quadrants and negative in the third and fourth quadrants.

Complete the following table of values of the cosecant function and compare your results with its graph in Figure 9.33.

θ	$g(\theta) = \sin \theta$	$f(\theta) = \dfrac{1}{\sin \theta} = \csc \theta$
0	0	$1/0 = \infty$
$\pi/6$	$\frac{1}{2}$	$1/\frac{1}{2} = 2$
$\pi/4$	$1/\sqrt{2}$	$\sqrt{2}/1 \approx 1.4$
$\pi/3$	$\sqrt{3}/2$	?
$\pi/2$	?	?
$3\pi/2$	?	?
$7\pi/4$	?	?
$5\pi/3$	?	?
$11\pi/6$	?	?

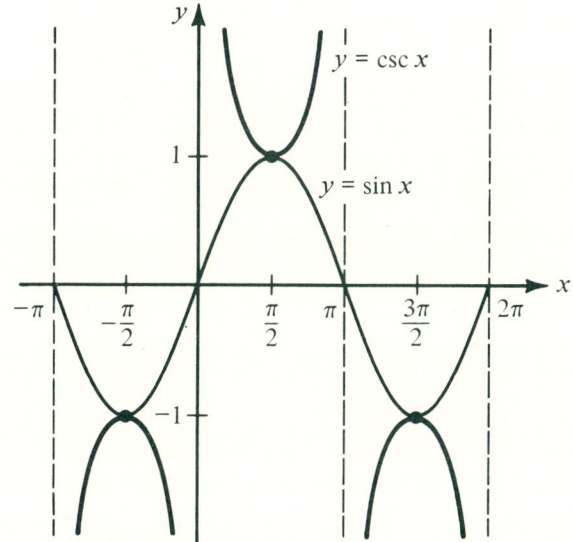

FIGURE 9.33.

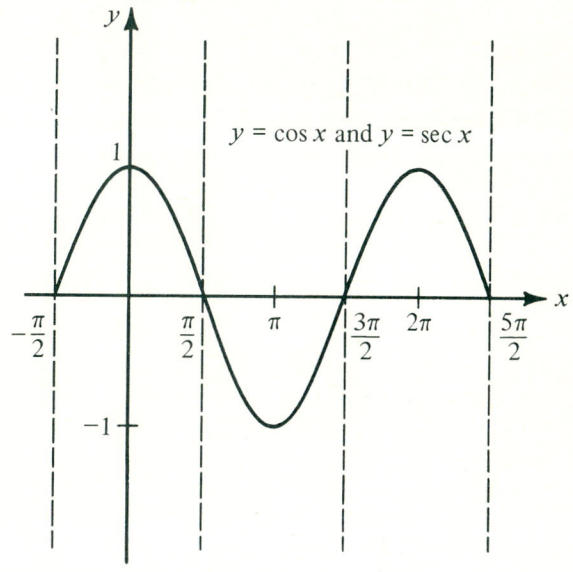

FIGURE 9.34.

Now use the same procedure to draw the graph of the secant function from that of the cosine by recalling that

$$\sec \theta = 1/(\cos \theta)$$

Since the period of the secant is 2π, the same as that of the cosine, we could graph one complete cycle of the secant function between $\theta = 0$ and $\theta = 2\pi$. But, because of the location of the vertical asymptotes, we will get a better graph of one complete cycle from

$$-\pi/2 < \theta < 3\pi/2, \qquad \theta \neq \pi/2$$

Complete the graph of the secant above that of the cosine in Figure 9.34. Remember that the signs of cos θ and sec θ are the same for each value of θ in their domains.

Your graph of the secant function above should touch the graph of the cosine function at $(0, 1)$, $(\pi, -1)$, and $(2\pi, 1)$. The dotted lines $\theta = -\pi/2$, $\pi/2$, $3\pi/2$, and $5\pi/2$ are its vertical asymptotes.

CHAPTER 9 The Circular Functions

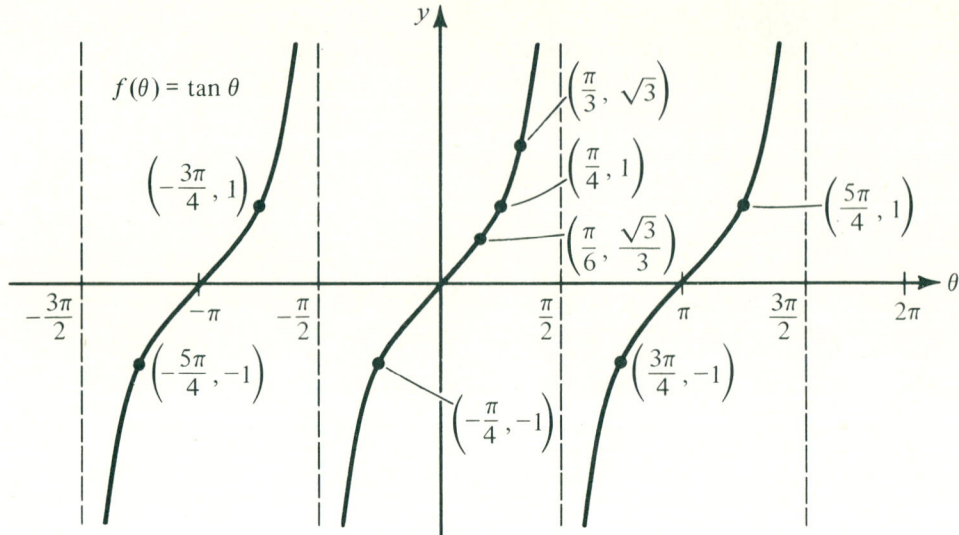

FIGURE 9.35.

The graphs of the tangent and cotangent functions are somewhat more complicated because their periods are not equal to 2π. Since

$$\tan\theta = (\sin\theta)/(\cos\theta)$$

and

$$\cot\theta = (\cos\theta)/(\sin\theta)$$

and the sine and cosine repeat themselves every 2π, the tangent and cotangent will also repeat every 2π. Their periods, however, are shorter than 2π.

By examining the changes in the values of the sine and cosine for $0 < \theta < \pi$ compared to $\pi < \theta < 2\pi$, we will see that the period of the tangent function is actually only π. For example,

$$\tan \pi/4 = \frac{\sin \pi/4}{\cos \pi/4} = \frac{\sqrt{2}/2}{\sqrt{2}/2} = +1$$

and

$$\tan(\pi/4 + \pi) = \frac{\sin 5\pi/4}{\cos 5\pi/4} = \frac{-\sqrt{2}/2}{-\sqrt{2}/2} = +1$$

Therefore,

$$\tan \pi/4 = \tan(\pi + \pi/4)$$

This illustrates the fact that the periodicity of the tangent function is π. To actually understand this you would need to try many more values than just $\pi/4$, and then construct a proof.

Now, let's find the intercepts and asymptotes for $f(\theta) = \tan\theta$ using the identity

$$\tan\theta = (\sin\theta)/(\cos\theta)$$

Since $\sin\theta = 0$ yields $\tan\theta = 0$, we know that the zeros for $f(\theta) = \tan\theta$ are $\theta = 0, \pm\pi, \pm 2\pi, \ldots \Leftrightarrow \theta = K\pi$, where $K \in$ integers. The asymptotes are located at the values of θ for which $\cos\theta = 0$,

$$\theta = \pm\pi/2, \pm 3\pi/2, \ldots \Leftrightarrow \theta = \pi/2 + K\pi$$

where $K \in$ integers. Some other values of $\tan\theta$ are also depicted in Figure 9.35.

The graph of the cotangent function is very similar to that of the tangent. In fact, since

$$\cot\theta = 1/(\tan\theta)$$

the cotangent function also has a period of π.

208

SECTION 9.5   More Circular Functions

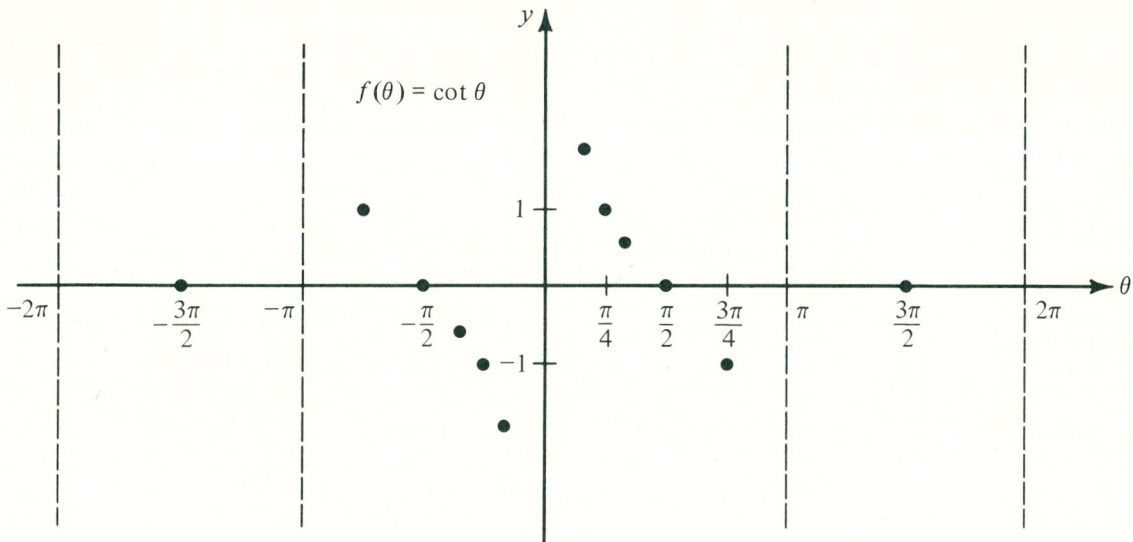

FIGURE 9.36.

Use the identity

$$\cot \theta = (\cos \theta)/(\sin \theta)$$

to determine the following information, then graph $f(\theta) = \cot \theta$.
Since

$$\cos \theta = 0 \Rightarrow$$
$$\theta = \underline{\quad ? \quad}, \underline{\quad ? \quad}, \ldots$$

we know that the graph of the cotangent function has \underline{\quad ? \quad} at

$$\theta = \underline{\quad ? \quad} \pm K\pi$$

where $K \in$ integers. Also, since

$$\sin \theta = 0 \Rightarrow$$
$$\theta = \underline{\quad ? \quad}, \underline{\quad ? \quad}, \ldots$$

the graph of $f(\theta) = \cot \theta$ will have \underline{\quad ? \quad}

at

$$\theta = \underline{\quad ? \quad} \pm K\pi$$

where $K \in$ integers.
  For $0 < \theta < \pi/2$, $\sin \theta$ and $\cos \theta$ are both positive. Therefore,

$$\cot \theta = (\cos \theta)/(\sin \theta)$$

is positive for $0 < \theta < \pi/2$.
  For $\pi/2 < \theta < \pi$, $\sin \theta > 0$ and $\cos \theta < 0$ show that $\cot \theta \underline{\quad ? \quad} 0$ for $\pi/2 < \theta < \pi$.
  Label the points in Figure 9.36, then draw the graph of the cotangent function. This graph is the graph of $f(\theta) = \tan \theta$ reflected about the horizontal axis then translated to the right $\pi/2$.

EXERCISE 9.5

A. FUNDAMENTALS
Give the equation that best describes each of the following graphs.

1. (a) 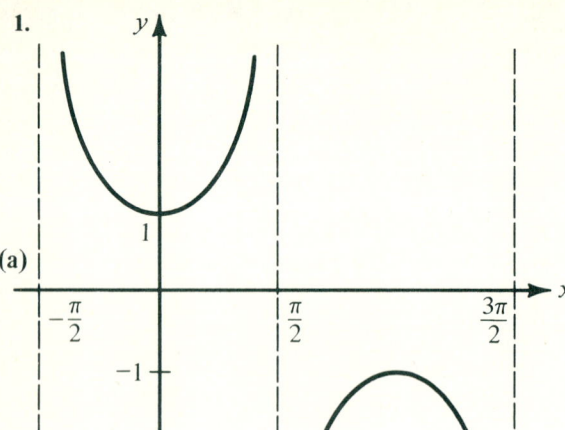 (b)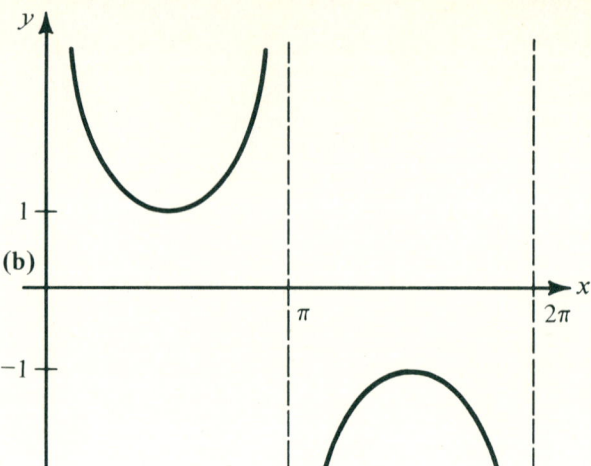

2. (a) 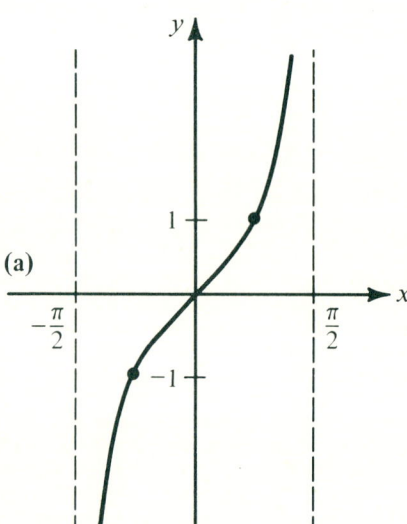 (b)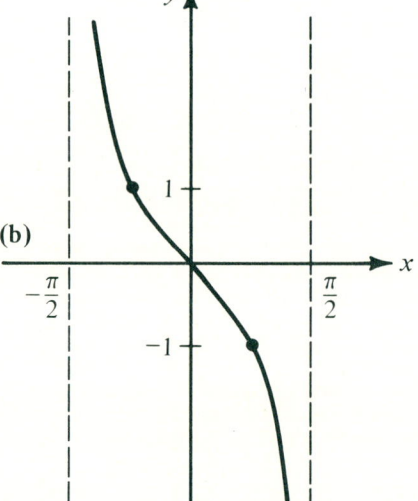

3. (a) 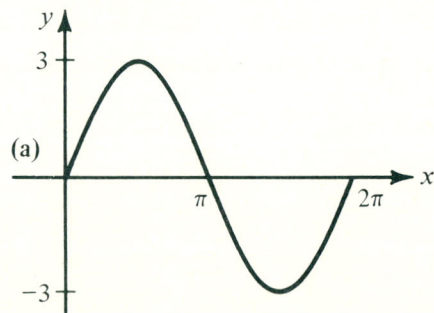 (b)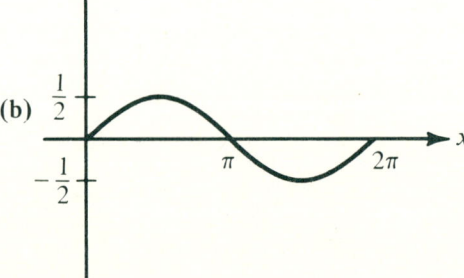

4.

(a) 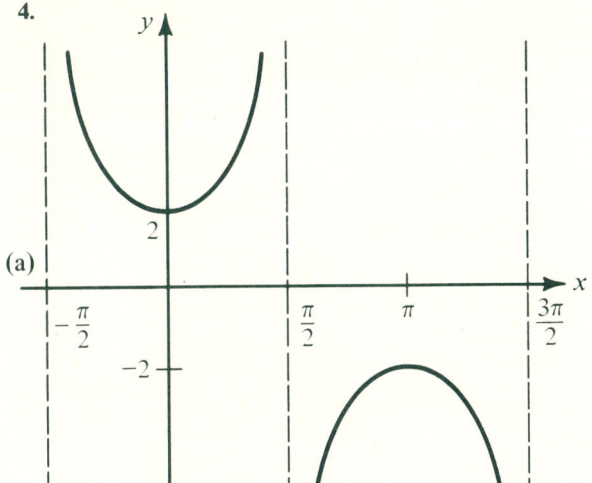 (b)

5.

(a) 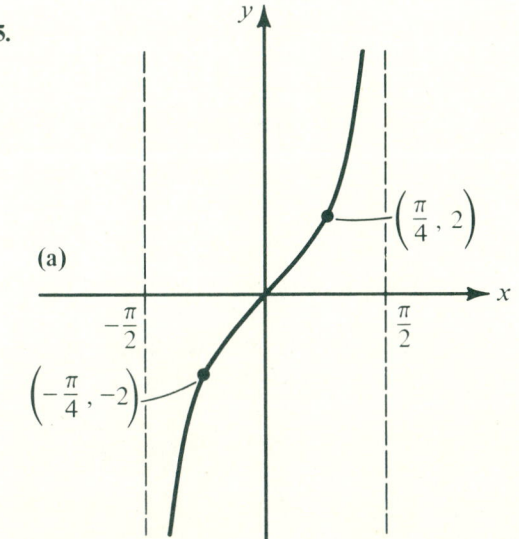 (b)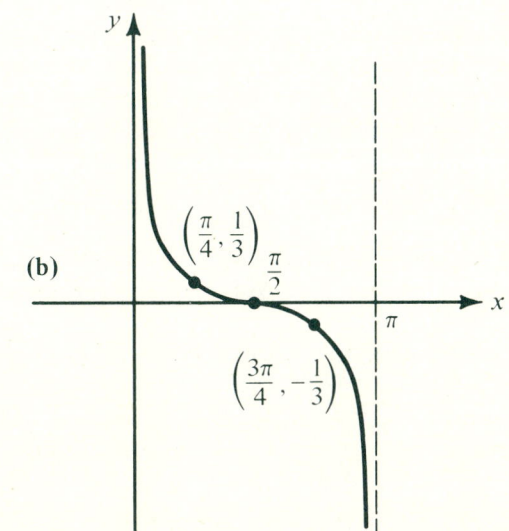

B. ESSENTIALS
Graph each of the following pairs of curves on the same coordinate system.

★6. (a) $f(\theta) = \tan \theta$ (b) $f(\theta) = \tan(\theta - \pi/4)$

7. (a) $f(\theta) = \tan \theta$ (b) $y = f(-\theta)$

8. (a) $f(\theta) = \sec 2\theta$ (b) $f(\theta) = 2 \sec \theta$

9. (a) $f(\theta) = \cot(\theta - \pi/2)$
    (b) $f(\theta) = \tan(-\theta)$

Graph each of the following, labeling all intercepts and asymptotes.

★10. $y = 2 \tan 3(\theta - \pi/4)$

11. $f(\theta) = \tfrac{1}{2} \tan 2(\theta + \pi/4)$

12. $f(\theta) = 2 \cot 3(\theta - \pi/4)$

13. $f(\theta) = \tfrac{1}{2} \sec 2\theta$ 14. $y = 2 \csc 2(\theta - \pi/3)$

★15. $y = \tfrac{1}{3} \csc(\theta/2 - \pi/4)$

C. APPLICATIONS

16. Prove that the period of $f(\theta) = \cot \theta$ is π.

★17. Decide whether $\tan \theta$, $\sec \theta$, $\csc \theta$, and $\cot \theta$ are even, odd, or neither. Prove your results.

18. Prove that the graph of $y = \cot \theta$ is the graph of $y = \tan \theta$ reflected about the x-axis, then translated to the right $\pi/2$.

• **19.** Graph $y = 2^x \tan x$.

• **20.** Graph $y = \sec x \tan x$.

9.6 THE INVERSES OF THE CIRCULAR FUNCTIONS

Before getting into this section you may find it helpful to review the general properties of inverses in Section 3.6.

By realizing that only one-to-one functions have inverses that are also functions, and that periodic functions are certainly not one to one (the value of the function is repeated periodically for different values of θ), we understand the first problem. If we wish the inverses of the circular functions to be functions, we must restrict their domains to values that will generate one-to-one functions.

For the sine function, we will consider a small section of the curve—not even one complete period—when considering its inverse. If we allow only values between $-\pi/2$ and $\pi/2$ for the independent variable, x, the resulting restricted sine function will be a one-to-one function. See Figure 9.37. There are other restricted domains that would yield a one-to-one function, but this is the best example (see Exercise 37).

Notice the capital letter on the new function, $y = \text{Sin } x$. This is interpreted as

$$y = \text{Sin } x \Leftrightarrow$$
$$y = \sin x \quad \text{and} \quad -\pi/2 \leq x \leq \pi/2$$

For the cosine function the customary restriction yielding a one-to-one function is illustrated in Figure 9.38.

$$y = \text{Cos } x \Leftrightarrow$$
$$y = \cos x \quad \text{and} \quad 0 \leq x \leq \pi$$

The tangent function is restricted to one period. Then,

$$y = \text{Tan } x \Leftrightarrow$$
$$y = \tan x \quad \text{and} \quad -\pi/2 < x < \pi/2$$

Now that we have three one-to-one functions defined, let's investigate their inverses.

Recall that if (a, b) is a point on the graph of a function, then (b, a) is a point on the graph of its inverse. This means that we can graph each inverse by reflecting the original graph about the line $y = x$.

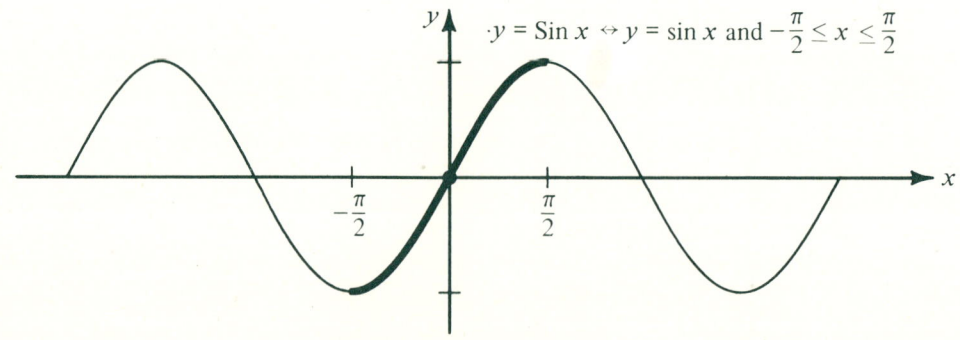

FIGURE 9.37.

SECTION 9.6 The Inverses of the Circular Functions

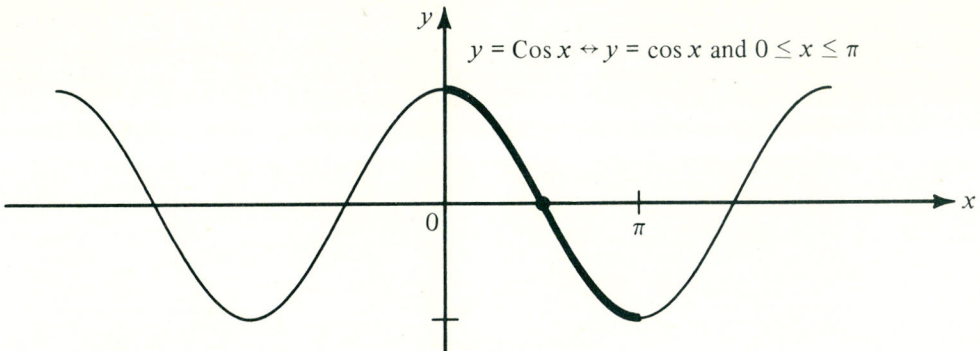

FIGURE 9.38.

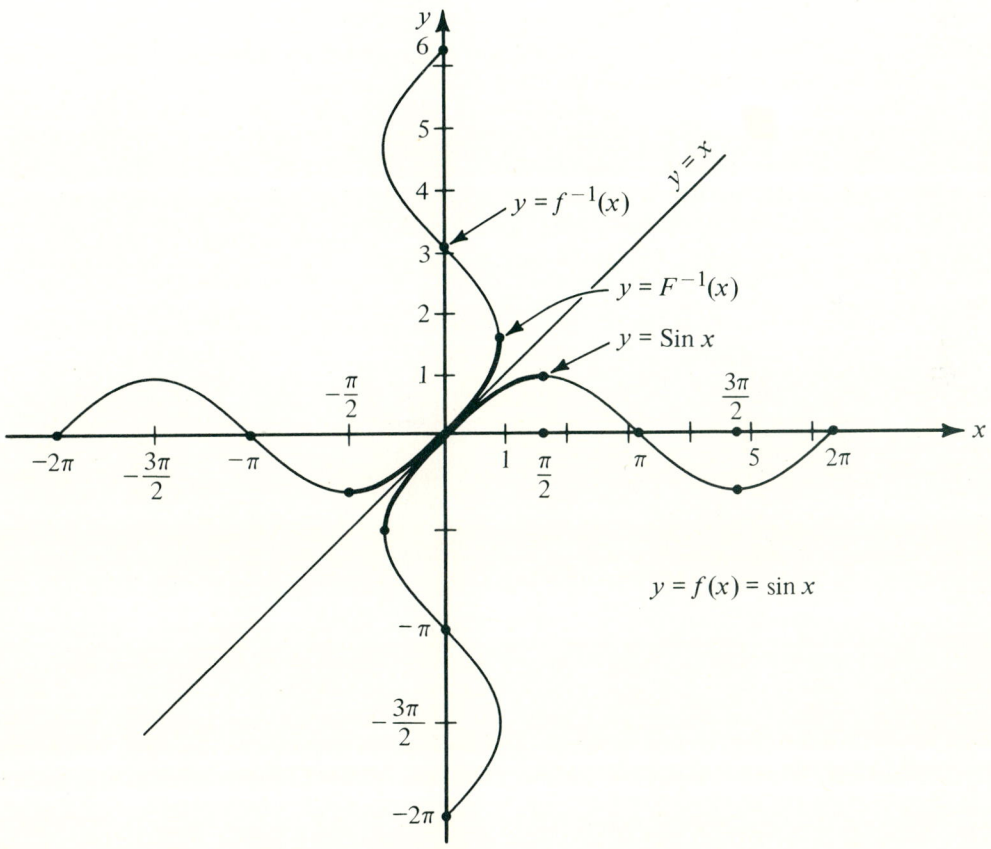

FIGURE 9.39.

CHAPTER 9   The Circular Functions

Let's reflect the graph of $y = \sin x$ about the line $y = x$ to obtain the graph of its inverse relation. See Figure 9.39.

Now, to find the equation of the inverse of a function, we solve for x, and then rename the variables, calling x, "y" and y, "x." Since there is no algebraic technique available to solve an equation like $y = \sin x$ for x, we will invent a new function, as we did for the inverses of exponential functions (logarithmic functions). This new function is called "Arcsine" and is defined as follows.

DEFINITION 9.5.

$$y = \text{Arcsin } x \Leftrightarrow x = \text{Sin } y$$

By renaming the variables we see that the inverse of $y = \text{Sin } x$ is $y = \text{Arcsin } x$. We will abbreviate $y = \text{Arcsin } x$ as $y = \text{Sin}^{-1} x$, which is read "y equals Arcsine of x" or "y equals the inverse of the sine of x." If we want to write $1/(\sin x)$ using exponential notation, we must write $(\sin x)^{-1}$, not $\sin^{-1} x$. To interpret an expression such as

$$y = \text{Arcsin } 1$$

change it into its equivalent statement about the sine function using the definition of the Arcsine.

$$y = \text{Arcsin } 1 \Leftrightarrow \text{Sin } y = 1$$

Then, since

$$\sin \pi/2 = 1$$

we have

$$\text{Arcsin } 1 = \pi/2$$

and, equivalently,

$$\text{Sin}^{-1} 1 = \pi/2$$

Complete the following:

1. $y = \text{Cos}^{-1} 0 \Leftrightarrow \text{Cos } y = 0$. Since $\cos \pi/2 = 0$,

$$\text{Cos}^{-1} 0 = \pi/2$$

(We often read the last statement as "the angle whose cosine is zero is $\pi/2$.")

2. $y = \text{Tan}^{-1} 1 \Leftrightarrow \text{Tan } y = 1$. Since $\tan \pi/4 = 1$,

$$\text{Tan}^{-1} 1 = \underline{\quad ? \quad}$$

3. $y = \text{Sin}^{-1} \sqrt{3}/2 \Leftrightarrow \underline{\quad ? \quad} = \sqrt{3}/2$. Since $\sin \pi/3 = \underline{\quad ? \quad}$,

$$\text{Sin}^{-1} \sqrt{3}/2 = \underline{\quad ? \quad}$$

4. $y = \text{Cos}^{-1} (-\tfrac{1}{2}) \Leftrightarrow \cos y = \underline{\quad ? \quad}$. Since $\cos \underline{\quad ? \quad} = -\tfrac{1}{2}$,

$$\text{Cos}^{-1} (-\tfrac{1}{2}) = \underline{\quad ? \quad}$$

5. $y = \text{Tan}^{-1} (-\sqrt{3}) \Leftrightarrow \underline{\quad ? \quad} = \underline{\quad ? \quad}$. Since $\tan \underline{\quad ? \quad} = -\sqrt{3}$,

$$\text{Tan}^{-1} (-\sqrt{3}) = \underline{\quad ? \quad}$$

6. $y = \text{Sin}^{-1} (-1) \Leftrightarrow \underline{\quad ? \quad} = \underline{\quad ? \quad}$. Since $\sin \underline{\quad ? \quad} = -1$,

$$\text{Sin}^{-1} (-1) = \underline{\quad ? \quad}$$

The correct values are $\text{Cos}^{-1} (-\tfrac{1}{2}) = -\pi/3$, $\text{Tan}^{-1} (-\sqrt{3}) = -\pi/3$, and $\text{Sin}^{-1} (-1) = -\pi/2$.

Now we are ready to graph the inverses of $y = \text{Cos } x$ and $y = \text{Tan } x$. Remember that if (a, b)

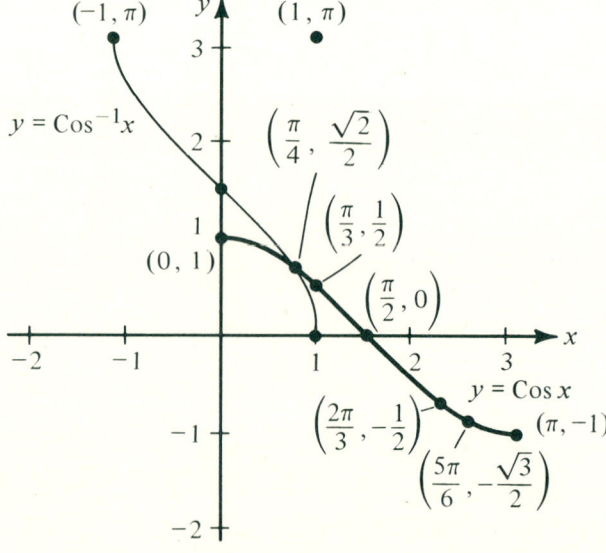

FIGURE 9.40.

SECTION 9.6 The Inverses of the Circular Functions

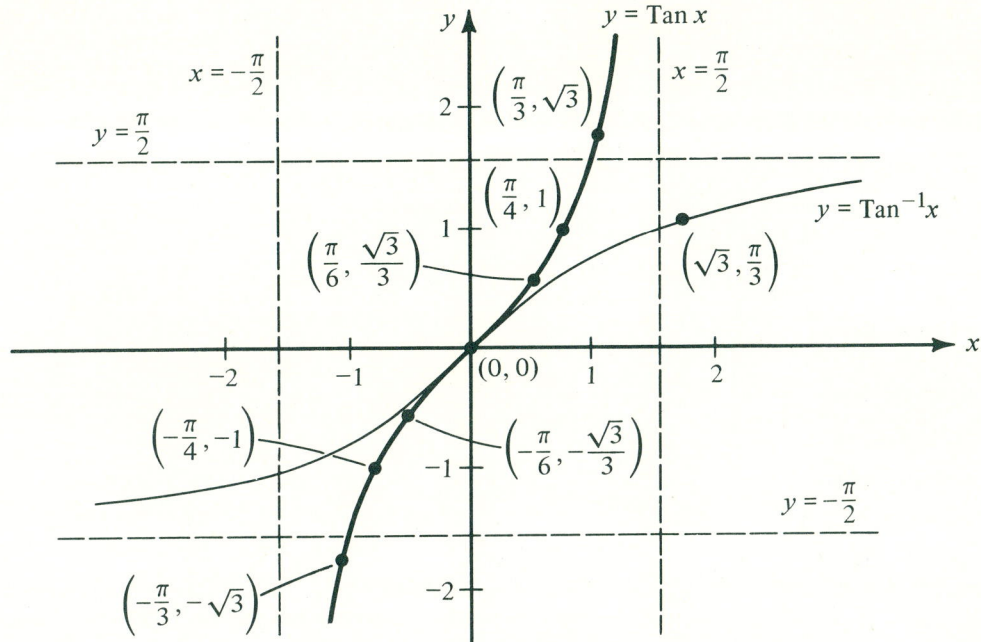

FIGURE 9.41.

is a point on one graph, (b, a) is on the graph of its inverse.

The graph of the Arctangent function has vertical asymptotes at $y = -\pi/2$ and $y = \pi/2$ and an x-intercept at $(0, 0)$.

EXERCISE 9.6

A. FUNDAMENTALS

Plot each of the given sets on a coordinate system. Then find the set of ordered pairs for the inverse function and plot these points. See if you recognize a trig function that would pass through the original set of points.

1. $f = \{(-\pi/2, 1), (-\pi/4, 0.7), (0, 0), (\pi/3, -0.9), (\pi/2, -1)\}$

2. $f = \{(-\pi/4, -1), (-\pi/6, -0.6), (0, 0), (\pi/3, 1.7), (\pi/4, 1)\}$

3. $f = \{(0, 1), (\pi/6, 0.9), (\pi/3, 0.5), (\pi/2, 0), (3\pi/4, -0.7)\}$

B. ESSENTIALS

Find the equations of the inverses of each of the following functions.

4. (a) $f(x) = \text{Sin } x$ (b) $g(x) = \text{Sin } 2x$
   (c) $h(x) = 2 \text{ Sin } x$

5. (a) $f(x) = \text{Cos } x$
   (b) $g(x) = \text{Cos } (x + \pi/4)$
   (c) $h(x) = \text{Cos } (x - \pi/4)$

6. (a) $f(x) = \text{Tan } (x + 3)$
   (b) $g(x) = 2 \text{ Tan } x$
   (c) $h(x) = 2 \text{ Tan } (x + 3)$

7. (a) $f(x) = \text{Sin } 3x$ (b) $g(x) = \text{Sin } (x - 2)$
   (c) $h(x) = \text{Sin } (3x - 6)$

Find the domains of each of the following one-to-one functions.

★8. (a) $f(x) = \text{Sin } x$
   (b) $g(x) = \text{Sin } (x - \pi/4)$
   (c) $h(x) = \text{Sin } (2x + \pi/3)$

CHAPTER 9   The Circular Functions

9. (a) $f(x) = \text{Cos } x$
   (b) $g(x) = \text{Cos } (x + \pi/3)$
   (c) $h(x) = 2 \text{ Cos } (x - \pi/4)$

10. (a) $f(x) = \text{Tan } x$ (b) $g(x) = f(x - \pi/4)$
    (c) $h(x) = f(x + \pi/2)$

Find the value of each of the following.

11. (a) $\text{Tan (Arcsin } \sqrt{2}/2)$
    (b) $\tan (\text{arcsin } \sqrt{2}/2)$
    (c) $\tan (\text{Arcsin } \sqrt{2}/2)$

12. (a) $\text{Tan (Cos}^{-1} \sqrt{2}/2)$
    (b) $\text{Tan (cos}^{-1} \sqrt{2}/2)$
    (c) $\tan (\cos^{-1} \sqrt{2}/2)$

13. (a) $\text{Cos}^{-1} (\text{Sin } \pi/6)$
    (b) $\text{Cos}^{-1} (\sin \pi/6)$
    (c) $\cos^{-1} (\sin \pi/6)$

14. $\text{Sin}^{-1} (\cos \pi/3)$ 15. $\text{Tan}^{-1} (\sin 3\pi/2)$

16. $\sin^{-1} (\cos 11\pi/3)$ 17. $\tan^{-1} (\sin \pi/2)$

18. $\sin^{-1} (\tfrac{1}{2} \tan \pi/4)$ 19. $\tan^{-1} (2 \sin 7\pi/6)$

20. (a) $\text{Sin (Sin}^{-1} \pi/6)$
    (b) $\text{Tan (Tan}^{-1} \pi/3)$
    (c) $\text{Cos}^{-1} (\text{Cos } 7\pi/6)$

Graph each of the following functions and relations and their inverses on the same coordinate system. State the domain and range for both. Find the equation of the inverse.

★21. $f(x) = \text{Sin } 3x$ 22. $y = 4 \text{ Cos } x$

23. $y = 3 \text{ Tan } x/2$ ★24. $y = \cos (x - \pi/4)$

25. $f(x) = 2 \sin (x + \pi/3)$

26. $y = \text{Tan}^{-1} (x/2) + \pi/4$

27. $y = \tfrac{1}{3} \text{Cos } 2(x - \pi/3)$

Solve each of the following equations.

★28. $\tfrac{1}{2} \text{Sin}^{-1} \sqrt{3}/2 = 2x$

29. $\tfrac{1}{4} \text{Cos}^{-1} (-\sqrt{3}/2) = x + 1$

★30. $\tfrac{1}{3} \tan^{-1} (-\sqrt{3}) = x - 1$

31. $\tfrac{1}{4} \sin^{-1} (-\sqrt{3}/2) = 2x$

★32. $\text{Sin}^{-1} x = \text{Tan}^{-1} x$

33. $\text{Cos}^{-1} x = \text{Sin}^{-1} 2x$

34. $\text{Cos}^{-1} x^2 = \pi/2 - \sin^{-1} x$

●35. $\pi/2 - \text{Sin}^{-1} x = \text{Tan}^{-1} 2x$

36. $\pi/2 - \text{Sin}^{-1} x = \text{Cot}^{-1} 2x$

C. APPLICATIONS

37. Find another restricted domain for $y = \cos x$ that will result in a one-to-one function. Then graph this restricted cosine function and its inverse on the same coordinate system.

38. Repeat Problem 37 for $y = \sin x$.

39. Repeat Problem 37 for $y = \tan x$.

ENDNOTES

1. The ancient Babylonians with their base-60 numbering system influenced the definition of the *degree* as 1/360 of a complete circle (revolution or rotation). Each degree contains 60 minutes and each minute contains 60 seconds.

2. The word *radian* (radial angle) was introduced by the physicist James T. Thompson in 1873. This scale for measuring angles was created by Thompson and Thomas Muir, a mathematician, to simplify certain mathematical and physical formulas.

CHAPTER 10

Trigonometry

10.1 INTRODUCTION

Trigonometry is the study of the geometrical properties of triangles and the functions that result from referring the ratios of pairs of sides of a triangle to the measure of its angles. We will begin by using right triangles, and then extend the study to obtuse triangles.

The relationship between the trigonometric functions and the circular functions in Chapter 9 will be apparent, since we will usually orient our triangles in an x,y-coordinate system as shown in Figure 10.1.

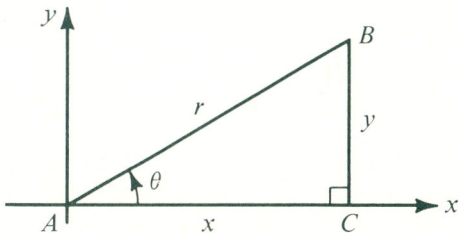

FIGURE 10.1.

The domains of the circular functions were real numbers (radians); the domains of the trig functions are degrees. Therefore, we will be working with degrees, minutes, and seconds in this chapter:

$$60 \text{ seconds} = 60'' = 1 \text{ minute}$$

$$60 \text{ minutes} = 60' = 1 \text{ degree}$$

$$360 \text{ degrees} = 360° = 2\pi \text{ radians}$$

In this chapter we often will refer to angle BAC as angle A instead of angle θ.

10.2 RIGHT TRIANGLE TRIGONOMETRY

First, let's define the six trigonometric functions in terms of the sides of a right triangle.

The placement of triangle ABC in a circle of radius r (Figure 10.2) demonstrates that the following definitions are the same as the previous definitions in Chapter 9.

CHAPTER 10   Trigonometry

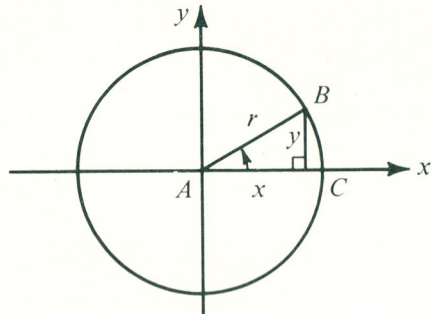

FIGURE 10.2.

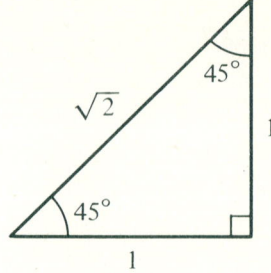

FIGURE 10.3.

DEFINITION 10.1.
In triangle *ABC* in Figure 10.2,

$$\sin A = \frac{\text{opposite side}}{\text{hypotenuse}} = \frac{y}{r}$$

$$\cos A = \frac{\text{adjacent side}}{\text{hypotenuse}} = \frac{x}{r}$$

$$\tan A = \frac{\text{opposite side}}{\text{adjacent side}} = \frac{y}{x}$$

$$\csc A = \frac{\text{hypotenuse}}{\text{opposite side}} = \frac{r}{y}$$

$$\sec A = \frac{\text{hypotenuse}}{\text{adjacent side}} = \frac{r}{x}$$

$$\cot A = \frac{\text{adjacent side}}{\text{opposite side}} = \frac{x}{y}$$

Therefore,

$$\sin 45° = \frac{\text{opposite}}{\text{hypotenuse}} = \frac{1}{\sqrt{2}} = \frac{\sqrt{2}}{2}$$

$$\cos 45° = \frac{\text{adjacent}}{\text{hypotenuse}} = \underline{}$$

and

$$\tan 45° = \frac{\text{opposite}}{\text{adjacent}} = \underline{}$$

If we had chosen *AC* to be of any length other than 1, we would have had a larger or smaller triangle *similar* to the one we used. (The angles would still be 45°, 45°, and 90°.) Since corresponding parts of similar triangles are proportional, the ratios of the sides of the other triangle would be exactly the same. For example, if *AC* = 10, the triangle is shown in Figure 10.4, where

$$\sin A = \frac{10}{\sqrt{200}} = \frac{10}{\sqrt{2}\sqrt{100}} = \frac{1}{\sqrt{2}} = \frac{\sqrt{2}}{2}$$

Now, let's find the values of the trig functions for angles that measure 30°, 45°, and 60°. These angles occur so frequently that it is beneficial to learn them. [See Section 9.3 for angles of measure $\pi/6$ (30°), $\pi/4$, and $\pi/3$.]

For 45°, the right triangle is isosceles, since the sum of the angles of any triangle must equal 180°. If we choose a length of 1 for side *AC*, then *BC* = 1 and, by the Pythagorean theorem, *AB* = $\sqrt{1^2 + 1^2} = \sqrt{2}$. See Figure 10.3.

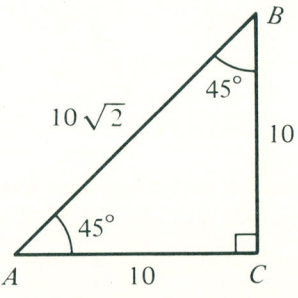

FIGURE 10.4.

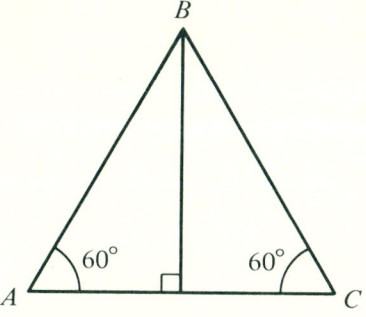

FIGURE 10.5.

For 30° and 60°, we will begin with the equilateral triangle ABD in Figure 10.5. If BC is the perpendicular line from B to AD, then angle $ABC = 30°$. Let's choose the smallest side of triangle ABC to be of unit length. Then,

$$AC = 1$$

Now, since $AB = AD$ and $AC = CD$, we have

$$AB = 2$$

Then, by the Pythagorean theorem,

$$BC = \sqrt{2^2 - 1^2} = \sqrt{3}$$

and we have the measure of all three sides. See Figure 10.6. Therefore,

$$\sin 60° = \frac{\text{opposite}}{\text{hypotenuse}} = \frac{\sqrt{3}}{2}$$

$$\sin 30° = \frac{\text{opposite}}{\text{hypotenuse}} = \frac{1}{2}$$

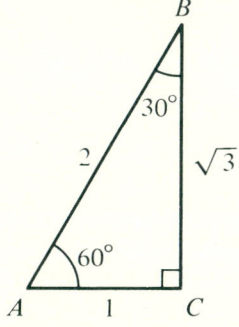

FIGURE 10.6.

Use the two triangles as references to complete the following table.

A	$\sin A$	$\cos A$	$\tan A$	$\csc A$	$\sec A$	$\cot A$
30°		$\sqrt{3}/2$				
45°		$\sqrt{2}/2$	1			
60°	$\sqrt{3}/2$		$\sqrt{3}$			

The last three columns are readily completed by using the following identities:

$$\csc A = \frac{1}{\sin A}$$

$$\sec A = \frac{1}{\csc A}$$

$$\cot A = \frac{1}{\tan A}$$

From these three equations, we see that the cosecant, secant, and cotangent are simply the reciprocals of the sine, cosine, and tangent, respectively.

For the values of the trig functions at 0°, 90°, 180°, and 270°, consider the point 1 unit from the origin on the terminal side of each of these angles (Figure 10.7). Then,

$$\sin 0° = \frac{0}{1} = 0 \qquad \csc 0° = \frac{1}{0} \text{ is undefined}$$

$$\cos 0° = \frac{1}{1} = 1 \qquad \sec 0° = \frac{1}{1} = 1$$

$$\tan 0° = \frac{0}{1} = 0 \qquad \cot 0° = \frac{1}{0} \text{ is undefined}$$

Since

$$\csc A = \frac{1}{\sin A}$$

and as A approaches 0, $\sin A$ approaches 0, as

$$A \to 0, \qquad \csc A \to \infty$$

which we abbreviate as

$$\csc 0° = \infty$$

CHAPTER 10   Trigonometry

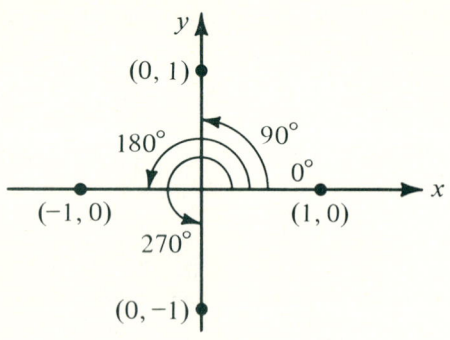

FIGURE 10.7.

$\sin 135° = \sin A = \dfrac{\sqrt{2}}{2}$

$\cos 135° = \cos A = -\dfrac{1}{\sqrt{2}} = -\dfrac{\sqrt{2}}{2}$

$\sin 210° = \sin A = -\tfrac{1}{2}$

$\cos 210° = -\dfrac{\sqrt{3}}{2}$

$\sin 300° = -\dfrac{\sqrt{3}}{2}$

$\cos 300° = \tfrac{1}{2}$

Complete the following table of values of the trig functions for angles that are multiples of 90°.

A	$\sin A$	$\cos A$	$\tan A$	$\csc A$	$\sec A$	$\cot A$
0°						
90°			∞	1	∞	
180°		−1				∞
270°	−1				∞	0

For angles greater than 90°, to find the reference triangle we draw a perpendicular line to the x-axis, as illustrated in Figure 10.8.

EXAMPLE I. Find the sine and cosine of the angles in Figure 10.8.

This process is formalized as the reduction formulas in Section 10.3.

EXERCISE 10.2

A. FUNDAMENTALS
Find the missing parts of the following right triangles.
1.

(a)

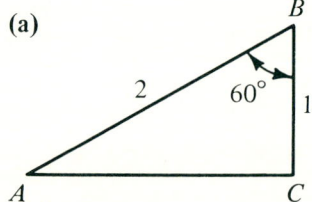

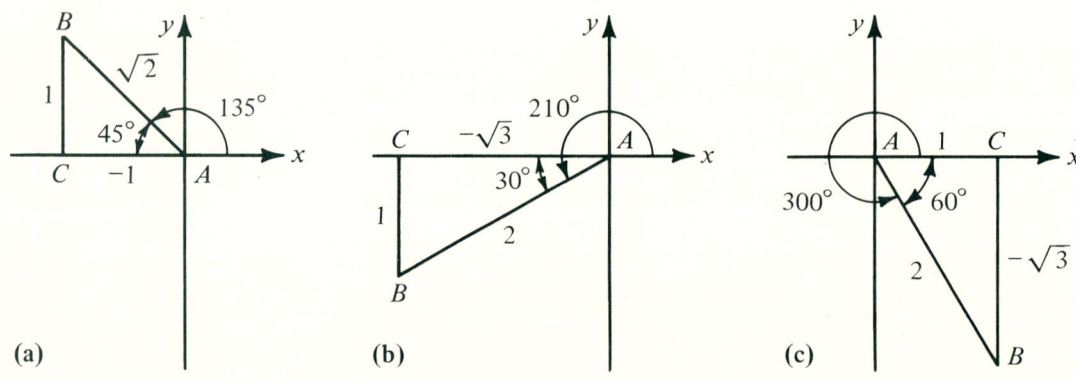

FIGURE 10.8.

SECTION 10.2 Right Triangle Trigonometry

(b)

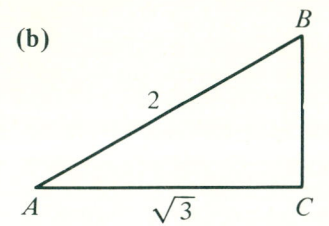

2.

(a)

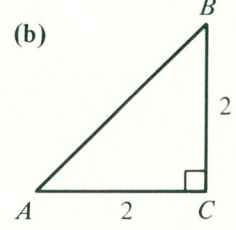
∠A = ∠B

(b)

Triangle with B at top right, right angle at C, AC = 2, BC = 2, vertices A, C at bottom.

3.

(a) Triangle with 30° at A, right angle at C, BC = 1.

(b) Triangle with 30° at A, right angle at C, BC = 4.

4.

(a)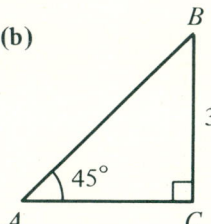
Triangle with 45° at A, right angle at C, BC = 1.

(b) Triangle with 45° at A, right angle at C, BC = 3.

Find the indicated trigonometric functions. Draw and label the reference triangle for each angle.

5. (a) $\sin 45°$ (b) $\cos 45°$ (c) $\tan 45°$
6. (a) $\sin 30°$ (b) $\sin 150°$ (c) $\sin 210°$

7. (a) $\cos 60°$ (b) $\cos 240°$
   (c) $\cos 300°$
8. (a) $\tan 45°$ (b) $\tan 225°$
   (c) $\tan(-45°)$

Evaluate the following for the given angles.

9. $A = 45°$ (a) $\sin A$ (b) $\csc A$
10. $A = 60°$ (a) $\cos A$ (b) $\sec A$
11. $A = 30°$ (a) $\tan A$ (b) $\cot A$
12. $A = 90°$
    (a) $\sin A$ (b) $\cos(90° - A)$
13. $A = 60°$
    (a) $\cos A$ (b) $\sin(90° - A)$
14. $A = 60°$
    (a) $\tan A$ (b) $\cot(90° - A)$
15. $A = 30°$
    (a) $(\sin A)^2$ (b) $(\cos A)^2$
16. $A = 45°$
    (a) $\sec^2 A$ (b) $1 + \tan^2 A$

B. ESSENTIALS

17. $A = 60°$ (a) $\cos \dfrac{A}{2}$ (b) $\dfrac{\cos A}{2}$
18. $A = 30°$ (a) $\sin 2A$ (b) $2 \sin A$
★19. $A = 45°$
    (a) $\sin 2A$ (b) $2 \sin A \cos A$
20. $A = 60°$
    (a) $\cos 2A$ (b) $\cos^2 A - \sin^2 A$
21. $A = 120°$
    (a) $\cos A$ (b) $1 - 2 \sin^2 \left(\dfrac{A}{2}\right)$
22. $A = 30°, B = 60°$
    (a) $\cos(B - A)$ (b) $\cos B \cos A + \sin B \sin A$
23. $A = 30°, B = 120°$
    (a) $\sin(A + B)$ (b) $\sin A \cos B + \cos A \sin B$
24. $A = 90°, B = 180°$
    (a) $\cos(A + B)$ (b) $\cos B \cos A - \sin B \sin A$
25. Find the values for the five other trig functions.
    (a) $\sin \theta = \frac{4}{5}$ (b) $\cos \theta = -\frac{4}{5}$
    $\cos \theta > 0$ $\sin \theta > 0$

221

(c) $\tan\theta = -\frac{5}{12}$
$\sin\theta < 0$

26. Find the values for the five other trig functions.

(a) $\sin\theta = 0.3$ (b) $\sin\theta = \frac{b}{5}$ (c) $\tan\theta = \frac{x}{h}$
$\tan\theta < 0$ $\cos\theta < 0$ $\sin\theta > 0$

10.3 THE FUNDAMENTAL IDENTITIES AND THE REDUCTION FORMULAS

Use the triangle in Figure 10.9 to complete the following. These equations hold for all values of A in the domain of the trig functions involved. These equations are called the *fundamental identities*.

1. $\sin^2 A + \cos^2 A = \left(\frac{y}{r}\right)^2 + (\underline{\ ?\ })^2$

$= \frac{y^2}{r^2} + \underline{\ ?\ } = \frac{?+?}{r^2} = \underline{\ ?\ }$

2. $\sec^2 A = (\underline{\ ?\ })^2 = \underline{\ ?\ }$

and

$1 + \tan^2 A = 1 + (\underline{\ ?\ })^2 = \frac{?}{x^2} = \underline{\ ?\ }$

3. $\csc^2 A = (\underline{\ ?\ })^2 = \underline{\ ?\ }$

and

$1 + \cot^2 A = 1 + (\underline{\ ?\ })^2 = \frac{?}{y^2} = \underline{\ ?\ }$

4. $\frac{\sin A}{\cos A} = \frac{?}{?} = \underline{\ ?\ }$

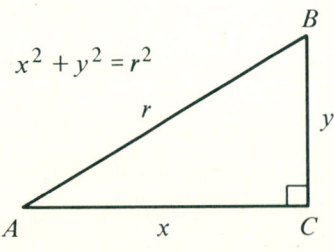

FIGURE 10.9.

and

$\tan A = \underline{\ ?\ }$

These three identities and the four reciprocal identities are summarized in the following.

THE BASIC TRIGONOMETRIC IDENTITIES.

$\sin^2 A + \cos^2 A = 1$

$\sec^2 A = 1 + \tan^2 A$

$\csc^2 A = 1 + \cot^2 A$

$\csc A = \dfrac{1}{\sin A}$

$\sec A = \dfrac{1}{\cos A}$

$\cot A = \dfrac{1}{\tan A}$

$\tan A = \dfrac{\sin A}{\cos A}$

In the last four identities, the denominator cannot equal zero.

Now, let's find out why the six trig functions are named in pairs—why three of them are "cofunctions" of the other three. Consider the following examples.

$\sin 60° = \underline{\ ?\ }$ and $\cos 30° = \underline{\ ?\ }$

$\sin 45° = \underline{\ ?\ }$ and $\cos 45° = \underline{\ ?\ }$

$\tan 60° = \underline{\ ?\ }$ and $\cot 30° = \underline{\ ?\ }$

$\sec 60° = \underline{\ ?\ }$ and $\csc \underline{\ ?\ } = 2$

Did you notice that the sums of the angles in each example are the same? These examples lead us to hypothesize that the value of a trig function at A always equals its cofunction evaluated at the complement of A, $90° - A$. We must now prove this statement, which is summarized in Theorem I.

SECTION 10.3 The Fundamental Identities and the Reduction Formulas

THEOREM I.
In a right triangle ABC, if

$$A + B = 90°$$

then

$$\sin A = \cos B$$
$$\tan A = \cot B$$

and

$$\sec A = \csc B$$

Proof: We will prove this theorem for sine and cosine and consider the other two proofs in the exercises. See Figure 10.10. Since $A + B + C = 180°$ and $C = 90°$, we have

$$A + B = 90°$$

Also, from Figure 10.10, we see that

$$\sin A = \frac{\text{opposite}}{\text{hypotenuse}} = \frac{y}{r}$$

and

$$\cos B = \frac{\text{adjacent}}{\text{hypotenuse}} = \frac{y}{r}$$

Therefore,

$$\sin A = \cos B$$

There are several identities that allow us to change trig functions such as $\sin 120°$, $\cos 330°$, and $\tan (-45°)$ to functions of positive acute angles:

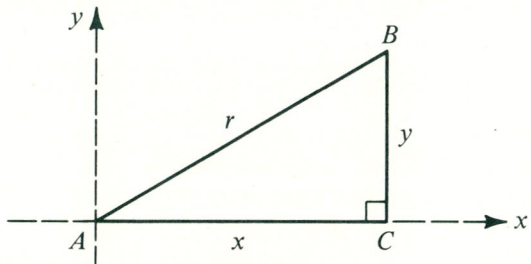

FIGURE 10.10.

$\sin 60°$, $\cos 30°$, and $-\tan 45°$, respectively. These identities, called *reduction formulas*, are summarized in the following theorems. In order to understand these very simple relationships you must draw the reference triangle for each angle mentioned. Use $A = 60°$ to illustrate each theorem.

THEOREM II.
For any angle A,

$$\sin (-A) = -\sin A$$
$$\cos (-A) = \cos A$$
$$\tan (-A) = -\tan A$$

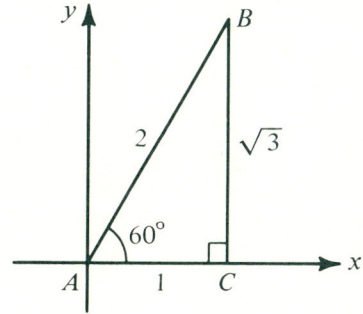

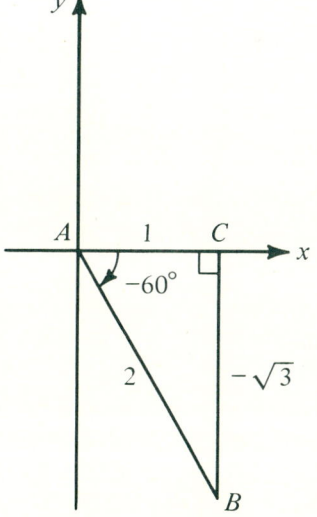

FIGURE T-II

223

CHAPTER 10   Trigonometry

$A = 60° \Rightarrow \sin A = $ __?__ , $\sin(-A) = $ __?__

$\cos A = $ __?__ , $\cos(-A) = $ __?__

$\tan A = $ __?__ , $\tan(-A) = $ __?__

> **Theorem III.**
> For any angle A,
> $$\sin(180° - A) = \sin A$$
> $$\cos(180° - A) = -\cos A$$
> $$\tan(180° - A) = -\tan A$$

$A = 60° \Rightarrow$

$\sin A = $ __?__ , $\sin(180° - A) = $ __?__

$\cos A = $ __?__ , $\cos(180° - A) = $ __?__

$\tan A = $ __?__ , $\tan(180° - A) = $ __?__

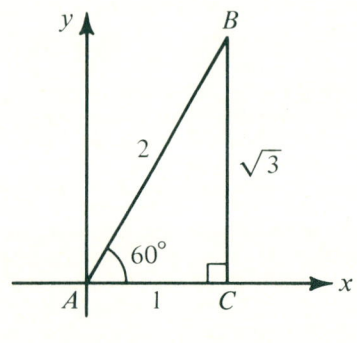

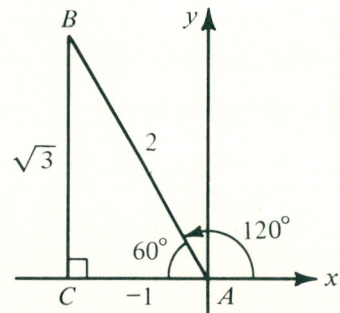

FIGURE T-III

> **Theorem IV.**
> For any angle A,
> $$\sin(180° + A) = -\sin A$$
> $$\cos(180° + A) = -\cos A$$
> $$\tan(180° + A) = \tan A$$

$A = 60° \Rightarrow \sin(180° + A) = $ __?__

$\cos(180° + A) = $ __?__

$\tan(180° + A) = $ __?__

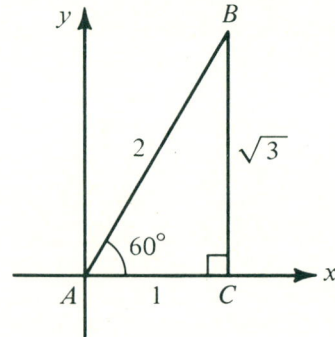

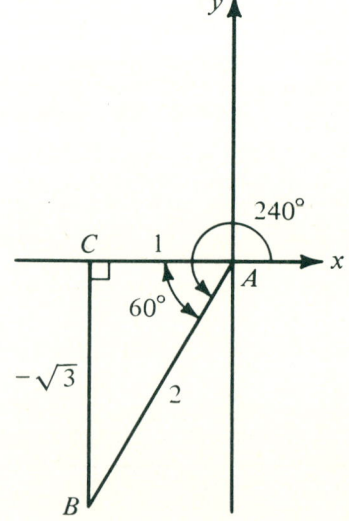

FIGURE T-IV

> **THEOREM V.**
> For any angle A,
> $$\sin(360° - A) = \sin(-A) = -\sin A$$
> $$\cos(360° - A) = \cos(-A) = \cos A$$
> $$\tan(360° - A) = \tan(-A) = -\tan A$$

If angle A is an acute angle, Theorems II and V both relate the functions of angles found in the fourth quadrant to the functional values of the first-quadrant angle A. Then, since the sine and the tangent are negative in the fourth quadrant and positive in the first quadrant, we have $-\sin A$ and $-\tan A$. Since the cosine is positive in both quadrants, there is no negative sign in front of $\cos A$.

EXERCISE 10.3

A. FUNDAMENTALS

Apply the reduction formulas to find the values of the following.

1. (a) $\sin 210° = \sin(180° + 30°) = -\sin 30°$ = ___?___
   (b) $\cos 210° = \cos(180° + 30°) =$ ___?___
   (c) $\tan 210° = \tan 30° =$ ___?___
2. (a) $\sin 150° = \sin(180° - 30°) = \sin 30°$ = ___?___
   (b) $\cos 150° =$ ___?___
   (c) $\tan 150° =$ ___?___
3. (a) $\sin(-45°)$ (b) $\cos(-30°)$
   (c) $\tan(-60°)$
4. (a) $\sin 330° = \sin(360° - 30°)$
   (b) $\cos 315°$ (c) $\tan 300°$

B. ESSENTIALS

Illustrate the basic identities by evaluating them at the values of A as indicated. Remember that identities are valid for all values of A in their domain.

5. $\sin^2 A + \cos^2 A = 1$ for
   (a) $A = 30°$ (b) $A = 60°$
6. $1 + \tan^2 A = \sec^2 A$ for
   (a) $A = 45°$ (b) $A = 30°$ (c) $A = 120°$
   (d) $A = 330°$
★7. $1 + \cot^2 A = \csc^2 A$ for
   (a) $A = 150°$ (b) $A = -30°$ (c) $A = 210°$
   (d) $A = 240°$
8. $\sec^2 A = 1 + \tan^2 A$ for
   (a) $A = 90°$ (b) $A = 270°$ (c) $A = 180°$
   (d) $A = 360°$

Check these equations for the values indicated.

★9. $\sin A \cdot \cot A = \cos A$ at $A = 45°$ and $A = 150°$.

10. $\tan^2 A + \sec^2 A = 1$ at $A = 0$, $A = 180°$, and $A = 45°$.

11. $\sin A \cdot \tan A + \cos A = \sec A$ for $A = 225°$ and $A = -30°$.

12. $\dfrac{\tan A \cdot \cot A}{\sin A} = \csc A$ for $A = 0$, $A = 30°$, and $A = 60°$.

13. $\tan A \cdot \sin A = \cos A$ for $A = 45°$ and $A = 150°$.

C. APPLICATIONS

14. Prove that for all angles A,
$$\tan(90° - A) = \cot A$$

15. Prove that for all angles A,
$$\sec(90° - A) = \csc A$$

16. Prove Theorem III.

17. Prove Theorem IV.

18. Use the results of Theorem III to prove whether the sine, cosine, and tangent functions are even, odd, or neither even nor odd functions.

10.4 BASIC EQUATIONS AND IDENTITIES

The fundamental identities introduced earlier will be helpful in solving equations involving trig functions as well as in deriving and proving more trig identities. To understand these applications, consider Examples I and II.

EXAMPLE I. Solve the following equation for all values of A, where $0 \leq A < 360°$:

$$3 \cos A = \cot A \sin A$$

Solution: First, let's use

$$\cot A = \frac{\cos A}{\sin A}$$

to express the entire equation in terms of sine and cosine. Then,

$$3 \cos A = \frac{\cos A}{\sin A} \cdot \sin A \Rightarrow$$

$$3 \cos A = \cos A$$

Now that only one trig function remains, we simply solve it by subtracting $\cos A$ from both sides. (All of the rules of algebra apply to these equations.) Then,

$$2 \cos A = 0 \Rightarrow$$

$$\cos A = 0 \Rightarrow$$

$$A = \frac{\pi}{2} \quad \text{or} \quad \frac{3\pi}{2}$$

EXAMPLE II. Solve the following equation for all values of A, where $0 \leq A < 360°$:

$$\sin A = \sqrt{3} \cos A$$

Solution: Divide both sides by $\cos A$ $\left[A \neq \frac{\pi}{2} + K\pi, \text{ since } \cos\left(\frac{\pi}{2} + K\pi\right) = 0 \right]$. Then,

$$\sin A = \sqrt{3} \cos A \Rightarrow$$

$$\frac{\sin A}{\cos A} = \sqrt{3} \Rightarrow$$

$$\tan A = \sqrt{3}$$

Then, from the 30°-60°-90° triangle in Figure 10.11, we see that

$$\tan 60° = \sqrt{3}$$

Since the tangent is positive in the first and third quadrants, we have

$$A = 60° \quad \text{or} \quad A = 240°$$

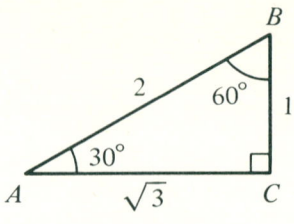

FIGURE 10.11

EXAMPLE III. Prove the following identity:

$$\sin^2 A + \sec^2 A - \tan^2 A \cos^2 A = 1 + \tan^2 A$$

Solution: We are going to use the fundamental identities

$$\sec^2 A = 1 + \tan^2 A \quad \text{and} \quad \tan A = \frac{\sin A}{\cos A}$$

where $A \neq 90° + K\pi$ and $K \in I$. Thus we have

$$\sin^2 A + \sec^2 A - \tan^2 A \cos^2 A$$

$$= \sin^2 A + \sec^2 A - \frac{\sin^2 A}{\cos^2 A} \cdot \cos^2 A$$

$$= \sin^2 A + \sec^2 A - \sin^2 A = \sec^2 A$$

$$= 1 + \tan^2 A$$

Then, according to the transitive law of equality, the first statement in our proof equals our last statement. Therefore,

$$\sin^2 A + \sec^2 A - \tan^2 A \cos^2 A = 1 + \tan^2 A$$

EXERCISE 10.4

A. FUNDAMENTALS
State the values of A between 0 and 360° that satisfy the following equations.

1. (a) $\tan A = -\sqrt{3}$ (b) $\tan A = \sqrt{3}$
2. (a) $\csc A = 2$ (b) $\sec A = 2$
3. (a) $\cot A = 1$ (b) $\cot A = -1$
4. (a) $\tan A = 1$ (b) $\sin A = \cos A$
5. (a) $\tan A = \cot A$ (b) $\sec A = \csc A$

6. (a) $\sin A \cdot \cos A = 0$
   (b) $\sin A(1 - \cos A) = 0$

B. ESSENTIALS

Solve the following equations for all values of A between 0 and 2π. Show your work.

★7. $\sin^2 A - \sin A = 0$

8. $\sin A \cdot \tan A = \sec A - \cos A$

9. $\cos A - \cos^2 A = 0$

★10. $\sec^2 A - \tan A = 1$

11. $\tan A - \sin A = 0$

12. $\csc^2 A - \cot A = 1$

13. $\cot A \cdot \sin A = \cos^2 A$

14. $\sin^2 A = \cos A + 1$

Prove that the following equations are identities by using the fundamental identities to transform the left side of the equation into the right. Note any restrictions on the variables.

★15. (a) $\sin A \cdot \tan A + \cos A = \sec A$
    (b) $\dfrac{\sec B \cdot \cot B \cdot \sin B}{\cos B \cdot \csc B} = \tan B$

16. (a) $\sin A \cdot \cot A = \cos A$
    (b) $\cos^2 A \cdot \tan^2 A = 1 - \cos^2 A$

17. $\dfrac{\tan B \cdot \cot B}{\sin B} = \csc B$

★18. $\dfrac{1 - \cos B}{\sin B} = \dfrac{\sin B}{1 + \cos B}$

19. $1 - \cos^2 A = (\tan A \cdot \cos A)^2$

20. (a) $\dfrac{1 + \sin A}{\cos A} = \dfrac{\cos A}{1 - \sin A}$
    (b) $\dfrac{\tan B}{\sec B - 1} = \dfrac{\sec B + 1}{\tan B}$

21. $\dfrac{\cos^2 A}{\sin A(1 + \csc A)} = 1 - \sin A$

10.5 HISTORICAL COMMENTS ON TRIGONOMETRY

Plane trigonometry (triangle measurement) requiring only the Pythagorean theorem and the concept of a right angle was probably known by the ancient Babylonians. The ancient Egyptians used trigonometry to survey after the yearly floods removed their property boundary markers. This trigonometry must have been very crude, for these people had no concept of angle measurement.

Many of the trig identities and laws are present in Euclid's work, but another Greek, Hipparchus (the "father of trigonometry"), is credited with compiling the first rough table of values of what we now call the sine.

The sine function was actually first introduced by the Siddhantas about 500 A.D.

The Arabs introduced unit circle trigonometry about 900 A.D. They were the first to use all six trig functions.

CHAPTER 11

Trigonometric Equations and Identities

11.1 TRIGONOMETRIC EQUATIONS

We have solved equations involving trigonometric functions before, but they were all carefully chosen. In this section, we will solve a wide variety of such "trigonometric equations." We will still have one major restriction on all of the equations in this section. In each equation, all the trig functions will have the same argument (all functions of x, or $2x$, etc., but not mixed). This final restriction will be removed in subsequent sections.

The periodicity of the trig functions requires that each equation that has a solution have an infinite number of solutions. We will refer to the solutions between 0 and 2π as the *particular solutions* and the expression that yields all possible solutions as the *general solutions*. This is illustrated in Example I.

EXAMPLE I. Find the particular and general solutions for

$$2 \sin x = \sqrt{2}$$

Solution:

$2 \sin x = \sqrt{2} \Rightarrow \sin x = \sqrt{2}/2 \Rightarrow x = \sin^{-1}(\sqrt{2}/2)$

The particular solutions ($0 \leq x \leq 2\pi$) are

$$x = \pi/4 \quad \text{and} \quad x = 3\pi/4$$

A partial list of the solutions,

$$-7\pi/4, -5\pi/4, \pi/4, 3\pi/4, 9\pi/4, 11\pi/4, 17\pi/4, 19\pi/4$$

can be expressed as the general solutions

$$x = \pi/4 + 2K\pi \quad \text{and} \quad x = 3\pi/4 + 2K\pi$$

where $K \in$ integers.

One way to learn the specific skills required to solve trig equations is by studying some typical examples. Remember that all the rules of algebra still apply.

EXAMPLE II. Find the general solution for

$$\sqrt{3} - \tan x = 0$$

Solution: First, let's solve for tan x, then for x.

$\sqrt{3} - \tan x = 0 \Rightarrow \tan x = \sqrt{3} \Rightarrow x = \tan^{-1} \sqrt{3}$

The particular solutions are

$$x = \pi/3 \quad \text{and} \quad x = 4\pi/3$$

which, when we add multiples of 2π, generate the list

$$\ldots, -2\pi/3, \pi/3, 4\pi/3, 7\pi/3, 10\pi/3, 13\pi/3, 16\pi/3, \ldots$$

This yields the general solution

$$x = \pi/3 + K\pi$$

where $K \in$ integers.

EXAMPLE III. Find the general solution for

$$2 \sin t - 1 = 0$$

Solution: First, we must solve for sin t, then for t.

$2 \sin t - 1 = 0 \Rightarrow 2 \sin t = 1 \Rightarrow$

$$\sin t = \tfrac{1}{2} \Rightarrow t = \sin^{-1} \tfrac{1}{2}$$

This yields the particular solutions

$$t = \pi/6 \quad \text{and} \quad t = 5\pi/6$$

and the general solutions

$$t = \pi/6 + 2K\pi \quad \text{and} \quad t = 5\pi/6 + 2K\pi$$

where $K \in$ integers.

EXAMPLE IV. Find the particular solutions for

$$2 \sin^2 x + \sin x - 1 = 0$$

Solution: Since this equation is a polynomial in sin x, let's substitute

$$u = \sin x$$

and solve for u. Then, we have

$2u^2 + u - 1 = 0 \Rightarrow$

$(2u - 1)(u + 1) = 0 \Rightarrow$

$2u - 1 = 0 \quad \text{or} \quad u + 1 = 0 \Rightarrow$

$u = \tfrac{1}{2} \quad \text{or} \quad u = -1$

Therefore,

$\sin x = u \Rightarrow$

$\sin x = \tfrac{1}{2} \quad \text{or} \quad \sin x = -1 \Rightarrow$

$x = \sin^{-1}(\tfrac{1}{2}) \quad \text{or} \quad x = \sin^{-1}(-1) \Rightarrow$

$x = \pi/6 \text{ or } 5\pi/6 \quad \text{or} \quad x = 3\pi/2$

To complete the solution of Example V, first use some fundamental identities to change all the trig functions to equivalent expressions that involve only one trig function, cos t. Then substitute

$$u = \cos t$$

and solve for u. Then, to find t we use

$$t = \cos^{-1} u$$

EXAMPLE V. Find the particular solutions for

$$2 \sin^2 t - \cos t = 2$$

Solution: First, if necessary, use $\sin^2 t + \cos^2 t = 1$ to eliminate $\sin^2 t$.

$2(\sin^2 t) - \cos t = 2 \Rightarrow$

$2(\underline{\quad ? \quad}) - \cos t = 2$

Then, let $u = \cos t$, yielding

$2(1 - u^2) - \underline{\quad ? \quad} = 2 \Rightarrow$

$2u^2 - u = 0 \Rightarrow$

$?(\underline{\quad ? \quad}) = 0 \Rightarrow$

$u = 0 \quad \text{or} \quad 2u - 1 = 0 \Rightarrow$

$u = 0 \quad \text{or} \quad u = \tfrac{1}{2} \Rightarrow$

$t = \cos^{-1} 0 \quad \text{or} \quad t = \cos^{-1}(\tfrac{1}{2}) \Rightarrow$

$t = \underline{\quad ? \quad} \quad \text{or} \quad t = \underline{\quad ? \quad} \text{ or } \underline{\quad ? \quad}$

since $\cos \pi/2 = 0$, $\cos \pi/3 = \tfrac{1}{2}$, and $\cos 5\pi/3 = \tfrac{1}{2}$.

Here are some general guidelines for solving trigonometric equations.

1. Use the fundamental identities to transform the equation into an equivalent equation involving only one trig function.

CHAPTER 11   Trigonometric Equations and Identities

2. Substitute a single letter (for example, u) for the remaining trig function.
3. Use the basic techniques of algebra (Chapter 1) to solve for u.
4. Use your knowledge of the values of the trig functions to find the particular solutions. Then, if necessary, specify the general solutions.

EXERCISE 11.1

A. FUNDAMENTALS

Find the solutions for each of the following equations and the particular solutions ($0 \leq x < 2\pi$ and $0 \leq t < 2\pi$) for the corresponding trig equations.

1. (a) $2u - 1 = 0$ (b) $2 \sin t - 1 = 0$
   (c) $2 \cos x - 1 = 0$

2. (a) $u(u - 1) = 0$ (b) $\sin t(\sin t - 1) = 0$
   (c) $\tan t(\tan t - 1) = 0$

3. (a) $(u - 1)(2u + 1) = 0$
   (b) $(\sin x - 1)(2 \sin x + 1) = 0$
   (c) $(\cos x - 1)(2 \cos x + 1) = 0$

4. (a) $u^2 - 3 = 0$ (b) $\tan^2 x - 3 = 0$
   (c) $\cot^2 x - 3 = 0$

5. (a) $2u^2 - 1 = 0$ (b) $2 \sin^2 t - 1 = 0$
   (c) $2 \cos^2 t - 1 = 0$

6. (a) $u^2 - 1 = 0$ (b) $\sin^2 t - 1 = 0$
   (c) $\tan^2 t - 1 = 0$

7. Write the general solutions for 5(b) and 5(c).
8. Write the general solutions for 6(b) and 6(c).

B. ESSENTIALS

Find all real numbers that satisfy the following equations.

9. (a) $2u^2 + u - 1 = 0$
   (b) $2 \sin^2 t + \sin t - 1 = 0$
   (c) $2 \sin^2 t - \cos t - 1 = 0$

10. (a) $2u^2 + 2u - 4 = 0$
    (b) $2 \cos^2 x + 2 \cos x - 4 = 0$
    (c) $2 \cos^2 x - 2 \sin x + 2 = 0$

11. (a) $6 \sec^2 x = 9 \sec x - 3$
    (b) $6 \cot^2 x = 9 \csc x - 9$

★12. $\tan^3 x - \tan^2 x = 3 \tan x - 3$

13. $\cot^3 x - \cot^2 x = 3 \cot x - 3$

★14. $16 \cos^4 x = 11 - 8 \sin^2 x$

15. $2 \cos^4 x = 2 - 3 \sin^2 x$

16. $2 \sin^4 t + \sin^3 t = 3 \sin^2 t + \sin t - 1$

17. (a) $\sqrt{u + 3} = 2u$
    (b) $\sqrt{\tan x + 3} = 2 \tan x$
    (c) $\sqrt{\sin x - 1} = 2 \cos x$

18. (a) $\tan x \sin x = 3/2$
    (b) $3 \sec x \tan x = 2$
    (c) $\dfrac{2 \sin^2 x + 3 \cos x}{\cos^2 x} = 0$

★19. $2 \sin x \cos x - \cos x = 0$

20. $4 \tan x \sin^2 x = 3 \tan x$

21. $2 \sin x \tan x - 2 \sin x + \tan x - 1 = 0$

22. $3 \sec^2 x \cot^2 x - 4 \cot^2 x = 6 \sec^2 x - 8$

C. APPLICATIONS

23. Graph $y = \sin x$ and $y = \cos x$ on the same coordinate system. Then, interpret the following general solution geometrically:

$$\sin x = \cos x$$

24. Graph $y = \sin 2x$ and $y = \cos x$ and then approximate the general solutions of the following equation:

$$\sin 2x - \cos x = 0$$

25. Illustrate geometrically the general solutions to the following equations. [HINT: Use $y = \sin^2 x$ and $y = 1$ for (a).]
    (a) $\sin^2 x - 1 = 0$ (b) $\tan^2 x - 1 = 0$

• 26. Use the quadratic formula and a calculator to find the particular solutions of the following.
    (a) $\sin^2 x - \sin x - 1 = 0$ (b) $\cos^2 x = \sin x$
    (c) $\sec x - \tan^2 x + 3 = 0$
    (d) $\csc x \cdot \cot x = \sec x + \cos x$

11.2 THE ADDITION FORMULAS

This section and the next will complete our study of the trigonometric identities. The identities in Section 11.3 are special cases of the addition formulas we will derive in this section.

When dealing with the cosine of the sum of two angles, it is very easy to proceed incorrectly. For example, consider

$$\cos(\alpha + \beta), \quad \text{where } \alpha = \pi/3 \text{ and } \beta = \pi/6$$

We know that

$$\cos(\pi/3 + \pi/6) = \cos(2\pi/6 + \pi/6)$$
$$= \cos(3\pi/6) = \cos \pi/2 = 0$$

But, what about $\cos \alpha + \cos \beta$?

$$\cos \alpha + \cos \beta = \cos \pi/3 + \cos \pi/6$$
$$= \frac{1}{2} + \frac{\sqrt{3}}{2} = \frac{1 + \sqrt{3}}{2} \neq 0$$

This proves that, in general,

$$\cos(\alpha + \beta) \neq \cos \alpha + \cos \beta$$

Now, let's determine how $\cos(\alpha + \beta)$ is related to the values of functions of α and β, $\cos \alpha$, $\sin \alpha$, $\cos \beta$, and $\sin \beta$.

Consider Figure 11.1, which depicts the terminal points of α, β, and $\alpha - \beta$ as P_1, P_2, and P_3, respectively.

$$x_1 = \cos \alpha$$
$$y_1 = \sin \alpha$$
$$x_2 = \cos \beta$$
$$y_2 = \sin \beta$$
$$x_3 = \cos(\alpha - \beta)$$
$$y_3 = \sin(\alpha - \beta)$$

The chords $P_1 P_2$ and $P_0 P_3$ are equal, since they

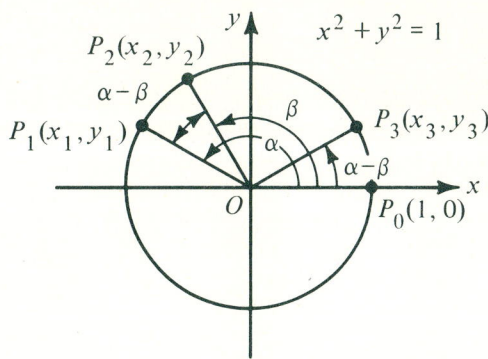

FIGURE 11.1.

cut off equal arcs of the circle. Then, by the distance formula, we have

$$P_1 P_2 = P_0 P_3 \Rightarrow$$
$$\sqrt{(x_1 - x_2)^2 + (y_2 - y_1)^2}$$
$$= \sqrt{(x_3 - 1)^2 + (y_3 - 0)^2}$$

which, by squaring both sides and collecting like terms, becomes

$$x_1{}^2 - 2x_1 x_2 + x_2{}^2 + y_2{}^2 - 2y_1 y_2 + y_1{}^2$$
$$= x_3{}^2 - 2x_3 + 1 + y_3{}^2 \Rightarrow$$
$$(x_1{}^2 + y_1{}^2) + (x_2{}^2 + y_2{}^2) - 2x_1 x_2$$
$$- 2y_1 y_2 - 1 - (x_3{}^2 + y_3{}^2) = -2x_3 \Rightarrow$$
$$1 + 1 - 2x_1 x_2 - 2y_1 y_2 - 1 - 1 = -2x_3 \Rightarrow$$
$$2x_3 = 2x_1 x_2 + 2y_1 y_2 \Rightarrow$$
$$x_3 = x_1 x_2 + y_1 y_2 \Rightarrow$$
$$\cos(\alpha - \beta) = \cos \alpha \cos \beta + \sin \alpha \sin \beta$$

Now let's test this formula (identity) for the values $\alpha = \pi/2$ and $\beta = \pi/3$. We know that

$$\cos(\alpha - \beta) = \cos(\pi/2 - \pi/3) = \cos \pi/6 = \sqrt{3}/2$$

Now,

$$\cos \alpha \cos \beta + \sin \alpha \sin \beta$$
$$= \cos \pi/2 \cos \pi/3 + \sin \pi/2 \sin \pi/3$$
$$= \underline{} \cdot \underline{} + \underline{} \cdot \underline{} = \underline{}$$

CHAPTER 11   Trigonometric Equations and Identities

EXAMPLE 1. Find the exact value of $\cos \pi/12$.

Solution: Since $\pi/12 = \pi/4 - \pi/6$, we can use the above formula for the cosine of the difference of two angles to find $\cos \pi/12$. We let $\alpha = \pi/4$ and $\beta = \pi/6$. Then,

$$\cos \pi/12 = \cos (\pi/4 - \pi/6)$$
$$= \cos \underline{\ ?\ } \cos \underline{\ ?\ } + \sin \underline{\ ?\ } \sin \underline{\ ?\ }$$
$$= \underline{\ ?\ } \cdot \underline{\ ?\ } + \underline{\ ?\ } \cdot \underline{\ ?\ }$$
$$= \underline{\ ?\ }$$

Your answer should be
$$\cos \pi/12 = \tfrac{1}{4}(\sqrt{6} + \sqrt{2}) \approx 0.9659$$

Now we can derive the other addition formulas from this one. We will use the identities

$$\cos(-\theta) = \cos \theta \quad \text{and} \quad \sin(-\theta) = -\sin \theta$$

in the next derivation.

$$\cos(\alpha + \beta) = \cos[\alpha - (-\beta)]$$
$$= \cos \alpha \cos(-\beta) + \sin \alpha \sin(-\beta)$$
$$= \cos \alpha \cdot \underline{\ ?\ } + \sin \alpha \cdot \underline{\ ?\ }$$

Therefore,

$$\cos(\alpha + \beta) = \cos \alpha \cos \beta - \sin \alpha \sin \beta \quad (1)$$

Now, for the sine function we will need the following result:

$$\sin \theta = \cos(\pi/2 - \theta) \Rightarrow$$
$$\sin(-\theta) = \cos(\pi/2 + \theta) \Leftrightarrow$$
$$-\sin \theta = \cos(\pi/2 + \theta) \Rightarrow$$
$$\sin \theta = -\cos(\pi/2 + \theta) \quad (2)$$

We will also need the result that

$$\sin(\pi/2 + \theta) = \cos(-\theta) = \cos \theta \quad (3)$$

Then, if we let $\theta = \alpha + \beta$ in Equation (2) in column 1, and then apply Equation (1), we have

$$\sin(\alpha + \beta) = -\cos[\pi/2 + (\alpha + \beta)]$$
$$= -\cos[(\pi/2 + \alpha) + \beta] \Rightarrow$$
$$\sin(\alpha + \beta) = -[\cos(\pi/2 + \alpha)\cos(\beta)$$
$$- \sin(\pi/2 + \alpha)\sin \beta]$$

which, from Equation (3), with $\theta = \alpha$, becomes

$$\sin(\alpha + \beta) = -\sin(-\alpha)\cos \beta + \cos(-\alpha)\sin \beta \Rightarrow$$
$$\sin(\alpha + \beta) = \sin \alpha \cos \beta + \cos \alpha \sin \beta$$

For $\sin(\alpha - \beta)$ we have

$$\sin(\alpha - \beta) = \sin[\alpha + (-\beta)]$$
$$= \sin \underline{\ ?\ } \cos \underline{\ ?\ } + \underline{\ ?\ } \cdot \underline{\ ?\ }$$
$$\cdot \underline{\ ?\ } \cdot \underline{\ ?\ } - \underline{\ ?\ } \cdot \underline{\ ?\ }$$

For the tangent of the sum of two angles we write

$$\tan(\alpha + \beta) = \frac{\sin(\alpha + \beta)}{\cos(\alpha + \beta)}$$

We can use the above identities to get

$$\tan(\alpha + \beta) = \frac{\sin \alpha \cos \beta + \cos \alpha \sin \beta}{\cos \alpha \cos \beta - \sin \alpha \sin \beta}$$

Then we divide the numerator and denominator by $\cos \alpha \cos \beta$ in order to write the equation as an expression involving only tangents:

$$\tan(\alpha + \beta) = \frac{\dfrac{\sin \alpha}{\cos \alpha} \cdot \dfrac{\cos \beta}{\cos \beta} + \dfrac{\cos \alpha}{\cos \alpha} \cdot \dfrac{\sin \beta}{\cos \beta}}{\dfrac{\cos \alpha}{\cos \alpha} \cdot \dfrac{\cos \beta}{\cos \beta} - \dfrac{\sin \alpha}{\cos \alpha} \cdot \dfrac{\sin \beta}{\cos \beta}} \Rightarrow$$

$$\tan(\alpha + \beta) = \frac{\tan \alpha + \tan \beta}{1 - \tan \alpha \tan \beta}$$

The formula for the tangent of the difference of two numbers $\alpha - \beta$ is

$$\tan(\alpha - \beta) = \frac{\tan \alpha - \tan \beta}{1 + \tan \alpha \tan \beta}$$

since
$$\tan(-\beta) = -\tan\beta$$

Theorem I summarizes the addition formulas. Since the other three trig functions can be readily expressed in terms of these formulas, separate formulas are seldom needed.

THEOREM I.
The addition formulas:

$$\sin(\alpha \pm \beta) = \sin\alpha\cos\beta \pm \cos\alpha\sin\beta$$

$$\cos(\alpha \pm \beta) = \cos\alpha\cos\beta \mp \sin\alpha\sin\beta$$

$$\tan(\alpha \pm \beta) = \frac{\tan\alpha \pm \tan\beta}{1 \mp \tan\alpha\tan\beta}$$

EXERCISE 11.2

A. FUNDAMENTALS

Use the addition formulas to evaluate the following.

1. (a) $\sin \pi/12$ (HINT: $\pi/12 = \pi/4 - \pi/6$.)
   (b) $\cos \pi/12$ (c) $\tan \pi/12$

2. (a) $\sin 5\pi/12$ (b) $\cos 5\pi/12$
   (c) $\tan 5\pi/12$

3. If $\sin u = \frac{3}{5}$ and $\sin v = \frac{4}{5}$, find
   (a) $\sin(u+v)$ (b) $\sin u + \sin v$

4. If $\tan \alpha = \frac{1}{2}$ and $\tan \beta = \frac{3}{4}$, find
   (a) $\tan(\alpha+\beta)$ (b) $\tan(\alpha-\beta)$

5. If $\sin x = \frac{1}{3}$ and $\sin y = 2\sqrt{2}/3$, find
   (a) $\sin(x+y)$ (b) $\cos(x+y)$
   (c) $\tan(x+y)$

B. ESSENTIALS

6. If $\tan \alpha = x$ and $\tan \beta = 1/x$, find
   (a) $\tan(\alpha+\beta)$ (b) $\tan(\alpha-\beta)$
   (c) $\sin(\alpha+\beta)$ (d) $\cos(\alpha-\beta)$

Use the addition formulas to prove the following identities. Interpret the results geometrically for α, an acute angle.

7. (a) $\sin(\pi/2 - \alpha) = \cos\alpha$
   (b) $\cos(\pi/2 - \alpha) = \sin\alpha$
   (c) $\sin(\pi/2 + \alpha) = \cos\alpha$

8. (a) $\sin(\pi - \alpha) = \sin\alpha$
   (b) $\cos(\pi - \alpha) = -\cos\alpha$
   (c) $\sin(\pi + \alpha) = -\sin\alpha$

Solve the following trig equations for their particular solutions.

★10. (a) $\sin^{-1}\frac{3}{5} + \sin^{-1}\frac{12}{13} = \sin^{-1} x$
   (b) $\cos^{-1}\frac{5}{13} + \sin^{-1}\frac{4}{5} = \cos^{-1} x$

11. (a) $\sin^{-1} x - \cos^{-1} x = \pi/6$
   (b) $\tan^{-1}(x-1) - \tan^{-1} 2x = \pi/4$

★12. $\cos 4x \cos 1 + \sin 4x \sin 1 = 1$

13. $\sin x^2 \cos x - \cos x^2 \sin x = 0$

★14. $\sin(x+y) + \sin(x-y) = \cos y$

15. $\cos(x+y) + \cos(x-y) = \sqrt{3}\cos y$

Prove or disprove that the following are trigonometric identities.

★16. $\sin 3\theta = 2\cos\theta - \cos^3\theta$

17. $\cos 3\theta = 4\cos^3\theta - 3\cos\theta$

18. $\dfrac{\cos(\alpha-\beta)}{\cos(\alpha+\beta)} = \dfrac{1 + \tan\alpha\tan\beta}{1 - \tan\alpha\tan\beta}$

19. $\cos(\alpha+\beta)\cos(\alpha-\beta) = \cos^2\alpha - \sin^2\beta$

20. $\cot(x-y) = \dfrac{\cot x \cot y + 1}{\cot y - \cot x}$

21. $\dfrac{\sin(\alpha+\beta)}{\sin(\alpha-\beta)} = \dfrac{\tan\alpha + \tan\beta}{\tan\alpha - \tan\beta}$

★22. $\tan^{-1} x - \tan^{-1} y = \tan^{-1}\dfrac{x-y}{1+xy}$

23. $\tan^{-1}\dfrac{x}{y} - \tan^{-1}\dfrac{x-y}{x+y} = \dfrac{\pi}{4}$

24. $\cos 2x = 1 - \sin^2 x$

25. $\sin 2x = 2\sin x \cos x$

C. APPLICATIONS

26. Derive an addition formula for $\sec(\alpha+\beta)$.

27. Derive an addition formula for $\csc(\alpha-\beta)$.

CHAPTER 11   Trigonometric Equations and Identities

11.3 THE HALF-ANGLE AND DOUBLE-ANGLE FORMULAS

The half-angle and double-angle formulas are *easily* derived from the addition formulas in Section 11.2. First, let's derive the double-angle formulas for sine and cosine.

We will discover how $\sin 2\alpha$ is related to functions of α, if we let

$$\sin 2\alpha = \sin(\alpha + \alpha)$$

and use the addition formula

$$\sin(\alpha + \beta) = \sin \alpha \cos \beta + \cos \alpha \sin \beta$$

We have

$$\sin 2\alpha = \sin(\alpha + \alpha) \Rightarrow$$
$$\sin 2\alpha = \underline{\ ?\ } \cdot \underline{\ ?\ } + \underline{\ ?\ } \cdot \underline{\ ?\ } \Rightarrow$$
$$\sin 2\alpha = 2\ \underline{\ ?\ } \cdot \underline{\ ?\ }$$

Therefore,

$$\sin 2\alpha = 2 \sin \alpha \cos \alpha$$

Now we will discover how $\cos 2\alpha$ is related to functions of α by using the addition formula for $\cos(\alpha + \beta)$:

$$\cos(\alpha + \beta) = \cos \alpha \cos \beta - \sin \alpha \sin \beta$$

That is,

$$\cos 2\alpha = \cos(\quad + \quad)$$
$$= \underline{\ ?\ } \cdot \underline{\ ?\ }$$
$$- \underline{\ ?\ } \cdot \underline{\ ?\ }$$
$$= \underline{\ ?\ } - \underline{\ ?\ }$$

Therefore,

$$\cos 2\alpha = \cos^2 \alpha - \sin^2 \alpha$$

This can be rewritten in terms of $\sin \alpha$ alone, using

$$\cos 2\alpha = \cos^2 \alpha - \sin^2 \alpha$$

and

$$\cos^2 \alpha + \sin^2 \alpha = 1$$

we have

$$\cos 2\alpha = (\underline{\ ?\ } - \underline{\ ?\ }) - \sin^2 \alpha$$
$$= \underline{\ ?\ } - \underline{\ ?\ } \cdot \sin^2 \alpha$$

Therefore,

$$\cos 2\alpha = 1 - 2 \sin^2 \alpha$$

We can also use $\sin^2 \alpha + \cos^2 \alpha = 1$ to eliminate the $\sin^2 \alpha$ from the double-angle formula for cosine:

$$\cos 2\alpha = \cos^2 \alpha - \sin^2 \alpha$$
$$= \cos^2 \alpha - (\underline{\ ?\ })$$
$$= \underline{\ ?\ }$$

Therefore,

$$\cos 2\alpha = 2 \cos^2 \alpha - 1$$

Now we can determine the half-angle formulas derived from the double-angle formulas for $\cos 2\alpha$ in terms of either $\sin^2 \alpha$ or $\cos^2 \alpha$. These formulas reduce squared trig functions to functions raised to the first power. They are essential in calculus, where linear (first-power) functions are easily manipulated.

Since

$$\cos 2\theta = 1 - 2 \sin^2 \theta$$

solving for $\sin^2 \theta$, we have

$$2 \sin^2 \theta = \underline{\ ?\ }$$

Therefore,

$$\sin^2 \theta = \frac{1 - \cos 2\theta}{2} \qquad (3)$$

This is called a half-angle formula because, if we let $2\theta = \alpha$, then $\theta = \alpha/2$, and Equation (3) becomes

$$\sin^2 \alpha/2 = \frac{1 - \cos \alpha}{2} \Leftrightarrow$$

$$\sin \alpha/2 = \pm \sqrt{\frac{1 - \cos \alpha}{2}}$$

Since

$$\cos 2\theta = 2 \cos^2 \theta - 1$$

234

if we solve for $\cos^2 \theta$, we have

$$2 \cos^2 \theta = \underline{\quad ? \quad} \Rightarrow$$

$$\cos^2 \theta = \frac{?}{?}$$

This is also a half-angle formula because the angle on the right side of the equation, θ, is half the angle on the right, 2θ. By letting $\alpha = 2\theta$ and $\alpha/2 = \underline{\quad ? \quad}$, we can rewrite this formula as

$$\cos \frac{\alpha}{2} = \pm \sqrt{\frac{1 + \cos \alpha}{2}}$$

Since the half- and double-angle formulas for the other four trig functions are relatively seldom needed, we will leave their derivation for the exercises.

In summary, we have Theorems II and III.

THEOREM II.
The double-angle formulas are

$$\sin 2\alpha = 2 \sin \alpha \cos \alpha$$

$$\cos 2\alpha = \cos^2 \alpha - \sin^2 \alpha$$

$$= 2 \cos^2 \alpha - 1$$

$$= 1 - 2 \sin^2 \alpha$$

THEOREM III.
The half-angle formulas are

$$\sin^2 \alpha = \frac{1 - \cos 2\alpha}{2}$$

or

$$\sin \frac{\alpha}{2} = \pm \sqrt{\frac{1 - \cos \alpha}{2}}$$

and

$$\cos^2 \alpha = \frac{1 + \cos 2\alpha}{2}$$

or

$$\cos \frac{\alpha}{2} = \pm \sqrt{\frac{1 + \cos \alpha}{2}}$$

The following are examples of equations and identities involving the half- and double-angle formulas.

EXAMPLE I. Solve the following equation for $0 \leq x < 2\pi$:

$$\sin 2x - \sin x = 0$$

Solution: Our first objective is to rewrite the problem in an equivalent form involving functions of the same variable, i.e., functions of x only, not of $2x$. Then, since

$$\sin 2x = 2 \sin x \cos x$$

we have

$$\sin 2x - \sin x = 0 \Leftrightarrow$$

$$2 \sin x \cos x - \sin x = 0 \Rightarrow$$

$$\sin x (2 \cos x - 1) = 0 \Rightarrow$$

$\sin x = 0 \quad$ or $\quad 2 \cos x - 1 = 0 \Rightarrow$

$x = \sin^{-1} 0 \quad$ or $\quad x = \cos^{-1}(\tfrac{1}{2}) \Rightarrow$

$x = 0$ or $\pi \quad$ or $\quad x = \pi/3$ or $5\pi/3$

EXAMPLE II. Solve the following for $0 \leq x < 2\pi$:

$$\cos 2x - \sin x = 0$$

Solution: The identity $\cos 2x = \cos^2 x - \sin^2 x$ would give us an equation whose functions are all of the same variable, but since it would not involve only one trig function, we would have to eliminate $\cos^2 x$ to solve the equation. Instead, let's use the identity

$$\cos 2x = 1 - 2 \sin^2 x$$

to obtain

$$\cos 2x - \sin x = 0 \Leftrightarrow$$

$$(\underline{\quad ? \quad}) - \sin x = 0 \Rightarrow$$

$$2 \sin^2 x + \sin x - 1 = 0$$

CHAPTER 11 Trigonometric Equations and Identities

Letting $u = \sin x$ yields $x = \sin^{-1} u$ and

$$2u^2 - u - 1 = 0 \Rightarrow$$

$$(\underline{\quad ? \quad})(\underline{\quad ? \quad}) = 0 \Rightarrow$$

$$\underline{\quad ? \quad} = 0 \quad \text{or} \quad \underline{\quad ? \quad} = 0 \Rightarrow$$

$$u = \underline{\quad ? \quad} \quad \text{or} \quad u = \underline{\quad ? \quad}$$

Therefore,

$$x = \sin^{-1}(\underline{\quad ? \quad}) \quad \text{or} \quad x = \sin^{-1}(\underline{\quad ? \quad}) \Rightarrow$$

$$x = \underline{\quad ? \quad} \quad \text{or} \quad x = \underline{\quad ? \quad}$$

These answers, $x = \pi/6$, $5\pi/6$, and $3\pi/2$, can be checked by substituting them into the original equation.

EXAMPLE III. Prove the following identity:

$$\frac{\sin 2x}{1 - \cos 2x} = \cot x$$

Solution:

$$\frac{\sin 2x}{1 - \cos 2x} = \frac{2 \sin x \cos x}{1 - (\cos^2 x - \sin^2 x)}$$

$$= \frac{2 \sin x \cos x}{(1 - \cos^2 x) + \sin^2 x}$$

$$= \frac{2 \sin x \cos x}{\sin^2 x + \sin^2 x} = \frac{2 \sin x \cos x}{2 \sin^2 x}$$

$$= \frac{\cos x}{\sin x} = \cot x$$

Therefore,

$$\frac{\sin 2x}{1 - \cos 2x} = \cot x$$

EXAMPLE IV. Prove the following identity:

$$\frac{\sin 2x}{1 + \cos 2x} = \tan x$$

Solution: We have

$$\frac{\sin 2x}{1 + \cos 2x} = \frac{?}{1 + (\underline{\quad ? \quad})}$$

$$= \frac{?}{(1 - \sin^2 x) + \underline{\quad ? \quad}}$$

$$= \frac{2 \cdot \sin x \cos x}{\underline{\quad ? \quad} + \cos^2 x \cdot ?} = \frac{?}{?} = \tan x$$

Therefore,

$$\frac{\sin 2x}{1 + \cos 2x} = \underline{\quad ? \quad}$$

EXERCISE 11.3

A. FUNDAMENTALS
Evaluate the following.
1. If $\cos \alpha = \frac{3}{5}$ and $\sin \alpha = \frac{4}{5}$, find
   (a) $\sin 2\alpha$ (b) $\cos 2\alpha$ (c) $\sin (\alpha/2)$
   (d) $\cos (\alpha/2)$

2. (a) Use $\sin^2\left(\frac{\alpha}{2}\right) = \frac{1 - \cos \alpha}{2}$ with $\alpha = \frac{\pi}{6}$ to find $\sin \frac{\pi}{12}$.

   (b) Find $\cos \frac{\pi}{12}$.

3. Show that if $\sin \alpha = \frac{3}{5}$ and $\sin \beta = \frac{4}{5}$, then
   (a) $\sin 2\alpha = \sin 2\beta$ (b) $\cos 2\alpha = \frac{7}{25}$
   (c) $\cos 2\alpha + \cos 2\beta = 0$ (d) $\sin 3\alpha = \frac{117}{125}$

B. ESSENTIALS
Solve the following trigonometric equations for the particular solutions.

★4. $\sin 2x = 2 \sin x$ 5. $\cos 2x = \cos x$
6. $\cos 2t = \cos^2 t$ 7. $\sin 2t = \sin^2 t$
8. $\sin 2\theta - \cos \theta = 0$
9. $\sin \theta \sin 2\theta = \cos \theta$ 10. $\cos 2x = \sin x$
★11. $\cos t \cos 2t = \sin 2t \sin t$
★12. $2 \sin^{-1} x = \cos^{-1} 1$

236

13. $2 \tan^{-1} x = \sin^{-1} \frac{4}{5}$

14. $2 \sin^{-1} x = \sin^{-1} 2x$

15. $2 \sin^{-1} \left(-\frac{\sqrt{3}}{2} \right) = \cos^{-1} x$

Prove that the following are trigonometric identities by transforming the left side of each equation into the right side.

★16. $\dfrac{\cos 2t}{1 - \sin 2t} = \dfrac{1 + \tan t}{1 - \tan t}$

17. $\sin 2t = \dfrac{2 \tan t}{1 + \tan^2 t}$

18. $\csc 2\theta + \cot 2\theta = \cot \theta$

19. $\dfrac{\sin 2x}{\sin x} - \dfrac{\cos 2x}{\cos x} = \sec x$

★20. $\dfrac{\sin 3x - \sin x}{\cos 3x + \cos x} = \tan x$

21. $2 \sin 2x \cos x - \sin 3x = \sin x$

22. $4 \sin \theta \cos^2 \theta - \sin 3\theta = \sin \theta$

23. $\dfrac{\sin 4\theta}{2 \sin 2\theta} = \cos 2\theta$

24. (a) $\tan 2x = \dfrac{2 \tan x}{1 - \tan^2 x}$

    (b) $\tan \dfrac{x}{2} = \dfrac{\sin x}{1 + \cos x} = \dfrac{\tan x}{\csc x + \tan x}$

    (c) $\tan x = \dfrac{1 - \cos 2x}{\sin 2x} = \dfrac{\sec 2x - 1}{\tan 2x}$

25. (a) $\cot \left(\dfrac{x}{2} \right) = \dfrac{\sin x}{1 - \cos x} = \dfrac{1}{\csc x - \cot x}$

    (b) $\cot x = \dfrac{1 + \cos 2x}{\sin 2x} = \dfrac{\csc 2x + 1}{\cot 2x}$

    (c) $\cot 2x = \dfrac{\cot^2 x - 1}{2 \cot x}$

In each of the following, prove (a), then use (a) to prove (b), and, finally, use (b) to prove (c).

26. (a) $\sin(\alpha + \beta) + \sin(\alpha - \beta) = 2 \sin \alpha \cos \beta$

    (b) $\sin x + \sin y$

$$= 2 \sin \left(\dfrac{x+y}{2} \right) \cos \left(\dfrac{x-y}{2} \right)$$

    (c) $\sin u - \sin v$

$$= 2 \cos \left(\dfrac{u+v}{2} \right) \sin \left(\dfrac{u-v}{2} \right)$$

★27. (a) $\cos(\alpha - \beta) + \cos(\alpha + \beta) = 2 \cos \alpha \cos \beta$

    (b) $\cos x + \cos y$

$$= 2 \cos \left(\dfrac{x+y}{2} \right) \cos \left(\dfrac{x-y}{2} \right)$$

    (c) $\cos u - \cos v$

$$= -2 \sin \left(\dfrac{u+v}{2} \right) \sin \left(\dfrac{u-v}{2} \right)$$

CHAPTER 12

Applications of Trigonometry

12.1 INTRODUCTION

Since applications very seldom use the familiar angles 30°, 45°, 60°, etc., we will need to use Table IV. Check the values for the above angles in the tables to make sure you understand how to use the tables.

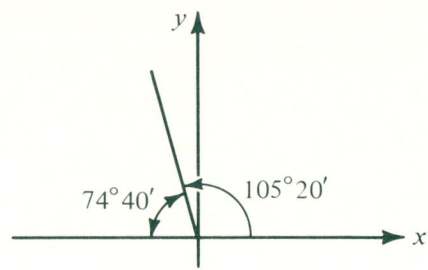

FIGURE 12.1.

EXAMPLE I. Find the approximate values of the following.
(a) $\sin 39°40' \approx 0.6383$
(b) $\cos 50°20' \approx 0.6383$
(c) $\tan 105°20' = -\tan(180° - 105°20')$
$= -\tan(74°40')$
≈ -3.647 (See Figure 12.1)
(d) $\sin 148° = \sin(180° - 148°)$
$= \sin 32°$
≈ 0.5299
(e) $\cos 148° = \cos(\quad - \quad)$
$= -\cos \underline{\quad ? \quad}$
≈ -0.8480

Notice that the signs of the trig functions still depend upon the quadrant in which the terminal side of the angle is found. In Example I, (c) and (d) are negative, because the tangent and cosine are negative in the second quadrant.

12.2 RIGHT-TRIANGLE APPLICATIONS

One way to study applications is to work through typical examples and then do similar problems. This next example illustrates the central procedure for right-triangle trigonometric applications and problems.

EXAMPLE I. Find the missing parts of the right triangles in Figure 12.2.

Solution: In part (a),

$$\sin A = \frac{BC}{AB} = \frac{6}{AB} \quad \text{and} \quad A = \pi/6 \Rightarrow$$

$$\frac{6}{AB} = \sin \frac{\pi}{6} = \frac{1}{2} \Rightarrow$$

$$\frac{6}{AB} = \frac{1}{2} \Rightarrow$$

$$AB = 12$$

Also,

$$\cos A = \frac{AC}{AB} \Rightarrow$$

$$\cos \frac{\pi}{6} = \frac{AC}{12} \Rightarrow$$

$$AC = 12 \cos \frac{\pi}{6} = 12\left(\frac{\sqrt{3}}{2}\right) = 6\sqrt{3}$$

Finally,

$$\angle A + \angle B = \frac{\pi}{2} \Rightarrow \angle B = \frac{\pi}{3}$$

In part (b),

$$\sin A = \frac{BC}{AB} \Rightarrow$$

$$\sin(?) = \frac{?}{AB} \Rightarrow$$

$$AB = \frac{?}{?} = ?$$

$$\cos A = \frac{AC}{AB} \Rightarrow$$

$$\cos(?) = \frac{AC}{4\sqrt{2}} \Rightarrow$$

Since $\quad AC = (4\sqrt{2})(?) = ?$

$$\angle A + \angle B = \frac{\pi}{2} \quad \text{and} \quad \angle A = \frac{\pi}{4}$$

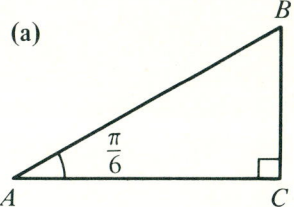 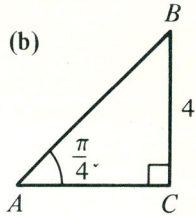

FIGURE 12.2.

we know that

$$\angle B = ?$$

This is a typical application. Notice the presence of a right triangle.

EXAMPLE II. A ten-meter ladder is leaning against a wall at an angle of 32°10′. How far up the wall does the ladder reach? See Figure 12.3.

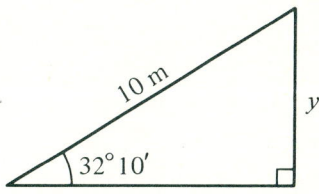

FIGURE 12.3.

Solution: We know that

$$\sin 32°10' = \frac{y}{10 \text{ m}} \Rightarrow$$

$$y = (10 \text{ m})(\quad ? \quad)$$

$$y = 5.02 \text{ m}$$

Trig is essential in surveying, where distances to be measured are given as the horizontal distances between 2 points, even when the points are located on a hill, as illustrated in Example III. First, we will need to define two concepts, the angle of elevation and the angle of depression. In Figure 12.4, an angle of elevation implies "looking up" and an angle of depression implies "looking down." Both are measured from the horizontal.

CHAPTER 12   Applications of Trigonometry

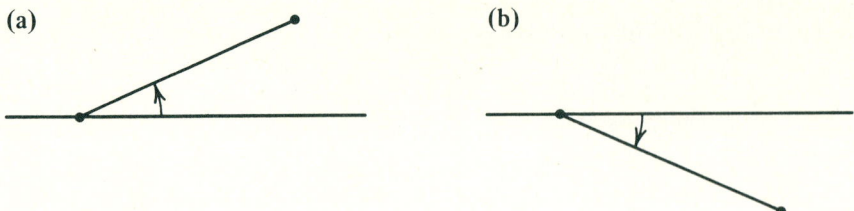

FIGURE 12.4. (a), an angle of elevation and (b), an angle of depression.

EXAMPLE III. The angle of elevation of a hill is measured as 12°10′, and the horizontal distance is given as 110.1 meters. How far up the hill should the surveying team go from point A before placing the stake for point B? See Figure 12.5.

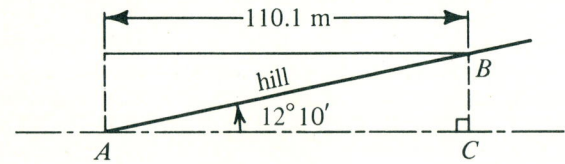

FIGURE 12.5.

Solution:

$$\cos 12°10' = \frac{AC}{AB} \Rightarrow$$

$$AB = \frac{(\quad ?\quad)}{\cos (12°10')} \Rightarrow$$

$$AB = \frac{(\quad ?\quad)}{(\quad ?\quad)} \Rightarrow$$

$$AB = \underline{\quad ?\quad}$$

Your solution for the distance up the hill should be 112.6 meters.

EXERCISE 12.2

A. FUNDAMENTALS
Find the missing parts of the following right triangles.

1.

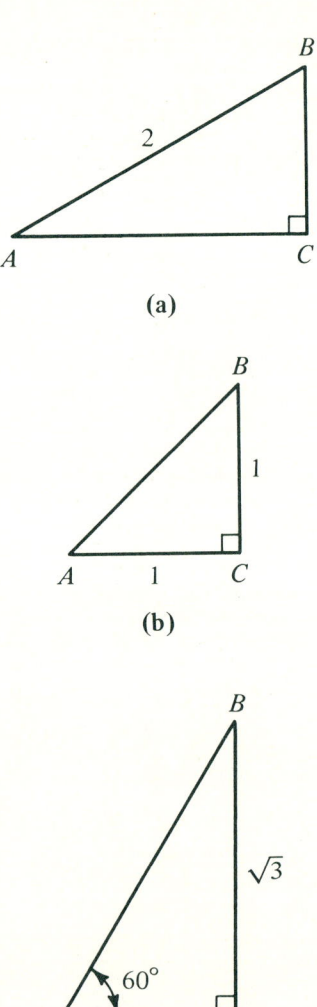

2.

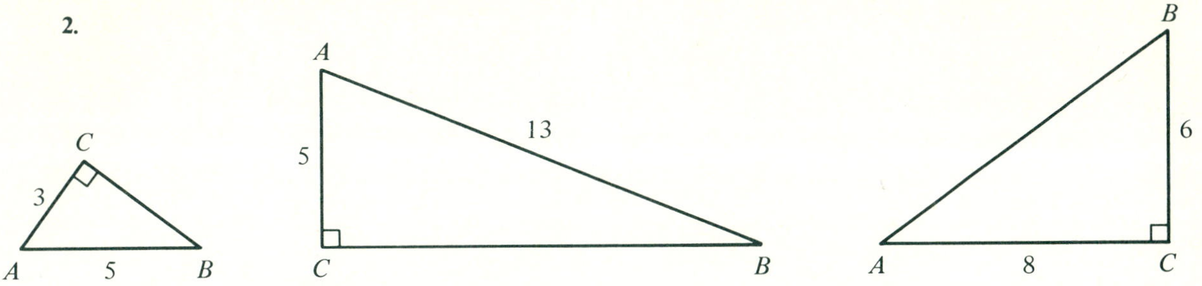

(a) Right triangle with right angle at C, $AC = 5$, CB (vertical leg labeled 3), hypotenuse from A to B.

(b) Right triangle with right angle at C, $AC = 5$, $AB = 13$.

(c) Right triangle with right angle at C, $AC = 8$, $BC = 6$.

3.

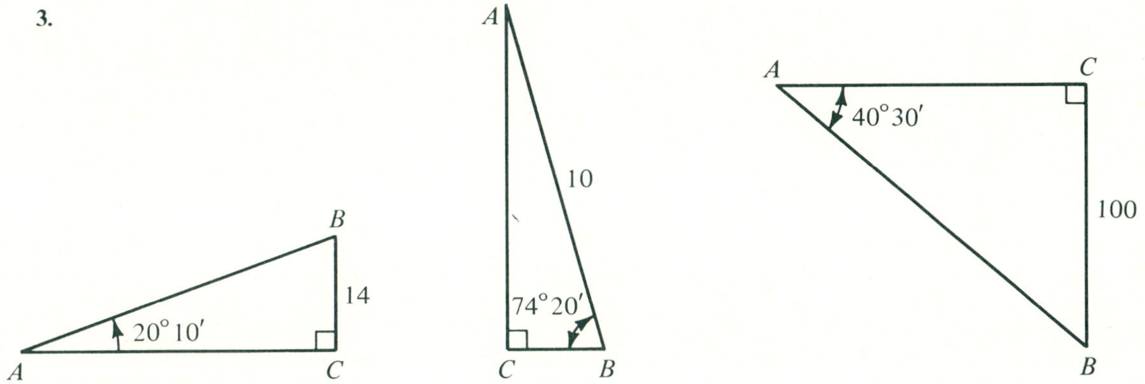

(a) Right triangle with right angle at C, angle at $A = 20°10'$, $BC = 14$.

(b) Right triangle with right angle at C, angle at $B = 74°20'$, $AB = 10$.

(c) Right triangle with right angle at C, angle at $A = 40°30'$, $CB = 100$.

B. ESSENTIALS

4. A ladder must reach a window 20 meters above the ground from the other side of a moat 10 meters wide. How long must the ladder be?

5. If the angle of elevation to the top of a Douglas fir tree from a point 20 meters from the base of the tree is $56°21'$, how tall is the tree?

6. The forest ranger on top of a lookout 10 meters high spots a fire at an angle of depression of $15°30'$. How far away from the base of the lookout is the fire?

7. How far away would the fire be if the angle of depression in Problem 6 were $4°20'$?

8. What would the angle of depression be if the fire were 1 kilometer away from the base of the lookout in Problem 6?

9. An ice cream cone contains a spherical scoop of ice cream. The center of the sphere is even with the top of the cone. If the radius of the ice cream is 2.5 centimeters and $\alpha = 15°$, what are the dimensions of the cone? See the following illustration.

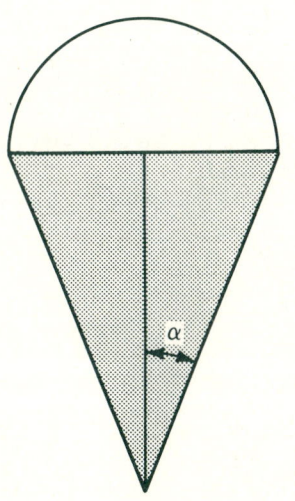

CHAPTER 12   Applications of Trigonometry

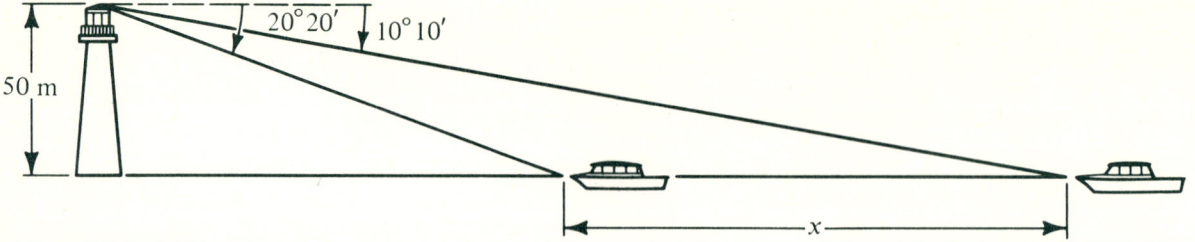

10. The angles of depression from the top of a lighthouse 50 meters above the sea to two boats are 10°10′ and 20°20′. How far apart are the two boats? How far are the boats from the lighthouse? See the above illustration.

11. A man 210 centimeters tall standing 15 meters from a lightpost casts a shadow 10 meters long. How tall is the lightpost? See the following illustration.

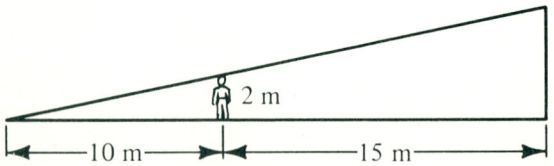

12. One night the moon was visible from a point in the Pacific for 11 hours and 50 minutes out of the 24 hours it took the moon to circle the earth. If the diameter of the earth is 12,742 kilometers, how high is the moon? See the following illustration.

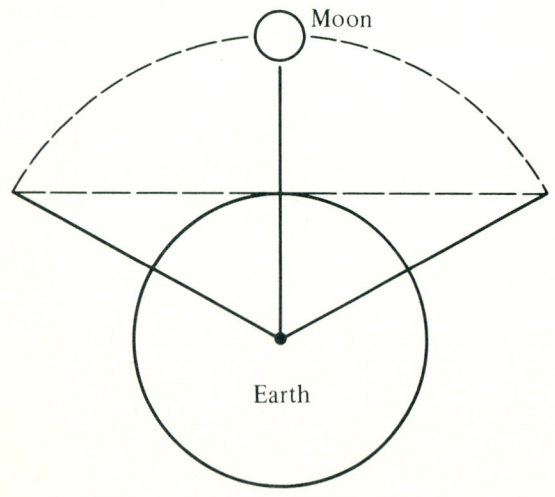

★13. A telecommunications satellite is locked in synchronous orbit about the earth, i.e., it completes one revolution in exactly the same amount of time that the earth completes one rotation. If it is traveling at a velocity of 1694 kilometers per hour, what is its altitude? (Assume it takes 24 hours to travel once around its circular path.) How many degrees of longitude, α, does the satellite cover? See the following illustration.

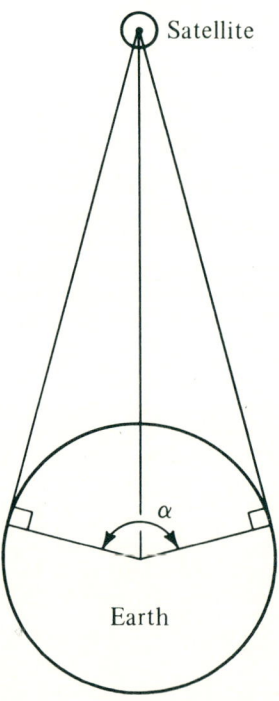

•14. Use Archimedes' method of exhaustion as a geometric model to define π as follows. The area of a circle of radius 1 is π. Consider circumscribed regular polygons of increasingly greater numbers of sides, as shown in the illustration. Find these areas. Then use inscribed regular polygons of increasingly greater number

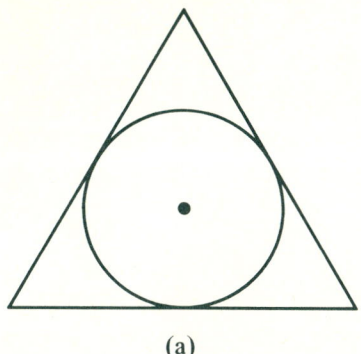
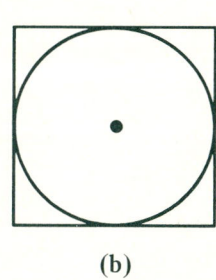
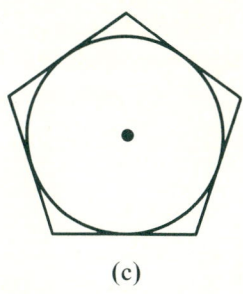

(a) (b) (c)

of sides. Find these areas. These sequences of areas converge on the area of the unit circle π. Use increasingly more sides on your polygons until you have calculated π accurately to two decimals, as Archimedes did without a calculator in about 240 B.C.!

12.3 AZIMUTH, BEARING, AND HEADING

This section concerns the most common methods of giving relative position and illustrates their usual applications.

Azimuth and heading are two names for exactly the same system of direction. Heading is used in navigation and azimuth in surveying. They are clockwise measures of the angle from a north–south line (the meridian). Some azimuths (headings) are shown in Figure 12.6. Example I illustrates their application.

EXAMPLE I. A ship is located 23 kilometers off a north–south shoreline. How long will it have to travel at 15 kilometers per hour to reach a port located at a heading of 13°20′ from the ship? See Figure 12.7 (page 244).

Solution: Angle B is also 13°20′, since BC is parallel to the north–south line through A. Therefore,

$$\sin B = \frac{23}{C} \Rightarrow$$

$$C = \frac{23 \text{ km}}{\sin 13°20'} \approx 100 \text{ km}$$

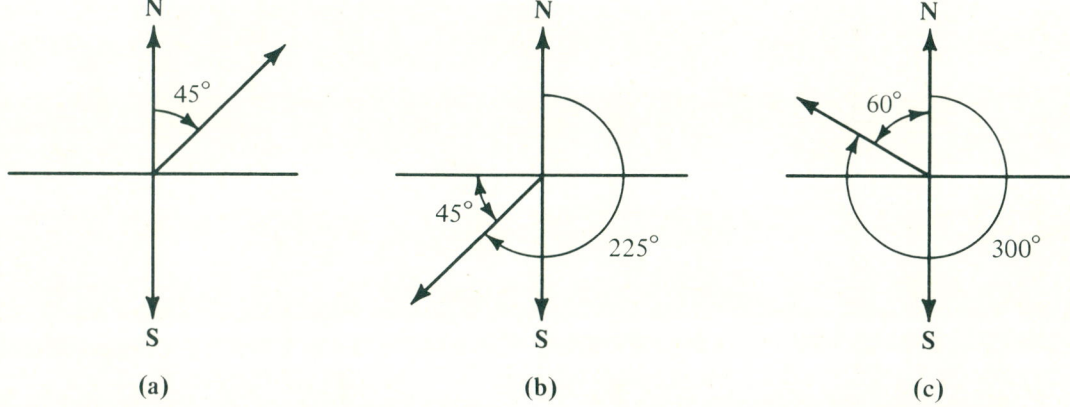

FIGURE 12.6. *(a), an azimuth of 45°; (b), an azimuth of 225°; (c), an azimuth of 300°.*

CHAPTER 12   Applications of Trigonometry

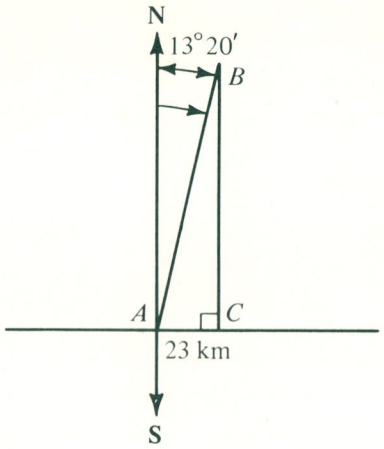

FIGURE 12.7.

Then, since $d = r \cdot t$, we have

$$t = \frac{d}{t} = \frac{100 \text{ km}}{15 \text{ km/hr}} = 6 \text{ hr } 40 \text{ min}$$

Trig is essential to surveying, where direction is given as *bearing*, as illustrated in the following example.

EXAMPLE II.   Consider Figure 12.8.

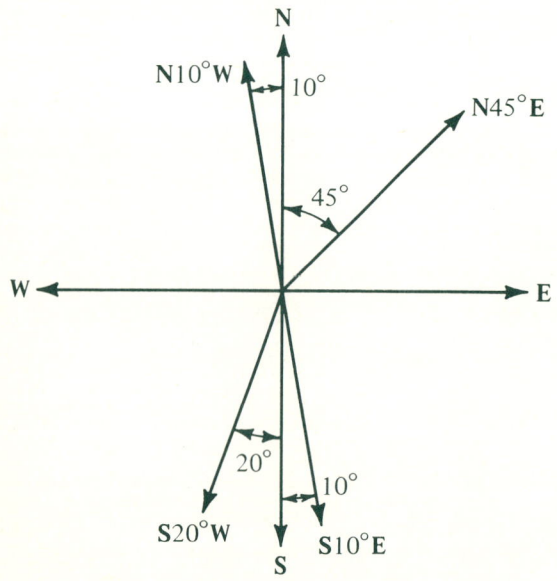

FIGURE 12.8.

In quadrants I and II the directions are east or west of north. In quadrants III and IV they are east or west of south. The direction straight up the *y*-axis is called "due north."

EXAMPLE III.   A surveying team is making a traverse over some open ground. By noon they have completed the following measurements. How far and at what bearing are they from their starting point? See Figure 12.9.

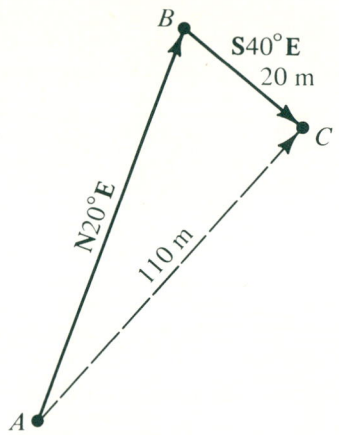

FIGURE 12.9.

Solution: To ensure that the coordinates of A, B, and C will be positive, we will position point A at (100, 100). See Figure 12.10.

To find the bearing and distance from A to C we will have to calculate the coordinates of points B and C. First, let's find x_1 and y_1 in order to find the coordinates of B.

$$\cos A = \frac{x_1}{110} \Rightarrow x_1 = 110 \cos 70° = 37.6 \text{ m}$$

and

$$\sin A = \frac{y_1}{110} \Rightarrow y_1 = 110 \sin 70° = 103.4 \text{ m}$$

Therefore, the coordinates of point B are

$$x = 100 + 37.6 = 137.6$$

and

$$y = 100 + \underline{} = \underline{}$$

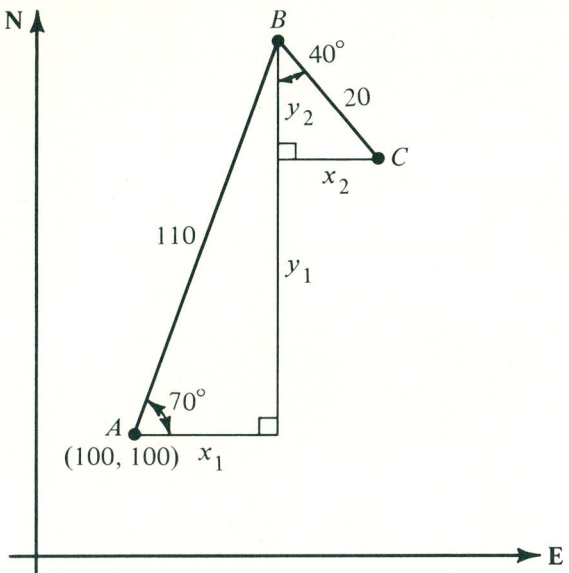

FIGURE 12.10.

Now we will find x_2 and y_2 in order to find the coordinates of C.

$$\sin B = \frac{x_2}{20} \Leftrightarrow x_2 = 20 \sin 40° = \underline{\quad ? \quad}$$

and

$$\cos B = \frac{?}{?} \Rightarrow y_2 = \underline{\quad ? \quad} \cdot \underline{\quad ? \quad} = \underline{\quad ? \quad}$$

Then, the coordinates of C are

$$x = 137.6 + x_2 = 150.4$$

and

$$y = 203.4 - y_2 = 218.7$$

The distance from A to C is found by the distance formula to be

$$d = \sqrt{(x - x')^2 + (y - y')^2}$$
$$= \sqrt{(50.4)^2 + (118.7)^2} \Rightarrow$$
$$d \approx 129 \text{ m}$$

The bearing from A to C is found in Figure 12.11.

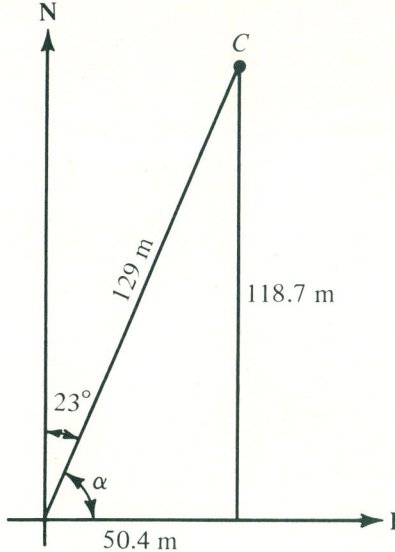

FIGURE 12.11.

We know that

$$\tan \alpha = \frac{118.7}{50.4} \Rightarrow \alpha \approx 67°$$

Therefore, the bearing is

N23°E

EXERCISE 12.3

A. FUNDAMENTALS

Make a drawing, similar to the one in Figure 12.6, to indicate the following directions. Use the origin as the starting point.

1. (a) an azimuth of 25°
   (b) a bearing of N25°E
   (c) a heading of 25°

2. (a) an azimuth of 30°
   (b) an azimuth of 150°
   (c) an azimuth of 330°

3. (a) a bearing of N40°E
   (b) a bearing of S40°E
   (c) a bearing of N40°W

CHAPTER 12   Applications of Trigonometry

4. (a) a heading of 110°
   (b) a bearing of S70°E
   (c) a heading of 250°

5. If a ship departs along a heading of 120°, what heading must it navigate to return?

6. If the bearing from point A to point B is S40°W, what is the bearing from B to A?

7. If the bearing from 0 to A is N45°E, and the bearing from 0 to B is S45°E, what is the bearing from A to B if they are the same distance from 0?

8. If two ships leave a port at the same time, ship A traveling at 10 knots on a heading of 135° and ship B traveling at 10 knots on a heading of 225°, what is the heading from A to B after 1 hour?

B. ESSENTIALS

9. If the bearing from 0 to A is N45°E, and from 0 to B is S45°E, how far apart are A and B, if each is 1 meter from the origin?

10. How far apart are the ships in Problem 8? (HINT: 1 knot = 1 nautical mile/hour ≈ 1.15 mph.)

11. If the ships in Problem 8 were traveling at 8 and 10 knots respectively, what would be the heading from B to A after 2 hours? How far apart would the ships be?

12. The azimuth from town A to town B is 15°. If a car leaves town B along an azimuth of 235°, and another car leaves town A along an azimuth of 325°, how fast must each car travel if they are to meet in 1 hour? Assume the towns are 90 km apart.

13. The survey crew of Example III has completed the following measurements. How far are they from their starting point, and at what heading?

From A to B: N40°E for 120 m
From B to C: N30°W for 30 m

14. Suppose the survey crew tried to follow a closed triangular path, but actually placed stakes along the plot described below. How far from closure are they? What is the bearing from A to D?

From A to B: 110 m at N50°E
From B to C: 120 m at S70°W
From C to D: 40 m at S30°E

12.4 APPLICATIONS OF THE LAW OF SINES AND THE LAW OF COSINES

This section extends the scope of the applications we can consider to triangles without a right angle. We will state the law of cosines and the law of sines, give an example, and then the proof.

THEOREM I.
The law of cosines is

$$a^2 = b^2 + c^2 - 2bc \cos A$$

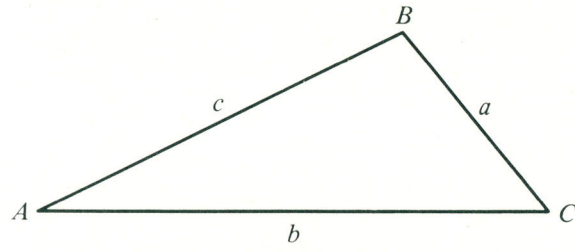

FIGURE TI.

This law is usually applied to find the length of one side of a triangle when the measure of the angle opposite this side and the other two sides are known. (Side, angle, side are given.)

EXAMPLE I. Find the measure of side a in Figure 12.12.

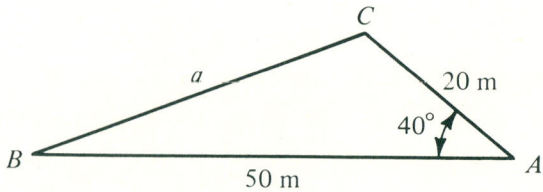

FIGURE 12.12.

Solution: By the law of cosines we see that

$$a^2 = b^2 + c^2 - 2bc \cos A$$

where

$b = 20$ m, $c = 50$ m,   and $A = 40°$

SECTION 12.4 Applications of the Law of Sines and the Law of Cosines

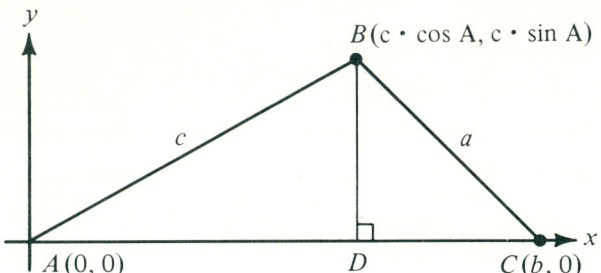

FIGURE 12.13.

Then

$$a^2 = (20)^2 + (50)^2 - 2(20)(50) \cos 40° \Rightarrow$$

$$a^2 = 1367.9 \Rightarrow$$

$$a \approx 36.99 \text{ m}$$

If you do not have access to a calculator, you may need to review Section 8.6 on computations involving logarithms in order to calculate $\sqrt{1367.9}$.

Now let's derive the law of cosines. This will require drawing auxiliary lines to transform the oblique triangle into more familiar ones having right angles. This is accomplished by drawing the perpendicular line BD as shown in Figure 12.13.

We want to prove that

$$a^2 = b^2 + c^2 - 2bc \cos A$$

Now, from the right triangle ABD, we have

$$\sin A = \frac{y}{c} \quad \text{and} \quad \cos A = \frac{x}{c}$$

Thus the coordinates of B are

$$x = c \cos A \quad \text{and} \quad y = c \sin A$$

Then, by the distance formula,

$$BC = \sqrt{(c \cos A - b)^2 + (c \sin A - 0)^2}$$

But,

$$BC = a \Rightarrow$$

$$a^2 = (c \cos A - b)^2 + (c \sin A)^2 \Rightarrow$$

$$a^2 = c^2 \cos^2 A - 2bc \cos A + b^2 + c^2 \sin^2 A \Rightarrow$$

$$a^2 = b^2 + c^2 \cos^2 A + c^2(1 - \cos^2 A) - 2bc \cos A \Rightarrow$$

$$a^2 = b^2 + c^2 - 2bc \cos A$$

EXAMPLE II. A forward observer for a gun emplacement is located 5 kilometers from the gun on an azimuth of 30°. If this forward observer located on a hill sights the target on an azimuth of 110° at a range of 8 kilometers, how far is the target from the gun, and along what azimuth should the gun fire? See Figure 12.14.

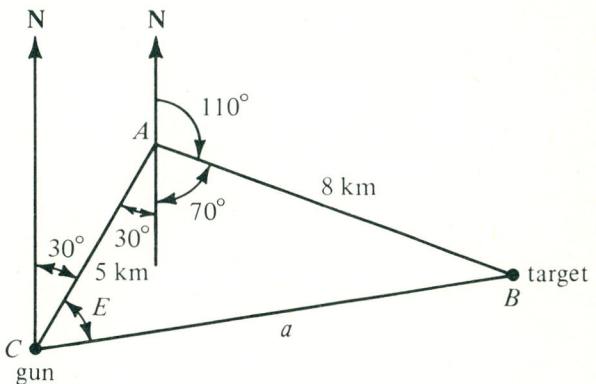

FIGURE 12.14.

Solution: Let a equal the distance we wish to find, $b = 5$ kilometers, and $c = 8$ kilometers. Then $\angle A = 100°$ and

$$a^2 = b^2 + c^2 - 2bc \cos A \Rightarrow$$

$$a^2 = 25 + 64 - 2(5)(8) \cos 100° \Rightarrow$$

$$a^2 = 89 - 80(-0.1736) \Rightarrow$$

$$a^2 = 102.89 \Rightarrow$$

$$a = 10.14 \text{ km}$$

The azimuth of the target is $\angle E + 30$. To find the measure of angle E, consider the following statement of the law of cosines using angle E as the included angle:

$$8^2 = 5^2 + a^2 - (2)(5)a \cos E$$

247

CHAPTER 12   Applications of Trigonometry

Since $a = 10.14$,

$$\cos E = \frac{5^2 + (10.14)^2 - 8^2}{(2)(5)(10.14)} \Rightarrow$$

$$E = \cos^{-1}(0.6294) \Rightarrow$$

$$E \approx 51°$$

Therefore, the azimuth from the gun to the target is

$$\angle E + 30° = 81°$$

Finding the azimuth is easier using the law of sines, as illustrated in the next example.

THEOREM II.
The law of sines is

$$\frac{\sin A}{a} = \frac{\sin B}{b} = \frac{\sin C}{c}$$

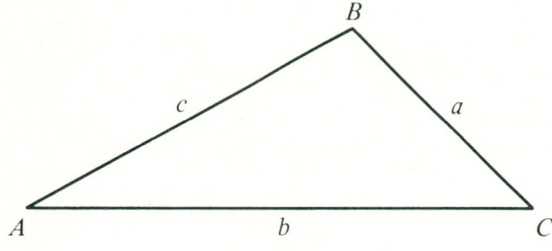

FIGURE TII.

EXAMPLE III.   Find angle E in Example II using the triangle in Figure 12.15, where we know an angle and two sides.

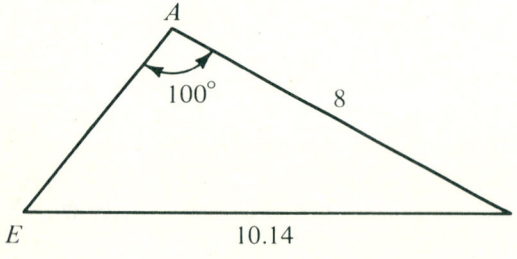

FIGURE 12.15.

Solution: We have

$$\frac{\sin E}{8} = \frac{\sin 100}{10.14} \Rightarrow$$

$$\sin E = \frac{8(0.9848)}{10.14} \Rightarrow$$

$$E = \sin^{-1}(0.7770) \Rightarrow$$

$$E \approx 51°$$

EXAMPLE IV.   Find the missing parts of the triangle in Figure 12.16, if $A = 30°$, $a = 2$, $B = 100°$, and $c = 8$, as shown in Figure 12.16.

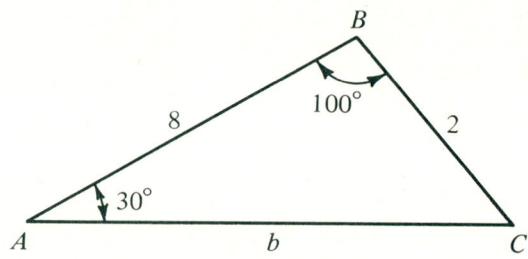

FIGURE 12.16.

Solution: From the law of sines,

$$\frac{\sin A}{a} = \frac{\sin B}{b} \Rightarrow$$

$$\frac{\sin 30°}{2} = \frac{\sin 100°}{b} \Rightarrow$$

$$b = \frac{2 \sin 100°}{\sin 30°} = 3.94$$

and

$$\frac{\sin A}{a} = \frac{\sin C}{c} \Rightarrow$$

$$\frac{\sin 30°}{2} = \frac{\sin C}{8} \Rightarrow$$

$$\sin C = 4 \sin 30° \Rightarrow$$

$$C = \sin^{-1} 2$$

Thus, angle C does not exist! Did you know this was going to happen? Try drawing the triangle to scale, and you will see why there is no

SECTION 12.4   Applications of the Law of Sines and the Law of Cosines

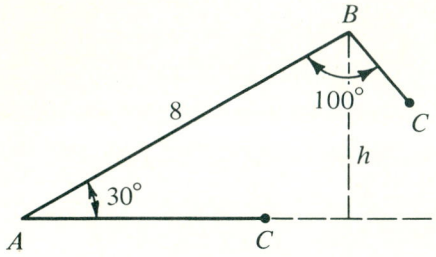

FIGURE 12.17.

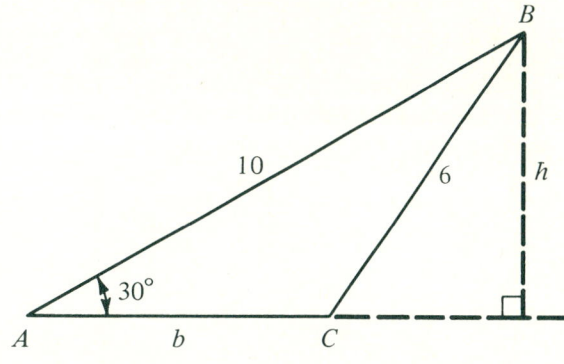

FIGURE 12.19.

value of angle C that will yield a triangle. This can be determined in advance by comparing BC with the altitude from angle B in Figure 12.17. Side BC must be greater than or equal to the altitude h. Do you see why b was too short? Since AB was not used in the calculation, it came from the

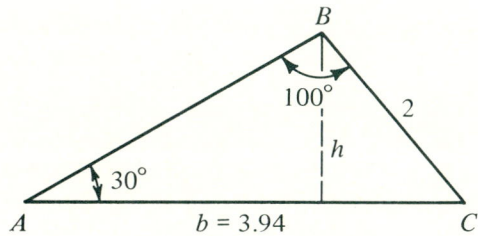

FIGURE 12.18.

triangle in Figure 12.18, which consistently employs all given values. In this triangle,

$$\angle C = 50° \Rightarrow$$

$$\sin C = \frac{h}{2} \Rightarrow$$

$$h = 2 \sin 50° = 1.53$$

which is a small enough altitude for a to reach side AC, since

$$a = 2 > h$$

These results are inconsistent with the original given values, where

$$AB = 8 \Rightarrow h = 4$$

and a is too short. Therefore, there is no solution to this problem, since no such triangle exists.

EXAMPLE V. Find the missing parts of the triangle in Figure 12.19.

Solution: First, let's compare $a = 6$ with the altitude h in Figure 12.19, to make sure we will have a triangle.

$$\sin 30° = \frac{h}{10} \Rightarrow$$

$$h = 5$$

and $a = 6 \Rightarrow a > h$ means that side a will reach AC and we will have a triangle.

Now, from the law of sines, we have

$$\frac{\sin A}{a} = \frac{\sin C}{c} \Rightarrow$$

$$\frac{\sin (?)}{(?)} = \frac{\sin C}{(?)} \Rightarrow$$

$$\sin C = \frac{(\quad) \sin 30°}{6} \Rightarrow$$

$$C = \sin^{-1} \tfrac{5}{6} \Rightarrow$$

$$C = 56°26' \Rightarrow$$

and

$$B = 180° - (A + C) = 93°34'$$

$$\frac{a}{\sin A} = \frac{b}{\sin B} \Rightarrow$$

$$\frac{(?)}{\sin (?)} = \frac{b}{\sin (?)} \Rightarrow$$

$$b = \underline{\quad ? \quad} = 11.98$$

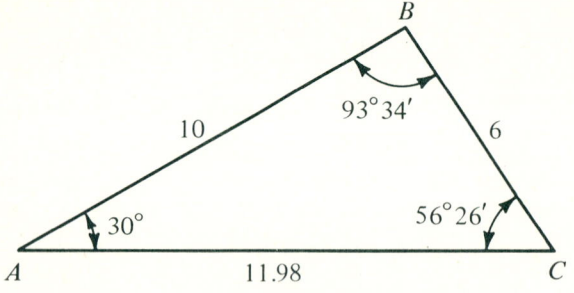

FIGURE 12.20.

This creates quite a different triangle from the one originally pictured. See Figure 12.20. It is possible to obtain an obtuse angle from these computations:

$$C = \sin^{-1} \tfrac{5}{6} \Rightarrow$$

$C = 56°26'$ or $C = 180° - 56°26' = 123°34'$

since the sine is positive in the first *and* second quadrants. The second value of angle C yields

$$b = 5.34$$

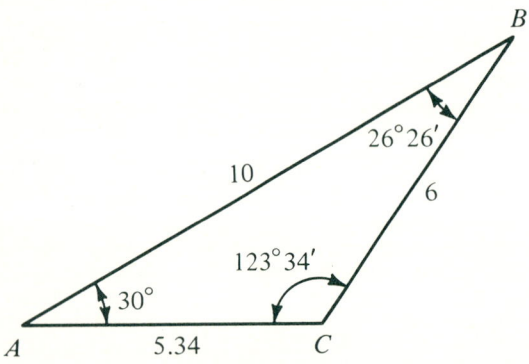

FIGURE 12.21.

and the triangle in Figure 12.21. This problem arose because the acute angle $\angle A = 30°$, being the only given angle, allowed either of the other two angles to be obtuse. (We have not had this problem before because we were always given one obtuse angle.) Such given information is ambiguous and the complete solution of such an ambiguous case consists of all the measurements for both triangles, or, if the application permits, the selection of the appropriate triangle using additional information.

EXAMPLE VI. A proposed road whose center line will have a bearing of S15°E is to cross an existing railroad track whose center line bearing is N70°E. The pictured coordinates having been established, it is necessary for the survey crew to find the distances a and b, in order to locate a stake at the proposed intersection. Find these distances. See Figure 12.22.

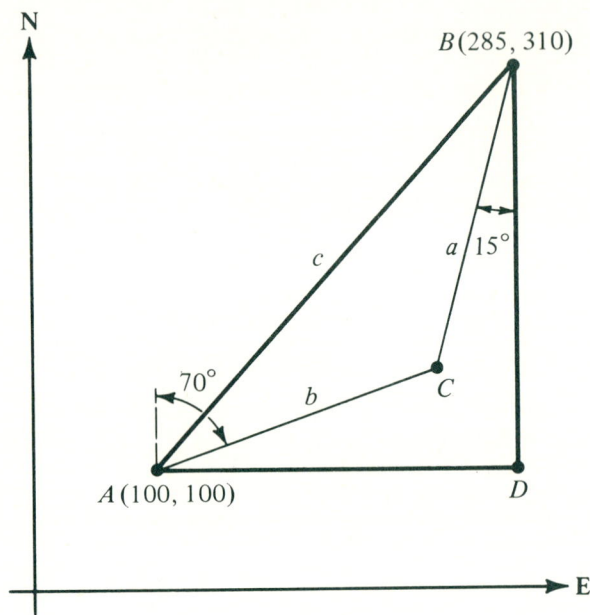

FIGURE 12.22.

Solution: From the right triangle ABD we can find the length of AB by the distance formula and the measure of $\angle BAC$ by finding the measure of $\angle BAD$. See Figure 12.23.

$$A = \tan^{-1}\left(\frac{210}{185}\right) = 48°37'$$

Now, we can find the measure of angle α from Figure 12.24:

$$\alpha = 70° - 48°37' \Rightarrow$$

$$\alpha = 21°23'$$

SECTION 12.4   Applications of the Law of Sines and the Law of Cosines

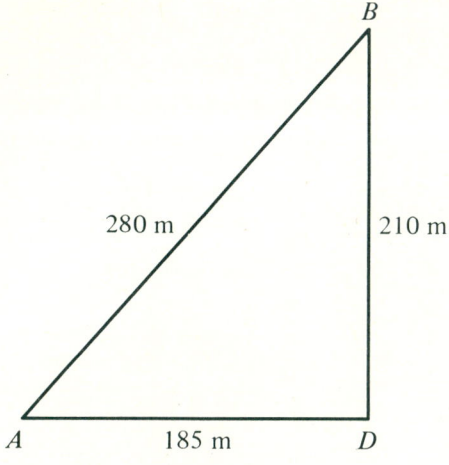

FIGURE 12.23.

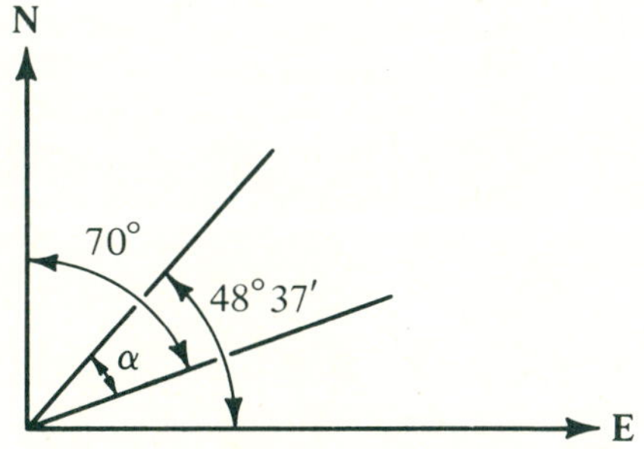

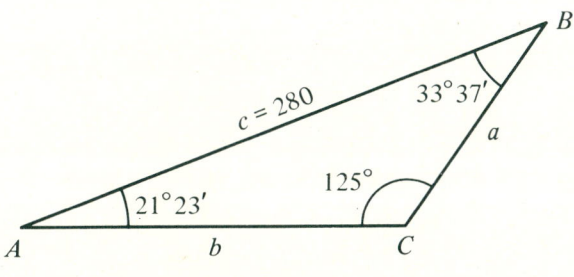

FIGURE 12.24.

The bearings of AC and BC lead us to the measure of $\angle C$ as $110° + 15° = 125°$. Now we know enough parts of the triangle to use the law of sines to find the measure of a and b.

FIGURE 12.25.

The law of sines yields:

$$\frac{c}{\sin C} = \frac{a}{\sin A} \Rightarrow$$

$$a = \frac{c \sin A}{\sin C} = \frac{(\ ?\ )(\ ?\ )}{(\ ?\ )} \Rightarrow$$

$$a = \underline{\qquad ?\qquad}$$

Also,

$$\frac{b}{\sin B} = \frac{c}{\sin C} \Rightarrow$$

$$b = \frac{c \sin B}{\sin C} = \frac{(\ ?\ )(\ ?\ )}{(\ ?\ )} \Rightarrow$$

$$b = \underline{\qquad ?\qquad}$$

Your answers should be $a \approx 124.6$ m and $b \approx 189.2$ m.

EXERCISE 12.4

A. Use the law of cosines to find the measure of the unknown sides of the following triangles. Sketch the triangles.

1. (a) $\angle A = 40°, b = 28, c = 13$
   (b) $\angle B = 25°, a = 28, c = 20$
   (c) $\angle C = 115°, a = 13, b = 20$

2. (a) $\angle B = 120°, a = c = 100$
   (b) $\angle A = 140°, b = c = 40$
   (c) $\angle B = 40°, c = 31.6, a = 50$

Use the law of sines to find the measure of the unknown angles in each of the following. Sketch the triangles.

3. (a) $\angle A = 10°, a = 100, b = 288$
   (b) $\angle C = 140°, a = 197, c = 288$

4. (a) $\angle A = 15°, b = 200, a = 151$
   (b) $\angle A = 20°, a = 200, b = 335$

B. Use the laws of sines and cosines to find all the unknown parts of the following triangles. If there is no possible triangle, write "no solution." If there are two possible triangles, give both solutions.

5. $\angle A = 40°, b = 28, c = 13$ (See Problem 1.)
6. $\angle B = 120°, a = c = 100$ (See Problem 2.)

7. $\angle A = 10°$, $a = 100$, $b = 288$ (See Problem 3.)

8. $\angle A = 15°$, $b = 200$, $a = 151$ (See Problem 4.)

9. (a) $\angle A = 30°$, $a = 60$, $c = 100$
   (b) $\angle A = 30°$, $a = 40$, $c = 100$

10. (a) $\angle B = 50°$, $b = 110$, $c = 140$
    (b) $\angle C = 50°$, $a = 140$, $c = 100$

C. APPLICATIONS

11. Consider the forward observer in Example II. If the observer is on a hill 4 kilometers from the gun emplacement, at a bearing of N40°E, and he locates the target at a bearing of S80°E, 10 kilometers away, what is the bearing from the gun to the target, and what is the range (distance)?

12. One forward observer on a hill 8 kilometers from a gun emplacement at a bearing of N30°W spots the target at a bearing of N20°E. Another observer 10 kilometers from the gun at a bearing of N40°E spots the target at a bearing of N50°W. What is the bearing from the gun to the target, and what is the range?

13. By standing at the west end of a lake, a surveying team is able to determine the placement of stakes on each side of the lake at its widest point. If one stake is located 1400 meters away, at an azimuth of 30°, and the other is 1200 meters away from the same spot, at an azimuth of 140°, how wide is the lake?

14. In the illustration below, points C and D are located across the river from A and B. Find the distance from C to D.

15. Two lighthouses 100 kilometers apart spot a ship at noon. From A its bearing is N50°W, and from B its bearing is N30°E. At 2:00 P.M. the new bearing from A is N70°W and from B is N10°W. Find the speed of the ship if B is due east of A.

12.5 INTRODUCTION TO POLAR COORDINATES

The location of a point in the plane can be established in other ways than by its horizontal (x) and vertical (y) distance from the origin. It is also common to identify the point P by giving its radial distance r and the angle θ, as shown in Figure 12.26. The ordered pair (r, θ) is called the

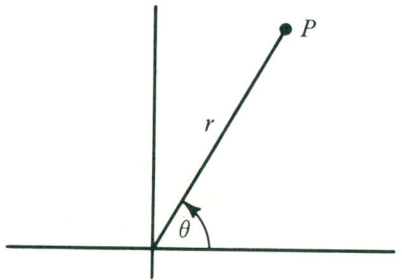

FIGURE 12.26.

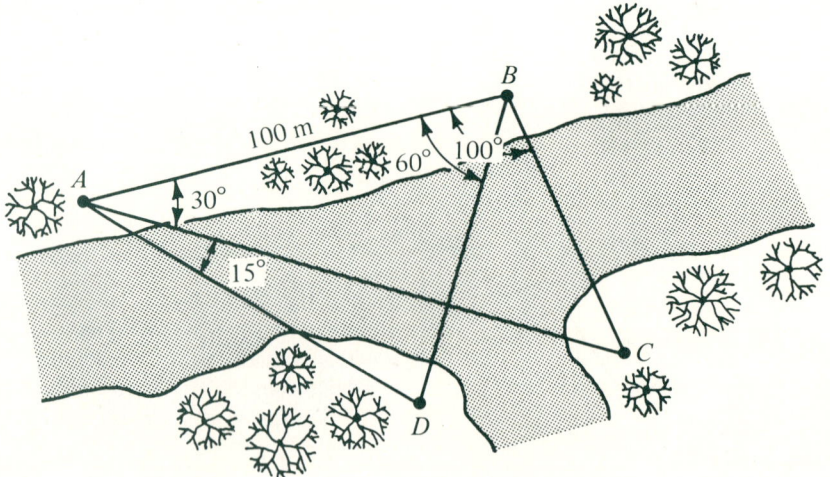

polar coordinates[1] for P. If the rectangular coordinates are (x, y), then the relationships between x and y and r and θ are established by the following illustration. These results are stated in Theorem I.

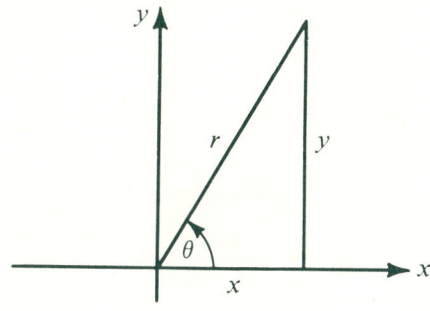

FIGURE TI.

THEOREM I.
If P has polar coordinates (r, θ) and Cartesian coordinates (x, y), then

$$x^2 + y^2 = r^2$$
$$x = r \cos \theta$$

and

$$y = r \sin \theta$$

Negative angles are measured clockwise and negative values of r are located on the opposite side of the origin from the terminal side of the angle θ. The two different names for the points P_1 and P_2 in Figure 12.27 illustrate these sign conventions. They also illustrate the ambiguity inherent in the polar coordinate system. There are many names for each point.

EXAMPLE I. Match each point in Figure 12.27 with its name.

P_3 _____
P_4 _____
P_5 _____
P_6 _____
P_7 _____

(a) $(1, -\pi/4)$
(b) $(-2, \pi/6)$
(c) $(3, 5\pi/6)$
(d) $(2, 7\pi/3)$
(e) $(2, \pi/6)$

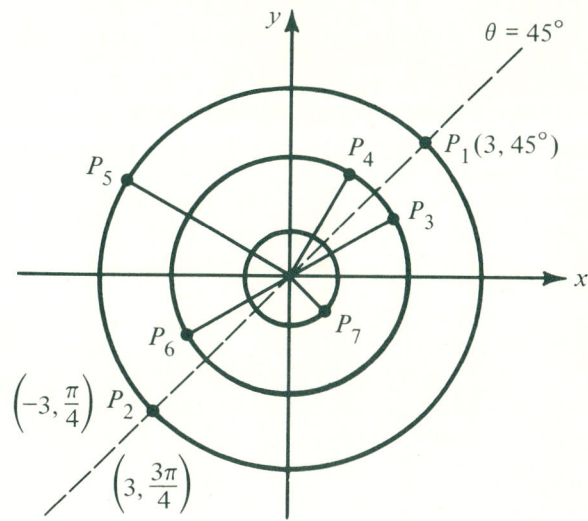

FIGURE 12.27.

The blanks should be "e, d, c, b, a."

EXAMPLE II. Change the following rectangular coordinates to polar coordinates: (a) $(4, 10)$ (b) $(-1, \sqrt{3})$. See Figure 12.28.

Solution: For (a),

$$x^2 + y^2 = r^2 \Rightarrow r = \sqrt{x^2 + y^2} \Rightarrow$$
$$r = \sqrt{16 + 100} \approx 10.8$$
$$\tan \theta = y/x \Rightarrow \tan \theta = 10/4 \Rightarrow$$
$$\theta = \tan^{-1} 2.5 \Rightarrow$$
$$\theta \approx 68°12'$$
$$(r, \theta) = (10.8, 68°12')$$

For (b),

$$r = \sqrt{x^2 + y^2} = \sqrt{\underline{\ ?\ }} \Rightarrow$$
$$r = \underline{\ ?\ }$$
$$\tan \theta = y/x \Rightarrow \tan \theta = \underline{\ ?\ } \Rightarrow$$
$$\theta = \tan^{-1}(\underline{\ ?\ }) \Rightarrow$$
$$\theta \approx \underline{\ ?\ } \Rightarrow$$
$$(r, \theta) = (2, 2\pi/3)$$

CHAPTER 12 Applications of Trigonometry

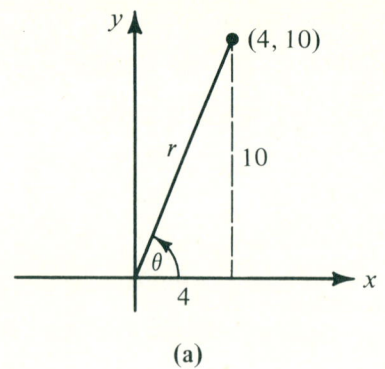

(a)

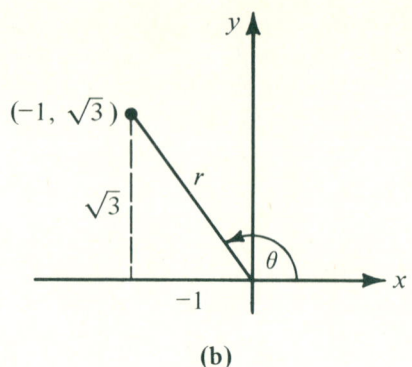

(b)

FIGURE 12.28.

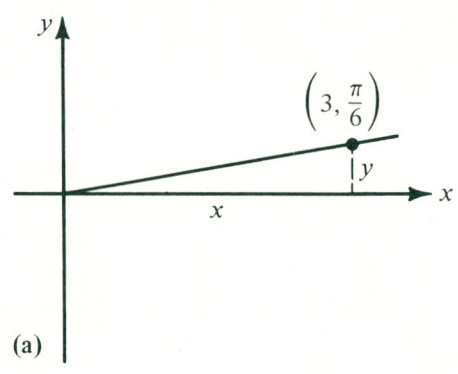

(a)

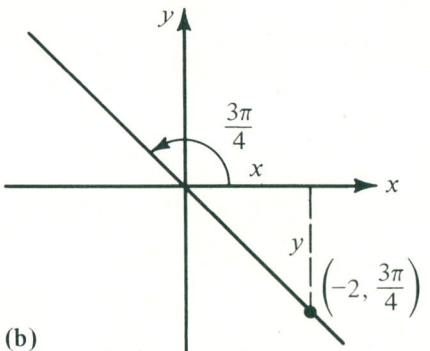

(b)

FIGURE 12.29.

EXAMPLE III. Change the following polar coordinates to Cartesian coordinates: (a) $(3, \pi/6)$ (b) $(-2, 3\pi/4)$. See Figure 12.29.

Solution: For (a),

$$x = r \cos \theta \Rightarrow$$
$$x = 3 \cos \pi/6 = 3\sqrt{3}/2$$
$$y = r \sin \theta \Rightarrow$$
$$y = 3 \sin \pi/6 = 3/2$$
$$(x, y) \approx (9.8, 1.5)$$

For (b),

$$x = r \cos \theta \Rightarrow$$
$$x = (\underline{\ ?\ }) \cos (\underline{\ ?\ }) = \underline{\ ?\ }$$
$$y = r \sin \theta \Rightarrow$$

$$y = (\underline{\ ?\ }) \sin (\underline{\ ?\ }) = \underline{\ ?\ }$$
$$(x, y) \approx (1.4, -1.4)$$

Now, let's use the equations relating polar and Cartesian coordinates to change equations given in one form to the other.

EXAMPLE IV. Straight lines passing through the origin have simple equations in both coordinate systems. Compare the equations depicted in Figure 12.30.

Solution: In general, polar equations of the form

$$r = \theta$$

are equivalent to Cartesian equations of the form

$$y = mx$$

254

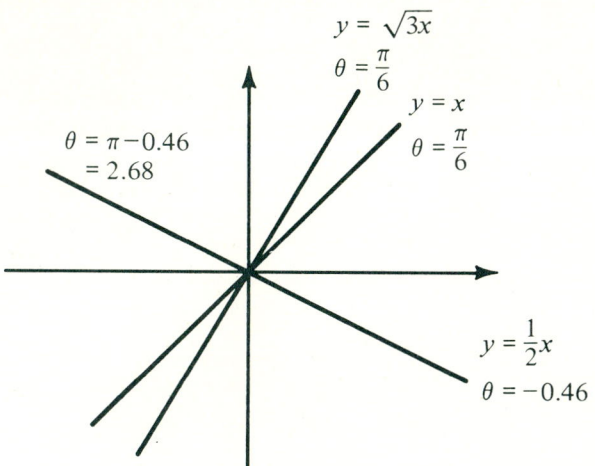

FIGURE 12.30

where $\theta = \tan^{-1}(y/x)$. This means that the slope of the line m is given by $\tan \theta$.

Now, let's examine equations of circles centered at the origin and on the x- or y-axis of Cartesian coordinate systems.

EXAMPLE V.
(a) Consider the equation $x^2 + y^2 = 4$.

$$x^2 + y^2 = r^2 \Rightarrow x^2 + y^2 = 4 \Leftrightarrow$$

$$r^2 = 4 \Rightarrow r = 2$$

The equation for a circle centered at the origin with radius a in polar coordinates is simply

$$r = a$$

The equations $r = -2$ represents the same circle as the equation $r = +2$.
(b) Consider $(x - 2)^2 + y^2 = 4$, a circle centered at $(2, 0)$ with $r = 2$:

$$x^2 + y^2 = r^2$$

and

$$x = r \cos \theta \Rightarrow (x - 2)^2 + y^2 = 4 \Rightarrow$$

$$(x^2 + y^2) - 4x = 0 \Leftrightarrow$$

$$r^2 - 4r \cos \theta = 0 \Rightarrow$$

$$r(r - 4 \cos \theta) = 0 \Rightarrow$$

$$r = 0 \quad \text{or} \quad r - 4 \cos \theta = 0 \Rightarrow$$

$$r = 4 \cos \theta$$

Since $\theta = \pi/2 \Rightarrow r = 0$ for $r = 4 \cos \theta$, the point $r = 0$ (the origin) is included in the cosine equation.
(c) Consider $(x + 3)^2 + y^2 = 9$:

$$x^2 + y^2 = r^2 \quad \text{and} \quad x = r \cos \theta$$

result in

$$(x + 3)^2 + y^2 = 9 \Rightarrow$$

$$(x^2 + y^2) + \underline{\quad ? \quad} = 0 \Leftrightarrow$$

$$\underline{\quad ? \quad} + 6(\underline{\quad ? \quad}) = 0 \Rightarrow$$

$$r(\underline{\quad ? \quad}) = 0 \Rightarrow$$

$$r = 0 \quad \text{or} \quad \underline{\quad ? \quad} = 0 \Rightarrow$$

$$r = -6 \cos \theta$$

From the original equation we see that this is a circle of radius 3 centered at $x = -3$, $y = 0$.
(d) Convert $r = 8 \sin \theta$ to rectangular coordinates and name its graph.

$$y = r \sin \theta \Rightarrow \sin \theta = \frac{?}{?} \Rightarrow$$

$$r = 8 \sin \theta \Leftrightarrow r = 8(\underline{\quad ? \quad}) \Rightarrow$$

$$r^2 = 8y$$

But

$$x^2 + y^2 = r^2 \Rightarrow$$

$$\underline{\quad ? \quad} = 8y \Rightarrow$$

$$x^2 + (y - \underline{\quad ? \quad})^2 = \underline{\quad ? \quad}$$

which represents a circle of radius 4 centered at $x = 0$, $y = 4$.

To help emphasize the difference between the two coordinate equations, take careful note of the similar-looking equations having entirely dissimilar graphs.

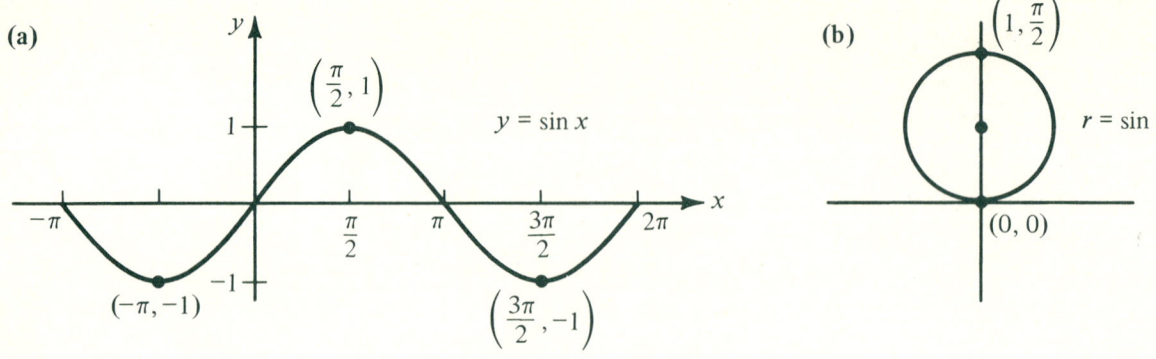

FIGURE 12.31.

EXAMPLE VI. Graph each of the following functions. (a) $y = \sin \theta$ and (b) $r = \sin \theta$.
See Figure 12.31.

Now let's continue our study of graphing in polar coordinates by examining some typical examples of polar relations. At this time we will graph these equations by creating extensive tables of ordered pairs (r, θ) by evaluating the relation at numerous arbitrarily selected values of θ. Then, by plotting these resulting points, we should be able to draw the graph. The more familiar we become with polar coordinates the fewer points we will need.

EXAMPLE VII. Plot the remaining points from the following table and complete the graph for $r = (\sin \theta)/(\cos \theta)$.

θ	0	$\pi/6$	$\pi/4$	$\pi/3$	$\pi/2$	$2\pi/3$	$3\pi/4$	$5\pi/6$	π
r	0	0.67	1.41	3.46	0	3.46	1.41	0.67	0

Do you recognize the graph in Figure 12.32 as a concave-up parabola? Let's change the equation to Cartesian coordinates.

$$r = \frac{\sin \theta}{\cos^2 \theta} \Rightarrow$$

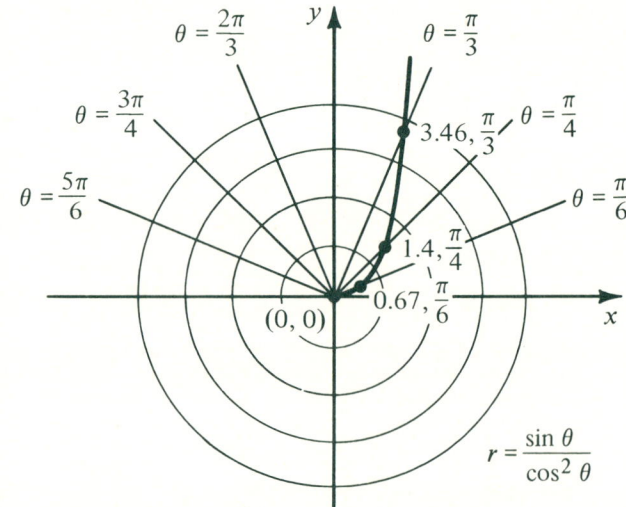

FIGURE 12.32.

$$r \cos^2 \theta = \sin \theta \Rightarrow$$
$$r^2 \cos^2 \theta = r \sin \theta \Rightarrow$$
$$x^2 = y$$

This example illustrates the fact that Cartesian coordinates are often easier to work with than polar coordinates. Equations with symmetry about a point (like a circle) are generally better suited to polar coordinates than those symmetric about a line (like $y = x^2$) or those with no symmetry.

Example VIII depicts a graph usually considered in polar coordinates. See Figure 12.33.

SECTION 12.5 Introduction to Polar Coordinates

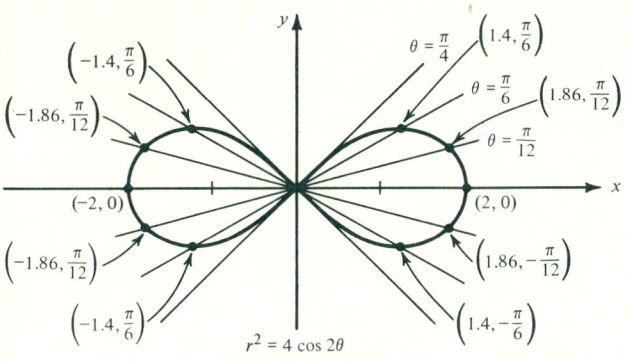

FIGURE 12.33.

EXAMPLE VIII. Graph $r^2 = 4\cos 2\theta$, called a *lemniscate*. See Figure 12.33.

θ	0	$3\pi/4$	$\pi/12$	$\pi/8$	$\pi/6$	$\pi/4$
r	± 2	0	± 1.86	± 1.68	± 1.4	0

EXAMPLE IX. Graph $r = 1 - 2\sin\theta$, called a *limaçon*. See Figure 12.34.

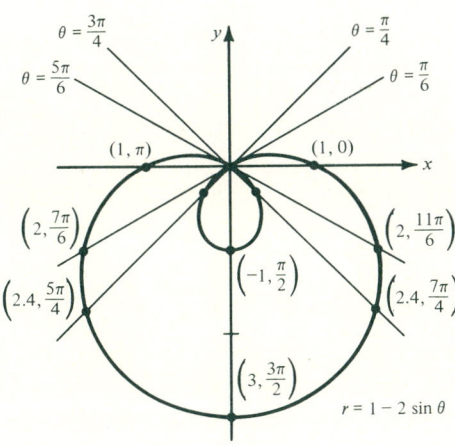

FIGURE 12.34.

θ	0	$\pi/6$	$\pi/4$	$\pi/2$	$3\pi/4$	$5\pi/6$	π	$7\pi/6$	$5\pi/4$	$3\pi/2$
r	1	0	-0.4	-1	-0.4	0	1	2	2.4	3

EXAMPLE X. Graph $r = \dfrac{4}{2 - \sin\theta}$. See Figure 12.35.

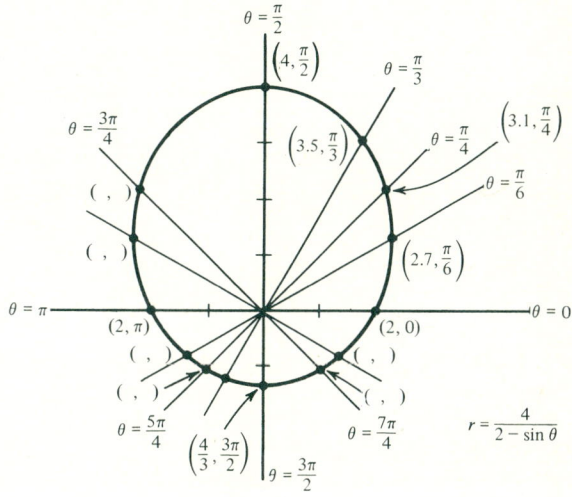

FIGURE 12.35.

r	0	$\pi/6$	$\pi/4$	$\pi/3$	$\pi/2$	$2\pi/3$	$3\pi/4$	$5\pi/6$	π	$5\pi/4$	$3\pi/2$	$7\pi/4$
θ	2	2.7	3.1	3.5	4	?	?	?	2	?	?	?

Now let's change the equation to Cartesian coordinates to see if it really is the equation of an ellipse.

$$r = \frac{4}{2 - \sin\theta} \Rightarrow$$
$$2r - r\sin\theta = 4 \Rightarrow$$
$$2\sqrt{x^2 + y^2} - y = 4 \Rightarrow$$
$$2\sqrt{x^2 + y^2} = y + 4 \Rightarrow$$
$$4(x^2 + y^2) = y^2 + 8y + 16 \Rightarrow 4x^2 + 3y^2 - 8y = 16$$

This is the equation of an ellipse since $4 \cdot 3 > 0$ and $4 \neq 3$. See Section 13.4.

CHAPTER 12   Applications of Trigonometry

EXERCISE 12.5

A. FUNDAMENTALS
Plot the following polar coordinates.
1. $(1, \pi/3), (-1, -\pi/3), (-1, \pi/3), (1, -\pi/3)$
2. $(3, \pi/6), (-2, \pi/6), (0, \pi/6), (7, \pi/6)$
3. $(2, \pi/4), (-2, 5\pi/4), (2, 9\pi/4), (-2, -3\pi/4)$
4. $(1, 0), (1, \pi/4), (1, \pi/2), (1, \pi), (1, 3\pi/2)$
5. $(0, 0), (1, \pi/4), (2, \pi/2), (3, 3\pi/4), (4, \pi), (5, 5\pi/4)$
6. Connect the points in Problems 3, 4, and 5 with a smooth curve. Can you give an equation for each graph?

Sketch the graphs of the following polar relations.
7. (a) $\theta = \pi/3$ (b) $\theta = 3\pi/4$
8. (a) $r = 1$ (b) $r = 5$
9. (a) $r = 2\cos\theta$ (b) $r = -2\cos\theta$
10. (a) $r = 4\sin\theta$ (b) $r = -4\sin\theta$
11. Change the polar equations in Problems 7–10 to Cartesian coordinates. The graphs will help.

B. ESSENTIALS
Use $x = r\cos\theta$, $y = r\sin\theta$, and $x^2 + y^2 = r^2$ to change the following to polar coordinates. Simplify completely. Name each graph.
12. (a) $x^2 + y^2 = 25$ (b) $y = \pm\sqrt{16 - x^2}$
13. (a) $x^2 + y^2 = 64$ (b) $y = +\sqrt{64 - x^2}$
14. (a) $(x - 1)^2 + y^2 = 25$
    (b) $x^2 + (y - 2)^2 = 25$
15. (a) $x^2 - y^2 = 1$ (b) $y^2 - x^2 = 1$
16. (a) $4x^2 + y^2 = 4$ (b) $x^2 + 4y^2 = 4$
17. (a) $y = x^2$ (b) $y = x^2 - 2x + 1$

Identify the following equations by translating them from polar coordinates to Cartesian coordinates.
18. ★(a) $r = 8\cos\theta$ (b) $r = -16\sin\theta$
19. (a) $r = 10\sin\theta$ (b) $r = -10\cos\theta$
20. (a) $r = \dfrac{6}{2 - 2\sin\theta}$ (b) $r = \dfrac{4}{1 - 2\cos\theta}$

Graph the following polar relations.
21. (a) $r = 1 + \cos\theta$ (b) $r = 2 + \cos\theta$
    (c) $r = 1 + 2\cos\theta$
22. (a) $r = \cos\theta$ (b) $r = \cos 3\theta$
    (c) $r = \cos 5\theta$
23. (a) $r = \sin 2\theta$ (b) $r = \sin 4\theta$
    (c) $r = \sin 6\theta$
24. (a) $r^2 = 2\cos 2\theta$ (b) $r^2 = 8\cos 2\theta$
25. (a) $r = \dfrac{1}{1 - \cos\theta}$ (b) $r = \dfrac{2}{1 + \cos\theta}$
26. (a) $r = \dfrac{1}{1 - \sin\theta}$ (b) $r = \dfrac{2}{1 + \sin\theta}$
27. (a) The spiral of Archimedes, $r = \theta$.
    (b) The logarithmic spiral, $\ln r = \theta \Leftrightarrow r = e^\theta$.
● 28. The lemniscate of Bernoulli, $r^2 = \cos 2\theta$.
● 29. The Ouija board curve, $r\theta = \sin\theta$.
● 30. The cissoid of Diocles, $r = \sin\theta \tan\theta$, $0 \le \theta \le \pi/2$.
● 31. A lituus, $r^2\theta = 1$.

12.6 PROPERTIES OF POLAR COORDINATES

In the last section we explored the basics of polar relations by studying examples. In this section we will examine general properties of polar relations and broad classifications of such equations based upon their graphs. This section, therefore, is a generalization of Section 12.5.

First, let's discuss symmetry and tests for symmetry for polar relations. If points (r_1, θ_1) and (r_2, θ_2) are on the graph of a curve that is symmetric about the x-axis, then the points $(r_1, -\theta_1)$ and $(r_2, -\theta_2)$ must be on the graph also. See Figure 12.36. This means that for a polar equation

$$r = f(\theta)$$

SECTION 12.6   Properties of Polar Coordinates

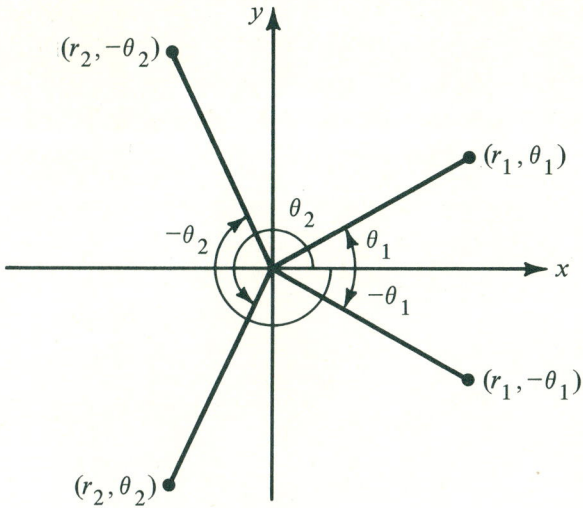

FIGURE 12.36.

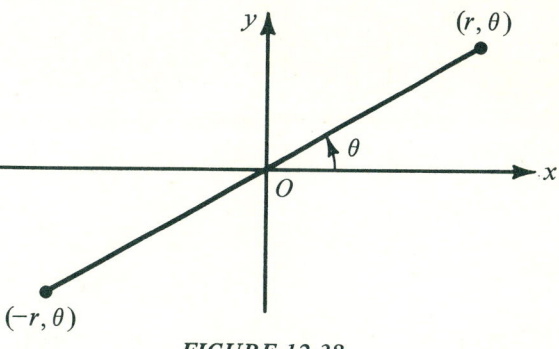

FIGURE 12.38.

to be symmetric about the x-axis,

$$f(-\theta) = f(\theta)$$

for all values of θ in the domain of f.

We know that

$$\cos(-\theta) = \cos\theta$$

for all θ. Therefore, one class of relations whose graphs must be symmetric about the x-axis consists of those involving only $\cos\theta$.

What can we say about equations involving only $\sin\theta$? We know that

$$\sin(\pi - \theta) = \sin\theta$$

for all θ. Geometrically, θ and $\pi - \theta$ are located on opposite sides of the y-axis. Therefore, points

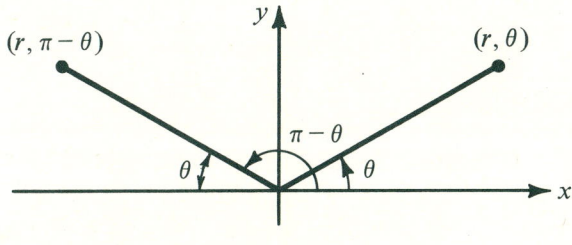

FIGURE 12.37.

like (r, θ) and $(r, -\theta)$ are symmetric about the y-axis. See Figure 12.37.

In general, if

$$f(\pi - \theta) = f(\theta)$$

for all values of θ in the domain of f, then its graph is symmetric about the y-axis.

For symmetry about the origin we need to have $(-r, \theta)$ on the graph whenever (r, θ) is on the graph. See Figure 12.38.

Equations involving even powers of r, such as

$$r^2 = \sin\theta$$

are examples of polar relations having origin symmetry. Notice that replacing r with $-r$ does not result in a different equation. Also notice the geometric consequences of evaluating the relation at some value of θ, for example, $\theta = \pi/6$. See Figure 12.39.

$$\theta = \frac{\pi}{6} \Rightarrow r^2 = \sin\frac{\pi}{6} = \frac{1}{2}$$

$$r = \frac{\pm\sqrt{2}}{2}$$

In summary, the following are the tests for symmetry where

$$r = f(\theta) \Leftrightarrow F(r, \theta) = f(\theta) - r = 0$$

259

CHAPTER 12   Applications of Trigonometry

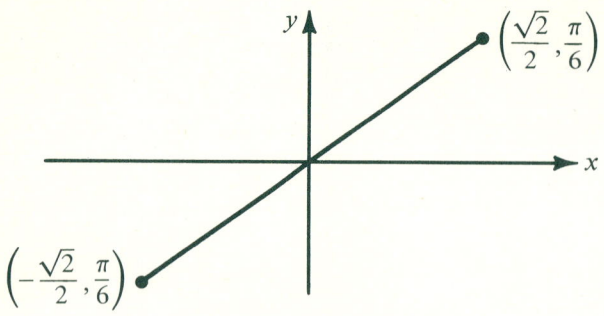

FIGURE 12.39.

TESTS FOR SYMMETRY.
1. Symmetric about the x-axis:
$$F(r, -\theta) = F(r, +\theta)$$
2. Symmetric about the y-axis:
$$F(r, \pi - \theta) = F(r, \theta)$$
3. Symmetric about the origin:
$$F(-r, \theta) = F(+r, \theta)$$
or
$$F(r, \pi + \theta) = F(r, \theta)$$

In Example I, each of the tests for symmetry is performed.

EXAMPLE I. Test the following polar relations for x-axis, y-axis, and origin symmetry.
(a) $r^2 = a^2 \cos 2\theta \Leftrightarrow F(r, \theta) = r^2 - a^2 \cos 2\theta = 0$
 1. $F(r, -\theta) = r^2 - a^2 \cos 2(-\theta) \Rightarrow$
    $F(r, -\theta) = r^2 - a^2 \cos 2\theta \Rightarrow$
    $F(r, -\theta) = F(r, \theta)$ (It is symmetric about the x-axis.)
 2. $F(r, \pi - \theta) = r^2 - a^2 \cos 2(\pi - \theta) \Rightarrow$
    $F(r, \pi - \theta) = r^2 - a^2 \cos 2\theta \Rightarrow$
    $F(r, \pi - \theta) = F(r, \theta)$ (It is symmetric about the y-axis.)
 3. $F(-r, \theta) = (-r)^2 - a^2 \cos 2\theta \Rightarrow$
    $F(-r, \theta) = r^2 - a^2 \cos 2\theta \Rightarrow$
    $F(-r, \theta) = F(r, \theta)$ (It is symmetric about the origin.)
See Example VIII for the graph.

(b) $r = 1 - 2 \sin \theta \Leftrightarrow F(r, \theta) = r + 2 \sin \theta - 1 = 0$
 1. $F(r, -\theta) = r + 2 \sin (\quad) - 1 \Rightarrow$
    $F(r, -\theta) = \underline{\quad ? \quad} \Rightarrow$
    $F(r, -\theta) = F(r, \theta)$
 2. $F(r, \pi - \theta) = r + 2 \sin (\quad) - 1 \Rightarrow$
    $F(r, \pi - \theta) = \underline{\quad ? \quad} \Rightarrow$
    $F(r, \pi - \theta) (= \text{or} \neq) F(r, \theta)$
 3. $F(-r, \theta) = (\quad) + 2 \sin \theta - 1 \Rightarrow$
    $F(-r, \theta) \neq F(r, \theta)$ (The test fails.)
See Example IX for the graph.

There is one drawback to these tests for symmetry of polar relations. They are not conclusive unless the test works. If the test fails, the graph may still be symmetric, as illustrated in Example II.

EXAMPLE II. Test the following polar relation for symmetry, then graph:
$$r = \sin \left(\frac{\theta}{2}\right)$$
1. $F(r, -\theta) = r - \sin \left(\frac{-\theta}{2}\right)$
   $= r + \sin \left(\frac{\theta}{2}\right)$
   $\neq F(r, \theta) \Rightarrow$
   $F(r, -\theta) = \text{or} \neq F(r, \theta)$
2. $F(r, \pi - \theta) = r - \sin (\quad)$
   $= r \underline{\quad ? \quad} \sin \left(\frac{\theta}{2}\right) (= \text{or} \neq) F(r, \theta)$
3. $F(-r, \theta) = (-r) - \sin \frac{\theta}{2} \neq F(r, \theta)$

None of the tests work, but look at the symmetry in its graph in Figure 12.40. The arrows indicate the sequence in which the curve is drawn as θ varies from 0 to 4π.

θ	0	$\frac{\pi}{2}$	π	$\frac{3\pi}{2}$	2π	$\frac{5\pi}{2}$	3π	$\frac{7\pi}{2}$	4π
r	0	0.7	1	0.7	0	-0.7	-1	-0.7	0

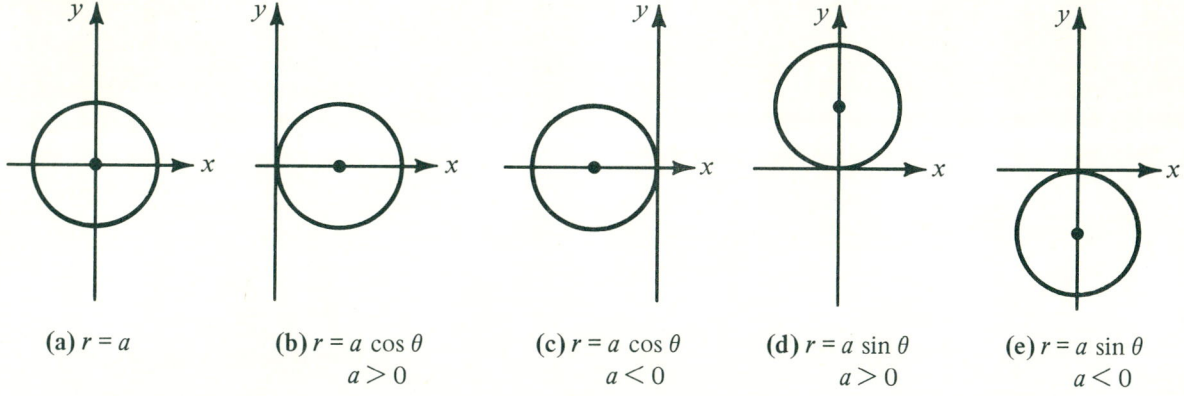

(a) $r = a$

(b) $r = a \cos \theta$
$a > 0$

(c) $r = a \cos \theta$
$a < 0$

(d) $r = a \sin \theta$
$a > 0$

(e) $r = a \sin \theta$
$a < 0$

FIGURE 12.40.

The reason the tests did not reveal the symmetry can be found by examining the polar coordinates for a pair of symmetric points and comparing them with the ones paired by the test. For example, $(0.7, \pi/2)$ and $(0.7, 3\pi/2)$ are on opposite sides of the x-axis. The test for symmetry about the x-axis says that if (r, θ) is there, then $(r, -\theta)$ will be there. The point $(0.7, -\pi/2)$ does not satisfy the original equation but it is the *same* point as $(0.7, 3\pi/2)$, which does satisfy the original equation. The absence of the uniqueness of the names of the points in the plane can cause difficulty unless you remember the ambiguity.

Below are some examples of the common classification of graphs in polar coordinates. You will find this list helpful when you are trying to graph any of these particular curves. Considering the tests for symmetry appropriate for the graph described will help you understand them. A table of values is necessary for complete understanding of why an equation has the indicated graph. This will be accomplished in the exercises.

CIRCLES: A circle of radius a centered at the origin has a polar equation $r = a$. See Figure 12.41(a). A circle of radius $a/2$ centered at $(a/2, 0)$ has a polar equation of the form $r = a \cos \theta$, and one centered at $(a/2, \pi/2)$ is of the form $r = a \sin \theta$.

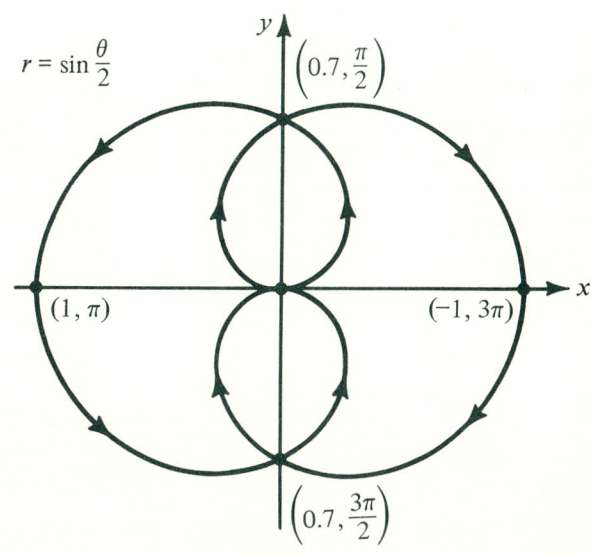

FIGURE 12.41.

ROSES: A rose has a polar equation of the form $r = a \cos n\theta$ or $r = a \sin n\theta$, where there are n pedals if n is odd or $2n$ pedals if n is even. See Fig. 12.42(a) and 12.42(b).

LIMAÇONS: A limaçon has an equation of the form $r = a \pm b \sin \theta$ or $r = a \pm b \cos \theta$. The graph

CHAPTER 12   Applications of Trigonometry

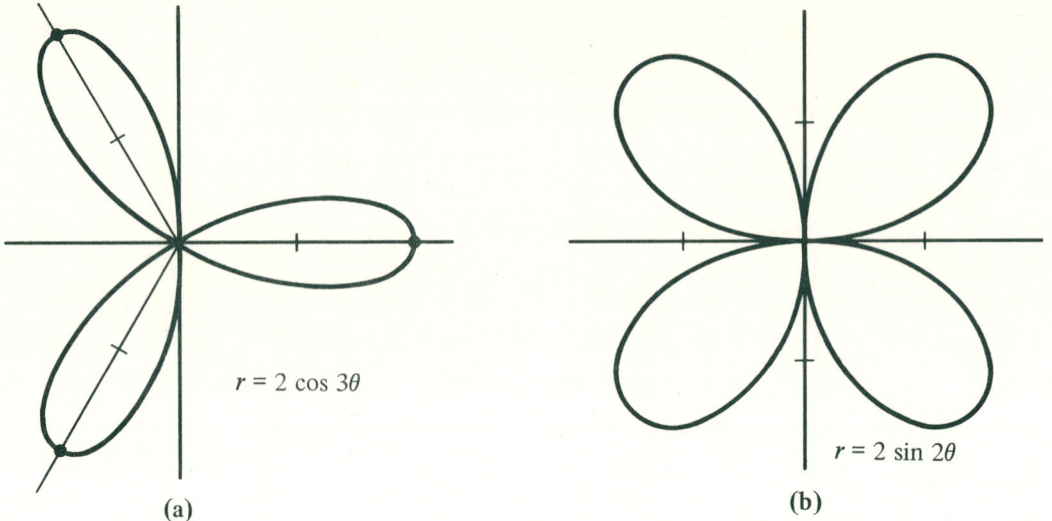

FIGURE 12.42.

can be any one of three basic shapes depending upon the relative sizes of a and b. Those shown in Figure 12.43 are the sine forms, with a and b positive. Those for cosine are similar except that they are symmetric about the x-axis. Those with negative a or b are rotated 180°.

LEMNISCATES: A lemniscate has an equation of the form $r^2 = a^2 \cos 2\theta$. See Example VIII. It is a figure eight.

SPIRALS: The spiral of Archimedes has the polar equation $r = a\theta$. See Figure 12.44.

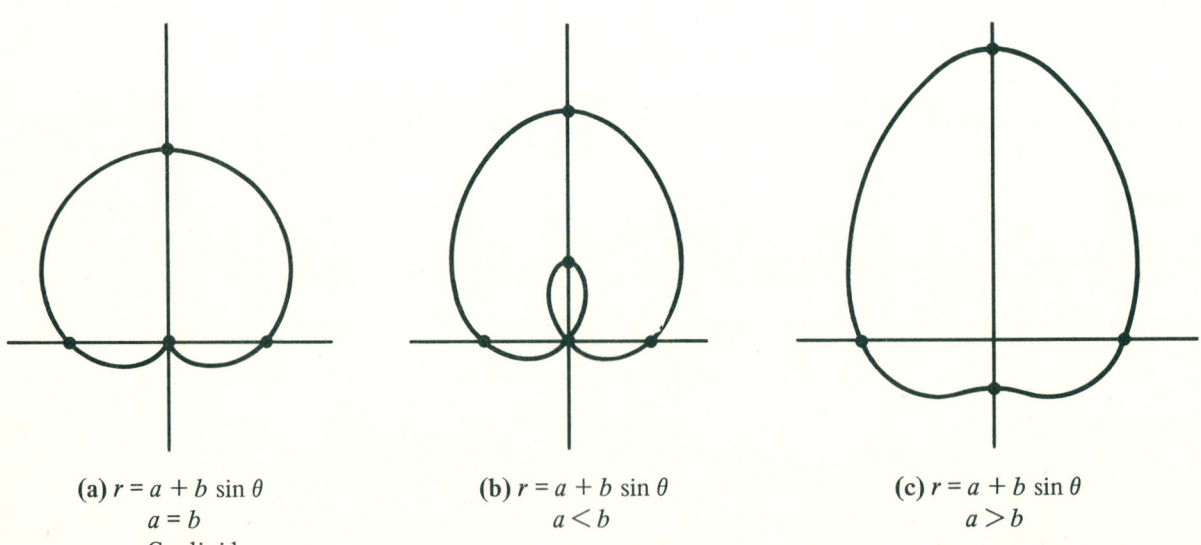

(a) $r = a + b \sin \theta$
   $a = b$
   Cardioid

(b) $r = a + b \sin \theta$
   $a < b$

(c) $r = a + b \sin \theta$
   $a > b$

FIGURE 12.43.

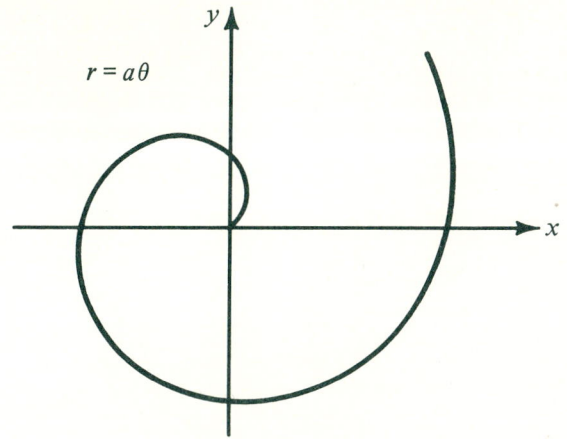

FIGURE 12.44.

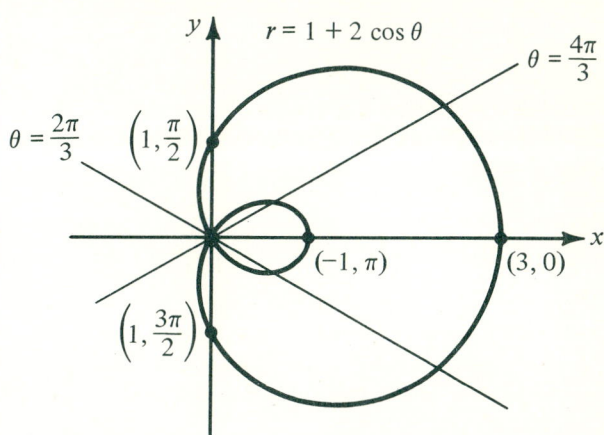

FIGURE 12.45. Tangents through the Origin.

Graphing many polar equations is simplified by finding the value(s) of θ for which $r = 0$. For example, if

$$r = 1 + 2 \cos \theta$$

then $r = 0$ yields

$$0 = 1 + 2 \cos \theta \Rightarrow$$
$$\theta = \cos^{-1}\left(-\tfrac{1}{2}\right) \Rightarrow$$
$$\theta = \frac{2\pi}{3} \quad \text{or} \quad \frac{4\pi}{3}$$

By examining Figure 12.45, we can see that these radial lines play important roles in the picture, since they tell us *exactly* when the curve passes through the origin. They are called *tangents through the origin*. In Figure 12.45 the r-values for $2\pi/3 < \theta < 4\pi/3$ are all negative.

In calculus it will be necessary to find the points of intersection of the graphs of two polar relations. Since a point lying on both curves may use different values of r and θ for each curve (remember, each point has many names), such algebra must be accompanied by a sound geometric appreciation of the graphs involved. Since the origin is a common point of intersection, which often goes by different names, we usually check this point separately.

EXAMPLE III. Find all points of intersection of the graphs of $r = 2 - 2 \cos \theta$ and $r = 2 \cos \theta$. Sketch the curves first.

Solution: See Figure 12.46.
$$r = 2 - 2 \cos \theta \quad \text{and} \quad r = 2 \cos \theta \Rightarrow$$
$$2 - 2 \cos \theta = 2 \cos \theta \Rightarrow$$
$$\cos \theta = \tfrac{1}{2} \Rightarrow$$
$$\theta = \frac{\pi}{3} \quad \text{or} \quad \frac{5\pi}{3}$$

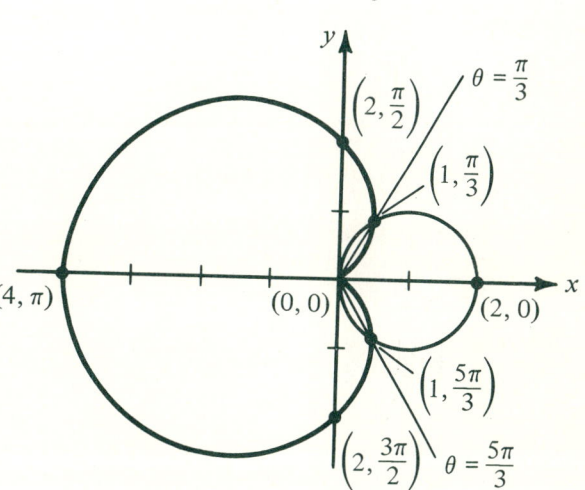

$r = 2 - 2 \cos \theta$ and $r = 2 \cos \theta$

FIGURE 12.46.

CHAPTER 12 Applications of Trigonometry

But notice that if $r = 0$, we have

$2 - 2\cos\theta = 0$ and $2\cos\theta = 0 \Rightarrow$

$\cos\theta = 1$ and $\cos\theta = 0 \Rightarrow$

$\theta = 0$ and $\theta = \dfrac{\pi}{2}$ or $\dfrac{3\pi}{2}$

Therefore, both graphs pass through the origin, but we did not discover this common point when we solved the equations simultaneously. Notice that $r = 0$ for different values of θ for each curve. Did you notice that the tangents through the origin are the x- and y-axes and that there is x-axis symmetry?

EXERCISE 12.6

A. FUNDAMENTALS

Test the following polar relations for symmetry as indicated. Draw the pair of points for each part.

1. $f(\theta) = \sin\theta$ is symmetric about the y-axis, since $f(\theta) = f(\pi - \theta)$ as illustrated below.

(a) $f\left(\dfrac{\pi}{3}\right) = \underline{\ ?\ }$ $f\left(\dfrac{2\pi}{3}\right) = \underline{\ ?\ }$

(b) $f\left(\dfrac{\pi}{4}\right) = \underline{\ ?\ }$ $f\left(\dfrac{3\pi}{4}\right) = \underline{\ ?\ }$

(c) $f\left(\dfrac{-\pi}{6}\right) = \underline{\ ?\ }$ $f\left(\dfrac{7\pi}{6}\right) = \underline{\ ?\ }$

2. $f(\theta) = \cos\theta$ is symmetric about the x-axis, since $f(\theta) = f(-\theta)$ as illustrated below. Draw the pair of points for each part.

(a) $f\left(\dfrac{\pi}{2}\right) = \underline{\ ?\ }$ $f\left(\dfrac{-\pi}{2}\right) = \underline{\ ?\ }$

(b) $f\left(\dfrac{3\pi}{4}\right) = \underline{\ ?\ }$ $f(-?) = \underline{\ ?\ }$

(c) $f\left(\dfrac{-5\pi}{6}\right) = \underline{\ ?\ }$ $f(?) = \underline{\ ?\ }$

3. $r^2 = \cos\theta \Leftrightarrow F(r, \theta) = \cos\theta - r^2 = 0$ is symmetric about the origin, since $F(r, \theta) = F(-r, \theta)$ as illustrated below. Draw the pair of points in each part.

(a) $F(1, 0) = 0$ $F(\underline{\ ?\ }, 0) = 0$

(b) $F\left(\underline{\ ?\ }, \dfrac{\pi}{3}\right) = 0$ $F(\underline{\ ?\ }, \underline{\ ?\ }) = 0$

(c) $F(1/\sqrt[4]{2}, \underline{\ ?\ }) = 0$ $F(\underline{\ ?\ }, \underline{\ ?\ }) = 0$

Name the symmetry present in each graph and see if the corresponding test for symmetry works.

4. $r = 4\cos\theta$

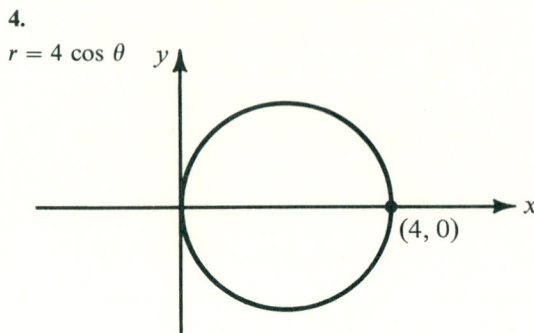

5. $r = 4\sin\theta$

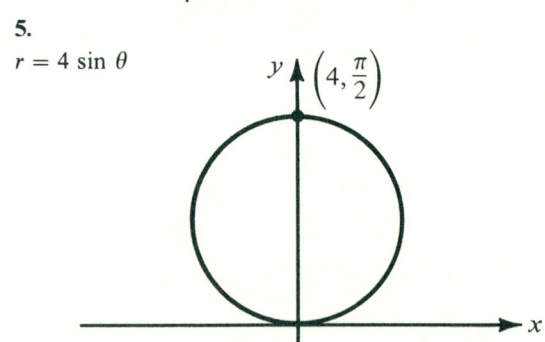

6. $r^2 = \sin 2\theta$

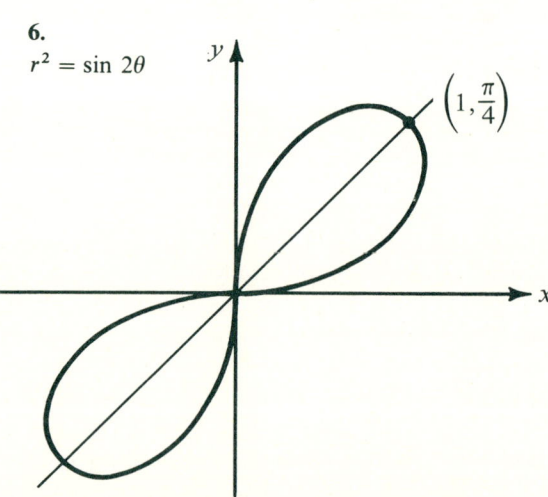

Solve the following equations simultaneously to find the point(s) of intersection of their graphs. Sketch the curves.

7. $r = 4\sin\theta$ $r = 4\cos\theta$

8. $r = 1 + 2\cos\theta$ $r = \cos\theta$

9. $r^2 = \sin 2\theta$ $r = 1$

B. ESSENTIALS

Test the following polar relations for symmetry, then sketch their graphs.

10. (a) $r^2 = \sin 2\theta$ (b) $r^2 = \cos 2\theta$

11. (a) $r^2 = \sin\theta$ (b) $r^2 = \sin 4\theta$
    (c) $r^2 = \sin 8\theta$

12. (a) $r = \cos 2\theta$ (b) $r = \cos 3\theta$
    (c) $r = \cos 4\theta$

13. (a) $r = 2 + 2\cos\theta$ (b) $r = 3 + 3\sin\theta$

14. (a) $r = 1 - 2\cos\theta$ (b) $r = 1 - 2\sin\theta$

15. (a) $r = 4 - 2\cos\theta$ (b) $r = 4 + 2\sin\theta$

Test each of the following pairs of polar relations for symmetry, find the point(s) of intersection, and sketch their graphs.

16. $r = \sin 2\theta$ $r = \sin\theta$

★17. $r = \cos 2\theta$ $r = -\cos\theta$

18. $r = \sqrt{2} + \sqrt{2}\cos\theta$ $r^2 = \cos 2\theta$

19. $r = 1 - 2\cos\theta$ $r = 4 + \cos\theta$

20. $r = 2\cos 2\theta$ $r = 2 + \cos\theta$

21. $r = 2\cos 2\theta$ $r = 2 + 4\sin\theta$

22. Find the tangents through the origin for the relations of Problems 18 and 20.

12.7 COMPLEX NUMBERS IN POLAR FORM

Before beginning this discussion of complex numbers, you may wish to review **Section 1.5**.

In this section we will compare the usual representation of a complex number

$$z = x + yi$$

where $i = \sqrt{-1}$, and its *polar form*, found by using the polar equations

$$x = r\cos\theta \quad \text{and} \quad y = r\sin\theta$$

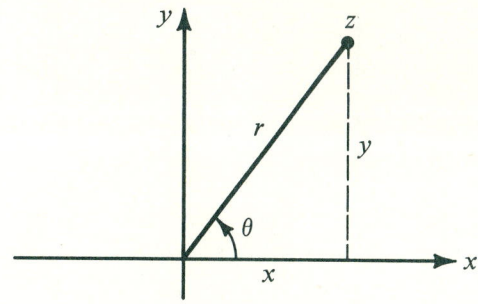

$$x + yi = r\cos\theta + r\sin\theta$$

FIGURE 12.47.

to write z as

$$z = r(\cos\theta + i\sin\theta)$$

We will often abbreviate this as

$$z = r \text{ cis } \theta$$

Consider the picture of a complex number $z = x + yi$ in Figure 12.47.

In this first example we will change from one form to the other.

EXAMPLE I. Write the following equations in polar form:
(a) $z = 2 + 2i$ (b) $z = -5\sqrt{3} + 5i$.

Solution: (a) To write the polar form for z, we need to find the values of r and θ in

$$z = r(\cos\theta + i\sin\theta)$$

Figure 12.48 depicts r, θ, x, and y.

$$\tan\theta = \frac{y}{x} = \frac{2}{2} \Rightarrow$$

$$\theta = \tan^{-1} 1 \Rightarrow$$

$$\theta = \frac{\pi}{4}$$

Then, by the Pythagorean theorem,

$$r^2 = 2^2 + 2^2 \Rightarrow$$

$$r = 2\sqrt{2}$$

CHAPTER 12 Applications of Trigonometry

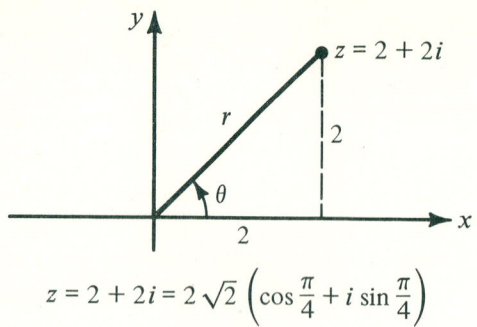

$$z = 2 + 2i = 2\sqrt{2}\left(\cos\frac{\pi}{4} + i\sin\frac{\pi}{4}\right)$$

FIGURE 12.48.

Therefore, the polar form for z is

$$z = 2\sqrt{2}\left(\cos\frac{\pi}{4} + i\sin\frac{\pi}{4}\right)$$

To change back to $z = x + yi$, we simply evaluate $\cos \pi/4$ and $\sin \pi/4$, and multiply through by $r = 2\sqrt{2}$ as follows. That is,

$$z = 2\sqrt{2}\left(\cos\frac{\pi}{4} + i\sin\frac{\pi}{4}\right) \Rightarrow$$

$$z = 2\sqrt{2}\left(\frac{\sqrt{2}}{2} + i\frac{\sqrt{2}}{2}\right) \Rightarrow$$

$$z = 2 + 2i$$

(b) To find the polar form for $z = -5\sqrt{3} + 5i$, see Figure 12.49.

$$\tan\theta = \frac{y}{x} = \frac{?}{?} \Rightarrow$$

$$\theta = \tan^{-1}\left(\frac{-\sqrt{3}}{3}\right) = \underline{\quad ? \quad}$$

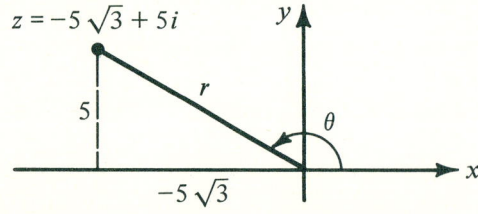

$$z = -5\sqrt{3} + 5i = 10 \text{ cis } \frac{5\pi}{6}$$

FIGURE 12.49.

and

$$r^2 = (x)^2 + (y)^2 = \underline{\quad ? \quad}$$

Therefore,

$$z = r(\cos\theta + i\sin\theta) \Rightarrow$$

$$z = 10\left(\cos\frac{5\pi}{6} + i\sin\frac{5\pi}{6}\right) \Rightarrow$$

$$z = 10 \text{ cis } \frac{5\pi}{6}$$

Now let's put the polar form of a complex number to work.

THEOREM I.
If z_1 and z_2 are two complex numbers,

$$z_1 = r_1(\cos\alpha + i\sin\alpha)$$

and

$$z_2 = r_2(\cos\beta + i\sin\beta)$$

Then, their product is

$$z_1 \cdot z_2 = r_1 \cdot r_2[\cos(\alpha + \beta) + i\sin(\alpha + \beta)]$$

Proof:

$$z_1 \cdot z_2 = r_1(\cos\alpha + i\sin\alpha)$$
$$\cdot r_2(\cos\beta + i\sin\beta)$$
$$= r_1 r_2(\cos\alpha\cos\beta + i\cos\alpha\sin\beta$$
$$+ i\sin\alpha\cos\beta + i^2\sin\alpha\sin\beta)$$
$$= r_1 r_2[(\cos\alpha\cos\beta - \sin\alpha\sin\beta)$$
$$+ (\cos\alpha\sin\beta + \sin\alpha\cos\beta)i]$$
$$= r_1 r_2[\cos(\alpha + \beta) + i\sin(\alpha + \beta)]$$

EXAMPLE II. Find the product $z_1 \cdot z_2$ for $z_1 = 2 + 2i$ and $z_2 = -5\sqrt{3} + 5i$ by simply multiplying them as usual and by using Theorem I.

Solution: We have

$$z_1 \cdot z_2 = (2 + 2i) \cdot (-5\sqrt{3} + 5i)$$
$$= \underline{\quad ? \quad} + \underline{\quad ? \quad} i$$

Then, from Example I,

$$z_1 = 2\sqrt{2}\left(\cos\frac{\pi}{4} + i\sin\frac{\pi}{4}\right)$$

and

$$z_2 = 10\left(\cos\frac{5\pi}{6} + i\sin\frac{5\pi}{6}\right)$$

Then

$$z_1 \cdot z_2 = r_1 \cdot r_2[\cos(\alpha + \beta) + i\sin(\alpha + \beta)] \Rightarrow$$

$$z_1 \cdot z_2 = (2\sqrt{2})(10)\cos\left(\frac{\pi}{4} + \frac{5\pi}{6}\right) + i\sin\left(\frac{\pi}{4} + \frac{5\pi}{6}\right)$$

$$= 20\sqrt{2}\left[\cos\left(\frac{13\pi}{12}\right) + i\sin\left(\frac{13\pi}{12}\right)\right]$$

$$\approx 20\sqrt{2}(-0.966) + i10\sqrt{2}(-0.259)$$

$$\approx -27.32 - 7.32i$$

Now let's examine z^n, where $n \in$ naturals. Use of the polar form is helpful at this stage. If

$$z = r(\cos\theta + i\sin\theta)$$

then, by Theorem I, with $\alpha = \beta = \theta$,

$$z^2 = r \cdot r[\cos(\theta + \theta) + i\sin(\theta + \theta)] \Rightarrow$$

$$z^2 = r^2(\cos 2\theta + i\sin 2\theta)$$

EXAMPLE III. Calculate z^2 for $z = -5\sqrt{3} + 5i$ by simply squaring z and also by changing z to its polar form and using the above result.

Solution: We have

$$z = -5\sqrt{3} + 5i \Rightarrow$$

$$z^2 = (-5\sqrt{3} + 5i)^2$$

$$= \underline{\quad ? \quad} - \underline{\quad ? \quad} + 25i^2$$

$$= \underline{\quad ? \quad} - \underline{\quad ? \quad} - \underline{\quad ? \quad}$$

$$= 50 - 50\sqrt{3}\,i$$

Then, from Example I,

$$z = -5\sqrt{3} + 5i \Rightarrow$$

$$z = 10\left(\cos\frac{5\pi}{6} + i\sin\frac{5\pi}{6}\right)$$

Then, since

$$z^2 = r^2(\cos 2\theta + i\sin 2\theta)$$

we have

$$z^2 = 100\left(\cos\frac{5\pi}{3} + i\sin\frac{5\pi}{3}\right)$$

$$= 100\left(+\frac{1}{2} - \frac{\sqrt{3}}{2}i\right)$$

$$= 50 - 50\sqrt{3}\,i$$

The following lemma is needed to prove the more general statement of powers of complex numbers in Theorem II. The proof of the lemma requires mathematical induction. See Section 2.12.

LEMMA I.
If z is a complex number whose polar form is

$$z = r(\cos\theta + i\sin\theta)$$

then

$$z^n = r^n(\cos n\theta + i\sin n\theta)$$

for all $n \in N$.

Proof:
1. $P(1)$: $z = r(\cos\theta + i\sin\theta) \Rightarrow$
   $z^1 = r^1(\cos 1 \cdot \theta + i\sin 1 \cdot \theta)$
2. $P(k) \Rightarrow P(k+1)$:
   $z^k = r^k(\cos k\theta + i\sin k\theta) \Rightarrow$
   $z^{k+1} = r^{k+1}[\cos(k+1)\theta + i\sin(k+1)\theta]$
   $z^{k+1} = z^k \cdot z = [r^k(\cos k\theta + i\sin k\theta)]$
   $\qquad \cdot [r(\cos\theta + i\sin\theta)] \Rightarrow$
   $z^{k+1} = r^{k+1}[(\cos k\theta \cos\theta - \sin k\theta \sin\theta)$
   $\qquad + (\sin k\theta \cos\theta + \sin\theta \cos k\theta)i] \Rightarrow$
   $z^{k+1} = r^{k+1}[\cos(k\theta + \theta)$
   $\qquad + i\sin(k\theta + \theta)] \Rightarrow$
   $z^{k+1} = r^{k+1}[\cos(k+1)\theta + i\sin(k+1)\theta]$

Now, let's apply this lemma to find the fourth power of a complex number.

CHAPTER 12   Applications of Trigonometry

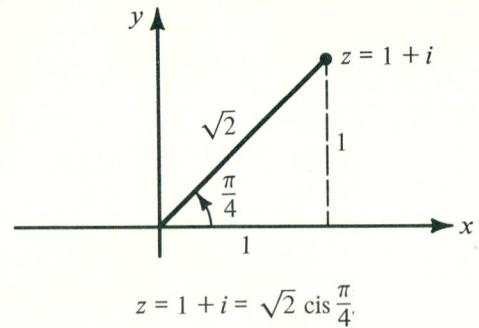

$z = 1 + i = \sqrt{2} \operatorname{cis} \frac{\pi}{4}.$

FIGURE 12.50.

EXAMPLE IV. For $z = 1 + 1i$, find z^4.

Solution: See Figure 12.50.

$$r = \sqrt{1^2 + 1^2} = \sqrt{2}$$

$$\theta = \tan^{-1}\left(\frac{1}{1}\right) = \frac{\pi}{4}$$

Then

$$z = \sqrt{2}\left(\cos\frac{\pi}{4} + i\sin\frac{\pi}{4}\right)$$

and

$$z^4 = (\sqrt{2})^4\left[\cos 4\left(\frac{\pi}{4}\right) + i\sin 4\left(\frac{\pi}{4}\right)\right] \Rightarrow$$

$$z^4 = 4(\cos\pi + i\sin\pi) \Rightarrow$$

$$z^4 = -4$$

Lemma I also allows us to find all the roots of a complex number as illustrated in Example V. Since the equation

$$x^3 - a^3 = 0$$

has exactly three solutions over the complexes, we will find three cube roots of a.

EXAMPLE V. Find the cube roots of $a = 4 + 4\sqrt{3}\,i$.

Solution: See Figure 12.51.

$$a^3 = 4 + 4\sqrt{3}\,i \Rightarrow$$

$$a^3 = 8\left[\cos\left(\frac{\pi}{3} + 2k\pi\right) + i\sin\left(\frac{\pi}{3} + 2k\pi\right)\right]$$

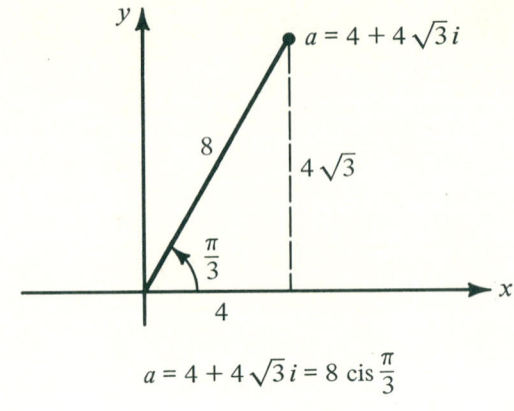

$a = 4 + 4\sqrt{3}\,i = 8 \operatorname{cis}\frac{\pi}{3}$

FIGURE 12.51.

where $k \in$ integers. Then, from Lemma I, we have

$$a^3 = r^3(\cos 3\theta + i\sin 3\theta)$$

Therefore,

$$r^3(\cos 3\theta + i\sin 3\theta) = 8\left(\cos\frac{\pi}{3} + i\sin\frac{\pi}{3}\right) \Rightarrow$$

$$r = 2$$

and

$$3\theta = \frac{\pi}{3} + 2k\pi$$

where $k \in$ integers. Then

$$\theta = \frac{\pi + 6k\pi}{9} = \begin{cases} \dfrac{\pi}{9} & \text{for } k = 0 \\[4pt] \dfrac{7\pi}{9} & \text{for } k = 1 \\[4pt] \dfrac{13\pi}{9} & \text{for } k = 2 \\[4pt] \dfrac{19\pi}{9} & \text{for } k = 3 \end{cases}$$

Notice that the sequence of values for θ begins to repeat after $k = 2$, since $\theta = (19\pi)/9$ is the same as $\theta = \pi/9$ in the polar form of z. We will only get three distinct values for the cube roots of a.

Now we have

$$a = r(\cos\theta + i\sin\theta) \Rightarrow$$

$$a = 2\left(\cos\frac{\pi}{9} + i\sin\frac{\pi}{9}\right) \approx 1.879 + 0.684i$$

$$a = 2\left(\cos\frac{7\pi}{9} + i\sin\frac{7\pi}{9}\right) \approx -1.532 + 1.286i$$

or

$$a = 2\left(\cos\frac{13\pi}{9} + i\sin\frac{13\pi}{9}\right) \approx -0.347 - 1.970i$$

Now let's extend Lemma I to all rational powers of z.

THEOREM II.
If z is a complex number whose polar form is

$$z = r(\cos\theta + i\sin\theta)$$

then

$$z^n = r^n(\cos n\theta + i\sin n\theta)$$

for any rational number n.

This theorem can also be extended to any real power of z, but the proof is beyond the scope of this text. The proof of Theorem II is based on Lemma I as follows.

Proof: First, let's consider the case where n is an integer, then we will use the fact that any rational number can be expressed as the quotient of two integers to include the rest of the rationals (the fractions).

CASE I. $n > 0$ is precisely Lemma I.

CASE II. $n = 0$, $n = 0 \Rightarrow z^0 = 1$, and $r^n(\cos n\theta + i\sin n\theta) = r^0(\cos 0 + i\sin 0) = 1 \cdot 1 = 1$.

CASE III. $n < 0$. If $n < 0$, then $-n > 0$ and we can apply Lemma I to the positive number $-n$. Then

$$z^{(-n)} = 1/z^n$$

and, by Lemma I,

$$z^{-n} = r^{(-n)}\cos(-n\theta) + i\sin(-n\theta) \Rightarrow$$

$$\frac{1}{z^n} = \frac{1}{r^n}\cos n\theta - i\sin n\theta \Rightarrow$$

$$z^n = \frac{1}{(1/r^n)\cos n\theta - i\sin n\theta} \Rightarrow$$

$$z^n = \frac{r^n}{\cos n\theta - i\sin n\theta} \cdot \frac{\cos n\theta + i\sin n\theta}{\cos n\theta + i\sin n\theta} \Rightarrow$$

$$z^n = \frac{r^n(\cos n\theta + i\sin n\theta)}{\cos^2 n\theta + \sin^2 n\theta} \Rightarrow$$

$$z^n = r^n(\cos n\theta + i\sin n\theta)$$

Now let's establish the theorem for all rational numbers.

CASE IV. $n = p/q$, where p and q are integers. First, let's assume $r = 1$ for convenience. Then, since p is an integer, we have

$$z^n = z^{p/q} \Rightarrow z^{nq} = z^p = \cos p\theta + i\sin p\theta$$

but

$$n = p/q \Rightarrow p = nq \Rightarrow$$

$$z^p = \cos nq\theta + i\sin nq\theta \Rightarrow$$

$$z^{p/q} = (\cos nq\theta + i\sin nq\theta)^{1/q}$$

However, since q is an integer,

$$(\cos n\theta + i\sin n\theta)^q = \cos nq\theta + i\sin nq\theta$$

This yields

$$(\cos nq\theta + i\sin nq\theta)^{1/q} = z^{p/q} = z^n$$

and

$$(\cos nq\theta + i\sin nq\theta)^{1/q}$$
$$= [(\cos n\theta + i\sin n\theta)^q]^{1/q}$$
$$= \cos n\theta + i\sin n\theta$$

Therefore,

$$z^n = \cos n\theta + i\sin n\theta$$

CHAPTER 12   Applications of Trigonometry

and, for $r \neq 1$, we have

$$z^n = r^n(\cos n\theta + i \sin n\theta)$$

for all $n \in$ rationals.

Theorem III is the generalization of Example V, and its proof is based upon the last theorem. It is called DeMoive's theorem.

THEOREM III.
DeMoive's theorem.
If $z = r(\cos \theta + i \sin \theta)$, then

$$z^{1/n} = r^{1/n}\left[\cos\left(\frac{\theta + 2\pi k}{n}\right) + i \sin\left(\frac{\theta + 2\pi k}{n}\right)\right]$$

where $k = 0, 1, \ldots, (n-1)$ are the n distinct nth roots of z.

Proof: We will use Theorem II to show that the proposed value of $z^{1/n}$ satisfies $(z^{1/n})^n = z$ as follows:

$$\left\{r^{1/n}\left[\cos\left(\frac{\theta + 2\pi k}{n}\right) + i \sin\left(\frac{\theta + 2\pi k}{n}\right)\right]\right\}^n$$

$$= (r^{1/n})^n\left[\cos\left(n \cdot \frac{\theta + 2\pi k}{n}\right) + i \sin\left(n \cdot \frac{\theta + 2\pi k}{n}\right)\right]$$

$$= r(\cos \theta + i \sin \theta)$$

$$= z$$

Now, let's use DeMoive's theorem to find the three cube roots of 1.

EXAMPLE VI. Find the cube roots of unity.

Solution: We know that

$$1 = 1(\cos 0 + i \sin 0) \Rightarrow$$

$$1^{1/3} = 1^{1/3}\left[\cos\left(\frac{0 + 2\pi k}{3}\right) + i \sin\left(\frac{0 + 2\pi k}{3}\right)\right] \Rightarrow$$

$$\sqrt[3]{1} = 1\left(\cos\frac{2\pi k}{3} + i \sin\frac{2\pi k}{3}\right)$$

$$k = 0 \Rightarrow \sqrt[3]{1} = 1$$

$$k = 1 \Rightarrow \sqrt[3]{1} = -\tfrac{1}{2} + \frac{\sqrt{3}}{2}i$$

$$k = 2 \Rightarrow \sqrt[3]{1} = -\tfrac{1}{2} - \frac{\sqrt{3}}{2}i$$

This can be checked as follows:

$$\sqrt[3]{1} = 1, \text{ since } 1^3 = 1$$

$$\sqrt[3]{1} = -\frac{1}{2} + \frac{\sqrt{3}}{2}i$$

since

$$\left(-\frac{1}{2} + \frac{\sqrt{3}}{2}i\right)^3 = \left(-\frac{1}{2} + \frac{\sqrt{3}}{2}i\right)\left(-\frac{1}{2} - \frac{\sqrt{3}}{2}i\right)$$

$$= 1$$

and

$$\sqrt[3]{1} = -\frac{1}{2} - \frac{\sqrt{3}}{2}i$$

since

$$\left(-\frac{1}{2} - \frac{\sqrt{3}}{2}i\right)^3 = \left(-\frac{1}{2} - \frac{\sqrt{3}}{2}i\right)(\underline{\quad ? \quad})$$

$$= \underline{\;?\;}$$

EXERCISE 12.7

A. FUNDAMENTALS
Express the following complex numbers in polar form. Remember that $x + yi = r \cos \theta + (r \sin \theta)i$, where $\theta = \tan^{-1}(y/x)$ and $r = \sqrt{x^2 + y^2}$.

1. (a) $z = 1 + i$ (b) $z = 1 - i$
2. (a) $z = 1$ (b) $z = i$
3. (a) $z = \sqrt{3} + i$ (b) $z = 1 - \sqrt{3}i$
4. Find both the Cartesian form ($x + yi$) and the polar form ($r \operatorname{cis} \theta$) of z^2 and z^3 for $z = 4 + 4i$.

B. ESSENTIALS

5. Find $z_1 \cdot z_2$ in Cartesian and polar form.
(a) $z_1 = 2\sqrt{3} + 2i$ and $z_2 = \overline{z_1}$
(b) $z_1 = 2\left[\cos\left(\frac{-\pi}{6}\right) + i \sin\left(\frac{-\pi}{6}\right)\right]$ and
$z_2 = 2\left(\cos\frac{\pi}{3} + i \sin\frac{\pi}{3}\right)$
(c) $z_1 = 5\left\{\cos\left[\tan^{-1}\left(\frac{-4}{3}\right)\right] + i \sin\left[\tan^{-1}\left(\frac{-4}{3}\right)\right]\right\}$
$z_2 = 2\sqrt{2}\left(\cos\frac{\pi}{4} + i \sin\frac{\pi}{4}\right)$
(d) $z_1 = -6 + 8i$ and $z_2 = -4 - 4i$

6. Find $\sqrt{z}$ for $z = 3 + 4i$ by solving the equation $a + bi = \sqrt{3 + 4i}$ for a and b and by using Theorem III.

7. Find the fourth roots of unity using DeMoive's theorem.

8. Solve the following using DeMoive's theorem.
(a) $z^3 = 8$ (b) $z^3 = 8 + 8i$ (c) $z^5 = i$
(d) $z^4 = 81i$ ★(e) $z^3 - 1 = \sqrt{3}i$
(f) $z^5 = -8 - 8i$ ★(g) $z^{1/3} = -27 - 27i$
(h) $z^4 = (2 + 2\sqrt{3}i)^3$ (i) $z^3 = (1 - i)^4$

9. (a) Interpret the geometric result of multiplying $z = 1 + \sqrt{3}i$ by i.
(b) Repeat for $z = r(\cos \theta + i \sin \theta)$ multiplied by
$$z = -i = \cos\frac{3\pi}{2} + i \sin\frac{3\pi}{2}$$

10. Express $z_1 = 3i$, $z_2 = \sqrt{3} + i$ and their quotient z_1/z_2 in polar form. What do you notice?

11. (a) Euler's formula, which takes calculus to derive, states that a complex number can be written as $x + yi = r(\cos \theta + i \sin \theta) = re^{i\theta}$. Write $z = 1 + i$ in exponential form.
(b) Extend the domain of $y = \ln x$ to the complex numbers by showing that
$$\ln z = \ln (re^{i\theta}) = \ln r + i\theta + 2k\pi i$$
(c) Find $\ln (1 + i)$.

12. (a) Show that for $z_1 = -4i$ and $z_2 = 3 - \sqrt{3}i$,
$$\frac{z_1}{z_2} = \frac{r_1(\cos \alpha + i \sin \alpha)}{r_2(\cos \beta + i \sin \beta)} = \frac{r_1}{r_2}[\cos(\alpha - \beta) + i \sin(\alpha - \beta)]$$

(b) Prove the equation stated in Problem 12(a).

13. Show that i^i is a real number. (HINT: Use logs and Euler's formula.)

12.8 THE TRIGONOMETRIC FUNCTIONS OF SMALL ANGLES (OPTIONAL)

Turn to Table IV and compare the value of x in radians with the corresponding values of sin x and tan x. Both values are small for very small values of x, but, of course, they are not equal. For example, if $x = 0.001$, then

$$\sin (0.001) \approx 0.0009999998$$

and

$$\tan (0.001) \approx 0.0010000003$$

But when the values are rounded off to four decimal places, the results are

$$\sin x = x$$

and

$$\tan x = x$$

for small values of x. For $x \leq 15°$ or $x \leq 0.25$ radians, sin x and tan x approximately equal x. Then, for small angles, we can replace tan x and sin x with x, resulting in a greatly simplified expression.

In order to justify replacing sin x with x for small values of x, we will prove that the ratio of x to sin x approaches 1 as x approaches zero. That is, we will show that

$$\lim_{x \to 0} \frac{x}{\sin x} = 1$$

This will prove that x and sin x become closer to being equal as x approaches zero. The proof is a rather clever geometric argument based on the three areas shown on the unit circle in the following illustration. We will show that the ratio $x/\sin x$ is

CHAPTER 12   Applications of Trigonometry

trapped between two expressions that approach 1 as x approaches 0, yielding

$$1 \le \underset{x \to 0}{\text{limit}} \frac{x}{\sin x} \le 1$$

which means the limit must equal 1.

> **THEOREM I.**
> $$\underset{x \to 0}{\text{limit}} \frac{x}{\sin x} = 1$$

Proof:

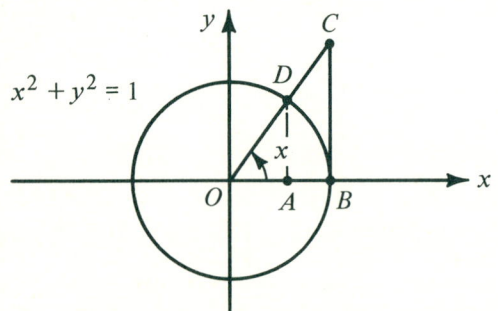

FIGURE TI.

The area of the sector of the unit circle cut off by angle x is larger than the area of triangle OAD and smaller than the area of triangle OBC. That is, since the area of triangle OAD is less than or equal to the area of sector OBD, which is less than or equal to the area of triangle OBC, we have

$$\tfrac{1}{2}b_1 h_1 \le \frac{x}{2\pi}(\pi r^2) \le \tfrac{1}{2} b_2 h_2 \Rightarrow$$

$$\tfrac{1}{2}(\cos x)(\sin x) \le \frac{xr^2}{2} \le \tfrac{1}{2}(1)\tan x \Rightarrow$$

$$\sin x \cos x \le x \le \frac{\sin x}{\cos x} \Rightarrow$$

for $\sin x > 0$

$$\cos x \le \frac{x}{\sin x} \le \frac{1}{\cos x}$$

As $x \to 0$, $\cos x \to 1$ and $1/\cos x \to 1$ means

$$\underset{x \to 0}{\text{limit}} \cos x \le \underset{x \to 0}{\text{limit}} \frac{x}{\sin x} \le \underset{x \to 0}{\text{limit}} \frac{1}{\cos x} \Rightarrow$$

$$1 \le \underset{x \to 0}{\text{limit}} \frac{x}{\sin x} \le 1 \Rightarrow$$

$$\underset{x \to 0}{\text{limit}} \frac{x}{\sin x} = 1$$

> **THEOREM II.**
> $$\underset{x \to 0}{\text{limit}} \frac{\sin x}{x} = 1$$

Since the ratio of x to $\sin x$ approaches 1 as x approaches 0, the ratio of $\sin x$ to x must also approach 1 as x approaches 0.

> **THEOREM III.**
> $$\underset{x \to 0}{\text{limit}} \frac{\tan x}{x} = 1$$

Proof:

$$\underset{x \to 0}{\text{limit}} \frac{\tan x}{x} = \underset{x \to 0}{\text{limit}} \frac{\sin x/\cos x}{x}$$

$$= \underset{x \to 0}{\text{limit}} \frac{\sin x}{x} \cdot \frac{1}{\cos x} = 1 \cdot 1 = 1$$

EXAMPLE I. The angle of depression of a section of storm drain is 17′. Find the drop per kilometer of pipe.

Solution: See Figure 12.52.

$$\tan 17' = \frac{y}{1 \text{ km}} \Rightarrow$$

$$y = 1 \text{ km}(\tan 17')$$

SECTION 12.8   The Trigonometric Functions of Small Angles (Optional)

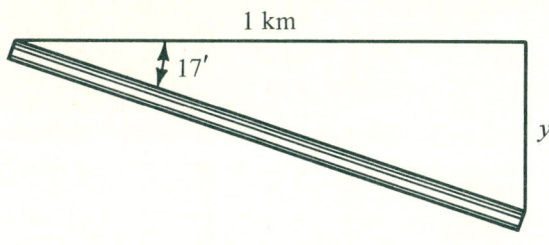

FIGURE 12.52.

But $17' = 0.005$ radians yields

$$y = 1 \text{ km}(\tan 0.005)$$

Then, since $\theta < 0.25$ radians,

$$\tan \theta = \theta \text{ yields}$$

$$y = 1 \text{ km}(0.005) \Rightarrow$$

$$y = 0.005 \text{ km}$$

For some applications, more accuracy may be required than approximating $\sin x$ or $\tan x$ with x will yield. Consider Example II.

EXAMPLE II. The angle of depression of another storm drain is $6°50'$. Find the drop per kilometer of pipe.

Solution: The "exact" value is

$$y = 1 \text{ km} \tan 6°50' = 0.1198 \text{ km} \Rightarrow$$

$$y = 119.8 \text{ m}$$

By using $\tan \theta = \theta$, we have

$$y = (1 \text{ km})(0.1193) = 9.1193 \text{ km} \Rightarrow$$

$$y = 119.3 \text{ m}$$

as the approximate value. The difference

$$\Delta y = 119.8 \text{ m} - 119.3 \text{ m} = 0.5 \text{ m}$$

may be too large to be ignored in some applications. The discrepancy arose because of the relatively large angle coupled with a fairly small distance (1 kilometer). The error per meter of pipe is very small, but the error per kilometer is rather large.

These concepts of small-angle trig are more appropriate in the field of astronomy, where the angles measured are extremely small and the distances are very large.

EXAMPLE III. The radius of the sun, which is 1.49×10^8 km from earth, subtends an angle of $16'$. Estimate the sun's diameter.

Solution: See Figure 12.53.

$$\tan \theta = \frac{r}{1.49 \times 10^8 + r}$$

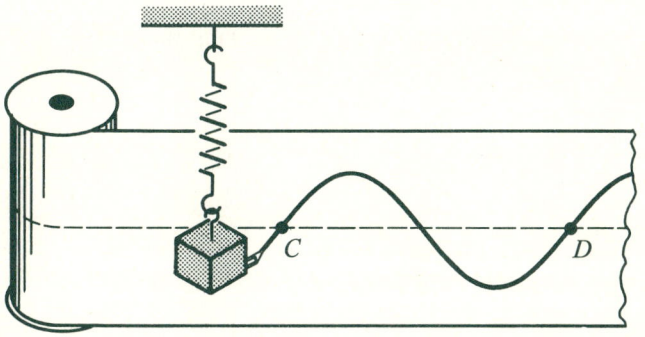
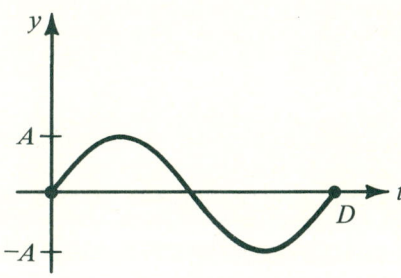

FIGURE 12.53. *A Weight on a Spring.*

but, since θ is very small,

$$\tan 16' = \tan 0.00465 \approx 0.00465 \Rightarrow$$

$$0.00465 = \frac{r}{1.49 \times 10^8 + r} \Rightarrow$$

$$6.928 \times 10^5 + 0.00465r = r \Rightarrow$$

$$r = 6.96 \times 10^5 \text{ km}$$

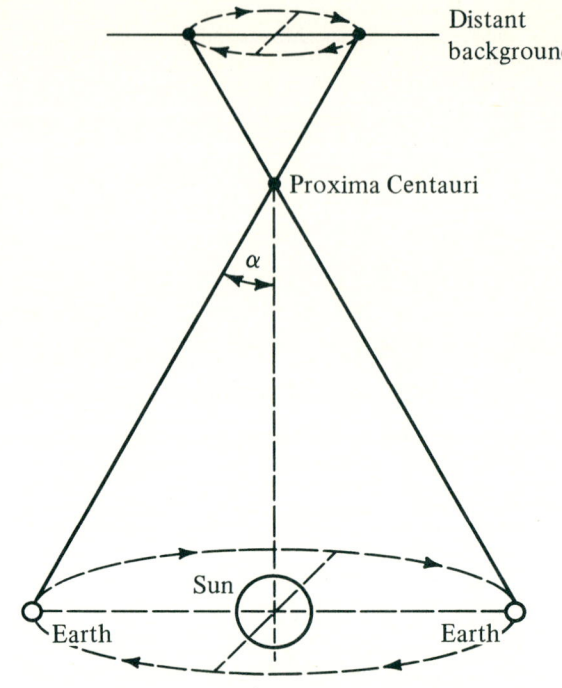

EXERCISE 12.8

A. Find the value of $\sin \theta$ and $\tan \theta$ for the following. If θ is in degrees, change it to radians.
  1. $\theta = 2$, $\theta = 0.2$, $\theta = 0.02$.
  2. $\theta = 10'$, $\theta = 10°$.
  3. $\theta = 6°$, $\theta = 8°$, $\theta = 10°$.

B. ESSENTIALS
  4. The angle of elevation of a railroad is 4°. Calculate the amount of rise in 20 kilometers using the approximation $\tan \theta \approx \theta$. Calculate the real rise in 20 kilometers.

  5. The radius of the moon subtends an angle of 31' on the earth. Estimate the moon's diameter if it is known to be 3.8×10^5 kilometers from earth.

  6. How large an angle does Mercury's diameter of 5,000 kilometers subtend on the sun when Mercury is 5.8×10^7 kilometers from the sun?

  7. The method of parallax is used to determine the distance from the earth to a star. If you sight an object against a distant background, as you move, the object changes position against that background. In the nineteenth century, Proxima Centauri, the nearest of all stars other than the sun, was found to be 4.25 light years from the earth. Find the degree that Proxima Centauri must have shifted against the background in observations made 6 months apart, as shown in the following illustration. We call α the parallactic shift. (HINT: The distance from the sun to the earth is 93 million miles and a light year is 5.886 million million miles, 5.886×10^{12} m.) See illustration above right.

  8. In 1838, the German astronomer Friedrich Bessel used the parallax method (see Problem 7) to find the distance to the star 61 Cygni. He observed a 6-month parallactic shift of about 0.3″. How far away did he find 61 Cygni to be? (It is actually about 10.7 light years away.)

C. APPLICATIONS
Use the facts that $\lim\limits_{x \to 0} \dfrac{\sin x}{x} = 1$ and $\lim\limits_{x \to 0} \dfrac{\tan x}{x} = 1$ to find the following limits.

  9. (a) $\lim\limits_{\theta \to 0} \dfrac{\sin 2\theta}{2\theta}$ (b) $\lim\limits_{\theta \to 0} \dfrac{\sin 2\theta}{\theta}$

  10. (a) $\lim\limits_{t \to -1} \dfrac{\tan(t+1)}{t+1}$

  (b) $\lim\limits_{t \to -2} \dfrac{\tan(t+2)}{2t+4}$

  ★11. (a) $\lim\limits_{x \to 0} \dfrac{x}{\sin 3x}$ (b) $\lim\limits_{x \to 0} \dfrac{3x}{\tan 2x}$

  12. $\lim\limits_{x \to 0} \sqrt{\dfrac{1 - \cos^2 x}{2x^2}}$

Use these facts to solve the following problems. If

$$\lim_{x \to a} f(x) = A \quad \text{and} \quad \lim_{x \to a} g(x) = B$$

then

$$\lim_{x \to a} f(x) \cdot g(x) = A \cdot B$$

13. $\displaystyle\lim_{x \to 0} \frac{\sin 2x}{\sin 3x}$
14. $\displaystyle\lim_{x \to 0} \frac{1 - \cos x}{x}$

12.9 SIMPLE HARMONIC MOTION (OPTIONAL)

Any periodic motion is said to be harmonic motion; simple harmonic motion is a special type of periodic motion. Some examples of harmonic motion include getting out of bed, the sun setting, a pendulum swinging, a piston moving, a weighted spring bouncing, and eating dinner. These are examples of periodically recurring events (motion), each of which has a *period* (the length of time in one cycle) and a *frequency* (the numer of cycles per unit time interval, for example, 365 sun settings per year).

If the periodic motion can be modeled by an equation of the form

$$y = A \sin (Bt + C)$$

it is said to be **simple harmonic motion**.

Simple harmonic motion results when the restoring force on the object is proportional to its displacement from equilibrium, that is, $F = kx$. Then, using Newton's second law, $F = ma$, the distance equation $y = A \sin (Bt + C)$ can be derived.

Some examples of simple harmonic motion are a swinging pendulum, the vibrating string of a musical instrument, electric current, waves in a lake, and the movement of a piston of an automobile engine.

Suppose we have a weight vibrating up and down on a spring suspended from above. If we draw a piece of paper past it at a constant speed, the marker records the path of the weight. See Figure 12.53 (page 273).

One complete cycle is depicted by the 2 points C and D. The time it takes for the weight to complete one cycle (repetition) is called its period. The number of cycles per unit time is its frequency, usually called cycles per second, cps, or vibrations per second, vps. The amplitude is given by A on the graph.

The corresponding Cartesian coordinate system graph is actually a sine curve that can be determined experimentally and verified mathematically using calculus.

Assume for some mass and particular spring the equation for the motion is

$$y = 4 \sin 2t$$

where y is in cps and t is in seconds. Then the amplitude of the motion would be 4, the period would be

$$t = \frac{2\pi}{B} = \frac{2\pi}{2} = \pi \text{ sec}$$

and the frequency would be

$$f = \frac{1}{t} = \frac{1}{\pi} \text{ cps}$$

A simple pendulum can be shown to behave similarly with an experiment similar to the one illustrated in Figure 12.54.

If the pendulum were that of a grandfather clock (having a very long pendulum), it might swing once every second (one complete cycle from C to D every second) for a total swinging distance of 10 centimeters (amplitude of 5 centimeters). Then its position at any time t is given by

$$y = 5 \sin 2\pi t$$

since

$$y = A \sin Bt$$

$$B = \frac{2\pi}{t}$$

and

$$t = 1 \text{ sec}$$

The fixed strings of a musical instrument like a piano or a guitar move according to harmonic motion. If we assume that this motion is simple harmonic motion, that it can be modeled by a sine curve, we can examine the "beat phenomena" often heard when two notes close to each other are sounded simultaneously. The result, used to tune musical instruments, pulsates (or beats). See below.

By international agreement the note A above middle C of a piano is tuned to vibrate at 440 vps. Often a guitar player will begin tuning his guitar by comparing the sound of his E string with that of the second E below middle C on a piano. This piano string is tuned to vibrate at about 80 vps. If the guitar is slightly out of tune, say its E string is vibrating at 70 vps, he will hear the beat. He then tunes out the beat by tightening his E string until the beat disappears. By examining the graphs of the two equations for the vibrating strings and

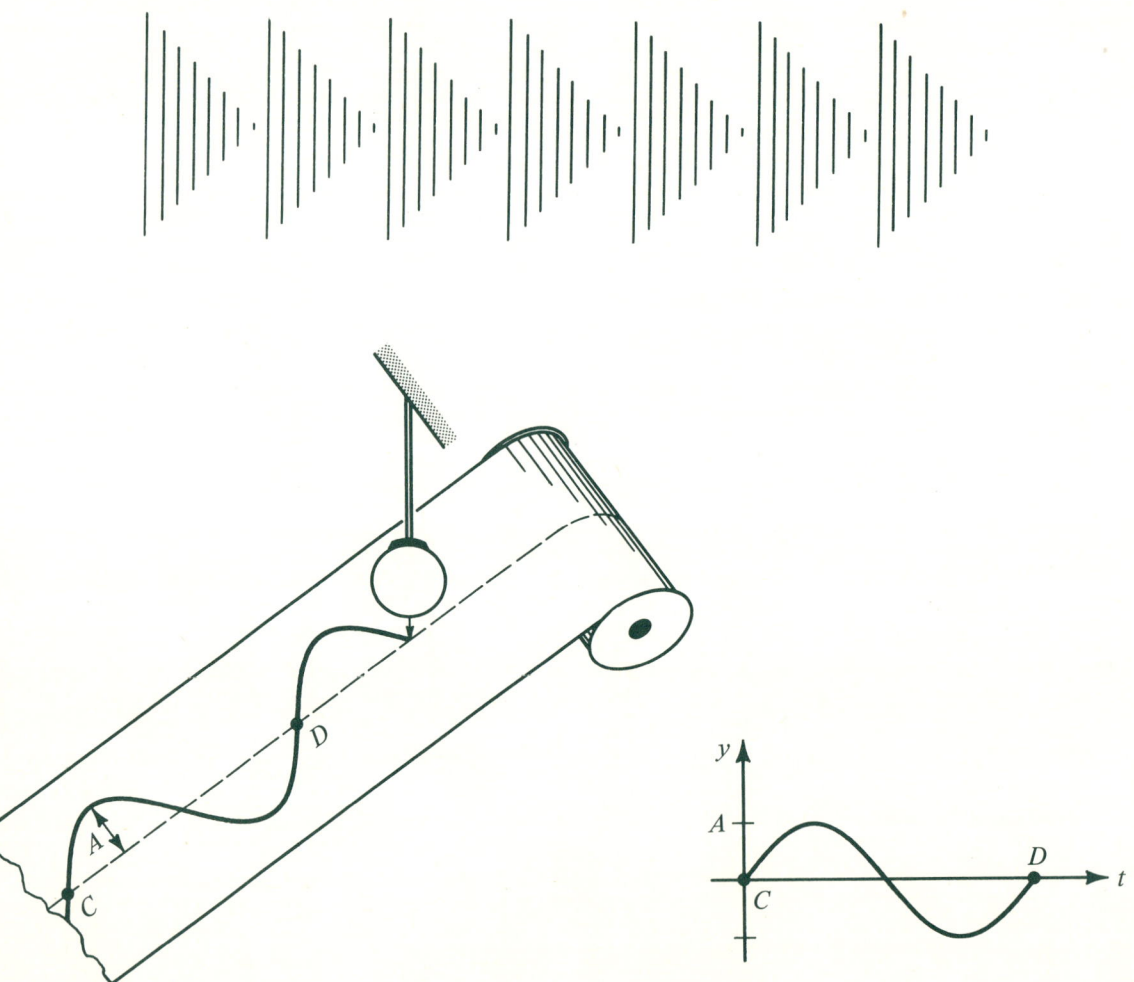

FIGURE 12.54. A Pendulum.

SECTION 12.9 Simple Harmonic Motion (Optional)

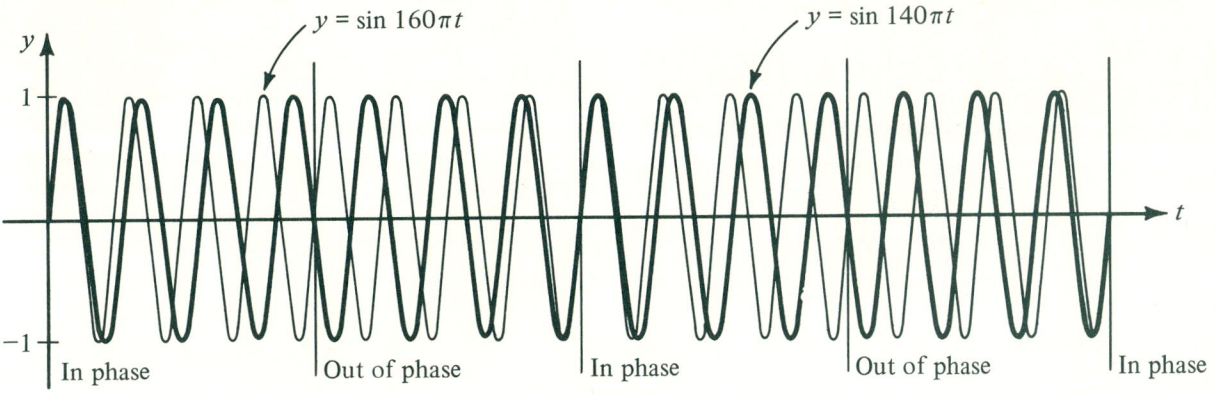

$y = \sin 160\pi t$ and $y = \sin 140\pi t$

FIGURE 12.55.

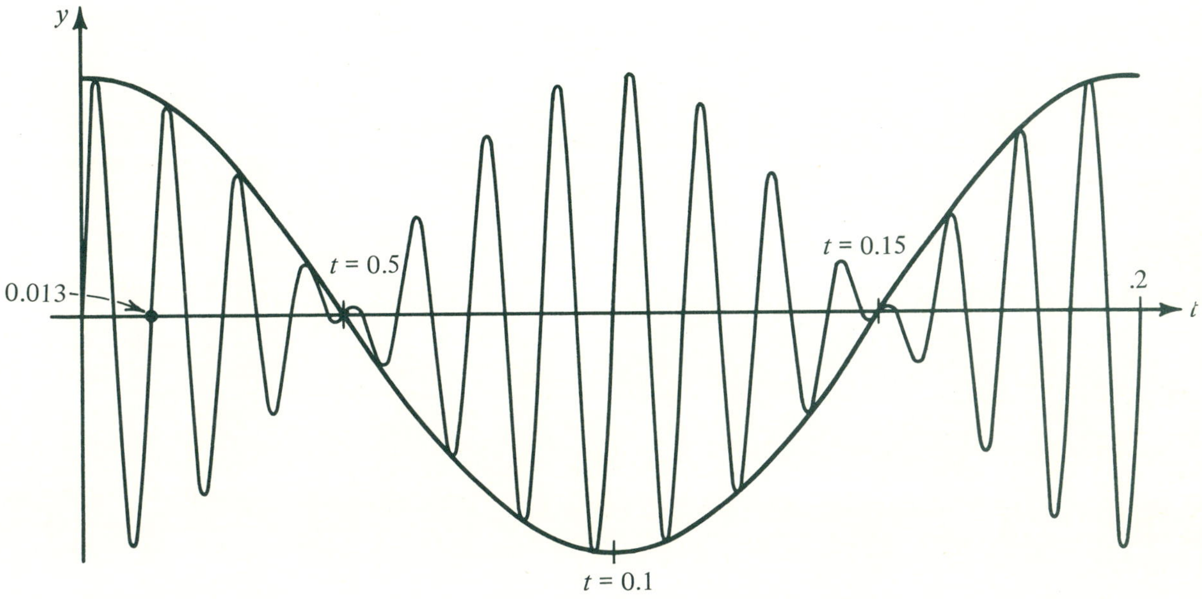

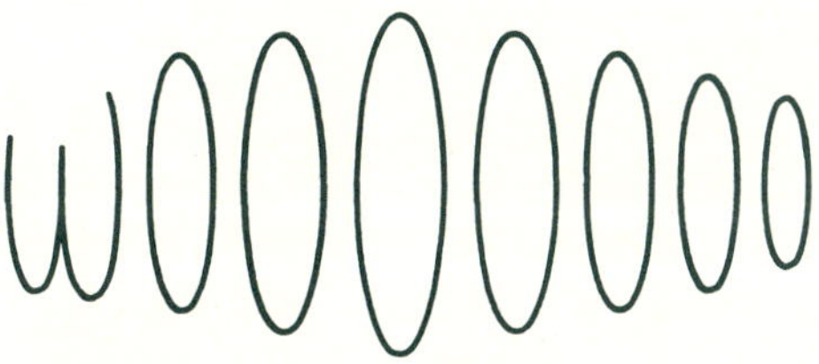

FIGURE 12.56. *The Beat Phenomenon.*

the resulting graph of their sum, we can see why two slightly out of tune vibrating strings cause a beat.

Let's assume that both equations have amplitudes of 1 for simplicity. Then, since in $y = A \sin Bt$,

$$B = 2\pi f$$

for $f = 80$ vps, we have

$$B = 2\pi(80) = 160\pi$$

and for $f = 70$ vps, we have

$$B = 2\pi(70) = 140\pi$$

Then the equations are

$$y = A \sin Bt \Rightarrow$$
$$y = 1 \cdot \sin 160\pi t$$

and

$$y = \sin 140\pi t$$

Their graphs are given in Figure 12.55, and the resulting graph of their sum in Figure 12.56. Notice the resulting amplitude of the combined sound of the piano and the "out of tune" guitar. This is a beat of 10 times per second. Notice that 80 cps minus 70 cps equals 10 cps.

The graph of the sum of the two sine curves in Figures 12.55 and 12.56 is very difficult to draw from the graphs of each of the curves individually. Instead, we use the identity

$$\sin A + \sin B = 2 \sin \tfrac{1}{2}(A + B) \cos \tfrac{1}{2}(A - B)$$

to express

$$y = \sin 160\pi t + \sin 140\pi t$$

as the product,

$$y = 2 \cos 10\pi t (\sin 150\pi t)$$

This can be thought of as a sine curve of varying amplitude. In electronics, $\sin 150\pi t$ is called the *carrier* and $2 \cos 10\pi t$ the *amplitude*. Notice that $\sin 150\pi t$ has a period of only $\tfrac{1}{75}$ of a second while $\cos 10\pi t$ has a period of $\tfrac{1}{5}$ of a second.

EXERCISE 12.9

A. FUNDAMENTALS

1. Build pendulums of the following lengths, and then time their periods to get an estimate of their frequencies. Is there any connection between the length of the pendulum and its frequency?
(a) 25 cm (b) 36 cm (c) 49 cm
(d) 64 cm

2. Build pendulums using several different weights of bobs, and time their periods. Keep their lengths equal. Is there any connection between the weight of a pendulum's bob and its frequency?

3. Construct two pendulums, each 1 meter long, and attach the end of one to the other's bob. Will they swing as one pendulum? Can you get the top one to swing without the bottom one swinging? Can you make the bottom one swing without the top?

B. ESSENTIALS

4. (a) Find the sine curve that models the sound made by striking middle C on a piano, if the corresponding string vibrates at 264 vps with an amplitude of 1. Graph the curve.
(b) Repeat (a) for G above middle C, if its frequency is 396 vps. Graph the curve.
(c) Express the sum of the sine curves from (a) and (b) above as a product and graph the result. These notes represent a fifth on our musical scale and $396 = \tfrac{3}{2}(264)$.

5. Repeat Problem 4 for middle C and the C below middle C. This interval represents 1 octave on our musical scale, and the frequency of the lower C is exactly one half that of middle C.

ENDNOTE

1. Polar coordinates were introduced by Jakob Bernoulli in 1691.

PART III

Algebra

CHAPTER 13

The Conic Sections

13.1 INTRODUCTION

The conic sections include lines, circles, parabolas, ellipses, and hyperbolas. Their equations are of the form

$$Ax^2 + Bxy + Cy^2 + Dx + Ey + F = 0$$

This chapter, therefore, is a complete discussion of all equations involving x, y, x^2, and y^2 terms.

The cones in Figure 13.1 depict these curves as cross-sections of cones, showing why they are called "conic sections." (See page 282.)

For the circle, the plane must be perpendicular to the axis of the cone, since tilting the plane slightly will generate an ellipse. A plane tilted at the same angle as the side of the cone cuts off a parabolic cross-section and a vertical plane, which cuts the cone twice and traces out a hyperbola.

We will now study each of these conic sections separately.[1]

13.2 THE CIRCLE

Consider the distances between the origin and the points in Figure 13.2 (page 282).

$$P_1 : S = \sqrt{(3-0)^2 + (4-0)^2}$$
$$S = \underline{}?\underline{}$$
$$P_2 : S = \sqrt{(?-0)^2 + (?-0)^2}$$
$$S = \underline{}?\underline{}$$
$$P_3 : S = \sqrt{}$$
$$S = \underline{}?\underline{}$$

Did you find that they are all located 5 units from the origin? The collection of all such points defines a circle of radius 5 centered at the origin.

For any circle, we have the following definition.

CHAPTER 13   The Conic Sections

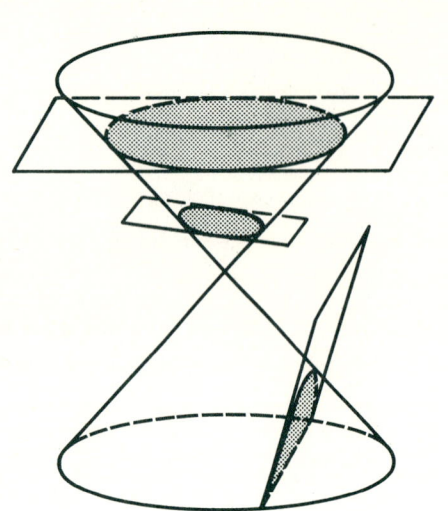
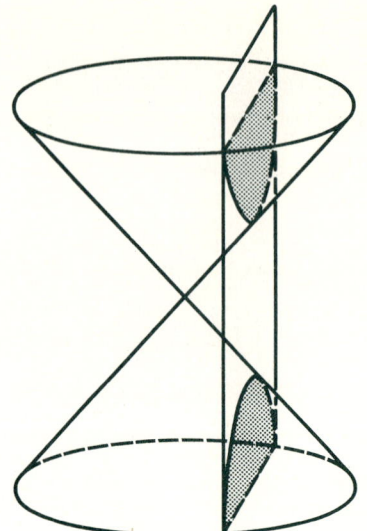

FIGURE 13.1.   *Cross-sections of a Cone.*

DEFINITION 13.1.
A circle is a set of points equidistant from a fixed point.

We can now derive the general equation of a circle centered at (h, k), with a radius of r.

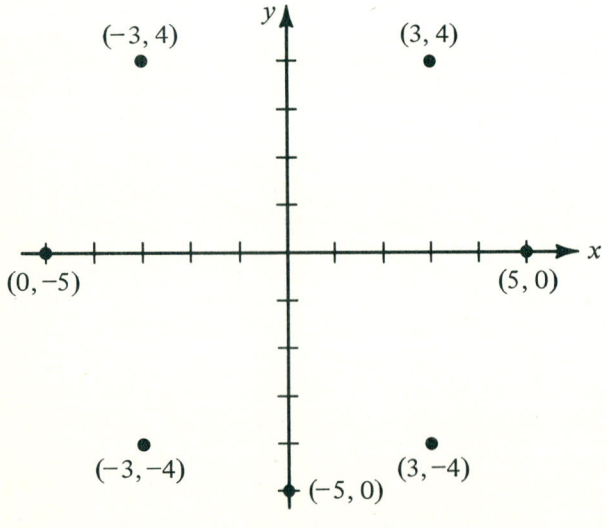

FIGURE 13.2.

Consider the set of all points (x, y), whose distances from a fixed point (h, k) equal r. See Figure 13.3. Then, by the distance formula, we know that if the distance from C to P is equal to r,

$$\sqrt{(x-h)^2 + (y-k)^2} = r \Rightarrow$$
$$(x-h)^2 + (y-k)^2 = r^2$$

THEOREM I.
The equation of a circle of radius r centered at (h, k) is

$$(x-h)^2 + (y-k)^2 = r^2$$

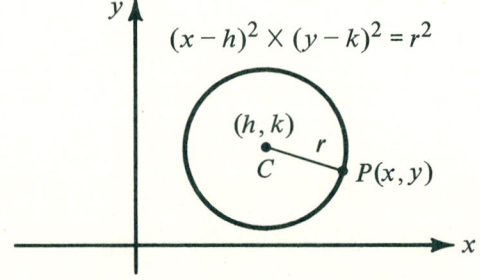

FIGURE 13.3.   Circle.

Notice that for every equation of the form

$$(x - h)^2 + (y - k)^2 = r^2$$

we also have

$$x^2 - 2xh + y^2 - 2ky - r^2 + h^2 + k^2 = 0$$

which, for $A \cdot C \neq 0$, is an equation of the form

$$Ax^2 + Cy^2 + Dx + Ey + F = 0$$

where

$$A = C$$

Usually the equations of circles are given in this second form. To draw the graph of such an equation, we "complete the square" to express it in the form

$$(x - h)^2 + (y - k)^2 = r^2$$

from which we can read the center (h, k) and the radius r. This is illustrated in Examples I and II.

EXAMPLE I. Graph

$$x^2 + y^2 + 4x - 2y = 0$$

Solution: Since $A = C = 1$, we know that this is the equation of a circle. (Remember that A and C are the coefficients of the squared terms.) To graph the circle, however, requires that we discover its center and radius. We can accomplish this by completing the square in order to rewrite the equation in the standard form

$$(x - h)^2 + (y - k)^2 = r^2$$
$$x^2 + y^2 + 4x - 2y - 20 = 0 \Rightarrow$$
$$(x^2 + 4x) + (y^2 - 2y) = 20 \Rightarrow$$
$$(x^2 + 4x + 4) + (y^2 - 2y + 1) = 20 + 4 + 1 \Rightarrow$$
$$(x + 2)^2 + (y - 1)^2 = 25$$

Hence, the center is at $(-2, 1)$ and the radius is $\sqrt{25} = 5$. The graph is shown in Figure 13.4.

EXAMPLE II. Graph

$$x^2 + y^2 = 25$$

Solution: This is of the form

$$(x - 0)^2 + (y - 0)^2 = 5^2$$

Hence, it is a circle, centered at $(0, 0)$ with a radius of 5. Its graph is shown in Figure 13.5. Example I is the translation of this curve to the left 2 and up 1.

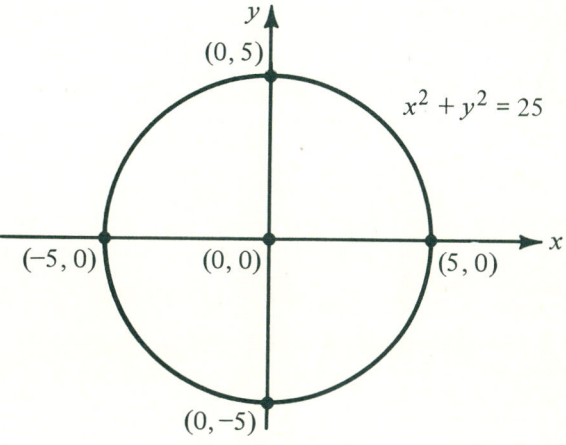

FIGURE 13.5.

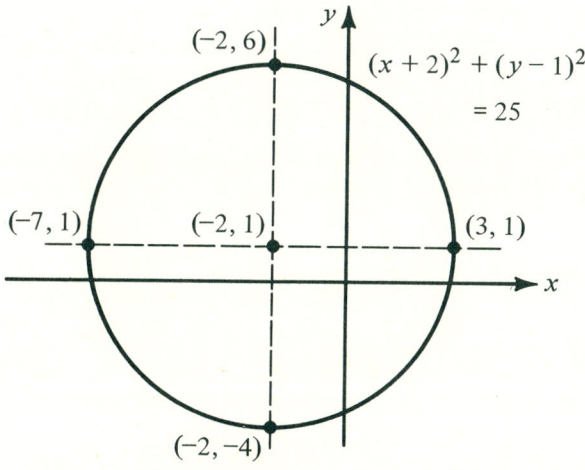

FIGURE 13.4.

EXERCISE 13.2

A. FUNDAMENTALS

Graph the following circles. It is important to label the center and indicate the radius.

1. (a) $x^2 + y^2 = 1$ (b) $x^2 + y^2 = 4$
2. (a) $x^2 + y^2 = 0$ (b) $x^2 - 16 = -y^2$
3. (a) $x^2 + y^2 = 5$ (b) $2x^2 + 2y^2 = 16$
4. (a) $(x - 1)^2 + (y - 3)^2 = 1$
   (b) $(x + 1)^2 + (y + 3)^2 = 1$
5. (a) $x^2 + (y^2 + 4y + 4) = 9$
   (b) $(x^2 + 2x + 1) + y^2 = 16$

6. (a) Find the equation that describes the set of all points that are 3 units from the origin.
(b) Find the equation for the set of points located 3 units from the point (1, 2).

B. ESSENTIALS
Graph and label the following.
7. (a) $x^2 + y^2 + 2x + 4y = 11$
   (b) $x^2 + y^2 - 2x - 4y = 11$
8. (a) $x^2 + y^2 + 2x - 4y = 11$
   (b) $x^2 + y^2 - 2x + 4y = 11$
9. (a) $x^2 + y^2 - 2x + 4y = 4$
   (b) $x^2 + y^2 - 2x + 4y = 11$
10. ★(a) $2x^2 + 2y^2 - 4x + 8y + 2 = 0$
    (b) $4x^2 + 4y^2 - 8x + 16y + 11 = 0$
11. (a) $y = \pm\sqrt{16 - x^2}$
    (b) $y = \sqrt{16 - x^2}$
12. (a) $y = -\sqrt{9 - x^2}$
    (b) $y = 2 - \sqrt{9 - x^2}$

★13. Find the range and domain for Problem 11.

14. Find the range and domain for Problem 12.

C. APPLICATIONS
15. (a) Find the equation of the tangent line to the circle $x^2 + y^2 = 16$ at (0, 4).
(b) Find the equation of the tangent line to the circle $x^2 + y^2 - 2x + 2y = 2$ at $(3, -1)$. Repeat for the tangent at (1, 1).
(c) Find the equation of the secant line through $(3, -1)$ and (1, 1) in (b).

16. Find the equation of the circle(s) tangent to the x and y-axis with radius of 4.

17. Find the number of turns of paper on a roll of toilet paper 8 centimeters in diameter, if each sheet is 0.02 centimeters thick. What assumptions must we make?

13.3 THE PARABOLA

Consider Figure 13.6. For each point there are two distances, one to the point $F: (0, 1)$, another to the line $y = -1$.

$P_1: d(P_1, F) = 2$

$d(P_1, L_1) = 2$

$P_2: d(P_2, F) = \sqrt{(4 - 0)^2 + (2 - 1)^2}$

= _____?_____

$d(P_2, L_2) = 2 - (-1) =$ _____?_____

$P_3: d(P_3, F) = \sqrt{\_\_\_\_\_?\_\_\_\_\_}$

= _____?_____

$d(P_3, L_3) =$ _____?_____

− _____?_____

= _____?_____

These pairs of distances are equal for each point. The collection of all points whose distance from $F(0, 1)$ equals the distance from the line $y = -1$ is called a *parabola*.

In general, a parabola is defined as follows.

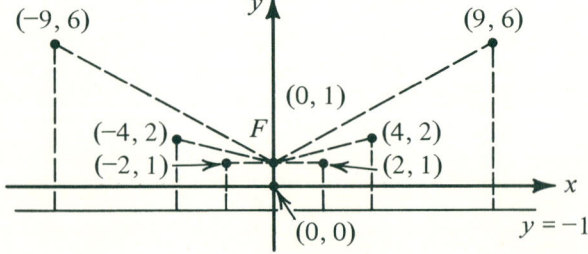

FIGURE 13.6.

SECTION 13.3 The Parabola

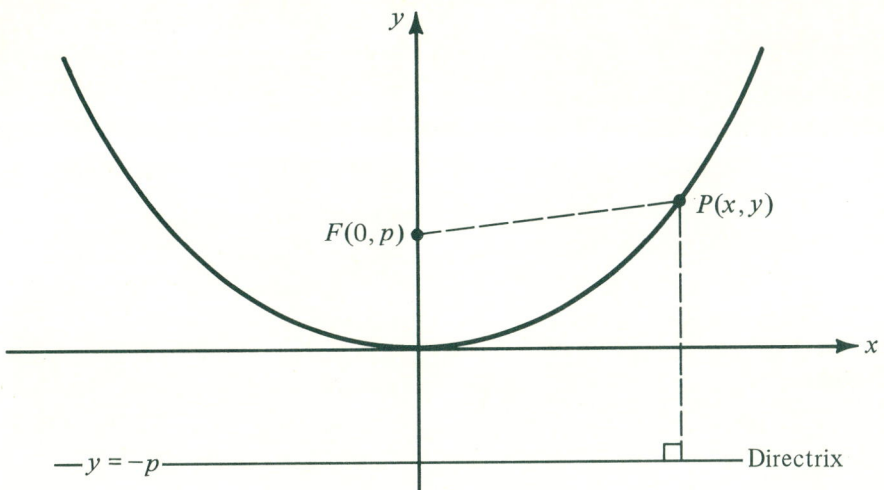

FIGURE 13.7. *Parabola.*

DEFINITION 13.2.
A parabola is the set of all points equidistant from a fixed point and a fixed line.

The fixed point is called the *focus* and the fixed line is called the *directrix*. The turning point on the parabola is called the *vertex*. (The focus cannot be located on the directrix.)

Now we can use the definition of a parabola to derive its equation. Consider Figure 13.7, which shows a fixed point F and a fixed line $y = -p$, with a typical point P given on the parabola. By definition, the distance from P to the line $y = -p$ must equal the distance from P to F. Therefore,

$$y + p = \sqrt{(x-0)^2 + (y-p)^2}$$

which, by squaring both sides, yields the general equation,

$$y^2 + 2py + p^2 = x^2 + y^2 - 2py + p^2$$
$$x^2 = 4py$$

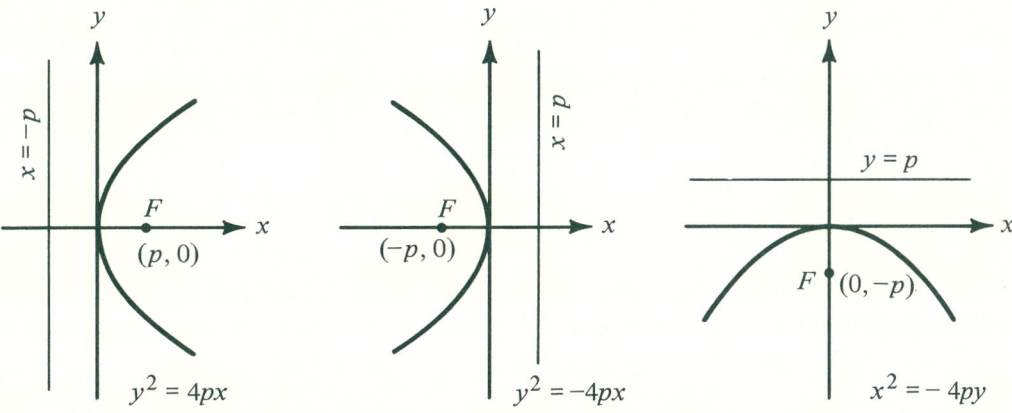

FIGURE 13.8. *Reflections and Inverses.*

CHAPTER 13   The Conic Sections

THEOREM II.
The equation of the parabola whose vertex is at (0, 0), whose focus is at (0, p), and whose directrix is $y = -p$ is $x^2 = 4py$.

The reflections and inverses of this parabola are shown in Figure 13.8, with their associated general equations.

These various orientations of the parabolas can easily be remembered by recalling that $y = x^2$ is symmetric about the y-axis and is concave up.

EXAMPLE I. Graph

$$y^2 = -8x \quad \text{and} \quad x^2 = 16y$$

Solution: $x^2 = 16y$ is a concave-up parabola in which $p = 4$ (since $4p = 16$); $y^2 = -8x$ opens to the left with $p = 2$. See Figure 13.9.

Notice how the shape of the curves can be determined by the relative size of p.

We can more easily graph parabolas whose vertices are not at the origin by considering them as translations of a parabola whose vertex is at the origin, exactly as we did for circles. In general, we have the following theorem.

THEOREM III.
The equation for a parabola whose vertex is at (h, k), whose focus is p units above its vertex, and whose directrix is p units below is given by $(x - h)^2 = 4p(y - k)$.

This theorem is illustrated by the next pair of parabolas, the second of which is simply a translation of the first.

EXAMPLE II. Graph $x^2 = 4py$ and $(x - h)^2 = 4p(y - k)$.

Solution: See Figure 13.10 (page 287).

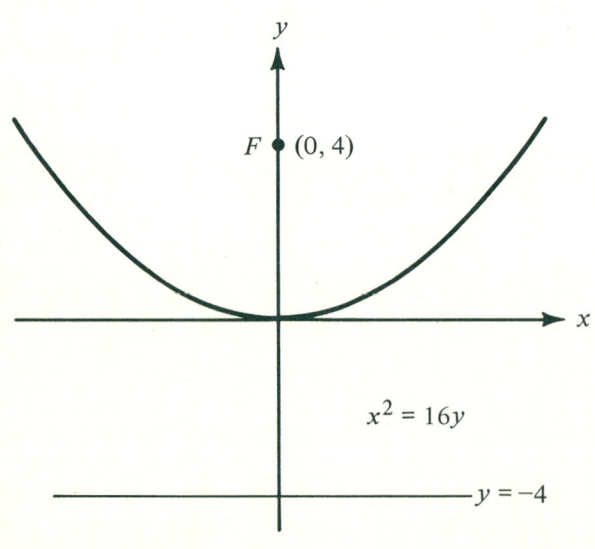
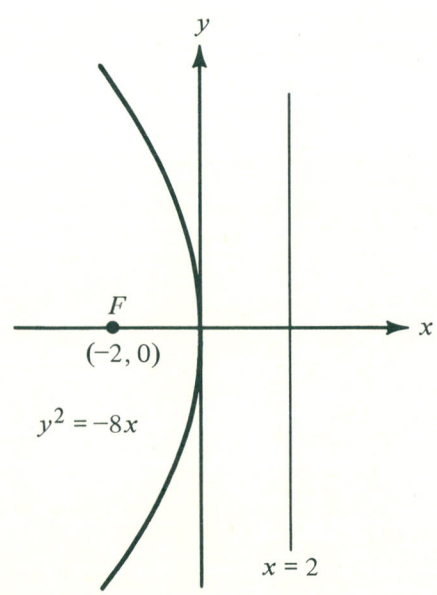

FIGURE 13.9.

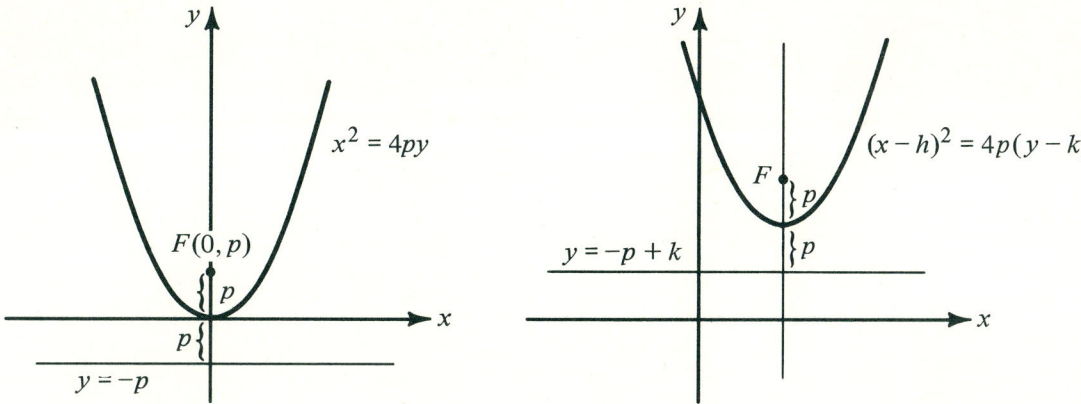

FIGURE 13.10. Parabolas.

There are other translations for parabolas that open down, to the left, or to the right. Notice that the equations of parabolas only have one squared term. They are represented by equations of the form

$$Ax^2 + Cy^2 + Dx + Ey + F = 0$$

where either

$$A = 0 \quad \text{or} \quad C = 0$$

but not both.

EXAMPLE III. Graph

$$x^2 + 2x + 4y - 7 = 0$$

Solution: We know this is a parabola because there is only one squared term. We must complete the square in order to find the vertex and the value of p.

$$x^2 + 2x = -4y + 7 \Rightarrow$$
$$x^2 + 2x + \underline{} = -4y + \underline{} \Rightarrow$$
$$(x + 1)^2 = -4(y - 2)$$

Therefore, the vertex is at $(-1, 2)$ and $p = 1$. This concave-down parabola is graphed in Figure 13.11.

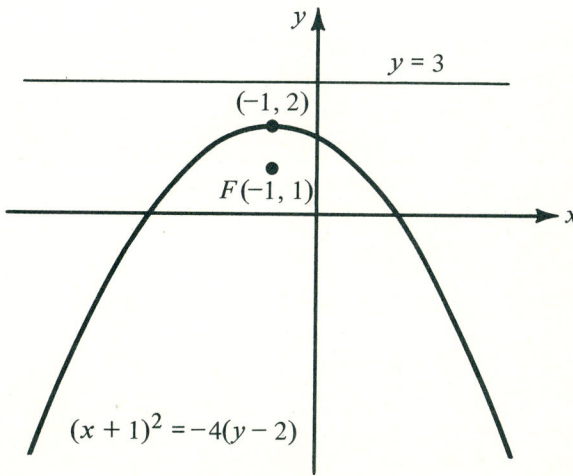

FIGURE 13.11.

EXERCISE 13.3

A. FUNDAMENTALS

Graph the following parabolas. Be sure to label the directrix and focus of each.

1. (a) $x^2 = 4y$ (b) $x^2 = 16y$
2. (a) $y^2 = 4x$ (b) $y^2 = -4x$

3. (a) $y^2 = x$ (b) $y^2 = 16x$

4. (a) $x^2 = -y$ (b) $x^2 = -64y$

5. (a) $x^2 = 8y$ (b) $(x-1)^2 = 8y$
   (c) $x^2 = 8(y-1)$

6. (a) $x^2 = -4y$ (b) $x^2 = -4(y+2)$
   (c) $x^2 = -4y + 8$

7. (a) $(y-1)^2 = -8(x+2)$
   (b) $(y+1)^2 = -8x - 16$
   (c) $y^2 + 2y + 1 = -4x + 8$

8. (a) What is the equation of the parabola whose vertex is at the origin and whose focus is at (4, 0)?
   (b) What is the equation of the parabola whose vertex is at $(-2, 1)$ and whose directrix is the y-axis?

B. ESSENTIALS
Graph and label the following parabolas.

9. (a) $y^2 - 2y - 4x = 7$
   (b) $x^2 - 2x - 4y = 7$

10. (a) $x^2 - 2x + 2y = 1$
    (b) $x^2 - 2x - 2y + 3 = 0$

11. (a) $y = 2x^2 - 4x + 1$
    (b) $y = 3x^2 - 4x + 1$

Find the equations of the following parabolas.

12. ★(a) directrix $x = 4$ and focus $(-4, 0)$.
    (b) directrix $y = -4$ and focus $(0, 4)$.

13. (a) directrix $y + 3 = 0$ and focus $(1, 4)$.
    (b) directrix $x + 2 = 0$ and focus $(3, 1)$.

C. APPLICATIONS

14. Find the equation of the parabola symmetric about the x-axis that has its vertex at the origin and that passes through (8, 4).

★15. Vertical highway curves are parabolic. Find the location of the culvert for the following valley.

16. The reflector of a headlight is parabolic. The light bulb must be located at the focus so that the light will travel parallel to the axis of symmetry. Locate the bulb for the following cross-section:

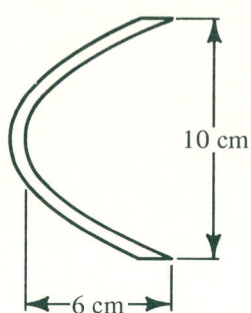

13.4 THE ELLIPSE

There are two equivalent forms for the equations of ellipses,

$$Ax^2 + Cy^2 + Dx + Ey + F = 0$$

where $A \cdot C > 0$ and $A \neq C$, and

$$\frac{(x-h)^2}{a^2} + \frac{(y-k)^2}{b^2} = 1$$

The restriction of $A \cdot C > 0$ in the first form forces A and C to not equal zero and to be of the same sign. The point (h, k) in the second form is the center of the ellipse, $2a$ is the length of the horizontal axis, and $2b$ is the length of the vertical axis, as shown in Figure 13.12.

As with the circle and the parabola, graphing ellipses whose equations are written in the first form will require completing the square in order to rewrite the equations in the second form. Then we can find the values of h, k, a, and b that are needed to complete the graph.

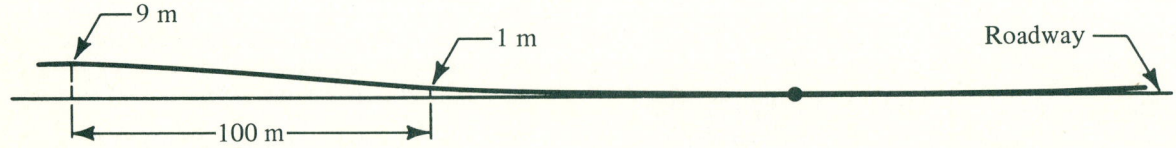

SECTION 13.4 The Ellipse

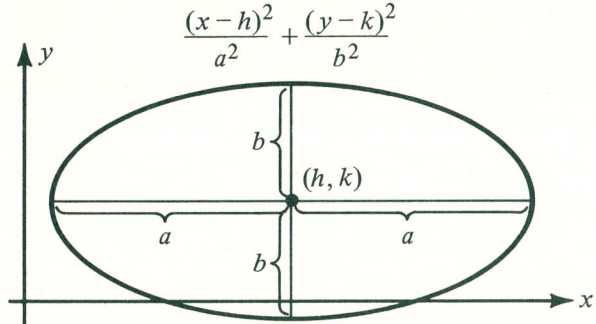

FIGURE 13.12. *Ellipse.*

Find the distances indicated for each of the points in Figure 13.13. As you might suspect, the collection of all the points whose distances conform to those of the examples will form an ellipse.

$P_1: d(P_1, F') = \sqrt{[0-(-4)]^2 + (3-0)^2}$

$= \underline{\quad ? \quad}$

$d(P_1, F)$

$= \sqrt{(\underline{\ ?\ } - \underline{\ ?\ })^2 + (\underline{\ ?\ } - \underline{\ ?\ })^2}$

$= \underline{\quad ? \quad}$

$P_2: d(P_2, F') = \sqrt{\left(\dfrac{10}{3} + 4\right)^2 + (\sqrt{5} - 0)^2}$

$= \underline{\quad ? \quad}$

$d(P_2, F)$

$= \sqrt{(\underline{\ ?\ } - \underline{\ ?\ })^2 + (\underline{\ ?\ } - \underline{\ ?\ })^2}$

$= \underline{\quad ? \quad}$

$P_3: d(P_3, F') = \underline{\quad ? \quad}, \quad d(P_3, F) = 1$

Do you see how these distances are related? If not, continue for P_4.

DEFINITION 13.3.
An **ellipse** is the set of all points $P(x, y)$ such that the distance from P to one fixed point F plus the distance from P to another fixed point F' equals a constant called $2a$.

Let's derive the formula for an ellipse centered at the origin ($h = k = 0$) using the above definition. Then we can graph any translated ellipse exactly as we graphed translated circles and parabolas.

Figure 13.14 shows an ellipse centered at the origin. Notice that F and F' are on the x-axis, equidistant from the origin, and the graph is symmetric about the x- and y-axes.

The distance from P to F plus the distance from P to F' must equal $2a$. Can you see why this "a" is the same as the "a" in the points $(a, 0)$ and

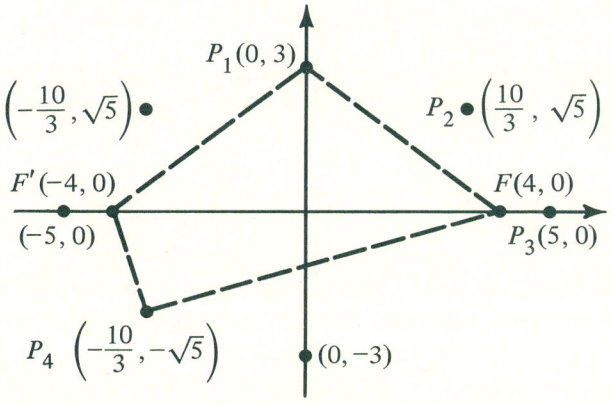

FIGURE 13.13.

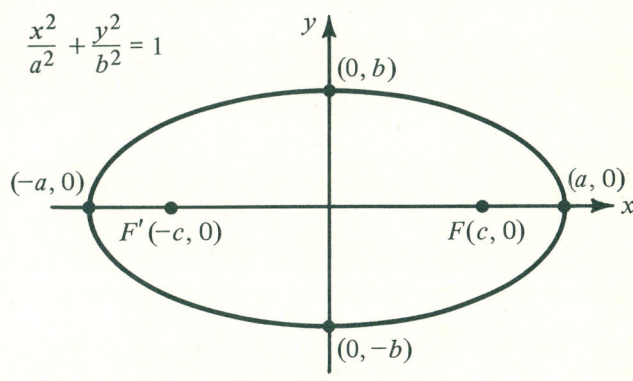

FIGURE 13.14. *Ellipse.*

CHAPTER 13   The Conic Sections

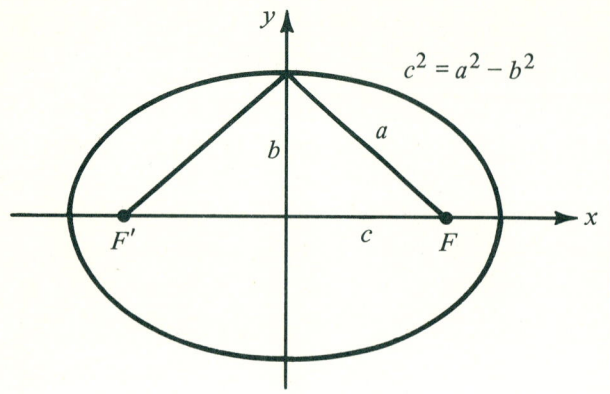

FIGURE 13.15.

$(-a, 0)$? [HINT: Let P be located at $(a, 0)$ and require that $PF + PF' = 2a$.] Now,

$$d(P, F) + d(P, F') = 2a \Rightarrow$$
$$\sqrt{(x + c)^2 + y^2} + \sqrt{(x - c)^2 + y^2} = 2a \Rightarrow$$
$$\sqrt{(x + c)^2 + y^2} = 2a - \sqrt{(x - c)^2 + y^2}$$

By squaring both sides we have

$$(x + c)^2 + y^2 =$$
$$4a^2 - 4a\sqrt{(x - c)^2 + y^2} + [(x - c)^2 + y^2] \Rightarrow$$
$$4a\sqrt{(x - c)^2 + y^2} = 4a^2 - 4cx \Rightarrow$$
$$a\sqrt{(x - c)^2 + y^2} = a^2 - cx \Rightarrow$$
$$a^2[(x - c)^2 + y^2] = (a^2 - cx)^2 \Rightarrow$$
$$a^2x^2 + a^2c^2 + a^2y^2 = a^4 + c^2x^2 \Rightarrow$$
$$a^2x^2 - c^2x^2 + a^2y^2 = a^4 - a^2c^2$$

This is the equation of an ellipse in terms of its x-intercepts, $(a, 0)$ and $(-a, 0)$, and its foci, $F(c, 0)$ and $F'(-c, 0)$. We can simplify the equation considerably by rewriting it in terms of only its x- and y-intercepts. This will also yield a better equation for graphing. Consider the point $(0, b)$. If we *define* this to be the y-intercept of the ellipse, then the

sum of the distances from there to F and F' must equal $2a$. Also, because of the symmetry, these distances are equal. Therefore, $c^2 = a^2 - b^2$ as shown in Figure 13.15.

Continuing, we now have

$$a^2x^2 - (a^2 - b^2)x^2 + a^2y^2 = a^4 - a^2(a^2 - b^2) \Rightarrow$$
$$b^2x^2 + a^2y^2 = a^2b^2 \Rightarrow$$
$$\frac{x^2}{a^2} + \frac{y^2}{b^2} = 1$$

which completes our derivation.

EXAMPLE I.   Graph and label $4x^2 + 9y^2 = 36$. First we divide both sides by 36 to rewrite the equation in the standard form. See Figure 13.16.

Solution: We have

$$\frac{4x^2}{36} + \frac{9y^2}{36} = \frac{36}{36} \Rightarrow$$
$$\frac{x^2}{9} + \frac{y^2}{4} = 1$$

Therefore,

$$a = 3, \quad b = 2, \quad \text{and} \quad c = \sqrt{a^2 - b^2} = \sqrt{5}$$

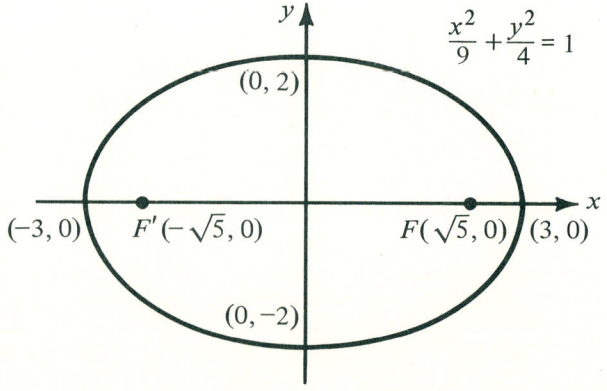

FIGURE 13.16.

290

EXAMPLE II. Graph and label $4x^2 + 9y^2 - 8x + 36y + 4 = 0$.

Solution: By completing the square, we have

$$4(x^2 - 2x) + 9(y^2 + 4y) = -4 \Rightarrow$$
$$4(x^2 - 2x + \underline{\quad ? \quad}) + 9(y^2 + 4y + \underline{\quad ? \quad})$$
$$= -4 + 4 + 36 \Rightarrow$$
$$4(x - 1)^2 + 9(y + 2)^2 = 36 \Rightarrow$$
$$\frac{4(x-1)^2}{36} + \frac{9(y+2)^2}{36} = \frac{36}{36} \Rightarrow$$
$$\frac{(x-1)^2}{9} + \frac{(y+2)^2}{4} = 1$$

Therefore, the center is at $(1, -2)$, $a = 3$, $b = 2$, and $c = \sqrt{5}$. See Figure 13.17. F' is located $\sqrt{5}$ units to the left of $(1, -2)$ at $(1 - \sqrt{5}, -2)$, and F is at $(1 + \sqrt{5}, -2)$.

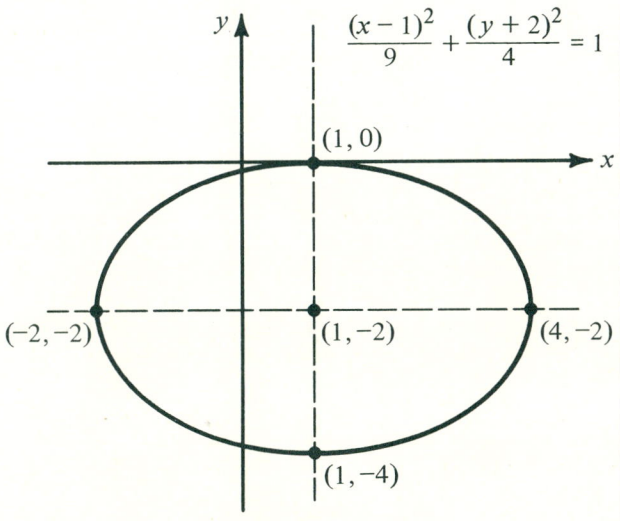

FIGURE 13.17.

The general equation of an ellipse centered at any point (h, k) is given in Theorem IV.

THEOREM IV.
The equation of the ellipse centered at (h, k), with a horizontal axis of length $2a$ and a vertical axis of length $2b$, whose foci are c units from the center on the horizontal axis is

$$\frac{(x-h)^2}{a^2} + \frac{(y-k)^2}{b^2} = 1$$

where

$$c^2 = a^2 - b^2, \quad a > b*.$$

EXERCISE 13.4

A. FUNDAMENTALS
Graph the following ellipses. It is necessary to label the foci and the end points of the *major* and *minor* axes, the long and short axes, respectively.

1. (a) $x^2 + 4y^2 = 4$ (b) $4x^2 + y^2 = 4$
2. (a) $4x^2 + 9y^2 = 36$ (b) $9x^2 + 4y^2 = 36$
3. (a) $25x^2 + 4y^2 = 100$
   (b) $36x^2 + 9y^2 = 324$
4. (a) $\dfrac{(x-1)^2}{4} + \dfrac{y^2}{9} = 1$
   (b) $\dfrac{(x-1)^2}{9} + \dfrac{y^2}{4} = 1$
5. (a) $\dfrac{x^2}{25} + \dfrac{(y-1)^2}{16} = 1$
   (b) $\dfrac{x^2}{9/4} + \dfrac{(y+1)^2}{4} = 1$
6. (a) $\dfrac{(x-1)^2}{4} + \dfrac{(y+2)^2}{9} = 1$
   (b) $\dfrac{(x+2)^2}{9} + \dfrac{(y-1)^2}{4} = 1$

---

* If $b > a$, then $c = \sqrt{b^2 - a^2}$ is the distance from the center to the foci along the vertical axis.

7. (a) What is the equation of the ellipse centered about the origin, with a major axis 3 centimeters long on the x-axis and a minor axis 1 centimeter long?

(b) What is the equation of the ellipse centered about $(2, -3)$, with a major axis 5 centimeters long on the x-axis and a minor axis 2 centimeters long?

B. ESSENTIALS
Graph and label the following ellipses.

8. ★(a) $9x^2 + 4y^2 + 18x - 27 = 0$
   (b) $4x^2 + 9y^2 + 8x - 32 = 0$

9. (a) $16x^2 + 9y^2 - 32x - 36y = 92$
   (b) $16x^2 + 9y^2 + 32x + 36y = 92$

10. $6x^2 + y^2 + 24x - 6y + 24 = 0$

11. $27x^2 + 20y^2 - 216x + 80y - 28 = 0$

12. (a) $y = 1 \pm \frac{1}{2}\sqrt{4 - x^2}$
    (b) $y = -2 + \frac{2}{3}\sqrt{9 - (x-1)^2}$

C. APPLICATIONS

13. If we set a sound or light source at one focus of an ellipse, the rays are reflected off the curve through the other focus. How far from the center of an elliptical room 15 meters by 8 meters should we locate a stereo speaker and a chair for listening to the stereo.

14. (a) Show that the length of a cord passing through the origin connecting 2 points (x_1, y_1) and (x_2, y_2) on the ellipse $\frac{x^2}{a^2} + \frac{y^2}{b^2} = 1$ is given by $S = 2\sqrt{x_1^2 + y_1^2}$.

    (b) Show that the length of a cord joining any two points (x_1, y_1) and (x_2, y_2) on the ellipse $\frac{x^2}{a^2} + \frac{y^2}{b^2} = 1$ is $S = \sqrt{1 + \left(\frac{\Delta y}{\Delta x}\right)^2} \, \Delta x$, where $\Delta x = x_2 - x_1$.

15. The earth travels around the sun in an elliptical orbit with the minimum distance from the sun (one of the foci) 147.5×10^6 kilometers and the maximum 152.5×10^6 kilometers. Find the equation of this orbit. These distances are called the Perigee and the Apogee, respectively, of the earth's orbit.

13.5 THE HYPERBOLA

The two forms for equations of hyperbolas are

$$Ax^2 + Cy^2 + Dx + Ey + F = 0$$

where $A \cdot C < 0$ and

$$\frac{(x - h)^2}{a^2} - \frac{(y - k)^2}{b^2} = 1$$

Here the restriction $A \cdot C < 0$ means neither A nor C is equal to zero, and A and C are of opposite signs. (Therefore, not equal.)

Let's begin with the nontranslated hyperbola ($h = k = 0$) in Figure 13.18.

The solid curve shown in Figure 13.18 is the hyperbola. The other lines are called asymptotes and are used as sketching aids. Hyperbolas have two asymptotes that we must sketch in order to obtain an accurate graph.

DEFINITION 13.4.
A **hyperbola** is the set of all points such that the difference of the distances to the two foci equals $2a$.

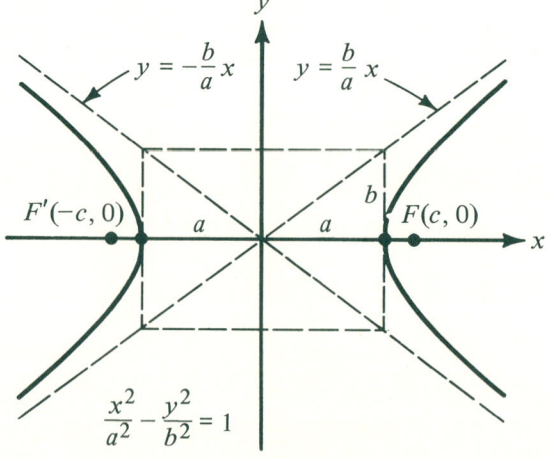

FIGURE 13.18. Hyperbola.

SECTION 13.5 The Hyperbola

Before we begin the derivation of the equation from the definition, let's establish the relationship between a, b, and c. We *define* b^2 as

$$b^2 = c^2 - a^2$$

The values of a and c will be determined by the equation of the hyperbola, and the value of b from a and c.

This definition of b results in the equations of these asymptotes being $y = \pm \dfrac{b}{a} x$. We will illustrate this in Example II.

Now, let's proceed with the derivation.

The fact that the distance from P to F minus the distance from P to F' equals $2a$ gives the equation:

$$\left| \sqrt{(x-c)^2 + y^2} - \sqrt{(x+c)^2 + y^2} \right| = 2a$$

where the absolute value symbols are used to guarantee that the number on the left will not be negative (see Figure 3.18).

By squaring both sides, we have

$$(x-c)^2 + y^2 - 2\sqrt{(x-c)^2+y^2}\sqrt{(x+c)^2+y^2}$$
$$+ (x+c)^2 + y^2 = 4a^2 \Rightarrow$$
$$x^2 + y^2 + c^2 - 2a^2$$
$$= \sqrt{(x-c)^2+y^2}\sqrt{(x+c)^2+y^2}$$

Then squaring again yields

$$x^4 + 2x^2y^2 + 2c^2x^2 - 4a^2x^2 + y^4 + 2c^2y^2$$
$$- 4a^2y^2 + c^4 - 4a^2c^2 + 4a^4$$
$$= x^4 - 2c^2x^2 + c^4 + x^2y^2 + 2cxy^2 + c^2y^2$$
$$+ x^2y^2 - 2cxy^2 + c^2y^2 + y^4$$

By collecting like terms we have

$$c^2x^2 - a^2x^2 - a^2y^2 - a^2c^2 + a^4 = 0$$

But $c^2 = a^2 + b^2$ yields

$$(a^2 + b^2)x^2 - a^2x^2 - a^2y^2$$
$$- a^2(a^2 + b^2) + a^4 = 0 \Rightarrow$$
$$b^2x^2 - a^2y^2 = a^2b^2 \Rightarrow$$
$$\frac{x^2}{a^2} - \frac{y^2}{b^2} = 1$$

EXAMPLE I. Graph and label $9x^2 - 16y^2 = 144$.

Solution: We have

$$\frac{x^2}{16} - \frac{y^2}{9} = 1 \Rightarrow$$
$$a = 4, \quad b = 3, \quad c = \sqrt{a^2 + b^2} = \sqrt{25} = 5$$

The asymptotes are $y = \pm \dfrac{b}{a} x = \pm \tfrac{3}{4}x$. See Figure 13.19.

EXAMPLE II. Show that the asymptotes to the hyperbola $9x^2 - 16y^2 = 144$ are $y = \pm \tfrac{3}{4}x$, as shown in Figure 13.19.

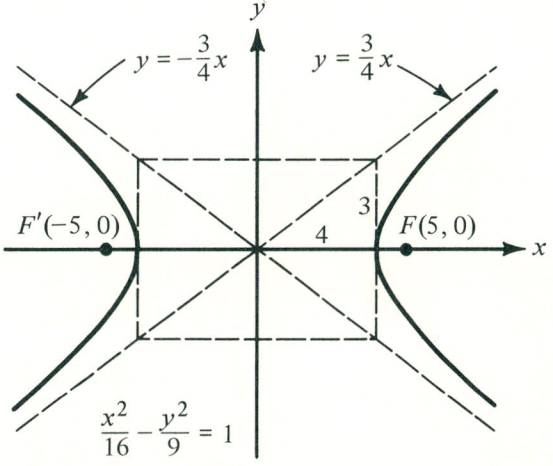

FIGURE 13.19.

Solution: We will show that the hyperbola approaches the line $y = \tfrac{3}{4}x$ as x becomes larger.

CHAPTER 13   The Conic Sections

If (x, y) is a point on the hyperbola, then its y-coordinate is

$$16y^2 = 144 - 9x^2 \Rightarrow$$

$$y = \sqrt{\frac{9x^2 - 144}{16}} = \sqrt{\frac{9(x^2 - 16)}{16}} \Rightarrow$$

$$y = \tfrac{3}{4}\sqrt{x^2 - 16}$$

We will use the positive square root for the point on the upper right side of the hyperbola in Figure 13.19. Notice that $x > 0$ also.)

We will show that the hyperbola approaches the line as x becomes larger by showing that as x approaches infinity

$$\tfrac{3}{4}\sqrt{x^2 - 16} \to \tfrac{3}{4}x$$

by showing that

$$\lim_{x \to \infty} (\tfrac{3}{4}\sqrt{x^2 - 16} - \tfrac{3}{4}x) = 0$$

Factoring the constant and using the conjugate to simplify the numerator, we have

$$\tfrac{3}{4} \lim_{x \to \infty} (\sqrt{x^2 - 16} - x)\left[\frac{\sqrt{x^2 - 16} + x}{\sqrt{x^2 - 16} + x}\right]$$

$$= \tfrac{3}{4} \lim_{x \to \infty} \left(\frac{(x^2 - 16) - x^2}{\sqrt{x^2 - 16} + x}\right)$$

$$= \tfrac{3}{4} \lim_{x \to \infty} \frac{-16}{\sqrt{x^2 - 16} + x} = 0$$

since dividing -16 by increasingly larger numbers results in quotients approaching zero.

Therefore, $y = \tfrac{3}{4}x$ is an asymptote to the hyperbola.

EXAMPLE III. Graph and label

$$9x^2 - 16y^2 - 18x - 64y - 199 = 0$$

Solution: We know that this equation represents a hyperbola, since $A = 9$ and $C = -16$ are opposite in sign. Since there are x^1 and y^1 terms in the equation, we know that the hyperbola is translated in both the x and the y directions. Again, we must complete the square in order to graph the curve. See Figure 13.20.

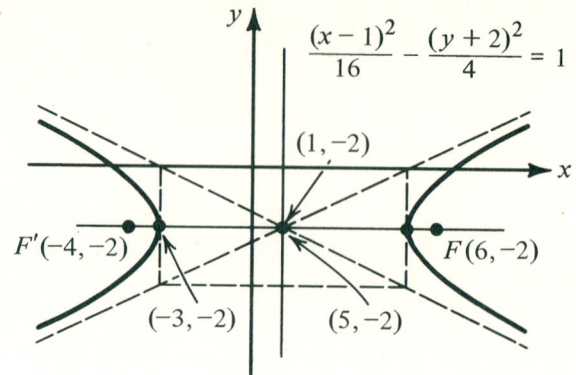

FIGURE 13.20.

$$9(x^2 - 2x + \underline{\ ?\ }) - 16(y^2 + 4y + \underline{\ ?\ })$$
$$= 199 + 9 - 64 = \underline{\ ?\ }$$

$$\frac{(x - 1)^2}{16} - \frac{(y + 2)^2}{9} = 1$$

Finding the equation of the asymptotes simply requires that we find the equation of the lines through $(1, -2)$ with slopes of $+\tfrac{3}{4}$ and $-\tfrac{3}{4}$.

$$y = m(x - x_0) + y_0 \Rightarrow$$

$$y = \underline{\ ?\ }(x - 1)\underline{\ ?\ } \quad \text{and} \quad y = \underline{\ ?\ }(x - 1)\underline{\ ?\ }$$

$$y = \tfrac{3}{4}x - \tfrac{11}{4} \quad \text{and} \quad y = -\tfrac{3}{4}x - \tfrac{5}{4}$$

The general equation for a hyperbola is summarized in Theorem V.

THEOREM V.
The equation of a **hyperbola** centered at (h, k), with vertices a and foci c units to the left and right of (h, k), with the slope of its asymptotes equaling $\pm \dfrac{b}{a}$ is given by

$$\frac{(x - h)^2}{a^2} - \frac{(y - k)^2}{b^2} = 1$$

where

$$a^2 + b^2 = c^2$$

294

If the coefficient of the y-term is positive and the coefficient of the x-term is negative, then the foci are on the y-axis.

Now, let's summarize the information about the conic sections.

$$Ax^2 + Cy^2 + Dx + Ey + F = 0$$

is the equation of

(i) a **line** if $A = C = 0$
(ii) a **circle** if $A = C \neq 0$
(iii) a **parabola** if either $A = 0$ or $C = 0$, but not both
(iv) an **ellipse** if $A \cdot C > 0$ and $A \neq C$
(v) a **hyperbola** if $A \cdot C < 0$

The general equations are

$(x-h)^2 + (y-k)^2 = r^2$ (circle)

$(x-h)^2 = 4p(y-k)$ (parabola)

$\dfrac{(x-h)^2}{a^2} + \dfrac{(y-k)^2}{b^2} = 1$ (ellipse)

$\dfrac{(x-h)^2}{a^2} - \dfrac{(y-k)^2}{b^2} = 1$ (hyperbola)

EXERCISE 13.5

A. FUNDAMENTALS

Graph the following hyperbolas, being sure to label the vertices, the asymptotes, and the foci.

1. (a) $x^2 - y^2 = 1$ (b) $y^2 - x^2 = 1$

2. (a) $\dfrac{x^2}{4} - \dfrac{y^2}{9} = 1$ (b) $\dfrac{y^2}{4} - \dfrac{x^2}{9} = 1$

3. (a) $\dfrac{x^2}{16} - \dfrac{y^2}{4} = 1$ (b) $\dfrac{x^2}{4} - \dfrac{y^2}{16} = 1$

Find the equations of the following asymptotes and graph the lines. Remember that the asymptotes pass through the center of the hyperbola.

★4. For $\dfrac{(x-1)^2}{1} - \dfrac{y^2}{1} = 1$

  (a) Through the point (2, 1)
  (b) Through the point (0, 1)

5. For $\dfrac{(x-2)^2}{4} - \dfrac{(y-1)^2}{1} = 1$

  (a) Through the point (4, 2)
  (b) Through the point (4, 0)

6. Complete the graphing of the two hyperbolas above if the vertices are (0, 0) and (2, 0) for number 4 and (0, 1) and (4, 1) for number 5.

B. ESSENTIALS

Graph and label.

7. (a) $\dfrac{(x+2)^2}{9} - \dfrac{(y-3)^2}{16} = 1$
   (b) $16(y-3)^2 - 9(x+2)^2 = 144$

8. (a) $4(x+1)^2 - 25(y-1)^2 = 100$
   (b) $4x^2 - 8x - 25y^2 = 96$

9. (a) $x^2 - 8x - y^2 = 9$
   (b) $x^2 - y^2 - 8y = 41$

10. $x^2 + 4x - y^2 + 6y = 14$

11. $4x^2 - 8x - 9y^2 - 36y = 68$

12. $25x^2 + 100x - 4y^2 + 24y \times 55 = 0$

Find the equation for each of the following hyperbolas.

13. Center at $(0,0)$, focus $(-5, 0)$, and vertex $(3, 0)$.

14. Center at $(0, 0)$, vertex $(2, 0)$, and asymptote $y = \tfrac{3}{2}x$.

15. Center at $(1, -2)$, focus $(6, -2)$, and asymptote $y = \tfrac{4}{3}x - \tfrac{10}{3}$.

Identify the following conic sections, then sketch and label each.

16. (a) $16x^2 - 9y^2 = 144$
    (b) $49x^2 + 24y^2 = 1176$

★17. (a) $25x^2 - 100x - 144y^2 + 288y = 3644$
    (b) $x^2 - 4x + y^2 - 2y = 164$

18. (a) $y^2 - 16x + 96 = 0$
    (b) $64x^2 - 36y^2 = 2304$

19. $3x^2 - 4y + 2 = 0$

20. $16x^2 - 32x + 4y^2 + 16y = 32$

21. $25x^2 - 200x - 9y^2 - 18y + 166 = 0$

22. $4x^2 + 25y^2 - 24x - 50y = 39$

23. $y^2 + 6y + 4x^2 + 8x - 51 = 0$

24. $x^2 - 6x - 8y + 17 = 0$

25. $12x + 2y + y^2 = 23$

26. $100x^2 - 36y^2 = 200x + 144y + 269$

27. $x^2 - y^2 = 0$

• **28.** $10y^2 - 5x^2 + 20x + 20y = 60$

• **29.** $1000x^2 + 2000x + 100y^2 - 400y = 98{,}600$

• **30.** $12x^2 - 112y - 36x + 91 = 0$

• **31.** $504y^2 - 336y - 675x^2 + 900x = 604$

C. APPLICATIONS

32. The vertices of the hyperbola $9x^2 - 7y^2 = 63$ are located at the foci of the ellipse $b^2x^2 + a^2y^2 = a^2b^2$ and vice versa. Find a^2 and b^2.

33. Use the reflective properties of hyperbolas and ellipses to prove that the two curves in Problem 16 are orthogonal, that is, that their tangents at the point of intersection are perpendicular.

34. The Long-Range Navigation System (Loran) uses hyperbolas to enable ships and planes to establish their locations. It uses two pairs of three stations transmitting pulses at two different frequencies. One station is called the master station and the other two are called slave stations. The master station and one slave are the foci for one family of hyperbolas; the master and the other slave are foci for another family of hyperbolas. Ships monitor the delay time between the signals sent from the master and those sent from the slaves to

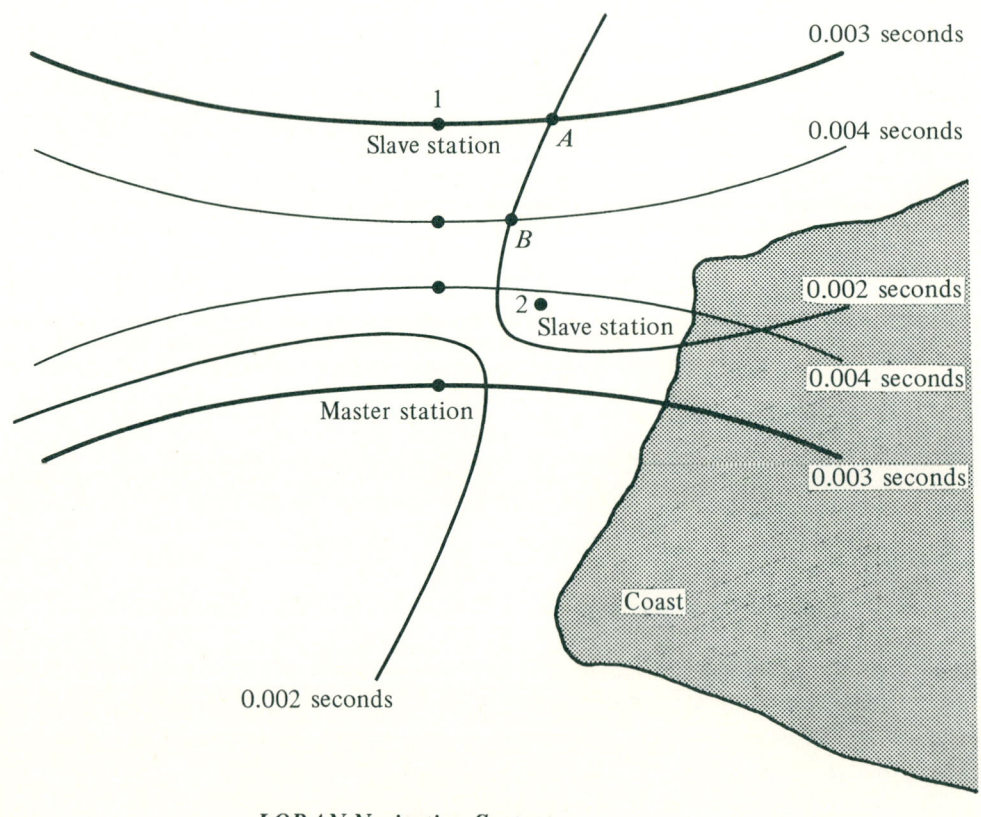

LORAN Navigation System

determine which pair of hyperbolas will give their location. (Do you remember the definition of a hyperbola?)

Ship A (shown in the following illustration) receives the signal from the first slave station 2000 microseconds (0.002 second) after receiving the signal from the master station and the signal from the second slave station 3000 microseconds after the master signal. This locates her position at the intersection of the pair of hyperbolas as shown.

(a) Explain how ship B determines her location.

(b) How does the fact that hyperbolas are not functions restrict the placement of the stations?

(c) Create a system of navigation based on ellipses analogous to the Loran system, and explain how it would work.

13.6 ROTATIONS AND THE CONIC SECTIONS (optional)

Now we are ready to discuss equations of the form

$$Ax^2 + Bxy + Cy^2 + Dx + Ey + F = 0$$

where

$$B \neq 0$$

Such equations represent rotated conics, such as those in Figures 13.21(a) and (b).

Once we rewrite the equations containing the $x \cdot y$-term as equivalent equations in u and v without a $u \cdot v$-term, graphing the conic section relative to the u- and v-axes will be the same as in the previous sections. The problem, then, is simply to find a change of variable that will eliminate the term containing the product of the variables x and y.

Let's find the relationship between the coordinates of a point given in the x, y-coordinate system and the coordinates of the same point in the u, v-coordinate system. Consider Figure 13.22, illustrating a rotation of θ to go from the original x, y-system to the new u, v-system. (See page 298.)

In the x, y-system we have

$$\sin \phi = \frac{y}{OP} \quad \text{and} \quad \cos \phi = \frac{x}{OP}.$$

In the u, v-system we have

$$\sin (\phi - \theta) = \frac{v}{OP} \quad \text{and} \quad \cos (\phi - \theta) = \frac{u}{OP}.$$

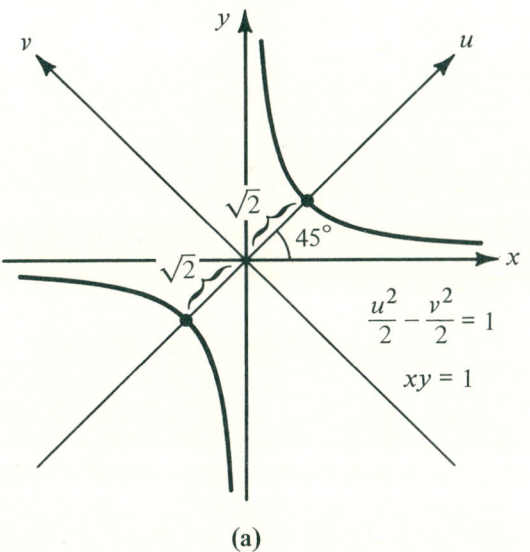

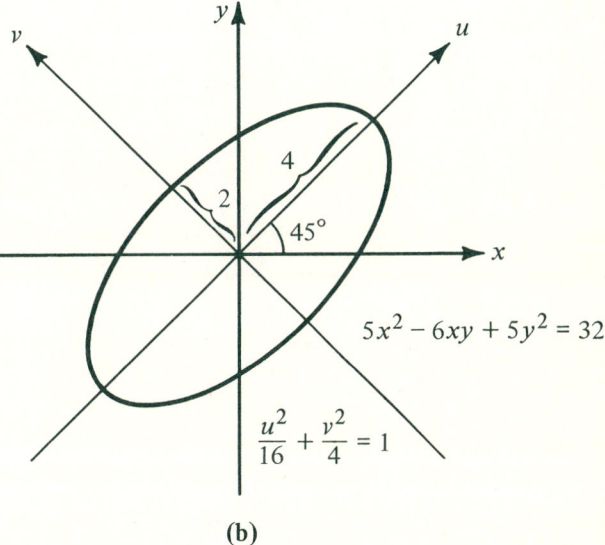

FIGURE 13.21.

CHAPTER 13 The Conic Sections

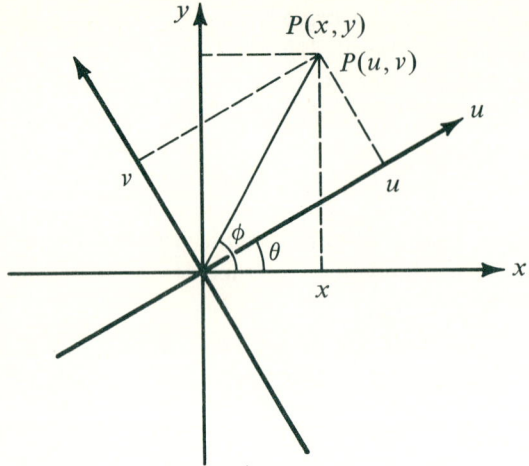

FIGURE 13.22.

Therefore,

$$v = OP \sin(\phi - \theta) = OP \sin\phi \cos\theta$$
$$- OP \cos\phi \sin\theta$$

But,

$$OP \sin\phi = y \quad \text{and} \quad OP \cos\phi = x \Rightarrow$$
$$v = y\cos\theta - x\sin\theta$$

Also,

$$u = OP \cos(\phi - \theta) = \underline{\quad ? \quad}$$
$$+ \underline{\quad ? \quad}$$

Therefore,

$$u = x\cos\theta + y\sin\theta$$

Now, if we solve these equations simultaneously for x and y, we have two pairs of equations relating the x, y- and the u, v-coordinate systems. The results are summarized in Theorem VI.

THEOREM VI.
If $P(x, y)$ are the coordinates of a point in the x, y-coordinate system, and $P(u, v)$ are the coordinates of the same point relative to the u, v-axis system formed by rotating the x, y-system through some angle θ, then

$$u = x \cos\theta + y \sin\theta$$
$$v = -x \sin\theta + y \cos\theta$$

and

$$x = u \cos\theta - v \sin\theta$$
$$y = u \sin\theta + v \cos\theta$$

Now we can find the coordinates relative to the x- and y-axes of the foci of the two conic sections whose graphs are in Figure 13.21.

EXAMPLE I. Find the foci of (a) $xy = 1$ and (b) $5x^2 - 6xy + 5y^2 = 32$.

Solution: (a) Since this equation is equivalent to

$$\frac{u^2}{2} - \frac{v^2}{2} = 1$$

relative to the u, v-axes, we have

$$c^2 = 2^2 + 2^2 = 8 \Rightarrow$$
$$c = 2\sqrt{2}$$

Therefore, the u, v-coordinates of foci are

$$(2\sqrt{2}, 0) \quad \text{and} \quad (-2\sqrt{2}, 0)$$

Now, since $\theta = 45°$, we have

$$x = u \cos 45° - v \sin 45°$$
$$= \pm 2\sqrt{2}\left(\frac{1}{\sqrt{2}}\right) - 0\left(\frac{1}{\sqrt{2}}\right) = \pm 2$$

and

$$y = u \sin 45° + v \cos 45°$$
$$= \pm 2\sqrt{2}\left(\frac{1}{\sqrt{2}}\right) + 0\left(\frac{1}{\sqrt{2}}\right) = \pm 2$$

The foci have coordinates $(2, 2)$ and $(-2, -2)$.
(b) This equation is equivalent to

$$\frac{u^2}{16} + \frac{v^2}{4} = 1$$

Therefore, the foci are c units from the origin on the u-axis, where

$$c^2 = 16 - 4 \Rightarrow$$
$$c = \underline{\ \ ?\ \ }$$

Therefore, the u, v-coordinates of the foci are

$$(\underline{\ \ ?\ \ }, 0) \quad \text{and} \quad (\underline{\ \ ?\ \ }, 0)$$

Then,

$x = u \cos 45° - v \sin 45°$

$= (\underline{\ \ ?\ \ })\left(\frac{1}{\sqrt{2}}\right) - (\underline{\ \ ?\ \ })\left(\frac{1}{\sqrt{2}}\right) = \underline{\ \ ?\ \ }$

and

$y = u \sin 45° + v \cos 45°$

$= (\underline{\ \ ?\ \ })\left(\frac{1}{\sqrt{2}}\right) - (\underline{\ \ ?\ \ })\left(\frac{1}{\sqrt{2}}\right) = \underline{\ \ ?\ \ }$

Therefore, relative to the original coordinate system, the vertices are

$$(2\sqrt{2}, 2\sqrt{2}) \quad \text{and} \quad (-2\sqrt{2}, -2\sqrt{2})$$

Now we are ready to learn how to derive the equation of a conic section relative to the u, v-coordinate system. Consider Example II.

EXAMPLE II. Find the angle through which the axes must be rotated to remove the x, y-term in the equation

$$5x^2 - 4xy + 8y^2 - 36 = 0$$

Solution: Substituting the equations from Theorem IV yields

$5(u \cos \theta - v \sin \theta)^2$
$\quad - 4(u \cos \theta - v \sin \theta)(u \sin \theta + v \cos \theta)$
$\quad + 8(u \sin \theta + v \cos \theta)^2 - 36 = 0$

By expanding and collecting like terms we find that the coefficient of the u, v-term is

$$6 \sin \theta \cos \theta - 4 \cos^2 \theta + 4 \sin^2 \theta$$

We wish to find the value of θ that will eliminate the u, v-term. Let's set its coefficient equal to zero and solve for θ.

$$6 \sin \theta \cos \theta - 4 \cos^2 \theta + 4 \sin^2 \theta = 0 \Rightarrow$$
$$3 \sin 2\theta - 4 \cos 2\theta = 0$$
$$\tan 2\theta = \tfrac{4}{3} \Rightarrow$$
$$2\theta = \tan^{-1} \tfrac{4}{3} \Rightarrow$$
$$\theta = \frac{1}{2} \tan^{-1} \tfrac{4}{3} \approx -0.46 \text{ radians} \Rightarrow$$
$$\theta \approx 26°$$

To complete the graph of this rotated conic section, we need to find the sine and cosine of θ. From Figure 13.23 we have

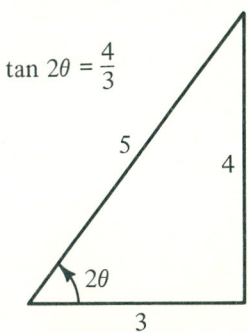

FIGURE 13.23.

$$\sin^2 \theta = \frac{1 - \cos 2\theta}{2} = \frac{1 - 3/5}{2} = \frac{1}{5}$$

and

$$\cos^2 \theta = \frac{1 + \cos 2\theta}{2} = \frac{1 + 3/5}{2} = \frac{4}{5}$$

Therefore,

$$\sin \theta = \frac{1}{\sqrt{5}} \quad \text{and} \quad \cos \theta = \frac{2}{\sqrt{5}}$$

CHAPTER 13 The Conic Sections

Now,
$$x = u\left(\frac{2}{\sqrt{5}}\right) - v\left(\frac{1}{\sqrt{5}}\right)$$
and
$$y = u\left(\frac{1}{\sqrt{5}}\right) + v\left(\frac{2}{\sqrt{5}}\right)$$

implies
$$5x^2 - 4xy + 8y^2 - 36 = 0 \Leftrightarrow$$
$$5\left(\frac{2}{\sqrt{5}}u - \frac{1}{\sqrt{5}}v\right)^2$$
$$- 4\left(\frac{2}{\sqrt{5}}u - \frac{1}{\sqrt{5}}v\right)\left(\frac{1}{\sqrt{5}}u + \frac{2}{\sqrt{5}}v\right)$$
$$+ 8\left(\frac{1}{\sqrt{5}}u + \frac{2}{\sqrt{5}}v\right)^2 - 36 = 0 \Rightarrow$$
$$4u^2 + 9v^2 - 36 = 0$$
$$5x^2 - 4xy + 8y^2 = 36 \Leftrightarrow \frac{u^2}{9} + \frac{v^2}{4} = 1$$

The graph is shown in Figure 13.24.

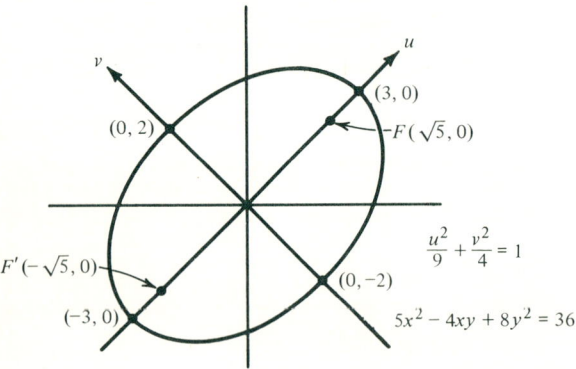

FIGURE 13.24.

By carrying out a process similar to the one used in the previous example, but using the general equation
$$Ax^2 + Bxy + Cy^2 + Dx + Ey + F = 0$$

we find that the angle of rotation is
$$\theta = \frac{1}{2}\tan^{-1}\left(\frac{B}{A-C}\right) \quad \text{for } A \neq C$$
or
$$\theta = 45° \quad \text{for } A = C$$

Then, we use the trig identities
$$\sin^2\theta = \frac{1 - \cos 2\theta}{2}$$
and
$$\cos^2\theta = \frac{1 + \cos 2\theta}{2}$$

to find the values of $\cos\theta$ and $\sin\theta$, allowing us to find expressions for x and y in terms of u and v from the equations
$$x = u\cos\theta - v\sin\theta$$
and
$$y = u\sin\theta + v\cos\theta$$

We will use the positive values of $\sin\theta$ and $\cos\theta$ from the identities, since we only need values of θ between $0°$ and $90°$.

Theorem VII enables us to recognize which conic section we have even before we rotate the coordinate axes.

THEOREM VII.
The equation $Ax^2 + Bxy + Cy^2 + Dx + Ey + F = 0$ may represent

(i) an **ellipse** if $B^2 - 4AC < 0$
(ii) a **parabola** if $B^2 - 4AC = 0$
(iii) a **hyperbola** if $B^2 - 4AC > 0$

If it does not represent one of these, then the equation represents either two straight lines, a point, or no ordered pairs of real numbers.

EXERCISE 13.6

A. FUNDAMENTALS
Use Theorem VII to determine which conic section is probably given by each of the following equations.

1. $x^2 - 2xy + y^2 - 1 = 0$
2. $3x^2 - 2xy - 4y^2 + 4y = 2$
3. $2x^2 + 5xy + 2y^2 - 3y = 4$
4. $x^2 + 4x + 3y^2 - 3xy = 1$

Determine the angle through which each of the following conic sections has been rotated.

5. $3x^2 - 4x - y^2 + 6y + 4xy - 3 = 0$
6. $2x^2 + 3y^2 + 6y - \sqrt{3}xy - 2\sqrt{3} = 0$
7. $2\sqrt{3}x^2 + \sqrt{3}x + 3\sqrt{3}y^2 + xy - \sqrt{3} = 0$
8. $5x^2 + 5y^2 + 6y + 6xy - 5 = 0$
9. $2x^2 - 4x + 2y^2 - 3y - 7 = 0$

Graph the following conic sections in the u, v-coordinate system, rotated through the indicated value of θ.

10. $\theta = \dfrac{\pi}{4}, \dfrac{u^2}{9} + \dfrac{v^2}{16} = 1$
11. $\theta = \dfrac{\pi}{3}, \dfrac{v^2}{4} - \dfrac{u^2}{16} = 1$
12. $\theta = \dfrac{\pi}{6}, (v-1)^2 = 4u$
13. $\theta = \dfrac{\pi}{4}, \dfrac{(u-1)^2}{4} + \dfrac{(v-2)^2}{9} = 1$

B. ESSENTIALS

14. Find the equation of the conic section in Problem 10, relative to the x, y-coordinate system.

15. Change the equation in Problem 11 to x, y-coordinates.

16. Change the equation in Problem 12 to x, y-coordinates.

Graph and label the following rotated conic sections.

★17. $x^2 + 2xy + y^2 = \sqrt{2}y - \sqrt{2}x$
18. $3x^2 + 13y^2 - 10xy = 36$
19. $10x^2 + 52xy + 10y^2 = 576$
20. $15x^2 + 40xy - 15y^2 + 625 = 0$
21. $101x^2 + 30xy + 29y^2 = 104$
22. $x^2 + 32\sqrt{3}x + 3y^2 + 32y - 2\sqrt{3}xy = 0$

Graph and label the following rotated and translated conic sections.

23. $4x^2 + 8xy + 4y^2 - 7\sqrt{2}x - 9\sqrt{2}y = -6$
24. $10x^2 - 12xy - 36\sqrt{2}x + 10y^2 - 28\sqrt{2}y + 4 = 0$
★25. $-6x^2 + 20xy - 20\sqrt{2}x - 6y^2 + 44\sqrt{2}y = 92$
26. $x^2 + 2\sqrt{3}xy + 2\sqrt{3}x - y^2 + 2y = 0$

13.7 SOLID ANALYTIC GEOMETRY (optional)

We are now ready to generalize the graphing of conic sections to three dimensions. That is, instead of drawing flat graphs in two-dimensional coordinate systems, we will draw solid graphs in three-dimensional systems. Find the missing coordinates (x, y, z) shown in the three-dimensional coordinate system in Figure 13.25. The solid axes represent positive reals, and the dotted axes represent negative reals. Notice that each point is found on the corner of a box instead of on the corner of a rectangle, as in the x, y-plane. This box is called a rectangular parallelepiped. Your coordinates should be $P_1(1, 2, 3)$, $P_2(3, -4, -2)$, $P_3(0, 0, -4)$, $P_4(0, 4, -4)$, and $P_5(-4, 4, -4)$. If you missed any of these, make sure you understand why before plotting the next points.

CHAPTER 13 The Conic Sections

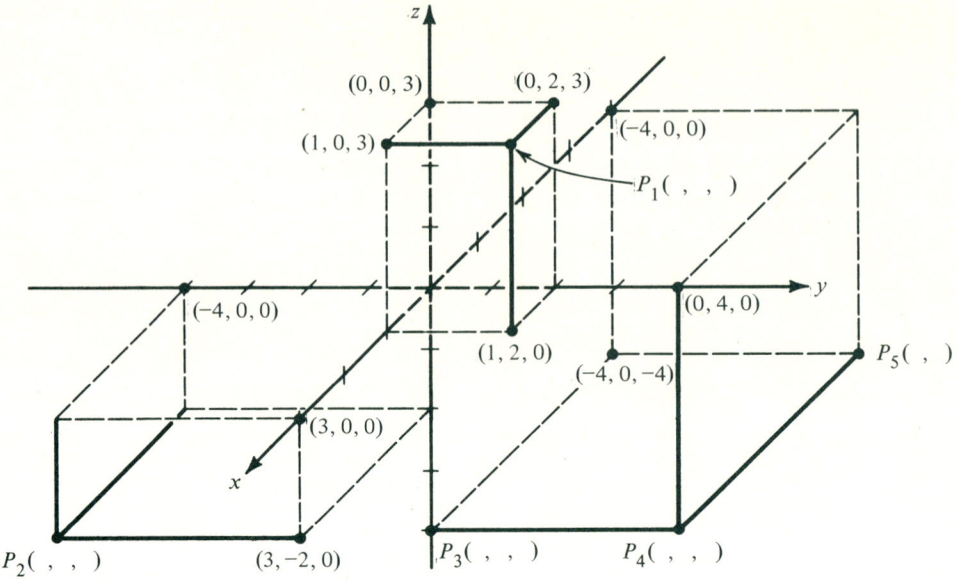

FIGURE 13.25.

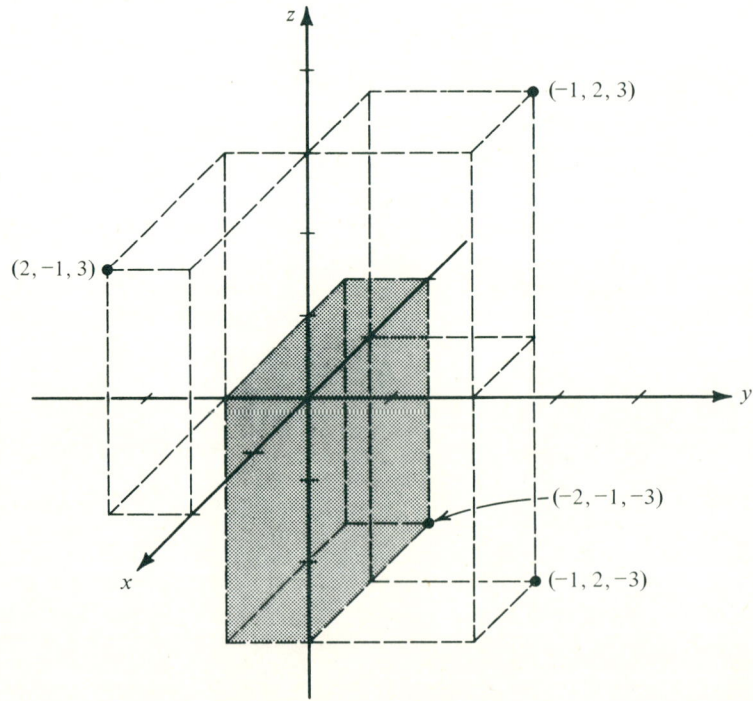

FIGURE 13.26.

EXAMPLE I. Plot the following in the axis system provided in Figure 13.26.

$$P_1(2, -1, 3)$$
$$P_2(-1, 2, 3)$$
$$P_3(-1, 2, -3)$$
$$P_4(2, -1, -3)$$

If you connect the 4 points in Example I, in order, they should form a rectangle. This rectangle is a portion of the plane $x + y = 1$. Check the coordinates of each point to see that they satisfy the equation. What do you notice about z? Notice that each of the following points satisfies the equation $x + y = 1$ and, therefore, lies on the plane illustrated in Figure 13.26.

$$\{(1, 0, 0), (1, 0, 1), (1, 0, 10), (1, 0, -100),$$
$$(1, 0, 1000)\}$$

Equations in two variables, like $x + y = 1$, $x + z^2 = 4$, and $y^2 + z = 10$, are easy to graph because parallel cross-sections taken along the axis of the missing variable are all the same. The graphs of such equations are called *cylinders*.

EXAMPLE II. Graph the three-dimensional solid represented by

$$x^2 + y^2 = 25$$

Solution: In the x, y-plane, this is the equation of a circle of radius 5, centered at the origin. Its graph for $z = 0$ is shown in Figure 13.27. The equations of this cross-section are

$$x^2 + y^2 = 25 \quad \text{and} \quad z = 0$$

Now, for any other value of z, the resulting cross-section will be the same two-dimensional graph. Study the graph of the **right circular cylinder** in Figure 13.28. Points satisfying the equation lie on the surface of the cylinder.

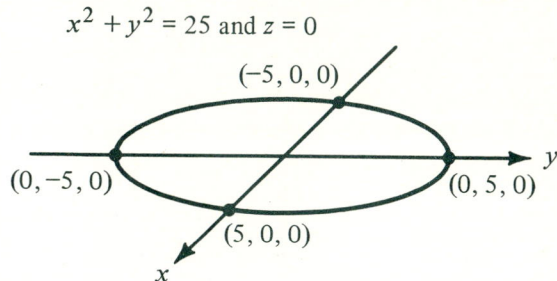

$x^2 + y^2 = 25$ and $z = 0$

FIGURE 13.27.

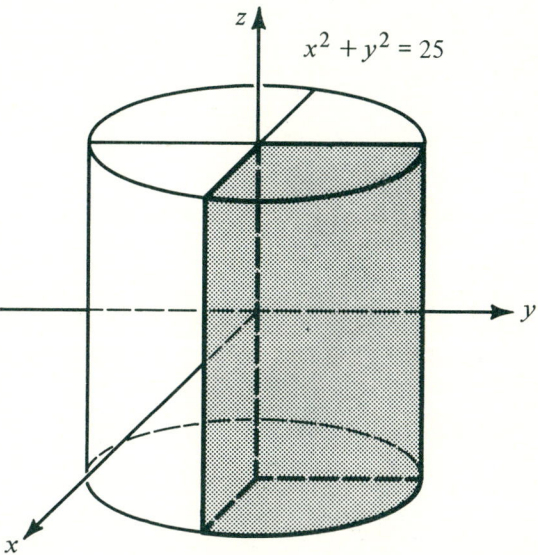

FIGURE 13.28.

EXAMPLE III. Complete the graph of the elliptic cylinder in Figure 13.29.

$$\frac{x^2}{9} + \frac{z^2}{1} = 1$$

Solution: Notice that since the equation is independent of y, the cross-section parallel to the x, z-plane will cut off identical pictures, called *traces*. See Figure 13.29. Because of the symmetry possessed by this graph, the first octant ($x \geq 0$, $y \geq 0, z \geq 0$) portion of the graph is often sufficient. See Figure 13.30.

CHAPTER 13 The Conic Sections

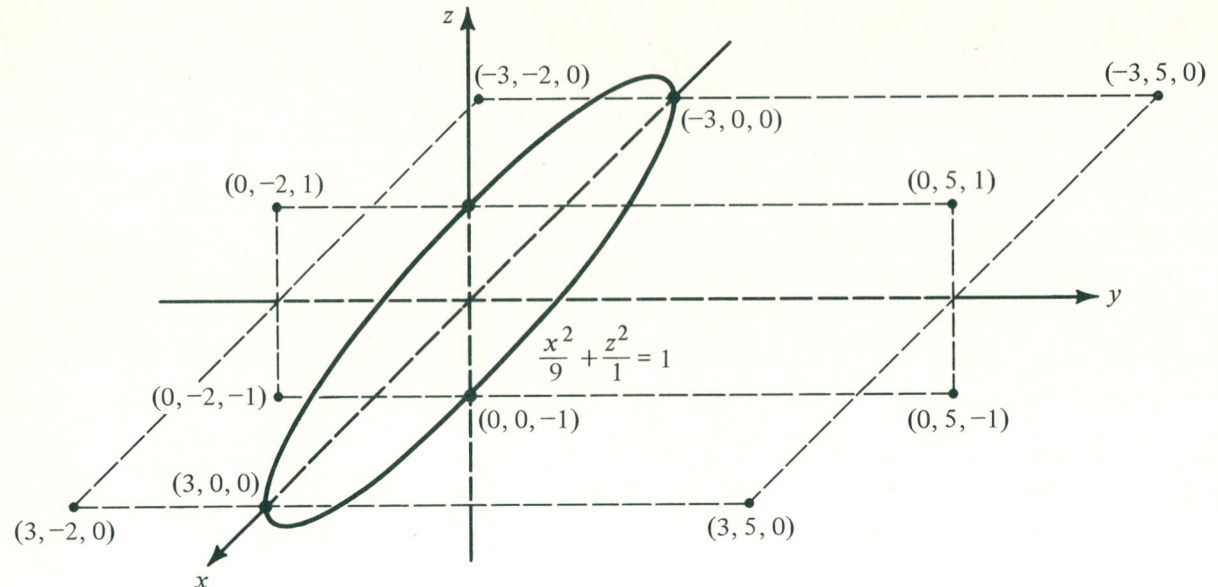

FIGURE 13.29.

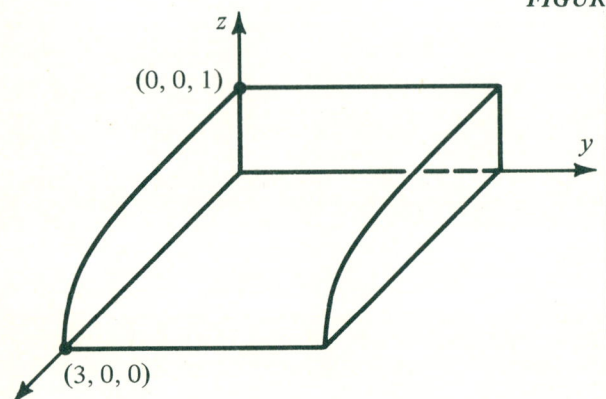

FIGURE 13.30.

EXAMPLE IV. Complete the positive portion of the graph of the hyperbolic cylinder

$$\frac{y^2}{9} - \frac{z^2}{1} = 1$$

Solution: First complete the graph of the y, z-trace ($x = 0$) in Figure 13.31. The asymptotes for this hyperbola are already pictured. Now we will copy this curve for the trace for some other positive value of x, for example, $x = 7$. Notice that we have planes for asymptotes in Figure 13.32.

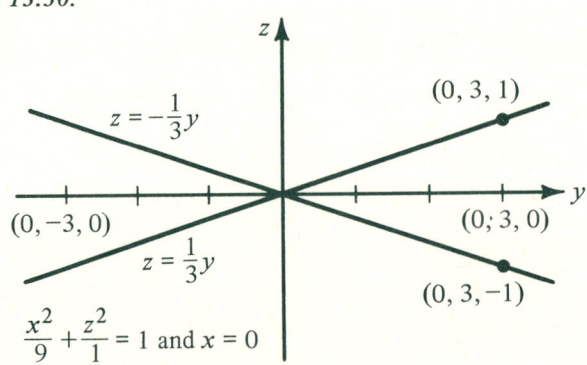

FIGURE 13.31.

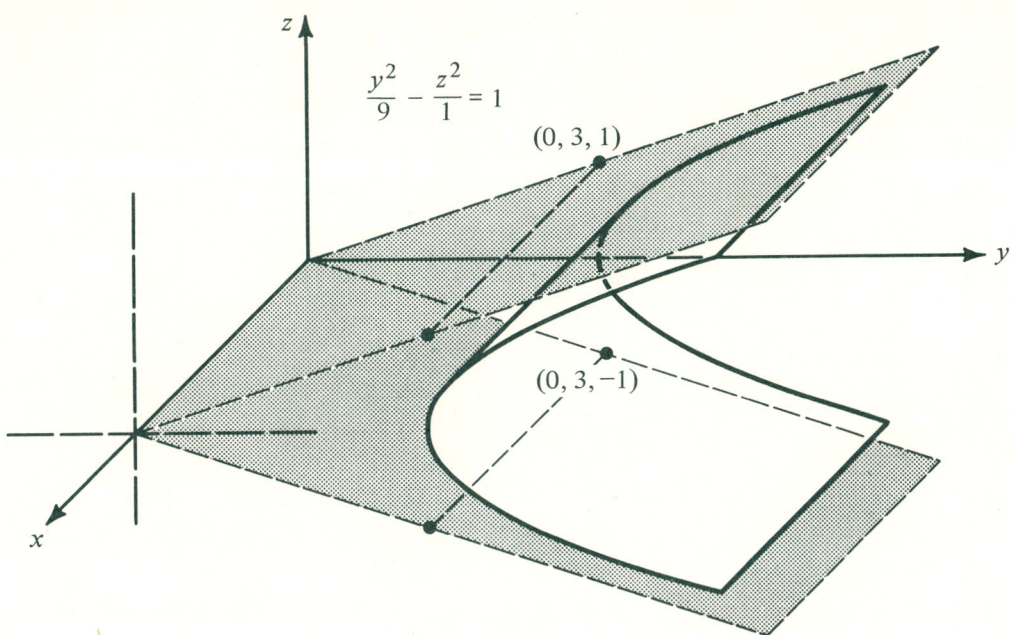

FIGURE 13.32.

Now let's draw the graph of an equation containing all 3 variables. We will need the cross-sections from each of the 3 planes, $x = 0$, $y = 0$, and $z = 0$, since they are all different. Then we will connect these traces.

EXAMPLE V. Graph the paraboloid

$$z = x^2 + y^2$$

Solution: First, study the traces.

x, y-trace ($z = 0$) is a point $(0, 0, 0)$

x, z-trace ($y = 0$) is the parabola $z = x^2$

y, z-trace ($x = 0$) is the parabola $z = y^2$

Notice that the trace $z = 4$ is the circle $x^2 + y^2 = 16$. Now, each of these traces is graphed and then connected to complete the picture as shown in Figure 13.33.

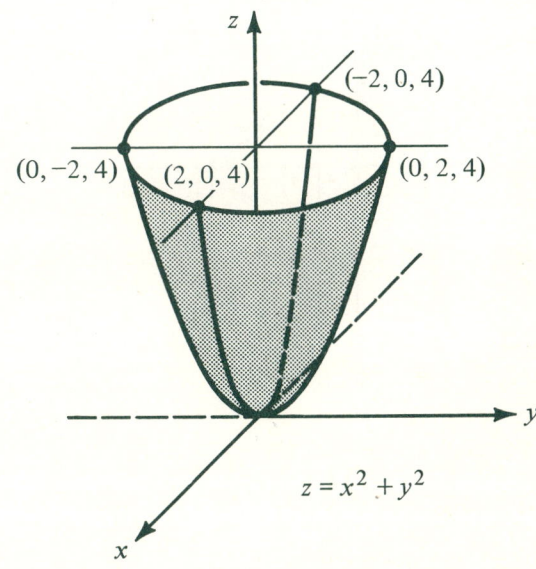

FIGURE 13.33.

CHAPTER 13 The Conic Sections

EXAMPLE VI. Describe the traces and then connect them to complete the graph of the ellipsoid in Figure 13.34.

$$\frac{x^2}{16} + \frac{y^2}{4} + \frac{z^2}{9} = 1$$

Solution: We have

$z = 0$ (x, y-trace) is the **ellipse**,

$$\frac{?}{\_\_} + \frac{?}{\_\_} = 1$$

$y = 0$ (x, z-trace) is the ___?___ ,

$$\frac{?}{\_\_} + \frac{?}{\_\_} = 1$$

$x = 0$ (y, z-trace) is the ___?___ ,

$$\frac{?}{\_\_} + \frac{?}{\_\_} = 1$$

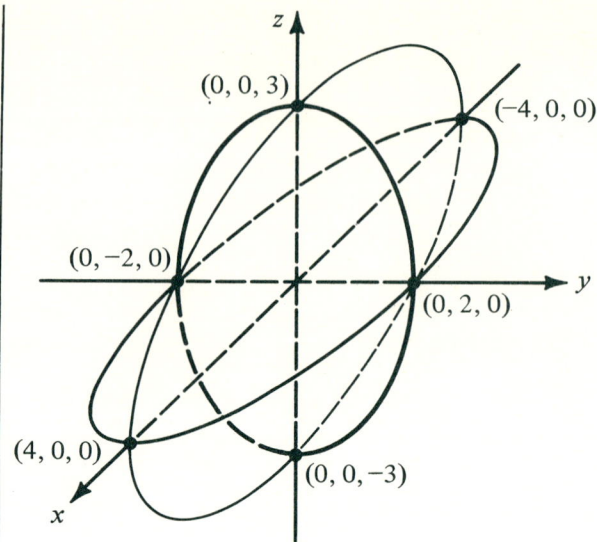

FIGURE 13.34.

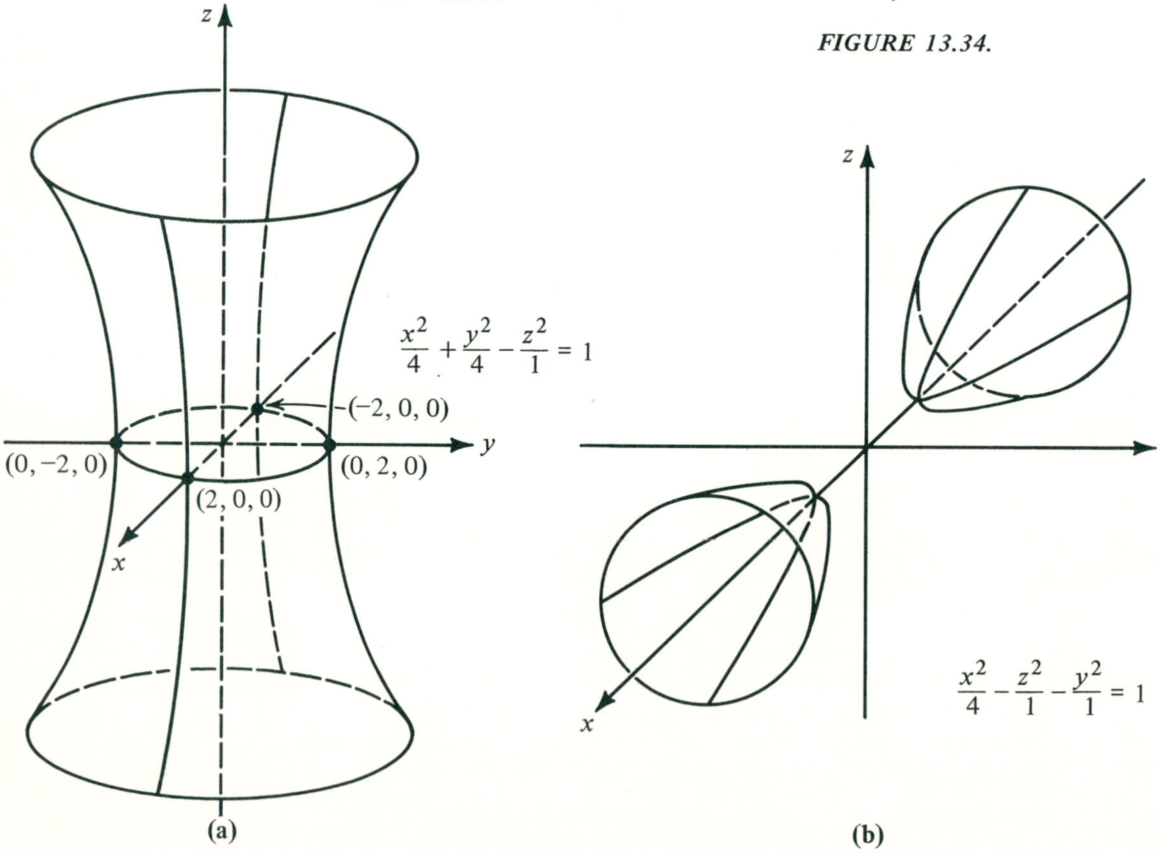

$$\frac{x^2}{4} + \frac{y^2}{4} - \frac{z^2}{1} = 1$$

(a)

$$\frac{x^2}{4} - \frac{z^2}{1} - \frac{y^2}{1} = 1$$

(b)

FIGURE 13.35. (a) A Hyperboloid of One Sheet, and (b) A Hyperboloid of Two Sheets.

They can be thought of as a two-dimensional hyperbola that revolves about either the z-axis or the x-axis. Examples of 2 hyperboloids are shown in Figures 13.35(a) and (b).

EXERCISE 13.7

A. FUNDAMENTALS
Draw each of the following graphs in the indicated plane.

1. In the y, z-plane $(x = 0)$.
(a) $y^2 + z^2 = 16$ (b) $y = z^2 + 1$
(c) $z = y^2 + 1$ (d) $\dfrac{y^2}{4} + \dfrac{z^2}{1} = 1$

2. In the x, y-plane $(z = 0)$.
(a) $x^2 + y^2 = 16$ (b) $\dfrac{x^2}{4} + y^2 = 1$
(c) $x^2 + \dfrac{y^2}{9} = 1$ (d) $y = (x - 2)^2$

3. In the x, z-plane $(y = 0)$
(a) $x^2 + z^2 = 16$ (b) $\dfrac{x^2}{9} + \dfrac{z^2}{1} = 1$
(c) $x = 4 - z^2$

B. ESSENTIALS
Sketch the indicated traces for each of the following solids. To find the equation of the trace, substitute the indicated value into the original equation.

4. Draw the traces of the sphere $x^2 + y^2 + z^2 = 169$ in the planes (a) $x = 0$ (b) $y = 0$ (c) $x = 12$ (d) $z = 5$.

5. Draw the traces of the ellipsoid $9x^2 + y^2 + 9z^2 = 169$ in the planes (a) $y = 5$ (b) $x = -4$ (c) $z = 4$.

6. Draw the traces of the elliptic paraboloid $4x^2 + y^2 = z$ in the planes (a) $z = 1$ (b) $z = 4$ (c) $z = 16$.

7. Draw the traces of the elliptic hyperboloid of 1 sheet $4x^2 + y^2 = z^2 + 4$ in the planes (a) $z = 0$ (b) $x = 0$ (c) $y = 0$.

Graph the following cylinders in the x, y, z-coordinate system. Label the intercepts.

8. (a) $x + y = 2$ (b) $y + z = 2$
(c) $x + z = 2$

9. (a) $4x + y = 8$ (b) $4z + y = 8$
(c) $4x + z = 8$

10. (a) $x^2 + y^2 = 25$ (b) $x^2 + z^2 = 25$
(c) $y^2 + z^2 = 25$

11. (a) $4x^2 + 9y^2 = 36$ (b) $4x^2 + 9z^2 = 36$
(c) $4y^2 + 9z^2 = 36$

12. (a) $y = x^2 - 9$ (b) $x = y^2 - 9$
(c) $z = x^2 - 9$

13. (a) $y = 16 - x^2$ (b) $z = 16 - x^2$
(c) $y = 16 - z^2$

14. (a) $xy = 1$ (b) $xz = 1$ (c) $yz = 1$

15. (a) $4x^2 - 16y^2 = 64$
(b) $16x^2 - 4z^2 = 64$
(c) $4y^2 - 16z^2 = 64$

Draw the following solids.

16. The spheres
(a) $x^2 + y^2 + z^2 = 4$ (b) $(x - 1)^2 + y^2 + z^2 = 9$
(c) $(x - 1)^2 + (y + 2)^2 + (z - 1)^2 = 9$

17. The ellipsoids
(a) $9x^2 + 4y^2 + z^2 = 36$
(b) $9(x - 1)^2 + y^2 + 4z^2 = 36$
(c) $4(x - 1)^2 + 9(y - 2)^2 + (z + 1)^2 = 36$

18. The paraboloids
(a) $x^2 + y^2 + z = 0$ (b) $4x^2 - y + 4z^2 - 16 = 0$
(c) $z = (x - 1)^2 + (y - 3)^2$ (try $z = 4$)
(d) $(x - 1)^2 + (y - 3)^2 - z - 16 = 0$

19. The elliptic paraboloids
(a) $x^2 + 4y^2 = z$ (b) $x + 4y^2 + 9z^2 = 36$
(c) $z = 4(x - 1)^2 + (y - 3)^2$ (try $z = 4$)
(d) $4(x - 1)^2 + 4(y - 5)^2 - z - 36 = 0$

Draw the three-dimensional graphs of the following cylinders.

20. $y = \log_2 x$ **21.** $z = 2^{y-1}$

22. $x = -\log_3 z$ **23.** $y = 1 - 3^x$

CHAPTER 13　　The Conic Sections

Graph each of the following solids. Identify the curve.
    24. $x^2 - 2x + y^2 + 4y + z^2 = 11$

★25. $4x^2 - 16x + 4y^2 + z^2 + 6z + 24 = 0$

★26. $z = 11 - x^2 + 2x - y^2 - 4y$

    27. $y = x^2 - 6x + z^2 - 8z + 17$

    28. $x = 8 - 4y^2 + 8y - z^2 + 4z$

    29. $4x^2 + 9z^2 - y^2 = 144$

    30. $9x^2 - y^2 - 16z^2 = 144$

    31. $y^2 - 2y - x^2 + z^2 = 0$

Graph the following hyperbolic paraboloids.
    32. $x^2 - y^2 + z = 1$ 33. $x + 4y^2 - z^2 = 4$

ENDNOTE

1. Eudoxus was one of the most brilliant mathematicians of the third century B.C., but it was one of his students, Menaechmus, who discovered the ellipse, hyperbola, and parabola from cross-sections of cones in about 360 B.C. He used three different cones, one acute-angled, one right-angled, and one obtuse-angled. The cross-sections were cut by a plane that in a two-dimensional drawing appears to be perpendicular to the sides of the cones.

These conic sections were originally called the "Menaechmian Triads," but Apollonius of Perga renamed them in about 225 B.C. in a treatise that was very close to our modern view. Apollonius' books on conic sections were comprehensive and excellently organized, often including his original work. He was the first to recognize that the hyperbola has two branches. Since his work preceded analytic geometry, it was done without the equations for the conic sections.

CHAPTER 14

Matrices and Determinants

14.1 INTRODUCTION

This chapter is a brief introduction to a vast area of study and applications called *linear algebra*. As we shall see, the solutions of systems of linear equations (Chapter 15) readily lend themselves to matrices and determinants. Most of the first part of this chapter concerns the basic definitions, notation, and vocabulary involved in linear algebra. Many of the more sophisticated concepts will be considered in the exercises.

14.2 MATRICES

First, let's define a matrix.[1]

> **DEFINITION 14.1.**
> A **matrix** is a rectangular array of elements having horizontal rows and vertical columns.

Some examples of matrices follow, where (a) is called a *row matrix*, (b) is a *column matrix*, (c) is a *matrix of order 3 × 2* (3 rows and 2 columns) read, "3 by 2," and (d) is a 2 × 3 matrix. The *order* of a matrix, the number of rows by the number of columns, is written $m \times n$.

(a) $(2 \quad 5 \quad \pi)$

(b) $\begin{pmatrix} 1 \\ -1 \\ 4 \end{pmatrix}$

(c) $\begin{pmatrix} 4 & -9 \\ 0 & 11 \\ 1 & 3 \end{pmatrix}$

(d) $\begin{pmatrix} 4 & -\frac{1}{2} & 2 \\ 7 & 11 & 0 \end{pmatrix}$

Definition 14.2 establishes the conditions for equality of matrices.

CHAPTER 14   Matrices and Determinants

DEFINITION 14.2.
Two matrices are equal if and only if they have the same order and their corresponding elements are equal.

The following illustrates matrix equality.

(a) $\begin{pmatrix} x \\ 2y \end{pmatrix} = \begin{pmatrix} 4 \\ 6 \end{pmatrix} \Leftrightarrow x = 4$ and $y = 3$

(b) $\begin{pmatrix} \ln x & 4 \\ y^2 & 1 \end{pmatrix} = \begin{pmatrix} 0 & 4 \\ 9 & 1 \end{pmatrix} \Leftrightarrow x = 1$ and $y = 3$

The definition of the addition of two matrices is quite natural. Add the following matrices by adding their corresponding elements.

(a) $\begin{pmatrix} 2 & 3 \\ 1 & 4 \end{pmatrix} + \begin{pmatrix} -1 & 0 \\ 2 & 5 \end{pmatrix} = $ _____?_____

(b) $\begin{pmatrix} 1 & 2 & -1 \\ 3 & 4 & 0 \end{pmatrix} + \begin{pmatrix} 2 & -1 & 0 \\ 4 & -3 & 1 \end{pmatrix}$

= _____?_____

(c) $\begin{pmatrix} 1 & 2 \\ 3 & 4 \\ 4 & 3 \end{pmatrix} + \begin{pmatrix} 2 & -1 & 3 \\ 4 & 0 & 1 \end{pmatrix}$

= _____?_____

The answers are $\begin{pmatrix} 1 & 3 \\ 3 & 9 \end{pmatrix}$, $\begin{pmatrix} 3 & 1 & -1 \\ 7 & 1 & 1 \end{pmatrix}$, and "cannot be added." Example (c) shows two matrices that are *not conformable to matrix addition* because they do not have the same order. The first is a 3 × 2 matrix and the second a 2 × 3 matrix.

There are two kinds of multiplications to consider for matrices. The first is called *scalar multiplication*. We will define a scalar as any real number. We will postpone the second type of multiplication, that of a matrix times a matrix, until Section 14.3.

EXAMPLE I. Find the indicated scalar products.

(a) $4 \begin{pmatrix} 2 & 3 \\ -1 & 2 \end{pmatrix} = \begin{pmatrix} 8 & 12 \\ -4 & ? \end{pmatrix}$

(b) $\frac{1}{3} \begin{pmatrix} 3 & 9 \\ 7 & 12 \end{pmatrix} = \begin{pmatrix} & ? & \end{pmatrix}$

The answer to (b) above is

$\begin{pmatrix} 1 & 3 \\ \frac{7}{3} & 4 \end{pmatrix}$

In general, scalar multiplication of any scalar, α, times a 3 × 3 matrix is defined as follows.

$\alpha \begin{pmatrix} a_1 & b_1 & c_1 \\ a_2 & b_2 & c_2 \\ a_3 & b_3 & c_3 \end{pmatrix} = \begin{pmatrix} \alpha a_1 & \alpha b_1 & \alpha c_1 \\ \alpha a_2 & \alpha b_2 & \alpha c_2 \\ \alpha a_3 & \alpha b_3 & \alpha c_3 \end{pmatrix}$

EXERCISE 14.2

A. FUNDAMENTALS

Perform the indicated operations whenever possible. If not possible, write, "not conformable."

1. (a) $\begin{pmatrix} 2 & -1 \\ 3 & 4 \end{pmatrix} + \begin{pmatrix} 1 & -3 \\ 2 & 7 \end{pmatrix}$

   (b) $\begin{pmatrix} 1 & -3 \\ 2 & 7 \end{pmatrix} + \begin{pmatrix} 2 & -1 \\ 3 & 4 \end{pmatrix}$

2. (a) $\left[\begin{pmatrix} 1 & 2 & 3 \\ -2 & 4 & 7 \end{pmatrix} + \begin{pmatrix} 1 & 3 & 7 \\ 2 & -1 & 4 \end{pmatrix} \right]$
   $+ \begin{pmatrix} 2 & 8 & -11 \\ 5 & -7 & 4 \end{pmatrix}$

   (b) $\begin{pmatrix} 1 & 2 & 3 \\ -2 & 4 & 7 \end{pmatrix}$
   $+ \left[\begin{pmatrix} 1 & 3 & 7 \\ 2 & -1 & 4 \end{pmatrix} + \begin{pmatrix} 2 & 8 & -11 \\ 5 & -7 & 4 \end{pmatrix} \right]$

3. $5 \begin{pmatrix} 1 \\ 0 \\ 0 \end{pmatrix} + 7 \begin{pmatrix} 0 \\ 1 \\ 0 \end{pmatrix} - 10 \begin{pmatrix} 0 \\ 0 \\ 1 \end{pmatrix}$

4. $\begin{pmatrix} 2 & 3 & 4 \\ 1 & -2 & 1 \end{pmatrix} + \begin{pmatrix} 2 & 1 \\ 3 & -2 \\ 4 & 1 \end{pmatrix}$

5. (a) $\begin{pmatrix} 4 & 7 & 1 \\ 3 & 2 & -2 \\ 1 & 5 & 7 \end{pmatrix} + \begin{pmatrix} 0 & 0 & 0 \\ 0 & 0 & 0 \\ 0 & 0 & 0 \end{pmatrix}$

(b) $\begin{pmatrix} -2 & 1 & 3 \\ 1 & -7 & 4 \\ -2 & 3 & -1 \end{pmatrix} + \begin{pmatrix} 2 & -1 & -3 \\ -1 & 7 & -4 \\ 2 & -3 & 1 \end{pmatrix}$

6. $\frac{1}{12}\begin{pmatrix} 4 & 2 \\ 6 & 6 \end{pmatrix}$ 7. $-\frac{1}{2}\begin{pmatrix} 2 & 3 \\ -2 & -4 \end{pmatrix}$

Compare each of the following.

8. (a) $[2+3]\begin{pmatrix} 1 & 2 & 3 \\ 4 & 5 & 6 \end{pmatrix}$

(b) $2\begin{pmatrix} 1 & 2 & 3 \\ 4 & 5 & 6 \end{pmatrix} + 3\begin{pmatrix} 1 & 2 & 3 \\ 4 & 5 & 6 \end{pmatrix}$

9. (a) $4\left[\begin{pmatrix} 1 & -2 \\ 3 & 1 \end{pmatrix} + \begin{pmatrix} 2 & 0 \\ -3 & 1 \end{pmatrix}\right]$

(b) $4\begin{pmatrix} 1 & -2 \\ 3 & 1 \end{pmatrix} + 4\begin{pmatrix} 2 & 0 \\ -3 & 1 \end{pmatrix}$

10. αA and $A\alpha$, where $A = \begin{pmatrix} 1 & 2 \\ 3 & 4 \end{pmatrix}$ and $\alpha = -3$.

11. $(\alpha \cdot \beta)A$ and $\alpha(\beta A)$, where $A = \begin{pmatrix} 1 & 2 \\ 3 & 4 \end{pmatrix}$, $\alpha = 3$, and $\beta = 8$

B. ESSENTIALS

Consider the set of all 2×2 matrices under normal matrix addition. You may wish to refer to the list of field properties in Section 1.1.

12. Is the set closed under matrix addition? That is, is the sum of two 2×2 matrices always a 2×2 matrix?

13. Is there an additive identity θ in the set, such that $A + \theta = \theta + A = A$, for all 2×2 matrices A? If so, what is it?

14. Does each matrix, A, have an additive inverse, $-A$, such that $A + (-A) = (-A) + A = \theta$? If so, what is the additive inverse of

$$\begin{pmatrix} a_1 & b_1 \\ a_2 & b_2 \end{pmatrix}$$

15. Illustrate that matrix addition of 2×2 matrices is commutative. (See Problem 1.)

16. Illustrate that matrix addition of 2×2 matrices is associative. (See Problem 2.)

17. Repeat the analysis described in Problems 12–16 for the set of all 3×3 matrices under matrix addition. A set that satisfies these properties under one operation is called a *commutative group*.

18. Repeat Problem 17 for the set

$$\left\{\begin{pmatrix} K & 0 \\ 0 & K \end{pmatrix} \middle| K \in \text{integers}\right\}$$

under normal matrix addition.

C. APPLICATIONS

Prove Problems 19–24 for scalars α and β and 2×2 matrices A, B, and C.

19. $(\alpha + \beta)A = \alpha A + \beta A$

20. $\alpha(A + B) = \alpha A + \alpha B$ 21. $\alpha A = A\alpha$

22. $\alpha(\beta A) = (\alpha \cdot \beta)A$

23. Repeat the proofs of Problems 19, 20, and 24 for 3×3 matrices.

24. Prove the following. (a) There exists a unique $m \times n$ matrix θ, such that

$$A + \theta = \theta + A = A$$

for all $m \times n$ matrices A.

(b) For any $m \times n$ matrix A, there exists a unique $m \times n$ matrix $-A$, such that

$$-A = (-1) \cdot A$$

and

$$A + (-A) = (-A) + A = \theta$$

CHAPTER 14 Matrices and Determinants

14.3 MATRIX MULTIPLICATION

The multiplication of two matrices is not comparable to the addition of two matrices. *If* the multiplication of two matrices were defined as

$$\begin{pmatrix} 1 & 2 \\ 3 & 4 \end{pmatrix} \begin{pmatrix} 2 & 6 \\ 3 & -5 \end{pmatrix} = \begin{pmatrix} 2 & 12 \\ 9 & -20 \end{pmatrix}$$

then any set of matrices would simply be the real number system in an array, having no more applicability than the reals themselves. That is *not* the way we multiply matrices.

Multiplication is illustrated in Example I. Notice that the product of the 2 × 3 and the 3 × 2 matrices is a 2 × 2 matrix.

EXAMPLE I.

$$\begin{pmatrix} 1 & 3 & 4 \\ 2 & 1 & 3 \end{pmatrix} \begin{pmatrix} 2 & 1 \\ -3 & 4 \\ 0 & 5 \end{pmatrix}$$

$$= \begin{pmatrix} 1(2) + 3(-3) + 4(0) & 1(1) + 3(4) + 4(5) \\ 2(2) + 1(-3) + 3(0) & 2(1) + 1(4) + 3(5) \end{pmatrix}$$

$$= \begin{pmatrix} -4 & 33 \\ 1 & 21 \end{pmatrix}$$

The first entry is the sum of the products of the elements in the first row of the first matrix and those in the first column of the second matrix. The position of an entry in the product matrix is determined by the *row* selected in the first matrix and the *column* selected in the second.

EXAMPLE II. Multiply the following matrices.

(a) $\begin{pmatrix} 2 & 4 & 3 \\ -2 & 1 & 5 \end{pmatrix} \begin{pmatrix} 1 & 2 \\ 3 & 0 \\ -4 & 1 \end{pmatrix} = \begin{pmatrix} 2 \cdot 1 + 4 \cdot 3 + 3(-4) & 2 \cdot 2 + 4 \cdot 0 + 3 \cdot 1 \\ -2 \cdot 1 + 1 \cdot 3 + 5(-4) & -2 \cdot 2 + 1 \cdot 0 + 5 \cdot 1 \end{pmatrix}$

$= \begin{pmatrix} \underline{?} & \underline{?} \\ \underline{?} & \underline{?} \end{pmatrix}$

(b) $\begin{pmatrix} 2 & 3 \\ 4 & -1 \end{pmatrix} \begin{pmatrix} 1 & 2 \\ 3 & 0 \end{pmatrix} = \begin{pmatrix} 2 \cdot 1 + 3 \cdot 3 & 2 \cdot \underline{} + 3 \cdot \underline{} \\ 4 \cdot \underline{} + (-1) \cdot \underline{} & \underline{} \cdot 2 + (\underline{}) \cdot 0 \end{pmatrix}$

$= \begin{pmatrix} \underline{} & 4 \\ 1 & \underline{} \end{pmatrix}$

(c) $\begin{pmatrix} 1 & 2 \\ 3 & 0 \\ -4 & 1 \end{pmatrix} \begin{pmatrix} 2 & 4 & 3 \\ -2 & 1 & 5 \end{pmatrix}$

$= \begin{pmatrix} 1 \cdot 2 + 2(-2) & 1 \cdot \underline{} + 2 \cdot \underline{} & 1 \cdot \underline{} + 2 \cdot 5 \\ \underline{} \cdot 2 + 0(-2) & \underline{} \cdot 4 + 0 \cdot 1 & \underline{} \cdot 3 + \underline{} \cdot \underline{} \\ -4 \cdot \underline{} + \underline{} \cdot \underline{} & (\underline{}) \cdot \underline{} + \underline{} \cdot \underline{} & (\underline{}) \cdot \underline{} + \underline{} \cdot \underline{} \end{pmatrix}$

$= \begin{pmatrix} \underline{} & \underline{} & \underline{} \\ \underline{} & \underline{} & \underline{} \\ \underline{} & \underline{} & \underline{} \end{pmatrix}$

Your answers should be

(a) $\begin{pmatrix} 2 & 7 \\ -19 & 1 \end{pmatrix}$ (b) $\begin{pmatrix} 11 & 4 \\ 1 & 8 \end{pmatrix}$

(c) $\begin{pmatrix} -2 & 6 & 13 \\ 6 & 12 & 9 \\ -10 & -15 & -7 \end{pmatrix}$

In matrix multiplication the matrices must have orders like

$$m \times p \quad \text{and} \quad p \times n$$

to be conformable. Their product will then have order $m \times n$. In the examples above, the number of columns of the first is always the same as the number of rows of the second. Their orders and the orders of the answers are

(a) a 2×3 times a 3×2 yielded a 2×2 matrix
(b) a 2×2 times a 2×2 yielded a 2×2 matrix
(c) a 3×2 times a 2×3 yielded a 3×3 matrix

Here is an example of nonconformable matrices. Notice how they simply do not "fit" as they should.

$$\begin{pmatrix} 2 & 3 & 1 \\ 1 & -3 & 2 \end{pmatrix} \begin{pmatrix} 4 & 5 \\ 8 & 7 \end{pmatrix}$$

$$= \begin{pmatrix} 2 \cdot 4 + 3 \cdot 8 + 1 \cdot \underline{?} & \\ 1 \cdot 4 + (-3) \cdot 8 + 2 \cdot \underline{?} & \\ & 2 \cdot 5 + 3 \cdot 7 + 1 \cdot \underline{?} \\ & 1 \cdot 5 + (-3) \cdot 7 + 2 \cdot \underline{?} \end{pmatrix}$$

A 2×3 and a 2×2 are not conformable to matrix multiplication because the number of columns of the first does not equal the number of rows of the second ($3 \neq 2$).

The next example illustrates an important consequence of this definition of multiplication.

EXAMPLE III. Find the products $A \cdot B$ and $B \cdot A$ for the following.

$$A = \begin{pmatrix} 2 & 1 \\ 3 & -1 \end{pmatrix} \quad \text{and} \quad B = \begin{pmatrix} 1 & 2 \\ 0 & -3 \end{pmatrix}$$

Solution:

$$AB = \begin{pmatrix} 2 & 1 \\ 3 & -1 \end{pmatrix} \begin{pmatrix} 1 & 2 \\ 0 & -3 \end{pmatrix}$$

$$= \begin{bmatrix} 2 \cdot 1 + 1 \cdot 0 & 2 \cdot 2 + 1 \cdot (-3) \\ 3 \cdot 1 + (-1) \cdot 0 & 3 \cdot 2 + (-1) \cdot (-3) \end{bmatrix}$$

$$= \begin{pmatrix} \underline{\quad} & \underline{\quad} \end{pmatrix}$$

$$BA = \begin{pmatrix} 1 & 2 \\ 0 & -3 \end{pmatrix} \begin{pmatrix} 2 & 1 \\ 3 & -1 \end{pmatrix} = \begin{pmatrix} 8 & \underline{\quad} \\ \underline{\quad} & \underline{\quad} \end{pmatrix}$$

The fact that these two products are not equal proves that matrix multiplication is not commutative. Notice Example III(a) and (c), where the commuted products are not even of the same order. This does not mean that the matrix product AB never equals BA, however. There are some special pairs that always commute (see Exercise 14.3, Problem 2).

EXERCISE 14.3

A. FUNDAMENTALS
Find the indicated products, if possible. If not possible, write "not conformable."

1. (a) $\begin{pmatrix} 2 & 3 \\ 1 & 4 \end{pmatrix} \begin{pmatrix} -2 & 3 \\ -2 & 1 \end{pmatrix}$

   (b) $\begin{pmatrix} -2 & 3 \\ -2 & 1 \end{pmatrix} \begin{pmatrix} 2 & 3 \\ 1 & 4 \end{pmatrix}$

2. (a) $\begin{pmatrix} 2 & 1 \\ 3 & 2 \end{pmatrix} \begin{pmatrix} 2 & -1 \\ -3 & 2 \end{pmatrix}$

   (b) $\begin{pmatrix} 2 & -1 \\ -3 & 2 \end{pmatrix} \begin{pmatrix} 2 & 1 \\ 3 & 2 \end{pmatrix}$

3. (a) $(1 \ 2 \ 3) \begin{pmatrix} 4 \\ 2 \\ -3 \end{pmatrix}$

   (b) $\begin{pmatrix} 4 \\ 2 \\ -3 \end{pmatrix} (1 \ 2 \ 3)$

4. (a) $\left[2 \begin{pmatrix} 3 & 1 \\ -2 & -1 \end{pmatrix} \right] \begin{pmatrix} 2 & 1 \\ -3 & 1 \end{pmatrix}$

CHAPTER 14 Matrices and Determinants

(b) $2\left[\begin{pmatrix} 3 & 1 \\ -2 & -1 \end{pmatrix}\begin{pmatrix} 2 & 1 \\ -3 & 1 \end{pmatrix}\right]$

5. (a) $\left[\begin{pmatrix} 1 & 3 & 4 \\ -1 & 0 & 2 \\ 1 & 4 & -1 \end{pmatrix}\begin{pmatrix} 0 & 3 & -2 \\ 1 & -1 & -2 \\ 2 & 2 & 1 \end{pmatrix}\right]$
$\times \begin{pmatrix} 1 & 2 & 0 \\ 1 & 3 & -1 \\ 2 & -2 & 2 \end{pmatrix}$

(b) $\begin{pmatrix} 1 & 3 & 4 \\ -1 & 0 & 2 \\ 1 & 4 & -1 \end{pmatrix}$
$\times \left[\begin{pmatrix} 0 & 3 & -2 \\ 1 & -1 & -2 \\ 2 & 2 & 1 \end{pmatrix}\begin{pmatrix} 1 & 2 & 0 \\ 1 & 3 & -1 \\ 1 & -2 & 2 \end{pmatrix}\right]$

6. (a) $\left[4\begin{pmatrix} 1 & 3 & 0 \\ -1 & 2 & 1 \end{pmatrix}\right]\left[2\begin{pmatrix} 1 & 2 \\ -1 & -2 \\ 3 & 4 \end{pmatrix}\right]$

(b) $8\left[\begin{pmatrix} 1 & 3 & 0 \\ -1 & 2 & 1 \end{pmatrix}\begin{pmatrix} 1 & 2 \\ -1 & -2 \\ 3 & 4 \end{pmatrix}\right]$

7. (a) $\begin{pmatrix} 2 & 3 \\ 4 & 7 \end{pmatrix}\begin{pmatrix} 1 & 0 \\ 0 & 1 \end{pmatrix}$

(b) $\begin{pmatrix} 3 & -1 & 2 \\ 4 & 7 & -11 \\ 10 & 9 & 8 \end{pmatrix}\begin{pmatrix} 1 & 0 & 0 \\ 0 & 1 & 0 \\ 0 & 0 & 1 \end{pmatrix}$

8. (a) $\begin{pmatrix} 1 & 3 \\ 2 & 7 \end{pmatrix}\begin{pmatrix} -7 & 3 \\ 2 & -1 \end{pmatrix}$

(b) $\begin{pmatrix} 3 & -2 & -3 \\ -2 & 1 & 0 \\ -4 & 2 & 1 \end{pmatrix}\begin{pmatrix} -1 & 4 & -3 \\ -2 & 9 & -6 \\ 0 & -2 & 1 \end{pmatrix}$

B. ESSENTIALS
Find the product of A and its transpose A^T, formed by changing the rows of A to columns.

9. $A = \begin{pmatrix} 1 & 2 \\ 3 & 4 \end{pmatrix}$ and $A^T = \begin{pmatrix} 1 & 3 \\ 2 & 4 \end{pmatrix}$

10. $A = \begin{pmatrix} 1 & 2 & -1 \\ 0 & 1 & 3 \\ 3 & -1 & 2 \end{pmatrix}$ and
$A^T = \begin{pmatrix} 1 & 0 & 3 \\ 2 & 1 & -1 \\ -1 & 3 & 2 \end{pmatrix}$

11. Does $A \cdot A^T = A^T \cdot A$ for the matrices in Problems 9 and 10?

12. Compare $A^T B^T$ and $B^T A^T$ with $(AB)^T$, and $(A + B)^T$ with $A^T + B^T$ for

(a) $A = \begin{pmatrix} 1 & 3 \\ 2 & -4 \end{pmatrix}$ and $B = \begin{pmatrix} 2 & 4 \\ -1 & 3 \end{pmatrix}$

(b) $A = \begin{pmatrix} 1 & 2 & -1 \\ 3 & -2 & 0 \\ 1 & 3 & 1 \end{pmatrix}$ and
$B = \begin{pmatrix} -1 & 2 & 2 \\ 3 & 4 & 5 \\ 2 & -1 & 0 \end{pmatrix}$

C. APPLICATIONS

13. Give one reason why the set of all 2×2 matrices is not a field under matrix addition and multiplication.

14. Prove that all matrices of the form $\begin{pmatrix} k & 0 \\ 0 & k \end{pmatrix}$, where $k \in$ reals, commute under matrix multiplication.

15. (a) Find A^2 and A^3, where $A^3 = A \cdot A \cdot A$ for
$A = \begin{pmatrix} 2 & 0 & 0 \\ 0 & 6 & 0 \\ 0 & 0 & 3 \end{pmatrix}$

(b) Find the expression for A^n.

14.4 DETERMINANTS

The **determinant** is a function that assigns a real-number to a square matrix.[2] That is, if A is a square $n \times n$ matrix, then the value of the determinants of A, written

$$\det A$$

is a *unique* real number. First, let's define the determinant of a 2×2 matrix and its value. We often refer to the array set off with straight vertical lines as a 2 by 2 (2×2) determinant or simply a second-order determinant.

SECTION 14.4 Determinants

DEFINITION 14.7.
If
$$A = \begin{pmatrix} a_1 & b_1 \\ a_2 & b_2 \end{pmatrix}$$
then
$$\det A = \begin{vmatrix} a_1 & b_1 \\ a_2 & b_2 \end{vmatrix} = a_1 b_2 - a_2 b_1$$

EXAMPLE I. Evaluate the determinants of the following 2×2 matrices.

(a) $A = \begin{pmatrix} 2 & 3 \\ -1 & 4 \end{pmatrix} \Rightarrow \det A = \begin{vmatrix} 2 & 3 \\ -1 & 4 \end{vmatrix}$
$= 2(4) - (-1)(3) = 11$

(b) $B = \begin{pmatrix} 3 & 2 \\ 4 & -1 \end{pmatrix} \Rightarrow \det B = \begin{vmatrix} 3 & 2 \\ 4 & -1 \end{vmatrix}$
$= 3(?) - 4(?) = -11$

(c) $C = \begin{pmatrix} -1 & 4 \\ 2 & 3 \end{pmatrix} \Rightarrow \det C = \begin{vmatrix} -1 & 4 \\ 2 & 3 \end{vmatrix}$
$= (?)(3) - (?)(4) = -11$

(d) $D = \begin{pmatrix} 2 & 6 \\ -1 & 8 \end{pmatrix} \Rightarrow \det D = \begin{vmatrix} & \\ & \end{vmatrix}$
$= (?)(?) - (?)(?) = 22$

Fortunately, it is not necessary to memorize a different procedure for evaluating larger-order determinants, since there is a technique that will reduce every determinant to sums of scalars times determinants of 2×2 matrices. We will explain this process for the determinant of a 3×3 matrix. It generalizes quite naturally to any other square matrix.

DEFINITION 14.8.
A **minor** of an element of a 3×3 determinant is the 2×2 determinant formed by eliminating the row and column containing that element.

EXAMPLE II. Find the indicated minors.
(a) The minor of 4 in

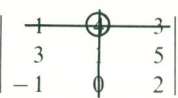

is
$$\begin{vmatrix} 3 & 5 \\ -1 & 2 \end{vmatrix}$$

Since 4 is in the first row, we eliminate the row, 1 4 3. Since 4 is in the second column, it is eliminated.

(b) The minor of 2 is
$$\begin{vmatrix} 1 & 4 \\ 3 & 1 \end{vmatrix}$$

(c) 5's minor is
$$\begin{vmatrix} 1 & 4 \\ -1 & 0 \end{vmatrix}$$

To evaluate the determinant of a 3×3 matrix, we reduce it to three 2×2 determinants, which are evaluated as before. These three 2×2 determinants are the three minors of the elements of *any* row or column. Here are the steps.

1. Choose a row or a column.
2. Find the minor for each element in your chosen row or column.
3. Multiply each element times the value of its minor.
4. Add or subtract these three products according to the following "checkerboard" pattern of signs.

$$\begin{pmatrix} + & - & + \\ - & + & - \\ + & - & + \end{pmatrix}$$

EXAMPLE III. Find the value of the determinant of the following 3×3 matrix.

$$A = \begin{pmatrix} 5 & -1 & 3 \\ 1 & 2 & 4 \\ 2 & 3 & -2 \end{pmatrix}$$

CHAPTER 14 Matrices and Determinants

(a) Choose the second row. Then the products of the elements and their minors follow

$$1 \cdot \begin{vmatrix} -1 & 3 \\ 3 & -2 \end{vmatrix}, \quad 2 \cdot \begin{vmatrix} 5 & 3 \\ 2 & -2 \end{vmatrix}, \quad 4 \cdot \begin{vmatrix} 5 & -1 \\ 2 & 3 \end{vmatrix}$$

These results,

$$1 \cdot (2 - 9) = -7, \quad 2(-10 - 6) = -32,$$

and

$$4(15 + 2) = 58$$

are then combined using the second row of signs, $-, +, -$, as follows:

$$-(-7) + (-32) - (58) = -93$$

(b) Choose the third row. Then

$$\det A = +3 \begin{vmatrix} 1 & 2 \\ 2 & 3 \end{vmatrix} - 4 \begin{vmatrix} 5 & -1 \\ 2 & 3 \end{vmatrix}$$

$$+ (-2) \begin{vmatrix} 5 & -1 \\ 1 & 2 \end{vmatrix}$$

$$= 3(\underline{\ ?\ } - \underline{\ ?\ }) - 4(\underline{\ ?\ } - \underline{\ ?\ })$$

$$- 2(\underline{\ ?\ } - \underline{\ ?\ })$$

$$= 3(\underline{\ ?\ }) - 4(\underline{\ ?\ }) - 2(\underline{\ ?\ })$$

$$= \underline{\ ?\ }$$

Notice the signs chosen were $+, -, +$, from the third column of the checkerboard pattern.

EXERCISE 14.4

A. FUNDAMENTALS
Find the determinants of the following matrices.

1. (a) $\begin{pmatrix} 1 & 2 \\ 3 & 4 \end{pmatrix}$ (b) $\begin{pmatrix} 2 & 2 \\ 6 & 4 \end{pmatrix}$
   (c) $\begin{pmatrix} -1 & 6 \\ 3 & 12 \end{pmatrix}$ (d) $\begin{pmatrix} -2 & 6 \\ 6 & 12 \end{pmatrix}$

2. (a) $\begin{pmatrix} 1 & 0 & 0 \\ 2 & 3 & 8 \\ 4 & 5 & 10 \end{pmatrix}$ (b) $\begin{pmatrix} -1 & 0 & 0 \\ -2 & 3 & 8 \\ -4 & 5 & 10 \end{pmatrix}$
   (c) $\begin{pmatrix} 1 & 0 & 0 \\ -2 & -3 & -8 \\ 4 & 5 & 10 \end{pmatrix}$

3. (a) $\begin{pmatrix} 0 & 2 & 3 \\ 0 & 1 & 1 \\ 3 & 4 & 2 \end{pmatrix}$ (b) $\begin{pmatrix} 0 & 2 & 3 \\ 3 & 4 & 2 \\ 0 & 1 & 1 \end{pmatrix}$
   (c) $\begin{pmatrix} 0 & 3 & 2 \\ 0 & 1 & 1 \\ 3 & 2 & 4 \end{pmatrix}$

4. (a) $\begin{pmatrix} 1 & 2 \\ -3 & -2 \end{pmatrix}$ (b) $\begin{pmatrix} 1 & -3 \\ 2 & -2 \end{pmatrix}$

5. (a) $\begin{pmatrix} 1 & 2 & -3 \\ -2 & 0 & 1 \\ 3 & 0 & 2 \end{pmatrix}$
   (b) $\begin{pmatrix} 1 & -2 & 3 \\ 2 & 0 & 0 \\ -3 & 1 & 2 \end{pmatrix}$

6. (a) $\begin{pmatrix} 1 & 3 \\ -2 & 4 \end{pmatrix}$ (b) $\begin{pmatrix} 1 & 5 \\ -2 & 0 \end{pmatrix}$
   (c) $\begin{pmatrix} 1 & 3 \\ 0 & 10 \end{pmatrix}$ (d) $\begin{pmatrix} 16 & 3 \\ 18 & 4 \end{pmatrix}$

7. (a) $\begin{pmatrix} 3 & 0 & 0 \\ -10 & 2 & 0 \\ 11 & 3 & 4 \end{pmatrix}$
   (b) $\begin{pmatrix} 3 & 5 & 9 \\ 0 & 2 & 10 \\ 0 & 0 & 4 \end{pmatrix}$
   (c) $\begin{pmatrix} 3 & 0 & 0 \\ 0 & 2 & 0 \\ 0 & 0 & 4 \end{pmatrix}$

8. (a) $\begin{pmatrix} 1 & 1 & 1 \\ 2 & 2 & 2 \\ 3 & 3 & 3 \end{pmatrix}$ (b) $\begin{pmatrix} 0 & 2 & 9 \\ 0 & 1 & 3 \\ 0 & -4 & 5 \end{pmatrix}$
   (c) $\begin{pmatrix} 2 & 3 & 2 \\ 4 & -9 & 4 \\ 6 & 4 & 6 \end{pmatrix}$

9. (a) $\begin{pmatrix} 1 & -1 & 2 \\ 3 & 1 & 4 \\ 2 & -3 & 1 \end{pmatrix}$ (b) $\begin{pmatrix} 4 & 0 & 6 \\ 3 & 1 & 4 \\ 2 & -3 & 1 \end{pmatrix}$
   (c) $\begin{pmatrix} 4 & 0 & 6 \\ 3 & 1 & 4 \\ 11 & 0 & 13 \end{pmatrix}$

B. ESSENTIALS
Find the indicated determinants for

$$A = \begin{pmatrix} 4 & -2 \\ 2 & -3 \end{pmatrix}, \quad B = \begin{pmatrix} 4 & 0 & -2 \\ 1 & 3 & 1 \\ 2 & 1 & 0 \end{pmatrix}$$

and
$$C = \begin{pmatrix} 3 & 1 & -2 \\ -10 & 0 & 4 \\ 2 & -1 & 2 \end{pmatrix}$$

10. (a) det A (b) det A^T

★11. (a) det $B \cdot$ det C (b) det $(B \cdot C)$

12. (a) det $(B + C)$ (b) det B + det C

13. (a) det $4A$ (b) 4 det A

★14. (a) det $2B$ (b) det $3B$

15. (a) det $[(3B)C]$ (b) 27 det B det C

Determine which matrices have determinants equal to zero. Such matrices are called *singular*.

16. $\begin{pmatrix} 2 & 3 & 4 \\ 1 & 4 & 2 \\ -3 & 7 & -6 \end{pmatrix}$

17. $\begin{pmatrix} 1 & x^2 & 4 - 2x \\ 3 & 3x^2 & 12 - 6x \\ -1 & 4x & 5 - 2x^2 \end{pmatrix}$

18. $\begin{pmatrix} 0 & 2 & 4 \\ 2 & 0 & 3 \\ 4 & 3 & 0 \end{pmatrix}$

19. Find the determinants of A, B, and AB for

(a) $A = \begin{pmatrix} 3 & 4 \\ 2 & 3 \end{pmatrix}$ and $B = \begin{pmatrix} 3 & -4 \\ -2 & 3 \end{pmatrix}$

(b) $A = \begin{pmatrix} 3 & -2 & -3 \\ -2 & 1 & 0 \\ -4 & 2 & 1 \end{pmatrix}$ and $B = \begin{pmatrix} -1 & 4 & -3 \\ -2 & 9 & -6 \\ 0 & -2 & 1 \end{pmatrix}$

C. APPLICATIONS

20. Prove that the determinant of a 3×3 matrix in lower triangular form,
$$\begin{pmatrix} a_1 & 0 & 0 \\ a_2 & b_2 & 0 \\ a_3 & b_3 & c_3 \end{pmatrix}$$
is the product of the diagonal elements,
$$a_1 \cdot b_2 \cdot c_3$$

21. Prove that the determinant of a 3×3 matrix in upper triangular form is the product of its diagonal elements.

22. Solve the equation (called the characteristic equation of A)
$$\det (A - \lambda I) = 0$$
for λ, where A is as defined below and I is the 3×3 multiplicative identity
$$I = \begin{pmatrix} 1 & 0 & 0 \\ 0 & 1 & 0 \\ 0 & 0 & 1 \end{pmatrix}$$

★(a) $A = \begin{pmatrix} 1 & 0 & 0 \\ 0 & 3 & 0 \\ 0 & 0 & 4 \end{pmatrix}$ (b) $A = \begin{pmatrix} 1 & 0 & 0 \\ 4 & -1 & 0 \\ 1 & 2 & 3 \end{pmatrix}$

(c) $A = \begin{pmatrix} 1 & 2 & 3 \\ 0 & 4 & 2 \\ 0 & 0 & 3 \end{pmatrix}$

23. Finding the coordinates of a point relative to a rotated coordinate system can be accomplished by matrix multiplication. Perform the indicated products and interpret geometrically.

★(a) $\begin{pmatrix} \dfrac{\sqrt{3}}{2} & \dfrac{1}{2} \\ -\dfrac{1}{2} & \dfrac{\sqrt{3}}{2} \end{pmatrix} \begin{pmatrix} 3 \\ 4 \end{pmatrix}$ (b) $\begin{pmatrix} \dfrac{\sqrt{3}}{2} & -\dfrac{1}{2} \\ \dfrac{1}{2} & \dfrac{\sqrt{3}}{2} \end{pmatrix} \begin{pmatrix} 3 \\ 4 \end{pmatrix}$

24. Multiply the 2×2 matrices (Problem 23) together. Interpret the significance of this result.

14.5 PROPERTIES OF DETERMINANTS

Fortunately, the process of evaluating a determinant can be greatly simplified by employing the following properties. We will apply each of these properties using 3×3 determinants, but the theorems hold for all determinants.

The first property tells us how to factor out a common factor of a row or column.

CHAPTER 14　　Matrices and Determinants

THEOREM II.
Multiplying a row or column by a constant (scalar) multiplies the value of the determinant by that constant. That is, for a row,

$$\begin{vmatrix} ka_1 & kb_1 & kc_1 \\ a_2 & b_2 & c_2 \\ a_3 & b_3 & c_3 \end{vmatrix} = k \begin{vmatrix} a_1 & b_1 & c_1 \\ a_2 & b_2 & c_2 \\ a_3 & b_3 & c_3 \end{vmatrix}$$

EXAMPLE I.

$$\begin{vmatrix} 2 & 1 & 1 \\ 4 & 3 & 4 \\ 2 & 4 & 1 \end{vmatrix} = 2 \begin{vmatrix} 1 & 1 & 1 \\ 2 & 3 & 4 \\ 3 & 4 & 1 \end{vmatrix}$$

since

$$\begin{vmatrix} 2 & 1 & 1 \\ 4 & 3 & 4 \\ 2 & 4 & 1 \end{vmatrix} = 2 \begin{vmatrix} 3 & 4 \\ 4 & 1 \end{vmatrix} - \begin{vmatrix} 4 & 4 \\ 6 & 1 \end{vmatrix} + \begin{vmatrix} 4 & 3 \\ 6 & 4 \end{vmatrix}$$

$$= -8$$

and

$$\begin{vmatrix} 1 & 1 & 1 \\ 2 & 3 & 4 \\ 3 & 4 & 1 \end{vmatrix} = \begin{vmatrix} 3 & 4 \\ 4 & 1 \end{vmatrix} - \begin{vmatrix} 2 & 4 \\ 3 & 1 \end{vmatrix} + \begin{vmatrix} 2 & 3 \\ 3 & 4 \end{vmatrix}$$

$$= -4$$

The next example illustrates the value of the following theorem, Theorem III.

EXAMPLE II. Evaluate the determinant of

$$A = \begin{pmatrix} 2 & 4 & 2 \\ 3 & 0 & -1 \\ 1 & 0 & 3 \end{pmatrix}$$

using the second column.

Solution: We have

$$\det A = -4 \begin{vmatrix} 3 & -1 \\ 1 & 3 \end{vmatrix} + 0 \begin{vmatrix} 2 & 2 \\ 1 & 3 \end{vmatrix} - 0 \begin{vmatrix} 2 & 2 \\ 3 & -1 \end{vmatrix}$$

$$= -4(9 + 1) + 0 + 0 = -40$$

The evaluation of the determinant in the above example was very easy because there was only one nonzero entry in column two. The next theorem tells us how we can create zeros in rows and columns of determinants.

THEOREM III.
Any constant multiple of a row (or column) may be added to any other row (or column) without changing the value of the determinant. That is, for a column,

$$\begin{vmatrix} a_1 & a_2 & a_3 \\ b_1 & b_2 & b_3 \\ c_1 & c_2 & c_3 \end{vmatrix} = \begin{vmatrix} a_1 & a_2 + ka_1 & a_3 \\ b_1 & b_2 + kb_1 & b_3 \\ c_1 & c_2 + kc_1 & c_3 \end{vmatrix}$$

EXAMPLE III. Show that the following three determinants are equal.

$$\begin{vmatrix} 2 & -1 & 2 \\ 3 & 2 & 0 \\ 1 & 5 & 3 \end{vmatrix} = \begin{vmatrix} 2 & -1 & 2 \\ 3 & 2 & 0 \\ 4 & 7 & 3 \end{vmatrix}$$

$$= \begin{vmatrix} 0 & -1 & 0 \\ 7 & 2 & 4 \\ 11 & 5 & 13 \end{vmatrix}$$

Solution: The second determinant is the first with the second row added to the third. The last determinant is the first with two times the second column added to both the first and third columns. Check the value of each determinant. You should get 47 for each. Use the first row to evaluate the third one.

The following, though not necessarily helpful in evaluating determinants, is an important property of determinants. Its proof is beyond the scope of this book.

THEOREM IV.
If A and B are both $n \times n$ matrices, then

$$\det (AB) = (\det A)(\det B)$$

EXERCISE 14.5

A. FUNDAMENTALS
Evaluate each of the following pairs of determinants to illustrate Theorem II.

SECTION 14.6 The Inverse of a Matrix

1. (a) $\begin{vmatrix} 1 & 2 \\ -1 & 3 \end{vmatrix}$ (b) $\begin{vmatrix} 4 & 2 \\ -4 & 3 \end{vmatrix}$

2. (a) $\begin{vmatrix} 1 & -4 \\ 2 & 3 \end{vmatrix}$ (b) $\begin{vmatrix} 3 & -12 \\ 2 & 3 \end{vmatrix}$

3. (a) $\begin{vmatrix} 1 & 0 & 0 \\ -1 & 2 & 2 \\ 4 & 1 & 3 \end{vmatrix}$

   (b) $\begin{vmatrix} 3 & 0 & 0 \\ -3 & 2 & 2 \\ 12 & 1 & 3 \end{vmatrix}$

4. (a) $\begin{vmatrix} 0 & 3 & 0 \\ 1 & 2 & 3 \\ -2 & 4 & 1 \end{vmatrix}$

   (b) $\begin{vmatrix} 0 & 3 & 0 \\ 1 & 2 & 3 \\ 4 & -8 & -2 \end{vmatrix}$

Evaluate each of the following determinants by first applying Theorem III to create two zeros in a row or column, then using that row or column in the evaluation.

5. $\begin{vmatrix} 2 & 1 & 1 \\ 3 & -1 & -1 \\ 3 & 4 & 2 \end{vmatrix}$ 6. $\begin{vmatrix} 3 & 6 & 3 \\ 2 & 4 & 2 \\ 4 & 3 & -1 \end{vmatrix}$

7. $\begin{vmatrix} 1 & 3 & 2 \\ 2 & -1 & -4 \\ -4 & 3 & 8 \end{vmatrix}$

B. ESSENTIALS

Verify each of the following without expanding the determinants.

8. $\begin{vmatrix} a_1 & b_1 \\ a_2 & b_2 \end{vmatrix} = \begin{vmatrix} a_1 + b_1 & b_1 \\ a_2 - a_1 + b_2 - b_1 & b_2 - b_1 \end{vmatrix}$

9. $\begin{vmatrix} a_1 & b_1 & c_1 \\ a_2 & b_2 & c_2 \\ a_3 & b_3 & c_3 \end{vmatrix} = \begin{vmatrix} a_1 - 2b_1 & \frac{1}{2}b_1 & b_1 + 2c_1 \\ a_2 - 2b_2 & \frac{1}{2}b_2 & b_2 + 2c_2 \\ a_3 - 2b_3 & \frac{1}{2}b_3 & b_3 + 2c_3 \end{vmatrix}$

10. $\begin{vmatrix} 3a_1 & 6b_1 & 3c_1 \\ a_2 & 2b_2 & c_2 \\ a_3 & 2b_3 & c_3 \end{vmatrix} = 6 \begin{vmatrix} a_1 & b_1 & c_1 \\ a_2 & b_2 & c_2 \\ a_3 & b_3 & c_3 \end{vmatrix}$

Evaluate the following determinants by applying Theorem III repeatedly until there is only one nonzero element in some row or column. Then use that row or column to evaluate the determinant.

★11. $\begin{vmatrix} 2 & 1 & 3 & 4 \\ -2 & -1 & 5 & 0 \\ 3 & 2 & -2 & 3 \\ 4 & 1 & 0 & 2 \end{vmatrix}$

12. $\begin{vmatrix} 2 & 4 & -3 & 1 \\ 3 & -2 & 1 & 0 \\ 4 & 2 & -1 & 3 \\ 2 & 5 & -2 & 4 \end{vmatrix}$

13. $\begin{vmatrix} 2 & 1 & 3 & 4 & -2 \\ -2 & 0 & 3 & -2 & 1 \\ 3 & -2 & -3 & 2 & 2 \\ 4 & -1 & 2 & -2 & 3 \\ 5 & 0 & 2 & 3 & 1 \end{vmatrix}$

14. $\begin{vmatrix} 3 & -1 & 2 & 4 & 3 \\ 0 & 3 & 1 & -2 & -3 \\ 2 & 4 & -1 & 3 & 2 \\ -1 & 0 & 2 & 1 & 3 \\ 4 & -2 & -1 & 3 & -2 \end{vmatrix}$

15. $\begin{vmatrix} 1 & 2 & 3 & 4 & 5 & 6 \\ -1 & -2 & -3 & 4 & -5 & 6 \\ 1 & -2 & 3 & -4 & 5 & -6 \\ -1 & -2 & 3 & 4 & -5 & -6 \\ -1 & -2 & -3 & 4 & 5 & 6 \\ 1 & -2 & -3 & -4 & -5 & 6 \end{vmatrix}$

C. APPLICATIONS

16. Compare the values of the determinants of the matrix

$$A = \begin{pmatrix} 1 & 3 & 6 \\ 2 & 1 & 1 \\ -1 & 2 & 3 \end{pmatrix}$$

and those of its *comatrix* A^c, formed by replacing each entry with its minor times $(-1)^{i+j}$, where i is its row and j is its column. For example, 6 is replaced with

$(-1)^{1+3} \begin{vmatrix} 2 & 1 \\ -1 & 2 \end{vmatrix} = 5.$

14.6 THE INVERSE OF A MATRIX

By "the inverse of a matrix," we mean its multiplicative inverse. Recall that in the real number field, the multiplicative inverse of 8 is $\frac{1}{8}$ since their product is the multiplicative identity, 1.

CHAPTER 14 Matrices and Determinants

A number must commute with its inverse so that

$$8 \cdot \tfrac{1}{8} = \tfrac{1}{8} \cdot 8 = 1$$

means 8 is the multiplicative inverse of $\tfrac{1}{8}$ and $\tfrac{1}{8}$ is the inverse of 8.

This restriction of having to commute with its inverse prevents any but square matrices from having (multiplicative) inverses. (What are the orders of the commuted products of a 2×3 and a 3×2 matrix?)

First, we will work with 2×2 matrices. The product of a matrix and its inverse must yield the multiplicative identity, which for 2×2 matrices is

$$I = \begin{pmatrix} 1 & 0 \\ 0 & 1 \end{pmatrix}$$

Therefore, if

$$A = \begin{pmatrix} 2 & 3 \\ 1 & 4 \end{pmatrix}$$

and

$$A^{-1} = \begin{pmatrix} a_1 & b_1 \\ a_2 & b_2 \end{pmatrix}$$

then

$$A \cdot A^{-1} = \begin{pmatrix} 1 & 0 \\ 0 & 1 \end{pmatrix} \Rightarrow$$

$$\begin{pmatrix} 2 & 3 \\ 1 & 4 \end{pmatrix} \begin{pmatrix} a_1 & b_1 \\ a_2 & b_2 \end{pmatrix} = \begin{pmatrix} 2a_1 + 3a_2 & 2b_1 + 3b_2 \\ a_1 + 4a_2 & b_1 + 4b_2 \end{pmatrix}$$

$$= \begin{pmatrix} 1 & 0 \\ 0 & 1 \end{pmatrix} \Rightarrow$$

$$2a_1 + 3a_2 = 1, \qquad 2b_1 + 3b_2 = 0$$
$$a_1 + 4a_2 = 0 \quad \text{and} \quad b_1 + 4b_2 = 1$$

These equations can now be solved to find the elements of A^{-1}. For example,

$$a_1 + 4a_2 = 0 \Rightarrow a_1 = -4a_2$$

and

$$2a_1 + 3a_2 = 1 \Rightarrow 2(-4a_2) + 3a_2 = 1$$

Therefore,

$$-8a_2 + 3a_2 = 1 \Rightarrow$$
$$a_2 = -\tfrac{1}{5}$$

Continuing similarly for a_1, a_3, and a_4 yields

$$a_1 = \tfrac{4}{5}, \qquad\qquad b_1 = -\tfrac{3}{5}$$
$$a_2 = -\tfrac{1}{5}, \quad \text{and} \quad b_2 = \tfrac{2}{5}$$

This can be checked by multiplying

$$A \cdot A^{-1} = \begin{pmatrix} 2 & 3 \\ 1 & 4 \end{pmatrix} \begin{pmatrix} \tfrac{4}{5} & -\tfrac{3}{5} \\ -\tfrac{1}{5} & \tfrac{2}{5} \end{pmatrix} = (\ ?\ )$$

Notice that det $A = 5$ and each of the numerators of the elements of A^{-1} can be found (except for perhaps a sign change) in A.

In general, if we were to repeat the process for

$$A = \begin{pmatrix} a_1 & b_1 \\ a_2 & b_2 \end{pmatrix}$$

we would find

$$A^{-1} = \begin{pmatrix} \dfrac{b_2}{\det A} & \dfrac{-b_1}{\det A} \\ \dfrac{-a_2}{\det A} & \dfrac{a_1}{\det A} \end{pmatrix}$$

for det $A \neq 0$. If det $A = 0$, then A is said to be singular and it does not possess a multiplicative inverse.

EXAMPLE 1. Find the multiplicative inverses for each of the following matrices and verify your results by showing that $A \cdot A^{-1} = I$.

(a) $A = \begin{pmatrix} 1 & 4 \\ 2 & 7 \end{pmatrix} \Rightarrow \det A = \begin{vmatrix} 1 & 4 \\ 2 & 7 \end{vmatrix} = -1$

and

$$A^{-1} = \begin{pmatrix} \dfrac{7}{-1} & \dfrac{-4}{-1} \\ \dfrac{-2}{-1} & \dfrac{1}{-1} \end{pmatrix} = \begin{pmatrix} \underline{?} & \underline{?} \\ \underline{?} & \underline{?} \end{pmatrix}$$

Check.

$$A \cdot A^{-1} = \begin{pmatrix} 1 & 4 \\ 2 & 7 \end{pmatrix} \begin{pmatrix} -7 & 4 \\ 2 & -1 \end{pmatrix}$$

$$= \begin{pmatrix} \underline{?} & \underline{?} \\ \underline{?} & \underline{?} \end{pmatrix} = I$$

(b) $A = \begin{pmatrix} 2 & 6 \\ 1 & 3 \end{pmatrix} \Rightarrow \det A = 0$ and A^{-1} does not exist.

(c) $A = \begin{pmatrix} 3 & 2 \\ 4 & 1 \end{pmatrix} \Rightarrow \det A = \underline{\quad ? \quad}$

and

$$A^{-1} = \begin{pmatrix} \frac{?}{-5} & \frac{?}{-5} \\ \frac{?}{-5} & \frac{?}{-5} \end{pmatrix} = \begin{pmatrix} & \\ & \end{pmatrix}$$

Check.

$$A \cdot A^{-1} = \begin{pmatrix} 3 & 2 \\ 4 & 1 \end{pmatrix} \cdot \begin{pmatrix} & \\ & \end{pmatrix} = \begin{pmatrix} 1 & 0 \\ 0 & 1 \end{pmatrix}$$

EXERCISE 14.6

A. FUNDAMENTALS

Find the multiplicative inverses, if any, of the following matrices.

1. $\begin{pmatrix} 1 & 2 \\ 3 & 4 \end{pmatrix}$ 2. $\begin{pmatrix} -1 & 0 \\ 2 & 3 \end{pmatrix}$ 3. $\begin{pmatrix} 4 & 6 \\ 2 & 3 \end{pmatrix}$

4. $\begin{pmatrix} 1 & 5 \\ -3 & -7 \end{pmatrix}$

Show that each of the following matrices commutes with its inverse. That is, show that $AA^{-1} = A^{-1}A$.

5. $\begin{pmatrix} 2 & 1 \\ 3 & 2 \end{pmatrix}$ 6. $\begin{pmatrix} 4 & 3 \\ -1 & 3 \end{pmatrix}$

7. $\begin{pmatrix} -2 & -4 \\ 2 & 4 \end{pmatrix}$

B. ESSENTIALS

Compare $A^{-1} \cdot B^{-1}$, $B^{-1} \cdot A^{-1}$, and $(A \cdot B)^{-1}$ for Problems 8 and 9.

★8. $A = \begin{pmatrix} 2 & 3 \\ 1 & 2 \end{pmatrix}$ and $B = \begin{pmatrix} 1 & 3 \\ 2 & 4 \end{pmatrix}$

9. $A = \begin{pmatrix} 1 & 3 \\ -1 & 2 \end{pmatrix}$ and $B = \begin{pmatrix} 2 & -4 \\ -1 & 2 \end{pmatrix}$

★10. Compare $\det A$ and $\det A^{-1}$ for

(a) $A = \begin{pmatrix} 3 & 1 \\ 2 & 1 \end{pmatrix}$ (b) $A = \begin{pmatrix} 3 & -1 \\ 4 & -2 \end{pmatrix}$

11. Compare $(A^{-1})^T$ and $(A^T)^{-1}$ for

$$A = \begin{pmatrix} 1 & 3 \\ 4 & 6 \end{pmatrix}$$

12. (a) The comatrix of a matrix is found by replacing each element with its minor times $(-1)^{i+j}$, where i is the row of the element and j is the column. Find the comatrix A^c for

$$A = \begin{pmatrix} -3 & 2 & 1 \\ -2 & 3 & 1 \\ 1 & -2 & 2 \end{pmatrix}$$

[HINT: The first entry, 3, is replaced with $(-1)^{1+1}(8) = 8$.]

(b) The adjoint A^a of a matrix A is the transpose of its comatrix $(A^c)^T$. Find the adjoint of A.

(c) The inverse of any $n \times n$ matrix is given by

$$A^{-1} = \frac{1}{\det A} \cdot A^a$$

for $\det A \neq 0$. Find the inverse for A in Part (a) and verify that $A \cdot A^{-1} = I$, where

$$I = \begin{pmatrix} 1 & 0 & 0 \\ 0 & 1 & 0 \\ 0 & 0 & 1 \end{pmatrix}$$

13. Use the following matrices to illustrate that

$$(AB)^{-1} = B^{-1}A^{-1}$$

$$A = \begin{pmatrix} 1 & 2 & 0 \\ 4 & 4 & 1 \\ 3 & 0 & 1 \end{pmatrix} \quad \text{and} \quad B = \begin{pmatrix} 4 & 4 & 1 \\ 4 & 5 & 1 \\ 5 & 6 & 2 \end{pmatrix}$$

C. APPLICATIONS

Prove the following theorems.

14. If $\det A = 0$, then A^{-1} does not exist. [HINT: Assume $p \Rightarrow q$ is false, and find a contradiction. Remember $\sim(p \Rightarrow q) \Leftrightarrow p \wedge \sim q$.]

15. The inverse of A is unique. (HINT: Assume there are two inverses and find a contradiction.)

CHAPTER 14   Matrices and Determinants

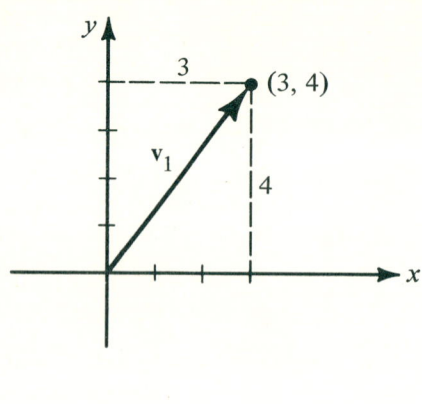

(a) $v_1 = (3, 4)$

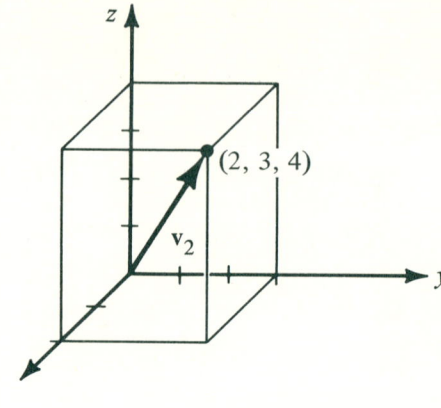

(b) $v_2 = (2, 3, 4)$

FIGURE 14.1.

★16. If $\det A \cdot \det B \neq 0$, then $(AB)^{-1} = B^{-1}A^{-1}$. [HINT: Use Problem 15 and the fact that $(AB) \cdot (AB)^{-1} = I$.]

17. If $\det A \cdot \det B \cdot \det C \neq 0$, then $(A \cdot B \cdot C)^{-1} = C^{-1} \cdot B^{-1} \cdot A^{-1}$.

18. If $\det A \neq 0$, then $(\alpha A)^{-1} = \dfrac{1}{\alpha} \cdot A^{-1}$ for any scalar, $\alpha \neq 0$.

14.7   VECTORS

The following definition of a vector reminds us that the properties of matrices also apply to vectors.

DEFINITION 14.11.
A **vector**, **v** (or $\vec{v}$), is a column or row matrix.

For two and three dimensions only, the following definition is common.

DEFINITION 14.12.
A **vector**, **v** (or $\vec{v}$), is a directed line segment having both magnitude and direction.

The second definition of a vector leads us to the definition of magnitude for a vector. Consider the two- and three-dimensional vectors in Figure 14.1. Their magnitudes (lengths) can be found using the distance formulas for two and three dimensions. Notice that we drew each vector with one end (its "tail") at the origin. This is always possible, since for any other vector there is an equal one (one of the same magnitude and direction) with its tail at the origin. There is a natural correspondence between the coordinates of the point located at the tip of a vector and its matrix representation. We often retain the commas when we write v_1 and v_2 in matrix form:

$$v_1 = (3, 4) \text{ or } v_1 = \begin{pmatrix} 3 \\ 4 \end{pmatrix}$$

and

$$v_2 = (2, 3, 4) \text{ or } v_2 = \begin{pmatrix} 2 \\ 3 \\ 4 \end{pmatrix}$$

For this reason, 3 is referred to as the x-component of v_1 and 4 is referred to as the y-component. Their respective magnitudes $\|v_1\|$ and $\|v_2\|$ are

$$\|v_1\| = \sqrt{(3-0)^2 + (4-0)^2} = \sqrt{3^2 + 4^2} = \sqrt{25} = 5$$

and
$$\|\mathbf{v}_2\| = \sqrt{(2-0)^2 + (3-0)^2 + (4-0)^2}$$
$$= \sqrt{2^2 + 3^2 + 4^2} = \sqrt{29}$$

DEFINITION 14.13.
If $\mathbf{v} = (x_1, x_2, \ldots, x_n)$, then
$$\|\mathbf{v}\| = \sqrt{x_1^2 + x_2^2 + \cdots + x_n^2}$$

Let's interpret the sum of two vectors geometrically. Since vectors are either row or column matrices, we already know how to add them.

EXAMPLE I. If $\mathbf{v}_1 = (5, 1)$ and $\mathbf{v}_2 = (2, 6)$, then interpret their sum, $\mathbf{v} = \mathbf{v}_1 + \mathbf{v}_2 = (7, 7)$, geometrically.

Solution: The sum $\mathbf{v}$, which is the diagonal of the parallelogram formed by $\mathbf{v}_1$ and $\mathbf{v}_2$, as shown in Figure 14.2, is called the *resultant* of $\mathbf{v}_1$ and $\mathbf{v}_2$.

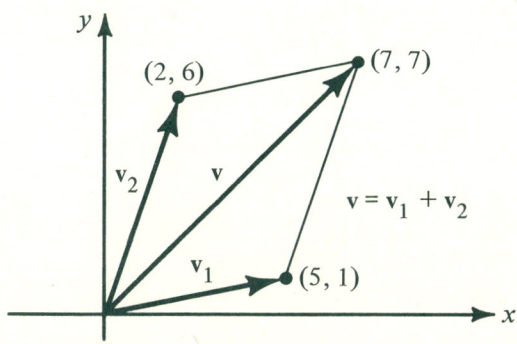

FIGURE 14.2.

EXAMPLE II. Compare the vectors $\mathbf{v} = (2, 2)$ and $3\mathbf{v} = (6, 6)$ geometrically. See Figure 14.3.

Solution: The vectors have the same direction but $3\mathbf{v}$ is 3 times as long as $\mathbf{v}$, since
$$\|\mathbf{v}\| = \sqrt{2^2 + 2^2} = \sqrt{8} = 2\sqrt{2}$$
and
$$\|3\mathbf{v}\| = \sqrt{6^2 + 6^2} = \sqrt{72} = 6\sqrt{2}$$

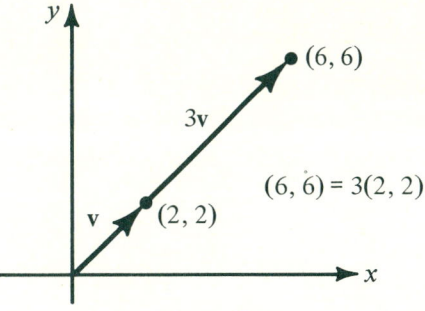

FIGURE 14.3.

THEOREM V.
For any vector $\mathbf{v}$ and a scalar λ,
$$\|\lambda \mathbf{v}\| = |\lambda|\|\mathbf{v}\|$$

Proof: Let $\mathbf{v} = (x_1, x_2, \ldots, x_n)$. Then
$$\|\lambda \mathbf{v}\| = \|\lambda x_1, \lambda x_2, \ldots, \lambda x_n\|$$
$$= \sqrt{(\lambda x_1)^2 + (\lambda x_2)^2 + \cdots + (\lambda x_n)^2}$$
$$= \sqrt{\lambda^2 (x_1 + x_2 + \cdots + x_n)}$$
$$= \sqrt{\lambda^2} \sqrt{x_1^2 + x_2^2 + \cdots + x_n^2}$$
$$= |\lambda|\|\mathbf{v}\|$$

Using Theorem V and Definition 14.13, we can find a vector of unit length in the direction of a given vector, as illustrated in Example III.

EXAMPLE III. Find a unit vector in the direction of each of the following vectors.
(a) $\mathbf{v} = (3, 4)$
Since
$$\|\mathbf{v}\| = \sqrt{3^2 + 4^2} = 5 \Rightarrow \|\tfrac{1}{5}\mathbf{v}\| = 1$$
our desired unit vector is
$$\mathbf{u} = \tfrac{1}{5}\mathbf{v} = (\tfrac{3}{5}, \tfrac{4}{5})$$
See Figure 14.4.
(b) $\mathbf{v} = (8, 0)$
$$\|\mathbf{v}\| = \sqrt{8^2 + 0^2} = 8 \Rightarrow \|\tfrac{1}{8}\mathbf{v}\| = 1$$

CHAPTER 14 Matrices and Determinants

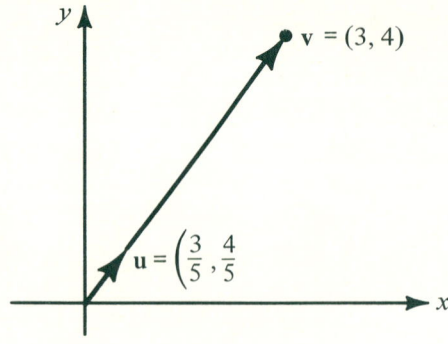

FIGURE 14.4.

and our unit vector, **i**, along the positive x-axis is

$$\mathbf{i} = (1, 0)$$

(c) $\mathbf{v} = (0, 4)$

$$\|\mathbf{v}\| = 4 \Rightarrow \|\tfrac{1}{4}\mathbf{v}\| = 1$$

and our unit vector, **j**, along the positive y-axis is

$$\mathbf{j} = (0, 1)$$

Notice that any two-dimensional vector can be expressed as a linear combination (sums of scalar products) of **i** and **j**. For

$$\mathbf{v} = (4, -7)$$

we can write

$$\mathbf{v} = 4(1, 0) + (-7)(0, 1)$$

or

$$\mathbf{v} = 4\mathbf{i} - 7\mathbf{j}$$

This is the reason 4 is often called the ith component of vector **v**, and -7 is called the jth component. For three space, we have

$$\mathbf{i} = (1, 0, 0)$$
$$\mathbf{j} = (0, 1, 0)$$

and

$$\mathbf{k} = (0, 0, 1)$$

as the unit vectors along each of the three positive axes. Now, any vector

$$\mathbf{v} = (\alpha_1, \alpha_2, \alpha_3)$$

can be written as a linear combination of **i**, **j**, and **k** as

$$\mathbf{v} = \alpha_1(1, 0, 0) + \alpha_2(0, 1, 0) + \alpha_3(0, 0, 1) \Leftrightarrow$$
$$\mathbf{v} = \alpha_1 \mathbf{i} + \alpha_2 \mathbf{j} + \alpha_3 \mathbf{k}$$

This is common notation.

THEOREM VI.
If $\mathbf{v} = (x_1, x_2, \ldots, x_n)$, then the unit vector in the direction of **v** is given by

$$\mathbf{u} = \frac{1}{\|\mathbf{v}\|} \mathbf{v}$$

where

$$\|\mathbf{v}\| = \sqrt{x_1^2 + x_2^2 + \cdots + x_n^2}$$

EXERCISE 14.7

A. FUNDAMENTALS

Perform the indicated vector operations for $\mathbf{v}_1 = (3, 4)$ and $\mathbf{v}_2 = (5, 12)$.

1. (a) $\mathbf{v}_1 + \mathbf{v}_2$ (b) $\mathbf{v}_2 + \mathbf{v}_1$
2. (a) $\|\mathbf{v}_1\|$ (b) $\|2\mathbf{v}_1\|$
3. (a) $\|\mathbf{v}_1\| + \|\mathbf{v}_2\|$ (b) $\|\mathbf{v}_1 + \mathbf{v}_2\|$

Perform the indicated vector operations for $\mathbf{v}_1 = (1, 2, 2)$ and $\mathbf{v}_2 = (5, 4, 5)$.

4. (a) $\|3\mathbf{v}_1\|$ (b) $\|-3\mathbf{v}_2\|$
5. $\mathbf{u}_1 = \dfrac{\mathbf{v}_1}{\|\mathbf{v}_1\|}$
6. (a) $\|\mathbf{v}_1\| + \|\mathbf{v}_2\|$ (b) $\|\mathbf{v}_1 + \mathbf{v}_2\|$

Interpret each of the following geometrically by drawing $\mathbf{v}_1 = (3, 4)$ and $\mathbf{v}_2 = (5, 12)$ and the result of each of the following in the x, y-plane.

7. $\mathbf{v}_1 + \mathbf{v}_2$ 8. $3\mathbf{v}_1$ 9. $-2\mathbf{v}_2$
10. $\|\mathbf{v}_1\|$

B. ESSENTIALS

Find the *scalar product*, $v_1 \cdot v_2$, for each of the following if $v_1 = (\alpha_1, \alpha_2)$ and $v_2 = (\beta_1, \beta_2) \Rightarrow v_1 \cdot v_2 = \alpha_1\beta_1 + \alpha_2\beta_2$

Sketch the vectors.

★11. $v_1 = (3, 4)$ $v_2 = (2, 1)$

12. $v_1 = (1, 0)$ $v_2 = (0, 1)$

13. $v_1 = (3, 15)$ $v_2 = (5, -1)$

The *scalar product* can also be found as follows:

$$v_1 \cdot v_2 = \|v_1\| \|v_2\| \cos \theta$$

where $(\|v_1\| \cos \theta)$ is the component of v_1 in v_2's direction, since θ is the angle between v_1 and v_2. See the following illustration. The two rules for finding scalar

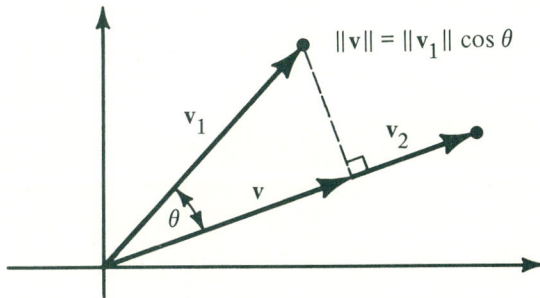

products are equivalent. Find the angle θ between the following pairs of vectors.

14. i and j ★15. $v_1 = (3, 2)$ and $v_2 = (4, -6)$

16. $v_1 = (1, 0)$ and $v_2 = \left(\frac{1}{\sqrt{2}}, \frac{1}{\sqrt{2}}\right)$

17. $v_1 = \left(\frac{\sqrt{3}}{2}, \frac{1}{2}\right)$ and $v_2 = \left(\frac{1}{2}, \frac{\sqrt{3}}{2}\right)$

The vector product $v_1 \times v_2$ is defined as follows for

$v_1 = (a_1, b_1, c_1)$ and $v_2 = (a_2, b_2, c_2)$

$$v_1 \times v_2 = \begin{vmatrix} i & j & k \\ a_1 & b_1 & c_1 \\ a_2 & b_2 & c_2 \end{vmatrix}$$

Geometrically, the vector $v_1 \times v_2$ is perpendicular to both v_1 and v_2 with magnitude equal in value to the area of the parallelogram $ABCD$ in the following illustration. Find the vector product, $v_1 \times v_2$, for each of the following pairs of vectors.

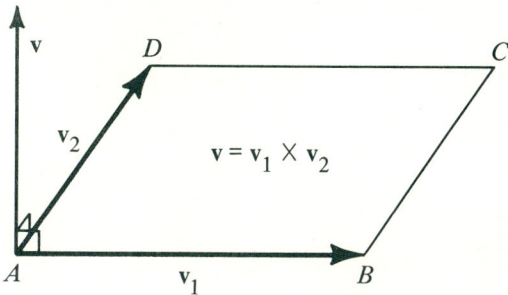

18. ★(a) $v_1 = (3, 2, -1)$, $v_2 = (-1, 3, -2)$
    (b) $v_1 = (4, 3, -2)$, $v_2 = (3, 2, -1)$

19. (a) $v_1 = (1, 2, 3)$, $v_2 = (-1, 3, -2)$
    (b) $v_1 = (2, 4, 6)$, $v_2 = (-1, 3, -2)$

20. (a) $v_1 = (2, 3, 1)$, $v_2 = (-4, -6, -2)$
    (b) $v_1 = (1, 2, -2)$, $v_2 = (2, -2, -1)$

C. APPLICATIONS

If we lift a weight of 10 pounds 5 feet, we say we have accomplished (10 lbs) · (5 ft) or 50 ft-lbs of *work*. In the metric system, the force required to lift the object is measured in Newtons and work is measured in Newton-meters. Vectors allow us to study work done by a force applied at an angle to the motion. The work done in moving an object against a constant force **F** along some distance **S** is defined as

$$W = -\mathbf{F} \cdot \mathbf{S} = -\|\mathbf{F}\| \|\mathbf{S}\| \cos \theta$$

where $(\|\mathbf{F}\| \cos \theta)$ is the component of the force along **S**.

Find the work done in the following instances.

21. A force $\mathbf{F} = (2, 3)$ is applied along the path $\mathbf{S} = (4, 6)$, (from $(0, 0)$ to $(4, 6)$).

22. $\mathbf{F} = (1, -1, 3)$ is applied along the path between $(1, -1, 2)$ and $(3, 4, 2)$. (HINT: $\mathbf{S} = (3 - 1, 4 - (-1), 2 - 2)$.)

★23. A boy pulls a wagon at an angle of 30° with the horizontal, with a force of 20 Newtons for a distance of 10 meters. See the following illustration.

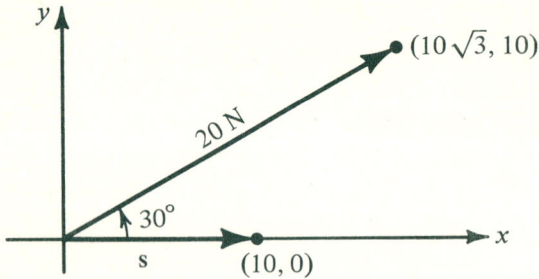

24. Three puppies are pulling on a rug with the forces and directions indicated in the following illustration. If the rug moves 12 meters, how much work have they done? Let **F** = mag (vector).

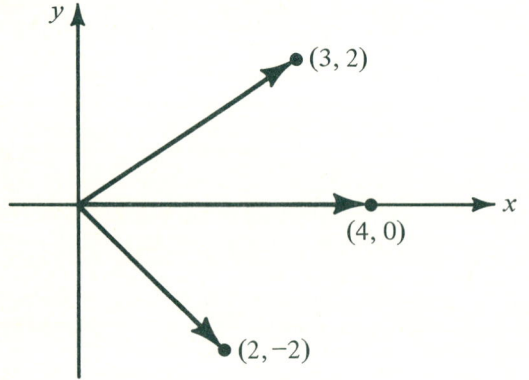

25. (a) Find the component of the force **F** heading down the inclined plane if $\theta = 30°$, if the mass (m) is 10 kilograms. (HINT: Since **F** = m**a**, where **a**, the acceleration due to gravity, is 9.8 m/sec², we have

$$\|\mathbf{F}\| = (10 \text{ kg})(9.8 \text{ m/sec}^2)$$
$$= 98 \text{ kg} \cdot \text{m/sec}^2 = 9.8 \text{ Newtons})$$

(b) Find the work done in pulling the block 10 meters up the inclined plane.

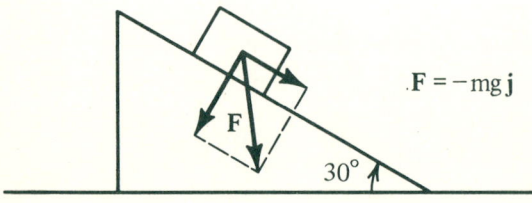

26. How much work is done in pulling a block, with a mass of 100 kilograms, up a height of 10 meters (vertical measurement), if $\theta = 45°$? See the following illustration.

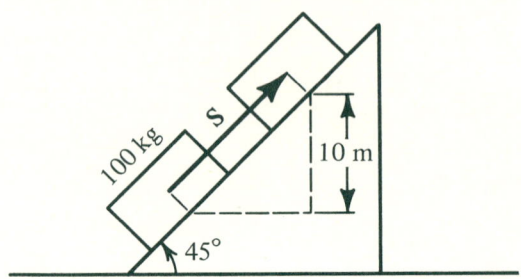

27. (a) Find the initial horizontal component, v_x, and vertical component, v_y, of the velocity of a bullet that was shot from a rifle with a muzzle velocity of 700 meters per second at an angle of inclination of 30°. (HINT: $\mathbf{v} = \mathbf{v}_x + \mathbf{v}_y$, where $\mathbf{v}_x$ is a constant (neglecting air resistance) varying only with different muzzle velocities and varying angles of inclination. $\mathbf{v}_y$ varies according to the constant rate of gravitational acceleration of -9.8 meters per second per second, as reflected in $v = -9.8t + v_0$, where v_0 is the initial velocity.)

(b) Find the horizontal distance the bullet travels if it is in the air 36 seconds.

(c) What is the magnitude of the horizontal component of its velocity at impact? Vertical component? What is its velocity at impact?

14.8 HISTORICAL COMMENTS ON VECTORS (optional)

Aristotle knew that forces can be represented as vectors and that combining forces can be analyzed using the parallelogram law, which was first explicitly stated by Galileo.

William Hamilton, through the intuitively based representation of complex numbers as directed line segments, lead the search for "three-dimensional complex numbers," later to be called *vectors*. Representing three-dimensional vectors as ordered triples (x, y, z) was naturally considered, but the algebra of complex numbers that went with two-dimensional vectors could not be extended to three dimensions. Even Gauss could not conceive of a suitable algebra for three-dimensional vectors.

Hamilton spent years trying to define the three-dimensional analogue of complex numbers because the algebra (vector algebra or linear algebra) was not to have the commutative law under multiplication. This was totally unprecedented. Finally, Hamilton defined *quaternions* as four-dimensional numbers of the form

$$3 + 4i + 6j + 2k$$

where $i = (1, 0, 0)$, $j = (0, 1, 0)$, and $k = (0, 0, 1)$. This resulting noncommutative algebra led to the algebra of vectors.

It was James Maxwell who began separating quaternions into the scalar (3) and vector ($4i + 6j + 2k$) components. The actual development of the algebra of vectors was finally accomplished by an American, J. W. Gibbs, in the 1880s and by Oliver Heaviside (1850–1925), an English mathematical physicist.

ENDNOTES

1. Arthur Cayley invented matrices in 1855 either directly from determinants or as a convenient mode of expressing systems of linear equations.

2. The theory of determinants is said to have originated with Leibniz in 1693 in dealing with systems of equations, although Kowa Seki considered determinants ten years earlier.

CHAPTER 15

Systems of Equations and Inequalities

15.1 LINEAR EQUATIONS IN TWO UNKNOWNS

At this point, we should have no difficulty graphing linear equations containing two variables. In this section, we will be concerned with finding the point of intersection of two such lines. This is not always possible, since the two equations may represent two parallel lines, in which case there will be no point of intersection, or they may be algebraically equivalent equations representing the same line, in which case there is no unique point of intersection. A system of parallel lines is called an *inconsistent* system, one of two equivalent equations is called *dependent*, and a system that yields a unique point of intersection is called *consistent*. The three cases are geometrically depicted in Figure 15.1.

There are several techniques used to find the point of intersection of the graphs of two linear equations. This point is the solution of the system of equations. One method of solving a system of two equations with two unknowns is by *substitution*. This is accomplished by solving one of the equations for one of its unknowns and then substituting this result into the other equation. This yields one equation having only one unknown, which is easily solved for the remaining variable. The result from this single equation is then substituted back into *either* original equation to find the value of the other variable. Let's use the consistent equations shown in Figure 15.1 as an example.

EXAMPLE 1. Solve the following system of equations.

$$x - y = 2$$
$$x + y = 4$$

Solution: Solving the second equation for y yields

$$y = 4 - x$$

This is now substituted for y in the first equation, which is then solved for the remaining variable x.

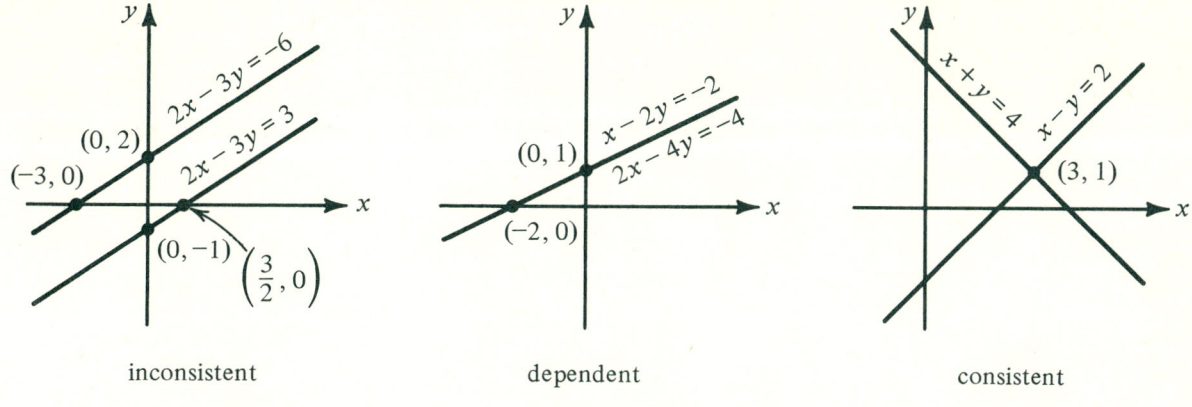

inconsistent   dependent   consistent

FIGURE 15.1.

Hence,
$$x - (4 - x) = 2 \Rightarrow$$
$$2x = 6 \Rightarrow$$
$$x = 3$$

This is a vertical line that passes through the point of intersection. To find the point on $x = 3$ that is the interesection of the two original lines, $x = 3$ is substituted back into either of the original equations or into $y = 4 - x$, which is equivalent to the second equation.

$$y = 4 - x \quad \text{and} \quad x = 3 \Rightarrow$$
$$y = 4 - 3 = 1$$

Therefore, the two lines intersect at the point (3, 1) and $x = 3$; $y = 1$ is the solution for the system of equations.

    The second method is called the *simultaneous solution* of the system, since both equations are used to find one equation with one unknown, which is then solved as before. In the following example, we can add the two equations to eliminate one of the variables.

    It is important to note, at this point, that adding two equations always results in an equation that is satisfied by the point(s) satisfying both of the original equations. That is, if (x_0, y_0) is the solution of a system of equations, then (x_0, y_0) will satisfy the sum of the two equations. Theorem 15.1 formalizes this result.

THEOREM 15.1.
If (x_0, y_0) satisfies
$$a_1 x + b_1 y = c_1$$
and
$$a_2 x + b_2 y = c_2$$
then (x_0, y_0) satisfies the sum of the two equations,
$$(a_1 + a_2)x + (b_1 + b_2)y = c_1 + c_2$$

Proof: If (x_0, y_0) satisfies both original equations, then
$$a_1 x_0 + b_1 y_0 = c_1$$
and
$$a_2 x_0 + b_2 y_0 = c_2$$
Therefore,
$$a_1 x_0 + a_2 x_0 + b_1 y_0 + b_2 y_0 = c_1 + c_2 \Rightarrow$$
$$(a_1 + a_2)x_0 + (b_1 + b_2)y_0 = c_1 + c_2$$
which says that (x_0, y_0) satisfies the sum of the two original equations, completing the proof.

CHAPTER 15 Systems of Equations and Inequalities

EXAMPLE II. Solve the following system.

$$2x - y = 3$$
$$x + y = 12$$

Solution: Adding the two equations eliminates y, and we have

$$3x = 15 \Rightarrow$$
$$x = 5$$

Then

$$x + y = 12 \quad \text{and} \quad x = 5 \Rightarrow$$
$$5 + y = 12 \Rightarrow$$
$$y = 7$$

and the solution is (5, 7).

Often the equations do not combine as easily as those in Example II, and we must change one or both equations before we add them if we wish to eliminate one of the variables.

EXAMPLE III. Solve the following system.

$$2x - 6y = 1$$
$$3x - 4y = 3$$

Solution: If we multiply both sides of the top equation by 2, and both sides of the bottom equation by -3, we have an equivalent system of equations in which the y-variable can be eliminated by adding the two equations. The resulting equivalent system is

$$4x - 12y = 2$$
$$-9x + 12y = -9$$

which, upon adding the two equations, yields a vertical line passing through the point of intersection.

$$-5x = -7 \Rightarrow$$
$$x = \tfrac{7}{5}$$

Since working with fractions is rather tedious, instead of substituting $x = \tfrac{7}{5}$ back into either original equation to find the y-coordinate, let's eliminate the x's *from the original system.* To accomplish this, we must multiply the first equation by __?__ and the second one by -2 before we add. Then the first equation becomes

$$6x - 18y = 3$$

and the second multiplied by -2 is

$$\underline{\ ?\ } + \underline{\ ?\ } = \underline{\ ?\ }$$

Then, by adding the two new equivalent equations, we have

$$-10y = -3 \Rightarrow$$
$$y = \tfrac{3}{10}$$

This horizontal line also passes through the point of intersection. Therefore, the graphs of the equations intersect at the point $(\tfrac{7}{5}, \tfrac{3}{10})$.

EXERCISE 15.1

A. FUNDAMENTALS

Determine whether the given point satisfies both of the corresponding linear equations. That is, determine if it is a solution of the system.

1. $2x - y = 1$
   $3x + 2y = 5$
   $(1, 1)$

2. $2x - 3y = 4$
   $x + 2y = 2$
   $(2, 0)$

3. $3x - 2y = 4$
   $3x + 7y = -1$
   $(-2, 1)$

4. $x = 2$
   $3x - 4y = 2$
   $(2, 1)$

5. $y = -5$
   $3x - 4y = 2$
   $(1, 2)$

6. $x + y = 3$
   $2x + 2y = 6$
   $(1, 2), (4, -1)$

B. ESSENTIALS

Solve the following systems by substitution.

7. $y = x - 3$
   $2x + y = 9$

8. $x = 4y - 3$
   $2x + 3y = 16$

9. $3x + y = 4$
   $4x - 3y = 27$

10. $x - 5y = 7$
    $2x + y = 3$

330

11. $3x + y = 11$
     $2x - y = 4$

12. $2x + 5y = 9$
     $5x - 5y = 5$

Solve the following systems simultaneously.

13. $x + 2y = 5$
     $3x - y = 8$

14. $3x - 4y = 7$
     $2x + 3y = -1$

15. $3y = 4x + 7$
     $2x = 3 + y$

16. $\dfrac{3x}{4} - \dfrac{2y}{3} = -1$
     $\dfrac{x}{2} + \dfrac{y}{6} = 3$

Use either technique to solve the following system. If there is no unique solution, determine if the lines are parallel lines having no common points, or one line having an infinite number of common points.

17. $3y = 2x - 5$
     $4x = 6y + 10$

18. $3x = 5 - 6y$
     $2x + 4y = 3$

19. $2x - 3y = 2$
     $9y - 6x = 1$

20. $x + 2y = 3$
     $3x + 2 = 5y$

21. $\dfrac{5x}{3} - \dfrac{2}{3}y = 13$
     $\dfrac{x}{9} + \dfrac{2y}{3} = 3$

22. $3x - 5y = 4$
     $10y - 6x = 3$

C. APPLICATIONS

Use either method to solve the system of two equations with two unknowns that models the following situations.

23. A swimmer who can travel at a rate of 5 kilometers per hour in still water spends 1 hour and 30 minutes swimming upstream 3 kilometers and back. What is the speed of the current? (HINT: Distance = rate × time.)

24. Two long trains pass each other going in opposite directions. The first engineer clocks the second train passing him at 90 kilometers per hour. The second engineer times the passing (from the time the engineer passed until the caboose passed) at 2 minutes. If one train is twice as long as the other, how long are the trains?

25. A bathtub holding 100 liters of water has a drain that leaks at a rate of 2 liters per minute. If twice as much hot water as cold is poured into the tub, it fills in 50 minutes, how fast is the hot water and the cold water entering the tub?

15.2 CRAMER'S RULE FOR TWO EQUATIONS WITH TWO UNKNOWNS

Cramer's Rule[1] is an excellent method of solving systems of n equations with n unknowns having unique solutions. Cramer's Rule also immediately tells us when the system does not have a unique solution.

Cramer's Rule for two equations with two unknowns requires that we memorize the general solution of the system

$$a_1 x + b_1 y = c_1$$
$$a_2 x + b_2 y = c_2$$

You will see that this is very easy once we find the solution, simultaneously. By multiplying the top equation by b_2 and the bottom by $-b_1$, we will eliminate the y-variable. The two new equations are

$$a_1 b_2 x + b_1 b_2 y = c_1 b_2$$

and

$$-a_2 b_1 x - b_1 b_2 y = -c_2 b_1$$

Now, let's add these two equations and solve for x.

$$a_1 b_2 x - a_2 b_1 x = c_1 b_2 - c_2 b_1$$
$$(a_1 b_2 - a_2 b_1) x = c_1 b_2 - c_2 b_1$$
$$x = \dfrac{c_1 b_2 - c_2 b_1}{a_1 b_2 - a_2 b_1}$$

Using a similar approach to eliminate the x's and solve for y yields

$$y = \dfrac{a_1 c_2 - a_2 c_1}{a_1 b_2 - a_2 b_1}$$

To simplify remembering this solution, we use a symbol called a **determinant**. Its definition is the difference of the two cross-products depicted below.

$$\begin{vmatrix} a & b \\ c & d \end{vmatrix} = ad - bc$$

Then, the equations
$$a_1 x + b_1 y = c_1$$
and
$$a_2 x + b_2 y = c_2$$
have solutions

$$x = \frac{\begin{vmatrix} c_1 & b_1 \\ c_2 & b_2 \end{vmatrix}}{\begin{vmatrix} a_1 & b_1 \\ a_2 & b_2 \end{vmatrix}} \quad \text{and} \quad y = \frac{\begin{vmatrix} a_1 & c_1 \\ a_2 & c_2 \end{vmatrix}}{\begin{vmatrix} a_1 & b_1 \\ a_2 & b_2 \end{vmatrix}}$$

because

$$\begin{vmatrix} c_1 & b_1 \\ c_2 & b_2 \end{vmatrix} = c_1 b_2 - c_2 b_1$$

$$\begin{vmatrix} a_1 & c_1 \\ a_2 & c_2 \end{vmatrix} = a_1 c_2 - \underline{\qquad ? \qquad}$$

and

$$\begin{vmatrix} a_1 & b_1 \\ a_2 & b_2 \end{vmatrix} = \underline{\qquad ? \qquad}$$

Notice that the determinant in the denominators of both have the coefficients of the variables written in exactly the array seen in the original system. This determinant is called the *coefficients' determinant*.

In solving for x, the determinant in the numerator is the coefficients' determinant, with the x-coefficients replaced by the constant terms.

In solving for y, the top determinant is the coefficients' determinant, with the y-coefficients replaced by the constant terms.

EXAMPLE I. Solve the following system using Cramer's Rule.

$$2x - 6y = 1$$
$$3x - 4y = 3$$

Solution: The coefficients' determinant is

$$D = \begin{vmatrix} 2 & -6 \\ 3 & -4 \end{vmatrix} = 2(-4) - (3)(-6) = 10$$

Therefore,

$$x = \frac{\begin{vmatrix} 1 & -6 \\ 3 & -4 \end{vmatrix}}{10} = \frac{(1)(-4) - (3)(-6)}{10}$$

$$= \frac{?}{} = \frac{?}{}$$

and

$$y = \frac{\begin{vmatrix} 2 & 1 \\ 3 & 3 \end{vmatrix}}{10} = \frac{(2)(3) - (3)(1)}{10} = \frac{3}{10}$$

The solution is

$$\left(\tfrac{7}{5}, \tfrac{3}{10}\right)$$

Notice where the constants 1 and 3 appear in the determinants in each solution.

Note that the equations should be written in the standard form

$$Ax + By = C$$

when using Cramer's Rule.

Thus far we have examined only consistent systems—ones with a unique solution. If we examine the two cases that do not generate unique solutions, we will see one of the great advantages of Cramer's Rule.

EXAMPLE II. Solve the following system using Cramer's Rule.

$$2x - 3y = 5$$
$$4x - 6y = 10$$

Solution: The coefficients' determinant is

$$D = \begin{vmatrix} 2 & -3 \\ 4 & -6 \end{vmatrix} = 2(-6) - 4(-3) = 0$$

We now know there is no unique solution, since we cannot divide by zero. Often this is sufficient, but if we wish to know whether there is no solution or an infinite number of solutions, we can continue.

$$x = \frac{\begin{vmatrix} 5 & -3 \\ 10 & -6 \end{vmatrix}}{0} = \frac{?}{0}$$

and

$$y = \frac{\begin{vmatrix} ? & ? \\ ? & ? \end{vmatrix}}{0} = \frac{?}{0}$$

This indeterminant form indicates that there may be infinitely many solutions. If x or y had turned out to be $c/0$, $c \neq 0$, there would be no solution.

THEOREM 15.2.
The system

$$a_1 x + b_1 y = c_1$$
$$a_2 x + b_2 y = c_2$$

has a unique solution

$$x = \frac{\begin{vmatrix} c_1 & b_1 \\ c_2 & b_2 \end{vmatrix}}{D} \quad \text{and} \quad y = \frac{\begin{vmatrix} a_1 & c_1 \\ a_2 & c_2 \end{vmatrix}}{D}$$

whenever

$$D = \begin{vmatrix} a_1 & b_1 \\ a_2 & b_2 \end{vmatrix} \neq 0$$

EXAMPLE III. Use Cramer's Rule to solve the following system.

$$6x - 4y = 1$$
$$3x - 2y = -2$$

Solution: The coefficients' determinant is zero, since

$$D = \begin{vmatrix} 6 & -4 \\ 3 & -2 \end{vmatrix} = \underline{\quad ?\quad}$$

Therefore, there is no unique solution. In this example, we see there is no solution at all, since

$$x = \frac{\begin{vmatrix} ? & ? \\ ? & ? \end{vmatrix}}{0} = \frac{-10}{0}$$

is undefined and

$$y = \frac{\begin{vmatrix} ? & ? \\ ? & ? \end{vmatrix}}{0} = \underline{\quad ?\quad}$$

Since these quotients are undefined not indeterminant, there are no values of x and y that satisfy both equations.

Theorem 15.2 summarizes this work.

EXERCISE 15.2

A. FUNDAMENTALS
Evaluate the following determinants.

1. (a) $\begin{vmatrix} 3 & 2 \\ 2 & 4 \end{vmatrix}$ (b) $\begin{vmatrix} 5 & 3 \\ 2 & 9 \end{vmatrix}$

2. (a) $\begin{vmatrix} 2 & 4 \\ 3 & 4 \end{vmatrix}$ (b) $\begin{vmatrix} 6 & 9 \\ 2 & 5 \end{vmatrix}$

3. (a) $\begin{vmatrix} 4 & 8 \\ 3 & 6 \end{vmatrix}$ (b) $\begin{vmatrix} 3 & -2 \\ 6 & -4 \end{vmatrix}$

4. (a) $\begin{vmatrix} 3 & 3 \\ 3 & 3 \end{vmatrix}$ (b) $\begin{vmatrix} 0 & 5 \\ 0 & 9 \end{vmatrix}$

5. (a) $\begin{vmatrix} 1 & 2 \\ 3 & 4 \end{vmatrix}$ (b) $\begin{vmatrix} 2 & 4 \\ 6 & 8 \end{vmatrix}$

B. ESSENTIALS
Use Cramer's Rule to find the solutions for the following. If there is no unique solution, decide whether it is because the equations represent parallel lines or because they represent the same line.

6. (a) $3x - y = 4$
       $x + 3y = 2$
   (b) $3x - y = -6$
       $x + 3y = 4$
   (c) $3x - y = 4$
       $3x - y = -6$

7. (a) $2x - 7y = 1$
       $3x - 2y = 10$
   (b) $2x - 7y = 1$
       $3x - 2y = -24$
   ★(c) $2x - 7y = 1$
       $3x - 2y = t$ (a fixed real number)

8. (a) $2x - 15y - 7 = 0$
       $y = \frac{2}{15}x + 1$
   (b) $x - 3y + 7 = 0$
       $\frac{2}{7}x = \frac{6}{7}y - 2$
   (c) $5x = 11y + 3$
       $22y = 10x + 9$

C. APPLICATIONS

9. In economics a competitive marketing situation is described by the two linear equations given below. The demand curve describes the consumers' willingness to buy at per unit price p, and the supply curve describes the businessmens' willingness to supply x units at price p. The intersection of these curves, where supply equals demand, is called *market equilibrium*. Find the per unit price p in dollars, and the quantity bought at market equilibrium x.

   ★(a) Demand: $x + 2p = 151$
       Supply: $3x - 4p = 448$
   (b) Demand: $2p + 3x = 44$
       Supply: $5x - 2p = 60$

10. The total expenditure (see Problem 9) is the per unit price times the quantity bought at market equilibrium. Find the total expenditure $p \cdot x$ if
    (a) Demand: $4p + x = 301$
        Supply: $3x - 8p = 888$
    (b) Demand: $3x + 5p = 55$
        Supply: $3x + 20 = 10p$

11. How much state and federal tax must you pay on $10,000.00 income if you must pay a state tax of 3% of the amount left after paying a federal tax of 20% of the amount left after paying the state tax?

12. A 10 liter radiator containing 30% antifreeze must contain 50% antifreeze to be safe. How much liquid should be replaced with pure antifreeze?

15.3 THREE-DIMENSIONAL SPACE

In this section we will examine three linear equations with three unknowns, x, y, and z. Each of these

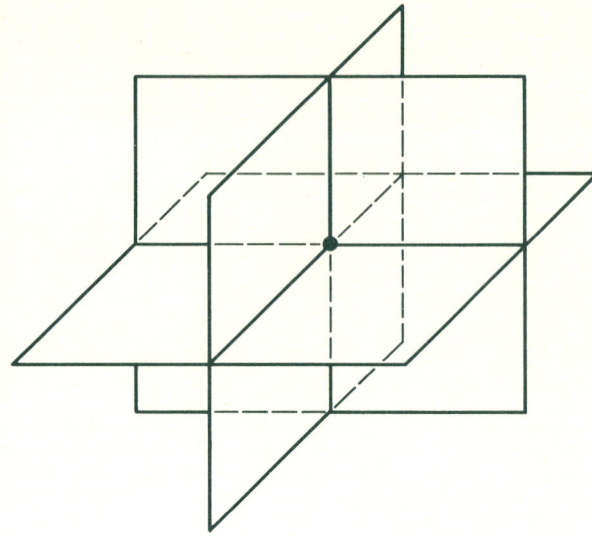

FIGURE 15.2. *Three Planes Intersecting in a Point.*

linear equations is the equation of a plane in three dimensions and the solutions, if any, are the points of intersection of the planes involved. We will begin our study with the most interesting possibility, a system having a unique solution. Geometrically, this will involve three planes intersecting in one point as pictured in Figure 15.2. See Section 13.7.

Since Cramer's Rule is often the fastest method for finding the unique solution when one exists, we will employ it first.

Since Cramer's Rule for three equations with three unknowns involves 3×3 determinants, you may wish to review Sections 14.4 and 14.5 before proceeding.

The generalization of Cramer's Rule from two equations with two unknowns is quite natural as the following example illustrates.

EXAMPLE I. Use Cramer's Rule to solve the system

$$x + 2y + z = 8$$
$$2x - 3y + z = -1$$
$$x + y - 2z = -3$$

Solution: First the coefficients' determinant is

$$D = \begin{vmatrix} 1 & 2 & 1 \\ 2 & -3 & 1 \\ 1 & 1 & -2 \end{vmatrix}$$

which by first multiplying the first column by -1 and adding it to the third column becomes

$$D = \begin{vmatrix} 1 & 2 & 0 \\ 2 & -3 & -1 \\ 1 & 1 & -3 \end{vmatrix}$$

Then let's add the second row to the third, yielding

$$D = \begin{vmatrix} 1 & 2 & 0 \\ 2 & -3 & -1 \\ -5 & 10 & 0 \end{vmatrix}$$

Now let's use the third column to evaluate the determinant.

$$D = \begin{vmatrix} 1 & 2 & 0 \\ 2 & -3 & -1 \\ -5 & 10 & 0 \end{vmatrix} = 0 \cdot \begin{vmatrix} 2 & -3 \\ -5 & 10 \end{vmatrix}$$

$$- (-1) \cdot \begin{vmatrix} 1 & 2 \\ -5 & 10 \end{vmatrix} + 0 \cdot \begin{vmatrix} 1 & 2 \\ 2 & -3 \end{vmatrix}$$

$$= 1 \cdot (10 - (-10)) = 20.$$

Then solving for x, y, and z, we have

$$x = \tfrac{1}{20} \begin{vmatrix} 8 & 2 & 1 \\ -1 & -3 & 1 \\ -3 & 1 & -2 \end{vmatrix} = \tfrac{1}{20} \begin{vmatrix} 8 & 2 & 9 \\ -1 & -3 & 0 \\ -3 & 1 & 5 \end{vmatrix}$$

$$= \tfrac{1}{20} \begin{vmatrix} 8 & -22 & 9 \\ -1 & 0 & 0 \\ -3 & 10 & -5 \end{vmatrix}$$

$$= \tfrac{1}{20}\left[-(-1) \begin{vmatrix} -22 & 9 \\ 10 & -5 \end{vmatrix} \right] \Rightarrow$$

$$x = \tfrac{1}{20}(110 - 90) = 1$$

$$y = \tfrac{1}{20} \begin{vmatrix} 1 & 8 & 1 \\ 2 & -1 & 1 \\ 1 & -3 & -2 \end{vmatrix} = \tfrac{1}{20} \begin{vmatrix} 1 & 8 & 9 \\ 2 & -1 & 0 \\ 1 & -3 & -5 \end{vmatrix}$$

$$= \tfrac{1}{20} \begin{vmatrix} 17 & 8 & 9 \\ 0 & -1 & 0 \\ -5 & 3 & -5 \end{vmatrix} \Rightarrow$$

$$y = \tfrac{1}{20}[(\underline{\ ?\ })|\underline{\ ?\ }|] = \underline{\ ?\ }$$

and

$$z = \tfrac{1}{20} \begin{vmatrix} 1 & 2 & 8 \\ 2 & -3 & -1 \\ 1 & 1 & -3 \end{vmatrix} = \tfrac{1}{20}|\underline{\quad ? \quad}|$$

$$= \tfrac{1}{20}|\underline{\quad ? \quad}| \Rightarrow$$

$$z = \tfrac{1}{20}[(\underline{\ ?\ })|\underline{\quad ?\quad}|] = \underline{\ ?\ }$$

The solution is

$$x = 1, y = 2, \text{ and } z = 3 \Leftrightarrow (1, 2, 3).$$

Figure 15.3 is the graph of the plane

$$x + 2y + z = 8$$

the lines, l_1, of intersection of this plane with

$$x + y - 2z = -3$$

the line, l_2, of intersection of the first plane

$$2x - 3y + z = -1$$

and their common point $(1, 2, 3)$.

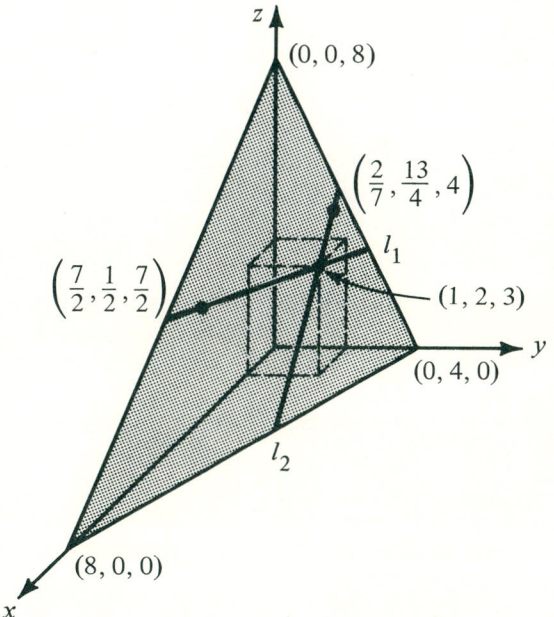

FIGURE 15.3.

All equations of the form

$$Ax + By + Cz = D$$

represent planes in three-dimensional space. A unique solution to a system of three such equations with three unknowns, x, y, and z, is the one point the three planes have in common, their point of intersection.

The concepts depicted in the previous examples are generalized in the following theorems on Cramer's Rule for systems of n equations with n unknowns.

THEOREM III.
The system of n equations with n unknowns

$$a_{11}x_1 + a_{12}x_2 + \cdots + a_{1n}x_n = c_1$$
$$a_{21}x_1 + a_{22}x_2 + \cdots + a_{2n}x_n = c_2$$
$$a_{n1}x_1 + a_{n2}x_2 + \cdots + a_{nn}x_n = c_n$$

has the solution $(b_1, b_2, \ldots, b_n)$ with

$$b_k = \frac{\begin{vmatrix} a_{11} & a_{12} & \cdots & c_1 & \cdots & a_{1n} \\ a_{21} & a_{22} & \cdots & c_2 & \cdots & a_{2n} \\ \vdots & \vdots & & \vdots & & \vdots \\ a_{n1} & a_{n2} & \cdots & c_n & \cdots & a_{nn} \end{vmatrix}}{D}$$

where $D = \begin{vmatrix} a_{11} & a_{12} & \cdots & a_{1n} \\ a_{21} & a_{22} & \cdots & a_{2n} \\ \vdots & \vdots & & \vdots \\ a_{n1} & a_{n2} & \cdots & a_{nn} \end{vmatrix} \neq 0$

is the coefficients' determinant, and in the numerator the constants $c_1, c_2, \ldots, c_n$ are in the kth column replacing the coefficients of x_k, $k = 1, 2, \ldots, n$.

THEOREM IV.
If the coefficients' determinant D equals zero, then the system does not have a unique solution.

EXERCISE 15.3

A. FUNDAMENTALS
Evaluate the following determinants.

1. (a) $\begin{vmatrix} 3 & 2 & -1 \\ 0 & 0 & 2 \\ 2 & 4 & 3 \end{vmatrix}$ (b) $\begin{vmatrix} 9 & 6 & -1 \\ 0 & 0 & 2 \\ 2 & 4 & 3 \end{vmatrix}$

(c) $\begin{vmatrix} 3 & 2 & -5 \\ 0 & 0 & 10 \\ 2 & 4 & 15 \end{vmatrix}$

2. (a) $\begin{vmatrix} 2 & 3 & 0 \\ -1 & 4 & 0 \\ 10 & 7 & 7 \end{vmatrix}$

(b) $\begin{vmatrix} 8 & 7 & -21 \\ 2 & 0 & 3 \\ -1 & 0 & 4 \end{vmatrix}$

(c) $\begin{vmatrix} 3 & -2 & 14 \\ 4 & 1 & 61 \\ 0 & 0 & 7 \end{vmatrix}$

3. (a) $\begin{vmatrix} 2 & -1 & 1 \\ 6 & -3 & 0 \\ 7 & -2 & -1 \end{vmatrix}$

(b) $\begin{vmatrix} 1 & -2 & 3 \\ 1 & 3 & 4 \\ -3 & 6 & -3 \end{vmatrix}$

(c) $\begin{vmatrix} 1 & 3 & 0 \\ 2 & -3 & 4 \\ 4 & -5 & 8 \end{vmatrix}$

B. ESSENTIALS
Find the solution for each of the following systems.

4. (a) $\begin{aligned} 2x + 2y + 3z &= 15 \\ x - y + 2z &= 5 \\ x - y - 2z &= -7 \end{aligned}$

(b) $\begin{aligned} x + y + z &= 6 \\ x + 2y + z &= 8 \\ 2x + 2y + z &= 9 \end{aligned}$

(c) $\begin{aligned} 2x - 2y + z &= 1 \\ x - 3y + 4z &= 7 \\ 2x - 4y + 3z &= 3 \end{aligned}$

5. (a) $\begin{aligned} 2x + 3y + z &= 4 \\ x - y + z &= 1 \\ -4x + y + 5z &= 20 \end{aligned}$

(b) $\begin{aligned} 2x + 3y + z &= 8 \\ x - y + z &= -5 \\ 4x - y - 5z &= 22 \end{aligned}$

(c) $3x - 2y - z = 11$
$x + y + z = -2$
$x + 4y - 5z = 49$

6. $4x + 3y - 3z = 5$
$6x + 6y - 5z = 7$
$6x + 3y - 4z = 8$

★7. $2x - 3y - z = 11$
$x - 2y - 4z = -4$
$3x - y - 2z = 8$

8. $2x + 2y - z = 3$
$4x + 4y - 2z = 1$
$4x + y - 2z = 3$

9. $3x - y + 3z - w = 6$
$x + y - 2z + w = 1$
$5x - y - 2z - 2w = -67$
$x + y + z - w = 2$

10. $x + y + z + w = 10$
$x + 2y - 2z + w = 3$
$2x + 2y - 3z + w = 1$
$x - 3y + 3z + 2w = 12$

C. APPLICATIONS

11. If you have a total of $1.00 in change in nickels, dimes, and quarters, with the total number of quarters and dimes being one more than the number of nickels, and there are 9 coins in all, how many of each coin do you have?

12. Find the solutions for 9 coins, nickels, dimes, and quarters, whose value is $1.00. There is no unique solution here, but since we only allow natural number solutions, there are a finite number of possible combinations. This is an example of a Diophantine system of equations—ones with integer solutions only.

13. Use three equations in three unknowns, a, b, and c, to find the equation of the parabola passing through the points $(-1, 1)$, $(1, -5)$, $(2, -2)$.

14. The refinement of 100 barrels of crude oil can be categorized into three products, heating oil, gasoline (all types), and petroleum by-products. If the number of barrels of gasoline always exceeds the sum of the number of barrels of heating oil and by-products by 50 barrels, and the number of barrels of oil used for production of the by-products is 5 percent of the total used for heating oil and gasoline combined, find what portion of the 100 barrels is used in each category.

15.4 MATRICES AND LINEAR SYSTEMS (optional)

Matrices lend themselves readily to solving systems of linear equations. We will begin applying matrices to systems of 2 equations and 2 unknowns by finding the inverse of the matrix composed of the coefficients of the unknowns.

$$3x + 5y = 13$$
$$7x - 2y = 3$$

The matrix equation

$$\begin{pmatrix} 3 & 5 \\ 7 & -2 \end{pmatrix} \begin{pmatrix} x \\ y \end{pmatrix} = \begin{pmatrix} 13 \\ 3 \end{pmatrix}$$

The equation is reasonable since the product of the 2×2 coefficient matrix and the 2×1 unknown matrix will yield a 2×1 matrix. We see that this matrix equation is, in fact, equivalent to the original system, since

$$\begin{pmatrix} 3 & 5 \\ 7 & -2 \end{pmatrix} \begin{pmatrix} x \\ y \end{pmatrix} = \begin{pmatrix} 3x + 5y \\ 7x - 2y \end{pmatrix} \Rightarrow$$

$$\begin{pmatrix} 3x + 5y \\ 7x - 2y \end{pmatrix} = \begin{pmatrix} 13 \\ 3 \end{pmatrix} \Rightarrow$$

$$3x + 5y = 13$$

and

$$7x - 2y = 3$$

Now we have a matrix equation of the form

$$AX = C$$

to solve. If det $A \neq 0$, then A^{-1} exists. Multiplying both sides by A^{-1} will lead to the solution.

$$A^{-1}(AX) = A^{-1}C \Rightarrow$$
$$(A^{-1}A)X = A^{-1}C \Rightarrow$$
$$IX = A^{-1}C \Rightarrow$$
$$X = A^{-1}C$$

Let's repeat this solution for our original system.

$$AX = C \Leftrightarrow \begin{pmatrix} 3 & 5 \\ 7 & -2 \end{pmatrix} \begin{pmatrix} x \\ y \end{pmatrix} = \begin{pmatrix} 13 \\ 3 \end{pmatrix}$$

$$A^{-1} = \begin{pmatrix} \frac{+2}{41} & \frac{+5}{41} \\ \frac{+7}{41} & \frac{-3}{41} \end{pmatrix}$$

since
$$\det A = -41$$

Therefore,

$$\begin{pmatrix} \frac{+2}{41} & \frac{+5}{41} \\ \frac{+7}{41} & \frac{-3}{41} \end{pmatrix} \begin{pmatrix} 3 & 5 \\ 7 & -2 \end{pmatrix} \begin{pmatrix} x \\ y \end{pmatrix} = \begin{pmatrix} \frac{+2}{41} & \frac{+5}{41} \\ \frac{+7}{41} & \frac{-3}{41} \end{pmatrix} \begin{pmatrix} 13 \\ 3 \end{pmatrix} \Rightarrow$$

$$\begin{pmatrix} 1 & 0 \\ 0 & 1 \end{pmatrix} \begin{pmatrix} x \\ y \end{pmatrix} = \begin{pmatrix} \frac{+26}{41} + \frac{+15}{41} \\ \frac{+91}{41} + \frac{-9}{41} \end{pmatrix} \Rightarrow$$

$$\begin{pmatrix} x \\ y \end{pmatrix} = \begin{pmatrix} 1 \\ 2 \end{pmatrix}$$

For a system of three equations with three unknowns, the process is identical except that finding the inverse of the coefficient matrix is more complicated. The process outlined in Exercise 14.6, Problem 12, for finding inverses applies to any $n \times n$ matrix whose determinant is not equal to zero. Theorem V completes this discussion.

Theorem V.
If the matrix equation for a system of n equations and n unknowns is
$$AX = C$$
then the unique solution for $\det A \neq 0$ is given by
$$X = A^{-1}C$$

EXERCISE 15.4

A. FUNDAMENTALS
Solve the following systems using Theorem V.
1. $x - 2y = 3$ 2. $3x = y + 2$
   $3x - 4y = 1$ $2x - 4y = -8$
3. $5x - 7y = 8$ 4. $7x = 3y - 2$
   $11x - 2y = 4$ $4y - x = 1$

Use Problem 12(c), Exercise 14.6, to find the inverses of the coefficient matrices for the B part of Exercise 15.3, then solve.

15.5 NON-UNIQUE SOLUTIONS AND GAUSSIAN REDUCTION

Gaussian reduction is simply a process for solving by the substitution technique in a set, formal pattern. It will always yield the solution, if one exists, even if there are more equations than unknowns, more unknowns than equations, or the determinant composed of the coefficients equals zero.

Gaussian reduction involves rewriting a system like

$$2x - 3y + 4z = 8$$
$$x - y + z = 2$$
$$2x + y + 3z = 1$$

as

$$x = y - z + 2$$
$$y = 2z - 4$$
$$z = 1$$

where each subsequent equation contains one less variable. This system can be readily solved from the bottom up, yielding the solution $(-1, -2, +1)$.

The next two examples illustrate Gaussian reduction on a system that is not $n \times n$ and on a system where the determinant of the coefficient matrix is zero. These are systems where Cramer's rule does not apply.

EXAMPLE I. Solve the following system using Gaussian reduction.

$$2x - 3y + 4z = 8$$
$$x - y + z = 2$$
$$2x + y + 3z = 1$$
$$4x + 2y - z = -9$$

Solution: Let's rearrange the equations so that the second equation is first. Then, solving it for x gives our first equation

$$x = y - z + 2 \qquad (1)$$

This is then substituted into the next equation to eliminate x.

$$2(y - z + 2) - 3y + 4z = 8$$

We then solve for y,

$$y = 2z - 4 \qquad (2)$$

Now equations (1) and (2) are substituted into the next equation to solve for z

$$2(y - z + 2) + (2z - 4) - 3z = -1 \Rightarrow$$
$$2[(2z - 4) - z + 2] + (2z - 4) - 3z = -1 \Rightarrow$$
$$z = 1 \qquad (3)$$

Now, equations (1), (2), and (3) are substituted into the last equation yielding

$$4(y - z + 2) + 2(2z - 4) - 1 = -9 \Rightarrow$$
$$4[(2z - 4) - 1 + 2] + 2[2(1) - 4] - 1 = -9 \Rightarrow$$
$$4[2(1) - 3] + 2(-2) - 1 = -9 \Rightarrow$$
$$0 = 0 \qquad (4)$$

Now, our system is reduced to

$$x = y - z + 2$$
$$y = 2z - 4$$
$$z = 1$$
$$0 = 0$$

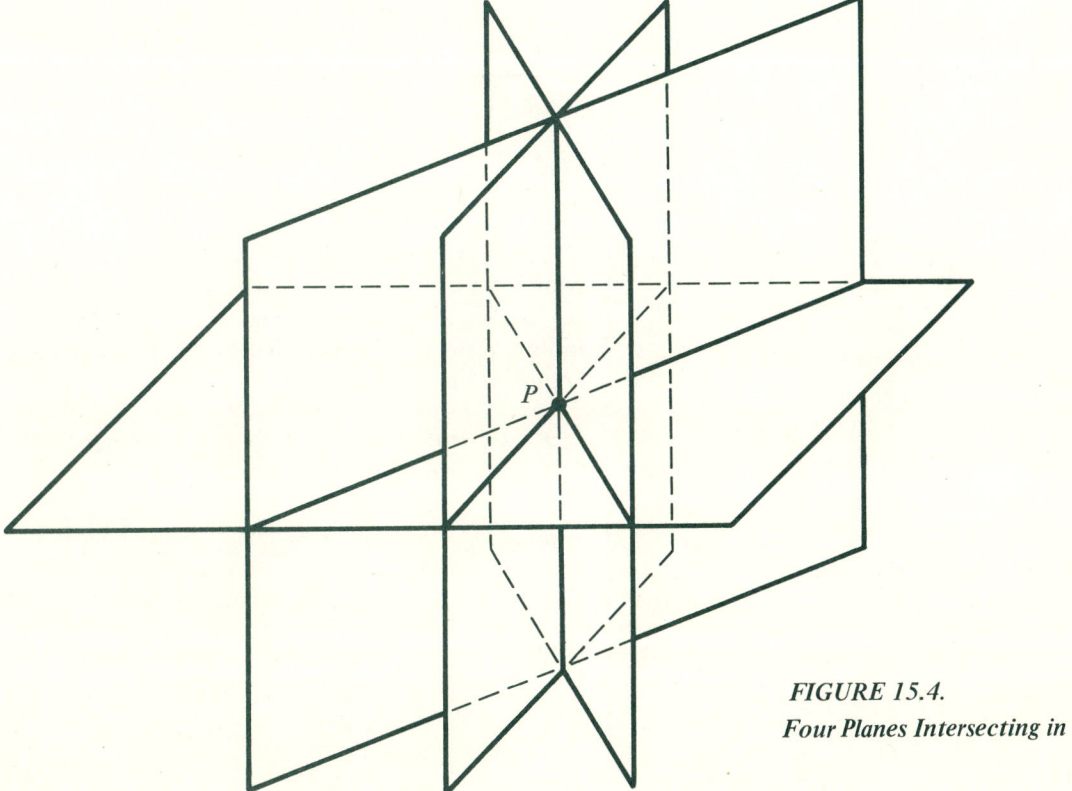

FIGURE 15.4.
Four Planes Intersecting in a Point.

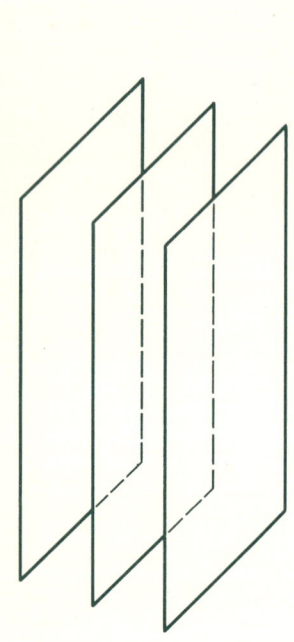
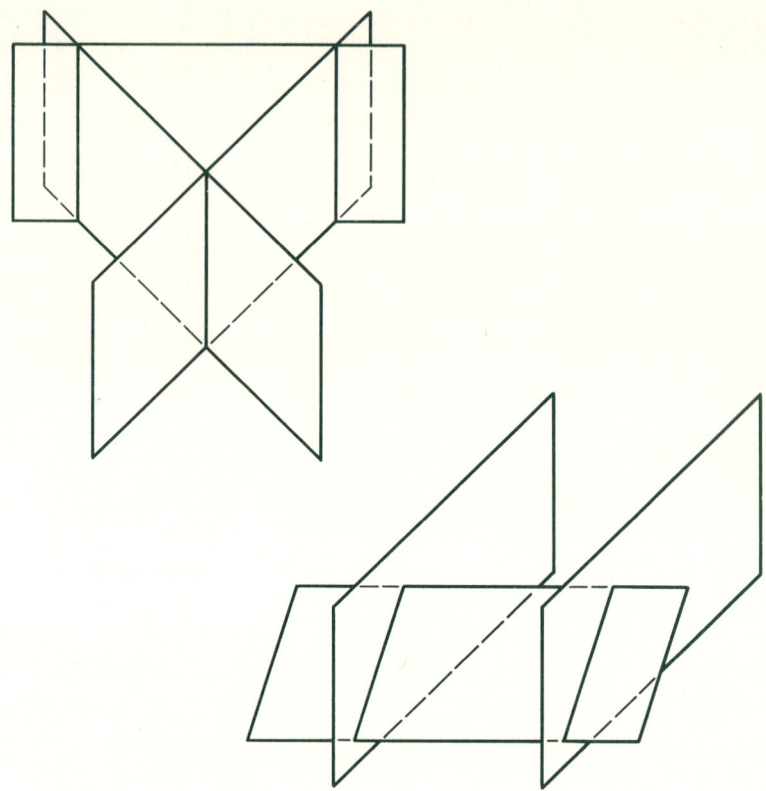

FIGURE 15.5.

Since $0 = 0$ is a consistent result, we now solve for y and x from $z = 1$. The four planes intersect in the point $(-1, -2, 1)$. Had the last equation proved to be inconsistent, like

$$3 = -9$$

then there would have been no solution to the system. This is quite likely when we have more equations than unknowns.

Figure 15.4 shows how four planes can intersect in a single point P. The bold lines show the line of intersection of each pair of planes.

Figure 15.5 shows how three equations with three unknowns may depict three planes with no single point in common, hence, no solution to the system.

The next example is a system of three equations with three unknowns whose corresponding planes intersect in a line. This system has an infinite number of solutions—all the points on the common line.

EXAMPLE II. Solve the following system by Gaussian reduction.

$$x + y - 5z = -7$$
$$2x - y - 4z = -5$$
$$x - 3y + 3z = 5$$

Solution: Since the coefficients' determinant,

$$D = \begin{vmatrix} 1 & 1 & -5 \\ 2 & -1 & -4 \\ 1 & -3 & 3 \end{vmatrix} = 0$$

(check this) we know the system does not have a unique solution. Either there is no solution or there are an infinite number of solutions.

Applying Gaussian reduction by first solving the first equation for x, then using this result to eliminate x from the second equation results in

$$x = -y + 5z - 7$$
$$y = 2z - 3$$

Substituting these results into the third equation yields the consistent result

$$0 = 0$$

Had this final result been a false statement, then there would have been no solution. Now, we know that there are an infinite number of solutions because our system cannot be solved for any specific value of an unknown. The planes, like those in Figure 15.6, intersect in a line.

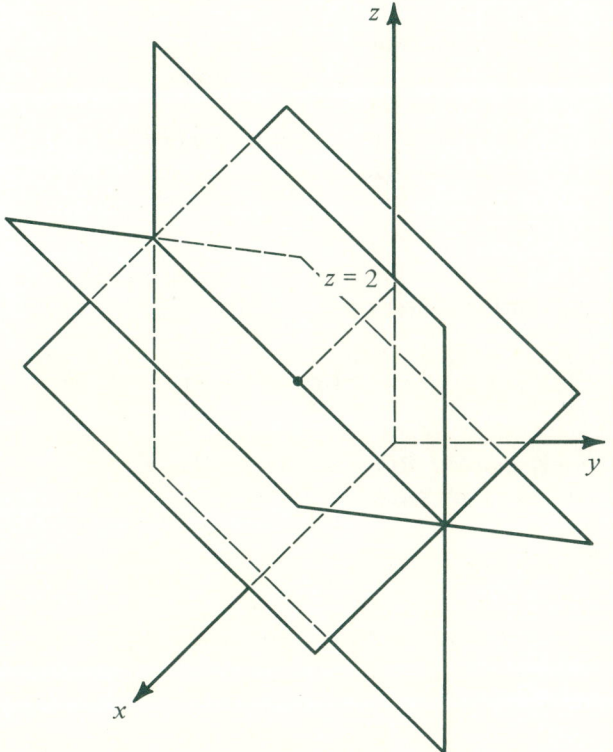

FIGURE 15.6.

The line of intersection of the three planes (in bold in Figure 15.6) gives an infinite number of points having x-, y-, and z-coordinates. If we fix the value of one of these unknowns, for example, z, we can then find the other two coordinates of a particular solution (one of the points on the bold line).

$$z = 2 \Rightarrow$$
$$y = 2z - 3 = 1$$

and

$$x = -y + 5z - 7 = 2$$

Therefore, one solution is (2, 1, 2). We could repeat this for as many different values of z as we wish. Each time we would find another solution. Instead, let's fix z at some specific but arbitrary real number, say $z = t$. Then

$$z = t \Rightarrow$$
$$y = 2z - 3 = 2t - 3$$

and

$$x = -y + 5z - 7 = 3t - 4$$

Our general solution generates all possible solutions, as t assumes different real values. It is written

$$(3t - 4, 2t - 3, t)$$

since

$$x = 3t - 4, \qquad y = 2t - 3, \qquad \text{and} \qquad z = t$$

EXERCISE 15.5

A. FUNDAMENTALS

Solve the following systems of linear equations using Gaussian reduction.

1. $3x - 2y = 4$
   $5x - 9y = 1$

2. $x - y + z = 4$
   $2x + y + 3z = 7$
   $3x + 4y + z = 1$

3. $x - 2y + 3z = -4$
   $2x + y - 2z = 6$
   $3x + 2y + z = -4$

4. $x - 2y + 3z = -7$
   $3x + y - 5z = 14$
   $2x - 3y + z = -6$

5. $\begin{aligned} x + 2y - z + w &= 2 \\ 2x + y - 3z + 2w &= -2 \\ 3x + 2y + z - 4w &= 7 \\ 2x - 3y - z + w &= -1 \end{aligned}$

6. $\begin{aligned} x + 2y - 3z + w &= -5 \\ 3x + y + z - w &= 4 \\ 2x - 3y + z - 3w &= 2 \\ x - 2y + 3z + w &= 3 \end{aligned}$

B. ESSENTIALS
Solve the following systems using Gaussian reduction.

7. $\begin{aligned} x + y - 3z &= 1 \\ x + 2y + z &= 1 \end{aligned}$

8. $\begin{aligned} x - 2y - 3z + 2w &= -4 \\ 3x + y - 4z + 2w &= 1 \\ 2x + 3y + z - 3w &= -4 \end{aligned}$

9. $\begin{aligned} x + 2y - 3z &= -2 \\ 2x - 3y - 3z &= -19 \\ 3x - 2y + z &= -10 \\ 2x + 3y + 4z &= 13 \end{aligned}$

10. $\begin{aligned} x - 3y + 2z &= 1 \\ x - 3y + 2z &= 3 \\ x - 3y + 2z &= 4 \end{aligned}$

11. $\begin{aligned} 6x + 2y - 2z &= -3 \\ 3x + 4y + 2z &= 3 \\ 12x - 6y - 14z &= -21 \end{aligned}$

12. $\begin{aligned} x + y + 2z - w &= -1 \\ 2x + y + 2z - 3w &= -4 \\ 2x - 3y + 2z + 9w &= -8 \\ 3x + 2y - 4z - 20w &= -9 \end{aligned}$

15.6 ECHELON FORM (optional)

Solving systems of linear equations by converting matrices to echelon form is simply Gaussian reduction in matrix form. This method has several advantages over Gaussian reduction. For example, it does not require variables and equal signs for organization.

For the system

$$\begin{aligned} 2x + 4y - z &= 0 \\ 3x - y + z &= 2 \\ x + y - 3z &= 6 \end{aligned}$$

we will convert the *augmented* matrix

$$\begin{pmatrix} 2 & 4 & -1 & 0 \\ 3 & -1 & 1 & 2 \\ 1 & 1 & -3 & 6 \end{pmatrix}$$

to the *echelon form*

$$\begin{pmatrix} 1 & a_1 & a_2 & a_4 \\ 0 & 1 & a_3 & a_5 \\ 0 & 0 & 1 & a_6 \end{pmatrix}$$

where $a_i \in$ reals. Theorem VI explains how this can be accomplished. As you read this theorem, remember that an augmented matrix is simply a system of linear equations.

THEOREM VI.
The following *row operations* (RO) result in a new augmented matrix representing an equivalent system of linear equations.

RO1. Interchanging two rows.
RO2. Adding a constant multiple of one row to another row.
RO3. Multiplying a row by a nonzero real number (scalar).

First, let's employ RO2 to change the entries in the third row first.

$$\begin{pmatrix} 2 & 4 & -1 & 0 \\ 3 & -1 & 1 & 2 \\ 1 & 1 & -3 & 6 \end{pmatrix} \sim \begin{pmatrix} 1 & 1 & -3 & 6 \\ 3 & -1 & 1 & 2 \\ 2 & 4 & -1 & 0 \end{pmatrix}$$

Now, employ RO2 to change the 3 and 2 in column 1 to zeros. Multiply row 1 by -3 and add it to row 2, and multiply row 1 by -2 and add to row 3.

$$\begin{pmatrix} 1 & 1 & -3 & 6 \\ 3 & -1 & 1 & 2 \\ 2 & 4 & -1 & 0 \end{pmatrix} \sim \begin{pmatrix} 1 & 1 & -3 & 6 \\ 0 & -4 & 10 & -16 \\ 2 & 4 & -1 & 0 \end{pmatrix} \sim$$

$$\begin{pmatrix} 1 & 1 & -3 & 6 \\ 0 & -4 & 10 & -16 \\ 0 & 2 & 5 & -12 \end{pmatrix}$$

Now, apply RO3 to change -4 in column 2 to 1. We multiply by $-\frac{1}{4}$.

$$\begin{pmatrix} 1 & 1 & -3 & 6 \\ 0 & -4 & 10 & -16 \\ 0 & 2 & 5 & -12 \end{pmatrix} \sim \begin{pmatrix} 1 & 1 & -3 & 6 \\ 0 & 1 & -\frac{5}{2} & 4 \\ 0 & 2 & 5 & -12 \end{pmatrix}$$

Now, employ RO2 to eliminate the 2 using row 2 times -2.

$$\begin{pmatrix} 1 & 1 & -3 & 6 \\ 0 & 1 & -\frac{5}{2} & 4 \\ ? & ? & ? & ? \end{pmatrix}$$

Now, divide the row 3 by 10 to get the echelon form

$$\begin{pmatrix} 1 & 1 & -3 & 6 \\ 0 & 1 & -\frac{5}{2} & 4 \\ 0 & 0 & 1 & -2 \end{pmatrix}$$

which represents the system

$$\begin{aligned} x + y - 3z &= 6 \\ y - \tfrac{5}{2}z &= 4 \\ z &= -2 \end{aligned}$$

which is readily solved for the solution

$$(1, -1, -2)$$

Now, let's see how echelon form matrices handle no solution and non-unique solutions in the next two examples.

EXAMPLE I. Solve the following system.

$$\begin{aligned} x + 3y + z &= 1 \\ x + 5y - z &= 5 \\ 2x + 9y - z &= 3 \end{aligned}$$

Solution: Write the augmented matrix with the constants 1, 5, 3 forming the fourth column.

$$\begin{pmatrix} ? \end{pmatrix}$$

Multiply the first row by -1, and add it to the second. Then, multiply the first row by -2, and add it to the third. This will change the first column to the final form 1, 0, 0.

$$\begin{pmatrix} 1 & 3 & 1 & 1 \\ ? & ? & ? & ? \\ 2 & 9 & -1 & 3 \end{pmatrix} \sim \begin{pmatrix} 1 & 3 & 1 & 1 \\ ? & ? & ? & ? \\ ? & ? & ? & ? \end{pmatrix}$$

Now divide row 2 by 2. Then eliminate the 3 from the third row.

$$\begin{pmatrix} 1 & 3 & 1 & 1 \\ ? & ? & ? & ? \\ 0 & 3 & -3 & 1 \end{pmatrix} \sim \begin{pmatrix} 1 & 3 & 1 & 1 \\ ? & ? & ? & ? \\ 0 & 3 & -3 & 1 \end{pmatrix} \sim \begin{pmatrix} 1 & 3 & 1 & 1 \\ ? & ? & ? & ? \\ 0 & ? & ? & ? \end{pmatrix}$$

Dividing row 3 by -5 yields the final echelon form matrix.

$$\begin{pmatrix} 1 & 3 & 1 & 1 \\ 0 & 1 & -1 & 2 \\ 0 & 0 & 0 & 1 \end{pmatrix}$$

Whenever the last row has only one nonzero entry in the third column, there is no solution to the system. This is readily seen by writing the system illustrated by the above matrix.

$$\begin{aligned} x + 3y + z &= 1 \\ y - z &= 2 \\ 0 &= 1 \end{aligned}$$

There is an inconsistent equation, $0 = 1$.

EXAMPLE II. Solve the following system.

$$\begin{aligned} x + y + 2z - w &= 2 \\ 2x + y + 3z - w &= 6 \\ 3x + 2y + 3z - w &= 6 \end{aligned}$$

CHAPTER 15   Systems of Equations and Inequalities

Solution: Since there are more unknowns than equations, we know there will not be a unique solution. Write the augmented matrix then convert the first column to 1, 0, 0.

$$\begin{pmatrix} ? & ? & ? & ? & ? \\ ? & ? & ? & ? & ? \\ ? & ? & ? & ? & ? \end{pmatrix}$$

$$\begin{pmatrix} 1 & ? & ? & ? & ? \\ 0 & ? & ? & ? & ? \\ 0 & ? & ? & ? & ? \end{pmatrix}$$

Now multiply row 2 by -1. Then change the second column to 1, 1, 0.

$$\begin{pmatrix} 1 & 1 & 2 & -1 & 2 \\ 0 & 1 & ? & ? & ? \\ 0 & -1 & -3 & 2 & 0 \end{pmatrix}$$

$$\begin{pmatrix} 1 & 1 & 2 & -1 & 2 \\ 0 & 1 & 1 & -1 & -2 \\ 0 & 0 & ? & ? & ? \end{pmatrix}$$

The final echelon form is found by dividing the third row by -2.

$$\begin{pmatrix} 1 & 1 & 2 & -1 & 2 \\ 0 & 1 & 1 & -1 & -2 \\ 0 & 0 & 1 & ? & ? \end{pmatrix}$$

This represents the system

$$x + y + 2z - w = 2$$
$$y + z - w = -2$$
$$z - \frac{1}{2}w = 1$$

which, by letting $w = t$ yields the solutions,

$$\left(\frac{t+6}{2}, \frac{3t-6}{2}, \frac{2-t}{2}, t \right)$$

EXERCISE 15.6

A. FUNDAMENTALS

Use the three row operations to transform the augmented matrix representation of the following systems into echelon form.

1. $x + 2y + z = 2$
    $2x - z = 3$
    $x + y = 3$

2. $3x - y + z = 0$
    $x + z = 0$
    $3x + y = 5$

3. $x - 2y + z = -8$
    $3x + y - z = 2$
    $4x = 2y + 3z - 17$

4. $x + y = -3$
    $2x - y + z = 11$
    $y - z + 2w = 3$
    $x + z - w = 5$

5. $x + 2y - w = 6$
    $3x - z + 2w = -4$
    $y - z + 3w = -1$
    $x + y + z = -2$

6. $2x + y - z + w = 3$
    $2x - y + 3z - w = 5$
    $x + 2y - z + 2w = 8$
    $2x + y + z - 3w = -12$

B. ESSENTIALS

Solve the following systems by any method available. Usually, the method of matrices in echelon form is better.

7. $x - 2y + z = -3$
    $3x - y - z = 8$
    $2x + 2y + 3z = -3$

8. $3x + 2y = 9$
    $x - 2z = 9$
    $3x + 4y = 15$
    $2y - z = 10$

9. $2x - y + z = 2$
    $3x + y - 2z = 3$

10. $3x - 2y + z - w = 1$
    $2x + y - z + 3w = 2$
    $x - 3y + 2z - w = 3$

11. $x - 3y - z + 2w = 6$
    $2x + 2y - 2z + w = 11$
    $3x - y - z + 2w - 10$
    $x + 3y + 3z - w = -5$
    $2x - y + z - 2w = -7$

15.7 SYSTEM OF NONLINEAR EQUATIONS

Systems of nonlinear equations are usually best solved by substitution. If you have two equations with two unknowns, solve for one of the variables in one equation, and substitute this result into the other equation. The next example shows how to find the points of intersection of an ellipse and a parabola.

EXAMPLE I. Solve the following system of non-linear equations.
$$2x^2 + 4y^2 = 17$$
$$x^2 + 2y = 1$$

Solution: First, solve for y in the second equation,
$$y = \frac{1 - x^2}{2}$$

Then substitute this result into the first equation to get one equation with one unknown.
$$2x^2 + 4\left(\frac{1-x^2}{2}\right)^2 = 17 \Rightarrow$$
$$2x^2 + 1 - 2x^2 + x^4 = 17 \Rightarrow$$
$$1 + x^4 = 17 \Rightarrow$$
$$x^4 = 16 \Rightarrow$$
$$x = \pm 2$$

Then,
$$y = \frac{1 - x^2}{2} \Rightarrow y = \frac{-3}{2}$$
$$\left(2, \frac{-3}{2}\right) \quad \text{and} \quad \left(-2, \frac{-3}{2}\right)$$

There are other techniques that are sometimes helpful in solving systems of nonlinear equations. Some of them are illustrated by the following examples.

EXAMPLE II. Solve the following system.
$$x^2 - y^2 = 1$$
$$2x^2 + y^2 = 2$$

Solution: This equation is solved exactly as two linear equations with two unknowns. By changing variables to u and v, where
$$u = x^2 \quad \text{and} \quad v = y^2$$

we have a linear system in u and v.
$$u - v = 1$$
$$2u + v = 2$$

whose solution is $u = 1$ and $v = 0$. Hence, $x = \pm\sqrt{u} = \pm 1$ and $y = \pm\sqrt{v} = 0$. Our solutions (1, 0) and (−1, 0) are the points of intersection of the above hyperbola and ellipse.

If the problem involves both first- and second-degree terms, sometimes using both equations simultaneously we can obtain an equation that is linear in one of the variables. We can then proceed as in Example I.

EXAMPLE III. Solve the following system.
$$x^2 - x + 2y + y^2 = 26$$
$$x^2 + 4x - 3y + y^2 = 21$$

Solution: Subtracting the two equations yields
$$-5x + 5y = 5 \Rightarrow$$
$$y = x + 5$$

Now, let's substitute this result into the first equation.
$$x^2 - x + 2(x + 5) + (x + 5)^2 = 26 \Rightarrow$$
$$x^2 - x + 2x + 10 + x^2 + 10x + 25 - 26 = 0 \Rightarrow$$
$$2x^2 + 11x + 29 = 0 \Rightarrow$$

$$(2x + 9)(x + 1) = 0 \Rightarrow$$
$$x = -\frac{9}{2} \quad \text{or} \quad x = -1.$$
$$x = -\frac{9}{2} \Rightarrow y = -\frac{9}{2} + 5 = \frac{1}{2}$$

and
$$x = -1 \Rightarrow y = -1 + 5 = 4$$

Hence, the solutions are $\left(-\frac{9}{2}, \frac{1}{2}\right)$, and $(-1, 4)$.

Consider a system of equations in which all terms containing unknowns are of degree two.

345

CHAPTER 15 Systems of Equations and Inequalities

Since this means that both curves must be symmetric about the origin (replacing x and y with $-x$ and $-y$ does not change the equations), the points of intersection must also lie on line(s) of the form

$$y = mx$$

Therefore, a substitution of $y = mx$ will yield the solutions. For example, consider first a circle and ellipse.

EXAMPLE IV. Solve the following system of nonlinear equations. See Figure 15.7.

$$9x^2 + y^2 = 25$$
$$x^2 + y^2 = 17$$

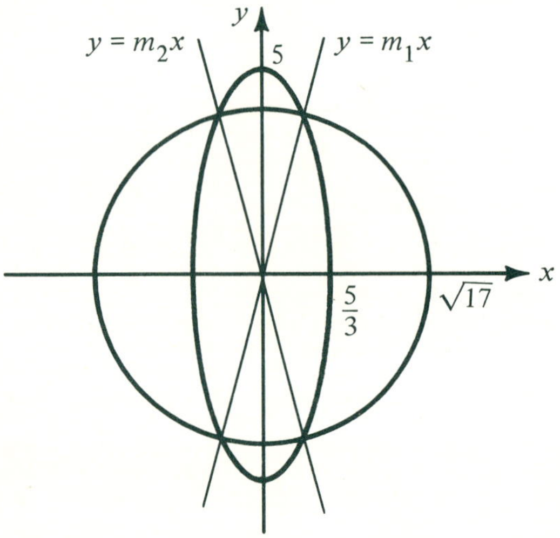

FIGURE 15.7.

Solution: Can you solve this system simultaneously? The four points of intersection lie on one of two lines passing through the origin. Therefore, the substitution of $y = mx$ leads to the solution. The equations become

$$9x^2 + (mx)^2 = 25$$
$$x^2 + (mx)^2 = 17$$

which, when solved for x^2 and equated, become

$$x^2 = \frac{25}{9 + m^2}$$

and

$$x^2 = \frac{17}{1 + m^2}$$

Since the fractions must be equal,

$$\frac{25}{9 + m^2} = \frac{17}{1 + m^2} \Rightarrow$$
$$25 + 25m^2 = 153 + 17m^2 \Rightarrow$$
$$8m^2 = 128 \Rightarrow$$
$$m^2 = 16 \Rightarrow$$
$$m = \pm 4$$

Therefore,

$$x^2 = \frac{17}{1 + m^2} \Rightarrow$$
$$x^2 = \frac{17}{1 + 16} \Rightarrow$$
$$x^2 = 1 \Rightarrow$$
$$x = \pm 1$$

and

$$y = mx \Rightarrow y = \pm 4(\pm 1) \Rightarrow$$
$$y = \pm 4$$

The points of intersection are

$$(1, 4), (1, -4), (-1, 4), (-1, -4)$$

The next system of nonlinear equations cannot readily be solved simultaneously. Since all terms are second degree, let's again substitute $y = mx$.

EXAMPLE V. Solve the following system.

$$x^2 + 2xy = 8$$
$$y^2 - xy = -1$$

Solution: Since all terms contain terms of degree two, a substitute of $y = mx$ will lead to a solution, if one exists.

$$y = mx \Rightarrow$$
$$x^2 + 2mx^2 = 8$$
$$m^2x^2 - mx^2 = -1$$

which, when solved for x^2 and equated, becomes

$$\frac{8}{1+2m} = \frac{-1}{m^2 - m}$$

Hence,

$$8m^2 - 8m = -1 - 2m \Rightarrow$$
$$8m^2 - 6m + 1 = 0 \Rightarrow$$
$$(4m - 1)(2m - 1) = 0 \Rightarrow$$
$$m = \frac{1}{4} \quad \text{or} \quad m = \frac{1}{2}$$

Since $y = mx$, we now have

$$y = \frac{1}{2}x \quad \text{or} \quad y = \frac{1}{4}x \Rightarrow$$

$$x^2 + 2x\left(\frac{1}{2}x\right) = 8 \quad \text{and} \quad x^2 + 2x\left(\frac{1}{4}x\right) = 8 \Rightarrow$$

$$x^2 + x^2 = 8 \Rightarrow \qquad x^2 + \frac{x^2}{2} = 8 \Rightarrow$$

$$x^2 = 4 \Rightarrow \qquad 3x^2 = 16 \Rightarrow$$

$$x = \pm 2 \qquad x = \pm \frac{4\sqrt{3}}{3}$$

Therefore, the solutions are

$$(2, 1), (-2, -1), \left(\frac{4\sqrt{3}}{3}, \frac{\sqrt{3}}{3}\right), \left(\frac{-4\sqrt{3}}{3}, \frac{-\sqrt{3}}{3}\right)$$

EXERCISE 15.7

A. FUNDAMENTALS
Solve the following equations by substitution. Identify each curve involved, i.e., circle, line, etc.

1. (a) $x^2 + y^2 = 16$ (b) $9x^2 + y^2 = 9$
   $y = 4$ $x = 1$

2. (a) $x^2 + y^2 = 25$ (b) $4x^2 + 9y^2 = 25$
   $y = \frac{4}{3}x$ $y = \frac{1}{2}x$

Solve the following equations simultaneously.

3. (a) $2x^2 - y^2 = -1$ (b) $x^2 - 3y^2 = 23$
   $x^2 + 2y^2 = 22$ $2x^2 - y^2 = 1$

4. (a) $2x^2 - y^2 = 1$ (b) $x^2 + y^2 = 20$
   $x^2 - 3y^2 = -2$ $4x^2 + 2y^2 = 48$

B. ESSENTIALS
Solve the following equations. Identify each curve.

5. $x^2 + y^2 = 45$ 6. $4x^2 + 9y^2 = 45$
   $y^2 = 4x$ $x^2 + y^2 = 10$

7. $4x^2 - y^2 = 12$
   $x^2 + 3y^2 = 16$

8. $x^2 - 4x + y^2 + 2y = 5$
   $3x + y = 5$

(You may not be able to identify some of these curves.)

9. $x^2 - xy + y^2 = 7$
   $2x^2 + 3xy + y^2 = 35$

10. $x^2 - 2xy + 2y^2 = 2$
    $x^2 + 4y^2 = 8$

11. $x^2 + 2xy - y^2 = 4$
    $x^2 + y^2 = 4$

12. $16x^2 - 9y^2 = 144$
    $49x^2 + 37y^2 = 1813$

C. APPLICATIONS

13. A boat with a two-member crew covers a 10 kilometer course (downstream and back) in 6 hours and 30 minutes. A five-member crew, which rows twice as fast, covers the same course in 2 hours and fifteen minutes. Find the speeds of the two boats and of the current.

14. Consider a rectangular window with a semicircle on top. If there is 12.96 square meters of glass in the semicircle part, and 40 square meters of glass in the rectangular portion, find the perimeter of the window.

15.8 SYSTEMS OF INEQUALITIES

We illustrate the solution of a system of inequalities graphically since the solution often contains an infinite number of points. Such a

CHAPTER 15   Systems of Equations and Inequalities

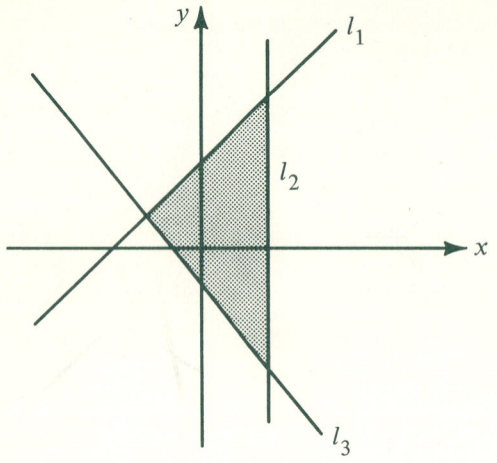

FIGURE 15.8.

graphical solution may look like that in Figure 15.8. The shaded region depicts the solution set for the system of inequalities

$$y \leq x + 4$$
$$x + y + 2 \geq 0$$
$$x \leq 3$$

where the boundaries of the region are the following lines,

$$l_1: y = x + 4$$
$$l_2: x = 3$$
$$l_3: x + y + 2 = 0$$

This suggests a reasonable way of progressing from a system of inequalities to its graphical solution. Begin by letting the inequalities be equalities and graphing the resulting equations. These equations will form the *boundaries* for the solution set. Since the solution may not be the region contained inside these boundaries, we must graph the solution set for each inequality separately, then the intersection of these regions will be our final solution.

EXAMPLE I. Solve the following system of inequalities graphically.

$$x + y \geq 4$$
$$y \leq x - 1$$
$$y \leq 4$$

Solution: Let $x + y = 4$, $y = x - 1$, and $y = 4$ and graph these equations. See Figure 15.9. For

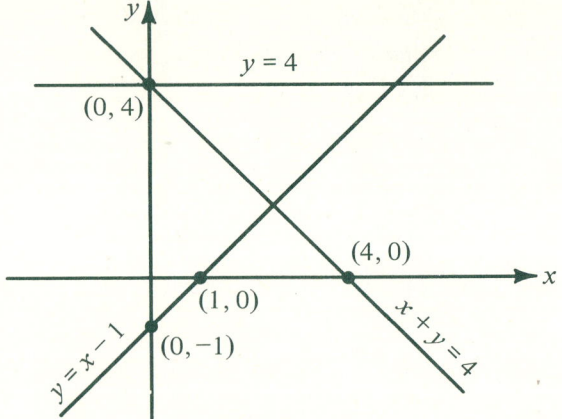

FIGURE 15.9.

$y \leq 4$, we want the region below the line $y = 4$. For $y \leq x - 1$, we want the region below the line $y = x - 1$, since all points on the line have their y-coordinate equal to $x - 1$, and the solution to the inequality contains points whose y-coordinates are less than the y-coordinates on the line. For $x + y \geq 4$, since that means $y \geq 4 - x$, we want the region above the line $y = 4 - x$, ($x + y = 4$). We can also determine which half of the plane divided by $x + y = 4$ represents the solution set for $x + y \geq 4$, by trial and error. If $x = 0$, $y = 0$ satisfies the inequality, then all points on the side of the line containing the origin $(0, 0)$ will satisfy the inequality, also. If $(0, 0)$ does not work, then the other side of the boundary is the one we want. Substituting $x = 0$, $y = 0$ into $x + y \geq 4$ yields $0 \geq 4$, which is false. Therefore, we shade the side of the boundary line $x + y = 4$ that does not contain the origin, that is, the region above $x + y = 4$. The final solution is indicated in Figure 15.10.

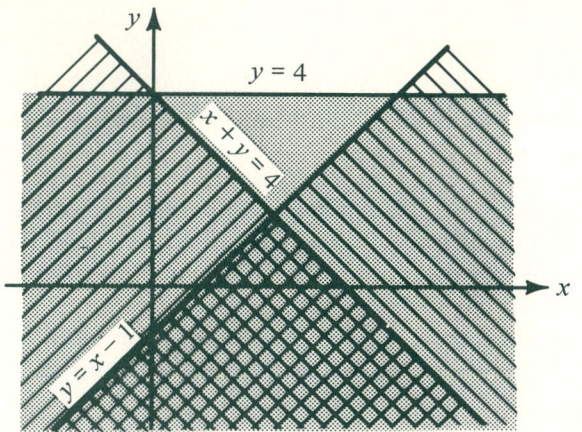

FIGURE 15.10.

The boundaries of systems of inequalities may not be linear equations. Consider the following system of nonlinear inequalities.

EXAMPLE II. Solve the following system of inequalities geometrically:

$$x^2 + y^2 < 16$$
$$y \geq 2 - |x|$$
$$y \geq x^2 - 9$$

Solution: Let $x^2 + y^2 = 16$, $y = 2 - |x|$, and $y = x^2 - 9$, then graph the resulting equations. Then, checking (0, 0) in each inequality yields

$x^2 - y^2 < 16 \Rightarrow 0^2 + 0^2 < 16 \Rightarrow 0 < 16$ (true)

$y \geq 2 - |x| \Rightarrow 0 \geq 2 - |0| \Rightarrow 0 \geq 2$ (false)

and

$y \geq x^2 - 9 \Rightarrow 0 \geq 0^2 - 9 \Rightarrow 0 \geq -9$ (true)

Therefore, we shade the interior of the circle $x^2 + y^2 = 16$, since that contains (0, 0). We shade the region above the absolute value function to exclude the origin located below the graph. Since the origin is above the parabola $y = x^2 - 9$, we shade this region. Notice that the boundary for the circle is a dotted curve indicating that the points on the circle, for which $x^2 - y^2 = 16$, are not included in solution set. See Figure 15.11.

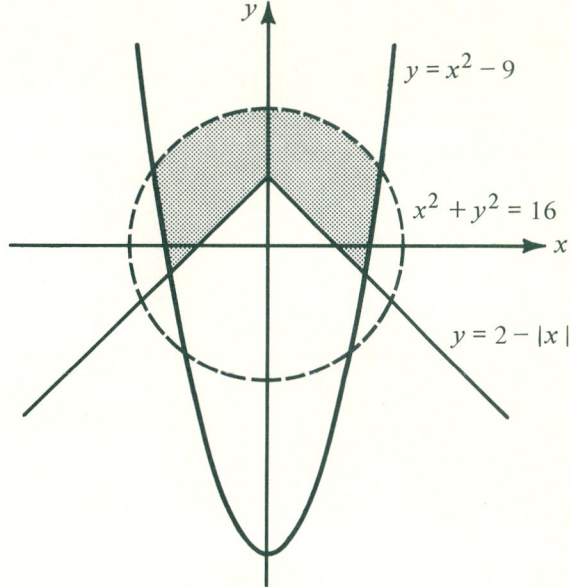

FIGURE 15.11.

EXERCISE 15.8

A. FUNDAMENTALS
Use shading to indicate the region of the plane containing points that satisfy the following inequalities.

1. (a) $y \leq 4$ (b) $y \geq -2$
2. (a) $x \leq -1$ (b) $x > 2$
3. (a) $2x - 3y < 4$ (b) $2x - 3y \geq 4$
4. (a) $y \geq \frac{3}{4}x - 2$ (b) $y < \frac{3}{4}x - 2$

Indicate that portion of the plane containing points that satisfy the following systems of inequalities, by shading.

5. (a) $x \geq 0$ (b) $x > 0$
   $y \leq 2x + 3$ $y > 2x + 3$
   $y \geq -2x + 1$ $y < -2x + 1$

6. (a) $y \geq 0$ (b) $x \geq 0$
   $y < 6x + 2$ $y \geq 3x + 2$
   $y < -x + 4$ $y < 4 - x$

7. $y < x$
   $x \leq 4$
   $2x + y \geq 8$

B. ESSENTIALS

Graph the solution set for the following systems.

8. (a) $x^2 + y^2 \leq 4$
    $y > x^2$
   (b) $x^2 + y^2 \geq 9$
    $y < 9 - x^2$

9. (a) $4x^2 + y^2 < 4$
    $x^2 + 4y^2 \geq 4$
   (b) $9x^2 + 4y^2 \leq 36$
    $4x^2 + 9y^2 \leq 36$

10. $y > \dfrac{1}{x}$
    $x^2 + y^2 \leq 16$

11. (a) $y \geq 4 - |x|$
    $x^2 + y^2 \leq 16$
    $y \geq 2$
   (b) $y < 4 - |x|$
    $y \leq 2$
    $x^2 + y^2 \leq 16$

12. $x^2 + 2x + y^2 \leq 0$
    $x^2 - 2x + y^2 \leq 0$
    $x^2 + (y - \sqrt{3})^2 \leq 1$

13. $x^2 + y^2 > 1$
    $y > \log x, x > 0$
    $y < -\log x$

15.9 LINEAR PROGRAMMING

Linear programming involves finding the maximum or minimum value of the object function $z = f(x, y)$, subject to linear constraints on the domain variables x and y. These linear constraints are linear inequalities. This subject does not concern computer programming, though computers are employed in solving such models for large numbers of variables with a large number of constraints.

The system of inequalities that make up the constraint will all include their boundary lines. That is, we will not use any strict inequalities. The solution to this system of inequalities is called the *feasible region*, since points in this region are feasible solutions to our problem. The point or points of this region that generate the maximum (or minimum) value of the dependent variable are called the *optimum* values.

Figure 15.12 depicts three examples of systems of linear constraints. The labeled vertices are called *nodes*. While problems involving unbounded feasible regions, such as (a), may or may not have optimum solutions, bounded ones, such as (b), always have, and infeasible ones, such as (c), never have optimum solutions.

Theorems VII, VIII, and IX indicate how linear programming problems are solved.

THEOREM VII.
A linear program having a bounded feasible region R has both minimum and maximum optimum values over R.

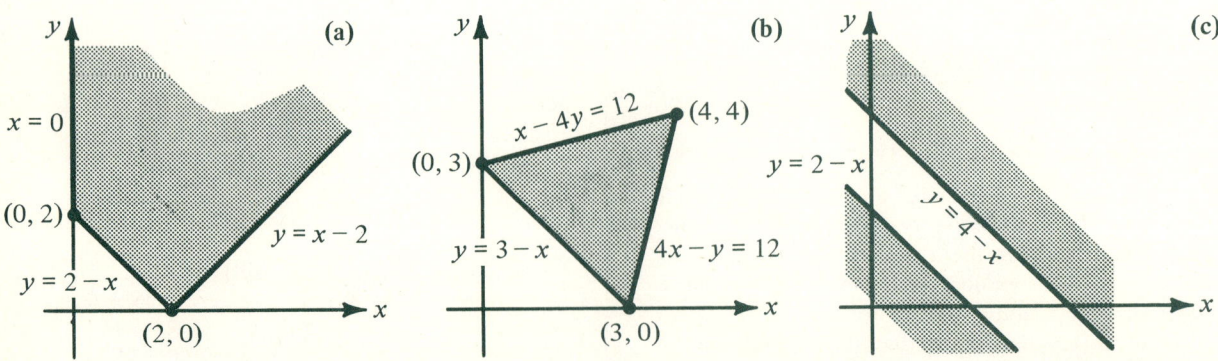

FIGURE 15.12. *(a) An Unbounded Feasible Region, (b) A Bounded Feasible Region, (c) An Infeasible Region.*

THEOREM VIII.
A constant object function,

$$Z = C, \quad C \in \text{reals}$$

has $Z = C$ as its optimum value.

THEOREM IX.
If a nonconstant object function attains its maximum (or minimum) over its feasible region R, then the optimum is attained at a node of R.

There is one other concept that concerns the uniqueness of the solution. The optimum value may be generated by all the points on a boundary, as well as by a point at a node. This does not contradict Theorem IX because the points at the nodes of this boundary are still optimum values whenever the boundary points are optimum values. The optimum values may not be unique. The next two examples illustrate a linear program for bounded feasible regions with a unique solution at a node and a nonunique solution on a boundary.

EXAMPLE I. Find the optimum (maximum and minimum) values of Z, where

$$Z = 6 - \frac{x}{2} - \frac{y}{2}$$

subject to the constraints

$$x \geq 0$$
$$y \geq 0$$
$$x - 5y + 20 \geq 0$$
$$25x - 15y \leq 30$$

Solution: The restrictions on x and y generate the feasible region R in Figure 15.13. All the nodes

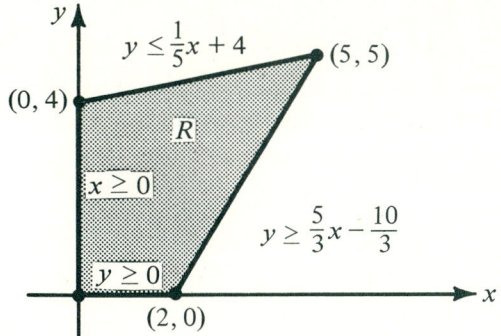

FIGURE 15.13.

are labeled. We now wish to find the minimum value on the plane

$$Z = 6 - \frac{1}{2}x - \frac{1}{2}y$$

over the feasible region graphed in Figure 15.13. The points on the plane above each of the nodes is labeled. The nodes are found by solving each pair of linear equations, simultaneously, for their point of intersection. Locate the section of the plane that is above R. Since this plane is strictly decreasing, it attains its minimum value over R. The minimum value is at one of the nodes. See Figure 15.14.

All the values on the plane above R are shaded. The smallest such value is $z = 1$, which occurs above the node (5, 5). The maximum value

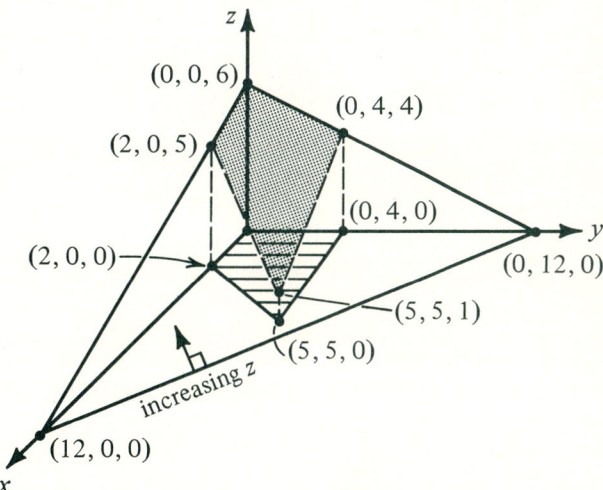

FIGURE 15.14.

of z subject to the linear constraints generating the feasible region R is z = 6. It occurs above the node (0, 0).

EXAMPLE II. Find the optimum values of z, where

$$z = 10 - \frac{3}{2}x - y$$

subject to the constraints

$$x \geq 0, \quad y \geq 0 \text{ (first quadrant)}$$
$$x + 2y \leq 4$$
$$3x + 2y \leq 6$$
$$3x - 2y \geq 3$$

Solution: The feasible region R in the x, y-plane is shown in Figure 15.15. The nodes are labeled. Transferring R to the x, y-plane in the x, y, z-coordinate system and sketching the first octant portion of the plane results in Figure 15.16.

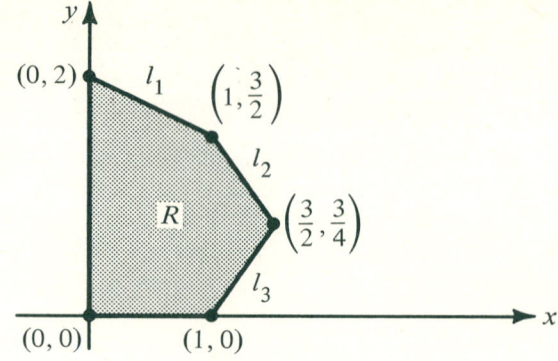

FIGURE 15.15.

The maximum value on the plane above R is z = 5 at the origin. There is a unique minimum value of z = 2, but it does not result from a unique ordered pair (x, y). Rather, every point on the line joining $(\frac{3}{2}, \frac{3}{4}, 0)$ and $(1, \frac{3}{2}, 0)$ will yield z = 2, the minimum. Therefore, our answer for the value of x and y minimizing z over R is

$$\{(x, y) \mid 3x + 2y = 6\}$$

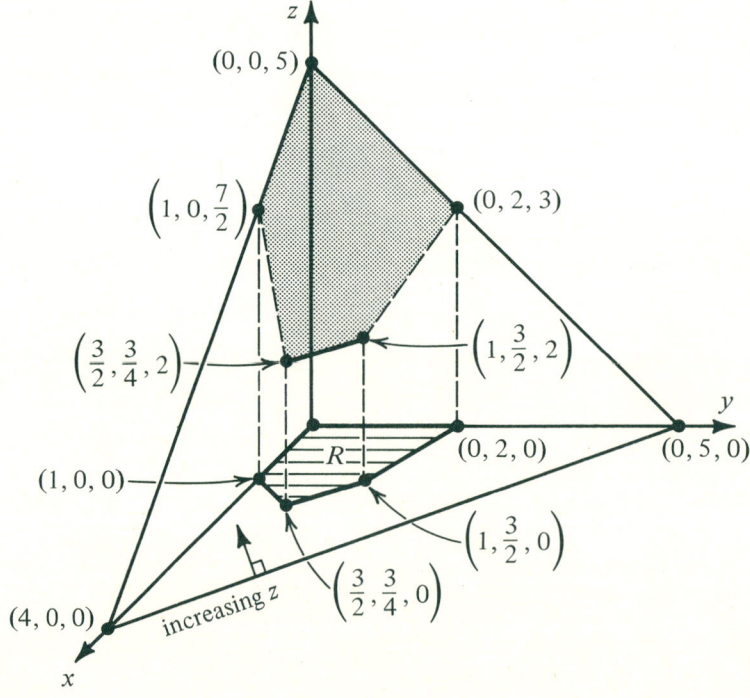

FIGURE 15.16.

SECTION 15.9 Linear Programming

The last "applied" example illustrates an unbounded feasible region.

EXAMPLE III. A company wishes to make a new brand of granola from whole wheat and nuts. Assume one gram of each commodity contains the following amounts of the indicated vitamins.

	vitamin A	vitamin B
whole wheat	12	8
assorted nuts	15	18

If whole wheat costs 10 cents per gram, and the nuts cost 20 cents per gram, minimize total cost

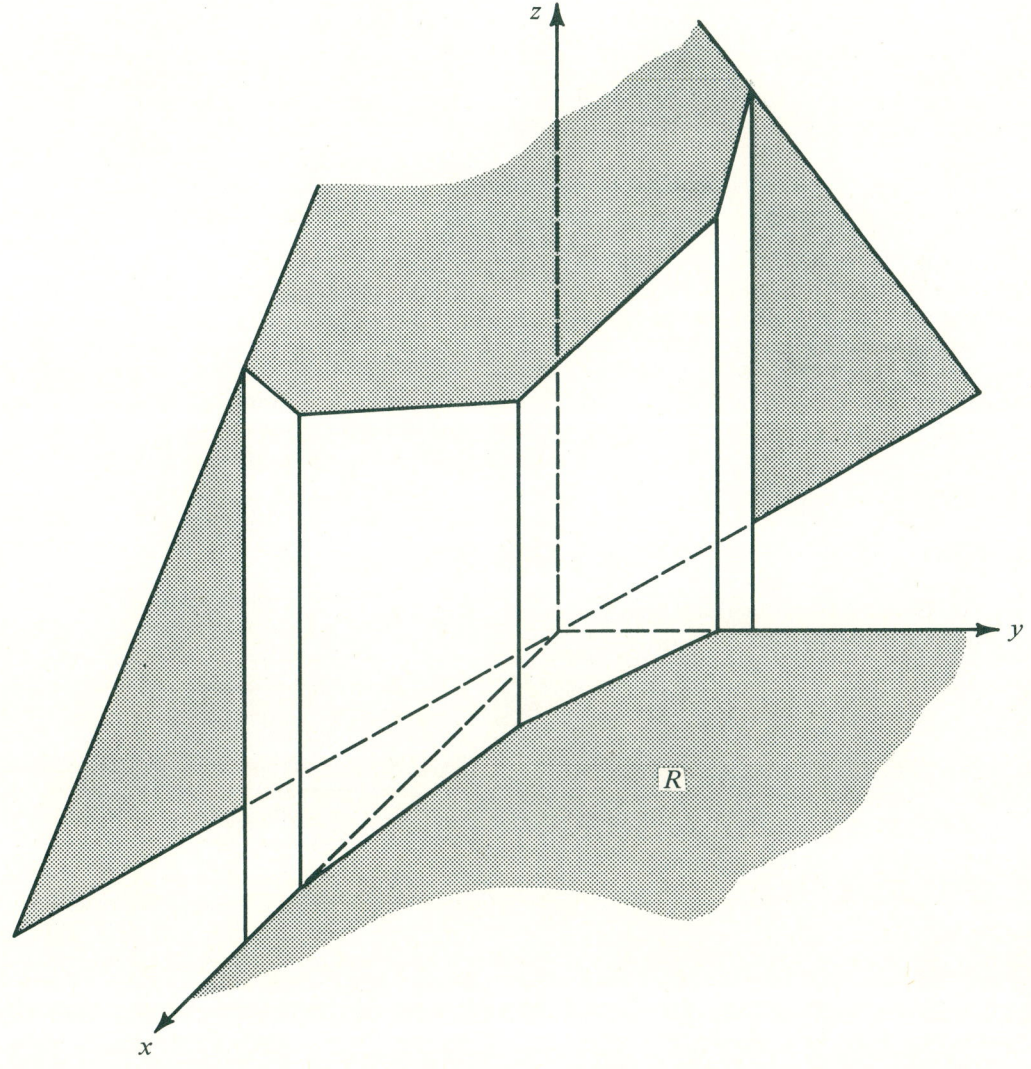

FIGURE 15.17.

353

subject to the following minimum requirements

15 units of vitamin A

18 units of vitamin B

Solution: Let x be the number of grams of wheat, and y the number of grams of nuts. Then, total cost z is

$$z = 0.10x + 0.20y$$

and the constraints are

$$x \geq 0, \quad y \geq 0$$
$$3x + 12y \geq 15 \quad \text{(for vitamin A)}$$

and

$$6x + 8y \geq 18 \quad \text{(for vitamin B)}$$

The geometric interpretation is given in Figure 15.17. The optimum value of z is the minimum cost of 35 cents for 2 grams of wheat and $\frac{3}{4}$ grams of nuts. There is, of course, no maximum cost. The more you use, the higher the cost. This is consistent with the model having an unbounded feasible region.

EXERCISE 15.9

A. FUNDAMENTALS

Draw the feasible region in the x, y-plane of the x, y, z-coordinate system and label all nodes. Classify each region as to *bounded* or *unbounded*.

1. $x \geq 0$
    $y \geq 0$
    $x - 4y \geq 12$
    $4x - y \leq 12$

2. $x \geq 0$
    $4y \geq 3x$
    $y \leq 2x - 5$

3. $y \geq 2 - x$
    $y \geq x - 2$
    $x + y \leq 6$
    $y \leq x + 2$

4. $y \leq 4x$
    $y \geq 2x$
    $3x - 2y + 5 \geq 0$

5. $2y \leq 3x + 7$
    $7x + y \leq 12$
    $x + 5y + 8 \geq 0$

6. $x \leq 4$
    $x + 3y + 2 \geq 0$
    $x + 4y \leq 16$
    $y \leq 2x + 4$

Evaluate the following functions at the nodes from the indicated exercise above. If possible, choose the optimum value.

7. $z = x + y$ (for Problem 1)
8. $z = 10x + 20y$ (for Problem 3)
9. $2x - 2y + z = 2$ (for Problem 3)

B. ESSENTIALS

Graph the feasible regions in the x, y-plane of the x, y, z-coordinate systems. Show the value of z associated with each node. Find any optimum values of z (maximums and minimums) present.

10. $3x + 4y + z = 24$
    $y \leq 3$
    $y \geq -x$
    $x \leq 4$

11. $x + y - z + 3 = 0$
    $x + y \geq 3$
    $5y \geq 6x - 18$
    $8y - 3x \leq 3$

12. $4x + 2y + z = 20$
    $2y + 3x \geq 8$
    $y \leq 4x + 4$
    $x - 4y + 2 \geq 0$

13. $z = 3x - 12y$
    $5y \leq x + 20$
    $x - y \leq 2$
    $4x + 3y \leq 8$
    $4y \leq x + 17$

C. APPLICATIONS

14. Assume you wish to minimize the cost of the granola in Example III if

(a) Wheat costs 15 cents per gram and nuts 15 cents per gram

(b) Wheat costs 10 cents per gram and nuts 1.00 dollar per gram

(c) Wheat costs 15 cents per gram and nuts 35 cents per gram

*15. An airline company has 707 and 747 airplanes with the following statistics:

plane	capacity	cost
707	100 people	$10/kilometer
747	200 people	$15/kilometer

Out of a total of 20 pilots available each week, 15 fly only the 707s and 5 fly both. The FCC says the airline must furnish at least 2000 seats per week between New York and Seattle. Find the number of each type of plane to use to minimize cost.

16. Repeat Problem 15 for at least 1000 seats but no more than 2000, and seating capacities of 200 for the 707s and 350 for the 747s.

17. Repeat Problem 15 for at least 4000 seats with carrying capacities of 200 and 300 people for the 707s and the 747s, respectively.

18. A creamery makes 5 dollars per kilogram profit on butter and 10 dollars per kilogram profit on yogurt. If total production must be less than or equal to 1000 kilograms per day, find the maximum profit subject to the constraint that yogurt production must be less than or equal to half the butter production.

• **19.**

	Calories	Grams of Protein	Milligrams of Calcium	Units of Vitamin A
Milk	0.676/gm	0.033/gm	1.67/gm	1.598/gm
Hamburger	2.88/gm	0.247/gm	0.106/gm	0.353/gm
Lettuce	0.13/gm	0	0.005/mg	3.0/gm

(a) If a typical young male requires at least 2900 calories, 70 milligrams of protein, and 800 milligrams of calcium per day, how much milk and hamburger should he take on a hike to supply these minimum needs with minimum total weight? (HINT: If x is the grams of milk and y is the grams of hamburger, then find the lowest point on the plane $z = x + y$, where z is total weight, above the region in the x, y-plane bounded by this system of inequalities: $x \geq 0$, $y \geq 0$, $0.676x + 2.88y \geq 2900$, $0.033x + 0.247y \geq 0.07$, $1.598x + 0.353y \geq 8$.)

(b) How many grams of hamburger and lettuce should you eat to obtain the above minimums and at least 5000 units of vitamin A, if you wish to minimize calories?

ENDNOTE

1. Cramer's Rule was actually created by Colin Maclaurin about 1729 but with poor notation. Cramer published the theory in 1750 in the context of solving a system of simultaneous equations to find the equation of the conic section passing through five points.

CHAPTER 16

Counting, Probability, and The Binomial Theorem

16.1 COUNTING

The action or process of counting refers to more than the procedure for determining a total amount. In this section we want to find the total amount without actually having to count. If we wanted to know how many possible outcomes there are to tossing a coin, two coins, a die, two dice, etc., we could list them all, and then count them, or we could employ the fundamental principle of counting, and establish the total much more quickly.

Carefully examine the following examples that illustrate the total number of ways things can happen in succession. They will help us understand how the counting principle works.

EXAMPLE I. Tossing a fair coin and one die.

A fair coin is one that is just as likely to land "heads" up as it is "tails" up. There are two possible outcomes (heads or tails) for the toss of the coin and six possible outcomes for the roll of the die. These outcomes are depicted below.

Coin:

tail   head

FIGURE 16.1.

Die:

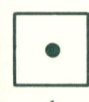   
1   2   3   4   5   6

FIGURE 16.2.

SECTION 16.1   Counting

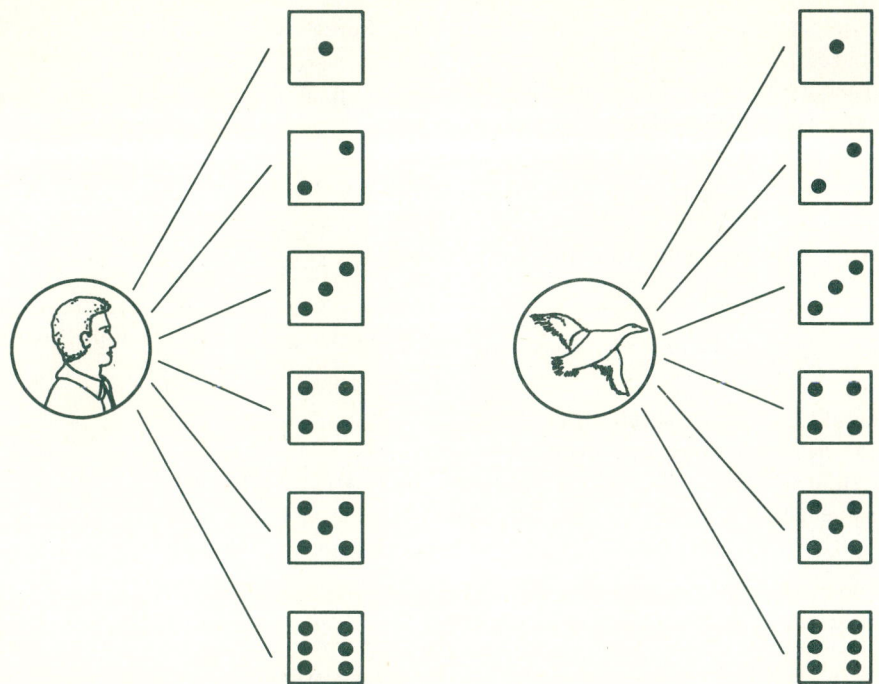

FIGURE 16.3.

The total possible different outcomes of this experiment of tossing a coin and rolling a die are shown in Figure 16.3.

Count the total number of different outcomes (head, 2) is one outcome, (tail, 5) is another, etc. Notice we have pictured two groups of six different outcomes and that

$$2 \times 6 = 12$$

EXAMPLE II. Roll two dice.

All possible outcomes are shown in Figure 16.4. Count them.

FIGURE 16.4.

357

There are six different outcomes for event *A* (rolling one of the die), and six different outcomes for event *B* (rolling the other die). There are thirty six total different outcomes to the rolling of two dice and

$$6 \times 6 = 36$$

These results are consistent with the following theorem.

THEOREM I.
The Fundamental Counting Principle: If event *A* can occur *M* different ways, and event *B* can occur *N* different ways, then the total number of ways event *A* followed by event *B* can occur is

$$M \times N$$

EXERCISE 16.1

A. FUNDAMENTALS
List the possible outcomes of the following experiments.
    1. (a) A toss of a penny.
(b) A toss of a penny and a dime.
(c) A toss of a penny, a dime, and a quarter.

    2. (a) Answering two true–false questions.
(b) Answering three true–false questions.
(c) Answering four true–false questions.

    3. (a) Creating subsets containing two elements from a set of two elements.
(b) Creating subsets containing two elements from a set of three elements.
(c) Creating subsets containing two elements from a set of four elements.

B. ESSENTIALS
Use the fundamental theorem of counting to "count" the number of different possible outcomes.
    4. (a) Subsets of one element from a set of four elements.
(b) Subsets of two elements from a set of four elements.
(c) Subsets of three elements from a set of four elements.
(d) Subsets of four elements from a set of four elements.

    5. Problem 1.

    6. Problem 2.

    7. The number of ways Fred, Jim, and Mike can date Sue and Sally.

    8. The number of ways Mom, Dad, and Junior (legal driving age) can ride in the front seat of a car.

16.2 PERMUTATIONS

Let's begin with a definition of the word permutation.

DEFINITION 16.1
A *permutation* is a change in position or order within a collection.

In this section we will employ the fundamental counting principle to count the total different possible permutations of a group.

EXPERIMENT I. Count the number of different ways three people can be seated in the front seat of a car.

If the people are named *A*, *B*, and *C*, either *A*, *B*, or *C* can drive. We will consider selecting the driver as one event. It can occur three ways. Now the person in the middle can only be selected two different ways because one of the three is behind the wheel. Therefore, our second event, selecting the middle person can occur two ways. This leaves only one person for the passenger side—one way for the third event to occur. The different seatings are depicted in Figure 16.5. Count them.

This can be calculated using the fundamental counting principle as

$$3 \cdot 2 \cdot 1 = 3! = 6$$

SECTION 16.2   Permutations

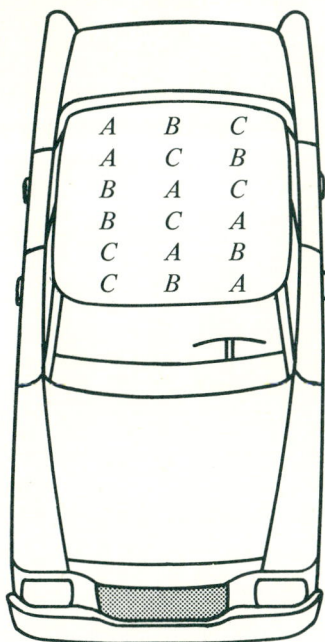

FIGURE 16.5.

since the first event can occur three ways, the second event, two ways, and the last event, one way. (See Appendix II for factorial notation!)

EXPERIMENT II. How many different ways can three people of a family of five ride in the front seat? Assume all can drive.

There are five possible drivers, leaving four for the middle, and three for the passenger side. Therefore, the total number of ways is

$$5 \cdot 4 \cdot 3 = 60$$

These experiments lead us to the next two theorems.

THEOREM II.
The total number of permutations of n objects is
$$n \cdot (n-1) \cdot (n-2) \cdots 1 = n!$$

THEOREM III.
The total number of permutations of n objects taking k at a time is
$$_nP_k = n \cdot (n-1) \cdot (n-2) \cdots [n - (k-1)]$$

EXPERIMENT III.
(a) How many different ways can a group of four people fill their offices of president, vice-president, secretary, and treasurer?

Since there are four people and four offices, this is the permutation of four things taken four at a time.

$$_4P_4 = 4! = 4 \cdot 3 \cdot 2 \cdot 1 = 24$$

(b) How many ways are there if secretary-treasurer is a combined office?

This is the permutation of four things taken three at a time.

$$_4P_3 = 4 \cdot 3 \cdot 2 = 24$$

EXAMPLE I. Evaluate the following.
(a) $_4P_2 = 4 \cdot 3 = 12$
(b) $_9P_3 = 9 \cdot 8 \cdot 7 = 504$
(c) $_{10}P_4 = 10 \cdot \underline{\ ?\ } \cdot \underline{\ ?\ } \cdot \underline{\ ?\ } = 5040$
(d) $_5P_5 = \underline{\ ?\ } \cdot \underline{\ ?\ } \cdot \underline{\ ?\ } \cdot \underline{\ ?\ } \cdot \underline{\ ?\ } = 120$

EXERCISE 16.2

A. FUNDAMENTALS
List the number of total possible permutations of the following. Stop if you get more than twenty five.

1. (a) The letters *ABC* using all three letters each time.

(b) The letters *ABCD* using all four letters each time.

2. (a) The letters *ABC* using only two letters each time.

(b) The letters *ABCD* using only two letters each time.

3. The different "words" that can be made from the letters of the following words if (i) all the letters are used each time, and (ii) if half the letters are used each time. Consider any combination of letters as a word.
(a) LIST   (b) BYTE   (c) AN

B. ESSENTIALS

Use the theorems on permutations to calculate the number of possible permutations for each of the following.

4. Problem 1.

5. Problem 2.

6. Problem 3.

★7. (a) Seating four people at a rectangular table.
(b) Seating six people at a rectangular table.
(c) Seating six people at a round table. (HINT: Since a round table has no "head," one person must be seated first to establish a starting point.)

8. One hundred cans of corn going to one of two different labeling machines.

9. (a) Ten buses going from Seattle to either New York or San Diego.
(b) Ten airplanes flying from Seattle to either New York, Dallas, or San Francisco.

10. A party of eight people arrive at a village inn one day. The food and lodgings are so good that they decide to stay at the inn for as long as they can be seated in a different order for every dinner. How long do they stay?

11. How many numbers can be made using three letters and four numbers?

12. (a) How many different phone numbers are there having two letters and five numbers?
(b) Having seven numbers?

13. Show why at least two people in Yakima, Washington, population 50,000, must have the same initials. Assume everyone has exactly three initials.

14. Solve the following for n algebraically.
(a) $_nP_3 = 60$ (b) $_nP_6 = 720$
(c) $_{(n-1)}P_4 = {_nP_3}$

15. ★(a) Derangements are the permutations (arrangements) where no element remains fixed. List the permutations of the letters ABC, and from that select the possible derangements. ACB is not a derangement, because A is in the same place. BCA is.
(b) List the derangements of the letters AB.
(c) List the derangements of the letters $ABCD$.
(d) Give the number of derangements of n objects in terms of the number of derangements of $n-1$ objects. This is called a recursive formula.
(e) Use the results of (c) to calculate the derangements of six objects.

16.3 COMBINATIONS

The following two examples illustrate the similarities and differences between permutations and combinations.

EXAMPLE I. How many different ways can a group of four people fill the offices of president, vice-president, and secretary-treasurer?

Solution: The election of the president can happen four ways, leaving three people for vice-president, and two for secretary-treasurer. Therefore, there are

$$4 \cdot 3 \cdot 2 = 24$$

different ways of filling the offices.

Notice that it does not make any difference which office we fill first. If we allow four for secretary-treasurer, then there are three left for president, and two for vice-president, or still $4 \cdot 3 \cdot 2 = 24$ different ways.

We call Example I the permutation of four things, taken 3 at a time, and write

$$_4P_3 = 4 \cdot 3 \cdot 2 = 4!$$

or

$$_4P_k = n \cdot (n-1) \cdot (n-2) \cdots (n-(k-1))$$

EXAMPLE II. How many different committees of three people can a group of ten people create?

Solution: In Example I an election of George as president, Sally as vice-president, and Henry as secretary–treasurer is a different outcome than Sally as president, George as vice-president, and Henry as secretary–treasurer. But, in this example, a committee consisting of George, Sally, and Henry is the same as one consisting of Sally, George, and Henry. If the four members were named George, Sally, Henry, and Fred, then the twenty four different possible election results are:

1. (G, S, H) (G, H, S) (H, S, G) (H, G, S)
   (S, G, H) (S, H, G)
2. (G, S, F) (G, F, S) (F, S, G) (F, G, S)
   (S, G, F) (S, F, G)
3. (G, H, F) (G, F, H) (H, F, G) (H, G, F)
   (F, G, H) (F, H, G)
4. (S, H, F) (S, F, H) (H, F, S) (H, S, F)
   (F, S, H) (F, H, S)

Note that in these twenty four possible results, each combination of people occurs six times—each in a different order. In order to solve Example II, we will only count each of these six permutations as one committee (since each involves the same combination of people). We see that there are only

$$\frac{24}{6} = 4$$

different combinations of people. We now know how many different combinations there are of four things taken three at a time. By noticing that $6 = 3!$, we have

$$_4C_3 = \frac{24}{6} = \frac{_4P_3}{6} = \frac{_4P_3}{3!}$$

In general,

$$_nC_k = \frac{_nP_k}{k!}$$

where the division by $k!$ eliminates the repetitions of each combination taken in a different order.

EXAMPLE III. Evaluate the following.
(a) $_5P_3 = \underline{} \cdot \underline{} \cdot \underline{} = 60$
(b) $_5C_3 = \frac{_5P_3}{3!} = \frac{?}{1 \cdot 2 \cdot 3} = 10$
(c) $_9P_3 = \underline{} \cdot \underline{} \cdot \underline{} = 504$
(d) $_9C_3 = \frac{_9P_3}{3!} = \frac{?}{6} = \underline{}$
(e) $_9C_6 = \underline{} = \frac{?}{?} = \underline{}$

(f) $_5C_2 = \underline{} = \frac{?}{?} = \underline{}$

The fact that (b) and (f) are both ten and (d) and (e) are both eighty-four is important. Can you see how the pairs are related? Consider five minus two for (b) and (f), and nine minus three for (d) and (e). These results are summarized in Theorem V. Theorem IV gives the more common formula for calculating combinations. It is then used to prove Theorem V.

THEOREM IV.
The total number of combinations of n things taken r at a time is given by

$$_nC_r = \frac{_nP_r}{r!} = \frac{n!}{r!(n-r)!}, \quad n \geq r$$

The derivation of the second formula from the first is accomplished by multiplying the top and bottom of

$$_nC_r = \frac{_nP_r}{r!} = \frac{n(n-1)(n-2)\cdots(n-(r-1))}{r!}$$

by $(n-r)!$.

THEOREM V.
The number of combinations of n things taken r at a time is the same as the number of combinations of n things taken $(n-r)$ at a time. In symbols,

$$_nC_r = {_nC_{(n-r)}}, \quad n \geq r$$

EXAMPLE IV. How many different poker hands are possible?

Solution: There are a total of fifty-two playing cards in a regular poker deck. A hand consists of any five cards. Therefore, the total number of different combinations of the fifty-two cards taken five at a time is the number of different poker hands.

$$_{52}C_5 = \frac{52!}{5!(52-5)!} = \frac{48 \cdot 49 \cdot 50 \cdot 51 \cdot 52}{1 \cdot 2 \cdot 3 \cdot 4 \cdot 5}$$

$$= 2 \cdot 49 \cdot 10 \cdot 51 \cdot 52 = 2{,}598{,}960$$

There are over two-and-one-half million different poker hands.

EXERCISE 16.3

A. FUNDAMENTALS

Compute the following combinations.

1. (a) $_8C_6$ (b) $_8C_2$
2. (a) $_{100}C_{97}$ (b) $_{100}C_0$
3. (a) $_7C_3$ (b) $_7C_7$
4. (a) $_{200}C_{198}$ (b) $_{200}C_2$

Classify each of the following either as a permutation or a combination.

5. The ways ten people can get into an elevator that holds only five people.

6. The ways five people can sit in the two front seats of a sports car.

7. The possible number of "words" using your seven letters in a Scrabble game.

8. The number of ways six people can line up to buy theater tickets.

9. The number of different ways three men and three women can use a men's and a women's restroom.

10. (a) List the possible permutations of the letters ABC taken two at a time ($_3P_2$).
(b) List the possible combinations of the letters ABC taken two at a time ($_3C_2$).

B. ESSENTIALS

11. Find the actual numbers involved in Problems 5-9 above.

Use an ordinary deck of fifty-two playing cards for the following problems. There are four suits, two red (hearts and diamonds) and two black (clubs and spades), each having the thirteen cards 2, 3, ..., 10, Jack, Queen, King, and Ace.

12. (a) How many different ways can you get a pair of Kings?
(b) Three Aces?
(c) The full house consisting of three Aces and two Kings?

13. How many different five-card heart flushes are possible (all hearts)?

14. How many possible straight flushes are there (five consecutive cards of one suit)?

15. How many different "words" can be formed from the following words?
(a) MISSILE   (b) ROLLER
(c) MISSISSIPPI

16. Eight people are waiting for three bank tellers. How many ways can they be served? How many possible lines are there, if there must be at least one person at each teller? How many lines are there, if there must not be more than four people at any window? How many lines are there, if there must be at least one, and no more than four, at any window?

17. Solve the following for n, algebraically.

$$_nC_3 = {_nC_1}$$

18. Prove that

$$_nC_k + {_nC_{k+1}} = {_{n+1}C_{k+1}} \quad \text{for all } n \in \text{naturals}$$

19. Prove Theorem V.

16.4 PROBABILITY

In this section we will be considering two common working definitions of probability. Probability is generally defined as the chance an event has of occurring.[1] Consider the following examples, which lead to our first definition.

EXAMPLE I. The probability of tossing a head with one toss of a *fair* coin is $\frac{1}{2}$.

EXAMPLE II. The probability of rolling a three on a *fair* die is $\frac{1}{6}$.

EXAMPLE III. The probability of drawing the Ace of spades from a *well-shuffled* ordinary deck of playing cards is $\frac{1}{52}$.

Did you detect an attempt to emphasize that in each example every outcome had an equal chance of occurring? We shall call the set of all possible outcomes the *sample space, SS.*

DEFINITION 16.2.
The *probability* of one event occurring out of a sample space of N equally likely events is

$$\frac{1}{N}$$

Definition 16.3 is a generalization of the one above. Consider these examples.

EXAMPLE IV. A red die and a green die are rolled. The probability that the numbers rolled will add up to four is $\frac{3}{36}$, or $\frac{1}{12}$.

The sample space in this example consists of the set of all possible combinations of the numbers on the two dice, which is found by the fundamental counting principle to be

$$6 \cdot 6 = 36$$

The total number of different ways the sum can equal four is shown in Figure 16.6.

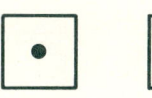

FIGURE 16.6.

EXAMPLE V. The probability of drawing an Ace from a regular deck of cards is $\frac{4}{52}$ or $\frac{1}{13}$, since there are four Aces in the deck of fifty-two cards.

EXAMPLE VI. The probability of drawing a heart from a regular deck of cards is $\frac{13}{52}$, or $\frac{1}{4}$, since there are thirteen hearts in the deck.

If we are interested in the probability of event A occurring, we will need to count the number of elements in set A, where set A contains all of the different ways A can happen. We will call the elements of set A *successes*. We will call the number of elements in the set of all possible outcomes the size of the sample space, SS. We will write the number of elements in set A as $n(A)$.

DEFINITION 16.3.
The *probability* of an event A is the ratio of the number of possible successes to the total number of elements in the sample space. We will write

$$P(A) = \frac{n(\text{Successes})}{n(\text{Sample Space})} = \frac{n(A)}{n(SS)}$$

EXAMPLE VII. What is the probability that the first three letters of your car's three letter–three number license plate will be ABC?

Solution: The number of total possible outcomes, the size of the sample space, is found using the fundamental counting principle as follows:

$$n(SS) = 26 \cdot 26 \cdot 26 \cdot 10 \cdot 10 \cdot 10 = 17{,}576{,}000$$

The number of successes is

$$n(A) = 1 \cdot 1 \cdot 1 \cdot 10 \cdot 10 \cdot 10 = 1000$$

Therefore,

$$P(A) = \frac{n(A)}{n(SS)} = \frac{?}{?} = 0.0006$$

This means the possibility of ABC is about

$$\frac{6}{100{,}000} \quad \text{or} \quad \frac{1}{16{,}667}$$

In all of these examples of calculating the probability of an outcome all possible outcomes were equally probable. Such events are called *unbiased*. In the examples cited, we calculated the probabilities for ideal unbiased experiments that could not be achieved in real life. There is no such thing as a fair coin, die, or deck of cards. Several tosses, rolls, and draws are probably not really random, since the possible outcomes are not really equally probable. The aspects of probability that a mathematician must strive to attain in any probability experiment, are random and unbiased outcomes. Faced with a specific coin, we would refuse to declare that the probability of tossing a head was really $\frac{1}{2}$. We would experiment extensively with the coin to see whether or not the ratio of heads to trials approached $\frac{1}{2}$. Even if the ratio never became exactly equal to $\frac{1}{2}$, we could call the coin fair if it were close to $\frac{1}{2}$.

Now let's examine what probability really means. Suppose you were tossing a fair coin and you tossed twenty heads in a row. Is this possible? Yes, but is it likely? What is the probability of your next toss being another head? Is it small since you have already tossed twenty heads in a row? No, the probability of a head on the twenty-first toss is still $\frac{1}{2}$. This discussion points out an underlying characteristic of probability. Simply because the probability of flipping a head is $\frac{1}{2}$, we do not know that exactly ten out of twenty tosses will be heads. We do know that if we continued tossing a fair coin long enough, the ratio of heads to tosses would *approach* 1 to 2.

It is important to understand how the fact that the probability of tossing a head is $\frac{1}{2}$ means that you are unlikely to toss several heads in a row, yet the probability of a head on the next toss is always $\frac{1}{2}$. Consider four tosses. The sample space is the set of all possible outcomes, which, since each event can occur two ways, contains

$$2 \cdot 2 \cdot 2 \cdot 2 = 16 \text{ elements}$$

Say we wanted exactly three heads in a row. The successes are depicted below as ordered quadruples. The first position is the outcome of the first toss, etc. There are two successes,

$$(H, H, H, T)$$

and

$$(T, H, H, H)$$

Hence, the probability of three heads in a row on three tosses is

$$P(3 \text{ Heads}) = \frac{n(\text{Successes})}{n(SS)} = \frac{2}{16} = \frac{1}{8}$$

Therefore, you have only about a 12% chance of tossing three heads in a row, even though there is a 50% chance of a head on any one toss.

Notice that the probability of a head and the probability of a tail add up to 1. The following definitions include four important properties of probabilities.

DEFINITION 16.4.
If an event is certain to occur, its probability is one.

DEFINITION 16.5.
If an event is not possible, its probability is zero.

DEFINITION 16.6.
The sum of the probabilities of all possible outcomes of an event equals one.

DEFINITION 16.7.
The probability of an event A occurring, $P(A)$, and event A not occurring, $P(A')$, must add up to one.

Some of the consequences of the above definitions are

$$P(SS) = 1$$
$$P(SS') = 0$$
$$P(A') = 1 - P(A)$$

Have you ever heard of the term "odds"? Someone might say the *odds for* a horse named Bilbo to win a race are 5 to 3, or that the *odds against* a heart attack are 3 to 2. In terms of probabilities, these simply mean

$$P(\text{Bilbo}) = \frac{5}{8} \quad \text{and} \quad P(\text{Bilbo}') = \frac{3}{8}$$

and

$$P(H') = \frac{3}{5} \quad \text{and} \quad P(H) = \frac{2}{5}$$

where $5 + 3 = 8$ and $3 + 2 = 5$ give the denominators.

EXERCISE 16.4

A. FUNDAMENTALS
Calculate the following probabilities.

1. (a) Two heads in a row, in four tosses of a fair coin.
(b) Not tossing two heads in a row, in four tosses of a fair coin.

2. (a) An Ace on one draw from an ordinary deck of playing cards.
(b) A spade. (c) An Ace of spades.

3. (a) No rain, if the chance of measurable precipitation is 84%.
(b) Rain or no rain, if there is a 96% chance of rain.
(c) Rain and no rain, if there is a 96% chance of rain.

4. (a) Winning a race, if the odds for you are 8 to 5.
(b) Losing the race in 4(a).

5. (a) Losing a race, if the odds against you are 4 to 1.
(b) Winning the race in 5(a).

6. (a) Rolling an even number on one roll of one die.
(b) Rolling an odd number.

7. (a) Rolling a sum that is a prime number on one roll of two dice.
(b) Rolling a sum that is a composite number.

B. ESSENTIALS
Calculate the following probabilities.

8. (a) Drawing a heart flush.
(b) Drawing one card to four hearts, completing a heart flush.
(c) Drawing two cards to three hearts, completing a heart flush.

9. (a) Drawing one card to the hand 4, 5, 6, 7, completing the straight.
(b) Drawing one card to the hand 4, 5, 7, 8, completing the straight.

10. (a) Rolling a six on one roll of a fair die.
(b) Rolling a sum of twelve (box cars) on a roll of two dice.

11. (a) Drawing a heart on one draw from a regular deck of cards, $P(H)$.
(b) Drawing an Ace, $P(A)$.
(c) Drawing an Ace and a heart.
(d) Drawing an Ace or a heart.

12. (a) Rolling a five on one roll of a fair die, $P(5)$.
(b) Tossing a head on one toss of a fair coin, $P(H)$.
(c) Getting a five and a head on the throw of one coin and one die.
(d) Getting a five or a head on the throw of one coin and one die.

13. Fred and Suzy are matching pennies. If they match (both heads or both tails) Fred pays Suzy $1.00. If they do not match, Suzy pays Fred $1.00. What is the probability that Suzy will have won $3.00 after four tosses? [HINT: One combination winning Suzy exactly $2.00 is (M, M, M, M'), where M means they match; (M, M', M, M') results in their breaking even.]

14. (a) In Problem 13, let M equal the number of matches and M' equal the number of mismatches in a sequence of four tosses. Then, in order for Suzy to win $2.00, $M + M' =$ __?__ and $M - M' =$ __?__ .

Solve for M and M'. Then, the one mismatch can occur on the first, second, third, or fourth toss. Therefore, $P(M') =$ __?__ and $P(M) = 1 - P(M') =$ __?__ .
(b) Calculate the probability that Suzy will be $4.00 ahead after eight tosses.
(c) Calculate the probability that Suzy will be $1.00 ahead after eight tosses.

15. Six people approach three bank tellers. Find the probability (a) that there is at least one person at each window, (b) that there are four people at window A, and (c) that there are no people at window A. [HINT: One of the sequences representing the sample space of six things distributed among the three different windows is (2, 4, 0), which indicates no one at window C.]

16. (a) Calculate the probability that no man dances with his own wife using derangements (Problem 15, Exercise 16.2) if there are eight couples. Derangements are arrangements where no element remains fixed.
(b) What is the probability that no cards will match when the cards of one deck are turned over one at a time with the cards of another deck.

17. The Mendelian theory of genetics is an example of applied probability. In his experiments with sweet peas, Mendel found that the color of the flowers of an offspring is governed by one of two possible forms (alleles) of the genes of the parents. We call the dominant gene A, and the recessive one a. The genotypes AA represent a dominant specimen, Aa represents a heterozygous specimen, and aa represents a recessive specimen. In peas, AA has red flowers, Aa has red flowers, and aa has white flowers. A parent contributes one gene, either A or a, to its offspring. A heterozygous parent can contribute either A or a with equal probability, while AA must contribute A.
(a) Create the sample space of the color of the peas resulting from two heterozygous peas.
(b) Peas can have either wrinkled or round seeds, round being dominant. Let R represent round and r represent wrinkled. What is the probability of plants that are heterozygous in both traits producing a white, wrinkled seeded pea?

16.5 THEOREMS ON PROBABILITY

Examine the following examples of the probability of successive events. This will help you understand *independent events*.

EXAMPLE I. Three consecutive tosses of a fair coin come up heads, but the probability of a head on the fourth toss is still $\frac{1}{2}$.

EXAMPLE II. You have rolled a seven as the sum of the numbers on two dice ten times in a row, but the probability of a seven on the next roll is still $\frac{6}{36} = \frac{1}{6}$.

EXAMPLE III. You draw two cards from a regular deck of playing cards. The probability of a heart on the first draw is $\frac{13}{52} = \frac{1}{4}$, but the probability of a heart on the second draw is dependent on the suit of the first card. If the first card was not a heart,

$$P(\text{second card being a heart}) = \frac{13}{52}$$

But, if the first card was a heart,

$$P(\text{second card being a heart}) = \frac{12}{52}$$

> **DEFINITION 16.8.**
> Two events are independent if the probability of one does not involve the outcome of the other. We write
> $$P(A|B) = P(A)$$
> meaning the probability of A, given that B has occurred, is simply the probability of A for independent events A and B.

Consider the following examples of determining probabilities of independent events first. Let's examine the probability of A and B, $P(A \cap B)$, to see how it is related to $P(A)$ and $P(B)$.

EXAMPLE IV. You toss a coin twice. The sample space of all possible outcomes of the two events is

(H, H), (H, T), (T, H), (T, T)

The probabilities of each outcome are

$$P(H) = \tfrac{1}{2}$$

and

$$P(T) = \tfrac{1}{2}$$

The probability of a head on the first toss and a tail on the second is $\tfrac{1}{4}$, since the second pair above represents the only success. How are the probabilities $P(H) = \tfrac{1}{2}$ and $P(T) = \tfrac{1}{2}$ related to $P(H \cap T) = \tfrac{1}{4}$?

EXAMPLE V. Two dice are rolled. (This is the same sample space as one die being rolled twice.) What is the probability of snake eyes (two 1s)?

We have

$$P(1 \text{ on the first die}) = \tfrac{1}{6}$$

and

$$P(1 \text{ on the second die}) = \tfrac{1}{6}$$

From the sample space we see that

$$P(\text{snake eyes}) = \tfrac{1}{36}$$

Then

$$P(1 \cap 1) = \tfrac{1}{6} \cdot \tfrac{1}{6} = \tfrac{1}{36}$$

In general, we have the following theorem.

THEOREM VI.
If $P(A|B) = P(A)$, then

$$P(A \cap B) = P(A) \cdot P(B)$$

This theorem is a direct consequence of the fundamental counting principle. If $P(A) = \tfrac{3}{5}$ and $P(B) = \tfrac{2}{3}$, then the total number of ways A followed by B can occur is

$$3 \cdot 2 = 6$$

and the total number of possible outcomes for possibilities associated with A, five, followed by those associated with B, three, is

$$5 \cdot 3 = 15$$

The sample space for A has five possible outcomes, and there are three in B's sample space. Then, the probability of A followed by B is simply the total number of successes, six, divided by the total number of possible outcomes, which is the size of the new sample space, fifteen. That is why

$$P(A \cap B) = P(A) \cdot P(B) = \frac{3}{5} \cdot \frac{2}{3} = \frac{6}{15}$$

Therefore, the probability of "A and B" also can be calculated by finding the number of elements in $A \cap B$. Then

$$P(A \cap B) = \frac{n(A \cap B)}{n(SS)}$$

where the sample space is composed of all possible outcomes in the sample space for A followed by the events from the sample space for B for independent events A and B.

EXAMPLE VI. A coin is tossed and a die is rolled. The sample space is in Section 16.1. What is the probability of a head and a six?

We see that

$$n(H \cap 6) = 2$$

and

$$n(SS) = 12$$

Therefore,

$$P(H \cap 6) = \frac{2}{12} = \frac{1}{6}$$

Also,

$$P(H) = \frac{1}{2} \quad \text{and} \quad P(6) = \frac{1}{6} \Rightarrow$$

$$P(H \cap 6) = P(H) \cdot P(6) = \frac{1}{2} \cdot \frac{1}{6} = \frac{1}{12}$$

What about the probability of a head or a six, $P(H \cup 6)$, in Example VI? From the sample space we see there are seven outcomes that have a head or a six. That is,

$$n(H \cup 6) = 7$$

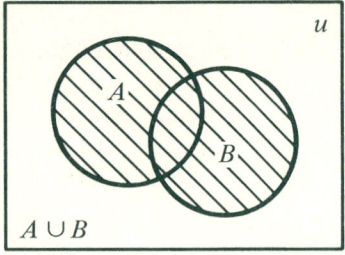 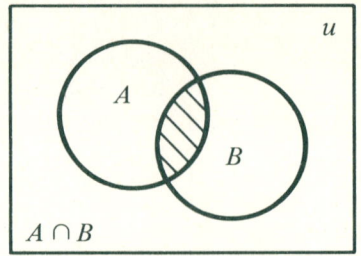 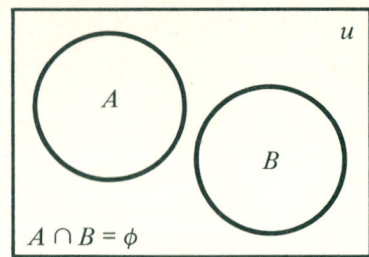

FIGURE 16.7.

Since there are still twelve possible outcomes,

$$P(H \cup 6) = \frac{7}{12}$$

Notice that

$$P(H) + P(6) = \frac{1}{2} + \frac{1}{6} = \frac{7}{12}$$

This is one more than above because

$$P(H) = \frac{1}{2}$$

counts the event (H, 6) as a success, as does

$$P(6) = \frac{1}{6}$$

This one success, (H, 6), was counted twice. In general, we have the following theorem.

THEOREM VII.
$$P(A \cup B) = P(A) + P(B) - P(A \cap B)$$

If two results cannot occur at the same time, for example, flipping a head and a tail on one toss of one fair coin, then we say that the events are **mutually exclusive**. For mutually exclusive events, we have the following corollary to Theorem VII.

COROLLARY I.
If A and B are mutually exclusive events, then
$$P(A \cap B) = 0$$
and
$$P(A \cup B) = P(A) + P(B)$$

The proof of Theorem VII and its corollary are direct results of a theorem from set theory whose proof is beyond the scope of this text, but it can be appreciated by examining the associated Venn diagrams. See Figure 16.7.

If $n(A)$ means the number of elements in set A, then

$$n(A \cup B) = n(A) + n(B) - n(A \cap B)$$

and if $A \cap B = \emptyset$, then

$$n(A \cup B) = n(A) + n(B)$$

When counting the number of elements in $A \cup B$, if $A \cap B \neq \emptyset$, then the elements in $A \cap B$ are included in the count of the elements of A and in the count of the elements of B.

When calculating probabilities we are often able to obtain additional information about the sample space. If such information results in a new value for the probability, we say that the probability was conditioned by the new information. Such **conditional probabilities** require careful consideration of the new, restricted sample space.

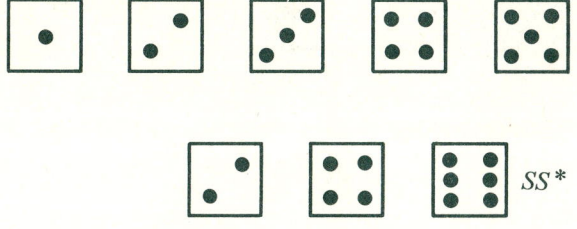

FIGURE 16.8.

EXAMPLE VII. If your five-card poker hand contains four spades and one heart, what is the probability of getting a flush (five cards of one suit) by discarding the heart and drawing one replacement card? The discard is not returned to the deck.

The number of spades not in your hand is

$$13 - 4 = 9$$

The number of cards not in your hand is

$$52 - 5 = 47$$

Therefore, the probability of a fifth spade is

$$\frac{n(\text{spades}^*)}{n(SS^*)} = \frac{9}{47} = 0.19$$

where SS^* is the new sample space. (The asterisk designates a revised set.)

EXAMPLE VIII. Calculate the following conditional probabilities. Consider a regular deck of playing cards and a die.
(a) Drawing an Ace given that you have drawn one card, an Ace of spades.

$$P(A|\text{Ace of spades}) = \frac{?}{?} = 0.06$$

(b) Drawing a second heart given that you have drawn the 3 of hearts.

$$P(H|3 \text{ of hearts}) = \frac{?}{51} = 0.24$$

(c) Drawing an Ace of spades if you are told that the card drawn was black.

$$P(\text{Ace of spades}|\text{card was black}) = \frac{?}{?} = 0.04$$

(d) The probability of rolling a six on one die if you are told that you have rolled an even number.

$$P(6|\text{It is even}) = \frac{?}{?} = 0.33$$

Conditional probability results from an alteration of the sample space. Consider the drawings for (d) above in Figure 16.8.

For (c) above, we have

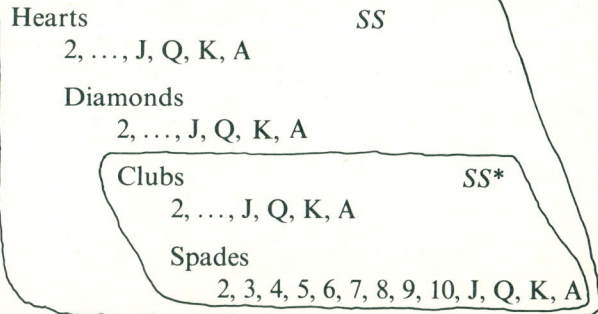

where SS^* contains only twenty-six cards.
Notice that for (d) we have

$$P(6 \cap \text{even}) = \frac{1}{6}$$

and

$$P(\text{even}) = \frac{3}{6}$$

where

$$P(6|\text{even}) = \frac{1}{3} = \frac{1/6}{3/6} = \frac{P(6 \cap \text{even})}{P(\text{even})}$$

For (c) we have

$$P(\text{Ace of spades} \cap \text{black}) = \frac{1}{52}$$

and

$$P(\text{black}) = \frac{26}{52}$$

where

$$P(\text{Ace of spades} | \text{black}) = \frac{1}{26} = \frac{1/52}{26/52}$$

$$= \frac{P(\text{Ace of spades} \cap \text{black})}{P(\text{black})}$$

THEOREM VIII.
The probability of A given B is

$$P(A|B) = \frac{P(A \cap B)}{P(B)}$$

This theorem can, of course, be used to find the probability of A and B, since

$$P(A \cap B) = P(A|B) \cdot P(B)$$

If A and B happen to be independent events, then

$$P(A|B) = P(A)$$

and

$$P(A \cap B) = P(A) \cdot P(B)$$

which is consistent with our previous results.

EXERCISE 16.5

A. FUNDAMENTALS

Classify the following events as independent or dependent.

1. (a) Tossing one coin twice.
(b) Tossing five coins, then tossing the heads again.

2. (a) Drawing a card, replacing it and drawing a second card.
(b) Drawing two cards at once.
(c) Dealing one card each to two people.

Decide whether or not the following are mutually exclusive outcomes ($P(A \cap B) = 0$).

3. (a) Event A is rolling a prime number on one die, and event B is rolling an even number.
(b) Event A is rolling a prime number, and event B is rolling an odd number.

4. (a) Event A is being a college graduate, and event B is earning more than $10,000.00 per year.
(b) Event A is being a college graduate, and event B is being married.

Use one of the words *independent*, *dependent*, or *mutually exclusive* to classify the following.

5. (a) Drawing an Ace and a spade on one draw from a regular deck of cards.
(b) Drawing an Ace and a Queen on one draw of one card.

6. (a) Being a high school graduate and being a college graduate.
(b) Not being a high school graduate and being a college graduate.

7. (a) Having blue eyes given that neither parent has blue eyes, if the characteristic of blue eyes is recessive.
(b) Having a blue-eyed child result from a blue-eyed father and a brown-eyed mother, if brown eyes are dominant.
(c) Having brown eyes when both parents have blue eyes.

Calculate the following probabilities.

8. $P(A) = \frac{1}{5}$, $P(B) = \frac{1}{3}$, and $P(A \cap B) = \frac{1}{10}$. Find (a) $P(A|B)$ (b) $P(B|A)$

9. A and B are mutually exclusive events with
(a) $P(A) = \frac{3}{10}$ and $P(B) = \frac{5}{10}$. Find $P(A \cap B)$ and $P(A \cup B)$.
(b) $P(A) = \frac{2}{5}$ and $P(B) = \frac{3}{5}$. Find $P(A \cap B)$ and $P(A \cup B)$.
(c) $P(A) = \frac{1}{3}$ and $P(A \cup B) = 1$. Find $P(B)$.
(d) $P(A) = \frac{1}{3}$ and $P(A \cup B) = \frac{3}{4}$. Find $P(B)$.
(e) $P(B) = \frac{1}{2}$ and $P(A \cup B) = \frac{3}{5}$. Find $P(A \cap B)$.

B. ESSENTIALS

10. Find the probability of the following.
(a) Tossing two heads in 1(b), if the first toss yielded four heads.
(b) Drawing an Ace for a second card, if your first card was an Ace.

11. Find the probabilities for Problem 5.

12. Find the probabilities for Problem 7, if bb is blue-eyed and bB and BB are brown-eyed, with b representing the recessive blue-eyed gene and B the dominant brown-eyed gene. A child receives two genes, like bB, one from each parent.

13. (a) Calculate the probability of drawing a Queen, if you draw from face cards only.
(b) Calculate the probability of drawing a heart, if you draw from only red cards.

14. Calculate the probabilities of being dealt the following poker hands in five cards from a well-shuffled deck of fifty-two regular playing cards.
(a) Two Kings (b) Three Kings
(c) Two Kings and three Queens (d) One pair
(e) Two pairs (f) Three of a kind
(g) A straight (h) A flush
(i) A straight flush (j) A full house
(k) A royal flush

★15. If $A_1 \cup A_2 \cup \cdots \cup A_n = SS$, where $A_i \cap A_j = \emptyset$ for $i = 1, \ldots, n$, $j = 1, \ldots, n$, and $i \neq j$, we say that the events A_i are mutually exclusive and that they partition SS. If B is some subset of SS, then Bayes' Theorem states

$$P(A_i|B) = \frac{P(A_i) \cdot P(B|A_i)}{\sum_{k=1}^{n} P(A_k)P(B|A_k)}, \quad 1 \leq i \leq n$$

Use Bayes' Theorem to calculate the following probabilities.
(a) A red urn contains four black marbles and two white ones, and a blue urn contains one black marble and five white ones. An urn is selected at random (say by the toss of a coin), and one marble extracted. If the marble is black, what is the probability that it came from the red urn. (HINT: $SS = \{$Draw a black marble from the red urn$\} \cup \{$Draw a black marble from the blue urn$\} = A_1 \cup A_2$.)

(b) Add a third urn containing two white and six black marbles to the collection in (a). If a black marble is drawn from an urn selected randomly, what is the probability that it was the third urn?

16. (a) The probabilities that two machines A and B will make a defective part are 0.03 and 0.10, respectively. If each machine does 50% of the work, what is the probability that a particular defective product came from A?
(b) If machine A contributes 30% of the production and B contributes 70%, what is the probability that a particular defective part came from machine B?

16.6 FINITE PROBABILITY DISTRIBUTIONS

This section is simply an introduction to probability distribution, a concept with many and varied applications in statistics. Since we will be concerned only with finite sample spaces, we will restrict our examples to binomial distributions.

Binomial distributions arise only under very restrictive conditions. One is that the possible outcomes of each event must be binary (one out of two possibilities), such as heads or tails, true or false, success or failure, and innoculated or not innoculated. Another restriction is based on the idea of generating outcomes as several trials, for example, several tosses of a coin. For each successive trial, the probabilities must remain fixed. These trials must be independent. The outcome of one is not affected by any previous outcome. Consider the following example.

EXAMPLE 1. Plot the probability distribution of heads in four tosses of a fair coin.

Solution: The $2^4 = 16$ possible outcomes are all listed below.

HHHH	HTHH	THHH	TTHH
HHHT	HTHT	THHT	TTHT
HHTH	HTTH	THTH	TTTH
HHTT	HTTT	THTT	TTTT

The total number for each of the following distributions of heads is given below. Compare numbers with the list.

$$4 \text{ heads: } {}_4C_4 = 1$$
$$3 \text{ heads: } {}_4C_3 = 4$$
$$2 \text{ heads: } {}_4C_2 = 6$$
$$1 \text{ head: } {}_4C_1 = 4$$
$$0 \text{ heads: } {}_4C_0 = 1$$

The probability of four heads is

$$P(H \cap H \cap H \cap H) = \frac{1}{2} \cdot \frac{1}{2} \cdot \frac{1}{2} \cdot \frac{1}{2} = \frac{1}{16}$$

This agrees with our list above, where only one out of sixteen outcomes contained four heads.

The probability of three heads is

$$P(H \cap H \cap H \cap T) + P(H \cap H \cap T \cap H)$$
$$+ P(H \cap T \cap H \cap H)$$
$$+ P(T \cap H \cap H \cap H)$$
$$= \frac{1}{2} \cdot \frac{1}{2} \cdot \frac{1}{2} \cdot \frac{1}{2} + \frac{1}{2} \cdot \frac{1}{2} \cdot \frac{1}{2} \cdot \frac{1}{2} + \frac{1}{2} \cdot \frac{1}{2} \cdot \frac{1}{2} \cdot \frac{1}{2}$$
$$+ \frac{1}{2} \cdot \frac{1}{2} \cdot \frac{1}{2} \cdot \frac{1}{2}$$
$$= 4\left(\frac{1}{2}\right)^3 \left(\frac{1}{2}\right)^1 = \frac{4}{16}$$

where $\frac{1}{2}$ is the probability of one tail occurring on each of the four tosses.

Completing the calculations of the probabilities we get the following:

$$P(4 \text{ heads}) = \frac{1}{16} = \frac{{}_4C_4}{16} = {}_4C_4 \left(\frac{1}{2}\right)^4 \left(\frac{1}{2}\right)^0$$

$$P(3 \text{ heads}) = \frac{4}{16} = \frac{{}_4C_3}{16} = {}_4C_3 \left(\frac{1}{2}\right)^3 \left(\frac{1}{2}\right)^1$$

$$P(2 \text{ heads}) = \frac{6}{16} = \frac{{}_4C_2}{16} = {}_4C_2 \left(\frac{1}{2}\right)^2 \left(\frac{1}{2}\right)^2$$

$$P(1 \text{ head}) = \frac{4}{16} = \frac{{}_4C_1}{16} = {}_4C_1 \left(\frac{1}{2}\right)^1 \left(\frac{1}{2}\right)^3$$

$$P(\text{no heads}) = \frac{1}{16} = \frac{{}_4C_0}{16} = {}_4C_0 \left(\frac{1}{2}\right)^0 \left(\frac{1}{2}\right)^4$$

The area of each rectangle is the probability of that outcome if we let the bases measure one unit.

Answer the following questions about the probable outcomes of tossing a coin four times. What is the probability of

(a) Tossing three heads? _____
(b) Not tossing three heads? _____
(c) Tossing at least two heads? _____
(d) Tossing more than three heads? _____
(e) Tossing no more than three heads? _____

Your answers should be 0.25, 0.75, 0.6875, 0.0625, and 0.9375.

We are now ready to consider another binomial distribution.

EXAMPLE II. Suppose you are playing a dice game with one die, where you win on a four and lose otherwise. Create the binomial distribution for three rolls.

Solution: On three rolls we could get any one of the following four results.

1. Three 4s:
$$P(4 \cap 4 \cap 4) = P(4) \cdot P(4) \cdot P(4)$$
$$= \tfrac{1}{6} \cdot \tfrac{1}{6} \cdot \tfrac{1}{6} = \left(\tfrac{1}{6}\right)^3$$

2. Two 4s:
$$P(4 \cap 4 \cap {\sim}4) + P(4 \cap {\sim}4 \cap 4)$$
$$+ P({\sim}4 \cap 4 \cap 4)$$
$$= \tfrac{1}{6} \cdot \tfrac{1}{6} \cdot \tfrac{5}{6} + \tfrac{1}{6} \cdot \tfrac{5}{6} \cdot \tfrac{1}{6} + \tfrac{5}{6} \cdot \tfrac{1}{6} \cdot \tfrac{1}{6} = 3\left(\tfrac{1}{6}\right)^2 \left(\tfrac{5}{6}\right)^1$$

3. One 4:
$$P(4 \cap {\sim}4 \cap {\sim}4) + P({\sim}4 \cap 4 \cap {\sim}4)$$
$$+ P({\sim}4 \cap {\sim}4 \cap 4)$$
$$= \tfrac{1}{6} \cdot \tfrac{5}{6} \cdot \tfrac{5}{6} + \tfrac{5}{6} \cdot \tfrac{1}{6} \cdot \tfrac{5}{6} + \tfrac{5}{6} \cdot \tfrac{5}{6} \cdot \tfrac{1}{6} = 3\left(\tfrac{1}{6}\right)^1 \left(\tfrac{5}{6}\right)^2$$

4. No 4s:
$$P({\sim}4 \cap {\sim}4 \cap {\sim}4) = \tfrac{5}{6} \cdot \tfrac{5}{6} \cdot \tfrac{5}{6} = \left(\tfrac{5}{6}\right)^3$$

Notice that

$${}_3C_3 = {}_3C_0 = 1 \quad \text{and} \quad {}_3C_2 = {}_3C_1 = 3$$

In general, we have

$$P(n \text{ fours in 3 tosses}) = P(n, 3) = {}_3C_n \left(\frac{1}{6}\right)^n \left(\frac{5}{6}\right)^{3-n}$$

when $\frac{1}{6}$ is the probability of rolling a 4 (a success) and $\frac{5}{6}$ is the probability of not rolling a 4 (a failure).

Answer the following questions about the outcome of rolling one die three times.

What is the probability of

1. Losing? _____?_____
2. Rolling at least one four? _____?_____
3. Rolling more than one four? _____?_____
4. Rolling less than two fours? _____?_____

Your answers should be 0.58, 0.345 + 0.07 + 0.005 = 0.42, 0.075, and 0.925.

In general, a binomial distribution will result after n trials, whenever the following conditions exist.

1. There are only two possible outcomes for each trial. They are generally called "success" and "failure."
2. The trials are independent.
3. The probabilities for success and failure remain the same for each trial.

The probability of x successes and $n - x$ failures in n trials is given by Theorem IX, where p is the probability of success and q is the probability of failure. Of course,

$$p + q = 1$$

THEOREM IX.

The probability of exactly x successes in n binomial trials is given by

$$p(x) = {}_nC_x p^x q^{n-x} \quad \text{for } 0 \leq x \leq n$$

EXERCISE 16.6

A. FUNDAMENTALS

1. Calculate the four values of

$$p(x) = {}_nC_x p^x q^{n-x}$$

for $n = 3$ and $x = 0, 1, 2,$ and 3 (the probabilities of getting 0, 1, 2, or 3 heads on three tosses of a fair coin).

2. Calculate the four values of $p(x)$ for $n = 4$ and $x = 0, 1, 2, 3,$ and 4 (the probability of drawing 0, 1, 2, 3, or 4 Aces, where the card is replaced and the deck reshuffled after each of the four draws). [HINT: $(\frac{1}{52})^0 = 1$, $(\frac{1}{52})^1 = 0.019$, $(\frac{1}{52})^2 = 0.0004 = 4 \times 10^{-4}$, $(\frac{1}{52})^3 = 0.000007 = 7 \times 10^{-6}$, and $(\frac{1}{52})^4 = 0.0000002 = 2 \times 10^{-7}$.] Graph the results.

B. ESSENTIALS

3. Construct the binomial distribution for five tosses of a coin. What is the probability that a coin will land heads on five tosses? Is it likely that such a coin is "fair"? What assumptions are you making in testing the hypothesis that the coin is fair?

The following applies only to successive trials *without* replacement.

4. If a population of size N can be divided into k successes and $n - k$ failures, then the hypergeometric probability distribution for the number of successes in a sample of size n is given by

$$p(x) = \frac{{}_kC_x \cdot {}_{(N-k)}C_{(n-x)}}{{}_NC_n}, \quad 0 \leq x \leq n$$

★(a) For a five-card poker hand, calculate the probability of four hearts using the hypergeometric distribution with $n = 5$, $N = 52$, $k = 13$, and $x = 4$.

(b) Find the six values of $p(x)$ for $x = 0, 1, 2, 3, 4,$ or 5 hearts and graph the results.

(c) If we have four red and five blue marbles in an urn, calculate the probability of drawing out three red marbles in six draws ($x = 3$, $k = 4$, $n = 6$, and $N = 9$). Interpret each of the numbers ${}_kC_x$, ${}_{N-k}C_{n-x}$, and ${}_NC_n$. What is the probability of drawing three blue marbles in four draws?

•5. Graph $p(x) = \dfrac{1}{\sqrt{2\pi}} e^{-x^2/2}$, the continuous normal probability distribution, where x can assume any real value. The total area under the curve is 1. The mean is zero, and about 68% of the area lies between $+1$ and -1.

16.7 THE BINOMIAL THEOREM

A binomial is a polynomial having two terms, such as

$$3x - 1, \quad x^2 + 2, \quad 4x^3 - \frac{3}{5}$$

The binomial theorem gives us a formula for expanding powers of binomials, such as

$$(3x - 1)^3, \quad (x^2 + 2)^{27}, \quad \left(4x^3 - \frac{3}{5}\right)^5$$

without performing 3, 27, or 5 multiplications. The theorem is a direct application of the technique used for counting combinations,

$$_nC_r = \frac{n!}{r!(n-r)!}$$

Before we discuss the binomial theorem let's develop a quick method for expanding binomials raised to small powers.

Using the above formula, let's find some possible combinations.

$$_1C_0 = 1, \quad _1C_1 = 1$$

$$_2C_0 = 1, \quad _2C_1 = \frac{2!}{1!(2-1)!} = 2, \quad _2C_2 = 1$$

$$_3C_0 = 1, \quad _3C_1 = \frac{3!}{1!2!} = 3, \quad _3C_2 = 3$$

$$_3C_3 = 1$$

$$_4C_0 = 1, \quad _4C_1 = \frac{4!}{1!3!} = 4, \quad _4C_2 = 6$$

$$_4C_3 = 4, \quad _4C_4 = 1$$

These results are tabulated and extended below. Do you see the pattern?

```
              1    1
           1    2    1
        1    3    3    1
     1    4    6    4    1
   1   5   10   10   5   1
 1   6  (15)  20   15   6   1
```

We know that the circled 15 is $_6C_2$, since it is in the 6th row and is the 3rd entry there.

The above array is called Pascal's triangle.[2] What is the second entry in the 100th row? What is the next to the last entry in the 100th row? Do you know why these entries are equal?

There is an obvious relationship between the numbers in Pascal's triangle, representing various combinations, and the coefficients of the following expanded binomials.

$$(a + b)^1 = a + b$$

$$(a + b)^2 = a^2 + 2ab + b^2$$

$$(a + b)^3 = a^3 + 3a^2b^1 + 3ab^2 + b^3$$

$$(a + b)^4 = a^4 + 4a^3b^1 + 6a^2b^2 + 4ab^3 + b^4$$

$$(a + b)^5 = a^5 + 5a^4b^1 + 10a^3b^2 + 10a^2b^3 + 5ab^4 + b^5$$

These coefficients also form Pascal's triangle! Now we can use Pascal's triangle to find the coefficients of the terms in the expansion of $(a + b)^n$, provided n is small enough to allow us to create enough rows of the triangle. The other factors in each term consist of a raised to some power and b raised to some power. Notice that the sum of the exponents of a and b equals the power on the corresponding binomial. The exponents of the first term, a, start at n and decrease to 0. The exponents on the second term, b, start at 0 and increase to n. Therefore, the powers of a and b are easily found.

EXAMPLE I. Use Pascal's triangle to expand the following.
(a) $(x + y)^3$ (b) $(x - 2y)^3$ (c) $(2x - 3)^4$

Solution: (a) From Pascal's triangle, we see that the coefficients are 1, 3, 3, 1. The powers of x are 3, 2, 1, 0, and the powers of y, second term, are 0, 1, 2, 3. Now we have

$$(x + y)^3 = 1 \cdot x^3 y^0 + 3x^2 y^1 + 3x^1 y^2 + 1 \cdot x^0 y^3$$

$$= x^3 + 3x^2 y + 3xy^2 + y^3$$

(b) $(x - 2y)^3 = 1x^3(-2y)^0 + 3x^2(-2y)^1$
$$+ 3x^1(-2y)^2 + 1x^0(-2y)^3$$
$$= x^3 - 6x^2 y + 12xy^2 - 8y^3$$

(c) $(2x - 3)^4 = ?(2x)^4(-3)^0 + ?(2x)^3(-3)^1$
$\qquad + ?(2x)^2(-3)^2 + ?(2x)^2(-3)^3$
$\qquad + ?(2x)^2(-3)^2$
$\qquad = 16x^4 - 96x^3 + 216x^2$
$\qquad - 216x + 81$

To understand why the coefficients of expanded binomial expressions can be found by using Pascal's triangle, which is composed of various combinations, examine the following closely. The circles $\bigcirc$ and triangles $\triangle$ represent any algebraic expression.

$(\bigcirc + \triangle)^3 = [(\bigcirc_1 + \triangle_1)(\bigcirc_2 + \triangle_2)](\bigcirc_3 + \triangle_3)$
$\qquad = [\bigcirc_1(\bigcirc_2 + \triangle_2) + \triangle_1(\bigcirc_2 + \triangle_2)]$
$\qquad\quad \times (\bigcirc_3 + \triangle_3)$
$\qquad = (\bigcirc_1\bigcirc_2 + \bigcirc_1\triangle_2 + \triangle_1\bigcirc_2 + \triangle_1\triangle_2)$
$\qquad\quad \times (\bigcirc_3 + \triangle_3)$
$\qquad = \bigcirc_1\bigcirc_2\bigcirc_3 + \bigcirc_1\triangle_2\bigcirc_3 + \triangle_1\bigcirc_2\bigcirc_3$
$\qquad\quad + \triangle_1\triangle_2\bigcirc_3 + \bigcirc_1\bigcirc_2\triangle_3$
$\qquad\quad + \bigcirc_1\triangle_2\triangle_3 + \triangle_1\bigcirc_2\triangle_3$
$\qquad\quad + \triangle_1\triangle_2\triangle_3$

From the first line, we see that there are three $\bigcirc$'s ($\bigcirc_1$, $\bigcirc_2$, $\bigcirc_3$) and three $\triangle$'s available. The term $\bigcirc_1 \cdot \bigcirc_2 \cdot \bigcirc_3$ is $_3C_3$, since all three $\bigcirc$s were used. There is only one such term since $_3C_3 = 1$. The terms $\bigcirc_1 \cdot \triangle_2 \cdot \bigcirc_3$, $\triangle_1 \cdot \bigcirc_2 \cdot \bigcirc_3$, and $\bigcirc_1 \cdot \bigcirc_2 \cdot \triangle_3$ are the three combinations of three $\bigcirc$s taken two at a time, $_3C_2$. There are three such combinations, since $_3C_2 = 3$. The terms $\triangle_1 \cdot \triangle_2 \cdot \bigcirc_3$, $\bigcirc_1 \cdot \triangle_1 \cdot \triangle_3$, and $\triangle_1 \cdot \bigcirc_2 \cdot \triangle_3$ depict the three ways of choosing $\bigcirc_1$, $\bigcirc_2$, or $\bigcirc_3$ one at a time, $_3C_1$. The last term containing no $\bigcirc$s represents $_3C_0 = 1$.

Now we are ready to state the binomial theorem, which explains how to expand $(a + b)^n$ for any natural number n.[3]

THEOREM X.
The Binomial Theorem:
$$(a + b)^n = \sum_{r=0}^{n} {}_nC_r\, a^{n-r}b^r$$

COROLLARY I.
In the expansion of the binomial $(a + b)^n$, the $(r + 1)$st term is given by
$${}_nC_r\, a^{n-r}b^r$$

As repeated applications of the above corollary amount to applying the binomial theorem for $r + 1 = 1, 2, 3, \ldots, n + 1$, we will illustrate the use of the binomial theorem by applying the corollary to find various terms.

EXAMPLE IV. Find the twenty-fifth term of $(2x - 1)^{27}$.

Solution:
$$r + 1 = 25 \Rightarrow r = 24$$
$${}_{27}C_{24}\, a^{27-24}b^{24} = \frac{27!}{24!\,(27-24)!}(2x)^3(-1)^{24}$$

where
$$\frac{27!}{24!\,(27-24)!} = \frac{24! \cdot 25 \cdot 26 \cdot 27}{24! \cdot 1 \cdot 2 \cdot 3}$$
$$= 25 \cdot 13 \cdot 9 = 2925 \Rightarrow$$
$${}_{27}C_{24}\, a^3b^{24} = (2925)(2x)^3(-1)^{24} = 23{,}400x^3$$

This would be very tedious to find by using Pascal's triangle, since we would need the twenty-fifth entry of the twenty-seventh row.

Now we will prove the binomial theorem. We will use mathematical induction, since we need to establish the theorem for all natural numbers n.

1. $p(1)$
$$p(n): (a + b)^n = \sum_{r=0}^{n} a^{n-r}b^r$$

CHAPTER 16 Counting, Probability, and the Binomial Theorem

and

$$n = 1 \Rightarrow \sum_{r=0}^{1} a^{1-r}b^r = a^{1-0}b^0 + a^{1-1}b^1 = (a+b)^1$$

2. $p(k) \Rightarrow p(k+1)$

$$(a+b)^k = \sum_{r=0}^{k} {}_kC_r a^{k-r}b^r \Rightarrow$$

$$(a+b)^{k+1} = \sum_{r=0}^{k+1} {}_{k+1}C_r a^{(k+1)-r}b^r$$

$(a+b)^{k+1} = (a+b)(a+b)^k \Rightarrow$

$$(a+b)^{k+1} = (a+b)\left[\sum_{r=0}^{k} {}_kC_r a^{k-r}b^r\right] \Rightarrow$$

$$(a+b)^{k+1} = (a+b)\left[a^k + \frac{k!}{(k-1)! \cdot 1!} a^{k-1}b \right.$$
$$+ \frac{k!}{(k-2)!2!} a^{k-2}b^2 + \cdots$$
$$\left. + \frac{k!}{1!(k-1)!} ab^{k-1} + b^k\right] \Rightarrow$$

$$(a+b)^{k+1} = \left(a^{k+1} + \frac{k!}{(k-1)!} a^k b + \frac{k!}{2!(k-2)!}\right.$$
$$\times a^{k-1}b^2 + \cdots + \frac{k!}{(k-1)!}$$
$$\left.\times a^2 b^{k-1} + ab^k\right) + \left(a^k b + \frac{k!}{(k-1)!}\right.$$
$$\times a^{k-1}b^2 + \frac{k!}{2!(k-2)!} a^{k-2}b^3 + \cdots$$
$$\left.+ \frac{k!}{(k-1)!} ab^k + b^{k+1}\right) \Rightarrow$$

$$(a+b)^{k+1} = a^{k+1} + \left[\frac{k!}{(k-1)!} + 1\right]a^k b$$
$$+ \left[\frac{k!}{2!(k-2)!} + \frac{k!}{(k-1)!}\right]a^{k-1}b^2$$
$$+ \cdots + b^{k+1}$$

where the $(r+1)$st term is

$$\left[\frac{k!}{r!(k-r)!} + \frac{k!}{(r-1)!(k-r+1)!}\right] a^{(k+1)-r}b^r$$

$$= \left[\frac{k-r+1}{k-r+1} \frac{k!}{r!(k-r)!}\right.$$
$$\left.+ \frac{r}{r} \frac{k!}{(r-1)!(k-r+1)!}\right] a^{(k+1)-r}b^r$$

$$= \left[\frac{[(k+1)-r]k!}{[(k+1)-r]r!(k-r)!}\right.$$
$$\left.+ \frac{r \cdot k!}{[r \cdot (r-1)!](k-r+1)!}\right] a^{(k+1)-r}b^r$$

$$= \frac{(k+1)! - r \cdot k! + rk!}{r!(k-r+1)!} a^{(k+1)-r}b^r$$

$$= \frac{(k+1)!}{r![(k+1)-r]!} a^{(k+1)-r}b^r$$

$$= {}_{k+1}C_r a^{(k+1)-r}b^r$$

Therefore,

$$(a+b)^{k+1} = \sum_{r=0}^{k+1} {}_{k+1}C_r a^{(k+1)-r}b^r$$

EXERCISE 16.7

A. FUNDAMENTALS

Expand the following binomials using Pascal's triangle.

1. (a) $(x+y)^5$ (b) $(x-y)^5$
2. (a) $(A-B)^6$ (b) $(2x-y)^6$
3. (a) $(x+1)^4$ (b) $(x^2+1)^4$
4. (a) $(2x-3)^4$ (b) $(2x-3)^5$
5. (a) $(2x-1)^3$ (b) $(2x^3-1)^3$
6. (a) $(x-\frac{1}{2})^7$ (b) $(3x-y)^7$

Use the binomial theorem to expand the following.

7. $(x-y)^7$ 8. $(2x+y)^6$

B. ESSENTIALS

Use the corollary to the binomial theorem to find the following.

9. (a) The first term of $(x - y)^{10}$
   (b) The last term of $(x - y)^{10}$

10. (a) The tenth term of $(a - b)^{12}$
    (b) The third term of $(a - b)^{12}$

11. (a) The eighth term of $(2x - y)^{10}$
    (b) The second term of $(2x - y)^{10}$

12. (a) The twenty-first and the eighth terms of $(a - b)^{27}$
    (b) The fourth and the twenty-fifth terms of $(a - b)^{27}$

Find the coefficient of the indicated term in each of the following.

13. x^5 in $(x - 3)^{10}$ 14. x^6 in $(x^2 - 2)^{15}$
15. x^6 in $(1 - x^3)^{10}$ 16. x^{-4} in $(1 - 1/x)^{15}$
17. x^2 in $(\sqrt{x} - 1/\sqrt{x})^{10}$
18. x^2 in $(x - 1/\sqrt{x})^{26}$

Approximate the following using the sum of the first three terms of the indicated binomial expansion. In Problem 20, the error can be shown to be less than the value of the fourth term. Find this upper bound for the error.

19. (a) $(1.03)^{12} = (1 + 0.03)^{12}$
    (b) $(\sqrt{2})^{10} \approx (1 + 0.41)^{10}$

20. (a) $(0.96)^{10} = (1 - 0.04)^{10}$ (b) $(0.83)^8$

21. Evaluate the following, where $i^2 = -1$.
    (a) $(1 + i)^4$ (b) $(2 - 3i)^6$

22. The binomial theorem can be extended to powers of polynomials with more than two terms. The extension to three terms for some power $n \in$ naturals is

$$(x + y + z)^n = \sum_{i+j+k=n} \binom{n}{i, j, k} x^i y^j z^k,$$

$i, j, k \in$ naturals or zero

where

$$\binom{n}{i, j, k} = \frac{n!}{i! j! k!}$$

and the sum is over all possible values of i, j, and k, whose sum is n.

(a) Expand $(x + y + z)^3$ by direct multiplication and by using the above theorem. The ten possible values of i, j, and k are (0, 0, 3), (0, 3, 0), (3, 0, 0), (1, 1, 1), (1, 2, 0), (1, 0, 2), (2, 1, 0), (2, 0, 1), (0, 1, 2), and (0, 2, 1).

(b) Expand $(x + y + z + w)^3$ by direct multiplication, extend the above to four terms, and compare the results. There are thirty-five terms in this expression.

23. The binomial theorem can be extended to rational exponents for $|x| < 1$ by the infinite series

$$(1 + x)^r = 1 + \frac{r}{1!} \cdot x + \frac{r(r-1)}{2!} \cdot x^2 + \frac{r(r-1)(r-2)}{3!} \cdot x^3 + \cdots + \frac{r(r-1)\cdots(r-n+1)}{n!} \cdot x^n + \cdots$$

Write the first four terms of the series expansion for the following

(a) $(\sqrt{2})^{1/2} = (1 + 0.4)^{1/2}$ (b) $\sqrt{1 - x}$
(c) $\sqrt{1.1}$ (d) $\sqrt[3]{0.98}$

ENDNOTES

1. In 1654 Pascal and Fermat wrote letters to each other about a particular problem that they attacked through the theory of combinations. This marked the beginning of the theory of probability. Jakob Bernoulli's book, *Ars Conjectandi*, published posthumously in 1713, was the first significant book on probability.

2. Pascal's triangle was first used by Pascal in 1654 to find the coefficients for binomial expansions, but it was known to many of his predecessors.

3. The binomial theorem was used by the thirteenth century Arabs and quite well known by the sixteenth century. It was Pascal who observed that the formula for combinations gives the binomial coefficients. In 1713 Jakob Bernoulli used combinations to prove the theorem for $(a + b)^n$, where n is a positive integer. It was extended to rational exponents by Newton and from binomial to multinomial by Leibniz.

BIBLIOGRAPHY

Bell, Eric T. *Development of Mathematics.* 2d ed. New York: McGraw-Hill Book Company, 1945.

Bell, Eric T. *Men of Mathematics.* New York: Simon & Schuster, Inc., 1961.

Bingham, Robert C. *Economics: Mathematically Speaking.* New York: McGraw-Hill Book Company, 1972.

Bishir, John W., and Drewes, Donald W. *Mathematics in the Behavioral and Social Sciences.* New York: Harcourt Brace Jovanovich, Inc., 1970.

Draper, Jean E., and Klingman, Jane S. *Mathematical Analysis: Business and Economic Applications.* 2d ed. New York: Harper & Row, 1972.

Eves, Howard. *Introduction to the History of Mathematics.* 3d ed. New York: Holt, Rinehart and Winston, Inc., 1969.

Kline, Morris. *Mathematical Thought from Ancient to Modern Times.* New York: Oxford University Press, 1972.

Kline, Morris. *Mathematics in Western Culture.* New York: Oxford University Press, 1964.

Placek, Ronald J. *Technical Mathematics with Calculus.* Englewood Cliffs, N.J.: Prentice-Hall, Inc., 1968.

Sears, Francis Weston, and Zemansky, A. W. *University Physics.* 4th ed. Reading, Mass.: Addison-Wesley Publishing Company, Inc., 1970.

Velz, Clarence J. *Applied Stream Sanitation.* New York: John Wiley & Sons, Inc., 1970.

Appendixes

APPENDIX I   A Chronology

APPENDIX II   Factorial Notation

APPENDIX III   Summation Notation

APPENDIX IV   Conversion Factors

APPENDIX V   The Meaning of *Limit*

APPENDIX VI   Tables
                      Exponential Functions
                      Logarithms Base 10
                      Logarithms Base *e*
                      Trigonometric Functions

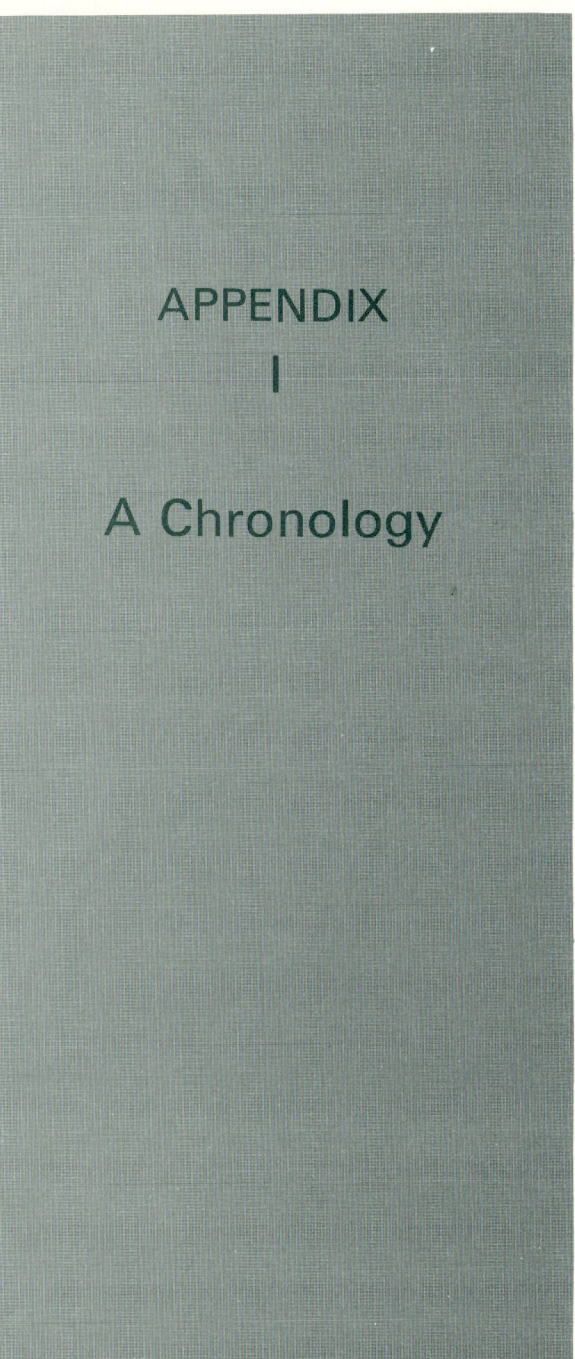

APPENDIX I

A Chronology

The following chronology is an incomplete, extremely oversimplified review of the historical development of the concepts of precalculus mathematics. It is interesting to note that completing this text really prepares you to begin studying seventeenth- and eighteenth-century mathematics.

2000 B.C. The Babylonians
    Arithmetic and quadratic equations.

1000 B.C. Early Egyptians
    Trigonometry; 3, 4, 5 right triangle.
Pythagoras of Samos (570?–500)
    The Pythagoreans made considerable advances in mathematics, astronomy, and music.

500 B.C. Zeno of Elea (480?–435)
    Formulated paradoxes of motion.
Hippocrates of Chios (450?)
    Created a logical presentation of the propositions of geometry.
Eudoxus of Cnidus (400?–350)
    Used the method of exhaustion; known chiefly for work in astronomy and mathematics.
Menaechmus (350?)
    Discovered the conic sections.
Aristotle of Stagira (384–322)
    Founded deductive logic.
Euclid (365?–300)
    His chief work Elements is a most definitive work on geometry.
Archimedes of Syracuse (287?–212)
    Greek mathematician and inventor.
Eratosthenes of Cyrene (275?–194)
    Generated prime numbers systematically.
Apollonius of Perga (262?–170)
    Treatise on conic sections.
Hipparchus (180?–120)
    Developed trigonometry; Greek astronomer.

100 B.C.

200 A.D. Diophantus
    Reputed inventor of algebra; chief work, Arithmetica.

APPENDIX 1

500 A.D. Zero introduced as a number by Hindus in India. Rules of signs for the integers traced to the Hindus.
Mohammed ibn Mûsâ al-Khowârizmi (?–850?)
  The first treatise on al-jebr (restoration and reduction) algebra.

1000 A.D. Omar Khayyám (?–1123?)
  Solved numerical cubics geometrically; he was an algebraic taxonomist and creative mathematician.
Leonardo Fibonacci (1180?–1250)
  Mathematician; discovered the "Fibonacci Sequence" in work with rabbits.
Nicole Oresme (1330?–1382)
  Developed coordinates and fractional exponents.
Johann Widmann (1460?–?)
  Possibly the first to introduce the signs + and − in arithmetic and algebra.
Francesco Maurolycus (1494–1575)
  Introduced mathematical induction.

1500 A.D. Jerome Cardan (1501–1576)
  First to investigate complex numbers. Extended the quadratic-type formula to cubics.
François Viète (1540–1603)
  Laid foundations for analytic geometry in his work in algebra and geometry.
John Napier (1550–1617)
  Invented logarithms.
Joost Bürgi (1552–1632)
  Complied the first table of antilogs.
Henry Briggs (1556?–1631)
  Calculated the first table of base 10 logarithms.
Galileo Galilei (1564–1642)
  Famous physicist and astronomer.
René Descartes (1596–1650)
  Invented analytic geometry.

1600 A.D. Pierre de Fermat (1601–1665)
  Worked in number theory and co-founded theory of probability.
John Wallis (1616–1703)
  Invented imaginary numbers.
Blaise Pascal (1623–1662)
  Treatise on conic sections; originated, with Fermat, the mathematical theory of probability.
Christian Huygens (1629–1695)
  Invented the pendulum clock, wrote first treatise on probability based on Pascal's and Fermat's letters.
Isaac Barrow (1630–1677)
  Stated the fundamental theorem of calculus.
Isaac Newton (1642–1727)
  Conceived idea of universal gravitation; credited with invention of calculus.
Gottfried Leibniz (1646–1716)
  Developed determinants, symbolic logic and discovered new notations of calculus.
Jakob Bernoulli (1654–1705)
  Introduced polar coordinates and contributed works in calculus and probability.
Guillaume de L'Hospital (1661–1704)
  Applied calculus to indeterminant forms and authored the first calculus book; studied with Jakob Bernoulli.
Johann Bernoulli (1667–1748)
  A prolific writer in analytic trig and optical phenomena; pioneer in exponential calculus. Fathered three sons who also became acclaimed mathematicians, Nicolas, Daniel, and Johann II.
Abraham DeMoivre (1667–1754)
  Author of works on annuities, probability, analytic trig (DeMoivre's Theorem) and series.
Brook Taylor (1685–1731)
  Developed most important technique for representing functions as infinite series.
Colin Maclaurin (1698–1746)
  He worked with applications of classical geometry and discovered Maclaurin's series.

1700 A.D. Leonhard Euler (1707–1783)
  The most prolific mathematician in history, even though blind the last fifteen years of his life. He introduced

384

these notations, $f(x)$, e, a, b, c, for triangle ABC, $\sum$, i, and Euler's formula.

Joseph Lagrange (1736–1813)
: First attempt to bring rigour to analysis. With Euler, one of the two greatest eighteenth century mathematicians, he worked with differential equations, number theory, and the theory of equations.

Pierre Simon de Laplace (1749–1827)
: His chief works include writings on differential equations, probability, celestial mechanics, and determinants.

Jean-Robert Argand (1768–1822)
: Introduced geometric representation for complex numbers.

Jean Fourier (1768–1830)
: Developed Fourier series—series involving trigonometric functions.

Karl Gauss (1777–1855)
: Propounded method of least squares at age eighteen, proved fundamental theorem of algebra, proposed an absolute system of magnetic units, devised solution for binomial equation.

Bernhard Bolzano (1781–1848)
: He laid the foundation for set theory.

Augustin Cauchy (1789–1857)
: Defined limit, continuity, and the derivative inspiring rigour in other mathematicians.

1800 A.D.

Niels Abel (1802–1829)
: Known for research in theory of elliptic functions. Proved that the quadratic formula could not be extended to equations of degree greater than four.

William Hamilton (1805–1865)
: His work on quaternions lead to vector algebra.

Évariste Galois (1811–1832)
: Determined which polynomial equations are solvable by algebraic operations.

Arthur Cayley (1821–1895)
: Introduced matrices as distinct entities.

First periodical on mathematics research (1826).

Pafnuti Tchebycheff (1821–1894)
: Influential in probability.

George Boole (1815–1864)
: Father of symbolic logic.

Karl Weierstrass (1815–1897)
: Developed modern theory of analytical functions.

Georg Riemann (1826–1866)
: Influential in generalizing mathematical concepts.

James Maxwell (1831–1879)
: Worked on theory of electromagnetism and vector algebra.

J. W. R. Dedekind (1831–1916)
: Defined real numbers and the fields of real and complex numbers; originated a theory of irrational numbers.

David Hilbert (1862–1943)
: The leading mathematician of the twentieth century, his work in integral equations is far beyond the treatment in this text.

Georg Cantor (1845–1918)
: Worked in number theory, using converging sequences to define irrational numbers; developed set theory.

Bertrand Russell (1872–1970)
: Mathematician and philosopher; established modern symbolic logic.

1900 A.D.

Kurt Gödel (1906–   )
: Showed that no axiomatic system is adequate for any significant branch of mathematics.

APPENDIX II

Factorial Notation

Factorial notation is particularly important in the study of the binomial theorem.

Five factorial, written 5!, means

$$1 \cdot 2 \cdot 3 \cdot 4 \cdot 5$$

Therefore,

$$5! = 120$$

By noticing that

$$4! = 1 \cdot 2 \cdot 3 \cdot 4$$

we see that

$$5! = 4! \cdot 5$$

In general,

$$n! = n \cdot (n-1) \cdot (n-2) \cdots 1$$

and

$$n! = n \cdot (n-1)!$$

Also, we must define

$$0! = 1$$

EXAMPLE 1. Simplify the following.

(a) $\dfrac{7!}{5!} = \dfrac{1 \cdot 2 \cdot 3 \cdot 4 \cdot 5 \cdot 6 \cdot 7}{1 \cdot 2 \cdot 3 \cdot 4 \cdot 5} = \dfrac{5! \cdot 6 \cdot 7}{5!} = 6 \cdot 7$

$= 42$

(b) $\dfrac{100!}{3! \cdot 97!} = \dfrac{97! \cdot 98 \cdot 99 \cdot 100}{1 \cdot 2 \cdot 3 \cdot 97!} = \dfrac{98 \cdot 99 \cdot 100}{1 \cdot 2 \cdot 3}$

$= 161{,}700$

APPENDIX III

Summation Notation

In this section we will encounter long summations, such as

$$1 + 2 + 3 + \cdots + 100$$

which we can write as

$$\sum_{i=1}^{100} i$$

The Greek letter sigma, $\sum$, stands for sum. The above expression is read, "the summation of i from $i = 1$ to $i = 100$." Then,

$$\sum_{i=1}^{4} 2i^2$$

is evaluated by first replacing i with the natural numbers between 1 and 4, one at a time, then summing the four results.

$$\sum_{i=1}^{4} 2i^2 = 2(1)^2 + 2(2)^2 + 2(3)^2 + 2(4)^2$$

$$= 2 + 8 + 18 + 32 = 60$$

There are several important theorems involving summation notation whose proofs rely on the field properties for addition and multiplication of real numbers.

How are the following three sums related?

$$\sum_{i=1}^{c} a_i, \quad \sum_{i=c+1}^{d} a_i, \quad \text{and} \quad \sum_{i=1}^{d} a_i$$

where $1 \leq c \leq d$. We have

$$(a_1 + a_2 + \cdots + a_c)$$
$$(a_{c+1} + a_{c+2} + \cdots + a_d)$$

and

$$(a_1 + a_2 + \cdots + a_d)$$

By the associative law,

$$\sum_{i=1}^{c} a_i + \sum_{i=c+1}^{d} a_i = \sum_{i=1}^{d} a_i$$

By the distributive law, $d(a_1 + a_2) = d \cdot a_1 + d \cdot a_2$ becomes

$$d \sum_{i=1}^{n} a_i = \sum_{i=1}^{n} da_i$$

APPENDIX III

We can regroup two summations over the same natural numbers to allow us to write one summation symbol as follows.

$$\sum_{i=1}^{n} a_i + \sum_{i=1}^{n} b_i = (a_1 + a_2 + \cdots + a_n)$$
$$+ (b_1 + b_2 + \cdots + b_n)$$
$$= (a_1 + b_1) + (a_2 + b_2) + \cdots + (a_n + b_n)$$
$$= \sum_{i=1}^{n} (a_i + b_i)$$

We have one more general equation,

$$\sum_{i=1}^{n} 1 = \underbrace{1 + 1 + \cdots + 1}_{n \text{ times}} = n$$

Following are some problems using summation notation.

EXAMPLE I. Evaluate the following sums.

(a) $\sum_{i=3}^{6} (2i^2 - 1) = (2 \cdot 3^2 - 1) + (2 \cdot 4^2 - 1)$
$$+ (2 \cdot 5^2 - 1) + (2 \cdot 6^2 - 1)$$
$$= 17 + 31 + 49 + 71 = 168$$

or

$$\sum_{i=3}^{6} (2i^2 - 1) = 2 \sum_{i=3}^{6} i^2 - \sum_{i=3}^{6} 1$$
$$= 2(3^2 + 4^2 + 5^2 + 6^2)$$
$$- (1 + 1 + 1 + 1)$$
$$= 2(86) - 4 = 168$$

(b) $\sum_{i=1}^{3} \left(\sum_{j=0}^{2} 2ij \right) = \sum_{i=1}^{3} (2i \cdot 0 + 2i \cdot 1 + 2i \cdot 2)$
$$= \sum_{i=1}^{3} 6i$$
$$= 6 \cdot 1 + 6 \cdot 2 + 6 \cdot 3 = 36$$

APPENDIX IV

Conversion Factors

Length

1 kilometer (km) = 1000 meters (m)
1 meter (m) = 100 centimeters (cm)
1 millimeter (mm) = 0.001 meters (m)
1 inch (in) = 2.54 centimeters (cm)
1 foot (ft) = 30.48 centimeters (cm)
1 yard (yd) = 0.9144 meter (m)
1 mile = 1.6093 kilometers (km)
1 centimeter (cm) = 0.3937 inch (in)
1 meter (m) = 39.37 inch (in)
1 kilometer (km) = 0.6214 mile (mi)

Volume

1 liter (l) = 1000 cubic centimeters (cm^3)
1 cubic meter (m^3) = 1000 liters (l)
1 liter (l) = 1.0567 quarts (qts)
1 cubic meter (m^3) = 35.32 cubic feet (ft^3)

Weight

1 kilogram (kg) = 1000 grams (g)
1 gram (g) = 1000 milligrams (mg)
1 kilogram (kg) = 2.205 pounds (lbs)
1 gram (g) = 0.0353 ounce (oz)
1 pound (lb) = 0.4536 kilogram (kg)
1 ounce (oz) = 28.35 grams (g)

APPENDIX V

The Meaning of *Limit*

To understand the limit concept intuitively, without studying the rigorous approach found in calculus, consider Figure A-1. We say

$$\lim_{x \to 2} \frac{(x+1)(x-2)}{x-2} = 3$$

because we can establish the value of the expression

$$\frac{(x+1)(x-2)}{x-2}$$

as close to 3 as we wish by simply choosing x close enough to $x = 2$. For example, suppose we were challenged to choose a value of x that would yield $f(x) = (x+1)(x-2)/(x-2)$ within 0.001 of 3. That is, $f(x)$ must be between 2.999 and 3.001. Since $f(1.999) = 2.999$ and $f(2.001) = 3.001$, we can show that any value of x such that

$$2.999 < x < 3.001$$

results in a corresponding value of $f(x)$ such that

$$2.999 < f(x) < 3.001$$

In general, we say

$$\lim_{x \to 2} f(x) = 3$$

because given *any challenge value* for the distance between $f(x)$ and 3 (no matter how small), we can find

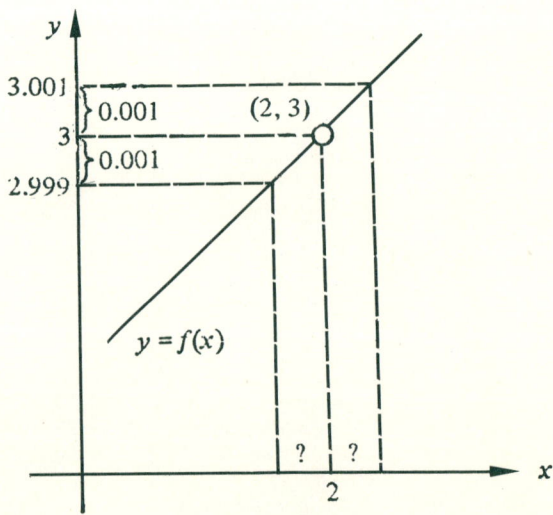

FIGURE A-1.

a corresponding interval about $x = 2$. As a result, every value of x within the interval about $x = 2$ has its corresponding y-value within the challenge distance of $y = 3$.

This concept of a limit is used in calculus to *prove* the following theorem on limits.

THEOREM I.
$\underset{x \to x_0}{\text{limit}} f(x) = A$ and $\underset{x \to x_0}{\text{limit}} g(x) = B$ implies

(a) $\underset{x \to x_0}{\text{limit}} f(x) \cdot g(x) = A \cdot B$

(b) $\underset{x \to x_0}{\text{limit}} \dfrac{f(x)}{g(x)} = \dfrac{A}{B}, \quad B \neq 0$

(c) $\underset{x \to x_0}{\text{limit}} \sqrt{f(x)} = \sqrt{A}$

These can be used in conjunction with

$$\underset{x \to \infty}{\text{limit}} \frac{1}{x^n} = 0, \qquad n > 1$$

to prove the following about horizontal asymptotes. Consider these specific examples.

EXAMPLE I. We know that $y = 0$ is an asymptote for

$$y = \frac{2x - 1}{3x^2 - 4x - 2}$$

since

$$\underset{x \to \infty}{\text{limit}} \frac{2x - 1}{3x^2 - 4x - 2} \cdot \frac{1/x^2}{1/x^2} = \underset{x \to \infty}{\text{limit}} \frac{2/x - 1/x^2}{3 - 4/x - 2/x^2}$$

$$= \frac{0 - 0}{3 - 0 - 0} = 0$$

Then we can make $y = (2x - 1)/(3x^2 - 4x - 2)$ assume values as close to $y = 0$ as we wish simply by choosing large enough values of x. As x approaches infinity, y approaches zero.

EXAMPLE II. We know that $y = \frac{2}{3}$ is an asymptote for

$$y = \frac{2x^2 - 1}{3x^2 + x - 2}$$

since

$$\underset{x \to \infty}{\text{limit}} \frac{2x^2 - 1}{3x^2 + x - 2} \cdot \frac{1/x^2}{1/x^2} = \underset{x \to \infty}{\text{limit}} \frac{2 - 1/x^2}{3 + 1/x - 2/x^2}$$

$$= \frac{2 - 0}{3 + 0 - 0} = \frac{2}{3}$$

EXAMPLE III. There is no horizontal asymptote for the function

$$y = \frac{x^3 - 1}{20x + 1}$$

since

$$\underset{x \to \infty}{\text{limit}} \frac{x^3 - 1}{20x + 1} \cdot \frac{1/x^3}{1/x^3} = \underset{x \to \infty}{\text{limit}} \frac{1 - 1/x^3}{20/x^2 + 1/x^3}$$

$$= \frac{1}{0} = \infty$$

We stated that a function is continuous if you can draw its graph without lifting your pencil from the paper. We can now use the limit concept to define continuity as in calculus.

A function $y = f(x)$ is continuous at $x = x_0$ if

(a) $f(x_0)$ exists

(b) $\underset{x \to x_0}{\text{limit}} f(x)$ exists

(c) $\underset{x \to x_0}{\text{limit}} f(x) = f(x_0)$

The third item states that we must be able to make $f(x)$ as close to $f(x_0)$ (make y as close to the *value* of y corresponding to x_0) as we wish (are challenged). This is accomplished by choosing values of x sufficiently close to $x = x_0$.

APPENDIX VI

Tables

TABLE I. *Exponential Functions*

TABLE II. *Logarithms Base 10*

TABLE III. *Logarithms Base* e

TABLE IV. *Trigonometric Functions*

TABLE I. *Exponential Functions*

x	e^x	e^{-x}	x	e^x	e^{-x}
0.0	1.00	1.00	3.1	22.2	.045
0.1	1.11	.905	3.2	24.5	.041
0.2	1.22	.819	3.3	27.1	.037
0.3	1.35	.741	3.4	30.0	.033
0.4	1.49	.670	3.5	33.1	.030
0.5	1.65	.607	3.6	36.6	.027
0.6	1.82	.549	3.7	40.4	.025
0.7	2.01	.497	3.8	44.7	.022
0.8	2.23	.449	3.9	49.4	.020
0.9	2.46	.407	4.0	54.6	.018
1.0	2.72	.368	4.1	60.3	.017
1.1	3.00	.333	4.2	66.7	.015
1.2	3.32	.301	4.3	73.7	.014
1.3	3.67	.273	4.4	81.5	.012
1.4	4.06	.247	4.5	90.0	.011
1.5	4.48	.223	4.6	99.5	.010
1.6	4.95	.202	4.7	110	.0091
1.7	5.47	.183	4.8	122	.0082
1.8	6.05	.165	4.9	134	.0074
1.9	6.69	.150	5.0	148	.0067
2.0	7.39	.135	5.1	164	.0061
2.1	8.17	.122	5.2	181	.0055
2.2	9.02	.111	5.3	200	.0050
2.3	9.97	.100	5.4	221	.0045
2.4	11.0	.091	5.5	245	.0041
2.5	12.2	.082	5.6	270	.0037
2.6	13.5	.074	5.7	299	.0033
2.7	14.9	.067	5.8	330	.0030
2.8	16.4	.061	5.9	365	.0027
2.9	18.2	.055	6.0	403	.0025
3.0	20.1	.050			

TABLE II. *Logarithms Base 10*

N	0	1	2	3	4	5	6	7	8	9
1.0	.0000	.0043	.0086	.0128	.0170	.0212	.0253	.0294	.0334	.0374
1.1	.0414	.0453	.0492	.0531	.0569	.0607	.0645	.0682	.0719	.0755
1.2	.0792	.0828	.0864	.0899	.0934	.0969	.1004	.1038	.1072	.1106
1.3	.1139	.1173	.1206	.1239	.1271	.1303	.1335	.1367	.1399	.1430
1.4	.1461	.1492	.1523	.1553	.1584	.1614	.1644	.1673	.1703	.1732
1.5	.1761	.1790	.1818	.1847	.1875	.1903	.1931	.1959	.1987	.2014
1.6	.2041	.2068	.2095	.2122	.2148	.2175	.2201	.2227	.2253	.2279
1.7	.2304	.2330	.2355	.2380	.2405	.2430	.2455	.2480	.2504	.2529
1.8	.2553	.2577	.2601	.2625	.2648	.2672	.2695	.2718	.2742	.2765
1.9	.2788	.2810	.2833	.2856	.2878	.2900	.2923	.2945	.2967	.2989
2.0	.3010	.3032	.3054	.3075	.3096	.3118	.3139	.3160	.3181	.3201
2.1	.3222	.3243	.3263	.3284	.3304	.3324	.3345	.3365	.3385	.3404
2.2	.3424	.3444	.3464	.3483	.3502	.3522	.3541	.3560	.3579	.3598
2.3	.3617	.3636	.3655	.3674	.3692	.3711	.3729	.3747	.3766	.3784
2.4	.3802	.3820	.3838	.3856	.3874	.3892	.3909	.3927	.3945	.3962
2.5	.3979	.3997	.4014	.4031	.4048	.4065	.4082	.4099	.4116	.4133
2.6	.4150	.4166	.4183	.4200	.4216	.4232	.4249	.4265	.4281	.4298
2.7	.4314	.4330	.4346	.4362	.4378	.4393	.4409	.4425	.4440	.4456
2.8	.4472	.4487	.4502	.4518	.4533	.4548	.4564	.4579	.4594	.4609
2.9	.4624	.4639	.4654	.4669	.4683	.4698	.4713	.4728	.4742	.4757
3.0	.4771	.4786	.4800	.4814	.4829	.4843	.4857	.4871	.4886	.4900
3.1	.4914	.4928	.4942	.4955	.4969	.4983	.4997	.5011	.5024	.5038
3.2	.5051	.5065	.5079	.5092	.5105	.5119	.5132	.5145	.5159	.5172
3.3	.5185	.5198	.5211	.5224	.5237	.5250	.5263	.5276	.5289	.5302
3.4	.5315	.5328	.5340	.5353	.5366	.5378	.5391	.5403	.5416	.5428
3.5	.5441	.5453	.5465	.5478	.5490	.5502	.5514	.5527	.5539	.5551
3.6	.5563	.5575	.5587	.5599	.5611	.5623	.5635	.5647	.5658	.5670
3.7	.5682	.5694	.5705	.5717	.5729	.5740	.5752	.5763	.5775	.5786
3.8	.5798	.5809	.5821	.5832	.5843	.5855	.5866	.5877	.5888	.5899
3.9	.5911	.5922	.5933	.5944	.5955	.5966	.5977	.5988	.5999	.6010
4.0	.6021	.6031	.6042	.6053	.6064	.6075	.6085	.6096	.6107	.6117
4.1	.6128	.6138	.6149	.6160	.6170	.6180	.6191	.6201	.6212	.6222
4.2	.6232	.6243	.6253	.6263	.6274	.6284	.6294	.6304	.6314	.6325
4.3	.6335	.6345	.6355	.6365	.6375	.6385	.6395	.6405	.6415	.6425
4.4	.6435	.6444	.6454	.6464	.6474	.6484	.6493	.6503	.6513	.6522
4.5	.6532	.6542	.6551	.6561	.6571	.6580	.6590	.6599	.6609	.6618
4.6	.6628	.6637	.6646	.6656	.6665	.6675	.6684	.6693	.6702	.6712
4.7	.6721	.6730	.6739	.6749	.6758	.6767	.6776	.6785	.6794	.6803
4.8	.6812	.6821	.6830	.6839	.6848	.6857	.6866	.6875	.6884	.6893
4.9	.6902	.6911	.6920	.6928	.6937	.6946	.6955	.6964	.6972	.6981
5.0	.6990	.6998	.7007	.7016	.7024	.7033	.7042	.7050	.7059	.7067
5.1	.7076	.7084	.7093	.7101	.7110	.7118	.7126	.7135	.7143	.7152
5.2	.7160	.7168	.7177	.7185	.7193	.7202	.7210	.7218	.7226	.7235
5.3	.7243	.7251	.7259	.7267	.7275	.7284	.7292	.7300	.7308	.7316
5.4	.7324	.7332	.7340	.7348	.7356	.7364	.7372	.7380	.7388	.7396
N	0	1	2	3	4	5	6	7	8	9

TABLE II. *Logarithms Base 10 (continued)*

N	0	1	2	3	4	5	6	7	8	9
5.5	.7404	.7412	.7419	.7427	.7435	.7443	.7451	.7459	.7466	.7474
5.6	.7482	.7490	.7497	.7505	.7513	.7520	.7528	.7536	.7543	.7551
5.7	.7559	.7566	.7574	.7582	.7589	.7597	.7604	.7612	.7619	.7627
5.8	.7634	.7642	.7649	.7657	.7664	.7672	.7679	.7689	.7694	.7701
5.9	.7709	.7716	.7723	.7731	.7738	.7745	.7752	.7760	.7767	.7774
6.0	.7782	.7789	.7796	.7803	.7810	.7818	.7825	.7832	.7839	.7846
6.1	.7853	.7860	.7868	.7875	.7882	.7889	.7896	.7903	.7910	.7917
6.2	.7924	.7931	.7938	.7945	.7952	.7959	.7966	.7973	.7980	.7987
6.3	.7993	.8000	.8007	.8014	.8021	.8028	.8035	.8041	.8048	.8055
6.4	.8062	.8069	.8075	.8082	.8089	.8096	.8102	.8109	.8116	.8122
6.5	.8129	.8136	.8142	.8149	.8156	.8162	.8169	.8176	.8182	.8189
6.6	.8195	.8202	.8209	.8215	.8222	.8228	.8235	.8241	.8248	.8254
6.7	.8261	.8267	.8274	.8280	.8287	.8293	.8299	.8306	.8312	.8319
6.8	.8325	.8331	.8338	.8344	.8351	.8357	.8363	.8370	.8376	.8328
6.9	.8388	.8395	.8401	.8407	.8414	.8420	.8426	.8432	.8439	.8445
7.0	.8451	.8457	.8463	.8470	.8476	.8482	.8488	.8494	.8500	.8506
7.1	.8513	.8519	.8525	.8531	.8537	.8543	.8549	.8555	.8561	.8567
7.2	.8573	.8579	.8585	.8591	.8597	.8603	.8609	.8615	.8621	.8627
7.3	.8633	.8639	.8645	.8651	.8657	.8663	.8669	.8675	.8681	.8686
7.4	.8692	.8698	.8704	.8710	.8716	.8722	.8727	.8733	.8739	.8745
7.5	.8751	.8756	.8762	.8768	.8774	.8779	.8785	.8791	.8797	.8802
7.6	.8808	.8814	.8820	.8825	.8831	.8837	.8842	.8848	.8854	.8859
7.7	.8865	.8871	.8876	.8882	.8887	.8893	.8899	.8904	.8910	.8915
7.8	.8921	.8927	.8932	.8938	.8943	.8949	.8954	.8960	.8965	.8971
7.9	.8976	.8982	.8987	.8993	.8998	.9004	.9009	.9015	.9020	.9025
8.0	.9031	.9036	.9042	.9047	.9053	.9058	.9063	.9069	.9074	.9079
8.1	.9085	.9090	.9096	.9101	.9106	.9112	.9117	.9122	.9128	.9133
8.2	.9138	.9143	.9149	.9154	.9159	.9165	.9170	.9175	.9180	.9186
8.3	.9191	.9196	.9201	.9206	.9212	.9217	.9222	.9227	.9232	.9238
8.4	.9243	.9248	.9253	.9258	.9263	.9269	.9274	.9279	.9284	.9289
8.5	.9294	.9299	.9304	.9309	.9315	.9320	.9325	.9330	.9335	.9340
8.6	.9345	.9350	.9355	.9360	.9365	.9370	.9375	.9380	.9385	.9390
8.7	.9395	.9400	.9405	.9410	.9415	.9420	.9425	.9430	.9435	.9440
8.8	.9445	.9450	.9455	.9460	.9465	.9469	.9474	.9479	.9484	.9489
8.9	.9494	.9499	.9504	.9509	.9513	.9518	.9523	.9528	.9533	.9538
9.0	.9542	.9547	.9552	.9557	.9562	.9566	.9571	.9576	.9581	.9586
9.1	.9590	.9595	.9600	.9605	9609	.9614	.9619	.9624	.9628	.9633
9.2	.9638	.9643	.9647	.9652	.9657	.9661	.9666	.9671	.9675	.9680
9.3	.9685	.9689	.9694	.9699	.9703	.9708	.9713	.9717	.9722	.9727
9.4	.9731	.9736	.9741	.9745	.9750	.9754	.9759	.9763	.9768	.9773
9.5	.9777	.9782	.9786	.9791	.9795	.9800	.9805	.9809	.9814	.9818
9.6	.9823	.9827	.9832	.9836	.9841	.9845	.9850	.9854	.9859	.9863
9.7	.9868	.9872	.9877	.9881	.9886	.9890	.9894	.9899	.9903	.9908
9.8	.9912	.9917	.9921	.9926	.9930	.9934	.9939	.9943	.9948	.9952
9.9	.9956	.9961	.9965	.9969	.9974	.9978	.9983	.9987	.9991	.9996
N	0	1	2	3	4	5	6	7	8	9

TABLE III. *Logarithms Base e*

N	.0	.1	.2	.3	.4	.5	.6	.7	.8	.9
1	0.000	0.095	0.182	0.262	0.336	0.405	0.470	0.531	0.588	0.642
2	0.693	0.742	0.788	0.833	0.875	0.916	0.956	0.993	1.030	1.065
3	1.099	1.131	1.163	1.194	1.224	1.253	1.281	1.308	1.335	1.361
4	1.386	1.411	1.435	1.459	1.482	1.504	1.526	1.548	1.569	1.589
5	1.609	1.629	1.649	1.668	1.686	1.705	1.723	1.740	1.758	1.775
6	1.792	1.808	1.825	1.841	1.856	1.872	1.887	1.902	1.917	1.932
7	1.946	1.960	1.974	1.988	2.001	2.015	2.028	2.041	2.054	2.067
8	2.079	2.092	2.104	2.116	2.128	2.140	2.152	2.163	2.175	2.186
9	2.197	2.208	2.219	2.230	2.241	2.251	2.262	2.272	2.282	2.293
10	2.303	2.313	2.322	2.332	2.342	2.351	2.361	2.370	2.380	2.389

TABLE IV. *Trigonometric Functions*

	Sin	Cos	Tan	Cot	
0° 00'	.0000	1.0000	.0000		**90° 00'**
10	029	000	029	343.8	50
20	058	000	058	171.9	40
30	.0087	1.0000	.0087	114.6	30
40	116	.9999	116	85.94	20
50	145	999	145	68.75	10
1° 00'	.0175	.9998	.0175	57.29	**89° 00'**
10	204	998	204	49.10	50
20	233	997	233	42.96	40
30	.0262	.9997	.0262	38.19	30
40	291	996	291	34.37	20
50	320	995	320	31.24	10
2° 00'	.0349	.9994	.0349	28.64	**88° 00'**
10	378	993	378	26.43	50
20	407	992	407	24.54	40
30	.0436	.9990	.0437	22.90	30
40	465	989	466	21.47	20
50	494	988	495	20.21	10
3° 00'	.0523	.9986	.0524	19.08	**87° 00'**
10	552	985	553	18.07	50
20	581	983	582	17.17	40
30	.0610	.9981	.0612	16.35	30
40	640	980	641	15.60	20
50	669	978	670	14.92	10
4° 00'	.0698	.9976	.0699	14.30	**86° 00'**
10	727	974	729	13.73	50
20	756	971	758	13.20	40
30	.0785	.9969	.0787	12.71	30
40	814	967	816	12.25	20
50	843	964	846	11.83	10
5° 00'	.0872	.9962	.0875	11.43	**85° 00'**
10	901	959	904	11.06	50
20	929	957	934	10.71	40
30	.0958	.9954	.0963	10.39	30
40	987	951	992	10.08	20
50	.1016	948	.1022	9.788	10
6° 00'	.1045	.9945	.1051	9.514	**84° 00'**
10	074	942	080	9.255	50
20	103	939	110	9.010	40
30	.1132	.9936	.1139	8.777	30
40	161	932	169	8.556	20
50	190	929	198	8.345	10
7° 00'	.1219	.9925	.1228	8.144	**83° 00'**
10	248	922	257	7.953	50
20	276	918	287	7.770	40
30	.1305	.9914	.1317	7.596	30
40	334	911	346	7.429	20
50	363	907	376	7.269	10
8° 00'	.1392	.9903	.1405	7.115	**82° 00'**
10	421	899	435	6.968	50
20	449	894	465	6.827	40
30	.1478	.9890	.1495	6.691	30
40	507	886	524	6.561	20
50	536	881	554	6.435	10
9° 00'	.1564	.9877	.1584	6.314	**81° 00'**
	Cos	Sin	Cot	Tan	

TABLE IV. *Trigonometric Functions (continued)*

	Sin	Cos	Tan	Cot	
9° 00'	.1564	.9877	.1584	6.314	**81° 00'**
10	593	872	614	197	50
20	622	868	644	084	40
30	.1650	.9863	.1673	5.976	30
40	679	858	703	871	20
50	708	853	733	769	10
10° 00'	.1736	.9848	.1763	5.671	**80° 00'**
10	765	843	793	576	50
20	794	838	823	485	40
30	.1822	.9833	.1853	5.396	30
40	851	827	883	309	20
50	880	822	914	226	10
11° 00'	.1908	.9816	.1944	5.145	**79° 00'**
10	937	811	974	066	50
20	965	805	.2004	4.989	40
30	.1994	.9799	.2035	4.915	30
40	.2022	793	065	843	20
50	051	787	095	773	10
12° 00'	.2079	.9781	.2126	4.705	**78° 00'**
10	108	775	156	638	50
20	136	769	186	574	40
30	.2164	.9763	.2217	4.511	30
40	193	757	247	449	20
50	221	750	278	390	10
13° 00'	.2250	.9744	.2309	4.331	**77° 00'**
10	278	737	339	275	50
20	306	730	370	219	40
30	.2334	.9724	.2401	4.165	30
40	363	717	432	113	20
50	391	710	462	061	10
14° 00'	.2419	.9703	.2493	4.011	**76° 00'**
10	447	696	524	3.962	50
20	476	689	555	914	40
30	.2504	.9681	.2586	3.867	30
40	532	674	617	821	20
50	560	667	648	776	10
15° 00'	.2588	.9659	.2679	3.732	**75° 00'**
10	616	652	711	689	50
20	644	644	742	647	40
30	.2672	.9636	.2773	3.606	30
40	700	628	805	566	20
50	728	621	836	526	10
16° 00'	.2756	.9613	.2867	3.487	**74° 00'**
10	784	605	899	450	50
20	812	596	931	412	40
30	.2840	.9588	.2962	3.376	30
40	868	580	994	340	20
50	896	572	.3026	305	10
17° 00'	.2924	.9563	.3057	3.271	**73° 00'**
10	952	555	089	237	50
20	979	546	121	204	40
30	.3007	.9537	.3153	3.172	30
40	035	528	185	140	20
50	062	520	217	108	10
18° 00'	.3090	.9511	.3249	3.078	**72° 00'**
	Cos	Sin	Cot	Tan	

TABLE IV. *Trigonometric Functions* (continued)

	Sin	Cos	Tan	Cot	
18° 00′	.3090	.9511	.3249	3.078	**72° 00′**
10	118	502	281	047	50
20	145	492	314	018	40
30	.3173	.9483	.3346	2.989	30
40	201	474	378	960	20
50	228	465	411	932	10
19° 00′	.3256	.9455	.3443	2.904	**71° 00′**
10	283	446	476	877	50
20	311	436	508	850	40
30	.3338	.9426	.3541	2.824	30
40	365	417	574	798	20
50	393	407	607	773	10
20° 00′	.3420	.9397	.3640	2.747	**70° 00′**
10	448	387	673	723	50
20	475	377	706	699	40
30	.3502	.9367	.3739	2.675	30
40	529	356	772	651	20
50	557	346	805	628	10
21° 00′	.3584	.9336	.3839	2.605	**69° 00′**
10	611	325	872	583	50
20	638	315	906	560	40
30	.3665	.9304	.3939	2.539	30
40	692	293	973	517	20
50	719	283	.4006	496	10
22° 00′	.3746	.9272	.4040	2.475	**68° 00′**
10	773	261	074	455	50
20	800	250	108	434	40
30	.3827	.9239	.4142	2.414	30
40	854	228	176	394	20
50	881	216	210	375	10
23° 00′	.3907	.9205	.4245	2.356	**67° 00′**
10	934	194	279	337	50
20	961	182	314	318	40
30	.3987	.9171	.4348	2.300	30
40	.4014	159	383	282	20
50	041	147	417	264	10
24° 00′	.4067	.9135	.4452	2.246	**66° 00′**
10	094	124	487	229	50
20	120	112	522	211	40
30	.4147	.9100	.4557	2.194	30
40	173	088	592	177	20
50	200	075	628	161	10
25° 00′	.4226	.9063	.4663	2.145	**65° 00′**
10	253	051	699	128	50
20	279	038	734	112	40
30	.4305	.9026	.4770	2.097	30
40	331	013	806	081	20
50	358	001	841	066	10
26° 00′	.4384	.8988	.4877	2.050	**64° 00′**
10	410	975	913	035	50
20	436	962	950	020	40
30	.4462	.8949	.4986	2.006	30
40	488	936	.5022	1.991	20
50	514	923	059	977	10
27° 00′	.4540	.8910	.5095	1.963	**63° 00′**
	Cos	Sin	Cot	Tan	

TABLE IV. *Trigonometric Functions* (continued)

	Sin	Cos	Tan	Cot	
27° 00'	.4540	.8910	.5095	1.963	**63° 00'**
10	566	897	132	949	50
20	592	884	169	935	40
30	.4617	.8870	.5206	1.921	30
40	643	857	243	907	20
50	669	843	280	894	10
28° 00'	.4695	.8829	.5317	1.881	**62° 00'**
10	720	816	354	868	50
20	746	802	392	855	40
30	.4772	.8788	.5430	1.842	30
40	797	774	467	829	20
50	823	760	505	816	10
29° 00'	.4848	.8746	.5543	1.804	**61° 00'**
10	874	732	581	792	50
20	899	718	619	780	40
30	.4924	.8704	.5658	1.767	30
40	950	689	696	756	20
50	975	675	735	744	10
30° 00'	.5000	.8660	.5774	1.732	**60° 00'**
10	025	646	812	720	50
20	050	631	851	709	40
30	.5075	.8616	.5890	1.698	30
40	100	601	930	686	20
50	125	587	969	675	10
31° 00'	.5150	.8572	.6009	1.664	**59° 00'**
10	175	557	048	653	50
20	200	542	088	643	40
30	.5225	.8526	.6128	1.632	30
40	250	511	168	621	20
50	275	496	208	611	10
32° 00'	.5299	.8480	.6249	1.600	**58° 00'**
10	324	465	289	590	50
20	348	450	330	580	40
30	.5373	.8434	.6371	1.570	30
40	398	418	412	560	20
50	422	403	453	550	10
33° 00'	.5446	.8387	.6494	1.540	**57° 00'**
10	471	371	536	530	50
20	495	355	577	520	40
30	.5519	.8339	.6619	1.511	30
40	544	323	661	501	20
50	568	307	703	1.492	10
34° 00'	.5592	.8290	.6745	1.483	**56° 00'**
10	616	274	787	473	50
20	640	258	830	464	40
30	.5664	.8241	.6873	1.455	30
40	688	225	916	446	20
50	712	208	959	437	10
35° 00'	.5736	.8192	.7002	1.428	**55° 00'**
10	760	175	046	419	50
20	783	158	089	411	40
30	.5807	.8141	.7133	1.402	30
40	831	124	177	393	20
50	854	107	221	385	10
36° 00'	.5878	.8090	.7265	1.376	**54° 00'**
	Cos	Sin	Cot	Tan	

TABLE IV. *Trigonometric Functions* (continued)

	Sin	Cos	Tan	Cot	
36° 00′	.5878	.8090	.7265	1.376	**54° 00′**
10	901	073	310	368	50
20	925	056	355	360	40
30	.5948	.8039	.7400	1.351	30
40	972	021	445	343	20
50	995	004	490	335	10
37° 00′	.6018	.7986	.7536	1.327	**53° 00′**
10	041	969	581	319	50
20	065	951	627	311	40
30	.6088	.7934	.7673	1.303	30
40	111	916	720	295	20
50	134	898	766	288	10
38° 00′	.6157	.7880	.7813	1.280	**52° 00′**
10	180	862	860	272	50
20	202	844	907	265	40
30	.6225	.7826	.7954	1.257	30
40	248	808	.8002	250	20
50	271	790	050	242	10
39° 00′	.6293	.7771	.8098	1.235	**51° 00′**
10	316	753	146	228	50
20	338	735	195	220	40
30	.6361	.7716	.8243	1.213	30
40	383	698	292	206	20
50	406	679	342	199	10
40° 00′	.6428	.7660	.8391	1.192	**50° 00′**
10	450	642	441	185	50
20	472	623	491	178	40
30	.6494	.7604	.8541	1.171	30
40	517	585	591	164	20
50	539	566	642	157	10
41° 00′	.6561	.7547	.8693	1.150	**49° 00′**
10	583	528	744	144	50
20	604	509	796	137	40
30	.6626	.7490	.8847	1.130	30
40	648	470	899	124	20
50	670	451	952	117	10
42° 00′	.6691	.7431	.9004	1.111	**48° 00′**
10	713	412	057	104	50
20	734	392	110	098	40
30	.6756	.7373	.9163	1.091	30
40	777	353	217	085	20
50	799	333	271	079	10
43° 00′	.6820	.7314	.9325	1.072	**47° 00′**
10	841	294	380	066	50
20	862	274	435	060	40
30	.6884	.7254	.9490	1.054	30
40	905	234	545	048	20
50	926	214	601	042	10
44° 00′	.6947	.7193	.9657	1.036	**46° 00′**
10	967	173	713	030	50
20	988	153	770	024	40
30	.7009	.7133	.9827	1.018	30
40	030	112	884	012	20
50	050	092	942	006	10
45° 00′	.7071	.7071	1.000	1.000	**45° 00′**
	Cos	Sin	Cot	Tan	

Solutions and Examples

CHAPTER 1

Exercise 1.1

1. (a) $-3/4, 2, 0$ (b) $1.4, 1.33\ldots, \sqrt[3]{8} = 2$
2. (a) $1, 0$ (b) $-2, 9/3 = 3, -4.0 = -4, -\sqrt{4} = -2$
3. (a) commutative   (b) associative and definition of $3 + 4$
4. (a) distributive and definition of $2 + 3$ (b) same as 4(a)
5. (a) associative   (b) commutative then associative $2 + (x + 5) = 2 + (5 + x) = (2 + 5) + x = 7 + x$
6. (a) distributive and multiplicative inverse for $4 \cdot 1/4 = 1$
   (b) distributive, associative, additive inverse, then additive identity
7. (a) associative   (b) associative twice
8. (a) distributive [here $(x + 2)$ is treated as "a" in $a(b + c) = ab + ac$]   (b) distributive
   (c) commutative   (d) associative   (e) distributive
9. (a) associative   (b) additive identity   (c) multiplicative law of equality   (d) associative
   (e) multiplicative inverse
10. (a) distributive   (b) associative   (c) additive identity   (d) additive law of equality
    (e) distributive   (f) associative   (g) additive inverses
11. $x + 2 = 4$
    $(x + 2) + (-2) = 4 + (-2)$ *additive law of equality*
    $x + [2 + (-2)] = 2$ *associative*
    $x + 0 = 2$ *additive inverses*
    $x = 2$ *additive identity*
17. $2x + 1 = x + 2$
    $(2x + 1) + (-1) + (-x) = (x + 2) + (-1) + (-x)$ *additive law of equality*
    $2x + [1 + (-1)] + (-x) = x + [2 + (-1)] + (-x)$ *associative*
    $(2x + 0) + (-x) = (x + 1) + (-x)$ *additive inverse, associative*
    $2x + (-x) = (1 + x) + (-x)$ *additive identity and commutative*
    $2x + (-x) = 1 + [x + (-x)]$ *associative*
    $[2 + (-1)]x = 1 + [1 + (-1)]x$ *distributive*
    $1 \cdot x = 1 + 0 \cdot x$ *additive inverse*
    $x = 1$ *multiplicative and additive identities*
19. $2(x + 3) = 3x + (-4)$
    $2x + 6 = 3x + (-4)$ *distributive*
    $-3x + (2x + 6) + (-6) = -3x + [3x + (-4)] + (-6)$ *additive law of equality*
    $(-3x + 2x) + [6 + (-6)] = (-3x + 3x) + [-4 + (-6)]$ *associative*
    $(-3 + 2)x + 0 = (-3 + 3)x + (-10)$ *distributive, additive inverse*
    $(-1)x + 0 = 0 \cdot x + (-10)$ *additive inverse*
    $(-1)x = -10$ *additive identity*
    $(-1)[(-1)x] = (-1)(-10)$ *multiplicative law of equality*
    $[(-1)(-1)]x = 10$ *associative*
    $1 \cdot x = 10$ *definition of multiplication*
    $x = 10$ *multiplicative identity*

20. If $a \cdot b = 0$, then $a = 0$ or $b = 0$.
    Proof.   Assume $a \neq 0$, then $1/a$ exists. If $a \cdot b = 0$, then

    $\dfrac{1}{a}(a \cdot b) = \dfrac{1}{a} \cdot 0$

    $\left(\dfrac{1}{a} \cdot a\right) \cdot b = 0$

    $1 \cdot b = 0$

    $b = 0$

    If $b \neq 0$, then $1/b$ exists, and a similar process leads to $a = 0$. Therefore, if $a \cdot b = 0$, then $a = 0$ or $b = 0$.

24. Assume there are two multiplicative inverses, x and y, for some real number, a. Then

    $x \cdot a = a \cdot x = 1$ and $y \cdot a = a \cdot y = 1$

    Therefore,

    $x \cdot a = y \cdot a$
    $(x \cdot a) \cdot x = (y \cdot a) \cdot x$
    $x \cdot (a \cdot x) = y \cdot (a \cdot x)$
    $x \cdot 1 = y \cdot 1$
    $x = y$

    Hence, the multiplicative inverse of any nonzero real number is unique.

26. (a) The naturals do not have additive or multiplicative inverses nor the additive identity 0. For 2, its additive inverse -2 and its multiplicative inverse $\frac{1}{2}$ are not natural numbers.
    (b) The integers do not have multiplicative inverses, except for ± 1.
    (c) The irrationals do not contain identities, 1 and 0, and they are not closed under addition or multiplication [$\sqrt{3} + (-\sqrt{3}) = 0$ and $\sqrt{2} \cdot \sqrt{8} = 4$].

Exercise 1.2

1. (a) 6   (b) 6   (c) 6
2. (a) -2 (b) -2 (c) -2
3. (a) 11   (b) 11   (c) 11
4. (a) 18   (b) 18 [(a) and (b) illustrate the distributive law]
5. 6   6. 6   7. -18
9. (a) If $a < b$ and $c < 0$, then $ac > bc$. If $c < 0$, then $-c > 0$ and

$a(-c) < b(-c)$	by Property IV
$ac - ac < ac - bc$	by Property III
$0 + bc < (ac - bc) + bc$	additive inverses and Property III
$bc < ac + (-bc + bc)$	associative and additive identity
$bc < ac$	additive inverses and identity
$ac > bc$	definition of ">"

11. (a) If $a \cdot b > 0$ and $a < 0$, then $b < 0$. If $a < 0$, then, since $a^2 > 0$,

$$\frac{1}{a^2} \cdot a < \frac{1}{a^2} \cdot 0 \qquad \text{by Property IV}$$

$$\frac{1}{a} < 0 \qquad \text{multiplicative inverses and identity}$$

and

$$\frac{1}{a}(a \cdot b) < \frac{1}{a} \cdot 0 \qquad \text{by Property IV}$$

$$\left(\frac{1}{a} \cdot a\right) \cdot b < 0 \qquad \text{associative}$$

$$1 \cdot b < 0 \qquad \text{multiplicative inverses}$$

$$b < 0 \qquad \text{multiplicative identity}$$

Exercise 1.3

1. (a) $x < 1$ (b) $x < 2$ (c) $x < 3$

 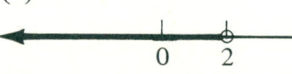 

2. (a) $x \geq 4$ (b) $x \geq 4/3$ (c) $x \geq -13$

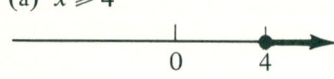 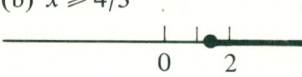 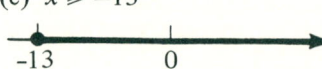

3. (a) $x \geq -4$ (b) $x \geq -10$ (c) $x \geq 4$

4. (a) $y < -1$ (b) $y > 0$ (c) $y < 28$

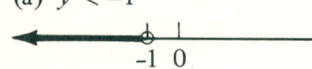  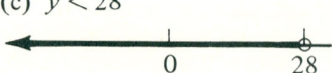

5. (a) $x \leq 17$ (b) $x \geq 13$ (c) $x \geq 18/5$

 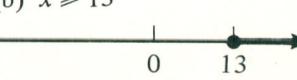 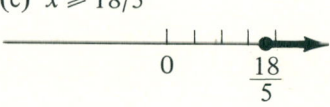

6. (a) All points x within four units of the origin

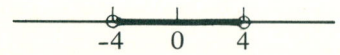

(b) All points x, three or more units from the origin

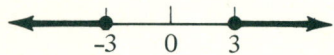

7. (a) All points x, four or less units from $x = 3$

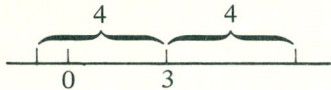

(b) All points x, three or more units from $x = 2$

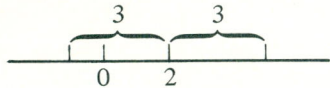

(c) All points x, two or less units from $x = -1$, since $|x + 1| = |x - (-1)|$

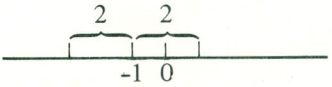

8. (a) $t = 7$ or $t = -7$
   (b) $t = 8$ or $t = -8$
   (c) No solution
9. (a) $x < -4$ or $x > 4$
   (b) $x \leq -4$ or $x \geq 4$
   (c) All reals
10. (a) $x = 3$ or $x = -3$
    (b) $x = 4$ or $x = -2$
    (c) $x = 2$ or $x = -4$
11. (a) $-3 < x < 3$
    (b) $-2 < x < 4$
    (c) $-4 < x < 2$
12. (a) $x < -4$ or $x > 4$
    (b) $x < -2$ or $x > 6$
    (c) $c < -6$ or $x > 2$
13. (a) $-3 < x < 5$
    (b) $-3 < x < 5$
14. (a) $x < -2$ or $x > 5$
    (b) $x < -2$ or $x > 5$
15. (a) $x = 5$ or $x = -1$
    (b) $x = 5$ or $x = -1$
16. (a) $x < -10$ or $x > 2$
    (b) $x < -10$ or $x > 2$
17. (a) $x < -1/4$ or $x > 5/4$
    (b) $x < -1/4$ or $x > 5/4$
18. (a) $|x| = 3$ means

    $x = -3$ or $x = 3$

    (b) $x^2 = 9$ means

    $\sqrt{x^2} = \sqrt{9}$

    $|x| = 3$

    $x = 3$ or $x = -3$
19. (a) $x = -2$ or $x = 2$
    (b) $x = -3$ or $x = 5$
21. Since $x^2 = (-x)^2$ and $2x - 1 = -(1 - 2x)$, we have
    (a) $x = -1/2$ or $x = 3/2$
    (b) $x = -1/2$ or $x = 3/2$
23. Since $4x^2 + 4x + 1 = (2x + 1)^2$
    (a) $x = -3$ or $x = 2$
    (b) $x = -3$ or $x = 2$

24. (a) If $|x+1| \leq |x|$, then
$$(x+1)^2 \leq x^2$$
$$x^2 + 2x + 1 \leq x^2$$
$$2x + 1 \leq 0$$
$$x \leq -1/2$$

25. (a) $x \leq 5/6$
    (b) $x \geq -1/2$ (Dividing by a negative reverses the order of the inequality.)

27. (a) $x \geq -7/2$
    (b) $x \leq 1/2$

29. $-1/2 < x < 1/2$

Exercise 1.4

1. (a) x^5
   (b) x^{15}
   (c) x^5

2. (a) x^3
   (b) x^3
   (c) x^3

3. (a) x^6
   (b) x^6
   (c) x^6

4. (a) x
   (b) x
   (c) x

5. (a) x^4/y^6
   (b) x^4/y^6
   (c) x^4/y^6

6. (a) x/y
   (b) x/y
   (c) x/y

7. (a) $2x$
   (b) $2x$
   (c) $2x$

8. (a) $2/3$
   (b) $8/27$
   (c) $8/27$

9. (a) $\sqrt{6}/6$
   (b) $2\sqrt{6}$

10. $(5\sqrt{3} - 6\sqrt{5})/15$

11. $(\sqrt{3} - 3)/3$

12. (a) $2\sqrt{3} + 2\sqrt{2}$
    (b) $(-\sqrt{5} - 5)/4$

13. (a) x^2
    (b) $1/x^2$
    (c) x

14. (a) x^2/y
    (b) y^3/x^2
    (c) x/y

15. (a) $4x^2z/y$
    (b) $y^3z^2/2x^2$
    (c) In division for like bases, the smaller power is subtracted from the larger power. For the x-factor, $-2 < -1$ means $x^{-2}/x^{-1} = 1/x^{-1-(-2)} = 1/x$.
$$\frac{2x^{-2}y^{-1}z^{-3}}{8x^{-1}y^{-2}z} = \frac{y^{-1-(-2)}}{4x^{-1-(-2)}z^{1-(-3)}} + \frac{y^{-1+2}}{4x^{-1+2}z^{1+3}} = \frac{y}{4xz^4}$$

The ys ended up in the numerator since $-1 > -2$ means we subtract the -2 from the -1. That is, we divided y^{-2} into y^{-1}. The divisor's exponent is subtracted from the dividend.

16. (a) b^3c/a^5
    (b) b^3c/a^5
    (c) b^3c/a^5

17. (a) x^3y^6/z^9
    (b) x^3y^6/z^9
    (c) y^9/x^6

18. (a) $x^2 + x$
    (b) $x + 1$
    (c) $x^3 + x^2$

19. $\sqrt{y} - \sqrt{x}$

20. $\dfrac{\sqrt{x}+1}{\sqrt{x}}$

21. Factor (using the distributive law) the common factor $(x + 1)$ to the lowest power to which it occurs, $-1/2$. Notice that

$$(x + 1)^{-1/2} \cdot (x + 1)^1 = (x + 1)^{+1/2}$$

$$(x + 1)^{1/2} - x(x + 1)^{-1/2} = (x + 1)^{-1/2}[(x + 1)^1 - x] = \dfrac{1}{\sqrt{x+1}} [1] = \dfrac{1}{\sqrt{x+1}}$$

Exercise 1.5

1. (a) $5 + 7i$
   (b) $-1 - 4i$
   (c) $(1 + \sqrt{2}) + 5i$

2. These products are real numbers because the factors are conjugates.
   (a) 13 (b) 13 (c) 3

3. (a) $-4 + 20i$
   (b) $9 - 45i$
   (c) $-16 + 80i$

4. (a) $2 + i$
   (b) $-2 + 3i$

5. (a) $-7/25 + (24/25)i$
   (b) $21/29 - (20/29)i$

6. (a) $3/2 + (9/8)i$
   (b) $32/13 - (43/13)i$

7. (a) $|z| = 5$ (d) $|z| = 13$
   (b) $|z| = 10$ (e) $|z| = 9$
   (c) $|z| = 13$ (f) $|z| = 3$

9. $|z_1 + z_2| = 10$ and $|z_1| + |z_2| = \sqrt{13} + \sqrt{41}$. They are not equal. In fact $|z_1 + z_1| \leq |z_1| + |z_2|$ for all complex numbers z_1 and z_2. In this case, $\sqrt{13} + \sqrt{41} \approx 10.01$. If $z_1 = 3$ and $z_2 = 4$, then $|z_1 + z_2| = |z_1| + |z_2|$

10. (d) $i^{27} = (i^2)^{13} \cdot i = (-1)^{13} \cdot i = -i$

11. $|2z| = 2|z|$. For $z = 6 - 8i$, both are 20. For $z = -3 + 4i$, both are 10. Notice that $6 - 8i = -2(-3 + 4i)$ but $|6 - 8i| \neq -2|-3 + 4i|$ that is, $|-2z| \neq -2|z|$.

12. (b) $x = \dfrac{1 + \sqrt{3}i}{2}$ is a cube root of -1, since

$$x^3 = \left(\dfrac{1 + \sqrt{3}i}{2}\right)\left(\dfrac{1 + \sqrt{3}i}{2}\right)\left(\dfrac{1 + \sqrt{3}i}{2}\right)$$

$$= \left(\dfrac{-2 + 2\sqrt{3}i}{4}\right)\left(\dfrac{1 + \sqrt{3}i}{2}\right) = \dfrac{-8}{8} = -1$$

13. $x = \pm 9i$ are square roots of -81.

15. Let $\sqrt{i} = a + bi$. Then,

$(\sqrt{i})^2 = (a + bi)^2$

$i = (a^2 - b^2) + 2abi$, where $i = 0 + 1 \cdot i$ means

$a^2 - b^2 = 0$ and $2ab = 1$

$a^2 = b^2$ and $a = \dfrac{1}{2b}$

$\left(\dfrac{1}{2b}\right)^2 = b^2$

A-7

$$\frac{1}{4} = b^4$$

$$b = \sqrt[4]{\frac{1}{4}} = \frac{1}{\sqrt{2}} = \frac{\sqrt{2}}{2}$$

and

$$a = \frac{1}{2b} = \frac{\sqrt{2}}{2}$$

Hence

$$\sqrt{i} = \sqrt{2}/2 + \sqrt{2}/2\, i.$$

There are two square roots of i. What is the other one?

Exercise 1.6

B. These are all perfect-square quadratics.
1. (a) 2 (b) 3 (c) 5
2. (a) −4 (b) −1 (c) −6
3. (a) 3/2 (b) −1/2 (c) 2/3
4. These are the difference of two squares.
   (a) $x = -3$ or $x = 3$
   (b) $x = -4$ or $x = 4$
   (c) $x = -1$ or $x = 1$
5. (a) $x = -5$ or $x = 5$
   (b) $x = -3/2$ or $x = 3/2$
   (c) $x = -4/3$ or $x = 4/3$
6. The sign on the last factor is +, so the solutions will have the same sign. Do you see why $x = 1$ (or $x = -1$) is a solution for each?
   (a) $x = 1$ or $x = 7$
   (b) $x = 1$ or $x = 10$
   (c) $x = 1$ or $x = 11$
7. (a) $x = -1$ or $x = -2$
   (b) $x = -1$ or $x = -6$
   (c) $x = -1$ or $x = -10$
8. Since these next two problems have a negative constant term, the solutions will have different signs.
   (a) $x = -1$ or $x = 3$
   (b) $x = -1$ or $x = 6$
   (c) $x = -1$ or $x = 9$
9. (a) $x = 1$ or $x = -12$
   (b) $x = 1$ or $x = -3$
   (c) $x = 1$ or $x = -6$
10. Notice how the solutions of Problems 10 and 11 are closely related. The slight change in sign in the quadratics is reflected in their solutions.
    (a) $x = -4$ or $x = -3$
    (b) $x = -6$ or $x = -2$
    (c) $x = -3$ or $x = -2$
11. (a) $x = 3$ or $x = 4$
    (b) $x = 2$ or $x = 6$
    (c) $x = 2$ or $x = 3$
12. (a) $x = -2/3$ or $x = 1/2$
    (b) $x = -1/4$ or $x = 3/2$
    (c) $x = 4/3$ or $x = 3/2$
13. (a) $x = \pm 2$
    (b) $x = -1$ or $x = 3$
    (c) $x = 1$ or $x = 7$
14. (a) $x = 1 \pm \sqrt{5}$
    (b) $x = 1 \pm \sqrt{3}$
    (c) $x = 2 \pm \sqrt{6}$
    (In Problem 14 these each have two solutions. For example, $x = 2 \pm \sqrt{6}$ means $x = 2 + \sqrt{6} \approx 4.4$ or $x = 2 - \sqrt{6} \approx 0.45$.)
15. (a) $x = 3 \pm \sqrt{13}$
    (b) $x = 1/2(3 \pm \sqrt{13})$

(c) $2x^2 - 4x - 1 = 0$

$x^2 - 2x = 1/2$, dividing both sides by 2

$x^2 - 2x + 1 = \dfrac{1}{2} + 1$, since $\left(\dfrac{-2}{2}\right)^2 = 1$

$(x - 1)^2 = 3/2$

$x - 1 = \pm\sqrt{\dfrac{3}{2}} = \pm\dfrac{\sqrt{6}}{2}$ yields $x = \dfrac{2 \pm \sqrt{6}}{2}$

16. $x = 1$ or $x = 5/2$
17. $x = \dfrac{1 \pm \sqrt{13}}{6}$
18. $x = -3/2$ or $x = 1$
19. $x = \dfrac{1 \pm \sqrt{33}}{8}$
20. $x = -2/3$ or $x = 1$
21. No real solutions, $x = 1/2(-1 \pm \sqrt{3}i)$ is the complex solution.
22. For $2x^2 - x - 1 = 0$, the discriminant

$b^2 - 4ac = (-1)^2 - 4(2)(-1)$
$= 1 + 8$
$= 9$

is positive. Therefore, there are two real roots. Since 9 is a perfect square, and a, b, and c are rational, there are two rational roots.

23. $b^2 - 4ac = 0$, one real
25. two real
26. no real solution, $b^2 - 4ac < 0$
27. For $3x^2 - 2x + k = 0$, the discriminant $b^2 - 4ac = (-2)^2 - 4(3)k = 4 - 12k$.

(i) For two real solutions, $b^2 - 4ac > 0$ and

$4 - 12k > 0$
$-12k > -4$
$k > \dfrac{1}{3}$

(ii) For no real solutions, $b^2 - 4ac < 0$ and

$4 - 12k < 0$
$-12k < -4$
$k > \dfrac{1}{3}$

(iii) For one real solution, $b^2 - 4ac = 0$ and

$4 - 12k = 0$
$k = \dfrac{1}{3}$

29. (i) $k > -9/8$, $k \neq 0$ (ii) $k < -9/8$ (iii) $k = -9/8$ or $k = 0$

31. $a = 64$ 33. $a = 15$ 34. $b^2 = 4c$

Exercise 1.7

1. yes, $u = x^2$
2. yes, $u = \sqrt{x}$
3. yes, $u = x^{1/3}$
4. yes, $u = x + 1$
5. no
6. no
7. (a) $x = \pm 1$
   (b) $x = \pm 3$
   (c) $x = \pm 4$
8. (a) $x = 1$
   (b) $x = 4$
   (c) $x = 1$ or $x = 9$
9. (a) $x^{2/3} + x^{1/3} = 2$. Let $u = x^{1/3}$. Then, $u^2 = x^{2/3}$ and $u^2 + u = 2$.

   $u^2 + u - 2 = 0$
   $(u + 2)(u - 1) = 0$
   $u + 2 = 0$ or $u - 1 = 0$
   $u = -2$ or $u = 1$

   Since $x^{1/3} = u$ means $x = u^3$, we have $x = (-2)^3 = -8$ or $x = (1)^3 = 1$.
   (b) $x = 4$ or $x = 9$
   (c) No solution

10. (a) $x = \pm 2$
    (b) $x = 1$ or $x = 2$
    (c) $x = 10$

11. $4/x + 3/\sqrt{x} = 1$
    $4x^{-1} + 3x^{-1/2} - 1 = 0$
    Let $u = x^{-1/2}$, then $u^2 = x^{-1}$ and $4u^2 + 3u - 1 = 0$.
    $(4u - 1)(u + 1) = 0$
    $4u - 1 = 0$ or $u + 1 = 0$

    $u = \dfrac{1}{4}$ or $u = -1$

    Since $u = x^{-1/2}$, $u^{-2} = (x^{-1/2})^{-2}$ yields $x = (1/4)^{-2} = (4)^2 = 16$. $u \neq -1$ since $u = 1/\sqrt{x}$, where $\sqrt{x}$ means the principal (positive) square root of x.

Exercise 1.8

1. (a) 12   (b) 27   (c) 20
2. (a) 18   (b) 84   (c) 60
3. (a) 231   (b) 90   (c) 1800
4. (a) 1260 (b) 525   (c) 900
5. (a) $2x(x + 2)$
   (b) $6x(x - 3)$
6. (a) $6x(x^2 - 4)(x^2 - 3)$
   (b) $2(x - 3)(x + 3)(x + 1)$
7. (a) 0   (b) 4   (c) 2
8. (a) $-1/16$ (b) 2   (c) 8

9. (a) $x = -1$ or $x = 6$
   (b) $x = 0$ or $x = 4$
   (c) $x = 1$

10. (a) $\dfrac{2}{3x-6} - \dfrac{5}{x-2} = \dfrac{-1}{3}$, LCM $= 3(x-2)$

   $3(x-2)\dfrac{2}{3(x-2)} - 3(x-2)\dfrac{5}{x-2} = 3(x-2)\dfrac{-1}{3}$

   $2 - 15 = -(x-2)$

   $x = 15$

11. (a) $x = 22/3$
    (b) $x = -12$

13. $x = 5/3$

15. $x = -2$

17. $x = -3$ or $x = 2$

18. (a) $x^2 + x = \dfrac{42}{x^2+x} - 1$. Let $u = x^2 + x$. Then

   $u = \dfrac{42}{u} - 1$

   $u \cdot u = u \cdot \dfrac{42}{u} - u \cdot 1$

   $u^2 = 42 - u$

   $u^2 + u - 42 = 0$

   $(u+7)(u-6) = 0$

   $u = -7$ or $u = 6$

   For $u = -7$, we have $x^2 + x = -7$ and $x = \dfrac{-1 \pm \sqrt{27}\,i}{2}$ (no real solution).

   For $u = 6$, we have $x^2 + x = 6$ and $x = -3$ or $x = 2$.

Exercise 1.9

1. (a) $x \leqslant 2$
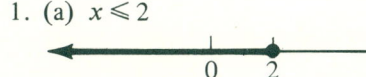
   (b) $x \leqslant 4$
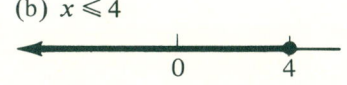
   (c) $x \leqslant 2$
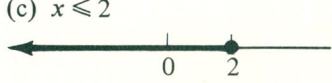

2. (a) $x \geqslant -2$
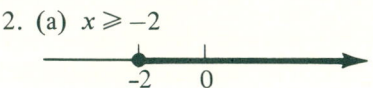
   (b) $x \geqslant -3$
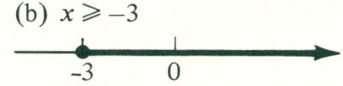
   (c) $x \geqslant -2$
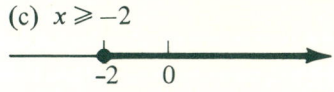

3. (a) $x > 3$
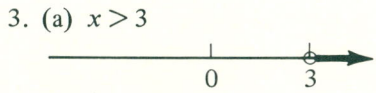
   (b) $x = 3$
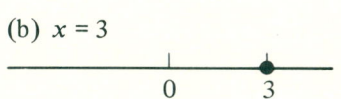
   (c) $x < 3$

4. (a) $x < -4$ 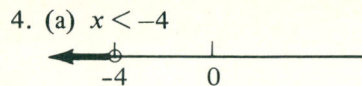 (b) $x < -6$ 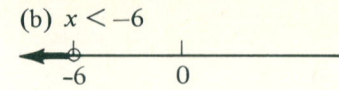 (c) $x < 2$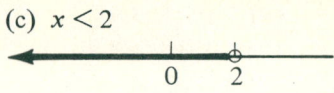

5. (a) $x \leqslant 2$  (b) $x < -5$ 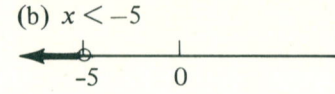 (c) $x < -11$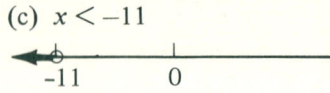

6. (a) $x < -4$ 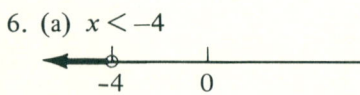 (b) $x > 3$ (c) $x > 5$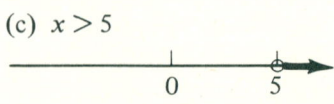

7. (a) $x \leqslant 3/5$ (b) $x \leqslant -3/4$ (c) $x > 0$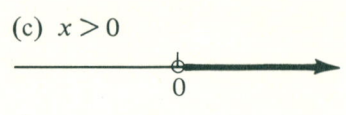

8. (a) $-2 \leqslant x \leqslant 2$ (b) $-2 < x < 2$ (c) $x < -2$ or $x > 2$

9. (a) $-2 < x < 1$ (b) $x \leqslant -2$ or $x \geqslant 1$

10. (a) $-1 \leqslant x \leqslant 0$ (b) $x \leqslant 0$ or $x \geqslant 2$ (c) $x \leqslant 0$ or $x \geqslant 2$

11. (a) $-1 < x \leqslant 5$ (b) $-1 \leqslant x < 4$ (c) $-2 < x < 3$

12. (a) $x \leqslant -1$ or $0 \leqslant x \leqslant 1$ (b) $-3 < x < -1$ or $x > 1$ (c) $-3 \leqslant x \leqslant -2$ or $x \geqslant 4$

13. $1 < x \leqslant 2$ (notice $x \neq 1$) 14. $-1 < x \leqslant 2$ 15. $-16 \leqslant x < -3$

16. (b) $4x^2 \leqslant 9$
    Let $4x^2 = 9$. Then $4x^2 - 9 = 0$.

    $(2x + 3)(2x - 3) = 0$
    $2x + 3 = 0$ or $2x - 3 = 0$

    $x = -\frac{3}{2}$ or $x = \frac{3}{2}$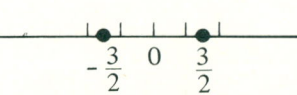

    Test $x = 0$ in the original inequality. If $x = 0$, then $4x^2 \leqslant 9$ becomes $0 \leqslant 9$, which is true. Therefore, $x = 0$, which is between our boundary points $x = -\frac{3}{2}$ and $x = \frac{3}{2}$, is a solution, so every point between $-\frac{3}{2}$ and $\frac{3}{2}$ must be solutions. Our answer, then, is

    $-\frac{3}{2} \leqslant x \leqslant \frac{3}{2}$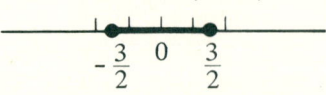

A-12

17. (a) $-2 \leqslant x \leqslant 4$
    (b) $-4 \leqslant x \leqslant 0$

19. (a) $-1/3 < x < 3/2$
    (b) $x \leqslant -1/3$ or $x \geqslant 3/2$

21. $3 \leqslant x \leqslant 4$

23. $x \leqslant -4/3$ or $x \geqslant 3/2$

24. The equation $x(6x + 7) \geqslant 3$ must be rewritten to make a statement about a positive or negative product.

$x(6x + 7) \geqslant 3$

$6x^2 + 7x - 3 \geqslant 0$

$(3x - 1)(2x + 3) \geqslant 0$

Case I. Both factors positive or zero.

$3x - 1 \geqslant 0$ and $2x + 3 \geqslant 0$ yields

$x \geqslant \frac{1}{3}$ and $x \geqslant -\frac{3}{2}$

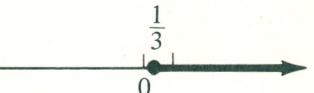

This means $x \geqslant \frac{1}{3}$, since any value of $x \geqslant \frac{1}{3}$ will also be greater than $-\frac{3}{2}$.

Case II. Both factors negative or zero.

$3x - 1 \leqslant 0$ and $2x + 3 \leqslant 0$ yields

$x \leqslant \frac{1}{3}$ and $x \leqslant -\frac{3}{2}$

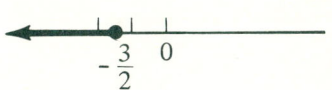

which means $x \leqslant -\frac{3}{2}$, since all values of $x \leqslant -\frac{3}{2}$ will also be less than $\frac{1}{3}$.

Any value of x making both factors positive or zero (Case I), *or* negative or zero (Case II), will cause the product $(3x - 1)(2x + 3)$ to be positive or zero. Therefore, our solution is

$x \leqslant \frac{3}{2}$ $\qquad$ $x \geqslant \frac{3}{2}$

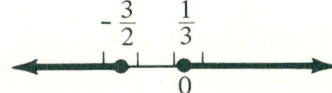

25. $-4 < x < 1$

27. $-3 \leqslant x < -1$ or $-1 < x < 4$

29. If $x^2 < a^2$, then $-a < x < a$. If $x^2 < a^2$, then $x^2 - a^2 < 0$ and $(x + a)(x - a) < 0$, where, without loss of generality, we can assume $a > 0$.

Case I. $x + a > 0$ and $x - a < 0$ means

$x > -a$ and $x < a$. Hence, $-a < x < a$.

Case II. $x + a < 0$ and $x - a > 0$, which is

$x < -a$ and $x > a$, has no solution. Therefore, $x^2 < a^2$ yields $-a < x < a$.

Exercise 1.10

1. (a) $x = 16$
   (b) No solution

2. (a) $x = 62$
   (b) No solution

3. (a) $x = 16$
   (b) $x = -16$

4. (a) $x = 4$
   (b) $x = 1$
5. (a) $x = 6$
   (b) $x = 1$
6. (a) $x = 1$
   (b) $x = 3$
7. (a) No solution
   (b) $x = -2$ or $x = -1$
9. $x = 1$
11. No solution

12. $\sqrt{4x - 3} - \sqrt{6 - 2x} = 3$ should be rewritten before squaring to yield the simplest result

$$\sqrt{4x - 3} = 3 + \sqrt{6 - 2x}$$
$$4x - 3 = 9 + 6\sqrt{6 - 2x} + (6 - 2x)$$
$$6x - 18 = 6\sqrt{6 - 2x}$$

Now divide both sides by 6 before squaring to keep our results simple.

$$x - 3 = \sqrt{6 - 2x}$$
$$x^2 - 6x + 9 = 6 - 2x$$
$$x^2 - 4x + 3 = 0$$
$$(x - 1)(x - 3) = 0 \text{ implies}$$

$x = 1$ or $x = 3$, but only $x = 3$ checks.

Check:
  If $x = 3$, then

$$\sqrt{4x - 3} - \sqrt{6 - 2x} = \sqrt{4 \cdot 3 - 3} - \sqrt{6 - 2 \cdot 3} = \sqrt{9} - 0 = 3$$

13. $x = -2$ or $x = 1$
15. $x = 3$.
   Since for $x = 3$,

$$\frac{2x - 3\sqrt{x}}{3} = \frac{6 - 3\sqrt{3}}{3} = 2 - \sqrt{3}$$

and

$$\frac{3}{2x + 3\sqrt{x}} = \frac{3}{6 + 3\sqrt{3}} = \frac{1}{2 + \sqrt{3}} \cdot \frac{2 - \sqrt{3}}{2 - \sqrt{3}} = \frac{2 - \sqrt{3}}{4 - 3} = 2 - \sqrt{3}$$

Exercise 1.11

1.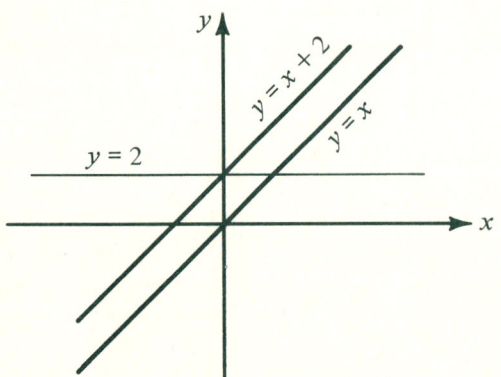

2.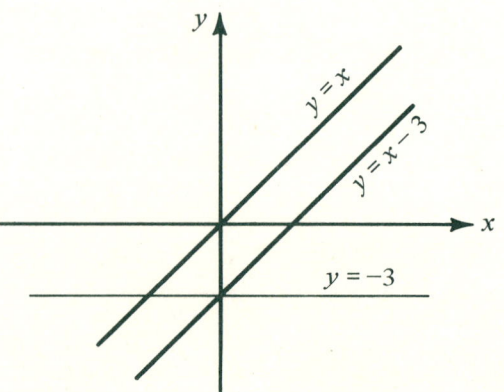

3.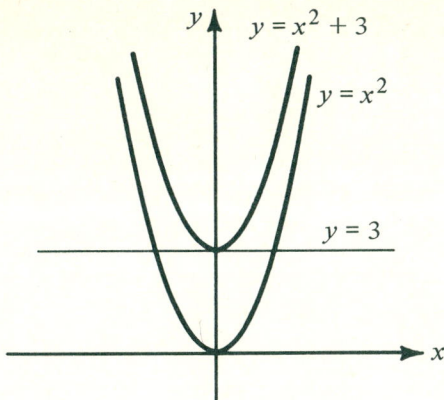

4.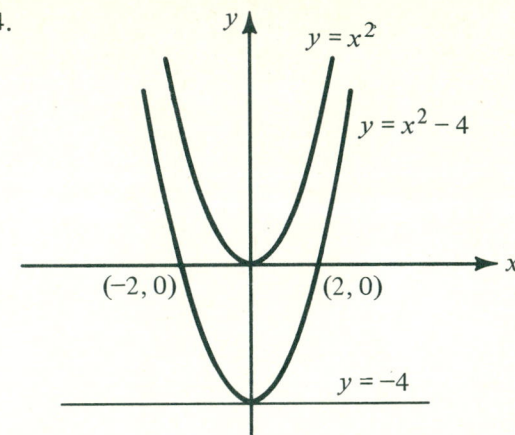

5.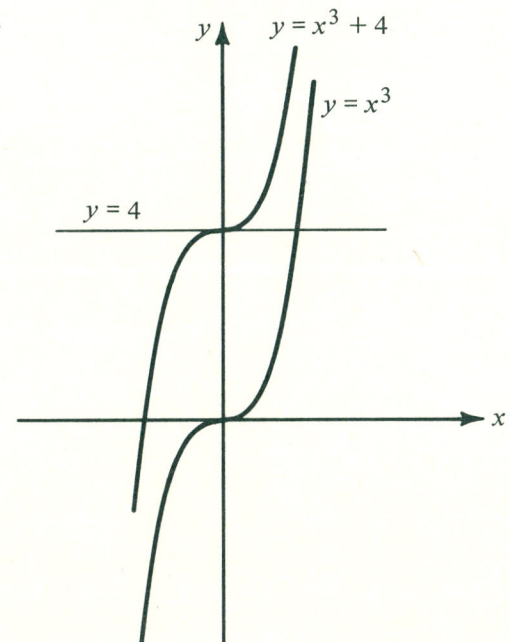

6.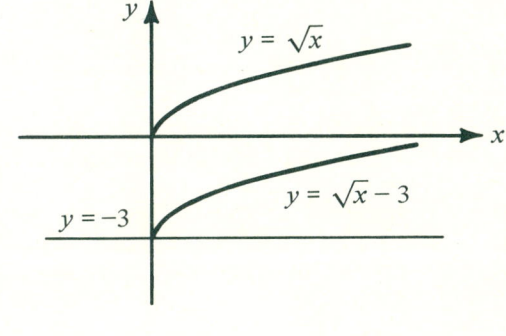

7. Below is a typical table of values.

x	$y = x$	$y = x^2$	$y = x^2 + x$
-3	-3	9	6
-2	-2	4	2
-1	-1	1	0
0	0	0	0
1	1	1	2
2	2	4	6

A–15

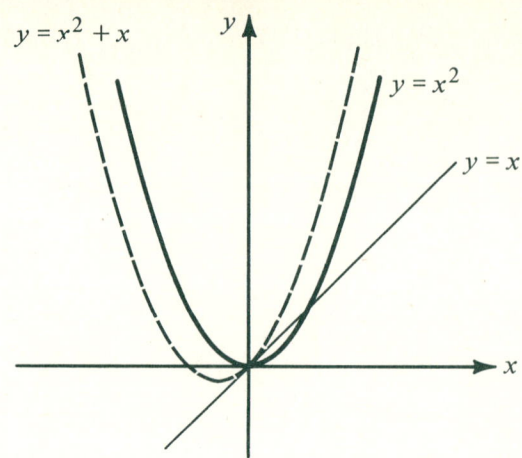

Then the points are plotted and the curves sketched. Notice that the basic shape of $y = x^2$ and $y = x^2 + x$ is the same. The vertical distance between the curves is always $|x|$. For small values of x, they are close. For large values the curves are far apart vertically. The curves cross only at $x = 0$.

8.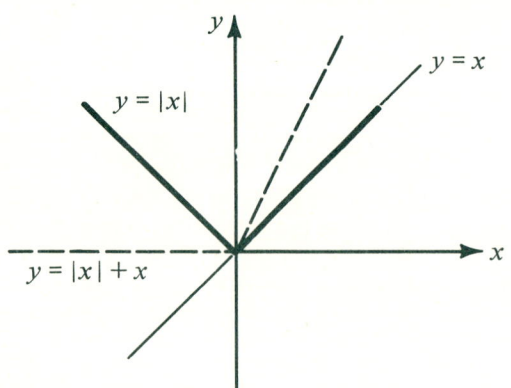

9.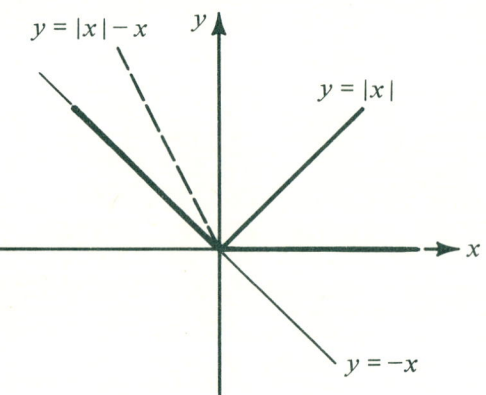

11.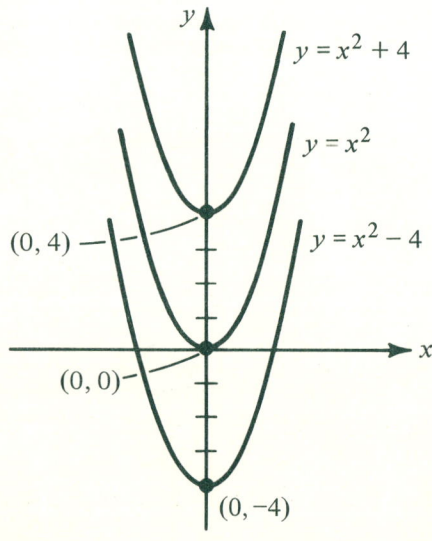

13.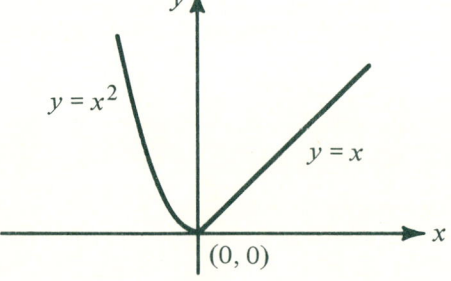

14. The table of values for the function might look like the following. Notice we evaluate $|x|$ for $x \geq 0$ and $[x]$ for $x < 0$.

x	$y = [x]$	$y = |x|$
-3	—	3
-2	—	2
-1	—	1
0	—	0
$\frac{1}{2}$	0	—
1	1	—
$\frac{5}{2}$	1	—
$\frac{5}{4}$	1	—
2	2	—
2.1	2	—

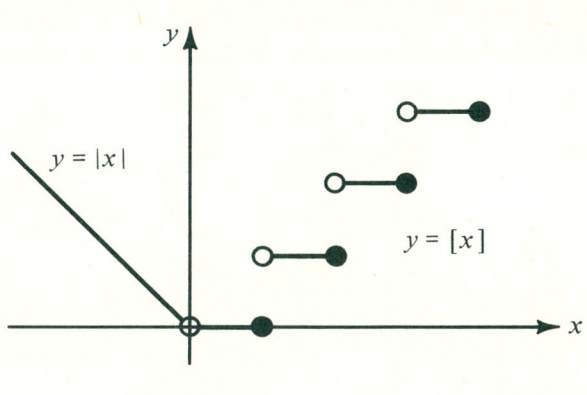

15.
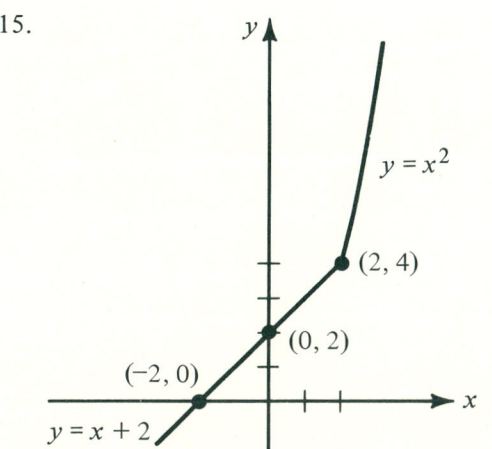

17.
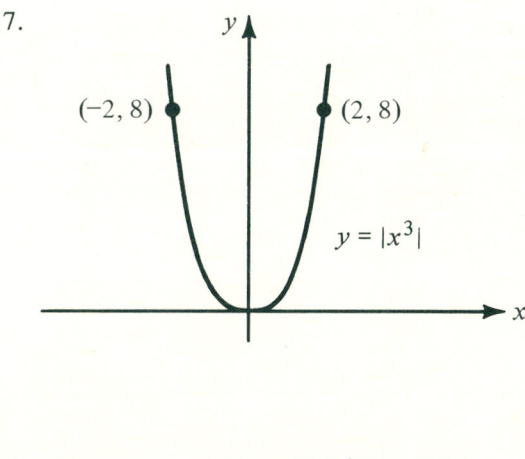

19.
(a)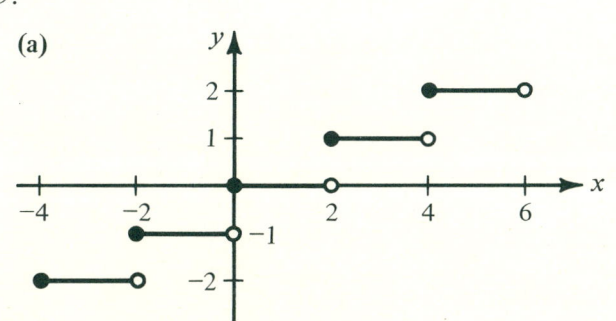

(b)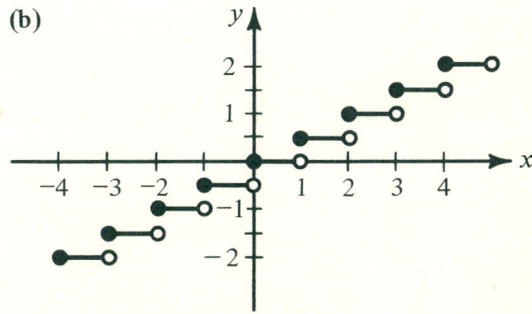

A-17

21.

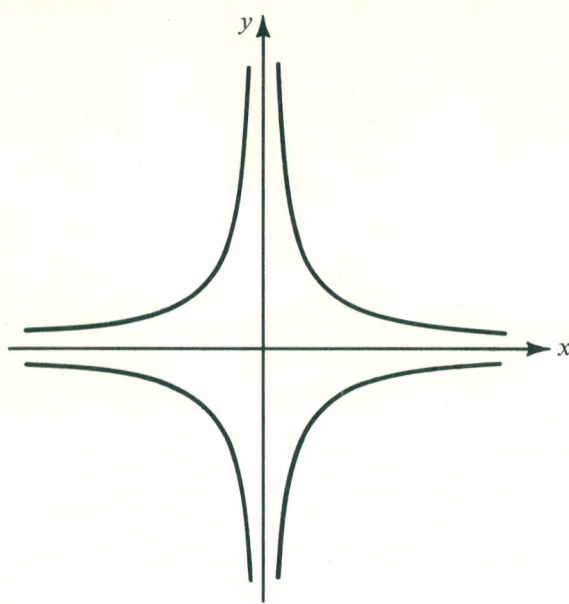

27. (a) $b^2 - 4ac = 0$ means that the quadratic equation $x^2 = 0$ has only one real root. This value of x for which $y = x^2$ is zero is the place where the graph of $y = x^2$ crosses the x-axis.
(b) $b^2 - 4ac < 0$ means that $x^2 + 1 = 0$ has no real roots. The graph of $y = x^2 + 1$ does not cross the x-axis. $y = x^2 + 1$ is never equal to zero for any real value of x.
(c) $b^2 - 4ac > 0$ means that $x^2 - 1 = 0$ has two real roots. The graph of $y = x^2 - 1$ has two x-intercepts $(-1, 0)$ and $(1, 0)$.

(a)
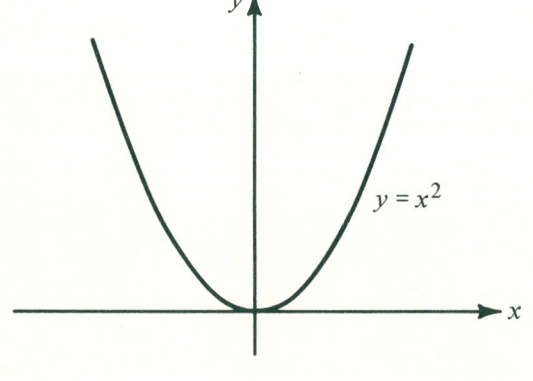
$y = x^2$

(b)
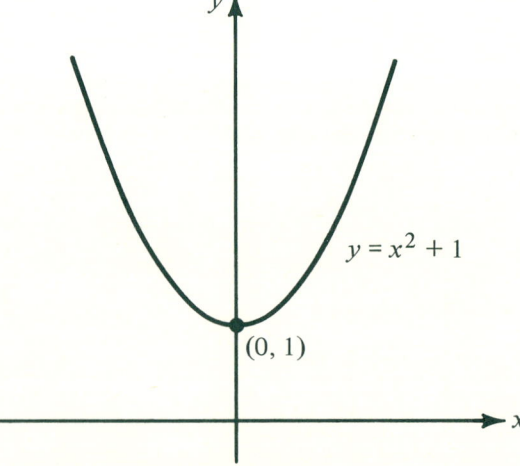
$y = x^2 + 1$

(c)

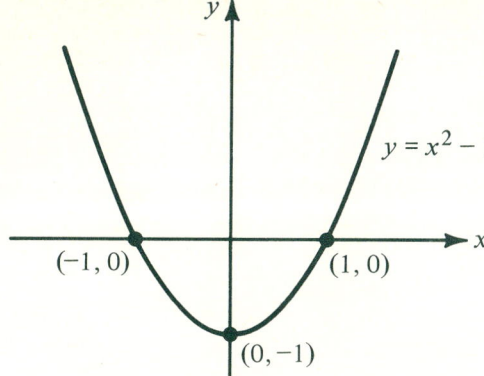

$y = x^2 - 1$

29.

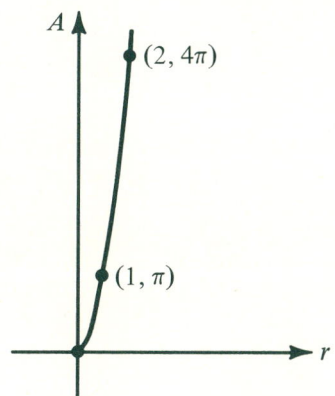

$A = \pi r^2, A \geq 0$ and $r \geq 0$

Exercise 1.12

1. Arithmetic, $d = 1$, $S = 3(11) = 33$
2. Geometric, $r = 2$, $S = [2 - 2(2)^5]/(1 - 2) = 62$
3. (a) Arithmetic, $d = 2$, $S = 10(22)/2 = 110$
   (b) Arithmetic, $d = 2$, $S = 10(3 + 21)/2 = 120$
4. (a) Arithmetic, $d = 4$, $S = 312$
   (b) Geometric, $r = 4$, $S = [4 - 4(4)^5]/(1 - 4) = 1364$
5. (a) Geometric, $r = 2$, $S = [6 - 6(2)^3]/(1 - 2) = 42$
   (b) Geometric, $r = 3$, $S = [6 - 6(3)^3]/(1 - 3) = 78$
6. (a) Arithmetic, $d = 2$, $S = 45$
   (b) Arithmetic, $d = 6$, $S = 90$

7. (a) $\sum_{i=1}^{100} 2i = 2 \sum_{i=1}^{100} i = 2(100)(101)/2 = 10{,}100$

(b) $\sum_{i=1}^{100}(2i-1) = 2\sum_{i=1}^{100} i - \sum_{i=1}^{100} 1 = 2(5050) - 100 = 10{,}000$

9. (a) 2156
   (b) 88,407

10. (c) $S = \sum_{i=1}^{10}(2i^3 - 3i^2)$

$= 2\sum_{i=1}^{10} i^3 - 3\sum_{i=1}^{10} i^2$

$= 2\dfrac{10^2(10+1)^2}{4} - 3\dfrac{10(10+1)(2 \cdot 10 + 1)}{6}$

$= 4895$

11. $S_6 = \dfrac{1}{2} + \dfrac{1}{4} + \dfrac{1}{8} + \dfrac{1}{16} + \dfrac{1}{32} + \dfrac{1}{64} = 0.9844$

$S_7 = S_6 + \dfrac{1}{128} = 0.9922$

$S_8 = S_7 + \dfrac{1}{256} = 0.9961$

$S_9 = S_8 + \dfrac{1}{512} = 0.9980$

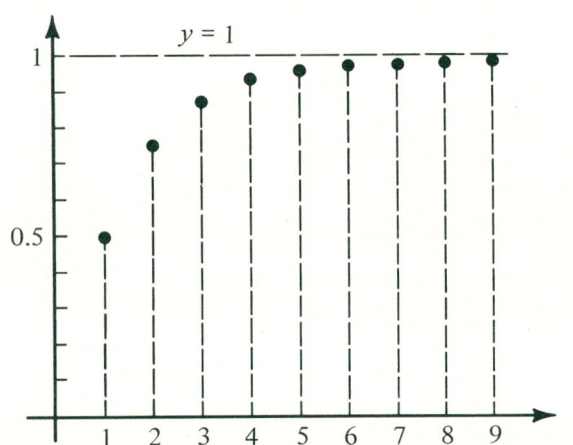

15. 3 meters

17. (a) 1/3   (b) 1

CHAPTER 2

Exercise 2.5

1. $\{0, 1, 2, 3, 4, 5, 6, 8\}$ 2. $\{\ \} = \emptyset$
3. $\{6\}$ 4. $\emptyset$
5. A, since $C \subseteq A$ 6. $\{0, 1, 2, 3, 4, 5, 6, 7, 8\} = U$, since $A' = E$
7. (a) $\{1, 3, 5, 7, 9\}$ (b) $\{0, 2, 4, 6, 8, 9\}$ (c) $A' \cup E' = \{9\}$ (d) $(A \cap E)' = \{9\}$
   This illustrates one of DeMorgan's Laws.

8. (a) $(A \cap B)' = \emptyset' = U$ (b) $(A \cup B)' = \{7, 9\}$ (c) $A' \cap B' = \{7, 9\}$ (d) $A' \cup B' = U$
   Parts (a) and (d), (b) and (c) depict DeMorgan's Law.

9. (a) U
   (b) $\{0, 2, 4, 9\}$
   (c) $\{0, 2, 4, 9\}$
   (d) U
   The equal results illustrate DeMorgan's Law.

10. (a) $\{2, 4, 6, 8\}$
    (b) $\{2, 4, 6, 7, 8\}$
    (c) $\{2, 4, 6, 8\}$
    (d) $\{2, 4, 6, 7, 8\}$
    The equal pairs illustrate the distributive property.

11. (a) $\{0, 2, 4, 6, 8\}$
    (b) $\{6, 8\}$
    (c) $\{0, 2, 4, 6, 8\}$
    Parts (a) and (c) depict correct usage of the distributive property.

12. (a) $\emptyset$
    (b) $\emptyset$
    The associative property.

13. (a) $\{0, 1, 2, 3, 4, 5, 6, 7, 8\}$
    (b) $\{0, 1, 2, 3, 4, 5, 6, 7, 8\}$
    The associative property.

14. False
15. True
16. True
17. True
18. False
19. True
20. False
21. True
22. False $A \cap E = \emptyset$ and $\emptyset \neq \{\emptyset\}$

23.

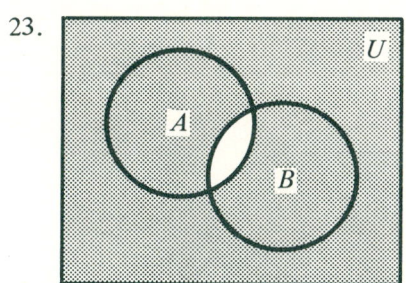

$(A \cap B)'$

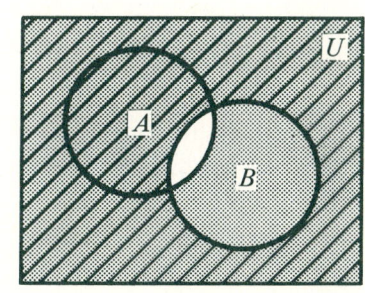

$A' \cup B'$

24.

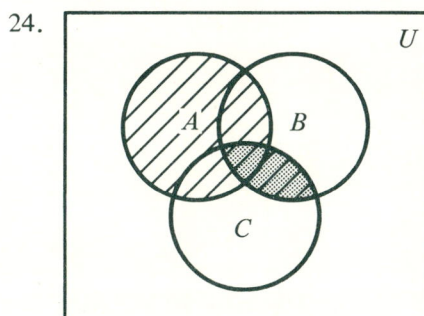

$A \cup (B \cap C)$

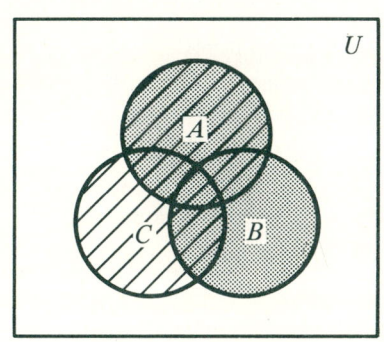

$(A \cup B) \cap (A \cup C)$

25.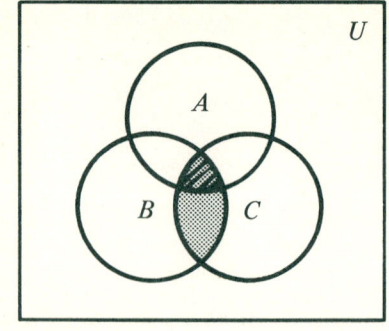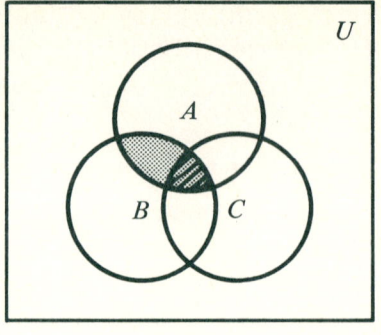

$A \cap (B \cap C)$ $(A \cap B) \cap C$

27. (a) A
    (b) B
29. (a) $\emptyset$
    (b) C
31. $\{x | x \leq -2\}$
33. $\{x | x \leq -1 \text{ or } x \geq 5\}$
35. $A \cup (A \cup B)' = A \cup B'$

$A \cup (A \cup B)' = A \cup (A' \cap B')$ DeMorgan's Law
$ = (A \cup A') \cap (A \cup B')$ Distributive
$ = U \cap (A \cup B')$ Since $A \cup A' = U$
$ = A \cup B'.$ Since $U \cap D = U$ for any set $D \subseteq U$.

Exercise 2.6

1. $A \times B = \{(a, p), (a, d), (a, q), (b, p), (b, d), (b, q), (c, p), (c, d), (c, q)\}$
2. $B \times A = \{(p, a), (d, a), (q, a), (p, b), (d, b), (q, b), (p, c), (d, c), (q, c)\}$
   Notice that from 1 and 2, $A \times B \neq B \times A$.
3. One possible subset is $\{(a, p), (b, d), (c, q)\}$
4. $\{(a, p), (b, d), (c, q)\}, \{(a, d), (b, p), (c, q)\}, \{(a, q), (b, p), (c, d)\}$
   $\{(a, p), (b, q), (c, d)\}, \{(a, d), (b, q), (c, p)\}, \{(a, q), (b, d), (c, p)\}$
5. (a) 9   (b) 6   (c) 6
6. $m \cdot n$
7. $\emptyset, \{(0, 2), (1, 2)\}, \{(0, 2)\}, \{(1, 2)\}$
   Such sets of ordered pairs in which each element of the first set appears with only one element from the second are called functions from A to B.
8. Since A contains two elements and B three, there cannot be any one-to-one pairing between their elements. Sets depicting a one-to-one relationship between a subset of A and B are $\emptyset, \{(0, 2)\}$ and $\{(1, 2)\}$. These sets are called one-to-one functions from A to B.
9. (a) $\emptyset, \{(2, 0)\}, \{(2, 1)\}$ (b) none
11. $(A \times B) \cup (A \times C) = A \times (B \cup C)$
    We will use the definition of equality ($A \subseteq B$ and $B \subseteq A$ means $A = B$) and the definition of subsets ($A \subseteq B$ means if $x \in A$, then $x \in B$).
    First, $(A \times B) \cup (A \times C) \subseteq A \times (B \cup C)$: Let $(x, y) \in [(A \times B) \cup (A \times C)]$, then $(x, y) \in (A \times B)$ or $(x, y) \in (A \times C)$ and $x \in A$ and $y \in B$, or $x \in A$ and $y \in C$. Hence, $x \in A$, and $y \in B$, or $y \in C$, and $x \in A$. Therefore $y \in (B \cup C)$, and $(x, y) \in [A \times (B \cup C)]$.

Second, $A \times (B \cup C) \quad (A \times B) \cup (A \times C)$: Let $(x, y) \in [A \times (B \cup C)]$, then $x \in A$ and $y \in (B \cup C)$. Then $x \in A$ and $y \in B$ or $y \in C$, and $x \in A$ and $y \in B$ or $x \in A$ and $y \in C$. Therefore, $(x, y) \in A \times B$ or $(x, y) \in A \times C$ and $(x, y) \in [(A \times B) \cup (A \times C)]$.

Exercise 2.8

1. (a) No
   (b) No
   (c) Yes
   (d) No
   (e) No

2. (a) p ∧ q
   (b) p ∨ q
   (c) p → q
   (d) p → q, if he is a politician, then he is a liar
   (e) (p ∧ q) → r
   (f) (p ∨ q) → r

3. (a) He is both honest and humble.
   (b) He is both.

4. (a) He is neither.
   (b) He is neither.
   (c) You are an honest politician.
   (d) You want to be heard but do not vote.

5.
p	q	p ∧ q	p → q	(p ∧ q) → q
T	T	T	T	T
T	F	F	F	T
F	T	F	T	T
F	F	F	T	T

7.
p	q	r	(q ∧ r)	p ∧ (q ∧ r)	(p ∧ q)	(p ∧ q) ∧ r
T	T	T	T	T	T	T
T	T	F	F	F	T	F
T	F	T	F	F	F	F
T	F	F	F	F	F	F
F	T	T	T	F	F	F
F	T	F	F	F	F	F
F	F	T	F	F	F	F
F	F	F	F	F	F	F

Since the truth values for p ∧ (q ∧ r) and (p ∧ q) ∧ r are identical, the propositions are equivalent.

9.
p	q	r	p ∨ (q ∧ r)			(p ∨ q) ∧ (p ∨ r)		
T	T	T	T	T	T	T	T	T
T	T	F	T	T	F	T	T	T
T	F	T	T	T	F	T	T	T
T	F	F	T	F	F	T	F	T
F	T	T	F	F	T	T	F	T
F	T	F	F	F	F	T	F	F
F	F	T	F	F	F	F	F	T
F	F	F	F	F	F	F	F	F
			1	↑	2	4	↑	5
				3			6	

The numbers indicate the order in which the truth values were found. Column 2, for example, gives the truth values for (q ∧ r). These are then compared with the truth values in column 1 to find the truth values for the entire disjunction. Since the columns with arrows are identical, the propositions are equivalent.

Exercise 2.9

1. John is not a student or not a fireman ($\sim p \lor \sim q$).

$$\sim(p \land q) \Leftrightarrow \sim p \lor \sim q$$

2. Mary is not a doctor or not a mother ($\sim p \lor \sim q$).

$$\sim(p \land q) \Leftrightarrow \sim p \lor \sim q$$

3. Henry is not a teacher or he is a researcher.

$$\sim(p \land \sim q) \Leftrightarrow \sim p \lor q$$

4. Joyce is not an author or Joyce is a poet.

$$\sim(p \land \sim q) \Leftrightarrow \sim p \lor q$$

5. It is raining and not pouring.

$$\sim(p \Rightarrow q) \Leftrightarrow p \land \sim q$$

6. You try and do not pass.

$$\sim(\sim p \Rightarrow q) \Leftrightarrow p \land \sim q$$

7. See above.

8. $p \lor \sim p$ is always true. Such propositions are called tautologies.

9. q must be false, since the hypothesis $p \lor \sim p$ is always true.

10. converse ($q \to p$)
    (a) If it is not raining, then the humidity is 100%.
    (b) If there is no dew forming, then the dew point is 28° and the temperature is 30°.
    contrapositive ($\sim q \to \sim p$)
    (a) If it is raining, then the humidity is not 100% (which is true, by the way).
    (b) If there is dew forming, then the dew point is not 28° or the temperature is not 30°.

11.

p	q	$p \lor q$	$\sim(p \lor q)$	$\sim p$	$\sim q$	$\sim p \lor \sim q$
T	T	T	F	F	F	F
T	F	T	F	F	T	F
F	T	T	F	T	F	F
F	F	F	T	T	T	T

13.

p	q	$[(p \Rightarrow q) \land p] \Rightarrow q$
T	T	T T T T T
T	F	F F T T F
F	T	T F F T T
F	F	T F F T F
		1 3 2 5 4

14. $[(p \Rightarrow q) \land \sim q] \Rightarrow \sim p$

T	F F T F
F	F T T F
T	F F T T
T	T T T T
1	3 2 5 4

The numbers indicate the order in which the truth values were determined. Column 3 was found from 1 and 2. Column 5 was found from 3 and 4 using the truth values for implications.

17. $\sim p \Rightarrow q$
    $r \Rightarrow \sim q$
    $\therefore r \Rightarrow p$

    By contraposition, $r \Rightarrow \sim q$ is equivalent to $q \Rightarrow \sim r$. Then $[(\sim p \Rightarrow q) \land (q \Rightarrow \sim r)] \Rightarrow \sim p \Rightarrow \sim r$ by the transitive law. This conclusion, again by contraposition, is equivalent to $r \Rightarrow p$. Therefore, the stated conclusion is valid.

19. $\sim(p \land q)$
    $r \Rightarrow p$
    $\therefore \sim r \Rightarrow q$

    This argument is not valid since the statement $[\sim(p \land q) \land (r \Rightarrow p)] \Rightarrow (\sim r \Rightarrow q)$ is false when p is true, q is false, and r is false. The easiest way to find this is to try to make the conclusion $(\sim r \Rightarrow q)$ false and all of the assumptions (the hypothesis) true. If this is possible then the argument is not valid.

21. $\quad p \vee q$ By Theorem II, $(p \vee q) \Leftrightarrow (\sim p \Rightarrow q)$ and $(s \vee r) \Leftrightarrow (\sim s \Rightarrow r)$. Then by contraposition we have
$\quad\quad q \Rightarrow \sim r$ $(\sim s \Rightarrow r) \Leftrightarrow (\sim r \Rightarrow s)$. Now, by the transitive law, $[(q \Rightarrow \sim r) \wedge (\sim r \Rightarrow s)] \Leftrightarrow (q \Rightarrow s)$ and
$\quad\quad \underline{s \vee r}$ $[(\sim p \Rightarrow q) \wedge (q \Rightarrow s)] \Leftrightarrow \sim p \Rightarrow s$. Therefore, the argument is valid.
$\quad\quad \therefore \sim p \Rightarrow s$ If we try to assign a truth value that will make the conclusion false and the assumptions all true, we will fail. That is, $\sim p \Rightarrow s$ is false if p is false and s is false. This forces q to be true and r true, which makes $q \Rightarrow \sim r$ false !

Exercise 2.10

1. $(\exists x)p(x)$
2. $(\forall x)q(x)$
3. $(\forall t)p(t)$
4. $(\exists x)p(x)$
5. (a) $(\exists x)p(x) \Rightarrow q(x)$, since $x = 4$ makes $p(x) \Rightarrow q(x)$ true and $x = -4$ makes it false, since $p(-4)$ is true but $q(-4)$, $(-4 = 4)$ is false.
   (b) $(\forall x)p(x) \Rightarrow q(x)$, since the only value that makes the hypothesis $p(x)$ true is 4 and $q(4)$ is also true. Remember, if the hypothesis of an implication is false, the implication is true.
6. $T_p = \{-1/3, 1\}$
7. $T_q = \{2\}$
8. $T_p = \{2, 3\}$
9. $T_q = \{2\}$
10. $T_p = \{-1/3\}$
11. $T_q = \{1/2\}$
12. $T_p = \{t | t \geq 4/3\}$
13. $T_q = \{t | -3 < t < 3\}$
15. (a) $T_p = \{x | x > 3\}$, $T_q = \{x | x < -3 \vee x > 3\}$, $p(x) \Rightarrow q(x)$ is true, since $T_p \subseteq T_q$.
    (b) $T_p = \{x | x < -3 \vee x > 3\}$, $T_q = \{x | x > 3\}$, $p(x) \Rightarrow q(x)$ is conditional (not always true), since $T_p \not\subseteq T_q$.
17. $p(x) : x/0 = 1$ is always false, even for $x = 0$, then, since $T_p = \emptyset$, $T_p \subseteq T_q$ because division by zero is undefined, and $p(x) \Rightarrow q(x)$ is true $\forall x$.
19. $T_p = \emptyset$ means $T_p \subseteq T_q$. Therefore, $p(x) \Rightarrow q(x)$ is true. Remember that absolute value is always positive or zero.
21. $T_p = \{x | x > 4\}$ $T_q = \{x | x < 1/2\}$
    $T_p \cup T_q = \{x | x < 1/2 \vee x > 4\} = T_{p \vee q}$
    $T_p \cap T_q = \emptyset = T_{p \wedge q}$
23. $T_p = \{x | x \leq 3/2\}$, $T_q = \{x | x < -7 \vee x > 1\}$
    $T_{p \vee q} = T_p \cup T_q = U$, where $U = \{x | x \in \text{reals}\}$
    $T_{p \wedge q} = T_p \cap T_q = \{x | x < -7 \vee 1 < x < 3/2\}$
25. $T_p = \emptyset$, $T_q = \{0\}$
    $T_{p \vee q} = T_p \cup T_q = \{0\}$
    $T_{p \wedge q} = T_p \cap T_q = \emptyset$

Exercise 2.11

1. If $a \cdot x = 0$ and $a \neq 0$, then $1/a$ exists, and $1/a(ax) = 1/a \cdot 0 \rightarrow$
$$\left(\frac{1}{a} \cdot a\right)x = 0 \rightarrow 1x = 0 \rightarrow$$
$$x = 0$$

3. If $a \vee b$ is divisible by 3, then $a = 3t_1$, $t \in$ integers $\vee$ $b = 3t_2$, $t_2 \in$ integers $\rightarrow a \cdot b = 3t_1 b \vee a \cdot b = a \cdot 3t_2 \rightarrow a \cdot b = 3k_1 \vee a \cdot b = 3 \cdot k_2$, where $k_1, k_2 \in$ integers. Therefore, $a \cdot b$ is divisible by 3.

5. Let $a \cdot b = 9$ and $a = b = 3$. Then, $a \cdot b$ is divisible by 9 (p is true) but neither a nor b is divisible by 3 (q is false). Therefore, the implication is false.

7. Let $x = -1/2$ and $y = 1/2$. Then $|x| + |y| = 1$ (p is true), but $x + y \neq 1$ (q is false). Hence, the implication is false.

9. $p \to q : a \cdot b \neq 0 \to a \neq 0 \wedge b \neq 0$ (implication)
   $\sim q \to \sim p : a = 0 \vee b = 0 \to a \cdot b = 0$ (contrapositive)
   If $a = 0 \vee b = 0$, then $a \cdot b = 0 \cdot b = 0 \vee a \cdot b = a \cdot 0 = 0 \to a \cdot b = 0$.

11. Assume it is false that the multiplicitive identity for the real numbers is unique. Then, $\exists x, y \in$ reals, $x \neq y$, such that $x \cdot a = a \wedge y \cdot a = a$, $\forall a \in$ reals $\to x \cdot a = y \cdot a$.
    Since this is true for all real numbers, it is true for $a = 4$. Then,

    $$x \cdot 4 = y \cdot 4 \to (x \cdot 4)\frac{1}{4} = (y \cdot 4)\frac{1}{4} \to x \left(4 \cdot \frac{1}{4}\right) = y \cdot \left(4 \cdot \frac{1}{4}\right) \to x \cdot 1 = y \cdot 1 \to x = y$$

    which is a contradiction of our assumption. Therefore, the assumption that the multiplicitive identity was not unique was incorrect. Hence, the multiplicitive identity is unique.

14. Assume the set of prime numbers is not infinite. Then the set of primes is finite, say $S = \{a_1, a_2, \ldots, a_n\}$. Let $x = a_1 \cdot a_2 \cdot \cdots \cdot a_n + 1$. Since x is larger than any prime in set S, by assumption, it cannot be prime. Therefore, it must have a prime divisor $a_1, a_2, \ldots$, or a_n. But, none of these primes divides

    $$x - a_1 \cdot a_2 \cdot a_2 \cdot \cdots \cdot a_n = 1$$

    (since none divides one). Therefore, none of these primes divides x. But, since x is composite, some prime must divide x which contradicts our assumptions that S contained all primes. Therefore, our assumption that the set of primes was finite must be wrong.

Exercise 2.12

1. $p(n) : \sum_{i=1}^{n} i = \frac{n(n+1)}{2}$

   $p(1): \sum_{i=1}^{1} i = 1$ and $n = 1 \Rightarrow \frac{n(n+1)}{2} = \frac{1(1+1)}{2} = 1$

   $p(2): \sum_{i=1}^{2} i = 1 + 2 = 3$ and $n = 2 \Rightarrow \frac{n(n+1)}{2} = \frac{2(2+1)}{2} = 3$

   $p(3): \sum_{i=1}^{3} i = 1 + 2 + 3 = 6$ and $n = 3 \Rightarrow \frac{n(n+1)}{2} = \frac{3(3+1)}{2} = 6$

   $p(4): \sum_{i=1}^{4} i = 1 + 2 + 3 + 4 = 10$ and $n = 4 \Rightarrow \frac{n(n+1)}{2} = \frac{4(4+1)}{2} = 10$

3. $p(n): \sum_{i=1}^{n} i^2 = \frac{n(n+1)(2n+1)}{6}$

   $p(2): \sum_{i=1}^{2} i^2 = 1^2 + 2^2 = 5$ and $n = 2 \Rightarrow \frac{n(n+1)(2n+1)}{6} = \frac{2(3)(5)}{6} = 5$

   $p(4): \sum_{i=1}^{4} i^2 = 1^2 + 2^2 + 3^2 + 4^2 = 30$ and $n = 4 \Rightarrow \frac{n(n+1)(2n+1)}{6} = \frac{4(5)(9)}{6} = 30$

5. $p(1) = 1, p(2) = 9, p(3) = 36, p(4) = 100$

6. Here is an alternate way to prove Example I.

$$p(n): \sum_{i=1}^{n} i = \frac{n(n+1)}{2}$$

I. $p(1): \sum_{i=1}^{1} i = 1$ and $\frac{n(n+1)}{2} = \frac{1(1+1)}{2} = 1$

II. $p(k) \Rightarrow p(k+1): \sum_{i=1}^{k} i = \frac{k(k+1)}{2} \Rightarrow \sum_{i=1}^{k+1} i = \frac{(k+1)[(k+1)+1]}{2}$

$$\sum_{i=1}^{k} i = \frac{k(k+1)}{2} \Rightarrow \sum_{i=1}^{k} i + (k+1) = \frac{k(k+1)}{2} + (k+1) \Rightarrow$$

$$\sum_{i=1}^{k+1} i = \frac{k(k+1) + 2(k+1)}{2} = \frac{(k+1)(k+2)}{2}$$

Therefore, if $\sum_{i=1}^{k} i = \frac{k(k+1)}{2}$, then $\sum_{i=1}^{k+1} i = \frac{(k+1)(k+2)}{2}$

8. $p(n): \sum_{i=1}^{n} i^2 = \frac{n(n+1)(2n+1)}{6}$

I. $p(1): \sum_{i=1}^{1} i^2 = 1^2 = 1$ and $\frac{n(n+1)(2n+1)}{6} = \frac{1(2)(3)}{6} = 1$

II. $p(k) \Rightarrow p(k+1): \sum_{i=1}^{k} i^2 = \frac{k(k+1)(2k+1)}{6} \Rightarrow \sum_{i=1}^{k+1} i^2 = \frac{(k+1)(k+2)(2k+3)}{6}$

$$\sum_{i=1}^{k+1} i^2 = \sum_{i=1}^{k} i^2 + (k+1)^2 = \frac{k(k+1)(2k+1)}{6} + (k+1)^2$$

$$= \frac{k(k+1)(2k+1) + 6(k+1)^2}{6} = \frac{(k+1)[k(2k+1) + 6(k+1)]}{6}$$

$$= \frac{(k+1)[2k^2 + 7k + 6]}{6} = \frac{(k+1)(k+2)(2k+3)}{6}$$

9. (a) $p(n): \sum_{i=1}^{n} \frac{1}{2^{i-1}} = \frac{2^n - 1}{2^{n-1}}$

I. $p(1): \sum_{i=1}^{1} \frac{1}{2^{i-1}} = \frac{1}{2^{1-1}} = \frac{1}{2^0} = \frac{1}{1} = 1$ and $\frac{2^n - 1}{2^{n-1}} = \frac{2^1 - 1}{2^{1-1}} = \frac{2-1}{2^0} = 1$

II. $p(k) \Rightarrow p(k+1)$: $\sum_{i=1}^{k} \frac{1}{2^{i-1}} = \frac{2^k - 1}{2^{k-1}} \Rightarrow \sum_{i=1}^{k+1} \frac{1}{2^{i-1}} = \frac{2^{k+1} - 1}{2^k}$

$$\sum_{i=1}^{k+1} \frac{1}{2^{i-1}} = \sum_{i=1}^{k} \frac{1}{2^{i-1}} + \frac{1}{2^k} \Rightarrow \sum_{i=1}^{k+1} \frac{1}{2^{i-1}} = \frac{2^k - 1}{2^{k-1}} + \frac{1}{2^k} \Rightarrow$$

$$\sum_{i=1}^{k+1} \frac{1}{2^{i-1}} = \frac{2}{2} \cdot \frac{2^k - 1}{2^{k-1}} + \frac{1}{2^k} = \frac{2^{k+1} - 2 + 1}{2^k} = \frac{2^{k+1} - 1}{2^k}$$

Hence, $\sum_{i=1}^{k+1} \frac{1}{2^{i-1}} = \frac{2^{k+1} - 1}{2^k}$, whenever $\sum_{i=1}^{k} \frac{1}{2^{i-1}} = \frac{2^k - 1}{2^{k-1}}$

14. (a) $p(n)$: 3 divides $n^3 - n + 3 \Leftrightarrow n^3 - n + 3 = 3t$ for some $t \in$ naturals.
   I. $p(1)$: $n = 1 \Rightarrow n^2 - n + 3 = 1^3 - 1 + 3 = 3 = 3 \cdot t$ for $t = 1$.
   II. $p(k) \Rightarrow p(k+1)$: $k^3 - k + 3 = 3t_1 \Rightarrow (k+1)^3 - (k+1) + 3 = 3t_2, t_1, t_2 \in N$
   $(k+1)^3 - (k+1) + 3 = k^3 + 3k^2 + 3k + 1 - k - 1 + 3$
   $\qquad = (k^3 - k + 3) + 3k^2 + 3k$
   $\qquad = 3 \cdot t_1 + 3k^2 + 3k$
   $\qquad = 3(t_1 + k^2 + k)$
   $\qquad = 3 \cdot t_2$, where $t_2 = t_1 + k^2 + k \in N$
   Therefore, if $k^3 - k + 3$ is divisible by 3, then $(k+1)^3 - (k+1) + 3$ is divisible by 3.

CHAPTER 3

Exercise 3.2

1. (a) $f(1) = 0$ (b) $f(0) = 0$
2. (a) $g(1) = 0$ (b) $g(-3) = 2$
3. (a) $h(s) = 2s^3 - s + 4$
   (b) $h(-t) = -2t^3 + t + 4$
4. (a) $f(4x) = 12x$
   (b) $f(5t) = 15t$
5. (a) $h(4x) = 48x^2$
   (b) $h(5t) = 75t^2$
   Did you notice that Problems 4 and 5 were evaluating their functions at the same values? In 4(a) $f(4x) = 4f(x)$, but for 5(a) $h(4x) \neq 4h(x)$. Few functions respond like 4(a). Functions of the form, $f(x) = mx$ have the property that $f(tx) = tf(x)$.
6. (a) $g(x + 2) = 2x^2 + 7x + 7$
   (b) $g(x - 1) = 2x^2 - 5x + 4$
7. (a) yes
   (b) yes
   (c) yes
8. (a) no   (b) no   (c) no
   8 (a) and (b) are equivalent.
9. (a) no   (b) no   (c) yes
   Inequalities are seldom single valued.
10. (a) no   (b) no
    Generally, with the presence of even powers of y, the relation is not a function.
11. (a) $y^2 = 1 - x$ is not a function, since for $x = -3, y^2 = 4 \Rightarrow y = 2$ or $y = -2$. Then, $x = -3$ is not paired with a unique value of y but rather with two values (b) $|y| = \sqrt{1 - x}$ is the same relation as $y^2 = 1 - x$, since $\sqrt{y^2} = |y|$.

12. (a) yes
    (b) yes
    (c) yes

13. (a) no
    (b) no
    (c) no

The power makes the difference. If the function contains only odd powers of x, such as those in 12, it will usually be one-to-one. If there are only even powers of x, it is not one-to-one because $f(x) = f(-x)$ for all x in the domain of $y = f(x)$. (Unless the domain contains only one element.)

14. (a) no
    (b) no

15. (a) $\{x \in \text{reals} \,|\, x \geq 0\}$
    (b) $\{x \,|\, x \geq -4\}$
    (c) $\{x \,|\, x \leq 3/2\}$

17. (a) $\{x \in \text{reals} \,|\, x \neq 1\}$
    (b) $\{x \in \text{reals} \,|\, x \neq 2\}$

18. (a) $\{t \in \text{reals} \,|\, t \neq 4 \wedge t \neq -1\} = \{-1, 4\}'$
    (b) $\{z \,|\, z \neq 4 \wedge z \neq -3\}$

19. (a) $\{x \,|\, x \geq 0\}$ (b) $\{x \,|\, x \neq 0\}$ (c) $\{x \,|\, x > 0\}$
    Notice that the restrictions on x in parts (a) and (b) are both included for the function in (c).

20. (a) $\{x \,|\, x \in \text{reals}\}$, since $x^2 + 4 > 0$ for all x.
    (b) $x^2 - x - 2 > 0$
        Let $x^2 - x - 2 = 0$
        Then,
        $(x - 2)(x + 1) = 0$
        $x - 2 = 0 \vee x + 1 = 0$
        $x = 2 \quad \vee \, x = -1$

    These are the boundary points for our solution set. Let's try a number in between them, for example, $x = 0$, to see if our solution is there or outside of them.
    $x = 0 \Rightarrow x^2 - x - 2 = -2 \not> 0 \Rightarrow x = 0$ is not in the solution set. Therefore,
    $x < -1$ or $x > 2$
    (c) $\sqrt{x + 1} - 2 \neq 0 \Rightarrow x \neq 3$
        $x + 1 \geq 0 \Rightarrow x \geq -1$
        $\{x \,|\, x \geq -1, \quad x \neq 3\}$

22. (c)

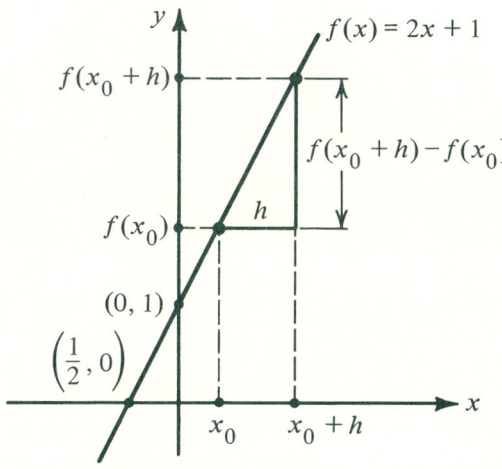

$$\frac{f(x_0 + h) - f(x_0)}{h}$$

$$= \frac{[2(x_0 + h) + 1] - [2x_0 + 1]}{h}$$

$$= \frac{2h}{h} = 2$$

In Chapter IV we will find that the ratio calculated above of change in height, $f(x_0 + h) - f(x_0)$, relative to a horizontal change in x, h, is called *slope*. The evaluation of the above difference quotient is a central concept of differential calculus.

A-29

23. (c) −5
25. (b) Part (i) is false, since, for $f(x) = x^2$, $f(2 + 3) = f(5) = 25$ but $f(2) + f(3) = 4 + 9 = 13 \neq 25$.
Any counterexample is a complete proof that 25(b) is false. Part (ii) is false also. Use $f(x) = x^2$ with $a = 2$ and $b = 5$ to evaluate $f(2) \cdot f(5)$ and $f(10)$.
26. $f(x) = x^2 \Rightarrow t \cdot f(x) = tx^2$ and $f(tx) = t^2x^2$. In general, $f(tx) \neq tf(x)$. But, for functions of the form $f(x) = mx$, such as $f(x) = -3x$, where $f(tx) = -3(tx) = tf(x) = t(-3x)$, $f(tx) = m(tx) = tf(x) = t(mx)$. See the following illustration.

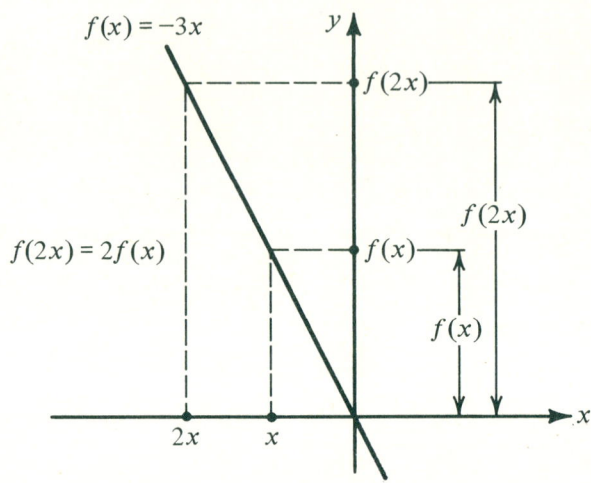

Notice that if you double the distance from the origin on the x-axis, you also double the corresponding distance on the y-axis.

Exercise 3.3

1. (a) $f(-x)$ is a reflection about the y-axis of $f(x)$.
   (b) $g(-x) = x^2 = g(x)$. Reflecting the graph of $y = x^2$ about the y-axis, does not result in a new picture.
   (c) $y = \sqrt{-x}$ has a domain of $\{x | x \leq 0\}$ since it is $y = \sqrt{x}$ reflected about the y-axis.
2. (a) $y = -f(x)$ reflects the graph about the x-axis. For $f(x) = x$, $f(-x) = -x$ and $-f(x) = -x$. This means we get the same result if we reflect $f(x) = x$ about either axis. This is very unusual.
   (b) $y = -g(x) = -x^2$ is $y = x^2$ reflected about the x-axis. The range for $y = -x^2$ is $\{y | y \leq 0\}$.
   (c) $y = -h(x) = -\sqrt{x}$ is $y = \sqrt{x}$ reflected about the x-axis. The domain of $y = -\sqrt{x}$ is still $\{x | x \geq 0\}$ but the range is $\{y | y \leq 0\}$.
3. These all illustrate reflections about the origin.
   (a) $-f(-x) = x = f(x)$ (b) $-g(-x) = -x^2$ (c) $-h(-x) = -\sqrt{-x}$
4. (a) $f(2) = 5$ 5. (a) $f(0) = 7$ 6. (a) $f(2) = 5$
   (b) $-f(2) = -5$ (b) $f(1) = 4$ (b) $f(-2) = 5$
7. (a) $-f(+1) = -4$ 9. (a) -3, since $g(2) = -f(2)$
   (b) $-f(-(-1)) = -4$ (b) -2, since $g(-2) = f(-(-2)) = f(2)$
                               (c) $g(-2) = -3$, since $g(-2) = -f(-(-2)) = -f(2)$
11. $g(x) = f(-x) \Rightarrow \{(1, 2), (-1, 2), (-2, 4), (2, 4)\} = g$, since, for example, $g(1) = f(-1) = 2$ gives us the first ordered pair of g.

13.
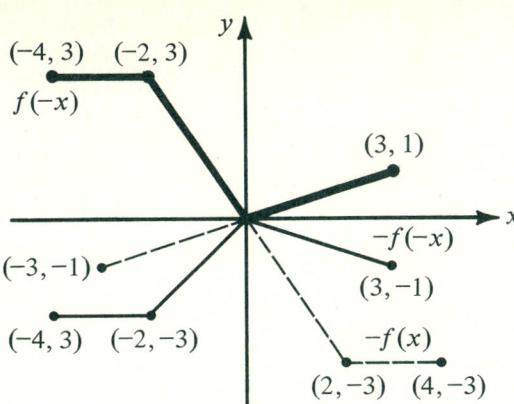

15.
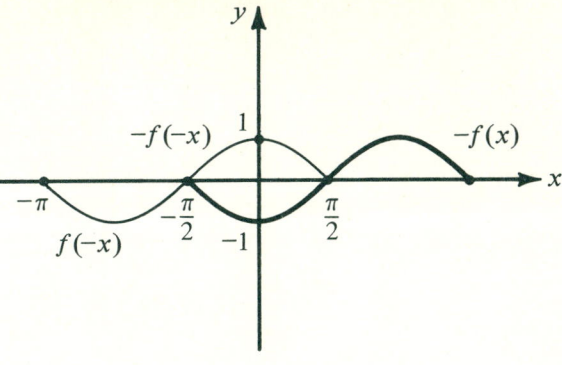

17.
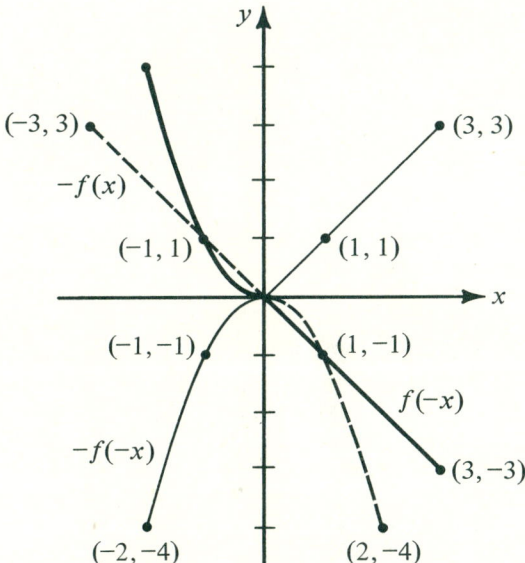

19.
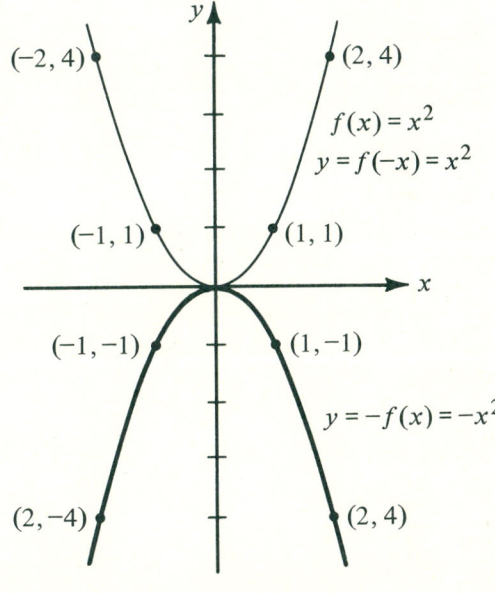

22. (b) $y = f(-x)$ is a reflection of (a) about the y-axis, since $f(-x)$ changes the signs of the x-coordinates of the ordered pairs of f.
  (c) $y = -f(-x)$ is a reflection of (a) about the origin, since $-f(-x)$ changes the signs of both coordinates of the ordered pairs of f.

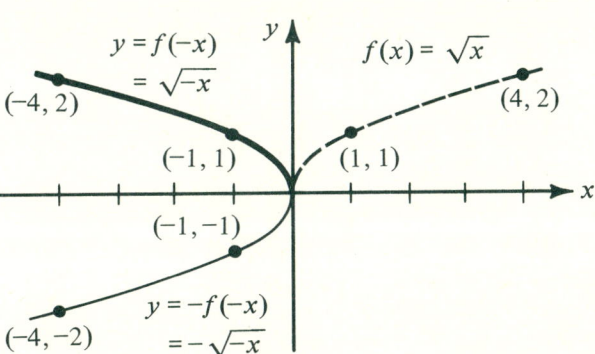

Exercise 3.4

1. (a) even
   (b) neither
   (c) neither

2. (a) odd
   (b) odd
   (c) neither

3. (a) neither
   (b) even
   (c) neither

4. (a) 3
   (b) $f(1) = -2$

5. (a) -2
   (b) $f(1) = 2$

6. (a) $F(x) = f(x) + g(x) = x^2 + |x|$ is even, since $F(-x) = (-x)^2 + |-x| = x^2 + |x| = F(x)$. (Remember that $|x| = |-x|$ like $|3| = |-3|$.)

   (b) $F(x) = x^3 + x$ is odd, since $F(-x) = (-x)^3 + (-x) = -x^3 - x = -(x^3 + x) = -F(x)$.

   (c) $F(x) = x^2 + x^3$ is neither, since $F(-x) = (-x)^2 + (-x)^3 = x^2 - x^3$, which is not either $F(x)$ or $-F(x) = -x^2 - x^3$.

   (d) $F(x) = |x| + x$ is neither, since $F(-x) = |-x| + (-x) = |x| - x$, which is not either $F(x)$ or $-F(x) = -|x| - x$.

7. (a) even   (b) even   (c) odd   (d) odd

8. (a) Let $F(x) = f(x) + g(x)$, where $f(x)$ and $g(x)$ are both even. Then $f(-x) = f(x)$ and $g(-x) = g(x)$. Therefore,

    $F(x) = f(x) + g(x) \Rightarrow$
    $F(-x) = f(-x) + g(-x) \Rightarrow$
    $F(-x) = f(x) + g(x) \Rightarrow$
    $F(-x) = F(x)$

    Hence, the sum of two even functions is even.

9. (b) Let $F(x) = f(x) \cdot g(x)$, where $f(x)$ and $g(x)$ are odd functions. Then $f(-x) = -f(x)$ and $g(-x) = -g(x)$. Therefore,

    $F(-x) = f(-x) \cdot g(-x) \Rightarrow$
    $F(-x) = (-f(x))(-g(x)) \Rightarrow$
    $F(-x) = f(x) \cdot g(x) \Rightarrow$
    $F(-x) = F(x)$

    Therefore, the product of two odd functions is even.

10. Consider the function $y = f(x)$ and this identity,

    $$f(x) = \frac{1}{2}(f(x) + f(-x)) + \frac{1}{2}(f(x) - f(-x))$$

    The function

    $$G(x) = \frac{1}{2}(f(x) + f(-x)) \text{ is even, since}$$

    $$G(-x) = \frac{1}{2}(f(-x) + f(x)) = G(x)$$

    and the function

    $$H(x) = \frac{1}{2}(f(x) - f(-x)) \text{ is odd, since}$$

    $$H(-x) = \frac{1}{2}(f(-x) - f(x)) = -H(x)$$

    Then, $f(x) = G(x) + H(x)$ is the sum of an even and an odd function as required.

11. Replacing x with $-x$ and y with $-y$ yields

$$(-x)^3(-y) + 2(-x)(-y) = 0 \rightarrow x^3y + 2xy = 0$$

the original equation. This means we have symmetry about the origin.

$$x^3y + 2xy = 0 \Leftrightarrow xy(x^2 + 2) = 0 \Leftrightarrow xy = 0 \Leftrightarrow$$

$$x = 0 \vee y = 0$$

this graph consists of the x- and y-axis.

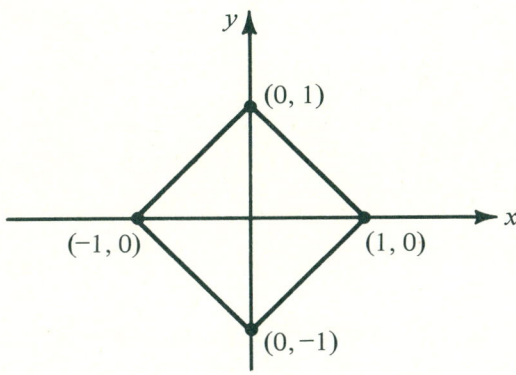

13. Replacing x and y with $-x$ and $-y$, respectively, yields the same equation, as does replacing either separately with its negative. Therefore, the graph of this relation is symmetric about the x-axis, the y-axis, and the origin.

Exercise 3.5

1.
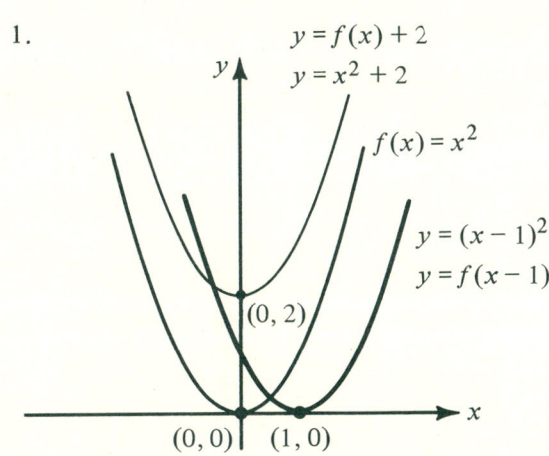

2.
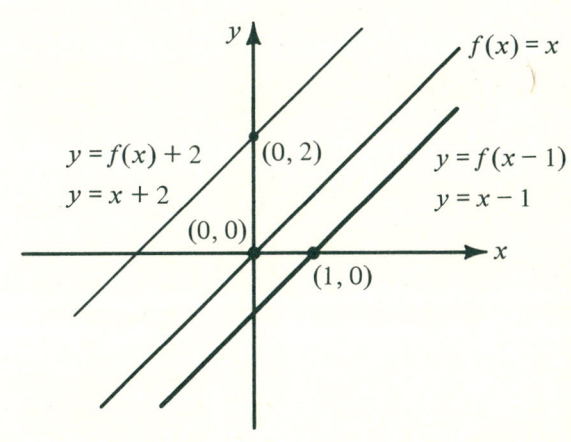

A-33

3.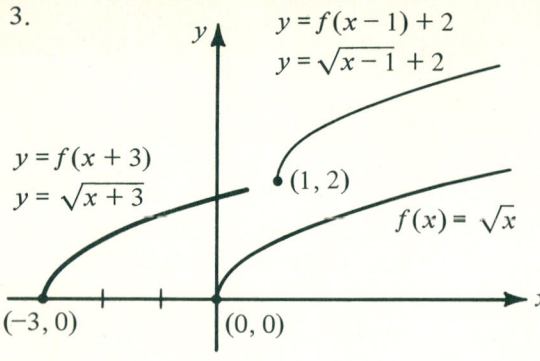

4.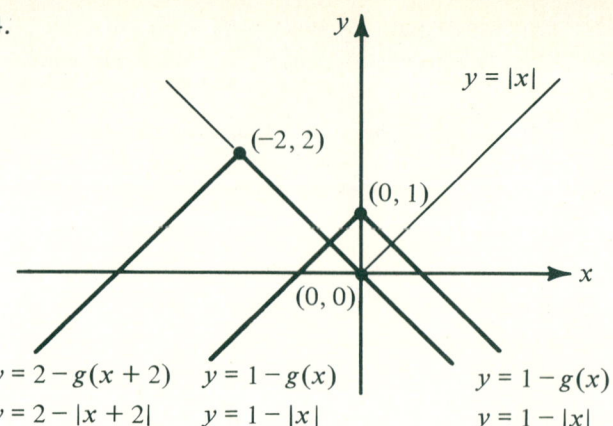

5.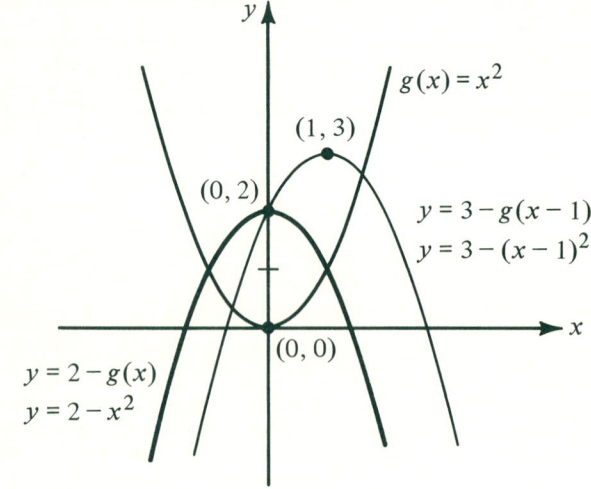

6. $y = \sqrt{-x} + 1$ is $y = \sqrt{x}$ reflected about the y-axis and then translated up 1. $y = \sqrt{-(x+2)} + 2$ is $y = \sqrt{x}$ reflected about the y-axis and then translated to the left 2 (from $(x + 2)$) and up 2. Notice how $y = f(-x) + 1$ changes the sign of the x-coordinate and adds 1 to the y-coordinate of each of the ordered pairs of f.

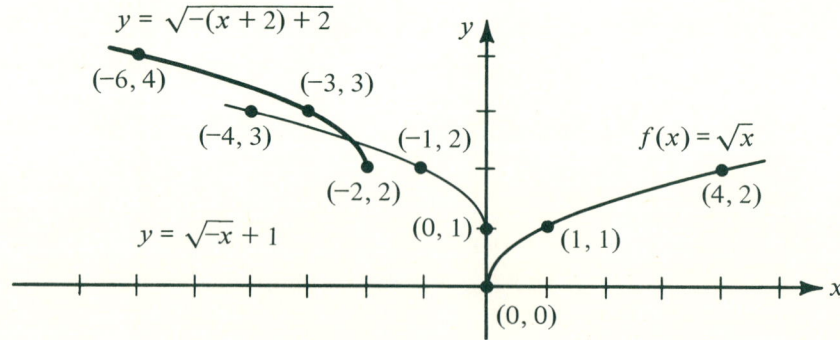

7. (a) Translation of $y = x^2$ to the right 1 and expansion *away* from the x-axis by a factor of 2.
   (b) Reflection about the x-axis, then translated to the left 2 and up 1. There is also a compression *towards* the x-axis by a factor of 2.
9. (a) Compression toward the y-axis by a factor of 3. Reflection about the y-axis (no real change, since $|-x| = |x|$). Translation up 1.
   (b) Expansion away from the x-axis, by a factor of 3. Reflection about the x-axis. Translation to the left 1.
11. $y = 2 - f(x - 1)$ is a reflection about the x-axis, then a translation to the right 1 and up 2.

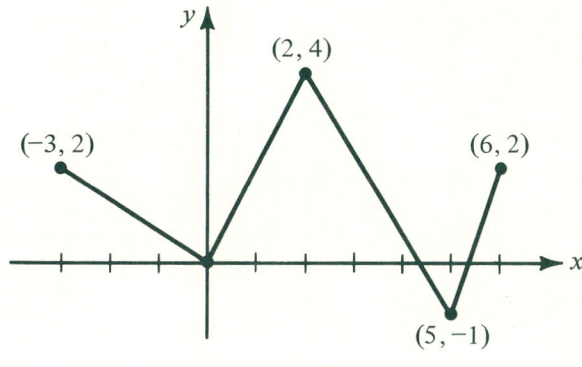

13. There is a compression towards the x-axis by a factor of 2, a reflection about the origin and a translation up 1.

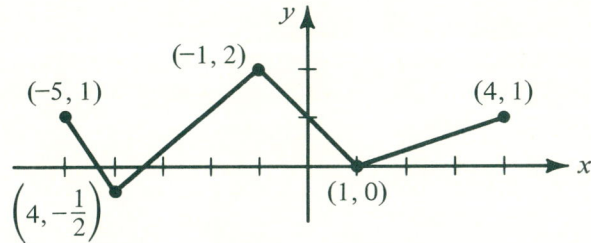

15. $f(4 - 2x) = f(-2(x - 2))$ means there is a compression towards the y-axis by a factor of 2, a reflection about the y-axis and a translation to the right 2.

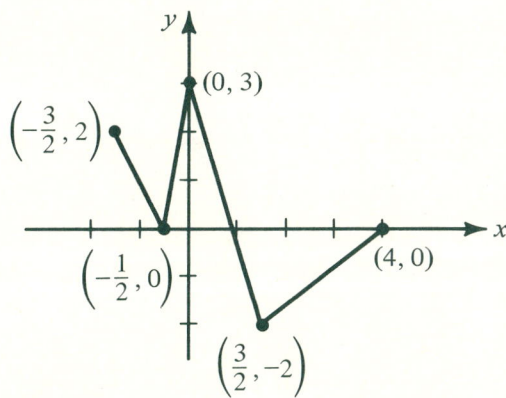

17. (a) A reflection about the y-axis and a translation up 2.
    (b) A reflection about the x-axis and a translation up 2.

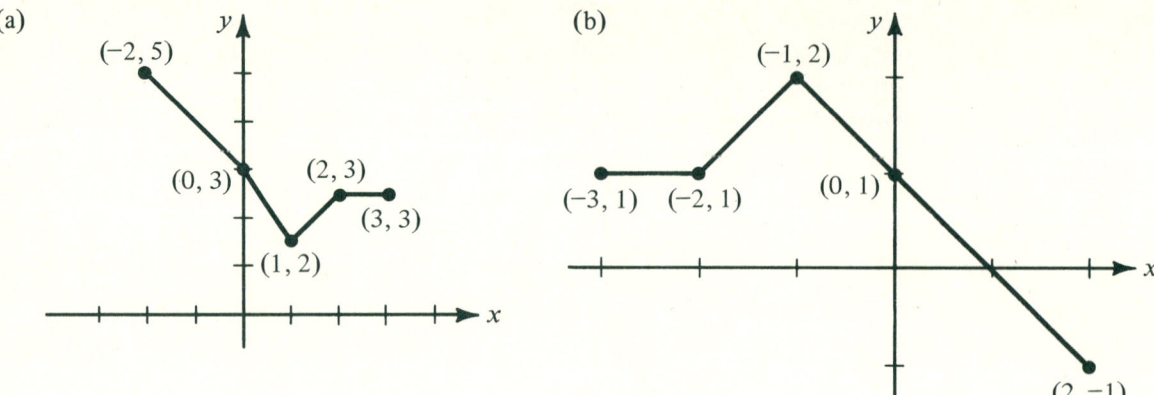

19. (a) There is a compression towards the x-axis by a factor of 2 and translations to the left 2 and down 1.
    (b) There is an expansion away from the x-axis by a factor of 2, a reflection about the y-axis and translation to the right 2 and up 3. Write $y = 2g(-(x-2)) + 3$.

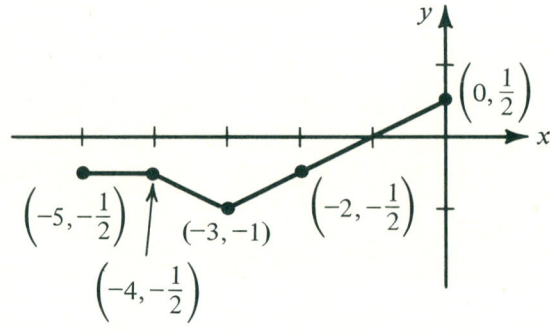

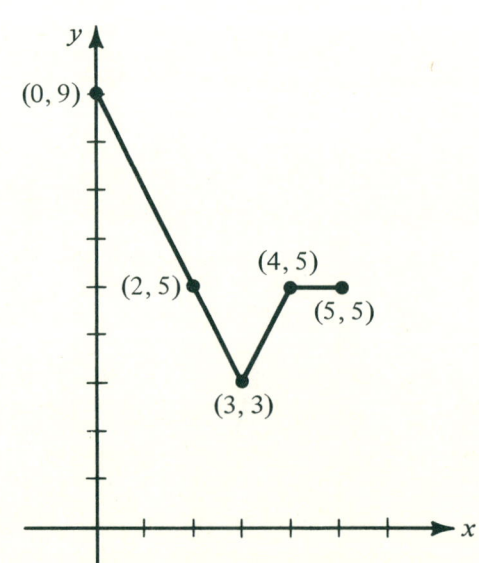

20. $f(x) = 2 - 3\sqrt{-(x-1)}$ is a reflection of $y = \sqrt{x}$ about the origin, then an expansion away from the x-axis of $y = -\sqrt{-x}$ by a factor of 3, then translated to the right 1 and up 2. The y-intercept is $f(0) = 2 - 3\sqrt{1} = -1$ and the x-intercept,

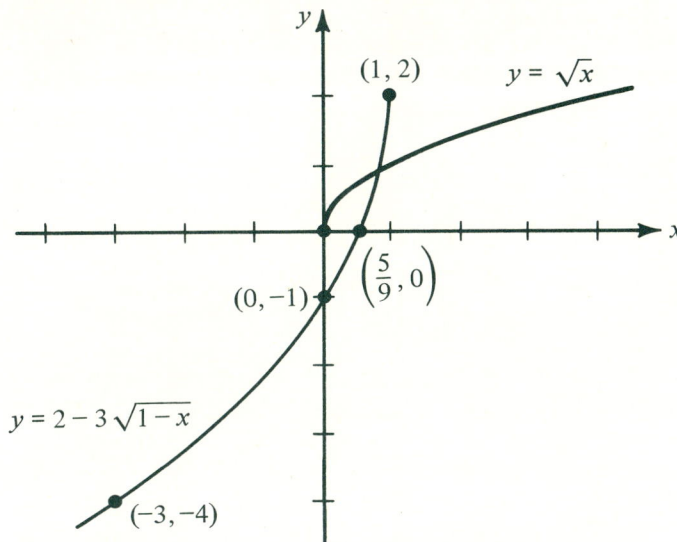

$f(x) = 0 \Rightarrow 2 - 3\sqrt{1-x} = 0$

$\sqrt{1-x} = \dfrac{2}{3} \Rightarrow 1 - x = \dfrac{4}{9} \Rightarrow x = \dfrac{5}{9}$

The domain is $1 - x \geq 0 \Leftrightarrow x \leq 1$ and the range is $y \leq 2$.

21. $f(x) = 3 + |x - 2|$

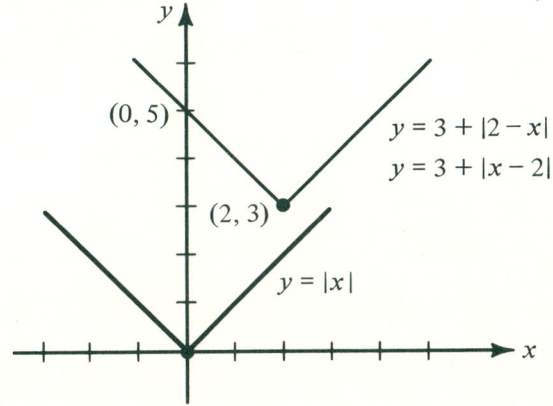

23. $g(x) = 2 - 3(x - 1)$

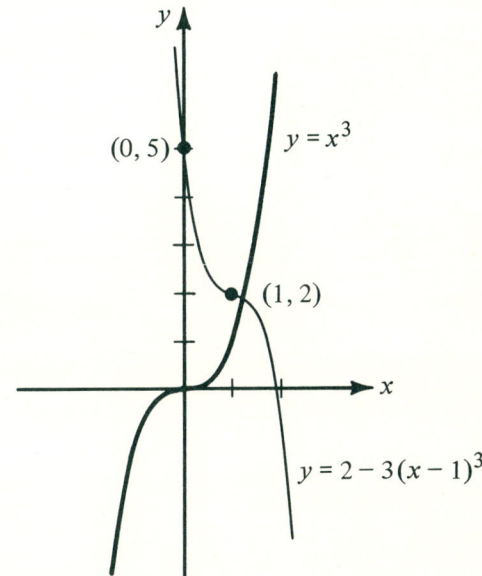

Exercise 3.6

1. (a) $f^{-1}(x) = x - 2$
   (b) $f^{-1}(x) = x + 2$

2. (a) $g^{-1}(x) = x/3$
   (b) $g^{-1}(x) = 3x$

A–37

3. (a) $y = (x - 3)/2$
   (b) $y = 5(x + 7)$

4. (a) $x \geq 0, y \geq 3$
   (b) $x \in$ reals, $y \geq 3$

5. (a) and (b) Domain and range are all real numbers.

6. (a) Domain is all reals except $x \neq 3$. Range is all reals except $y = 0$, since $1/(x - 3)$ can be positive or negative but not zero. Solve for x and you will see.
   (b) Domain is all reals except $x \neq 0$. Range is all reals except $y - 3$. This is more easily seen by solving for x. Then,

$$y = \frac{3x + 1}{x} \Rightarrow xy = 3x + 1 \Rightarrow$$

$$xy - 3x = 1 \Rightarrow$$

$$(y - 3)x = 1 \Rightarrow$$

$$x = \frac{1}{y - 3} \Rightarrow$$

$$y \neq 3$$

7. (a) All real numbers
   (b) Domain and range are all nonzero real numbers.

9. $f(x) = 3x - 5 \Rightarrow f^{-1}(x) = \dfrac{x + 5}{3}$

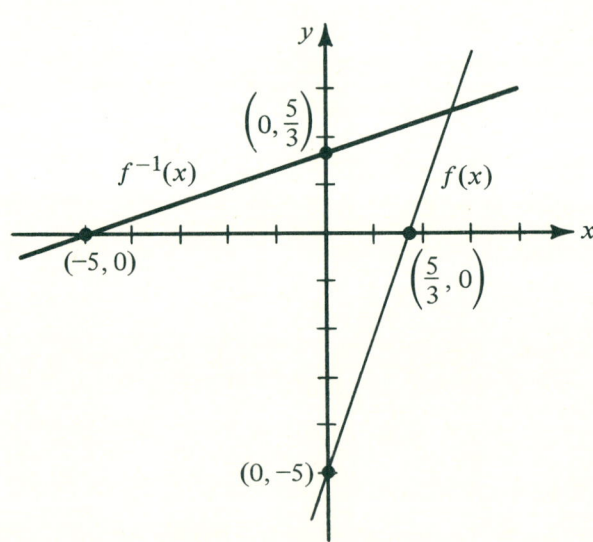

11. $f(x) = -x^3 \Rightarrow f^{-1}(x) = -\sqrt[3]{x}$

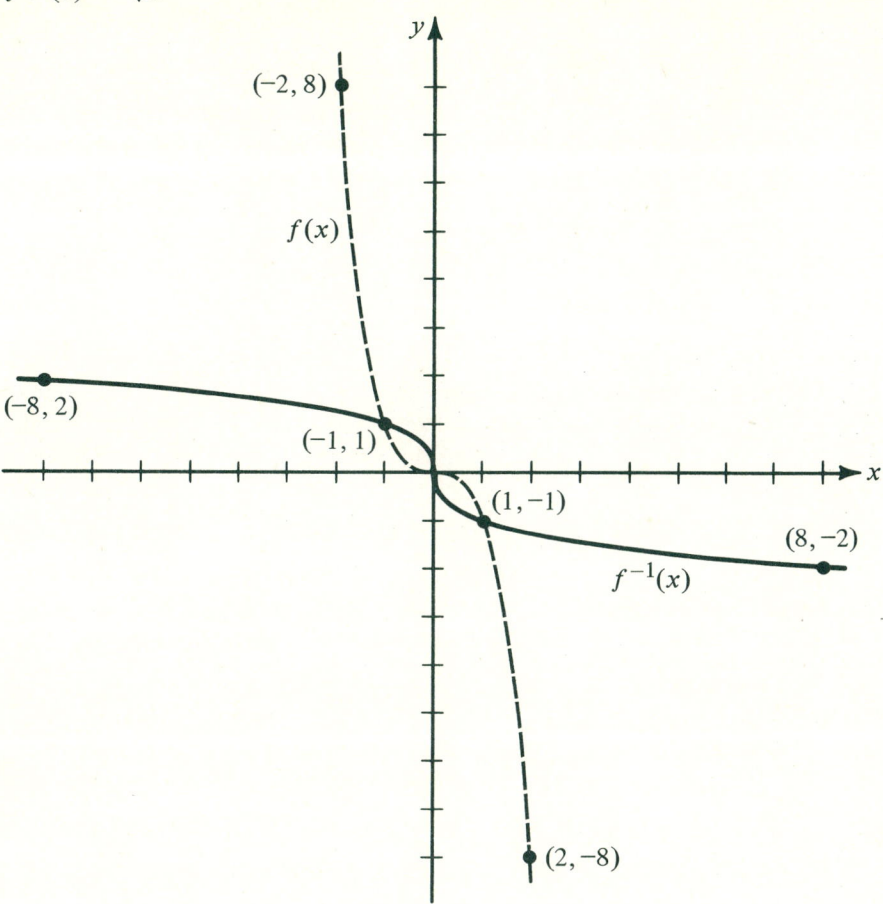

13. $y = f(-x)$ is the reflection about the y-axis.

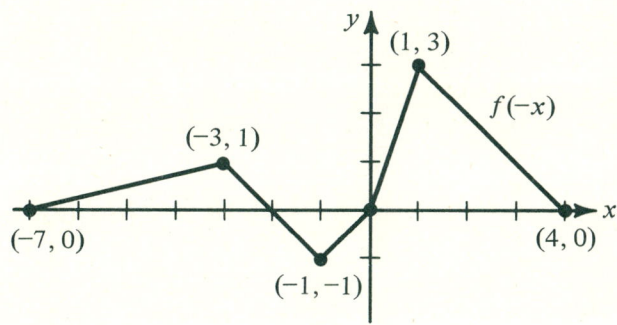

15. $y = -f(-x)$ is the reflection about the origin.

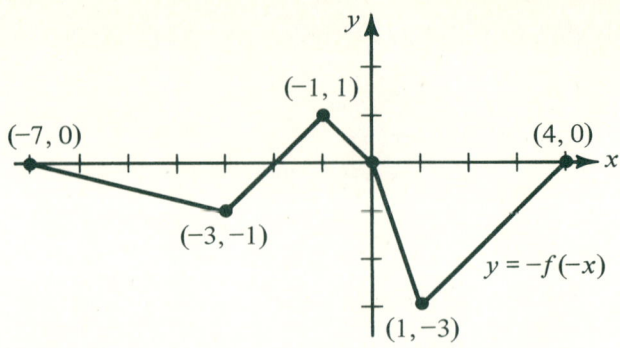

17. In creating the new functions from $y = f(x)$, as required for each of these, the order in which the changes are made are dictated by the parentheses, as always. In $y = f^{-1}(-x)$, the inverse relation of $y = f(x)$ is evaluated at $-x$, causing the graph of $y = f(x)$ to change to that of $y = f^{-1}(x)$ [by changing each ordered pair (a, b) to (b, a)], which is then reflected about the y-axis. In $y = (-f(x))^{-1}$, the reflection of $y = f(x)$ about the x-axis is accomplished first, then we find the inverse of this new function. We must closely examine $y = (f(-x))^{-1}$ to understand $y = f^{-1}(-x)$ and $y = (-f(x))^{-1}$ for $y = -f^{-1}(x)$. If finding the inverse is to be done last, it must be written explicitly.

Notice that the final graphs in 16 (a) and (b) are identical. See Problem 20.

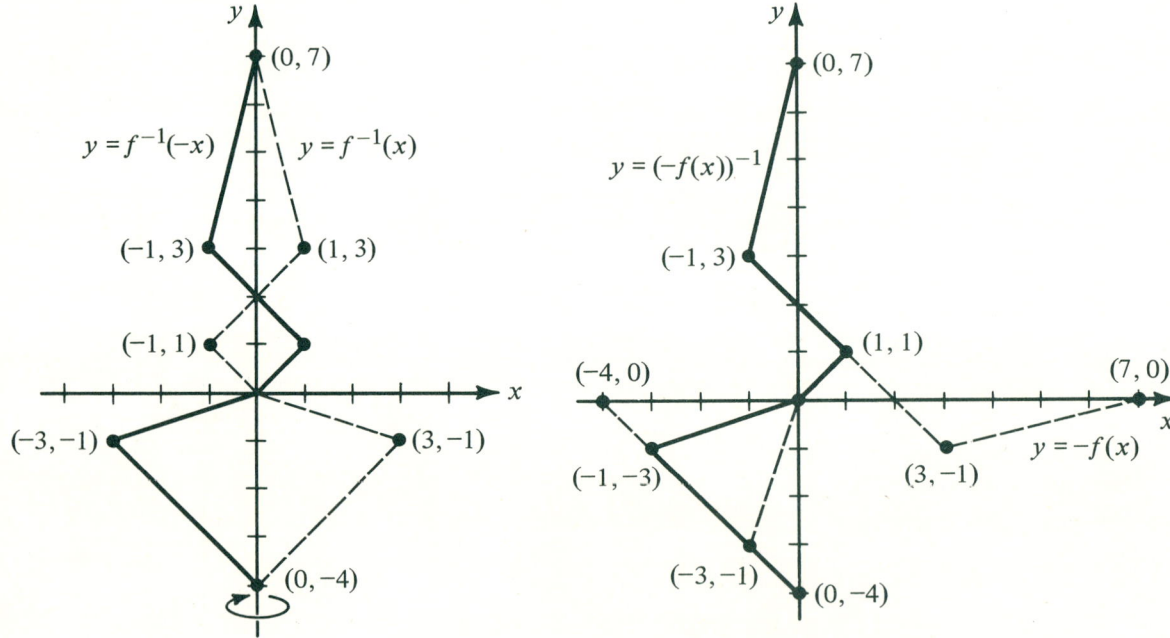

19. (a) $y = -f^{-1}(x)$
$y = -(x^2), x \geq 0$
$y = -x^2, x \geq 0$

(b) $y = (f(-x))^{-1}$ is the inverse of $y = f(-x) = \sqrt{-x}$ which is $y = -x^2, x \leq 0$.

$y = -f^{-1}(x)$ is different from $y = (-f(x))^{-1}$. The first is the reflection of $y = f^{-1}(x)$ about the x-axis. The second is the inverse of $y = -f(x)$. The restrictions on x, $x \geq 0$, come from $y = \sqrt{x}$. Notice $y \geq 0$ as well as $x \geq 0$.

Exercise 3.7

1.

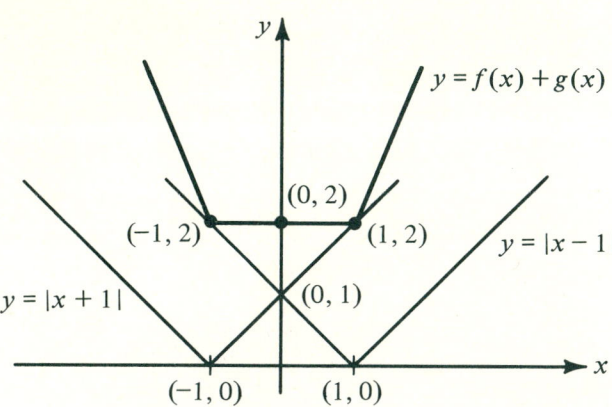

2.

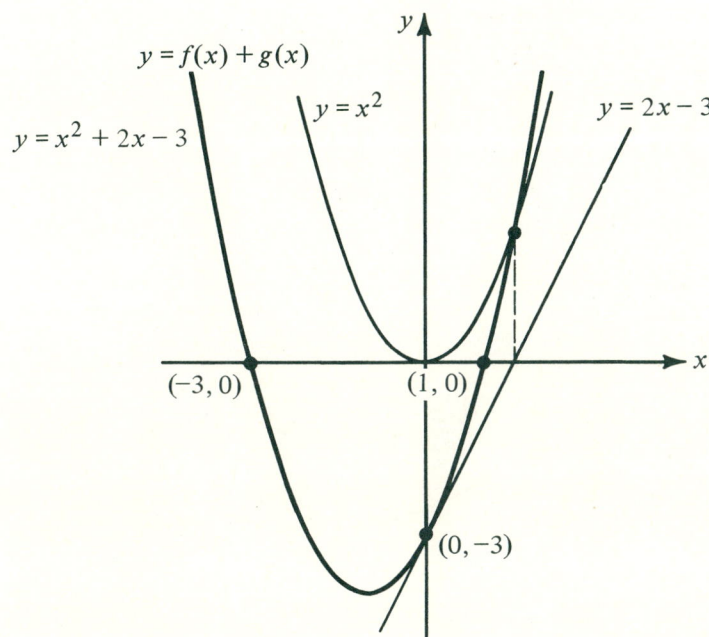

3.

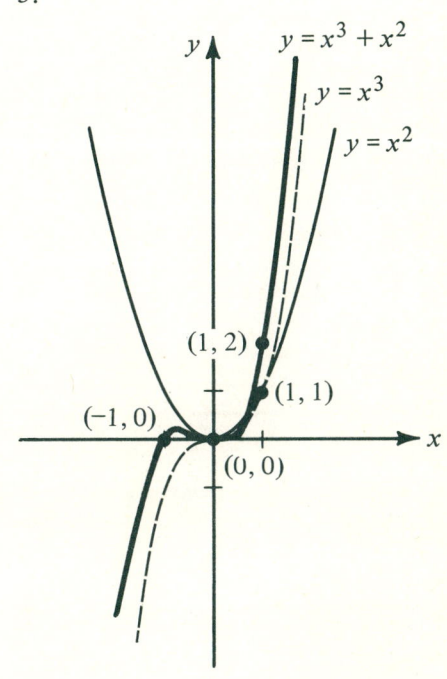

4. (a) 9
   (b) 9
5. (a) 3
   (b) not real
6. (a) 27
   (b) 15
7. (a) not real
   (b) 1

8.

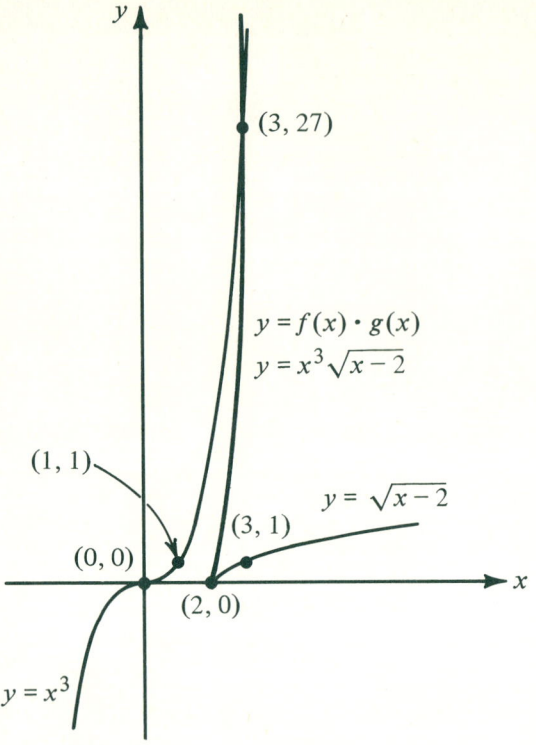

9. (c) $y = |x + 3|/x$

It is important to notice the signs and the relative size of each factor, $1/x$ and $|x + 3|$. Since $|x + 3|$ is always positive or zero, the product $1/x \, |x + 3|$ is negative whenever $1/x$ is negative. There is an x-intercept at $x = -3$, since $-\frac{1}{3} \cdot 0 = 0$. As x nears 0, $1/x$ increases greatly in size, causing the product to increase greatly in size.

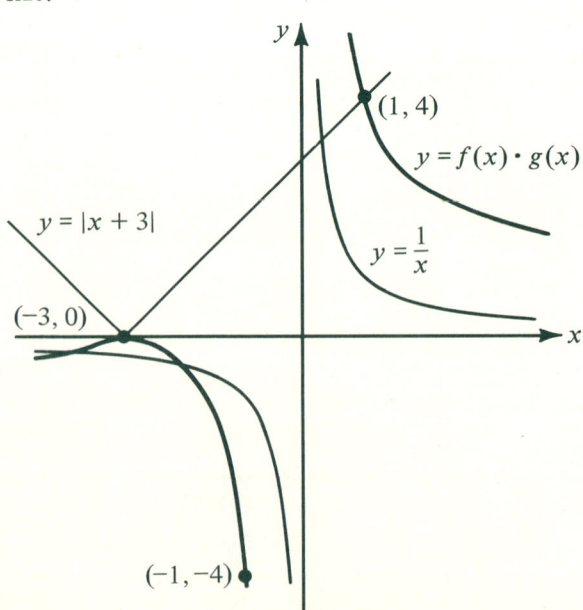

11. $y = g(f(x)) = g(\sqrt{x-1}) = \sqrt{x-1} + 4$

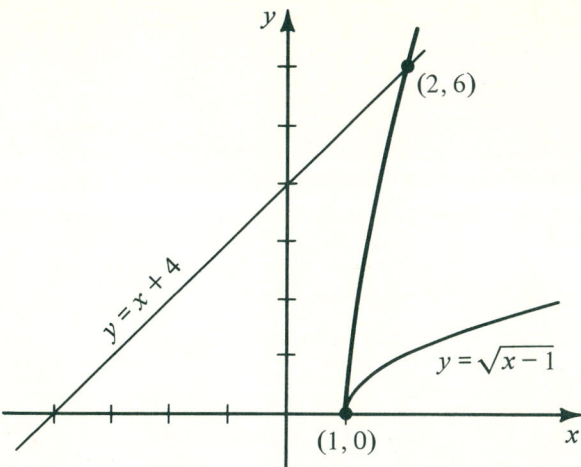

12. $y = g(f(x)) = g(x^2 - 9) = \sqrt{x^2 - 9}$

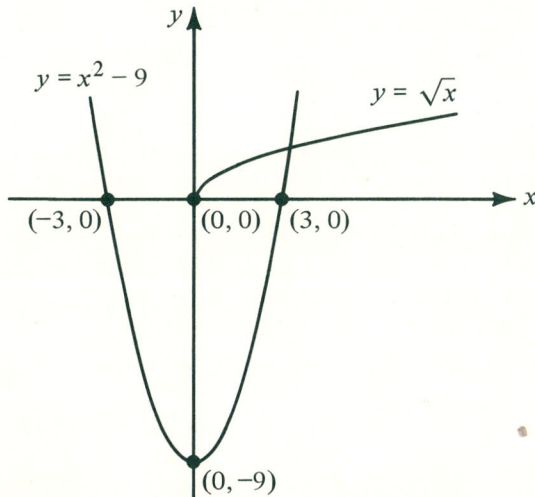

Here the domains and ranges are of central importance. We can use only values of x for which $y = x^2 - 9$ is positive, since in $\sqrt{x^2 - 9}$ we are going to be taking the square root of these y values. The graph of (a) shows positive y values for

$$x < -3 \quad \text{or} \quad x > 3$$

and

$$y = 0 \text{ at } x = \pm 3$$

The graph of (c) follows.

(c)

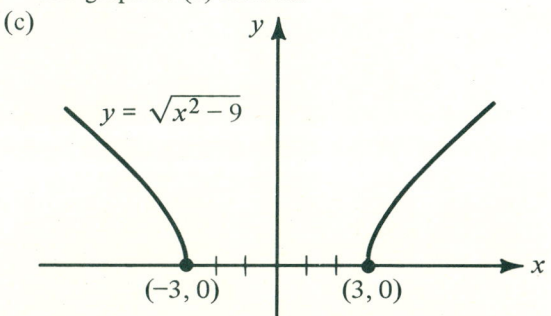

13. (a) $f(x) = (x - 2)^2$
    (b) domain: $\{x | x \geq +1\}$
    (c) $f(g(x)) = f(\sqrt{x-1}) = x + 3 - 4\sqrt{x-1}$

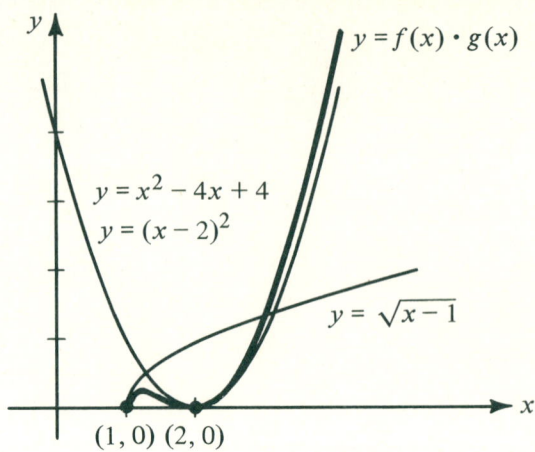

15. $f(x) = 1/3\, x + 5$ and $f(f^{-1}(x)) = x \Rightarrow f(f^{-1}(x)) = 1/3\, f^{-1}(x) + 5$ must equal x. Therefore, $1/3\, f^{-1}(x) = x - 5 \Rightarrow f^{-1}(x) = 3x - 15$.

17. Consider $g(x) = \sqrt{x+1}$ and $f(x) = x^2 - 1$. Then, $f(g(x)) = (\sqrt{x+1})^2 - 1 = (x+1) - 1 = x$ but $g(f(x)) = \sqrt{(x^2 - 1) + 1} = \sqrt{x^2} = |x|$.

19. (a) Let $h(x) = x + 1$ and $k(x) = x^2$. Then, $h(k(x)) = x^2 + 1$.
    (b) Let $h(x) = x + 1$ and $k(x) = x^2$. Then, $k(h(x)) = (x + 1)^2$.

21. $f(x) = (x + 3)^2$. Let $h(x) = x + 3$ and $k(x) = x^2$. Then, $f(x) = k(h(x))$.

22. If this function is to be written as a composite of only two simpler functions, it first must be rewritten as follows.

$f(x) = \sqrt{x^2 - 8x + 1} \Rightarrow$
$f(x) = \sqrt{(x^2 - 8x + 16) - 15} \Rightarrow$
$f(x) = \sqrt{(x - 4)^2 - 15}$

Then, since
$$f(x) = \sqrt{u^2 - 15}$$
where
$$u = x - 4$$
we say that $f(x)$ is composed of the two functions
$$h(x) = \sqrt{x^2 - 15} \quad \text{and} \quad k(x) = x - 4$$

23. $h(x) = \sqrt{x^2}$ and $k(x) = x + 3 \Rightarrow f(x) = h(k(x))$. (This could be rewritten $f(x) = |x + 3|$, since $\sqrt{x^2} = |x|$.)

25. $f(x)$ is translated to the left 1.

27. $f(x)$ is compressed toward the y-axis by a factor of 2.

29. The dimensions of the product is gallons/hour.

31. x, quantity must be non-negative and the ranges of $F(x)$ and $C(x)$ are 28 and non-negative reals, respectively.

33. $t = 1 \Rightarrow v = 290.2$ m/sec $\Rightarrow$ KE $= 421{,}080$ gm-m/sec.
    $t = 20 \Rightarrow v = 104$ m/sec $\Rightarrow$ KE $= 54{,}080$ gm-m/sec.
    $v = 0 \Rightarrow$ KE $= 0$ (at $5 = 30.6$ sec).

35. at $t = 0, p \approx 9$ ft. When $t = 10, p \approx 4$ ft.

CHAPTER 4

Exercise 4.1

1. These lines are parallel to the *x*-axis, hence their slopes are 0.
2. These lines are vertical (parallel to the *y*-axis), hence their slopes are infinitely large (undefined). We write $m = \infty$.
3. (a) $m = 1$ (b) $m = 1$ 4. (a) $m = 3$ (b) $m = 3$

   The following points lie on perpendicular lines (5 and 6).
5. (a) $m = 4$ (b) $m = -1/4$ 6. (a) $m = 3$ (b) $m = -1/3$
7. $B(0, 6)$ 8. $B(8, 0)$
9. (a) $B(4, 6)$ (b) $B(0, 4)$ These lines are perpendicular.
10. (a) $B(-2, 6)$ (b) $(-1, 7)$ These lines are perpendicular.

    These following lines are perpendicular.
11. (a) $m = 1$ (b) $m = -1$ 12. (a) $m = 1/2$ (b) $m = -2$
13. These lines are parallel with $m = 2/3$.
15. $m = -19.6$ m/sec/sec. The velocity, measured in meters per second is decreasing at a rate of -19.6 meters per second every second.
17. (a) *Marginal Revenue* is the rate at which the revenue, $R(x)$, is changing relative to changes in quantity sold, x. It is often described as the revenue produced by selling one more unit, that is, change in revenue per unit. $R(0) = -1000$, measured in dollars, is the initial cost. Since it is the cost incurred from producing nothing, it is called the *fixed cost*. It is not a function of quantity.
    (b) The slope of the cost function, $m = 0.8$, is the rate at which cost, $C(x)$, varies relative to changes in quantity, x, produced. Marginal cost is often described as the cost of producing one more, that is, change in cost per unit. $C(0) = 100$ gives the initial (fixed) cost. Marginal Revenue, MR, and marginal cost, MC, are equal at the quantity level that produces maximum profit. Suppose $MR < MC$. Then the revenue received from producing one more would be less than the cost incurred by producing one more. Decrease production. Suppose $MR > MC$. Then revenue from one more exceeds the cost of one more. Increase production. If $MR = MC$, it will not improve profit (Revenue − Cost) by increasing or decreasing production, so profit must be at a maximum.
19. Let $y = m_1 x + b_1$ be parallel to $y = m_2 x + b_2$ as shown below.

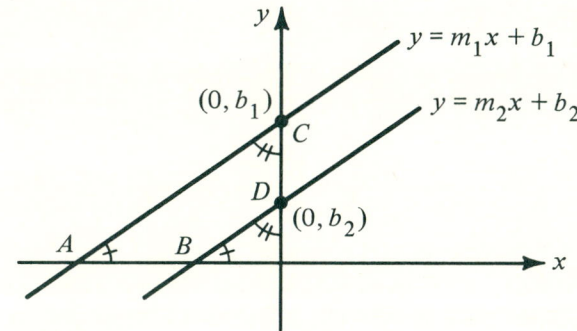

Since the *x*- and *y*-axis are transversals of the parallel lines, the indicated pairs of angles are equal. Therefore, $\triangle AOC$ is similar to $\triangle BOD$. Since corresponding sides of similar triangles are proportional,

$$m_1 = \frac{OC}{AO} = \frac{OD}{BO} = m_2$$

21. Let $m_1 \cdot m_2 = -1$ for the lines shown below.

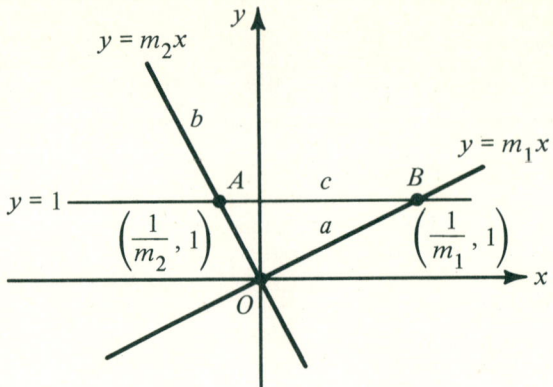

If we can show that $a^2 + b^2 = c^2$, then $\triangle ABO$ is a right triangle by the Pythagorean theorem.

$$c = \frac{1}{m_1} - \frac{1}{m_2} \Rightarrow$$

$$c^2 = \left(\frac{1}{m_1} - \frac{1}{m_2}\right)^2 \Rightarrow$$

$$c^2 = \frac{1}{m_1^2} - \frac{2}{m_1 m_2} + \frac{1}{m_2^2}$$

But

$$m_1 m_2 = -1 \Rightarrow$$

$$c^2 = \frac{1}{m_1^2} - \frac{2}{-1} + \frac{1}{m_2^2} = \frac{1}{m_1^2} + \frac{1}{m_2^2} + 2$$

From the triangles below, we can find the measure of a and b using the Pythagorean theorem.

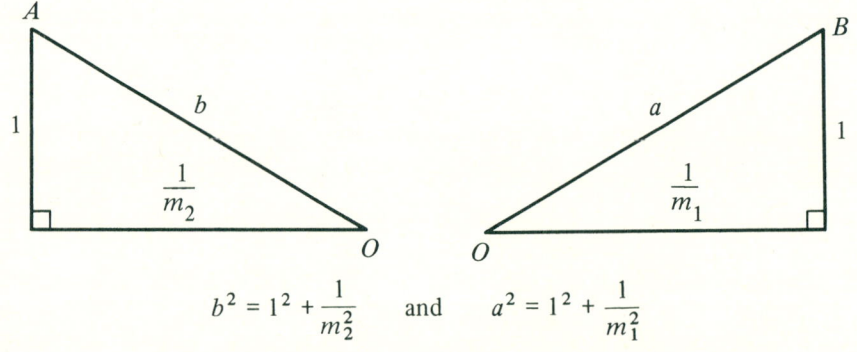

$$b^2 = 1^2 + \frac{1}{m_2^2} \quad \text{and} \quad a^2 = 1^2 + \frac{1}{m_1^2}$$

Therefore,

$$a^2 + b^2 = \frac{1}{m_1^2} + \frac{1}{m_2^2} + 2 \Rightarrow c^2 = a^2 + b^2$$

as required.

25. $m = 6$.

Exercise 4.2

1. (a) 5   (b) 5   2. (a) 10   (b) 13
3. (a) $4\sqrt{2}$ (b) $2\sqrt{2}$ 2   5. No
7. Yes, $y \neq -3$ 9. Yes
12. (a) The triangle inequality states that the sum of the lengths of any two sides of a triangle is greater than the length of the third side. We will show that $d(AC) + d(CB) > d(AB)$ in the following.

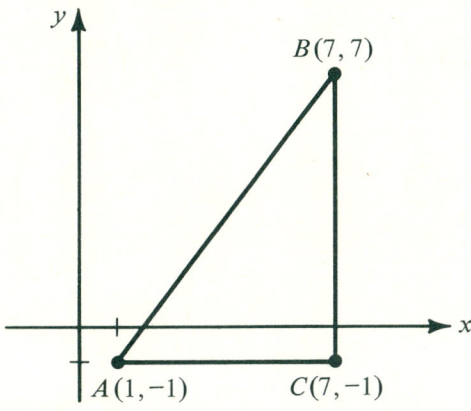

$$d(AC) = \sqrt{(x_1 - x_2)^2 + (y_1 - y_2)^2} = \sqrt{(1-7)^2 + (-1-(-1))^2}$$
$$= \sqrt{36} = 6$$
$$d(CB) = \sqrt{(7-7)^2 + (-1-7)^2} = \sqrt{64} = 8$$
$$d(AB) = \sqrt{(1-7)^2 + (-1-7)^2} = \sqrt{36+64} = 10$$
$$6 + 8 > 10 \Rightarrow d(AC) + d(CB) > d(AB)$$

The only time the distances are equal ($d(AC) + d(CB) = d(AB)$) is when ACB is a straight line.

13. In general, the midpoint between (x_1, y_1) and (x_2, y_2) is $[(x_1 + x_2)/2, (y_1 + y_2)/2]$, as shown below.

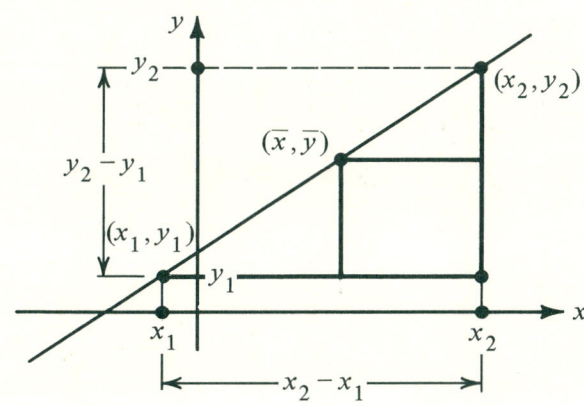

$$\bar{x} = x_1 + (x_2 - x_1)/2 = (2x_1 + x_2 - x_1)/2 = (x_1 + x_2)/2$$

and

$$\bar{y} = y_1 + (y_2 - y_1)/2 = (2y_1 + y_2 - y_1)/2 = (y_1 + y_2)/2$$

Then, for $A(2, -4)$ and $B(8, -2)$,
$$(\bar{x}, \bar{y}) = [(2+8)/2, -4 - 2/2] = (5, -3)$$

15. These are circles.
    (a) $x^2 + y^2 = 16$ (b) $(x - 3)^2 + (y + 4)^2 = 25$
16. The set of all such points is an ellipse.

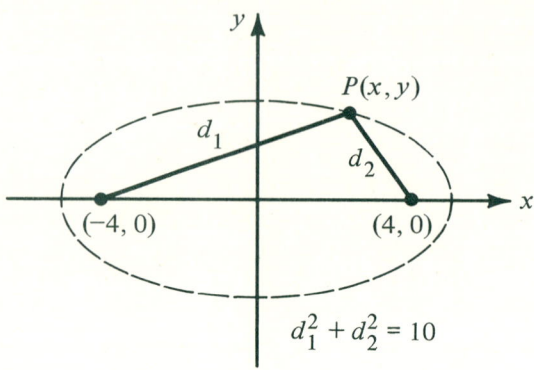

$$\sqrt{(x-4)^2 + (y-0)^2} + \sqrt{(x+4)^2 + (y-0)^2} = 10 \Rightarrow$$
$$\sqrt{x^2 - 8x + 16 + y^2} = 10 - \sqrt{x^2 + 8x + 16 + y^2}$$

which by squaring both sides becomes,

$$x^2 - 8x + 16 + y^2 = 100 - 20\sqrt{x^2 + 8x + 16 + y^2} + x^2 + 8x + 16 + y^2 \Rightarrow$$
$$5\sqrt{x^2 + 8x + 16 + y^2} = 4x + 25$$

which, after squaring both sides again yields,

$$25(x^2 + 8x + 16 + y^2) = 16x^2 + 200x + 625 \Rightarrow$$
$$9x^2 + 25y^2 = 225$$

17. This is the hyperbola, $9x^2 - 16y^2 = 1$.

Exercise 4.3

1. (a) $y = 3x$
   (b) $y = 3x - 11$
2. (a) $y = -1/3\, x$
   (b) $y = -x/3 - 10/3$
3. (a) $y = 3x/2 - 1/2$
   (b) $y = 3x/2 - 17/2$
4. (a) $y = 3x - 6$
   (b) $y = -x/3 + 4$
5. (a) $y = -x - 4$
   (b) $y = x - 1$
6. First we must rewrite the equation in the form $y = mx + b$ to determine its slope.
$$2x - 3y = 5 \Rightarrow$$
$$3y = 2x - 5 \Rightarrow$$
$$y = \frac{2}{3}x - \frac{5}{3}$$

Then, $m = 2/3$ and any parallel line will also have a slope of $2/3$. Then, translate $y = 2/3\, x$ to the left 1 and down 3, so it will pass through $(-1, -3)$.

$y = \frac{2}{3}(x + 1) - 3$

$y = 2/3\, x - 7/3$

7. $3x + 4y - 7 = 0 \Rightarrow$

$4y = -3x + 7 \Rightarrow$

$y = -\dfrac{3}{4}x + \dfrac{7}{4}$

This line has a slope of $-3/4$, so any perpendicular line must have a slope of $+4/3$, the negative reciprocal of $-3/4$. Then, we want to translate $y = \tfrac{4}{3}x$ to the right 2 and down 1 so it will pass through $(2, -1)$.

$$y = \dfrac{4}{3}(x - 2) - 1 \Rightarrow$$

$$y = \dfrac{4}{3}x - \dfrac{11}{3}$$

9. $y = 3/2\, x - 35/24$ or $36x - 24y = 35$

11. (a) $y = -3x$ becomes $y = -3(x - 3)$ (b) $y = 3(x - 3)$ becomes $y = -3(x - 3)$

These results are the same for linear functions of the form $y = mx$. That is, reflection then translation is the same as translation then reflection. This is not true for most functions.

13. $AC = 3x + 5$ is translated up 2 by simply adding 2 to $3x + 5$. Then the new average cost function, AC^*, is $AC^* = 3x + 7$.

15. $AC^* = 3(x - 4) + 5$ 17. $y = 6x$

Exercise 4.4

1. (a)

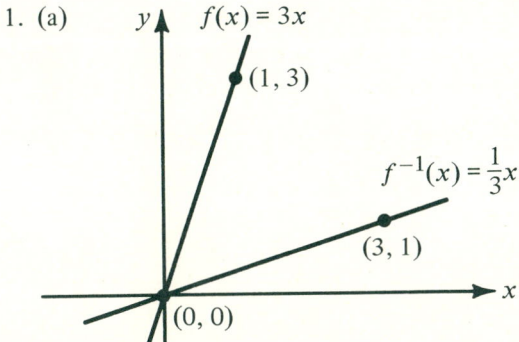

(b)

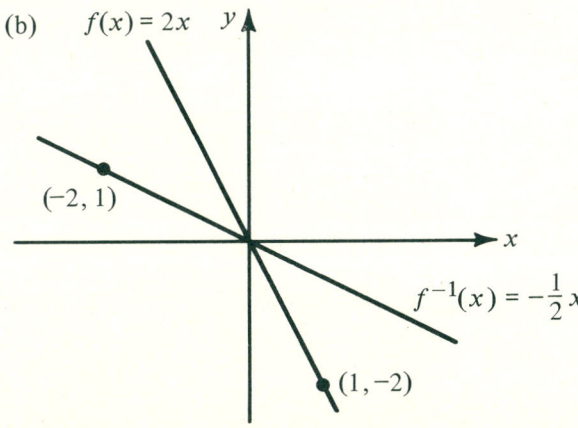

A–49

2.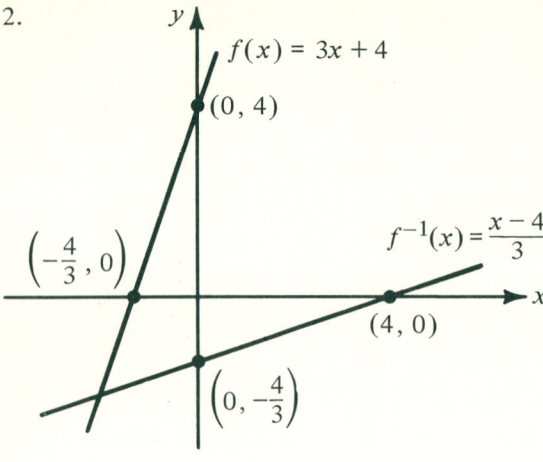

3.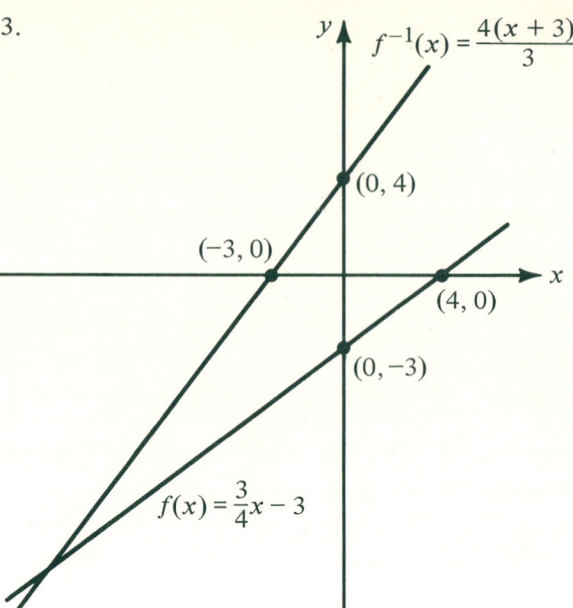

4. (a) $y = 3$ has as its inverse $x = 3$.
   (b) $x = -2$ has as its inverse $y = -2$.
5. $y = x$ is its own inverse.
6. (a) $x = 3$
   (b) $y = -4$
7. (a) $y = x + 4/3$
   (b) $y = 3(x + 5)$

9. $f^{-1}(x) = x + 3/5$ and $f(f^{-1}(x)) = 5(x + 3)/5 - 3 = x$ and $f^{-1}(f(x)) = \dfrac{5(x + 3)}{5} - 3 = x$.

11. Draw line AB whose slope is -1.

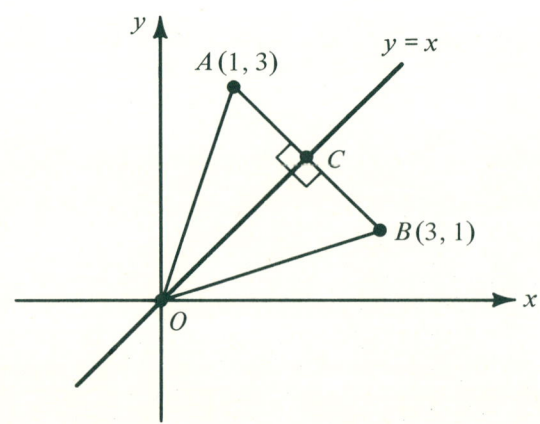

Since $d(OA) = \sqrt{10}$ and $d(OB) = \sqrt{10}$, triangles OAC and OBC are congruent. Therefore, $AC = BC$ and the symmetry is established.

13. $F = 9/5C + 32°$ is the formula for changing a Celsius temperature into its equivalent Fahrenheit temperature. Its inverse, found by solving for C, is

$$\frac{9}{5}C = F - 32 \Rightarrow C = \frac{5}{9}(F - 32)$$

is the formula for changing from Fahrenheit to Celsius. For the derivation of these formulas, see Problem 16, Exercise 4.3.

15. (a) $D = 5F$ (b) $D = 2.78M$

CHAPTER 5

Exercise 5.2

These first two represent parabolas that are translated up from *a* to *b* to *c*.
1. (a) $(0, 3), (-3, 0), (-1, 0)$
   (b) $(0, 4), (-2, 0)$
   (c) $(0, 5)$ no x-intercepts
2. (a) $(0, 3), (3, 0), (-1, 0)$
   (b) $(0, -1), (1, 0)$
   (c) $(0, -2)$ no x-intercepts
3. (a) $(0, -4), (1 + \sqrt{17}/2, 0), (1 - \sqrt{17}/2, 0)$
   (b) $(0, 4), (1 + \sqrt{17}/2, 0), (1 - \sqrt{17}/2, 0)$
   Since $h(x) = -k(x)$, $k(x)$ is the reflection about the x-axis of $h(x)$.
4. (a) $(2, 1)$
   (b) $(-3, -3)$
   (c) $(-4, -11)$
5. (a) $(4, -17)$
   (b) $(4, 17)$, (b) is a reflection about the x-axis of (a).
6. (a) $(+2, 4)$ (b) $(-2, 4)$, (b) is a reflection about the y-axis of (a).
7. (a) $(-3, -8)$ (b) $(-3, -16)$, (b) is an expansion away from the x-axis by a factor of 2 of (a).
8. (a) $(1, -3)$ (b) $(1/2), -3)$, (b) is a compression of (a) towards the y-axis by a factor of 1/2.
9.

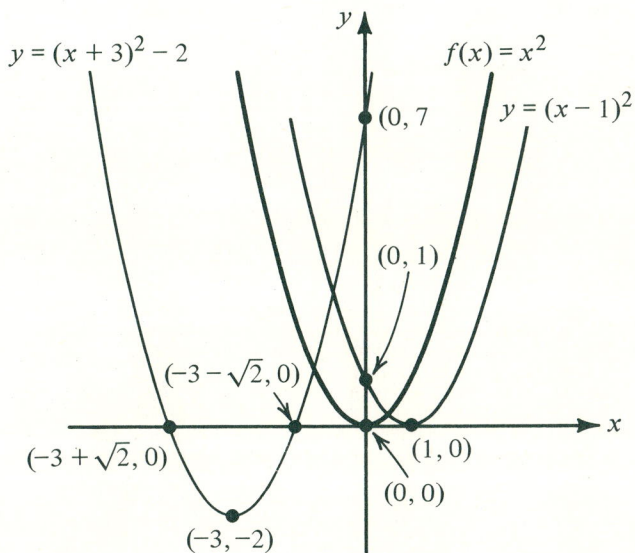

A-51

10.

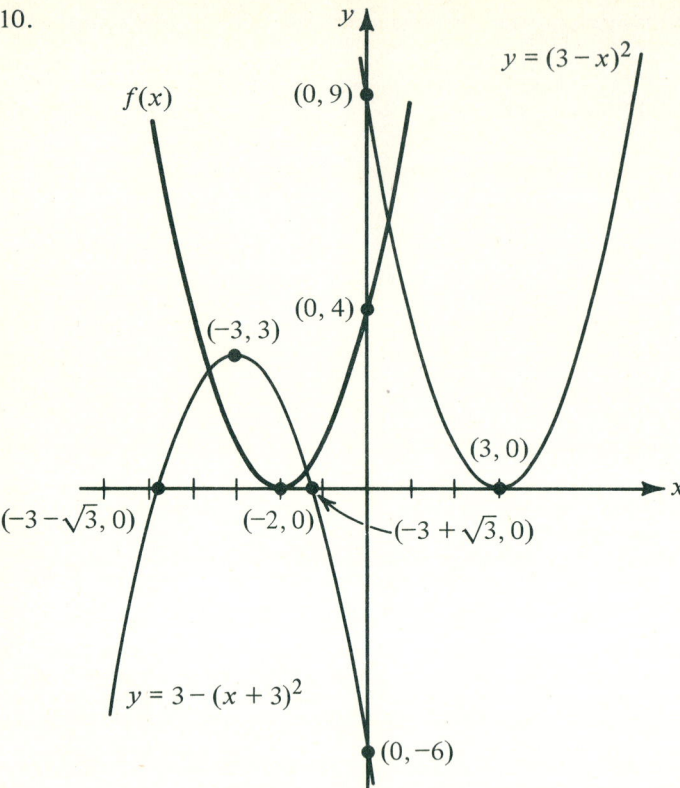

11.

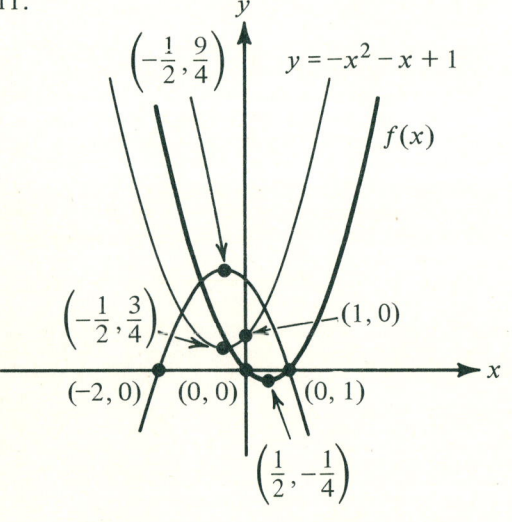

13.

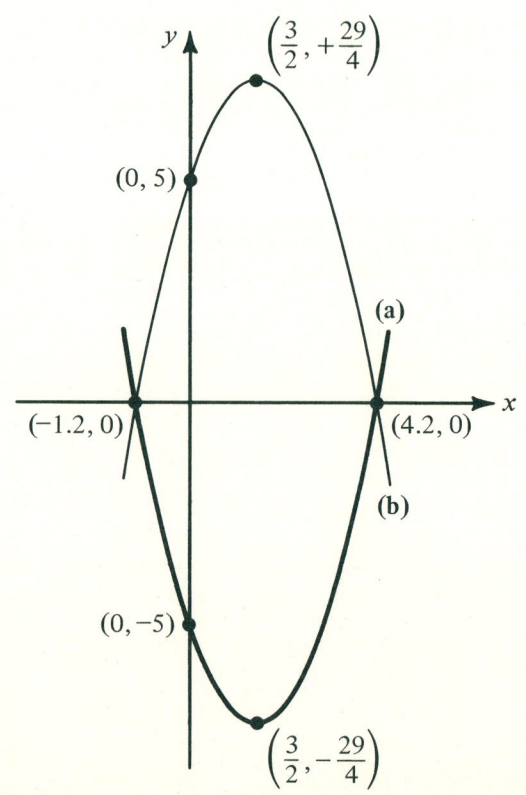

14. (a) $y = 6x^2 + 5x - 6$

$x = 0 \Rightarrow y = -6$

$y = 0 \Rightarrow 6x^2 + 5x - 6 = 0$

$(3x - 2)(2x + 3) = 0$

$3x - 2 = 0$ or $2x + 3 = 0$

$x = -\dfrac{b}{2a} = \dfrac{-5}{2(6)} = -\dfrac{5}{12} \Rightarrow$

Vertex:

$x = -\dfrac{b}{2a} = \dfrac{-5}{2(6)} = -\dfrac{5}{12} \Rightarrow$

$y = 6\left(\dfrac{-5}{12}\right)^2 + 5\left(\dfrac{-5}{12}\right) - 6 \approx -7.04$

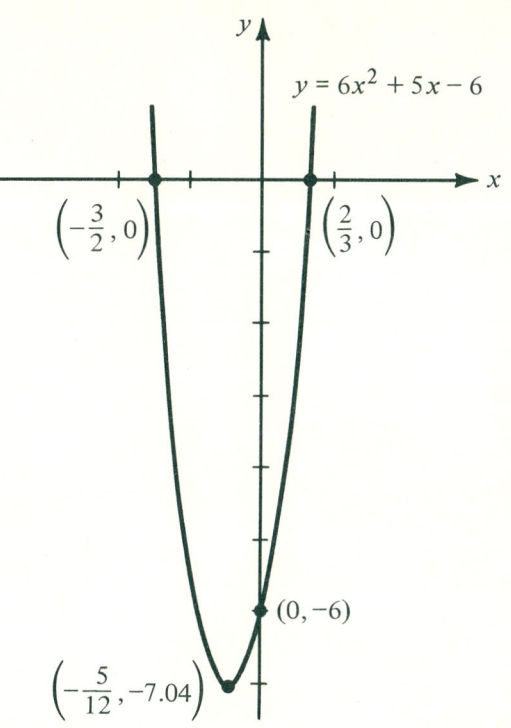

15. 9(a) $D = \{x|x \in \text{reals}\}$, $R = \{y|y \geq 0\}$
    9(b) $D = \{x|x \in \text{reals}\}$, $R = \{y|y \geq 0\}$
    9(c) $D = \{x|x \in \text{reals}\}$, $R = \{y|y \geq -2\}$

17. $f(x) = ax^2 + bx + c \Rightarrow$

$f\left(\dfrac{-b}{2a} + x\right) = a\left(\dfrac{-b}{2a} + x\right)^2 + b\left(\dfrac{-b}{2a} + x\right) + c$

$= a\left(\dfrac{b^2}{4a^2} - \dfrac{b}{a}x + x^2\right) - \dfrac{b^2}{2a} + bx + c$

$= \dfrac{b^2}{4a} - bx + ax^2 - \dfrac{b^2}{2a} + bx + c$

$= ax^2 - \dfrac{b^2}{4a} + c$

$f\left(\dfrac{-b}{2a} - x\right) = a\left(\dfrac{-b}{2a} - x\right)^2 + b\left(\dfrac{-b}{2a} - x\right) + c$

$= a\left(\dfrac{b^2}{4a^2} + \dfrac{b}{a}x + x^2\right) - \dfrac{b^2}{2a} - bx + c$

$= \dfrac{b^2}{4a} + bx + ax^2 - \dfrac{b^2}{2a} - bx + c$

$= ax^2 - \dfrac{b^2}{4a} + c$

Therefore, $f(-b/2a + x) = f(-b/2a - x)$ for all real numbers, x. Hence, the parabola is symmetric about the line through its vertex, $x = -b/2a$.

Exercise 5.3

1. (a) $y = \pm\sqrt{x}$
   (b) $y = \pm\sqrt{x} - 1$

2. (a) $y = \pm\sqrt{x-4}$
   (b) $y = \pm\sqrt{4-x}$

3. (a) $y = \pm\sqrt{x} - 3$
   (b) $y = \pm\sqrt{x} - 4$

4. The domains and ranges for 3(a) and 3(b) are the same since both originals are perfect square quadratics.

$$\text{Domain of } f = \text{Range of } f^{-1} = \{t | t \in \text{reals}\}$$
$$\text{Range of } f = \text{Domain of } f^{-1} = \{s | s \geq 0\}$$

5. $f(x) = x^2 - x - 12$

$$f^{-1}(x) = \frac{1}{2}(1 \pm \sqrt{4x + 49})$$

$$D_f = R_{f-1} = \{s | s \in \text{reals}\}$$

$$R_f = D_{f-1} = \{t | t \geq 12\tfrac{3}{16}\}$$

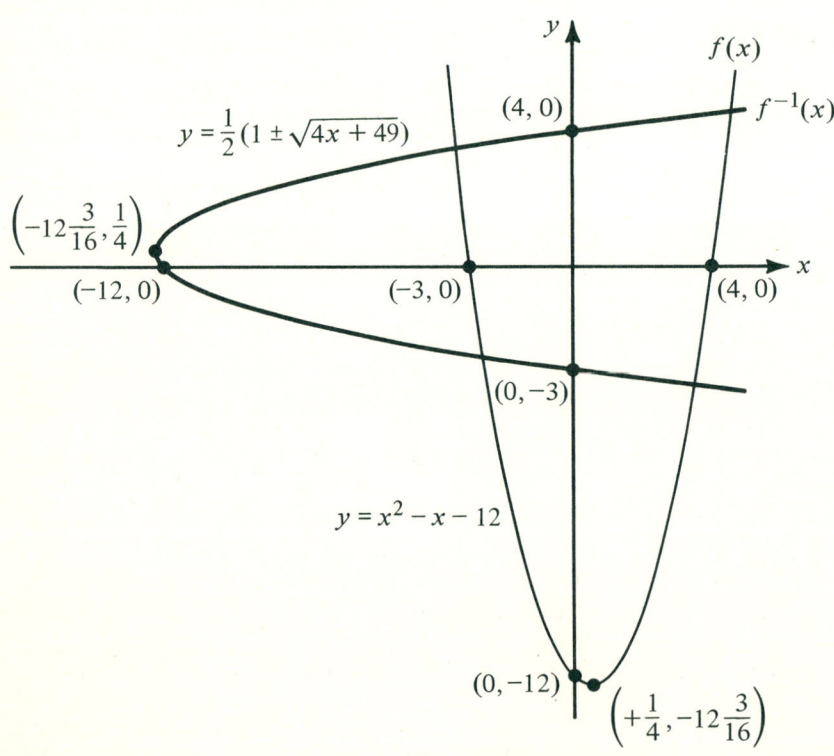

A-54

7. $h(x) = 2x^2 - 3x - 5$

$h^{-1}(x) = \dfrac{1}{4}(3 \pm \sqrt{8x + 49})$

$D_f = R_{f-1} = \{s | s \in \text{reals}\}$

$R_f = D_{f-1} = \left(t \middle| t \geqslant -6\dfrac{1}{8}\right)$

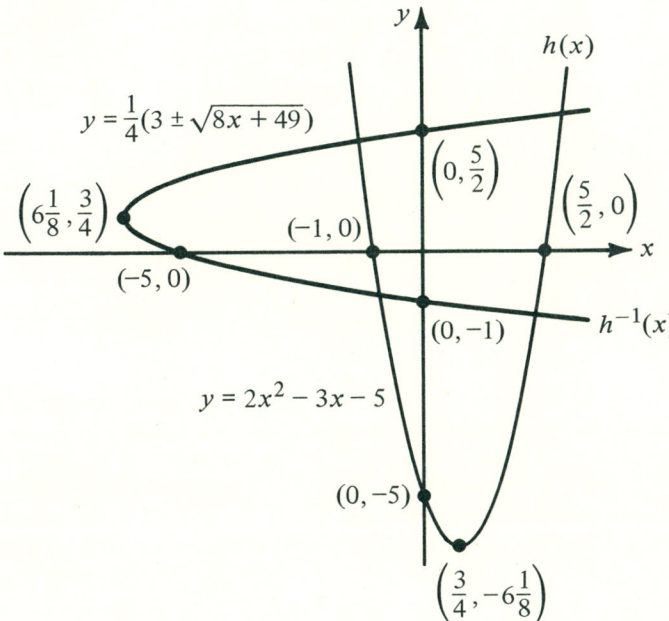

9. $f^{-1}(x) = x^2 - 9$

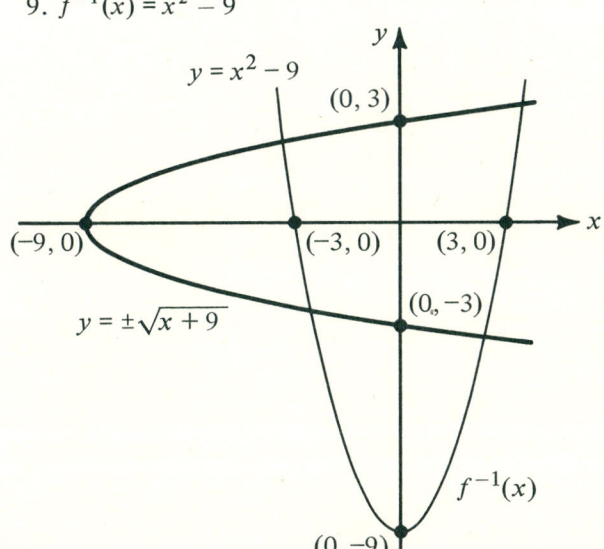

11. $f^{-1}(x) = x^2 + 2x - 15$

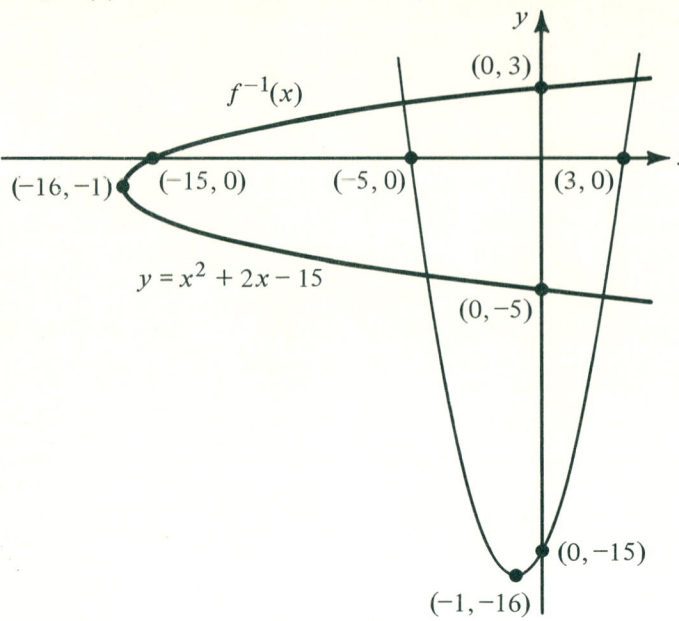

Exercise 5.4

These are all minimums because $a > 0$ means the parabolas are concave up.
1. (a) $y = 0$ (occurs at $x = -3$)
   (b) $y = 0$ (at $x = -4$)
2. (a) $y = -9$ (at $x = 0$)
   (b) $y = -16$ (at $x = 0$)
3. (a) $y = -25/4$ (at $x = 3/2$)
   (b) $y = -25/4$ (at $x = 3/2 + 2 = 7/4$)

These are all maximums because $a < 0$ means the parabolas are concave down.

4. (a) $y = 9$ (at $x = 0$)   5. (a) $y = 0$ (at $x = 5$)
   (b) $y = 16$ (at $x = 0$)   (b) $y = 0$ (at $x = 6$)
6. (a) $y = 9$ (at $x = 1$)
   (b) $y = 9 + 1 = 10$ (at $x = 1$)

Problems 1 and 5 depict perfect-square quadratics whose vertices are on the x-axis. Problems 2 and 4 depict even quadratics (only even powers of x) that are symmetric about the y-axis. This means that their vertices are on the y-axis.

7. $t = -b/2a = -19.6/-9.8 = 2$ sec $\Rightarrow S = 78.8$ m
9. $s = 104.9$ m at $t = 0.1$ sec
11. (a) $y = -25$ is the minimum at $x = 4$
    (b) $y = -25$ is the minimum at $x = -2$ and $x = 2$
13. 25m x 50m and maximum area is 1250m^2
15. Minimum AC is $15.75 per radio. It occurs at a production level of $x = 2$ radios. (This number could mean 2,000 or 2 carloads of radios.)

17. Profit = Total Revenue − Total Cost ⇔
$$P(x) = R(x) - C(x) \Rightarrow$$
$$P(x) = -5.00x^2 + 75.00x$$
which is concave-down parabola whose maximum is located at its vertex,
$$x = -\frac{b}{2a} = -\frac{75}{2}(-5) = 7.5$$
which yields the maximum profit of
$$P(7.5) = -5(7.5)^2 + 75(7.5) = \$281.25$$

19. After some arbitrary time t has passed, the ships are located as shown in the following. The object is to find the value of t for which s is minimum. During this time t, the ship A has traveled
$$D = R \cdot t = 20t \text{ km}$$
and ship B has traveled $15t$ km.
Then, by the Pythagorean theorem we have
$$s^2 = (200 - 20t)^2 + (15t)^2$$
$$s^2 = 625t^2 - 8000t + 40,000$$

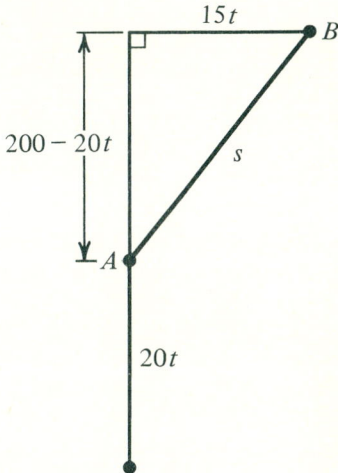

If we find the value of t that minimizes s^2, then we will also have the value of t that minimizes s. Since the above equation represents a concave-up parabola, its minimum value occurs at its vertex
$$t = -\frac{b}{2a} = +\frac{8000}{2(625)} = \frac{8000}{1250} = 6.4 \text{ hours}$$
The actual minimum distance is $s = 120$ km.

21. $t = 16$ hours.
23. The best fitting line in this case is $y = \bar{y}$, where $\bar{y}$ is the arithmetic average (mean) of the temperatures, that is, $y = 100$.
24. $y = -59/3x$.

Exercise 5.5

1. (a) $y = x^2 - 9$
   (b) $y = x^2 - 25$
2. (a) $y = (x - 4)^2 = x^2 - 8x + 16$
   (b) $y = (x + 1)^2 = x^2 + 2x + 1$
3. (a) $y = x^2 - 4x + 3$
   (b) $y = -x^2 + 4x - 3$
4. (a) $y = -2/9 \, x^2 + 4/3 \, x + 1$
   (b) $y = -2/9 \, x^2 - 4/3 \, x + 1$
5. (a) $y = 1/9 \, (2x^2 - 8x - 10)$
   (b) $y = -1/9 \, (2x^2 - 8x - 10)$
6. (a) $a_0 = 0, a_1 = 1, a_2 = 1$
   (b) $y = -x^2$

A-57

7. (a) Let $(x_0, y_0) = (1, 5)$, $(x_1, y_1) = (-2, 8)$ and $(x_2, y_2) = (3, 13)$
Then,
$$a_0 = y_0 \Rightarrow a_0 = 5$$
$$a_1 = \frac{y_1 - y_0}{x_1 - x_0} = \frac{8 - 5}{-2 - 1} = -1 \quad \text{and} \quad a_2 = \frac{y_2 - y_0 - (x_2 - x_0)a_1}{(x_2 - x_0)(x_2 - x_1)} \Rightarrow$$
$$a_2 = \frac{13 - 5 - (3 - 1)(-1)}{(3 - 1)(3 + 2)} = \frac{10}{10} = 1$$

Then,
$$f(x) = a_0 + (x - x_0)a_1 + (x - x_1)(x - x_0)a_2 \Rightarrow$$
$$f(x) = (+5) + (x - 1)(-1) + (x + 2)(x - 1)(1) \Rightarrow$$
$$f(x) = x^2 + 4$$

(b) $f(x) = 1/10\,(3x^2 - 39x - 14)$

9. (a) $y = x^2 - 2x - 8$
   (b) $y = -2x^2 + 4x + 30$

11. (a) $y = x^2 + x + 1$
    (b) $y = 2x^2 + 2x + 2$

13. $s = -4.9t^2 + 20t + 200$
15. $s = -4.9t^2 + 50$
    and
    $s_0 = 50\text{m}\ (t = 0 \Rightarrow s = 50\text{m}) \Rightarrow$
    $s = -4.9t^2 + v_0 t + 50$
    $(t = 2 \Rightarrow s = 69.6) \Rightarrow 69.6 = -19.6 + 2v_0 + 5 \Rightarrow v_0 = 0$
    Then, the equation is
    $$s = -4.9t^2 + 50$$
    Now, let's check to make sure the data point $(1, 45.1)$ satisfies this equation.
    $$t = 1 \Rightarrow s = -4.9(1^2) + 50 = 45.1$$

17. Minimum cost is $4.00 per radio (this is the average cost) at a production level of $x = 2000$ radios.
19. $y = 2x^2 + 3x + 1$
21. $y = 3x^2 - 2x + 5$

CHAPTER 6

Exercise 6.2

1. (a) $2x^2 - 3x + 1$
   (b) $2x^2 + x - 1$
2. (a) $3x^2 - 11x + 32 + 132/x + 4$
   (b) $3x^2 - 14x + 58 + -294/x + 5$
3. (a) $x^3 + 3x^2 + 2x - 2 + -4/x - 1$
   (b) $x^3 - x - 2 + 2/x - 2$
4. (a) $a_n = 3$, degree is 4
   (b) $a_n = 3$, degree is 4

5. (a) $a_n = 1$, degree is 2
   (b) not a polynomial
6. (a) $a_n = 4$, degree is 5
   (b) not a polynomial
7. (a) $a_n = 2$, degree $= 3$
   (b) $a_n = 4$, degree $= 3$
   (c) $a_n = 6$, degree $= 3$
   (d) $a_n = 8$, degree $= 6$
8. (a) $a_n = 2$, degree $= 4$
   (b) $a_n = 4$, degree $= 3$
   (c) $a_n = 2$, degree $= 4$
   (d) $a_n = 8$, degree $= 7$
9. (a) $a_n = 3$, degree $= 4$
   (b) $a_n = -3$, degree $= 4$
   (c) $a_n = 1$, degree $= 3$
   (d) $a_n = -9$, degree $= 8$
10. (a) $\underline{1\rfloor}$ 1   4   1   −6
              1   5   6
         1   5   6   0

    $$\frac{x^3 + 4x^2 + x - 6}{x - 1} = x^2 + 5x + 6 \Rightarrow$$

    $x^3 + 4x^2 + x - 6 = (x - 1)(x^2 + 5x + 6) + 0$

11. (a) $(x^3 + 7x^2 + 15x + 9)(x + 3) + 0$
    (b) $(x^3 + 9x^2 + 27x + 27)(x + 1) + 0$
13. (a) $(2x^3 - 3x^2 - 8x - 3)(x + 3) + 0$
    (b) $(2x^3 - 5x^2 + 3x - 39)(x + 4) + 147$
15. (a) $(x^3 + 3x^2 + 4x + 12)(x - 3) + 40$
    (b) $(x^3 + 4x^2 + 11x + 44)(x - 4) + 180$
17. Let $f(x) = a_n x^n + \cdots + a_0$ and $g(x) = b_m x^m + \cdots + b_0$, where $n > m$. Then,
    $$f(x) + g(x) = a_n x^n + \cdots + (a_m + b_m)x^m + (a_{m-1} + b_{m-1})x^{m-1} + \cdots + (a_0 + b_0)$$
    and the degree of $f(x) + g(x)$ equals n, the higher of the degrees of $f(x)$ and $g(x)$.

Exercise 6.3

1. (a) The possible values of the rational zeros are $p/q = \pm 1, \pm 1/2$, and $f(x) = (x + 1)(x - 1)(2x - 1)$
   (b) $p/q = \pm 1, \pm 1/2$, and $g(x) = -(x + 1)(x - 1)(2x - 1)$
2. (a) $p/q = \pm 1, \pm 2, \pm 4, \pm 1/3, \pm 2/3, \pm 4/3$, and $f(x) = (x + 2)(x - 2)(x + 1)$
   (b) $g(x) = f(-x)$ (a reflection about the y-axis); $g(x) = (x - 2)(x + 2)(x - 1)$
3. (a) $p/q = \pm 1, \pm 5, \pm 1/2, \pm 5/2$, and $y = (2x^2 + x + 5)(x + 1)(x - 1)$
   (b) This is a reflection of 3(a) about the y-axis. $y = (2x^2 - x + 5)(x - 1)(x + 1)$
4. (a) $(x + 3)^3(x + 1)$
   (b) $y = (x - 3)^3(x - 1)$

5. $y = 9x^4 + 36x^3 + 35x^2 - 4x - 4$ means $p = \pm 1, \pm 2, \pm 4$, and $q = \pm 1, \pm 3, \pm 9$

$$\frac{p}{q} = \pm 1, \pm 2, \pm 4, \pm \frac{1}{3}, \pm \frac{2}{3}, \pm \frac{4}{3}, \pm \frac{1}{9}, \pm \frac{2}{9}, \pm \frac{4}{9}$$

Since there is only one sign change, there will be, at most, 1 positive real root. $f(-x) = 9x^4 - 36x^3 + 35x^2 + 4x - 4$ has three sign changes. There will be 3, or 1 negative root.

```
2 | 9   36    35   -4    -4
        18   108  286   564
    ─────────────────────────
      9   54   143  282   560
```

We know $f(2) = 560$ and, since we divided by a positive and all the signs were positive, no value larger than $x = 2$ will be a zero.

```
-2 | 9   36    35   -4   -4
        -18   -36    2    4
     ────────────────────────
       9   18    -1   -2    0
```

Now we have
$$y = (x + 2)(9x^3 + 18x^2 - x - 2)$$
where the possible rational roots of the second factor are $\pm 1, \pm 2, \pm 1/3, \pm 2/3, \pm 1/9$, or $\pm 2/9$.

```
-2 | 9    18   -1   -2
        -18    0    2
     ────────────────────
       9     0   -1    0
```

Now
$$9x^3 + 18x^2 - x - 2 = (x + 2)(9x^2 + 0x - 1)$$
and
$$y = (x + 2)^2(9x^2 - 1) = (x + 2)^2(3x + 1)(3x - 1)$$

6. $y = (x - 3)(x + 2)(3x - 1)(2x + 3)$

7. For Problem 2(a): $f(x)$ has one sign change $\Rightarrow$ 1 or no positive roots; $f(-x) = -3x^3 + x^2 + 12x - 4$ also has one $\Rightarrow$ 1 or no negative roots.
   For Problem 3(a): $f(x)$ has one sign change $\Rightarrow$ 1 or 0 positive roots.

9. (a) Since the theorems do not allow us to find negative upper bounds on the set of zeros of a polynomial, even though the possible rational zeros include -1, we will start with $p = 1$. Then we will simply try successively larger divisors until we finally have all positive numbers for the coefficients of $q(x)$ and the remainder r in $p(x) = (x - a)q(x) + r$.

```
1 | 1   -2   -1   -1        2 | 1   -2   -1   -1
        1   -1   -2                2    0   -2
    ───────────────────         ───────────────────
      1   -1   -2   -3            1    0   -1   -3

                    3 | 1   -2   -1   -1
                            3    3    6
                        ───────────────────
                          1    1    2    5
```

Success! $x = 3$ is an upper bound for the set of zero of $f(x)$.

(b) $x = 1$

10. (a) $x = 4$ ($x = 3$) is the LUB, but $r = 0 \not> 0$)
    (b) $x = 2$ ($x = 1$ is the LUB)
11. 9(a) $x = -1$, 9(b) $x = -3$

Exercise 6.4

1. (a)

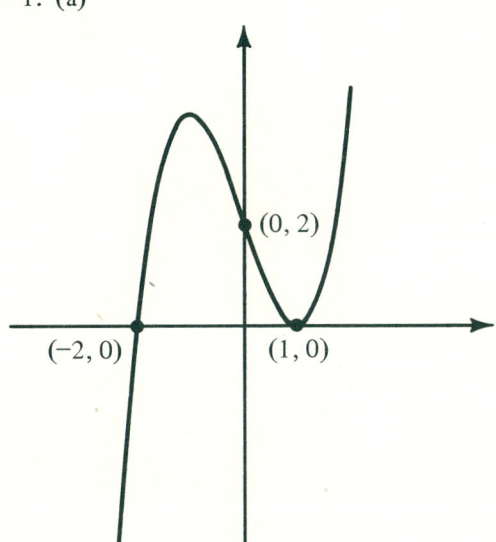

(b)

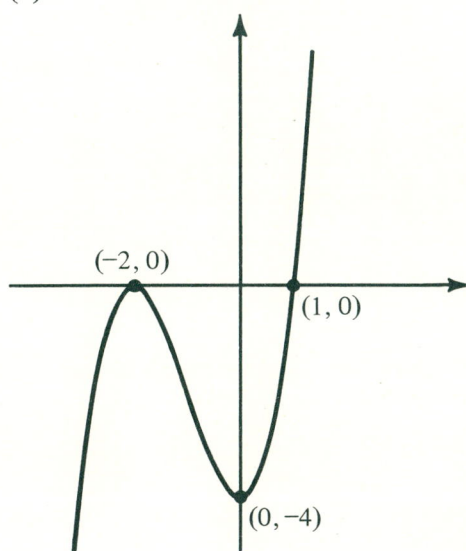

3. (a)

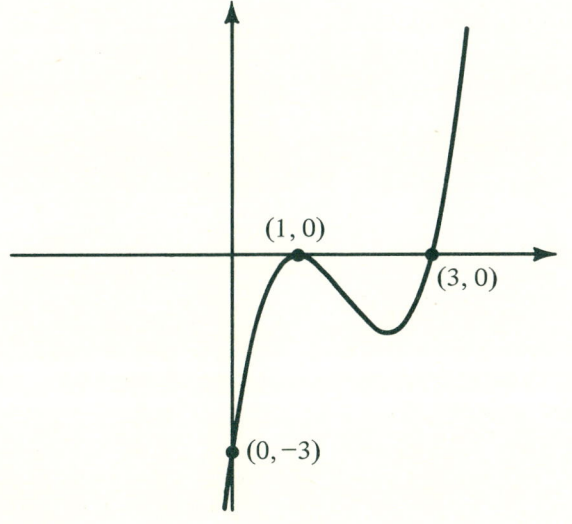

(b)

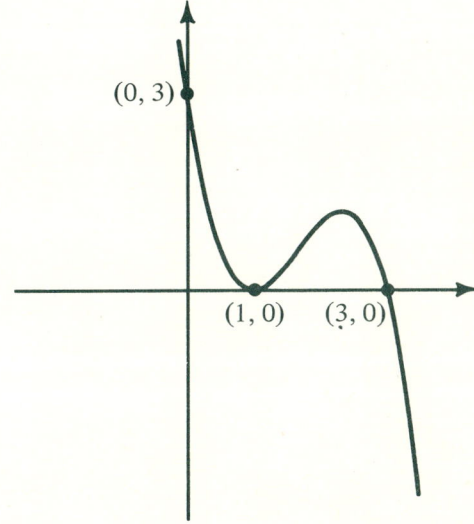

5. (a)

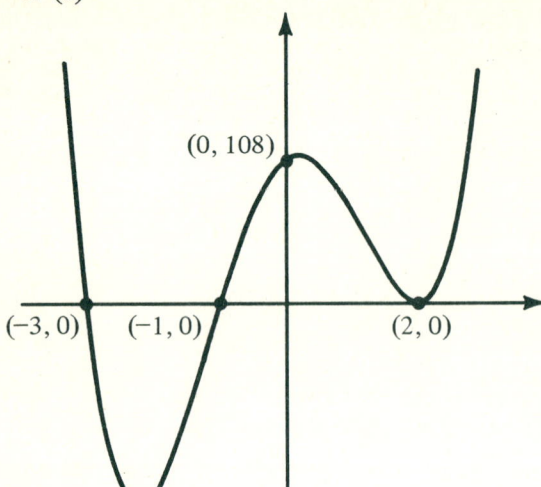

(b)

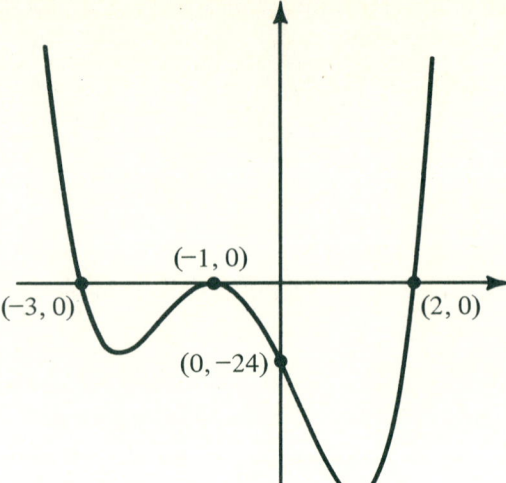

7. (a)

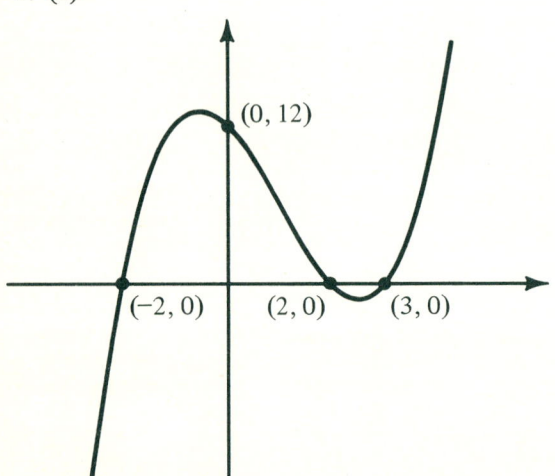

(b)

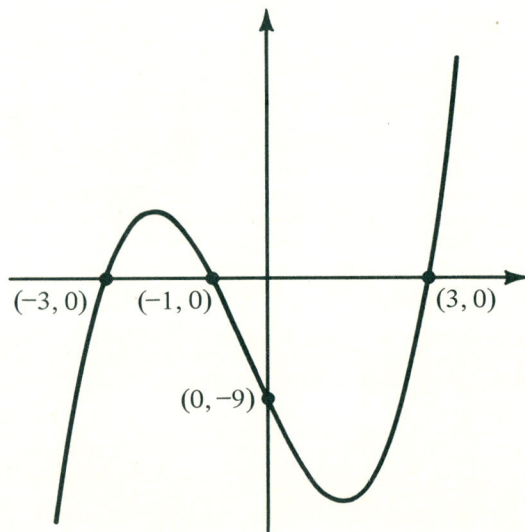

8. (a) $p(x) = 2x^3 + x^2 - 6x - 3$. The candidates for rational zeros are $p/q = \pm 1, \pm 3, \pm 1/2, \pm 3/2$, which are tested using synthetic division. The resulting factorization of $p(x)$ is $p(x) = (x + 1/2)(2x^2 - 6)$ $= (2x + 1)(x^2 - 3)$. The intercepts, therefore, are

$$\left(-\frac{1}{2}, 0\right), (\sqrt{3}, 0), (-\sqrt{3}, 0), \text{ and } (0, -3)$$

Graphed below is this odd-degree polynomial with leading coefficient positive. It goes up to the right and down to the left.

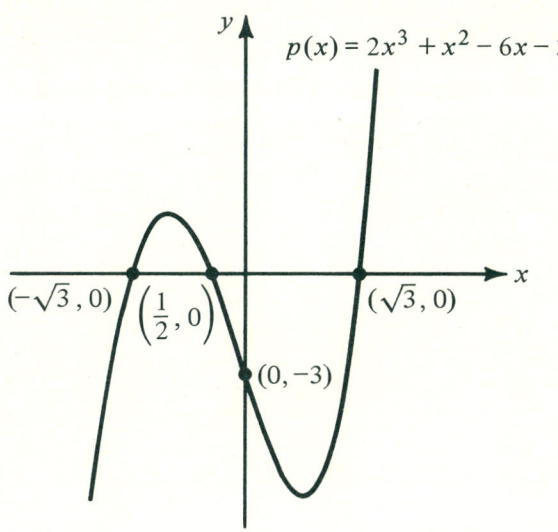

9. (a)   (b), (c)

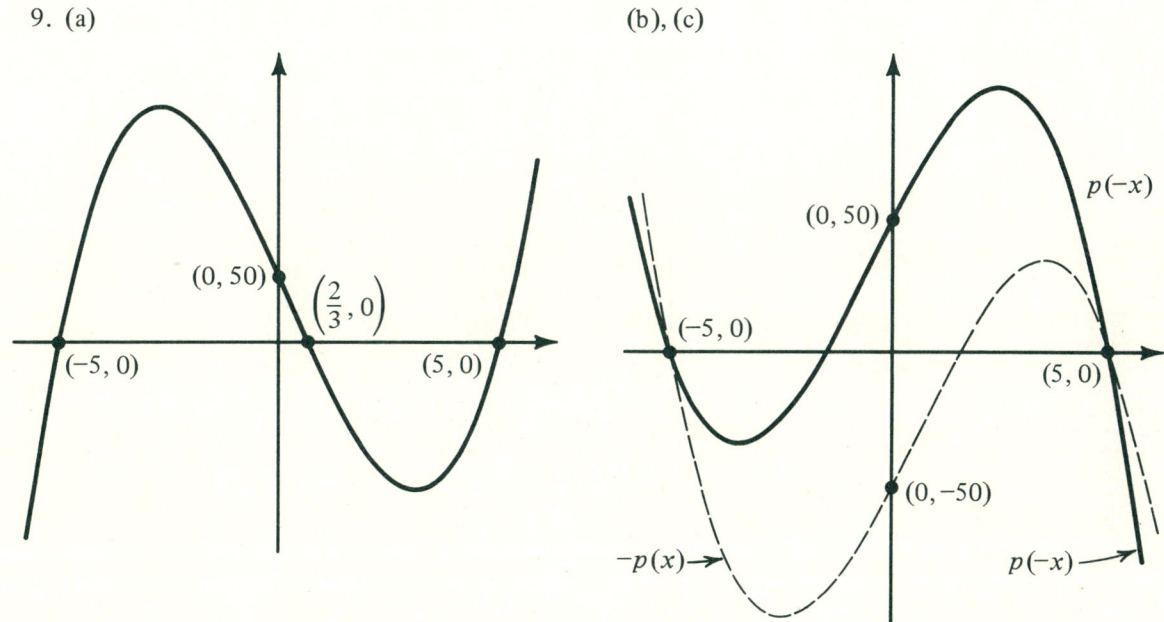

A-63

11.

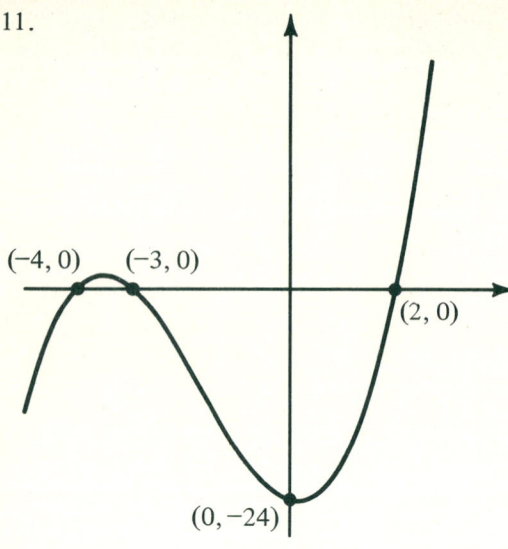

12. $y = x^4 - 4x^3 + 3x^2 + 4x - 4$. The candidates for rational zeros are only $p/q = \pm 1, \pm 2, \pm 4$.

$$
\begin{array}{r|rrrrr}
\underline{1|} & 1 & -4 & 3 & 4 & -4 \\
 & & 1 & -3 & 0 & 4 \\
\hline
\underline{-1|} & 1 & -3 & 0 & 4 & 0 \Rightarrow y = (x-1)(x^3 - 3x^2 + 4) \\
 & & -1 & 4 & -4 & \\
\hline
\underline{2|} & 1 & -4 & 4 & 0 \Rightarrow y = (x-1)(x+1)(x^2 - 4x + 4) \\
 & & 2 & -4 & & \\
\hline
 & 1 & -2 & 0 \Rightarrow y = (x-1)(x+1)(x-2)(x-2)
\end{array}
$$

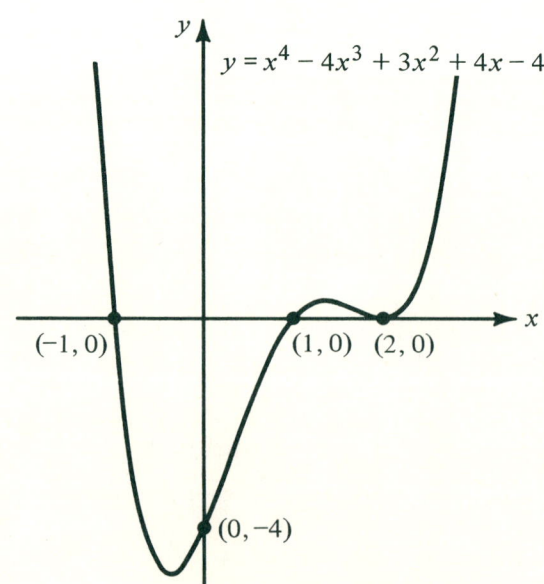

A-64

Notice that $x = 2$ is a zero of multiplicity two. The curve will touch the x-axis there like $y = (x - 2)^2$ does. The y-intercept is $(0, -4)$. This even-degree polynomial with leading coefficient positive goes up on both ends.

13.
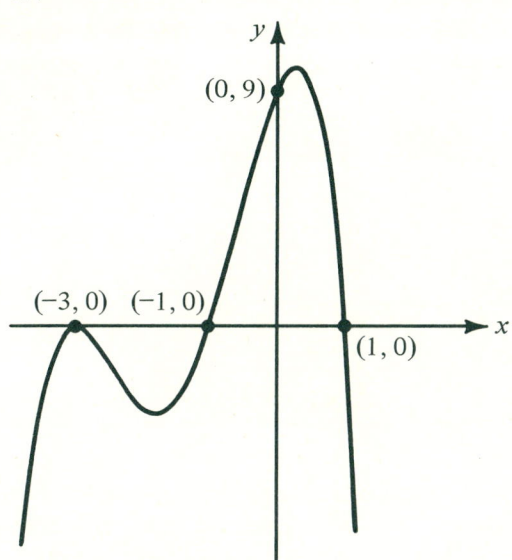

15.
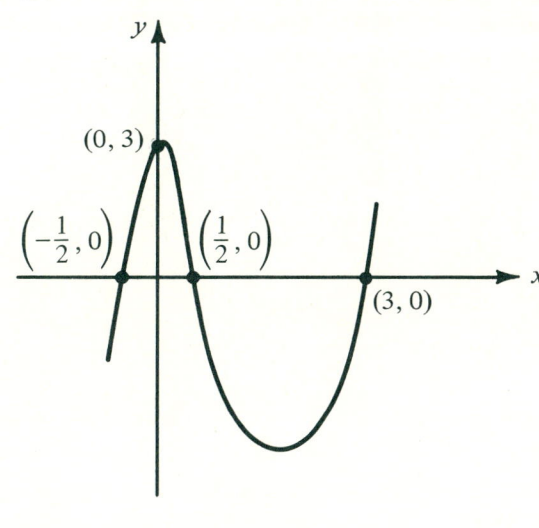

16. $y = -3x^3 - x^2 + 20x - 12 = -(x - 2)(3x - 2)(x + 3)$
This is an odd-degree polynomial with a negative leading coefficient so it goes down to the right and up to the left, like $y = -x$.

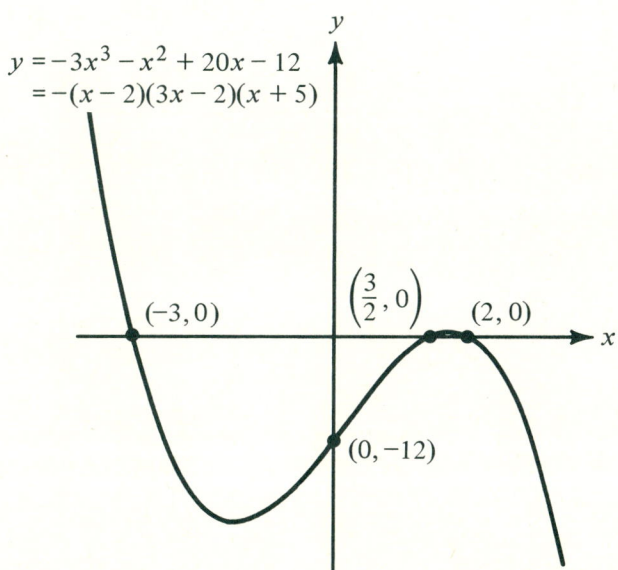

17.

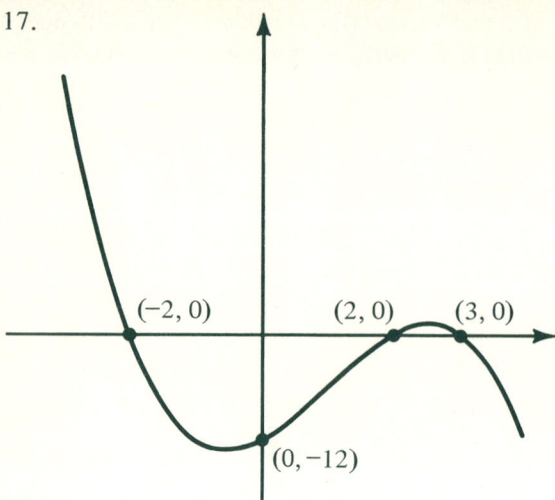

19.

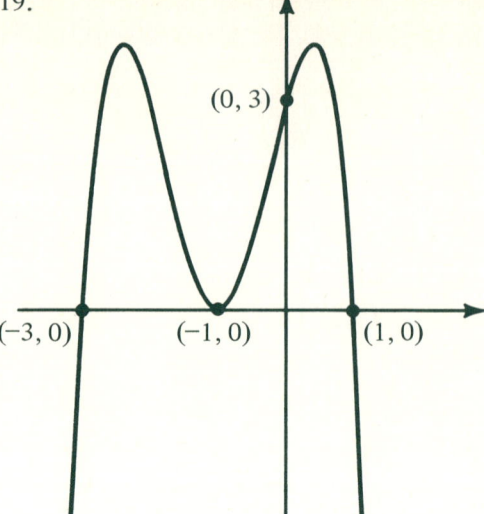

21.

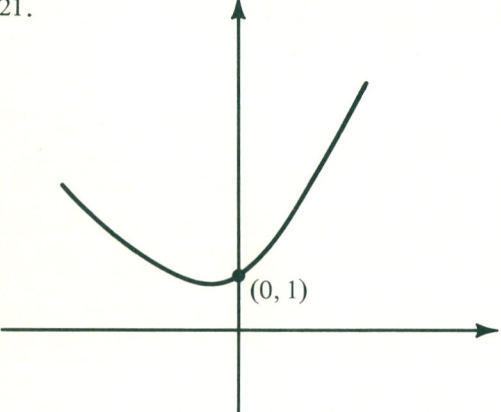

23. (a) $y = 1/96\,(x + 4)(x + 1)^2(x - 4)(x - 6)$
(b) $y = cx^2(x + 5)(x - 3)^3$, $c \in$ reals

(a)

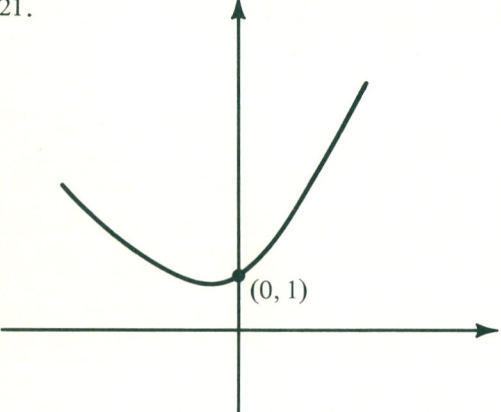

(b)

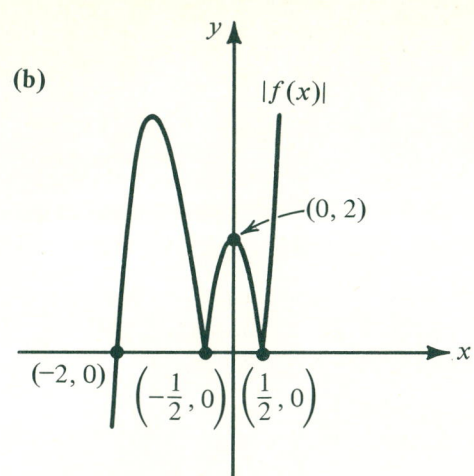

(c)

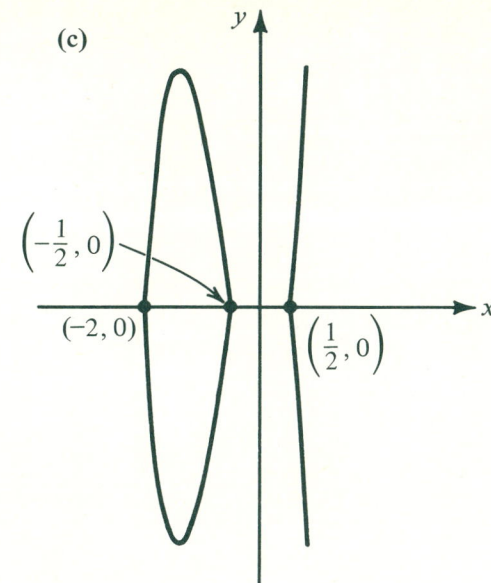

26. (b) $y = 7/4\,x^3 + 17/2\,x^2 - 9/2\,x + 2$ 27. (b) $y = 2x^3 - 4x^2 + x + 3$

Exercise 6.5

1. $x \approx 1.673$
3. $x \approx -3.632$
5. $x \approx 1.097$
7. $x = 2.7690625 \pm 0.0003125$
   $2.76875 \leqslant x \leqslant 2.769375$
9. $2.8898 \leqslant x \leqslant 2.88985$
   $x = 2.889825 \pm 0.000025$
10. $p(x) = (x + 1)(x^2 - 3x - 5)$
11. $p(x) = -(x + 2)(2x - 3)(x^2 - x - 3)$
15. $\sqrt[3]{21} \approx 2.7589$
17. $x \approx 1.6266$

Exercise 6.6

1. (a) $x = 1, x = -1/2 + \sqrt{3}/2\,i$, or $x = -1/2 - \sqrt{3}/2\,i$
   (b) $x = \pm 1$ or $x = \pm i$
3. (a) $x = \pm 1$ or $x = -1/2 \pm \sqrt{3}/2\,i$
   (b) $x = \pm 2$ or $x = -1 \pm 2i$
5. $y = z^4 - 2z^3 + 6z^2 - 8z + 8$

$1-i$	1	-2	6	-8	8
		$1-i$	-2	$4-4i$	-8
$1+i$	1	$-1-i$	4	$-4-4i$	0
		$1+i$	0	$4+4i$	
	1	0	4	0	

$y = [z - (1 - i)][z - (1 + i)][(z^2 + 4)]$

Therefore, the zeros are

$$z = 1 - i, \quad z = 1 + i, \quad \text{or} \quad z = \pm 2i$$

CHAPTER 7

Exercise 7.3

1. Asymptote at $x = -2$, hole at $x = -1$, since $y = (x + 1)^2/(x + 1)(x + 2) = (x + 1)/(x + 2)$, $x \neq -1$.
2. $x = -1/2$ and $x = -1$ are asymptotes.
3. $x = +2$ and $x = -2$ are asymptotes, and $(-1, 8/3)$ is a hole.
4. $x = +2$ and $x = -2$ are asymptotes.   5. $x = +2$, $x = -2$, and $x = 3$ are asymptotes.
6. $x = -3$, $x = 1$, and $x = 2$ are asymptotes, $(-1, -2/3)$ is a hole.
7. All those in which the degree of the polynomial in the denominator is greater than the degree of the polynomial in the numerator; Problems 5 and 6 have the x-axis as an asymptote.
9. (a) $y = 1/(x - 3)$ (b) $y = 4/(x - 3)$

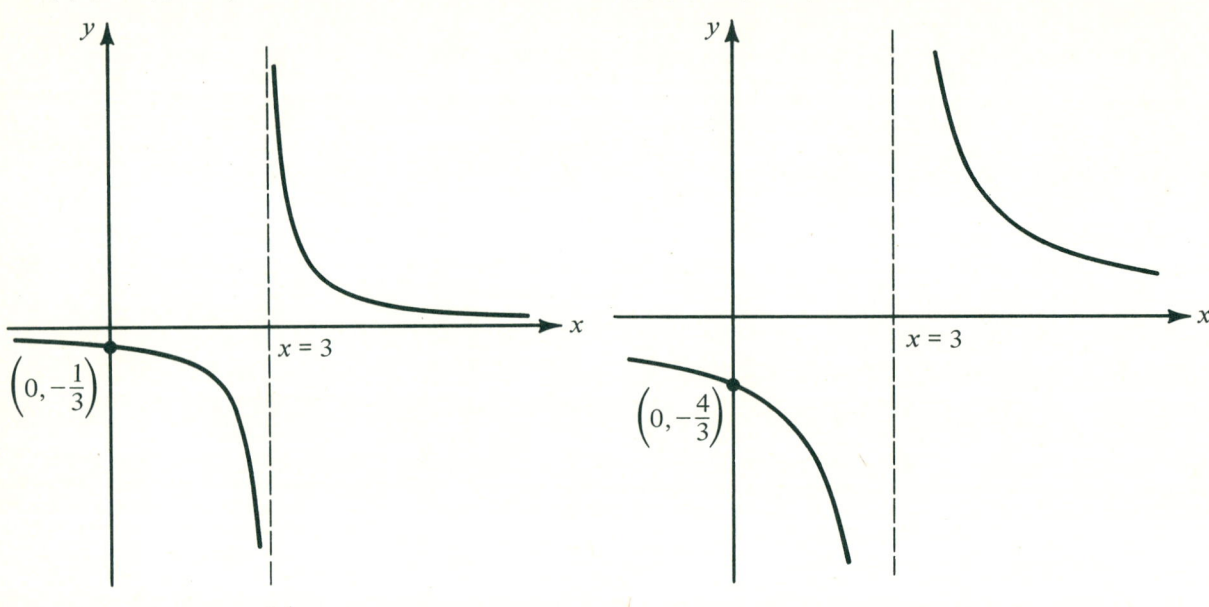

(c) $y = 4/(5x - 3)$

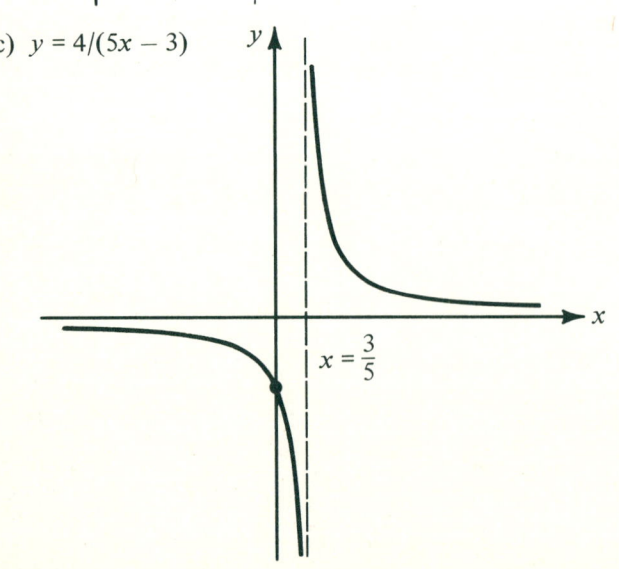

11. (a) $y = 1/(x + 1)^2$ (b) $y = 2/(x + 1)^2$

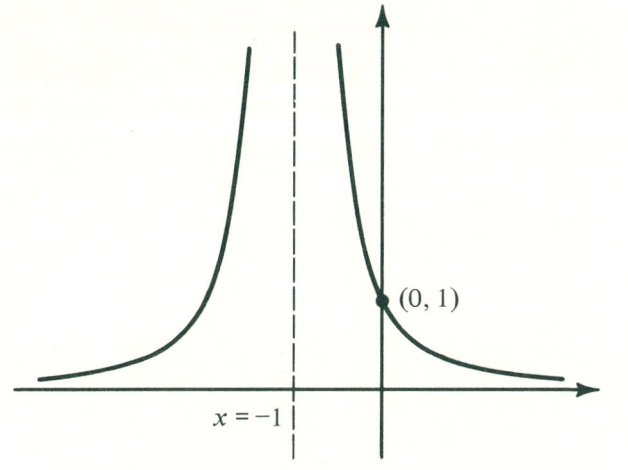
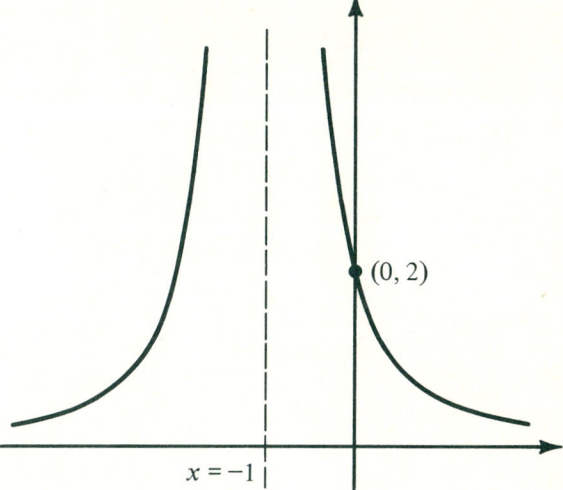

Compared with the even function $y = 1/x^2$, which is symmetric about the y axis with $y \geq 0$ for all $x \neq 0$ and has a vertical asymptote at $x = 0$ (see Figure 7.5), $y = 2/(x + 1)^2$ is translated to the left 1 and expanded away from the x-axis by a factor of 2. This yields $x = -1$ as the line of symmetry and the vertical asymptote.

(c) $y = 1/(2x + 1)^2$ (d) $y = 3/(4x + 1)^2$

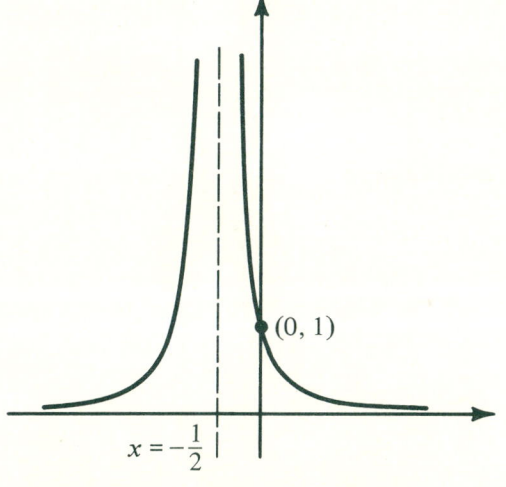
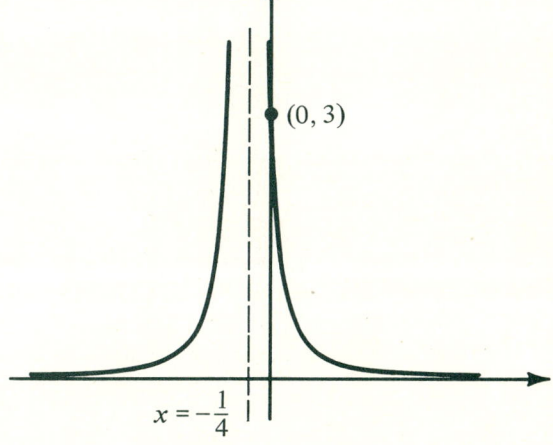

13. (a) $y = -1/x^2$

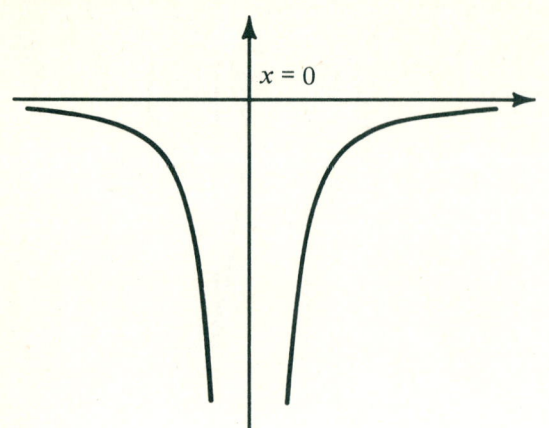

(b) $y = -1/(x-3)^2$

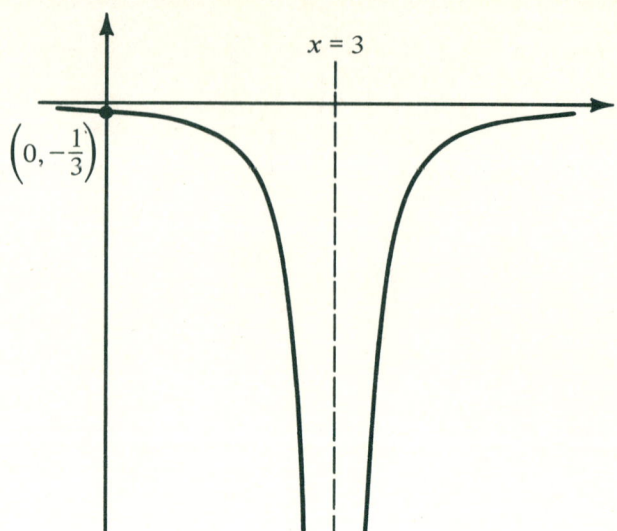

(c) $y = -2/(4-3x)^2$

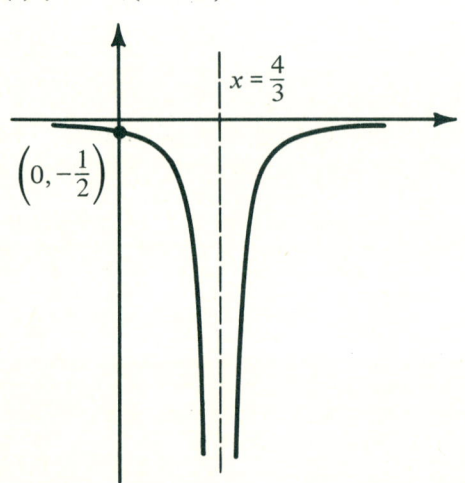

14. (c) $y = 1/(x^2 + 2x + 3)$

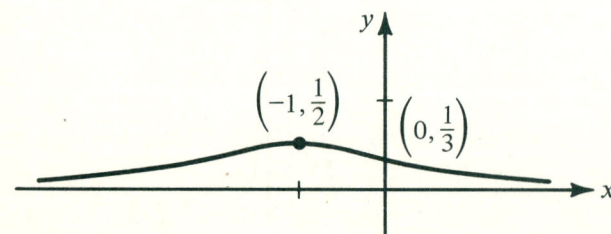

Since $x^2 + 2x + 3 \neq 0$ for any real value of x ($b^2 - 4ac = -8 < 0$), there are no vertical asymptotes. By rewriting this function by completing the square in the denominator we can recognize it as a translation of the graph in Figure 7.6.

$$y = \frac{1}{x^2 + 2x + 3} = \frac{1}{(x^2 + 2x + 1) + 2} = \frac{1}{(x + 1)^2 + 2}$$

Now, since the minimum value of $y = (x + 1)^2 + 2$ is $(-1, 2)$ (the vertex of this concave-up parabola, the maximum value of $y = 1/(x + 1)^2 + 2$ is $(-1, 1/2)$ because of the translation to the left 1 indicated by the $(x + 1)^2$ term. The x-intercept is $(0, 1/3)$ and the x-axis is the horizontal asymptote since the degree of the polynomial in the numerator (0) is smaller than the degree of that in the denominator (2).

15. (a) $y = \dfrac{1}{x^2 + 1}$

(b) $y = \dfrac{-2}{x^2 + 1}$

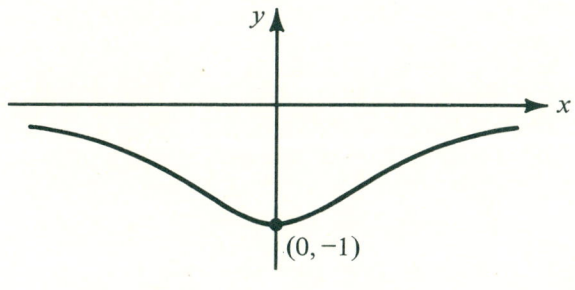
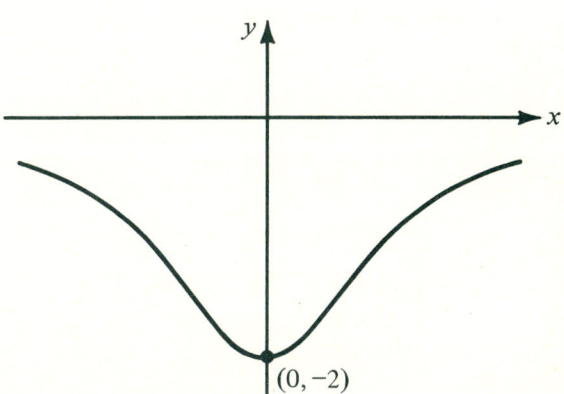

(c) $y = \dfrac{-1}{x^2 + 4}$

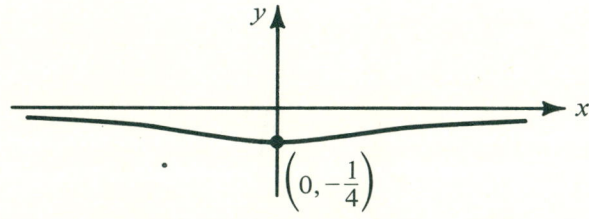

A-71

23. (a)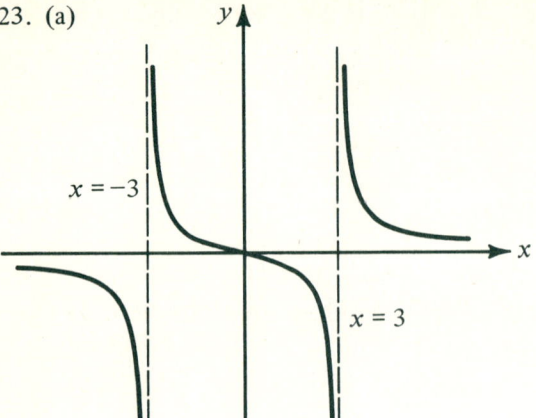

(b)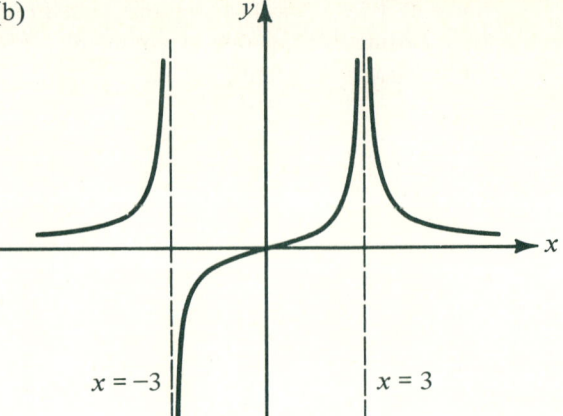

(c)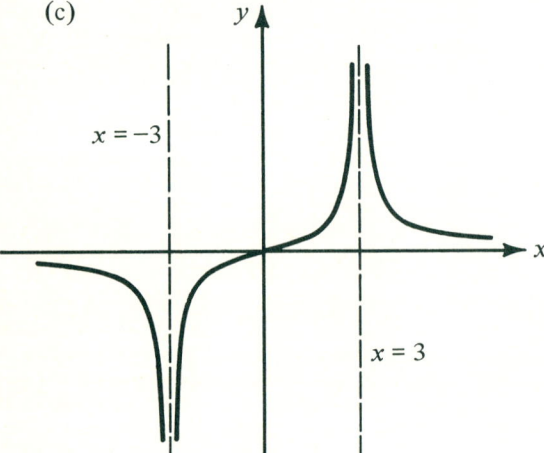

Exercise 7.4

1. (a)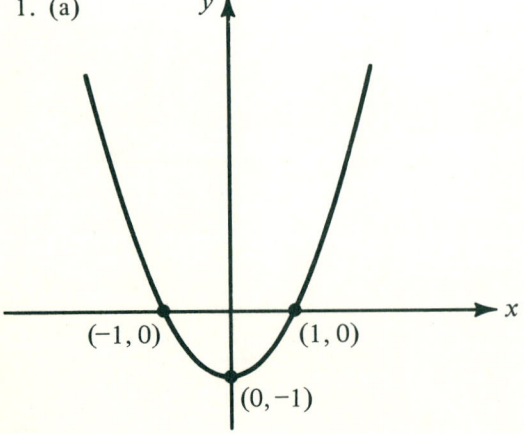

(b)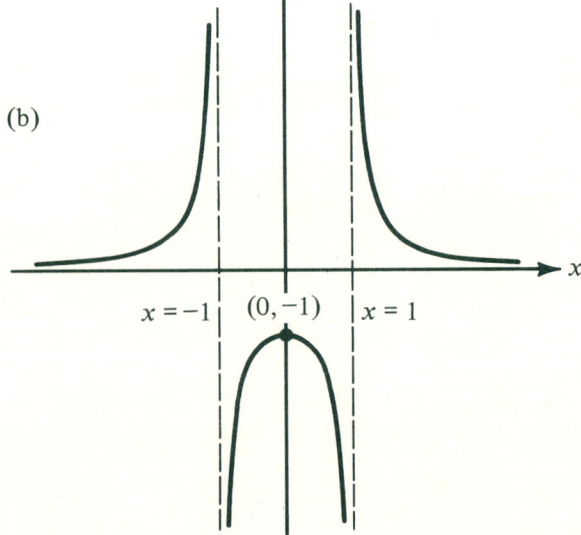

A-72

2. (b)

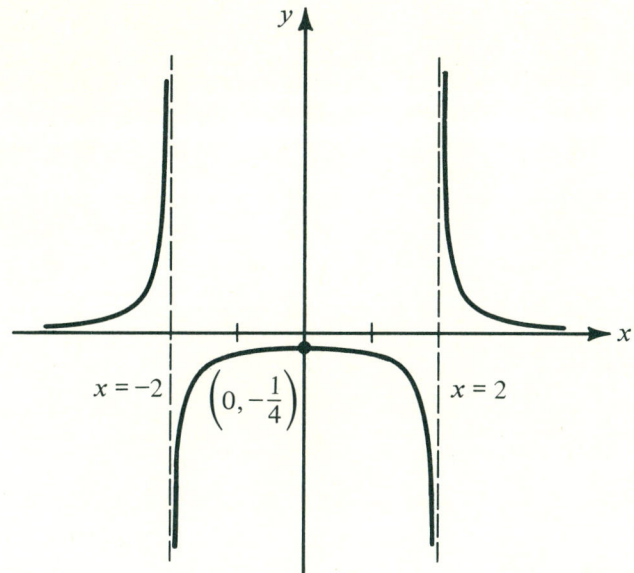

3. (a)

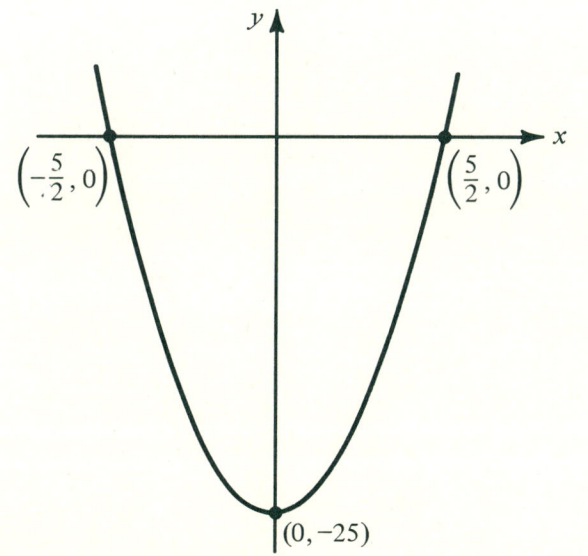

(b)

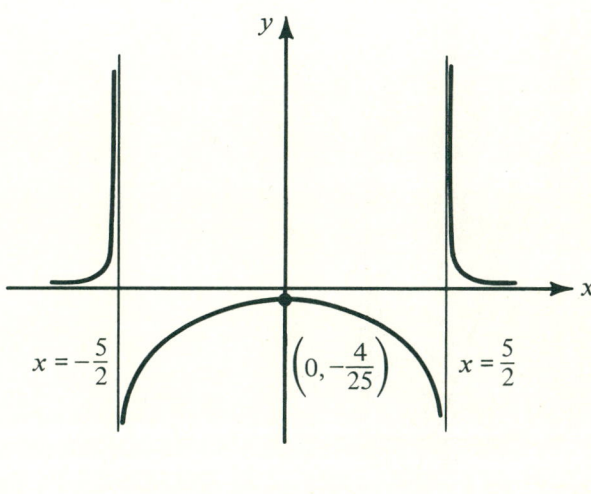

4. (b)
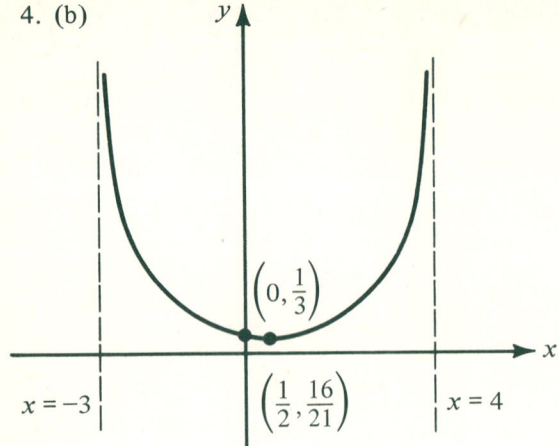

5. (a)
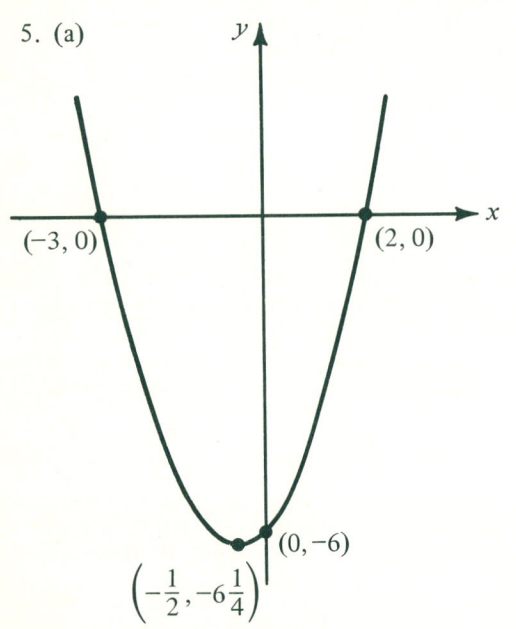

(b)
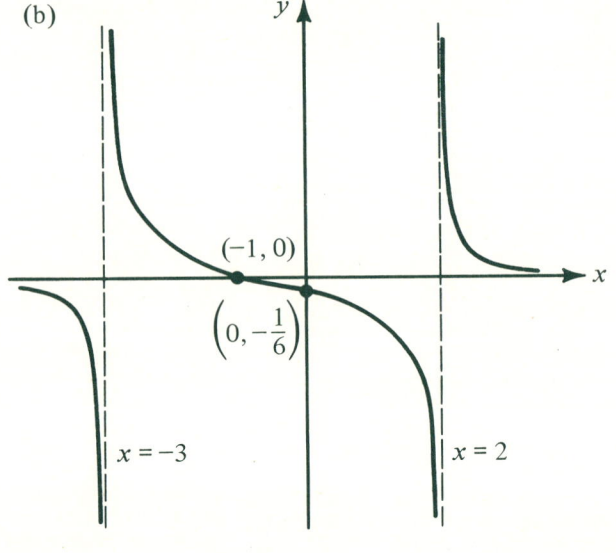

6. (b)
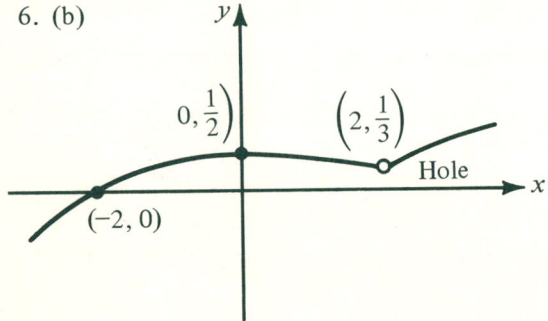

7. $y = \dfrac{2}{x^2 - 1} + \dfrac{2}{x + 1} = \dfrac{2}{(x + 1)(x - 1)} + \dfrac{2}{x + 1} \cdot \dfrac{x - 1}{x - 1} = \dfrac{2x + 1}{(x + 1)(x - 1)}$

Since the degree of the denominator, 2, is greater than that of the numerator, 1, the x-axis is the horizontal asymptote. Since $x = 1$ and $x = -1$ cause division by 0, but not 0/0, they are vertical asymptotes. Now, for the intercepts.

$y = 0 \Rightarrow \dfrac{2x + 1}{x^2 - 1} = 0 \Rightarrow 2x + 1 = 0 \Rightarrow x = -\dfrac{1}{2}$

$x = 0 \Rightarrow y = \dfrac{1}{-1} = -1$

$y = \dfrac{2x}{x^2} - 1$

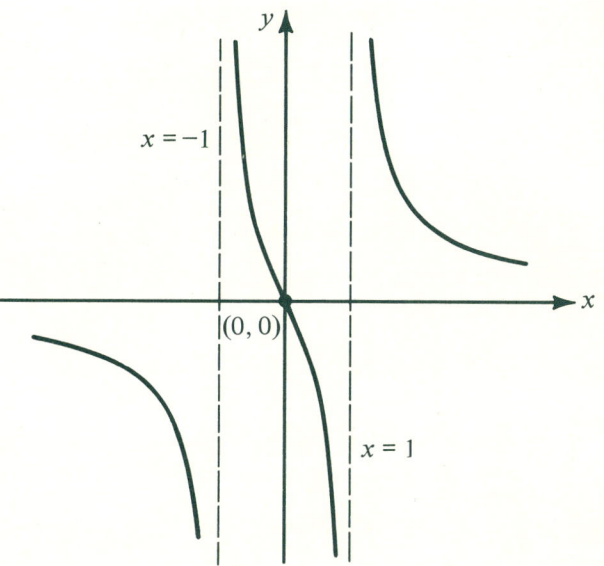

9. $y = \dfrac{-4}{x^2 - 2x - 3}$

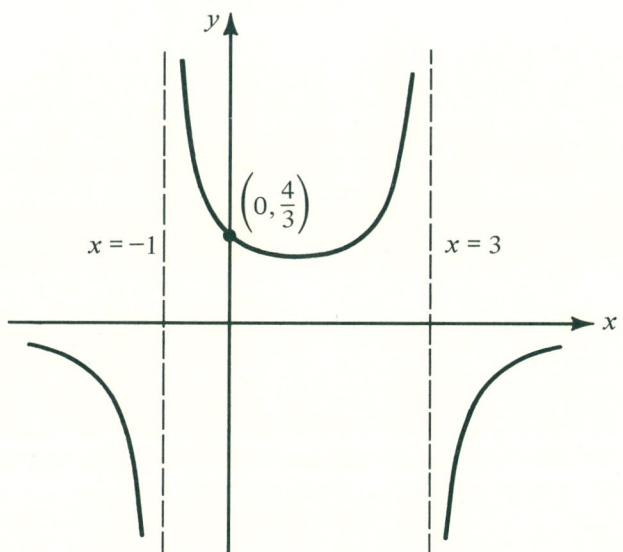

11. $y = \dfrac{-7x}{2x^2 - 9x + 4}$

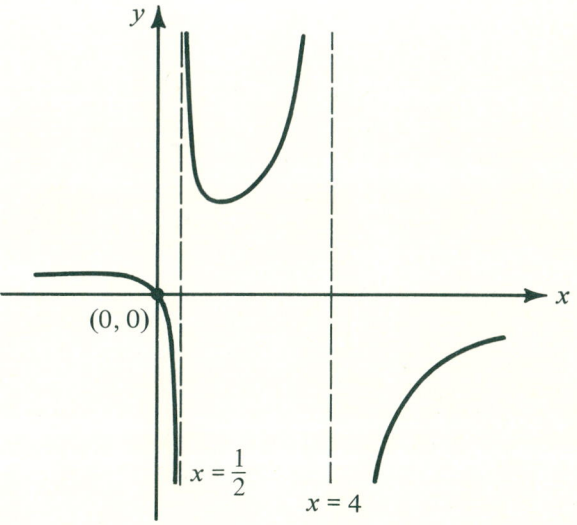

A-75

13. $g(x) = \dfrac{5}{x^3 - 13x + 12}$

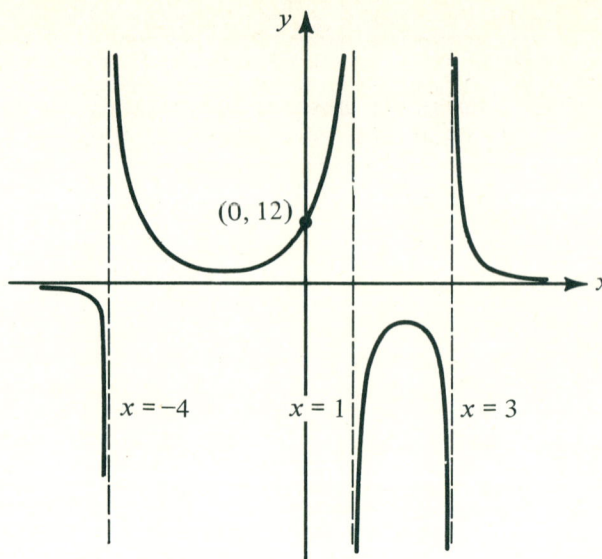

15. $h(x) = \dfrac{2x - 2}{x^2 + 4x + 3}$

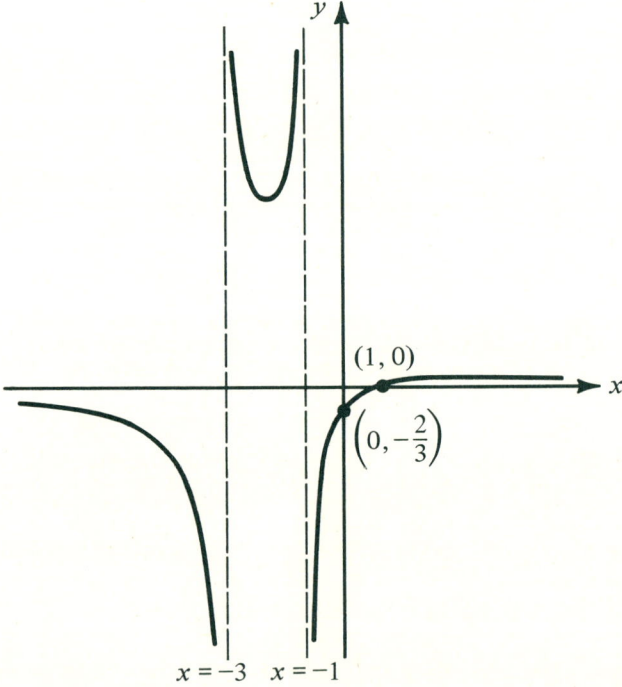

17. $\dfrac{x^2 + x - 2}{x^4 - 3x^2 - 4} = \dfrac{(x - 1)(x + 2)}{(x - 2)(x + 2)(x^2 + 1)}$

$(x-1)(x+2)$

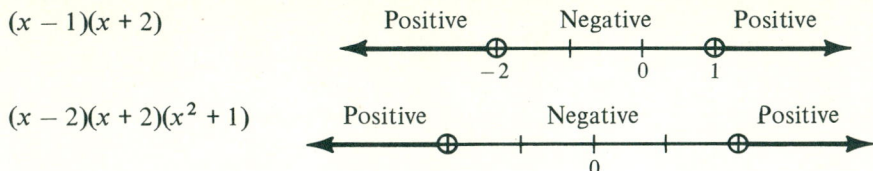

$(x-2)(x+2)(x^2+1)$

The quotient will be positive or zero when the numerator is zero (and the denominator is not) or when the signs of the numerator and denominator are the same, both positive or both negative. From the chart above, we see that these values are

$$x < -2 \quad \text{or} \quad -2 < x \leq 1 \quad \text{or} \quad x > 2$$

Notice that $x = -2$ is not allowable because 0/0 is undefined.

The original question asks for the values of x that make the quotient positive. To determine these; we will examine the signs of the numerator and denominator for various values of x, recalling that the quotient of two numbers is positive when they have the same sign, both positive or both negative.

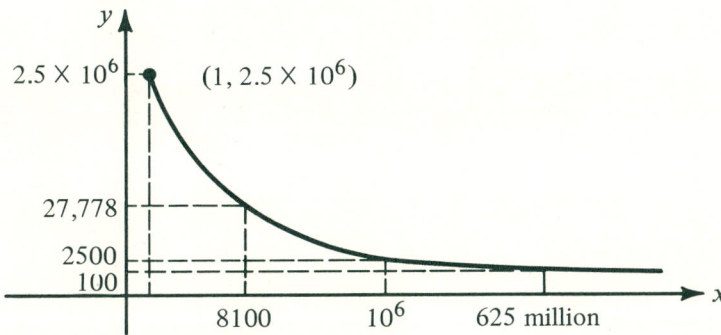

Hence, the signs agree for $\{x \mid x \leq 1 \text{ or } x > 2\}$ at $x = 1$, the quotient is zero but at $x = 2$ it is undefined.

19. $-3 < x < -2 \quad \text{or} \quad -1 \leq x \leq \dfrac{1}{2} \quad \text{or} \quad 1 < x < 2 \quad \text{or} \quad x > 3$

Notice that zeros and asymptotes both act as boundary points for our solution set.

20. (a) (i)

[Graph showing y vs x with point $(1, 2.5 \times 10^6)$, and dashed lines at $y = 27{,}778$, 2500, 100 corresponding to $x = 8100$, 10^6, 625 million]

(ii) 2,500.
(iii) 625 million dollars.
(iv) 27,778 have greater than $8100.00, therefore, 2,472,222 have less.

A-77

21. (a) The largest amount of butter that can be produced is 2,997 kilograms (at this level there is no yogurt being produced. The most yogurt that can be produced is 1,988 kilograms. Only use the first quadrant portion of the sketch, where x and y are both positive.

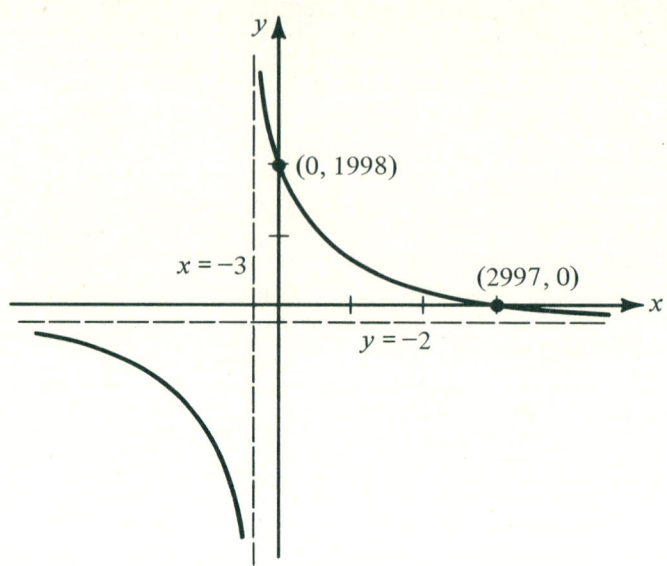

(b) Maximum gas is 15,996 barrels; maximum stove oil is 3,999 barrels.

Exercise 7.5

1. (a) $y = 2$
   (b) $y = 4$
2. (a) $y = -2$
   (b) $y = -3$
3. (a) $y = x$
   (b) $y = x$
4. $y = 3$
5. $y = 3x + 4$
6. $y = 4$
7.
9.

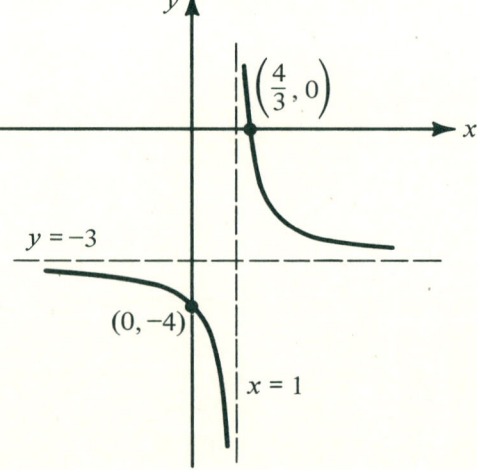

11.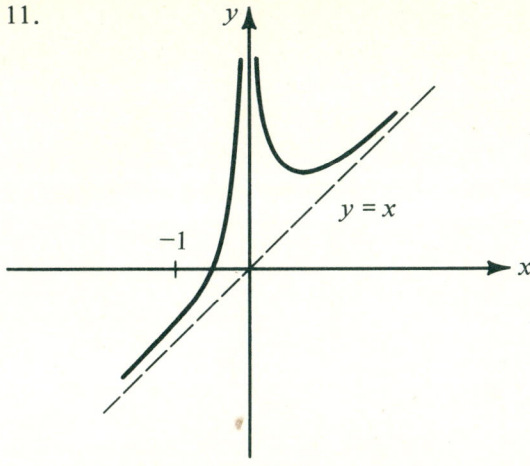

12. $y = 2x - 1 + (x - 3)/x^2 - 4 = 2x^3 - x^2 - 7x + 1/x^2 - 4$

There is an oblique asymptote, $y = 2x - 1$, since

$$\lim_{x \to \infty} \frac{x - 3}{x^2 - 4} = 0$$

There are vertical asymptotes at $x = \pm 2$, since

$$\lim_{x \to \pm 2} y = \pm \infty$$

The curve crosses its oblique asymptote at $x = 3$, $y = 5$. The y-intercept is $(0, -\frac{1}{4})$ but the x-intercepts are irrational. By watching the sign of the remainder we can see that there are three, one between -1 and -2, one between 0 and $\frac{1}{2}$, and the other between 2 and 3.

$$y = 0 \rightarrow 2x^3 - x^2 - 7x + 1 = 0$$

Then

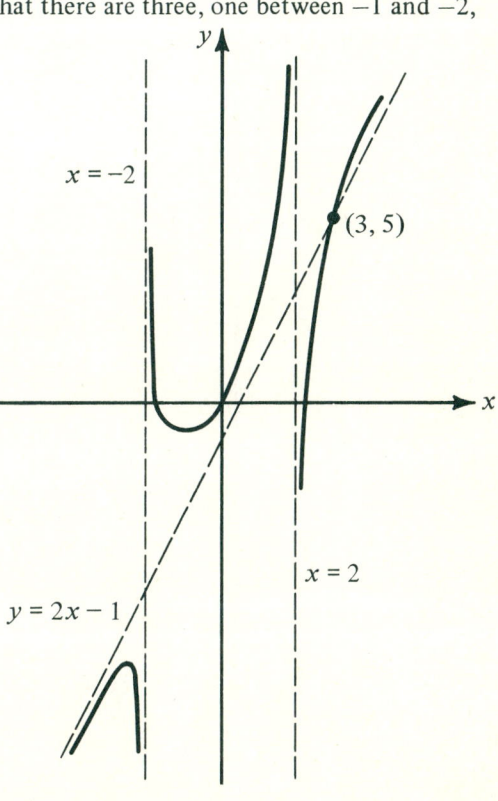

$$\begin{array}{r|rrrr} -2 & 2 & -1 & -7 & 1 \\ & & -4 & 10 & -6 \\ \hline & 2 & -5 & 3 & -5 \end{array}$$

$$\begin{array}{r|rrrr} -1 & 2 & -1 & -7 & 1 \\ & & -2 & 3 & 4 \\ \hline & 2 & -3 & -4 & 5 \end{array}$$

$$x = 0 \rightarrow 2x^3 - x^2 - 7x + 1 = 1$$

$$\begin{array}{r|rrrr} \frac{1}{2} & 2 & -1 & -7 & 1 \\ & & 1 & 0 & -\frac{7}{2} \\ \hline & 2 & 0 & -7 & -\frac{5}{2} \end{array}$$

$$\begin{array}{r|rrrr} 3 & 2 & -1 & -7 & 1 \\ & & 6 & 15 & 24 \\ \hline & 2 & 5 & 8 & 25 \end{array}$$

13.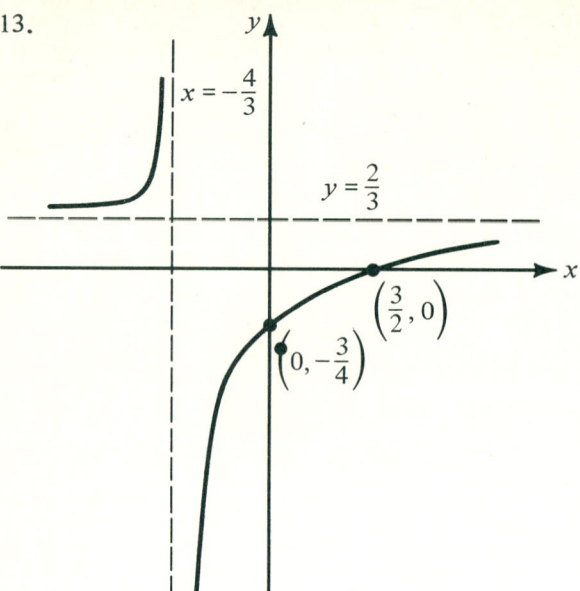

15.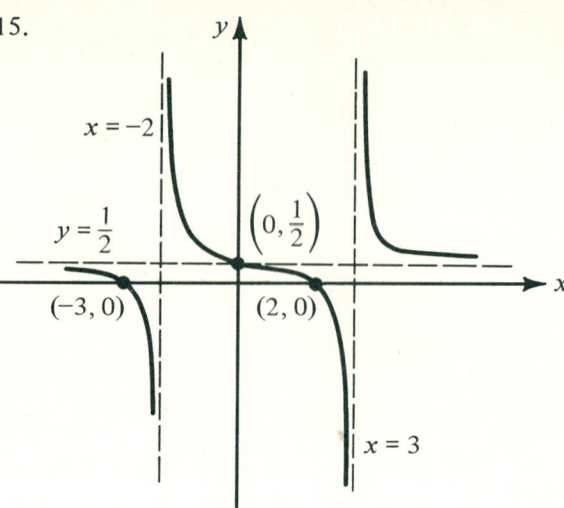

17.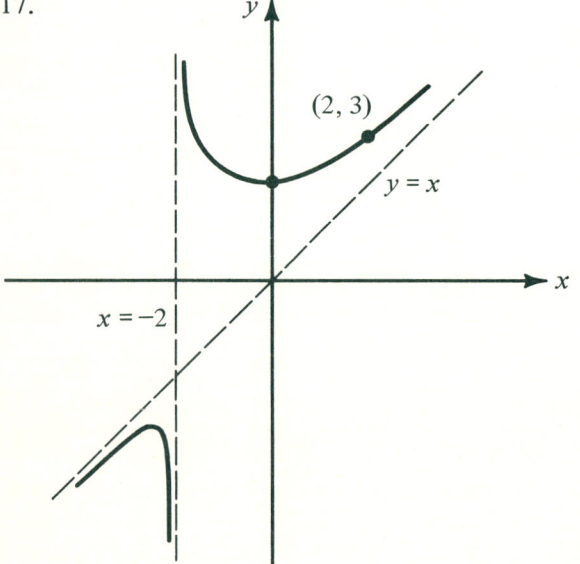

18.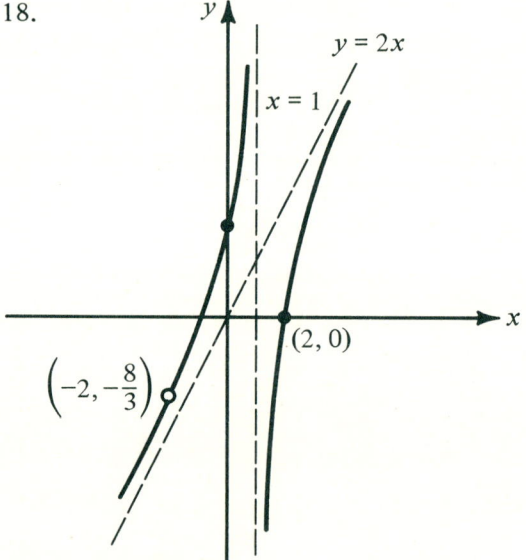

19. $f(x) = (2x^4 - 10^2 + 8)/(x^4 - 2x^3 - 15x^2)$
    Intercepts: $f(0)$ undefined

$$f(x) = 0 \Rightarrow 2x^4 - 10x^2 + 8 = 0$$
$$x^4 - 5x^2 + 4 = 0$$
$$(x + 1)(x - 1)(x + 2)(x - 2) = 0$$

Asymptotes: $x = \pm 1$ or $x = \pm 2$

$$\lim_{x\to\infty} \frac{2x^4 - 10x^2 + 8}{x^4 - 2x^3 - 15x^2} \cdot \frac{1/x^4}{1/x^4} = \lim_{x\to\infty} \frac{2 - 10/x^2 + 8/x^4}{1 - 2/x - 15/x^2} = 2$$

Therefore, $y = 2$ is a horizontal asymptote. To find the vertical asymptotes, set the denominator equal to 0 and solve for x.

$$x^4 - 2x^3 - 15x^2 = 0 \Rightarrow$$
$$x^2(x + 3)(x - 5) = 0 \Rightarrow$$
$$x = 0, x = -3 \lor x = 5$$

Notice that $x = 0$ is a zero of multiplicity 2. Compare the behavior of the function near $x = 0$ with its behavior near $x = -3$ or $x = 5$.

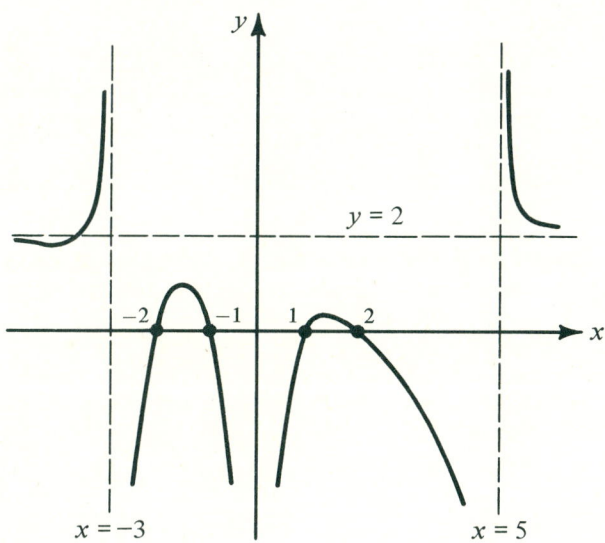

21. (a)   (b)

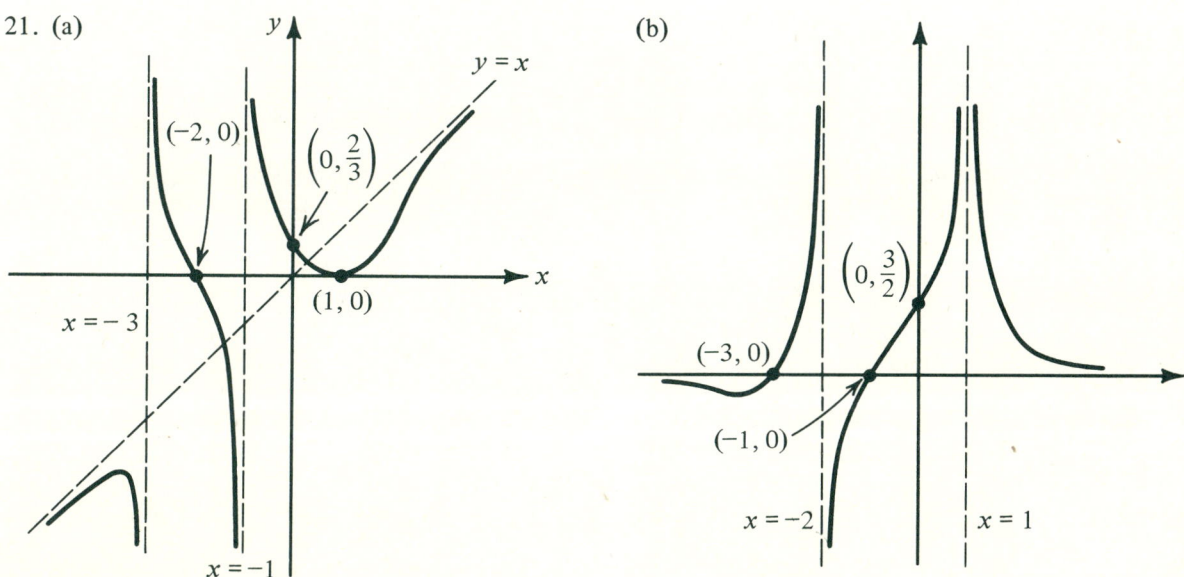

23. The asymptote $t = 1$ represents the minimum number of months in which the project can be completed. Notice that as the time decreases to 1 month, the cost C grows fast. If more time were allowed for the project, the cost would decrease for a while, but eventually, due to inflation, the costs would start climbing. This is depicted by the oblique asymptote, $C = \frac{1}{2}t - 12$, whose slope is 5%. The domain is $t > 1$ and the range is $C \geqslant 0$.

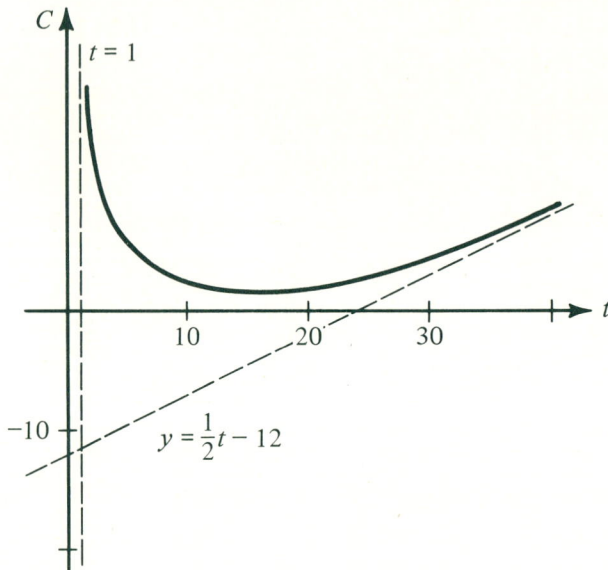

25. The domain is $t \geqslant 0$. The range is all real numbers—positive v representing motion away from earth, negative v representing motion towards earth. The horizontal asymptote, $u = c$, represents the limit to speed, the speed of light.

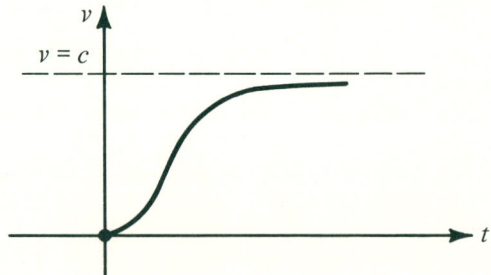

27.

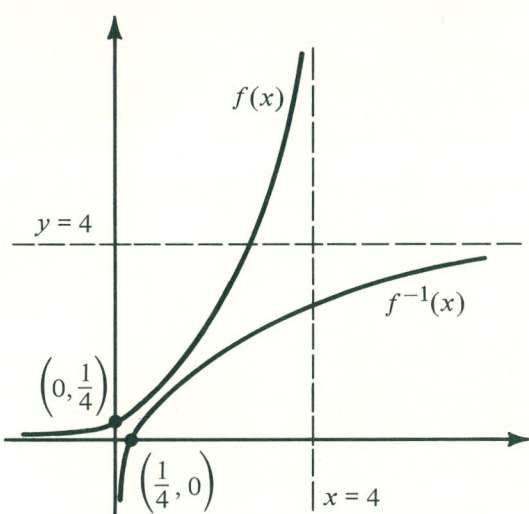

29.

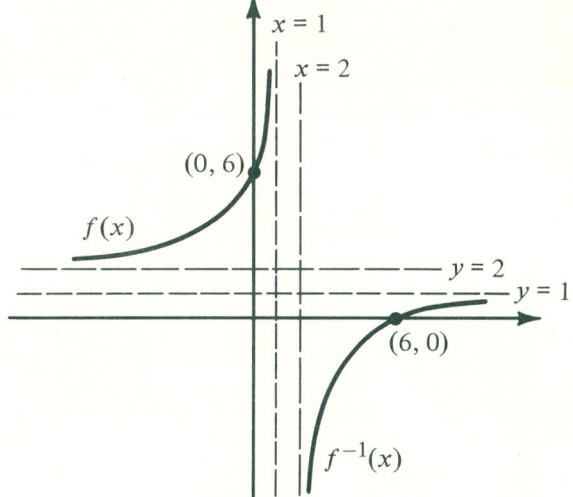

31.

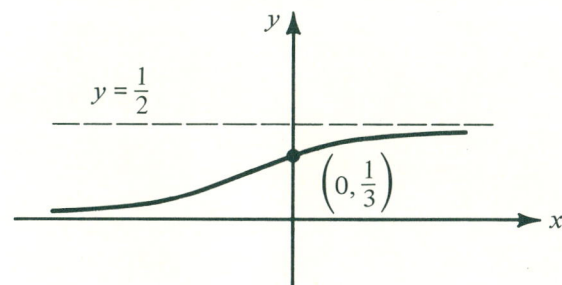

Exercise 7.6

1. $\dfrac{3/7}{x-4} + \dfrac{-3/7}{x+3}$

2. $\dfrac{2}{x-4} + \dfrac{2}{3x+9}$

3. $3 + \dfrac{2}{x+1} - \dfrac{1}{x-1}$

4. $\dfrac{2x+3}{x^2+1} + \dfrac{1}{x}$

5. $\dfrac{2x-1}{(x+1)^2} + \dfrac{1}{x+1}$

6. $\dfrac{2x-3}{(x+1)^2} + \dfrac{3}{x-1}$

7. $\dfrac{2x}{(x-2)^2} + \dfrac{1}{x-2} + \dfrac{3}{x+3}$

8. $2x + 1 + \dfrac{1}{x^2+x+1} + \dfrac{2}{x}$

9. $2x + 1 + \dfrac{1}{x} + \dfrac{x+2}{x^2+x+1}$

A-83

11. $y = \dfrac{2x^2 + 2}{x^4 - 2x^2 + 1} = \dfrac{2x^2 + 2}{(x-1)^2(x+1)^2} = \dfrac{A}{x-1} + \dfrac{B}{(x-1)^2} + \dfrac{C}{x+1} + \dfrac{D}{(x+1)^2}$

$2x^2 + 2 = A(x-1)(x+1)^2 + B(x+1)^2 + C(x+1)(x-1)^2 + D(x-1)^2$

$x = 1 \Rightarrow 4 = 4B \Rightarrow B = 1$

$x = -1 \Rightarrow 4 = 4D \Rightarrow D = 1$

$x = 0 \Rightarrow 2 = -A + B + C + D$ and $D = B = 1 \Rightarrow 2 = 2 - A + C \Rightarrow A = C$

$x = 2 \Rightarrow 10 = 9A + 9B + 3C + D$ and $D = B = 1 \Rightarrow 9A + 3C = 0$ and $A = C \Rightarrow$

$A = 0$ and $C = 0$

Hence,

$$y = \dfrac{1}{(x-1)^2} + \dfrac{1}{(x+1)^2}$$

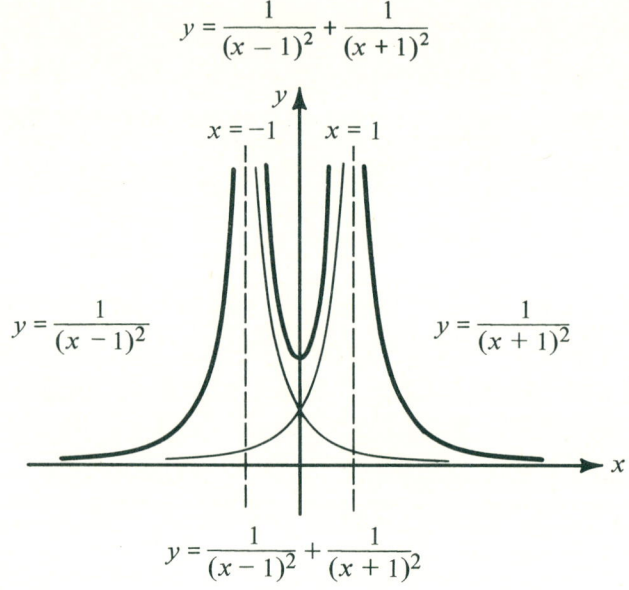

CHAPTER 8

Exercise 8.2

1. (a) 1/4, 1, 4
   (b) 1/9, 1, 9
   (c) 1/25, 1, 25
3. (a) −2, −4, −8
   (b) −2, 4, −8
   (c) 2, −4, 8

5. (a) 1/10, 1/100, 1/1000
   (b) 1, 1/10, 1/100
   (c) 1, 1/10, 1/100 [5 (b) and (c) are equivalent since $10^1 \cdot 10^x = 10^{x+1}$.]

2. (a) 1/6, 6, 36
   (b) 2/9, 2/3, 2
   (c) 6, 12, 24
4. (a) 1/4, 1/2, 1
   (b) 1, 1/2, 1/4
   (c) 1, 1/2, 1/4 [4 (b) and (c) are equivalent since $2^{-x} = 1/2x$.]

6. (a) 1/3, 1, 3
   (b) 1/3, 1, 3
   (c) 1/4, 1/2, 1 [6 (a) and (b) are equivalent since $1/3^1 \cdot 3^x = 3^{x-1}$. For (c), $1/4 \cdot 2^x = 1/2^2 \cdot 2^x = 2^{x-2}$.]

7. (a) $x = 3$
   (b) $x = 4$
8. (a) $x = -1$
   (b) $x = -2$
9. (a) $x = -3$
   (b) $x = -1$
10. (a) $x = 1$
    (b) $x = -1$ from $4^{x-1} = 4^{-2}$
11.

13.

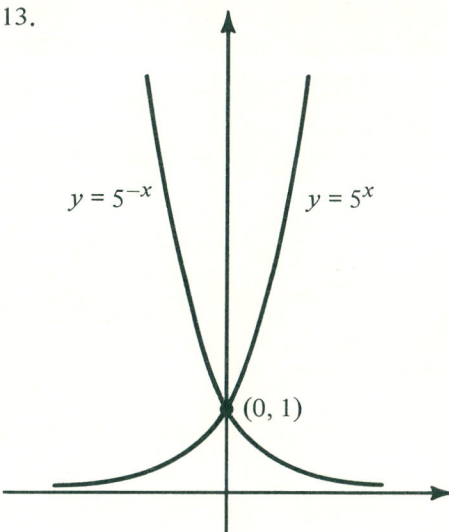

14. These graphs have the same basic shape since $(1/3)^x = 1/3^x = 3^{-x}$. Both graphs have the positive x-axis as their horizontal asymptote and $(0, 1)$ as their y-intercept. To the left, $y = 7^{-x}$ increases more rapidly because of its larger base. The larger base also causes it to approach its asymptote faster than $y = 3^{-x}$.

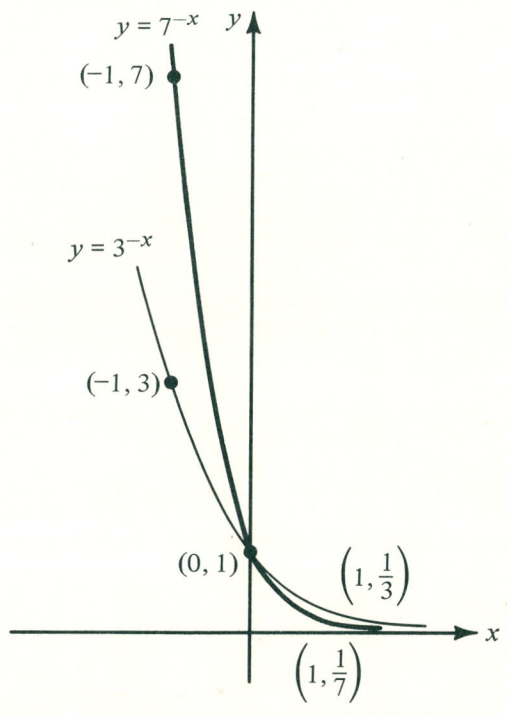

15.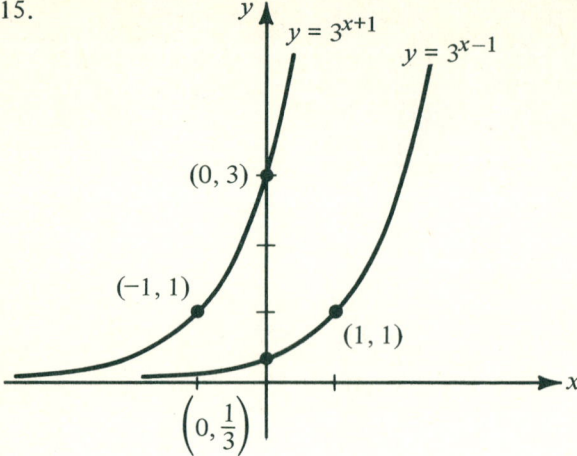

17.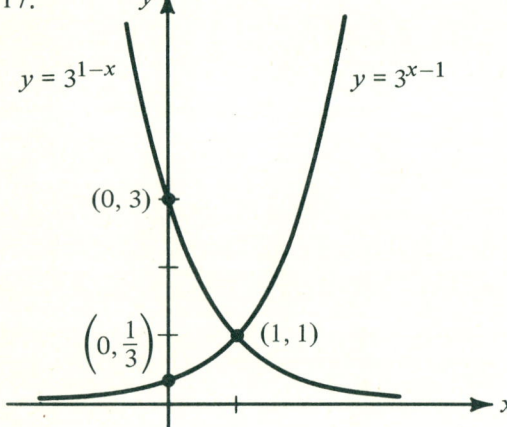

19.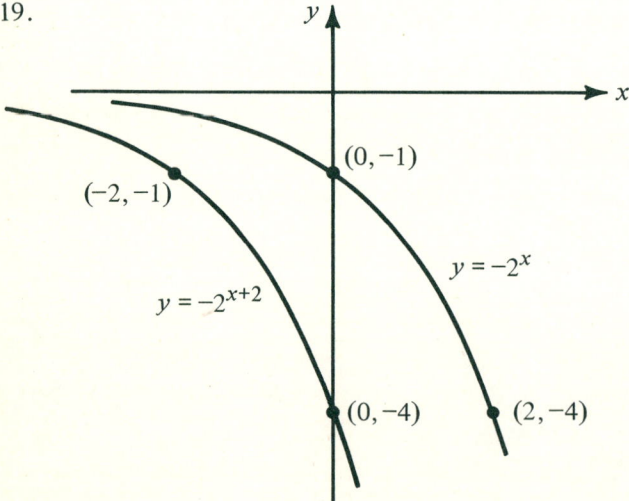

20. (b) $y = 1 - f(x+2) = 1 - 3^{-(x+2)}$.

This is a reflection of 20 (a) about the x-axis then a translation to the left 2 and up 1. Since 20 (a) has the x-axis as its asymptote, the translation up 1 will result in a new asymptote of $y = 1$.

Intercepts: $x = 0 \Rightarrow$
$$y = 1 - 3^{-(0+2)}$$
$$y = 1 - 3^{-2}$$
$$y = 1 - \frac{1}{3^2}$$
$$y = 1 - \frac{1}{9} = \frac{8}{9}$$

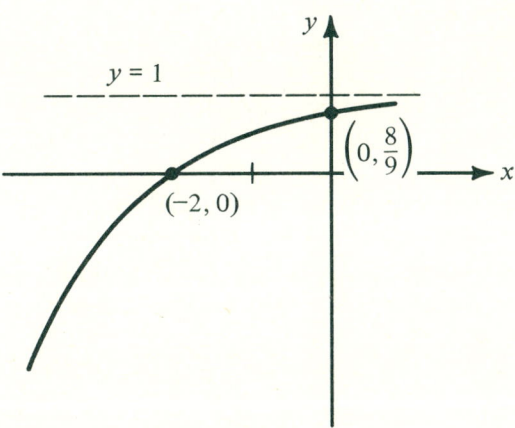

$y = 0 \Rightarrow$
$1 - 3^{-(x+2)} = 0 \Rightarrow$
$3^{-(x+2)} = 1 \Rightarrow$
$-(x+2) = 0 \Rightarrow$
$x = -2$

21.

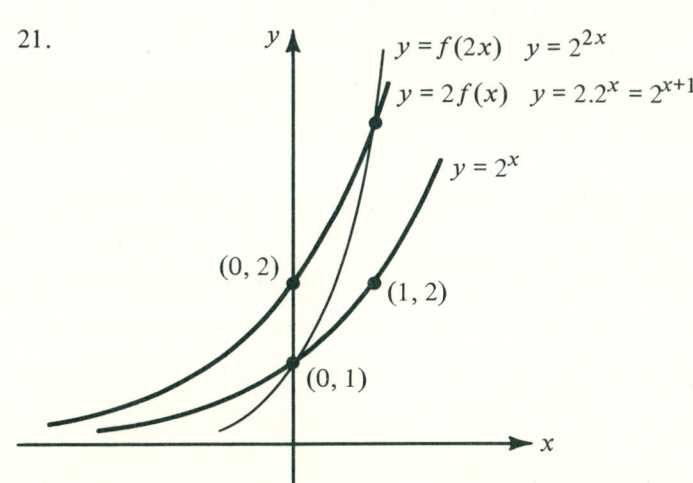

$y = f(2x) \quad y = 2^{2x}$
$y = 2f(x) \quad y = 2 \cdot 2^x = 2^{x+1}$

A-87

22. Since
$$\lim_{x \to \infty} 3^{-x} = 0$$
the product $2x \cdot 3^{-x}$ will approach zero as x increases in size. The exponential factor grows much more rapidly than $2x$ as x approaches $-\infty$, so that the only real contribution made by $2x$ is the sign, which makes the product negative. Near $x = 0$, the factor $2x$ does dominate, however, and at $x = 0$, $y = 2x \cdot 3^{-x} = 0$.

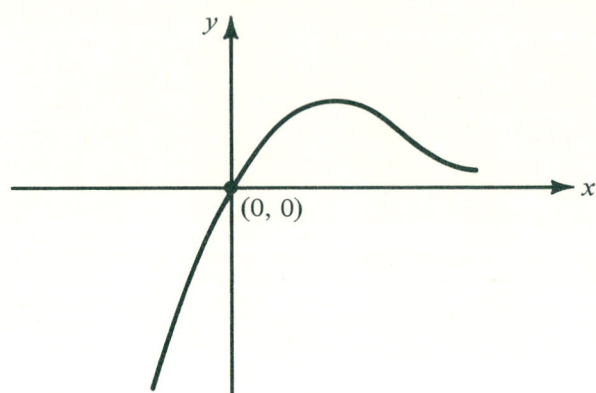

23.

25.

29. $f(x) = 4^x$

$$0.5 \left[0.2 \left[\begin{matrix} f\left(\frac{3}{2}\right) = 8 \\ f(\sqrt{3}) = y \\ f(2) = 16 \end{matrix} \right] \Delta y \right] 8$$

$\dfrac{\Delta y}{8} = \dfrac{0.2}{0.5}$

$\Delta y = 3.2$

$y \approx 8 + \Delta y = 11.2$

30. $10^{\sqrt{2}} \approx 36$

A-88

Exercise 8.3

1. (a) 40,000
   (b) 33,746
   (c) 66,612

2. (a) 17.7 grams
   (b) 10.98 grams

3. (a) 76 centimeters of Hg
   (b) 44.68 centimeters of Hg
   (c) 62.22 centimeters of Hg

4. 10 minutes

5. Over 1 kilometer high

7. $I = I_1 10^{-0.4M}$
   For the sun,
   $$I_S = I_1 10^{-0.4(-27)} = I_1 10^{10.8}$$
   and for Alpha Centauri
   $$I_A = I_1 10^{-0.4(-0.3)} = I_1 10^{0.12}$$
   The ratio of their apparent intensities is
   $$\frac{I_S}{I_A} = \frac{I_1 10^{10.8}}{I_1 10^{0.12}} = 10^{10.68} = 4.8 \times 10^{10}$$
   This means that the sun's apparent brightness is about 50 billion times that of Alpha Centauri.

8. We know that the model for this culture will be of the form
   $$N = N_0 e^{+kt}$$
   where $N_0 = 10,000$ and $t = 2 \Rightarrow N = 15,000$.
   Therefore,
   $$N = 10,000 e^{+kt} \quad \text{and} \quad t = 2 \Rightarrow N = 15,000 \Rightarrow$$
   $$15,000 = 10,000 e^{+2k} \Rightarrow 1.5 = e^{+2k}$$
   From the exponential table, we see that
   $$e^{0.4} \approx 1.5$$
   therefore,
   $$+2k = 0.4 \Rightarrow k = 0.2$$
   Hence, the exponential growth model for this bacteria culture is
   $$N = 10,000 e^{0.2t}$$
   where t is in days. In one week ($t = 7$) there will be
   $$N = 10,000 e^{0.2(7)}$$
   $$= 10,000 e^{1.4}$$
   $$= (10,000)(4.06)$$
   $$= 40,600 \text{ bacteria}$$

9. $A_1 = \$93,200.00$ and $A_2 = \$186,340.00$

11. $A = 2P \Rightarrow (1.06)^n = 2 \Rightarrow n \approx 12$

13. $N = 1000a^{0.20t}$, $(t = 0 \to N = 100) \Rightarrow a = 0.1$
    Hence, $N = 1000(1/10)^{0.20t} = 10^3 \cdot 10^{-0.20t}$.

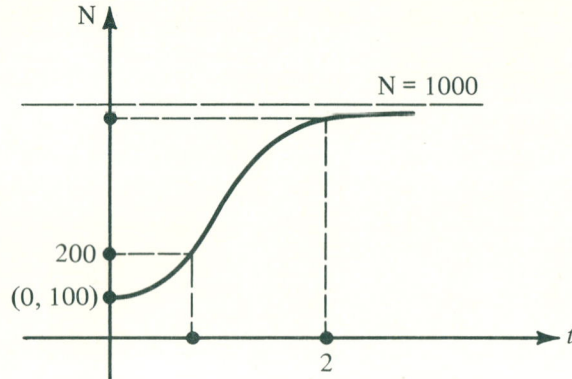

$t = 2 \Rightarrow N =$ and $N = 200 \Rightarrow t =$

15. (a) 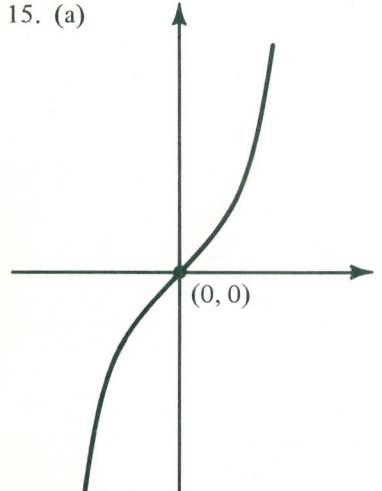   (b)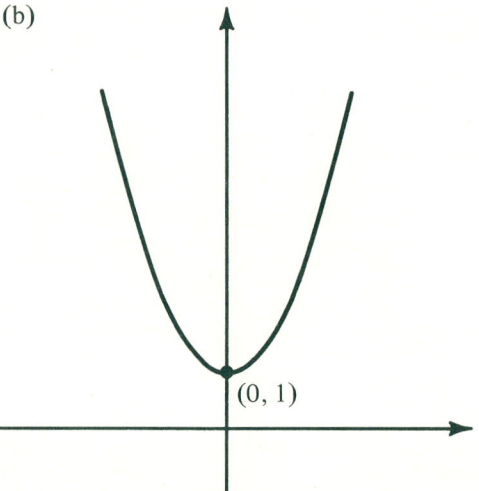

Exercise 8.4

1. (a) $\log_3 1 = 0$ (b) $\log_{10} 1 = 0$ (c) $\log_e 1 = 0$
2. (a) $\log_5 5 = 1$ (b) $\log_{10} 10 = 1$ (c) $\log_e e = 1$
3. (a) $\log_2(1/2) = -1$ (b) $\log_{10} 1/100 = -2$ (c) $\log_{10} 0.001 = -3$
4. (a) $\log_2 8 = 3$ (b) $\log_3 81 = 4$ (c) $\log_4 64 = 3$

The following pairs of graphs (5–10) are inverses of each other. Notice the symmetry about the line $y = x$. Also notice that the base for the exponential function is always the same as the base for its inverse logarithmic function. Notice the domains and ranges of the pairs of functions. The domain of one is the range of the other and vice versa.

5.
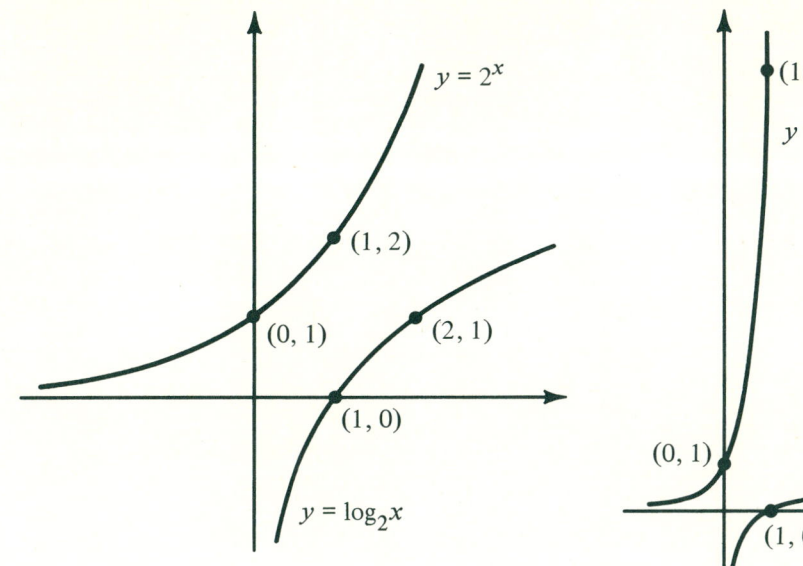

6.
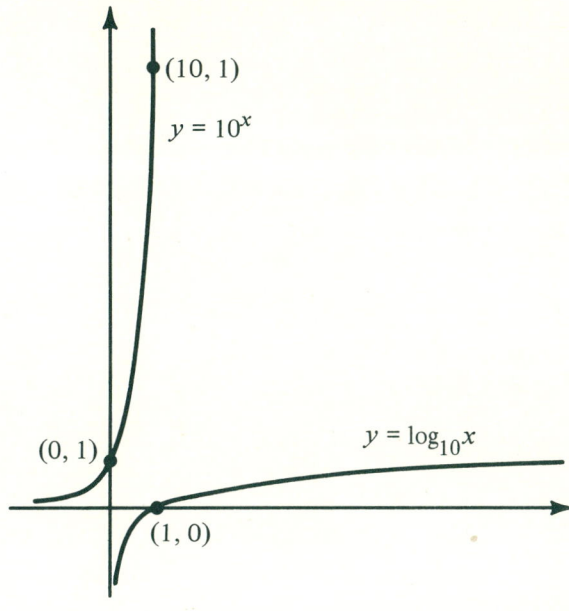

7.
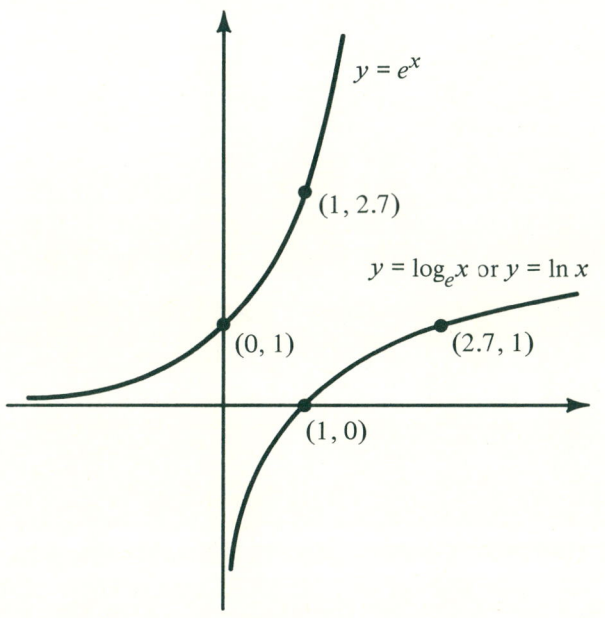

8.
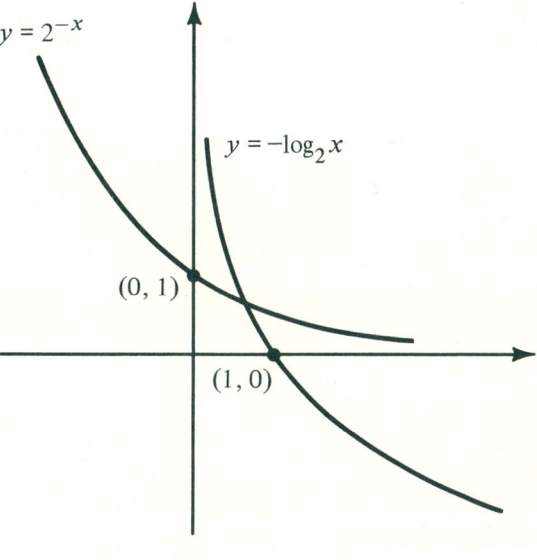

9.

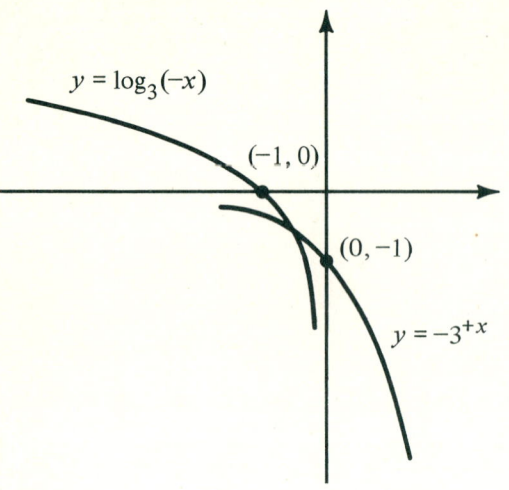

10.

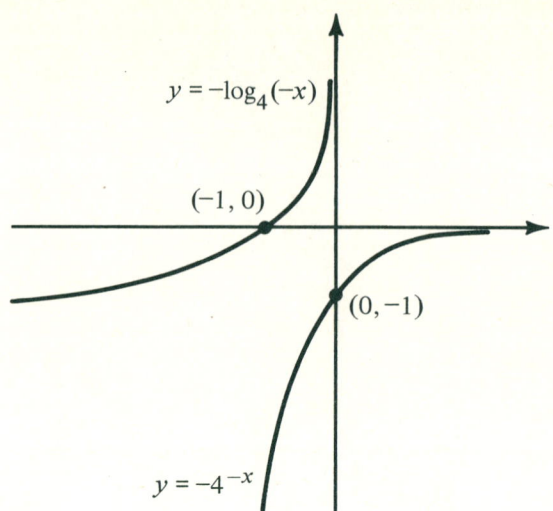

11. (a)

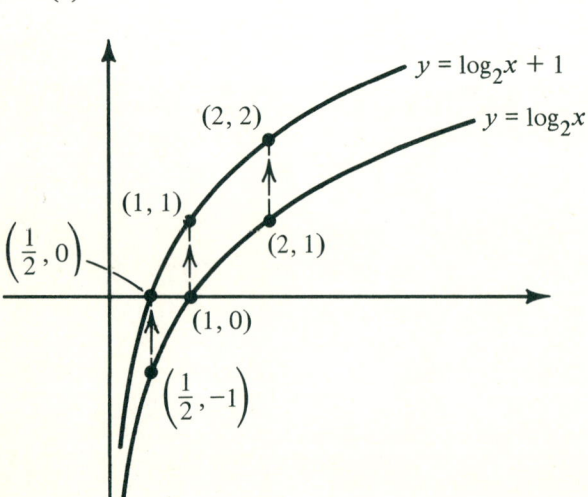

(b)

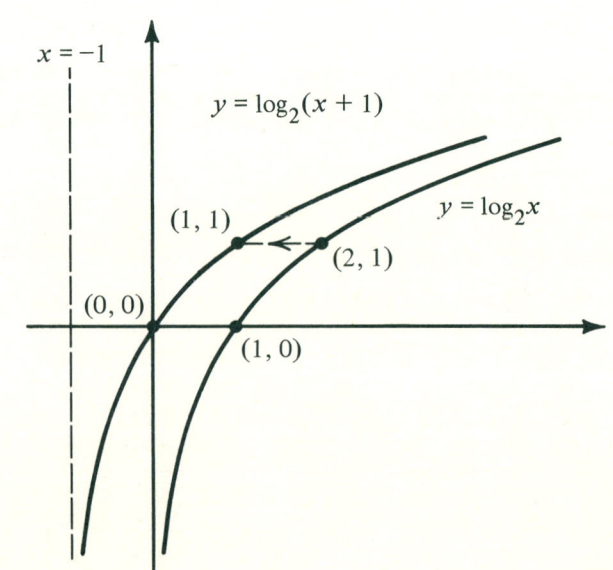

13. $y = 2^{x+1} \Rightarrow f^{-1}(x) = \log_2 x - 1$
    $D_f = \{x | x \in \text{reals}\}$
    $D_{f^{-1}} = \{x | x > 0\}$

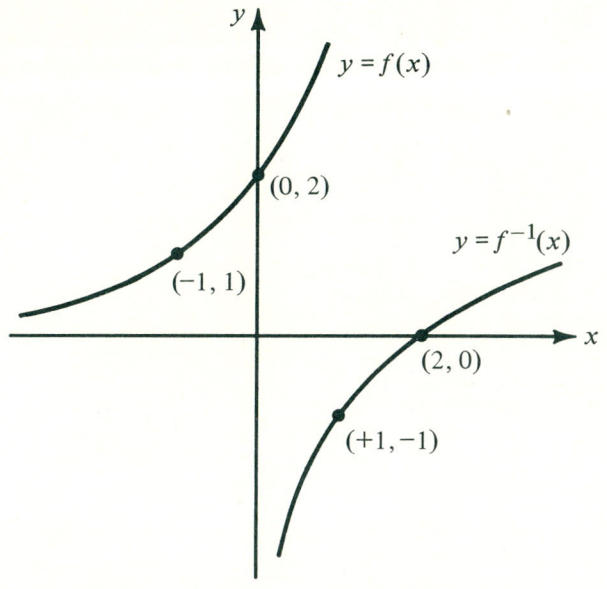

15. $y = 1 - 2^x \Rightarrow f^{-1}(x) = \log_2(1 - x)$
    $D_f = \{x | x \in \text{reals}\}$
    $D_{f^{-1}} = \{x | x < 1\}$

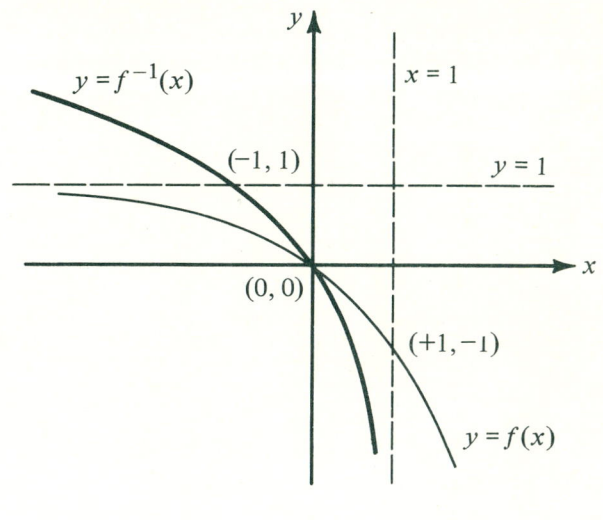

17. $y = 3^{1-x} + 2 \Rightarrow f^{-1}(x) = 1 - \log_3(x - 2)$
    $D_f = \{x | x \in \text{reals}\}$
    $D_{f^{-1}} = \{x | x > 2\}$

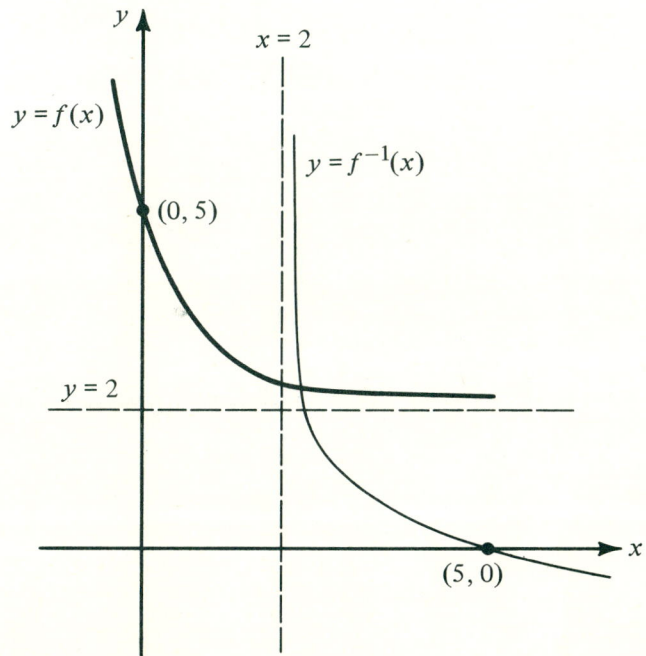

A-93

18. $y = e^{x+1} - 1$
    Equation of the inverse:
    1. Solve for x:
       $y + 1 = e^{x+1} \Leftrightarrow$
       $\log_e(y + 1) = x + 1 \Rightarrow$
       $x = \log_e(y + 1) - 1$
    2. Rename the variables:
       $y = \log_e(x + 1) - 1 \Leftrightarrow$
       $y = \ln(x + 1) - 1$
       $D_f = \{x | x \in \text{reals}\}$
       $D_{f^{-1}} = \{x | x > -1\}$ so that $x + 1 > 0$

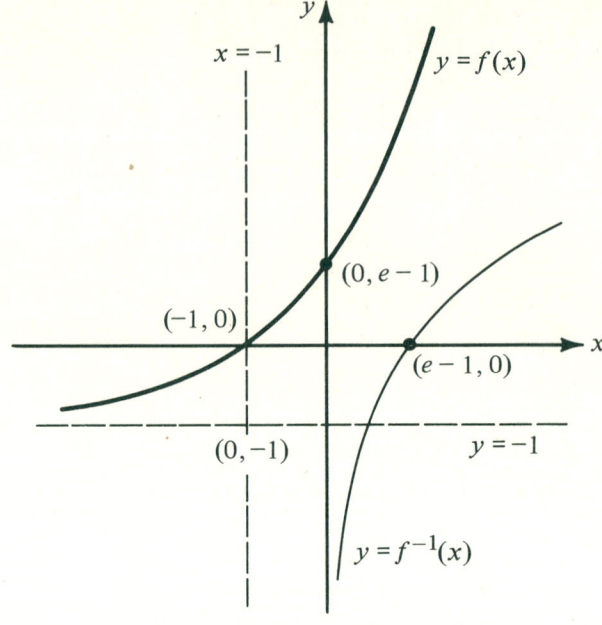

19. $y = \log_3 x \Rightarrow f^{-1}(x) = 3^x$
    $D_f = \{x | x > 0\}$
    $D_{f^{-1}} = \{x | x \in \text{reals}\}$

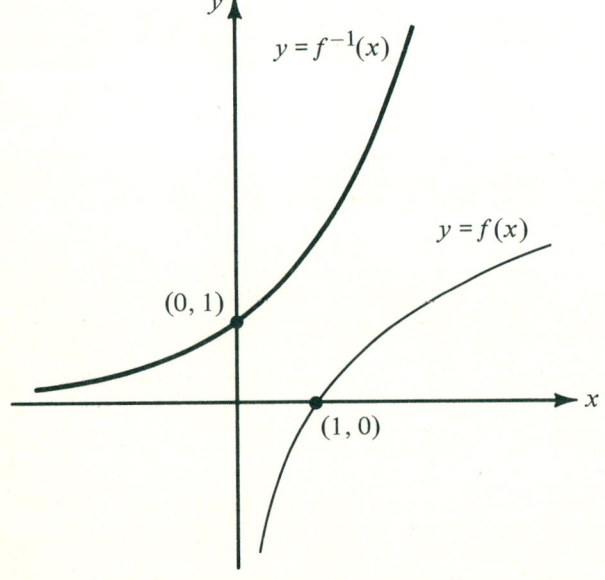

21. $y = 3 - \log_2 x \Rightarrow f^{-1}(x) = 2^{3-x}$
    $D_f = \{x | x > 0\}$
    $D_{f^{-1}} = \{x | x \in \text{reals}\}$

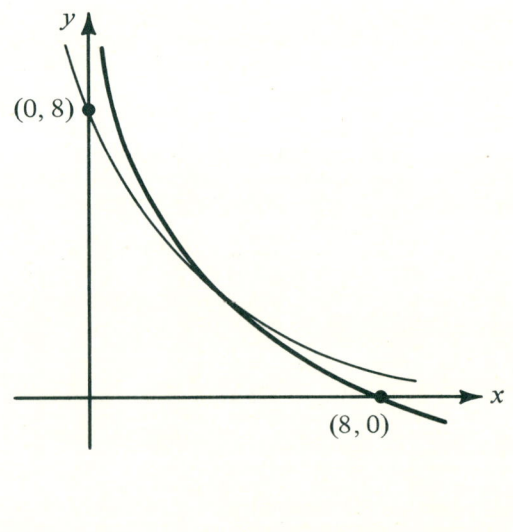

A-94

23. $y = 4 - \log_5(x + 3)$
　　For the domain,
　　$x + 3 > 0 \Rightarrow x > -3$

　　For the equation of its inverse:
　　1. Solve for x:
　　　　$\log_5(x + 3) = 4 - y \Leftrightarrow$
　　　　$x + 3 = 5^{4-y} \Rightarrow$
　　　　$x = -3 + 5^{4-y}$
　　2. Rename the variables:
　　　　$y = -3 + 5^{4-x}$
　　　　$D_{f^{-1}} = \{x | x \in \text{reals}\}$
　　　Intercepts:
　　　　$y = 0 \Rightarrow 4 - \log_5(x + 3) = 0$
　　　　　　　$\log_5(x + 3) = 4$
　　　　　　　$x + 3 = 5^4$
　　　　　　　$x = 622$
　　　　$x = 0 \Rightarrow y = 4 - \log_5 3$
　　　where

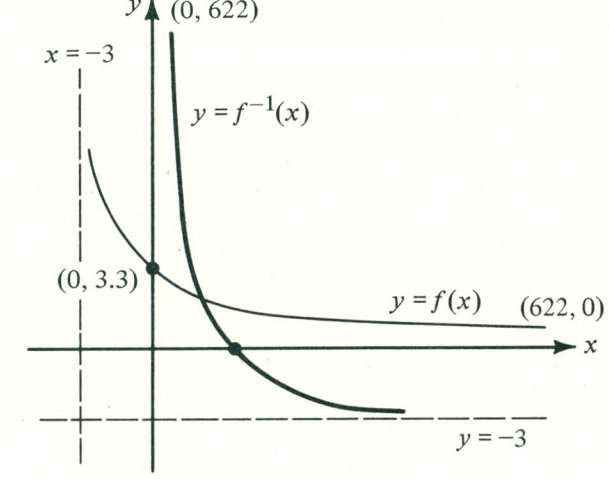

$$\log_5 3 = \frac{\log_{10} 3}{\log_{10} 5} \approx 0.68$$

　　and

$$y = 4 - 0.68 = 3.32$$

25. $y = 4 - \log_2(2x + 3)$
　　$D_f = \{x | x > -\frac{3}{2}\}$
　　$D_{f^{-1}} = \{x | x \in \text{reals}\}$
　　$f^{-1}(x) = \frac{2^{4-x}}{2} - \frac{3}{2} = 2^{3-x} - \frac{3}{2}$

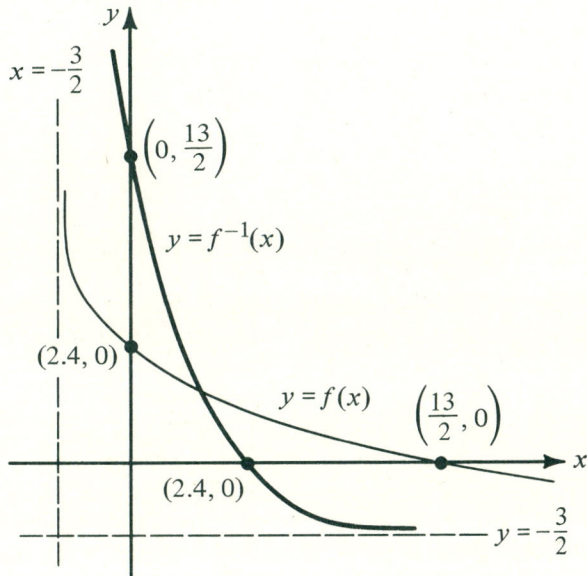

A-95

27.

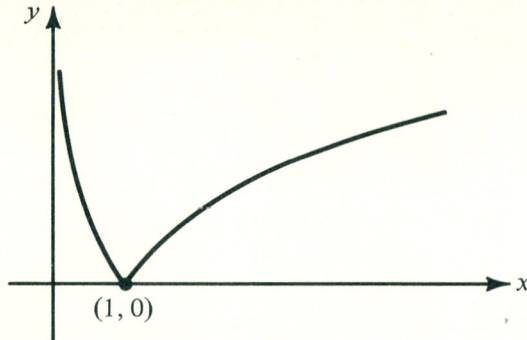

29. The squared x term in 29(b) allows negative values of x. It is an even function, symmetric about the y-axis. The right side of the curves are identical.

(a)

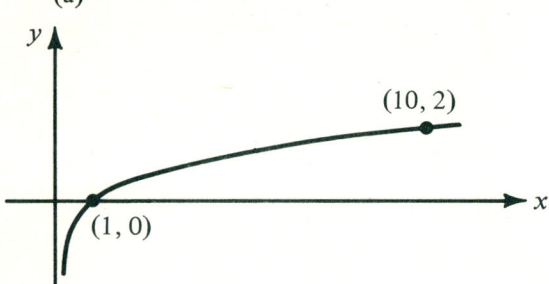

(b)

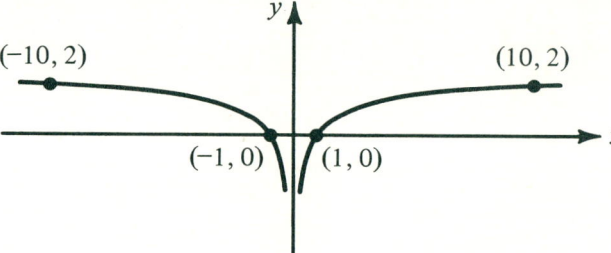

31. These are exactly the same graphs, since
$$y = \log_3 3x = \log_3 3 + \log_3 x = 1 + \log_3 x$$

Geometrically, that means for $\log_3 x$, a translation up 1 is the same as a compression toward the y-axis by a factor of 3.

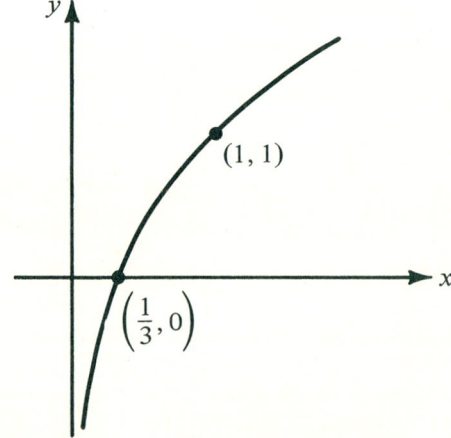

33. $y = \log_1 x \Leftrightarrow x = 1^y$ is the constant *relation*, $x = 1$

A-96

Exercise 8.5

1. (a) $x = 2$ (b) $x = 10$ (c) $x = e$ 2. (a) $x = 100$ (b) $x = 0.01$ (c) $x = 10$
3. (a) $x = 8$ (b) $x = -8$ (c) $x = 7$ 4. (a) $x = 9$ (b) $x = -8$
5. (a) $x = 1/10$ (b) $x = -1.9$ 6. (a) $x = 2$ or $x = -1$ (b) $x = -0$ or $x = 1$
7. (a) $x = 3$ (b) $x = 1$
  (c) $\log(x + 2) = -1 - \log x \Rightarrow$
    $\log(x + 2) + \log x = -1 \Rightarrow$
    $\log x(x + 2) = -1 \Rightarrow$
    $x^2 + 2x = 10^{-1} \Rightarrow$
    $10x^2 + 20x - 1 = 0 \Rightarrow$

$$x = \frac{-20 \pm \sqrt{400 - 4(10)(-1)}}{2(10)}$$

  $x \approx 0.05$ (Exclude the negative root since it is not in the domain of $\log x$.)
9. (a) $x = 2$ (b) $x = 3$ (c) $x = 5$
11. $\ln x = \ln(x - 6) - \ln(x - 4) \Rightarrow$
    $\ln x = \ln(x - 6/(x - 4))$
    Then, by taking the antilog of both sides, we get an equation where the log of both sides equals the above equation.

$$x = \frac{x - 6}{x - 4}$$

$$x^2 - 4x = x - 6$$
$$x^2 - 5x + 6 = 0$$
$$(x - 3)(x - 2) = 0$$
$$x - 3 = 0 \quad \text{or} \quad x - 2 = 0$$
$$x = 3 \quad \text{or} \quad x = 2$$

  But neither is in the domain. Therefore, there is no solution.

13. $x = 1$ or $x = 3/2$ 15. $x = 3/2$
17. $\log_2\sqrt{x + 6} - \log_2(x - 1) = 3/2 \Rightarrow$
    $\log_2\sqrt{x + 6}/(x - 1) = 3/2 \Rightarrow$
    $\sqrt{x + 6}/(x - 1) = 2^{3/2} = \sqrt{8} \Rightarrow$
    $x + 6/(x^2 - 2x + 1) = 8 \Rightarrow$
    $x + 6 = 8x^2 - 16x + 8 \Rightarrow$
    $8x^2 - 17x + 2 = 0 \Rightarrow$
    $(8x - 1)(x - 2) = 0 \Rightarrow$
    $8x - 1 = 0$ or $x - 2 = 0 \Rightarrow$
    $x = 1/8$ or $x = 2 \Rightarrow$
    $x = 2$
19. 0.94   21. 0.17
23. 0.15   25. 1.47

27. (a) $x = 2$
    (b) $x = 1/100$
    (c) (a googol) $x = 10^{100}$

29. (a) $\log_3 5 = 1.465$
    (b) $1 - \log_4 3 = 0.2075$

Exercise 8.6

1. (a) 0.3997
   (b) 2.3997
2. (a) 0.5694
   (b) −1.4306
3. (a) 3.4330
   (b) −3.5670
4. (a) 3.37
   (b) 337
5. (a) 5.49
   (b) 0.00549
6. (a) 0.000940
   (b) 0.00940
7. (a) 2.3365
   (b) 4.419
8. (a) $y = 37.5$
   (b) $y = 29.19$
9. (a) $y = 4$
   (b) $y = 4$
10. (a) $y = 15$
    (b) $y = 5$
11. (a) 0.4448
    (b) 2.6699
12. (a) 0.1427
13. (a) 3.575
    (b) 3.6707
15. (a) 1.658
    (b) 1.453
17. (a) 1.96×10^{10}
    (b) 1268
19. $L = 0.375$ meters

Exercise 8.7

1. 10,000 times
3. $M_1 = 6.2, M_2 = 7.6, M_3 = 7$

$$M = \log A - \log A_0 \Rightarrow$$

$$M_1 - M_2 = \log A_1 - \log A_2 = \log \frac{A_1}{A_2} \Rightarrow$$

$$6.2 - 7.6 = -1.4 = \log \frac{A_1}{A_2} \Rightarrow$$

$$1.4 = \log \frac{A_2}{A_1} \Rightarrow$$

$$\frac{A_2}{A_1} \approx 25.1 \Rightarrow$$

$$A_2 \approx 25.1 \, A_1$$

A_2 is over 25 times stronger than A_1.
Similarly, Managua's quake was approximately 6.31 times stronger than Alaska's.

5. $S = 10^9 z \Rightarrow A = 10 \log S/z = 10 \log 10^9 z/z = 90$db.

7. $L = L_0 10^{-kt}$
   $(t = 0 \Rightarrow L = 200 \text{ mg/l}) \Rightarrow L = 200 \cdot 10^{-kt}$
   $(t = 2 \Rightarrow L = 150 \text{ mg/l}) \Rightarrow 150 = 200 \cdot 10^{-2k} \Rightarrow$
   $\quad 10^{-2k} = 0.75 \Rightarrow$
   $\quad -2k \quad = \log 0.75$
   $\quad -2k \quad = 0.8751 - 1$
   $\quad k \quad\; = 0.0624$
   For $L = 0.5$ mg/l, we have
   $$L = 200 \cdot 10^{-0.0624t} \Rightarrow 0.5 = 200 \cdot 10^{-0.0624t}$$
   $$-0.0624t = \log 0.025 = 0.3979 - 2$$
   $$t = 25.7 \text{ days}$$

Exercise 8.8

1. (a) $y = 10x^3 \Leftrightarrow v = 3u + 1$ \qquad (b) $y = 100x^3 \Leftrightarrow v = 3u + 2$
2. (a) $y = 100x^{-1} \Leftrightarrow v = -u + 2$ \qquad (b) $y = 100x^{-4} \Leftrightarrow v = -4u + 2$

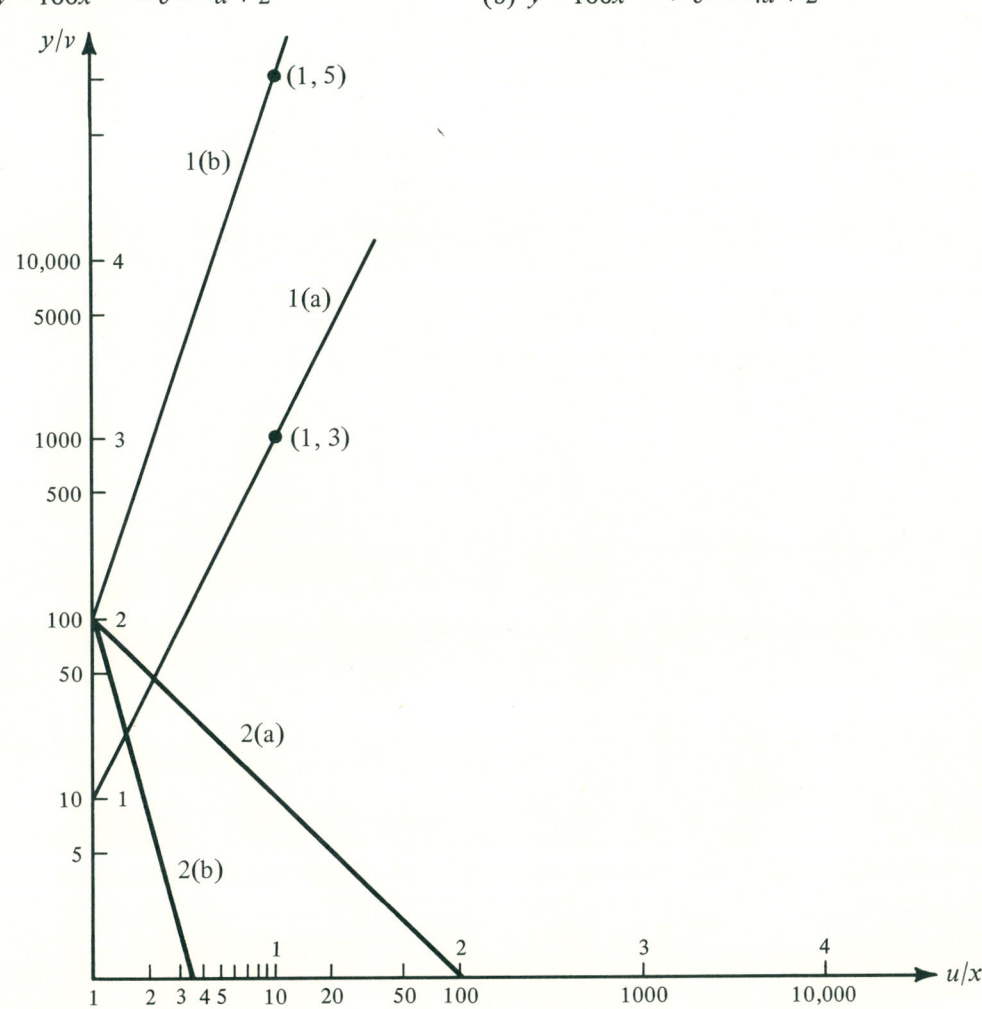

3. (a) $y = x^2 \Leftrightarrow v = 2u$ (b) $y = 100x^2 \Leftrightarrow v = 2u + 2$

4. (a) $y = 80x^{-2.3} \Leftrightarrow v = -2.3u + \log 80$ (b) $y = 8x^{2.3} \Leftrightarrow v = 2.3u + 0.9031$

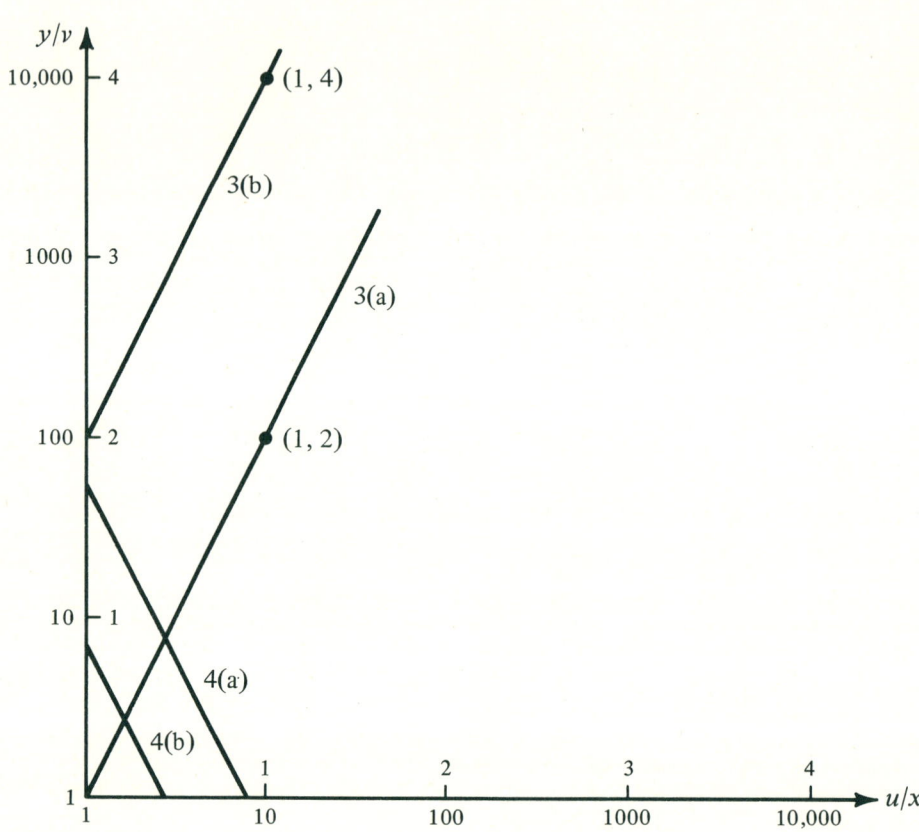

5. (a) $y = 25(10^{-x}) \Leftrightarrow v = -x + 1.4$ (b) $y = 100(10^{-x}) \Leftrightarrow v = -x + 2$
7. (a) $y = 90(10^{-3.1x}) \Leftrightarrow v = -3.1x + 1.954$ (b) $y = 9(10^{3.1x}) \Leftrightarrow v = 3.1x + 0.954$

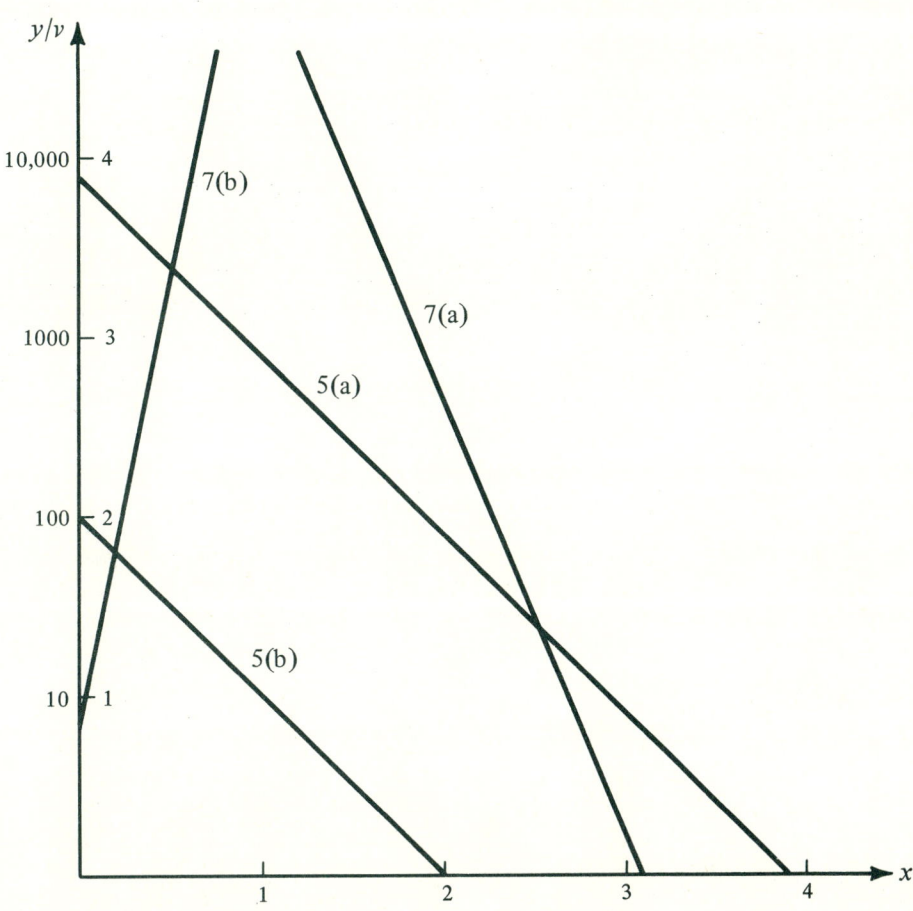

9. Plotting these data points on a log–log system yields the straight line $v = 5u + 3.6$, which represents the power function $s = 4000x^5$.

11. Plotting these points on a semilog coordinate system yields the straight line $v = M + \log A_0$, which represents the exponential function.
$$A = A_0 10^M$$

13. $m = \dfrac{\Delta v}{\Delta 1/T} = \dfrac{-II}{4.58} \Rightarrow H \approx 10{,}048$ calories

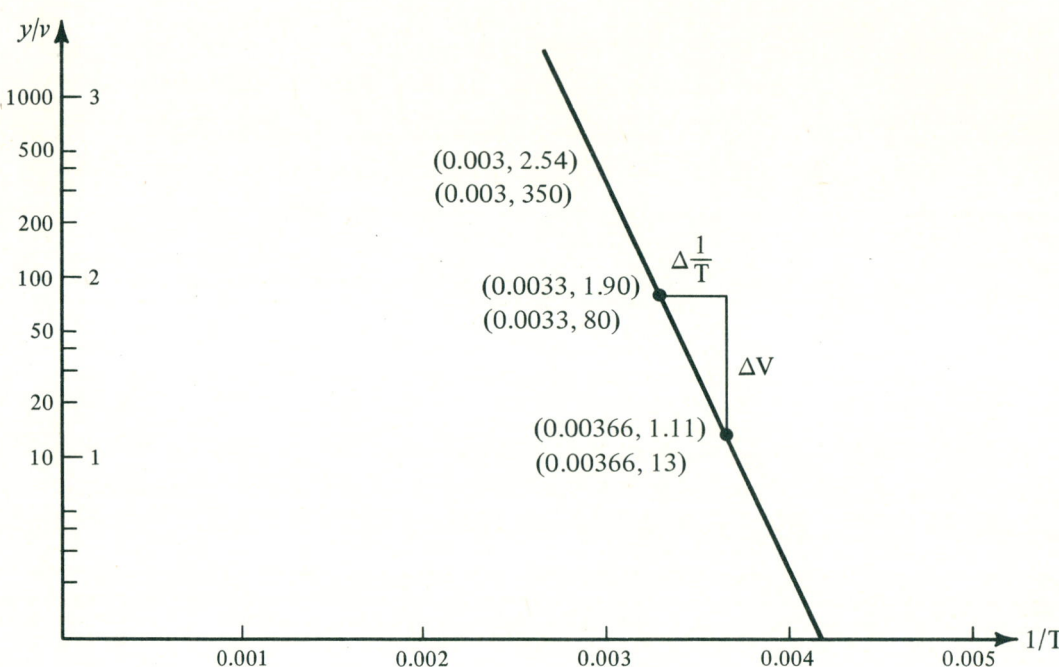

CHAPTER 9

Exercise 9.2

1. (a) 0
   (b) 2π
   (c) 4π
3. (a) $\pi/4$
   (b) $3\pi/4$
   (c) $5\pi/4$
5. (a) $\pi/3$
   (b) $2\pi/3$
   (c) $4\pi/3$
7. (a) 90°
   (b) 270°
   (c) 450°

2. (a) π
   (b) $-\pi$
   (c) 3π
4. (a) $\pi/6$
   (b) $5\pi/6$
   (c) $7\pi/6$
6. (a) 180°
   (b) 360°
   (c) 540°
8. (a) 30°
   (b) 150°
   (c) −30°

9. (a) 45°
   (b) 315°
   (c) −495°

10. (a) 120°
    (b) −300°
    (c) −420°

11.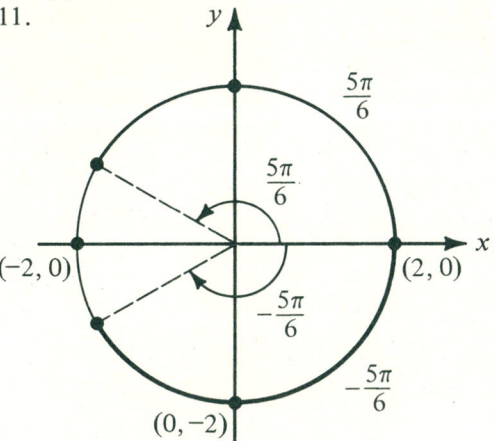

12.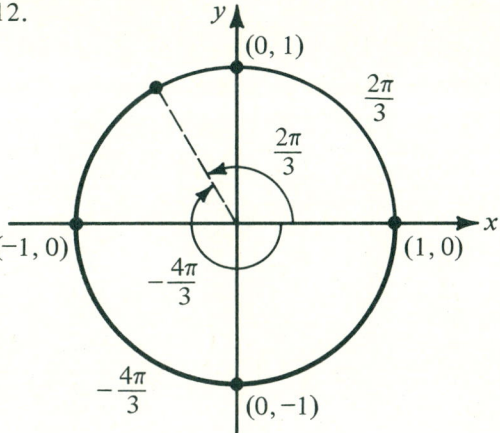

13.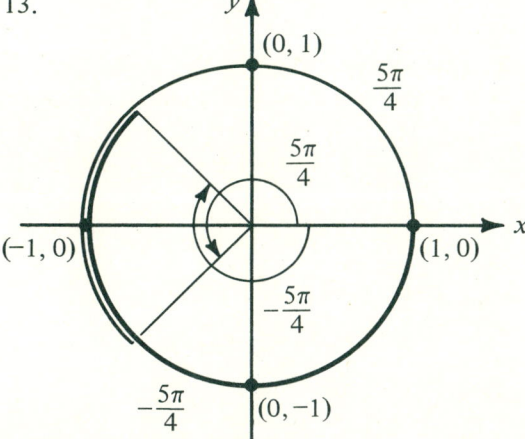

15. (a) (−1, 0)   (b) (−1, 0)

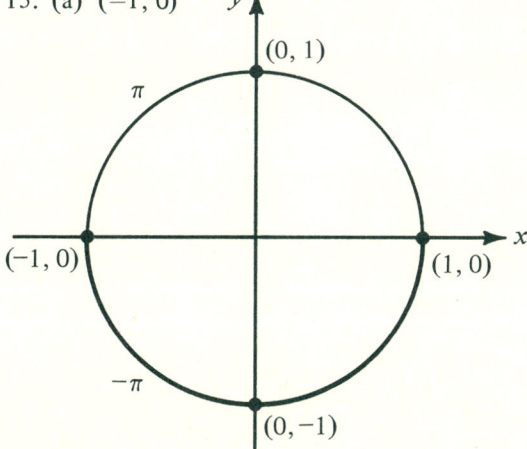

17. (a) $7\pi/2 = 2\pi + 3\pi/2 (0, -1)$ (b) $11\pi/2 = 4\pi + 3\pi/2 (0, -1)$

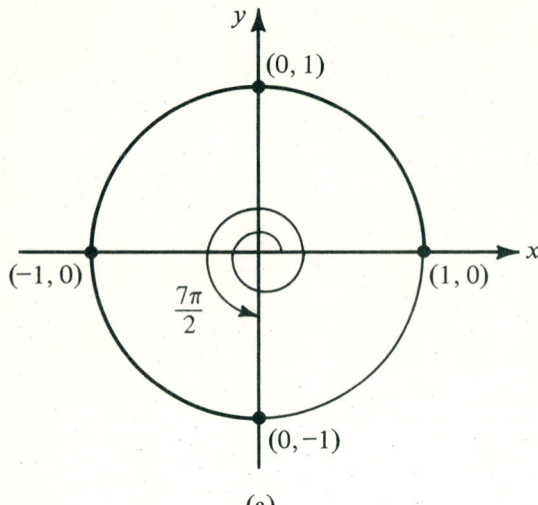

(a)

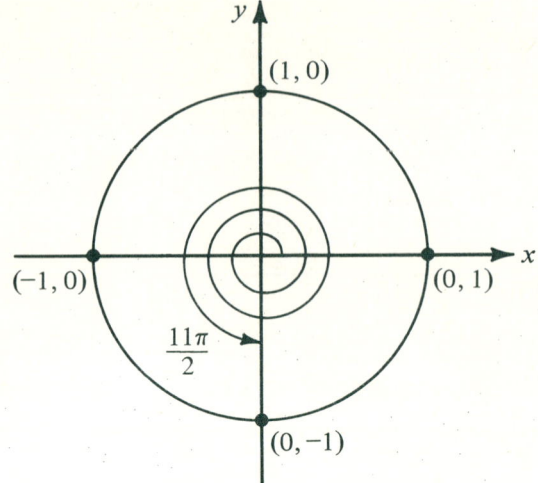

18. $25°16' = 25(16°/60) \approx 25.267° \cdot \dfrac{\pi}{180°} \approx 0.14\pi \approx 0.441$

19. 2.068 21. 4.317
23. -4.891

Exercise 9.3

1. (a) $1/2$
   (b) $1/2$
   (c) $1/2$
3. (a) $\sqrt{2}/2$
   (b) $-\sqrt{2}/2$
5. (a) $1/2$
   (b) $1/2$
   (c) $\sqrt{3}/2$
   (d) $\sqrt{3}/2$
   (e) $-1/2$
   (f) $-1/2$
   (g) $\sqrt{3}/2$
   (h) $\sqrt{3}/2$
7. (a) $\sqrt{2}/2$
   (b) $\sqrt{2}/2$
9. (a) $-1/2$
   (b) $-1/2$
11. (a) $1, \sqrt{2}$
    (b) $\sqrt{3}/2, 1$

2. (a) $-1/2$
   (b) $-1/2$
   (c) $-1/2$
4. (a) $\sqrt{2}/2$
   (b) $-\sqrt{2}/2$
6. (a) $\sqrt{3}/2$
   (b) $-\sqrt{3}/2$
   (c) $1/2$
   (d) $-1/2$
   (e) $1/2$
   (f) $-1/2$
8. (a) $-\sqrt{2}/2$
   (b) $\sqrt{2}/2$
10. (a) $1/2$
    (b) $1/2$
13. (a) 1
    (b) 1
    (c) 1

15. (a) $-1/2, -1/2$
    (b) $0, 0$
    (c) $1/2, 1/2$
16. (a) $\theta = \pi/2 \Rightarrow \cos^2 \theta/2 = (\cos \pi/4)^2 = (\sqrt{2}/2)^2 = 1/2$ and $1 + \cos \theta = 1 + \cos \pi/2 = 1$. Therefore, $\cos^2 \theta/2 = 1/2(1 + \cos \theta)$ for $\theta = \pi/2$.
    (b) $\cos^2 \pi/12 = 1/2(1 + \cos \pi/6) \Rightarrow \cos \pi/12 = \sqrt{(1 + \sqrt{3}/2)/2}$
    $= 1/2\sqrt{2 + \sqrt{3}} \approx 0.966$
17. (a) $\pi/6, 5\pi/6$
    (b) $7\pi/6, 11\pi/6$
19. (a) $\pi/4, 3\pi/4$
    (b) $5\pi/4, 7\pi/4$
21. $\alpha = \beta/2$
    and
    $\alpha = \pi/6$ or $11\pi/6 \Rightarrow \beta/2 = \pi/6$ or $\beta/2 = 11\pi/6 \Rightarrow$
    $\beta = \pi/3$ or $\beta = 11\pi/3$
23. (a) $\alpha = 3\pi/4$ or $\alpha = 5\pi/4$
    (b) $\alpha = -\beta/2 \Rightarrow \beta = -3\pi/2$ or $-5\pi/2$
25. (a) $\theta = \pi/4$
    (b) $\pi/4 < \theta < 5\pi/4$
27. Since $\angle OP'P$ is an inscribed angle, its measure is half that of its intercepted arc,
$$OP'P = 1/2(2\pi/3) = \pi/3$$
Then, since an equiangular triangle is equilateral, $\triangle OPP'$ is equilateral and $PP' = 1$.
Now, by the Pythagorean Theorem, we have
$$y_0 = \sqrt{1 - x_0^2} \text{ and } x_0 = \frac{1}{2}(OP') = \frac{1}{2} \Rightarrow$$
$$y_0 = \sqrt{3}/2$$

29.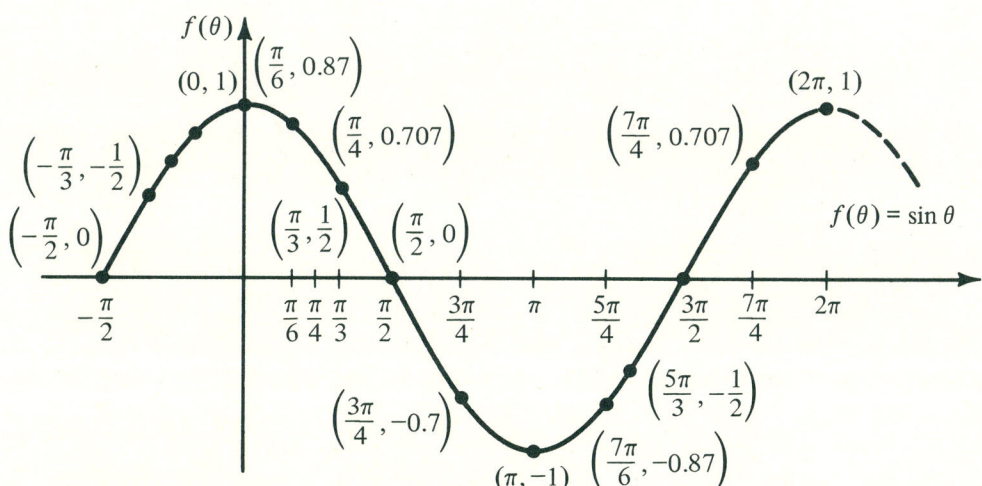

Exercise 9.4

1. (a) (ii)
   (b) (i)
2. (a) (ii)
   (b) (i)
3. (a) (ii)
   (b) (i)
4. (a) (ii)
   (b) (i)
5. (a) (i)
   (b) (ii)

7.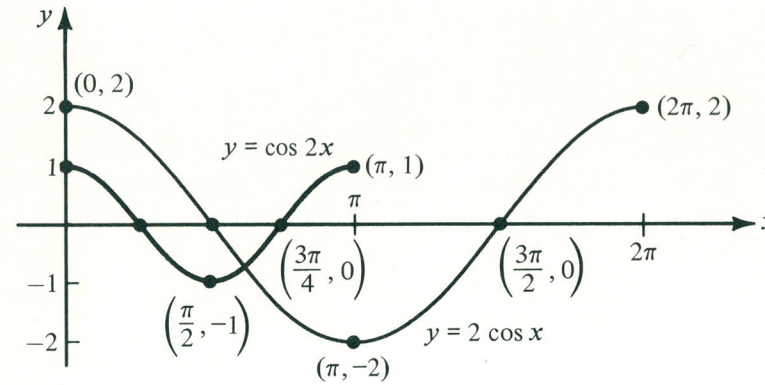

9.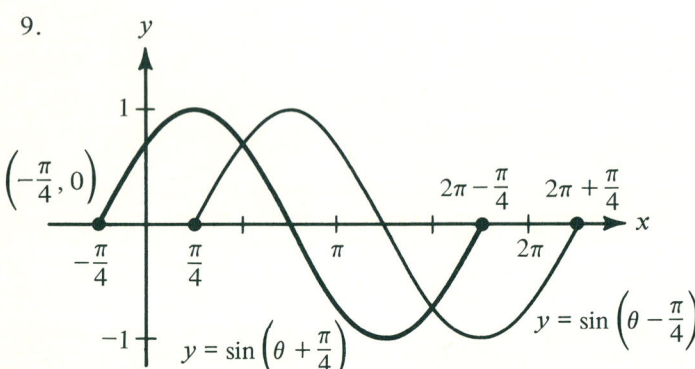

11.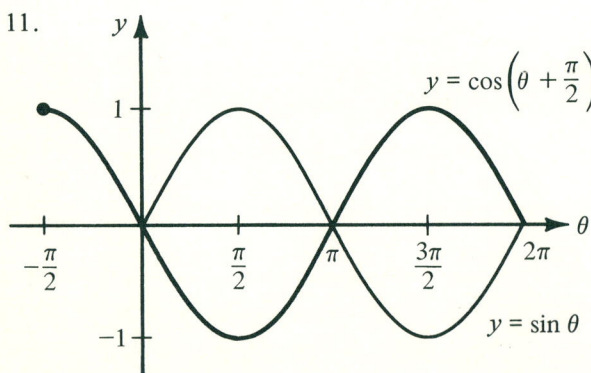

13.

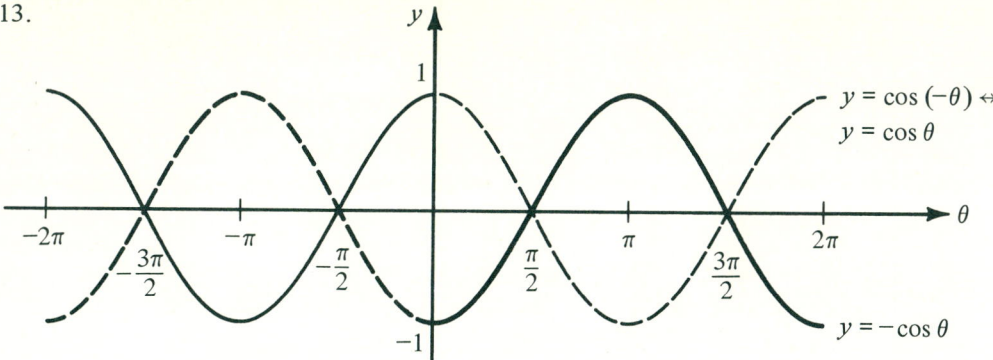

15.

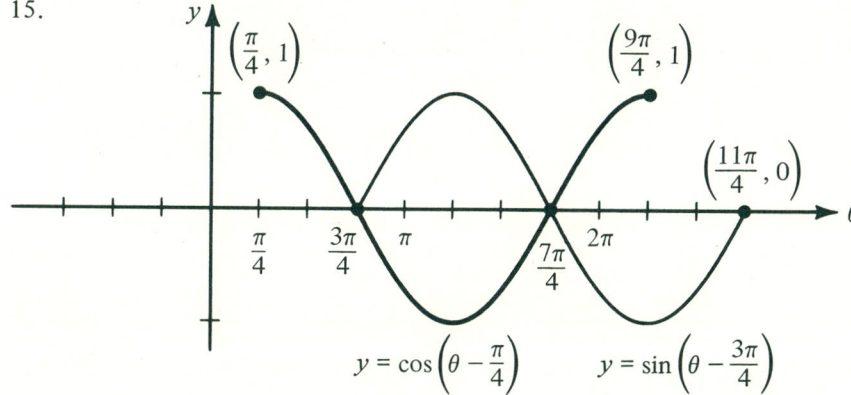

16. $f(\theta) = 2 \sin 3(\theta - \pi/6)$ has an amplitude of 2 and a phase shift of $\pi/6$ to the right. One complete period will require the argument $3(\theta - \pi/6)$ to assume all values between 0 and 2π. That is,

$$0 \leq 3(\theta - \pi/6) \leq 2\pi \Rightarrow$$
$$0 \leq \theta - \pi/6 \leq 2\pi/3 \Rightarrow$$
$$\pi/6 \leq \theta \leq 5\pi/6$$

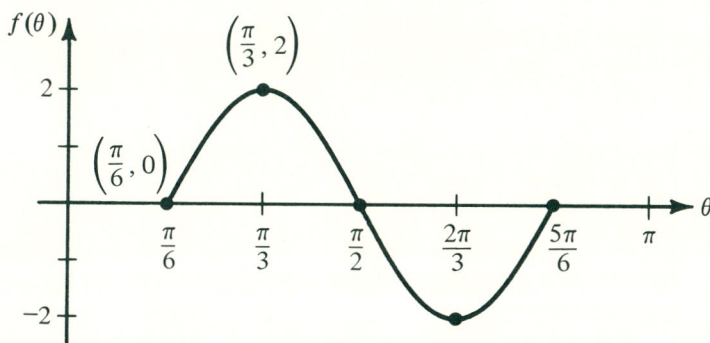

A-107

17.

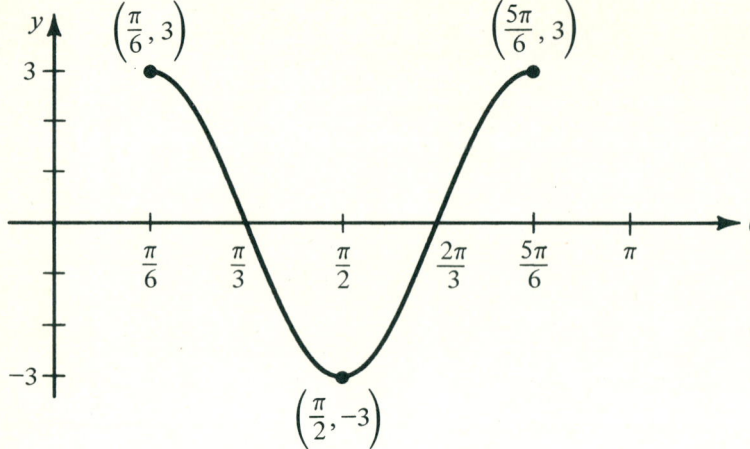

19.

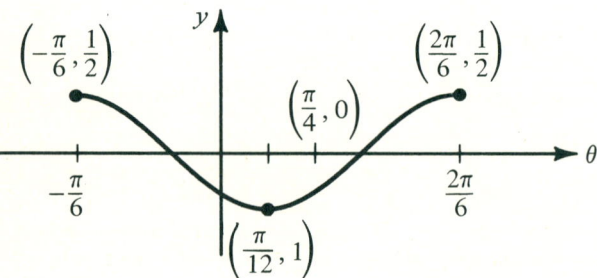

21.

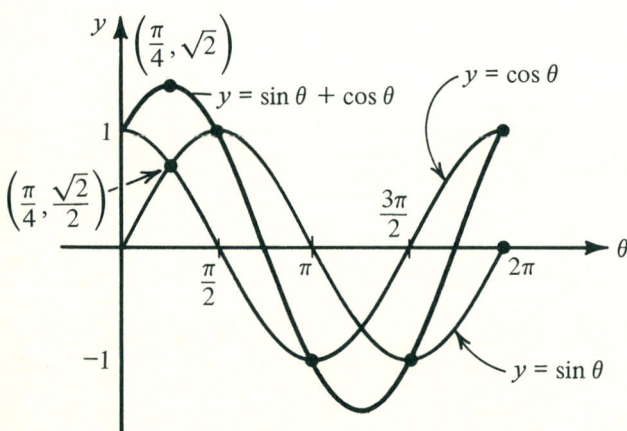

23.

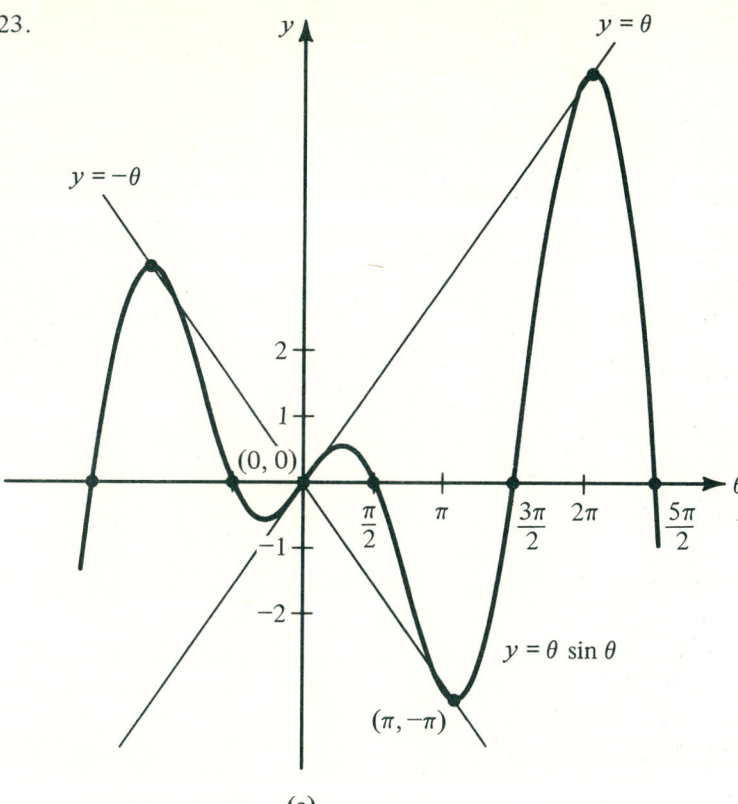

(a)

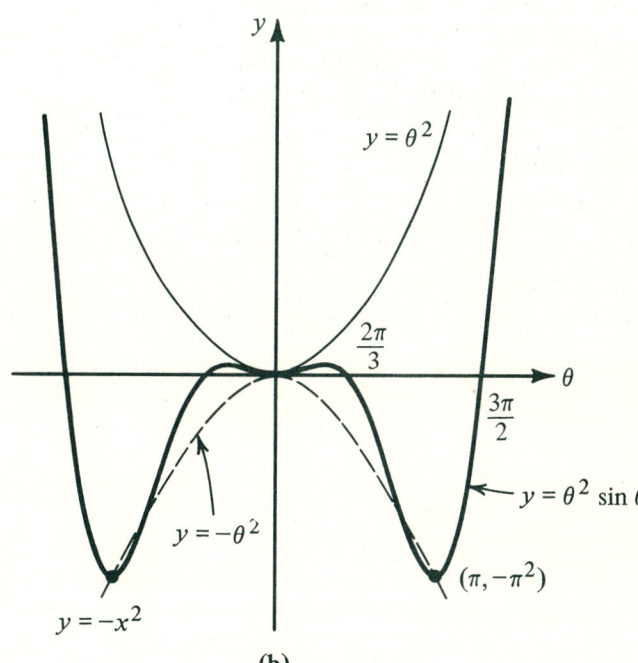

(b)

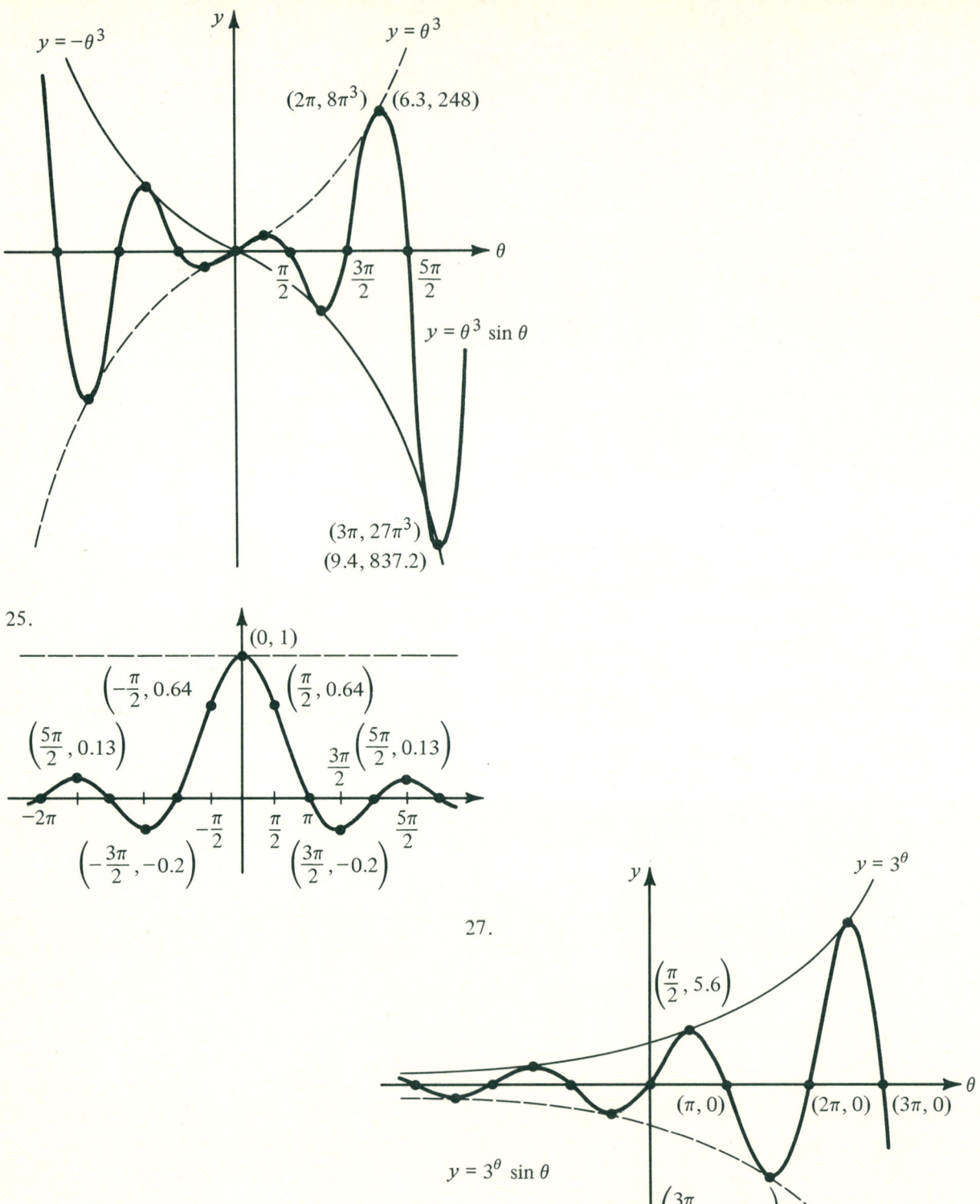

29.
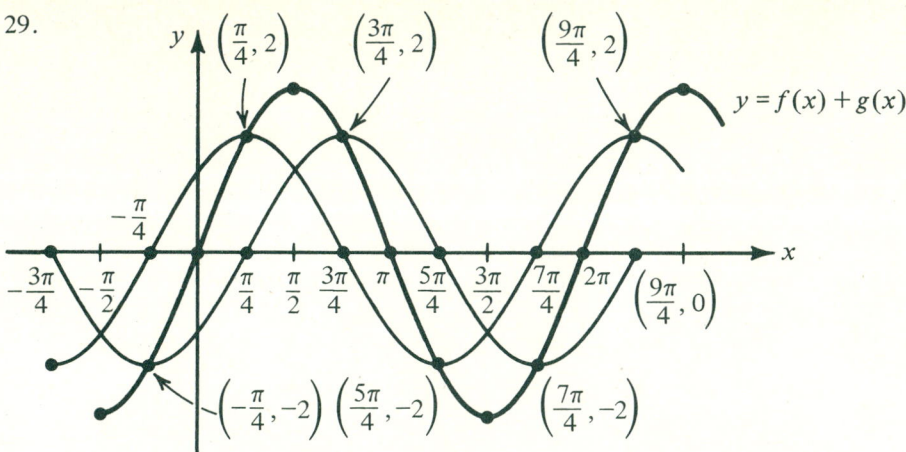

31.
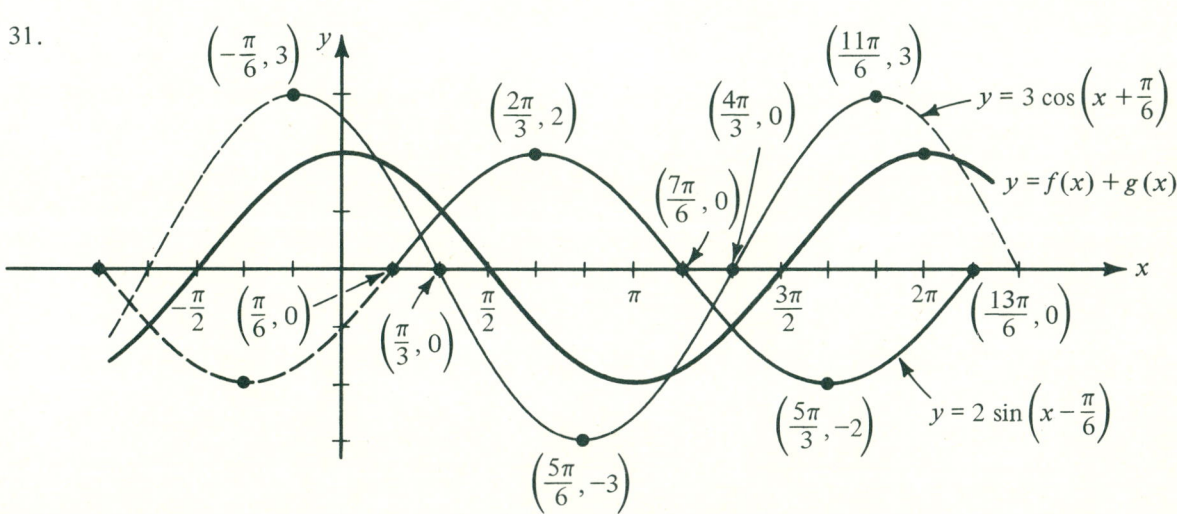

Exercise 9.5

1. (a) $f(\theta) = \sec \theta$
   (b) $y = f(\theta - \pi/2)$
   $= \sec(\theta - \pi/2)$

2. (a) $f(\theta) = \tan \theta$
   (b) $y = f(-\theta) = -f(\theta)$
   $= -\tan \theta$

3. (a) $f(\theta) = 3 \sin (\theta/2)$
   (b) $g(\theta) = 1/2 \cos (\theta/2)$

4. (a) $f(\theta) = 2 \sec \theta$
   (b) $f(\theta) = 1/2 \sec (\theta - \pi/2)$

5. (a) $f(\theta) = 2 \tan \theta$
   (b) $y = 1/3 \, f(\pi/2 - \theta)$
   $= 1/3 \tan (\pi/2 - \theta)$

A–111

6. Notice that (b) is a translation to the right of (a) giving (b) a phase shift of $\pi/4$. The vertical asymptotes shift also.

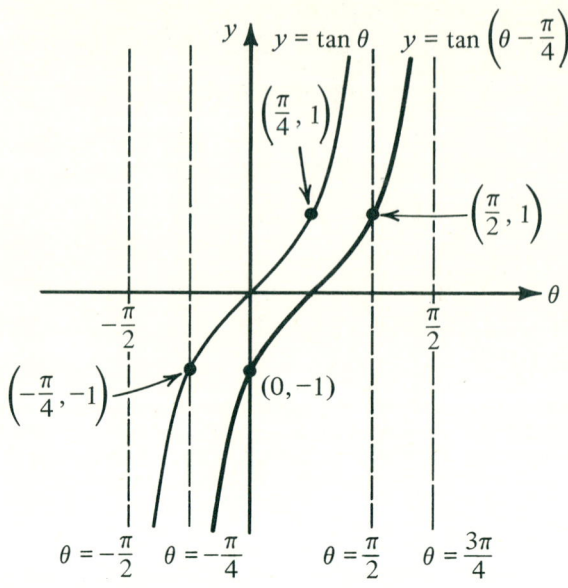

7.

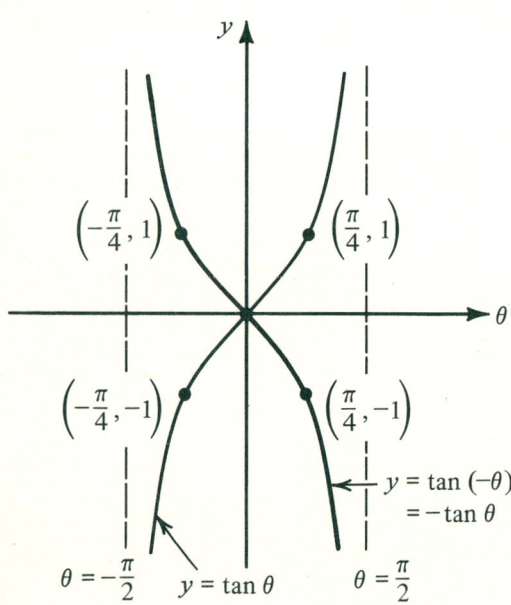

9.

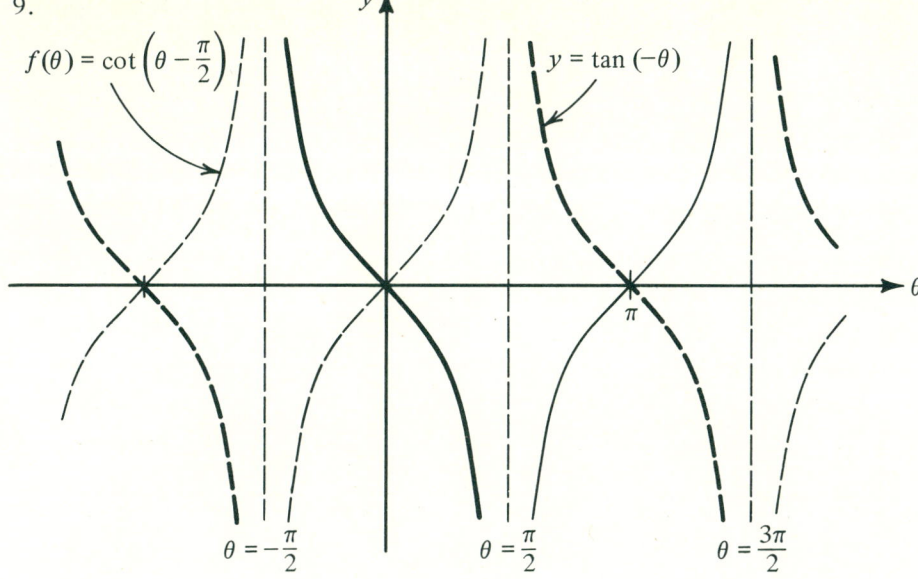

10. $y = 2 \tan 3(\theta - \pi/4)$ is expanded away from the x-axis by a factor of 2, compressed toward the y-axis by a factor of 3 and translated to the right $\pi/4$. In order to complete one period, the argument of this tangent function, $3(\theta - \pi/4)$, must assume all values between $-\pi/2$ and $\pi/2$. That is,

$$-\pi/2 < 3(\theta - \pi/4) < \pi/2 \Rightarrow$$
$$-\pi/6 < \theta - \pi/4 < \pi/6 \Rightarrow$$
$$\pi/12 < \theta < 5\pi/12$$

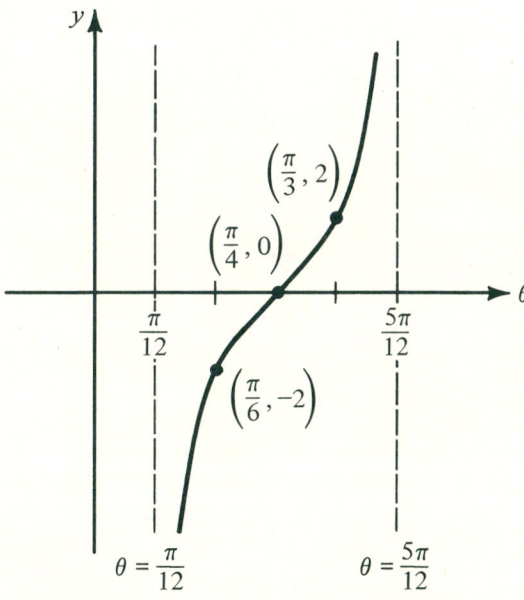

11.

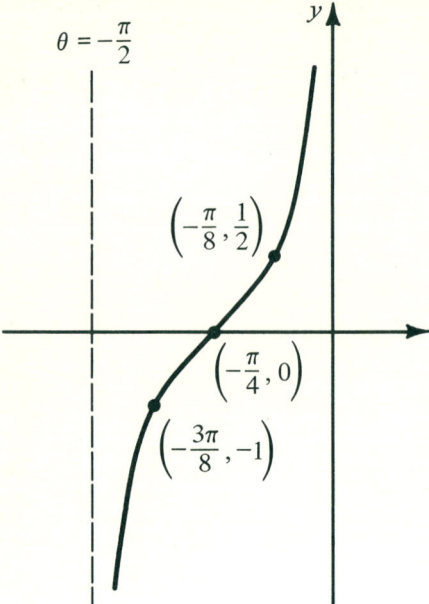

13.

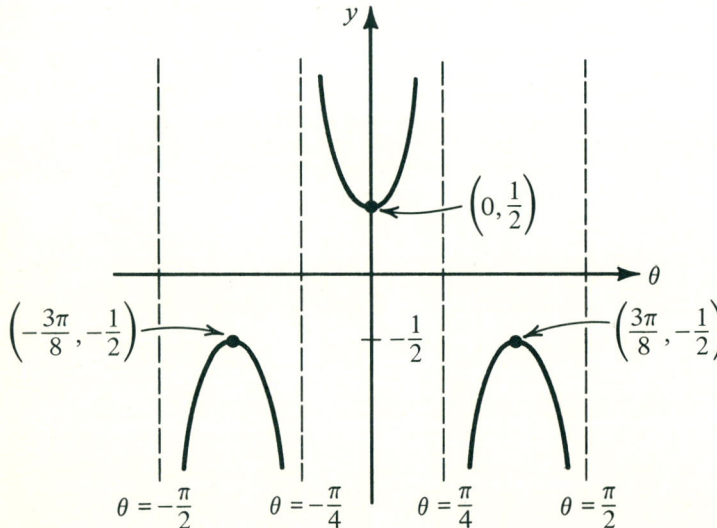

15. $y = 1/3 \csc(\theta/2 - \pi/4) = 1/3 \csc 1/2 (\theta - \pi/2)$ is compressed toward the x-axis by a factor of 3, expanded away from the y-axis by a factor of 2, and has a phase shift of $+\pi/2$. In order to complete one period, the argument of this cosecant function, $\theta/2 - \pi/4$, must assume all values between $-\pi$ and π. That is,

$$-\pi < \frac{\theta}{2} - \frac{\pi}{4} < \pi \quad \text{and} \quad \frac{\theta}{2} - \frac{\pi}{4} \neq 0$$

$$\frac{-3\pi}{4} < \frac{\theta}{2} < \frac{5\pi}{4} \quad \text{and} \quad \frac{\theta}{2} \neq \frac{\pi}{4}$$

$$\frac{-3\pi}{2} < \theta < \frac{5\pi}{2} \quad \text{and} \quad \theta \neq \frac{\pi}{2}$$

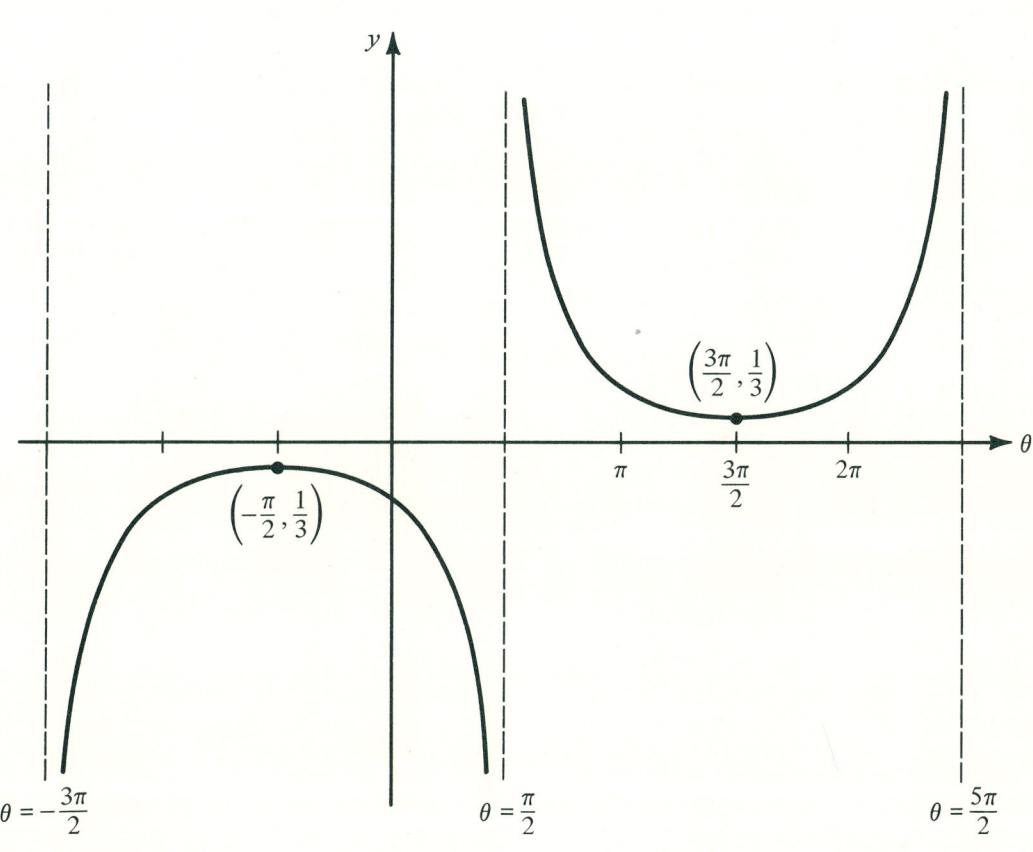

17. Consider $f(-\theta)$ for each of the functions. Remember that $f(-\theta) = f(\theta) \Leftrightarrow f(\theta)$ is even and $f(-\theta) = -f(\theta) \Leftrightarrow f(\theta)$ is odd.

$$f(\theta) = \tan \theta \Rightarrow f(-\theta) = \tan(-\theta) = -\tan \theta = -f(\theta)$$
$$f(\theta) = \sec \theta \Rightarrow f(-\theta) = \sec(-\theta) = \sec \theta = f(\theta)$$
$$f(\theta) = \csc \theta \Rightarrow f(-\theta) = \csc(-\theta) = -\csc \theta = -f(\theta)$$
$$f(\theta) = \cot \theta \Rightarrow f(-\theta) = \cot(-\theta) = -\cot \theta = -f(\theta)$$

19.

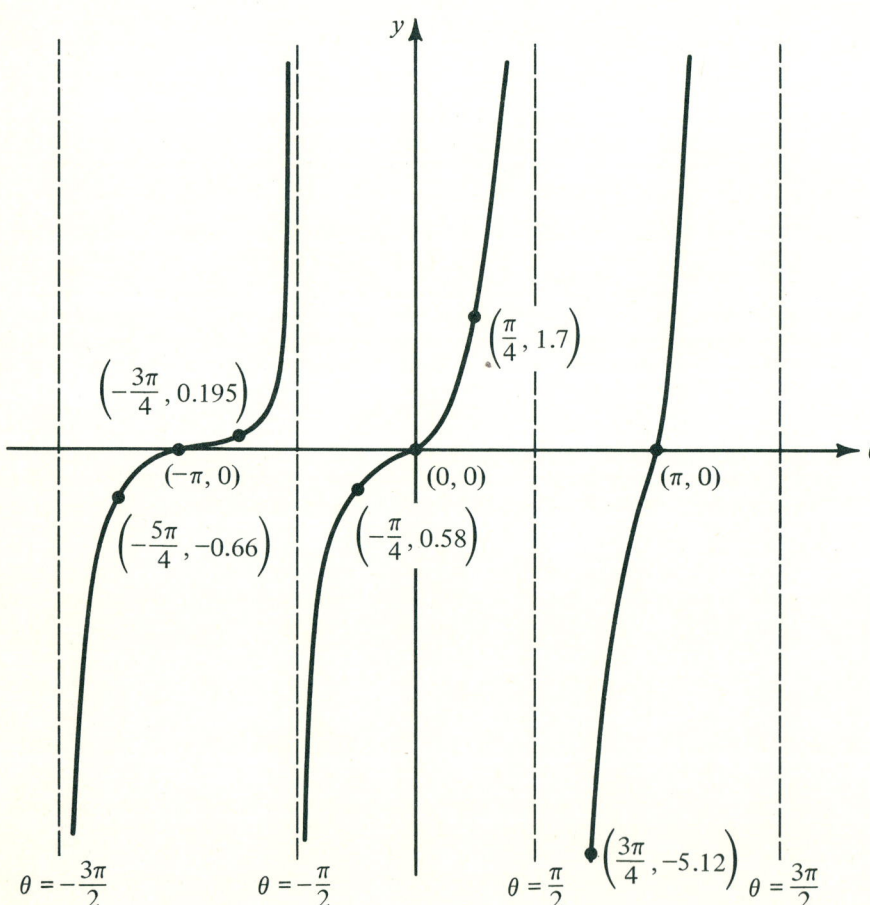

Exercise 9.6

1.

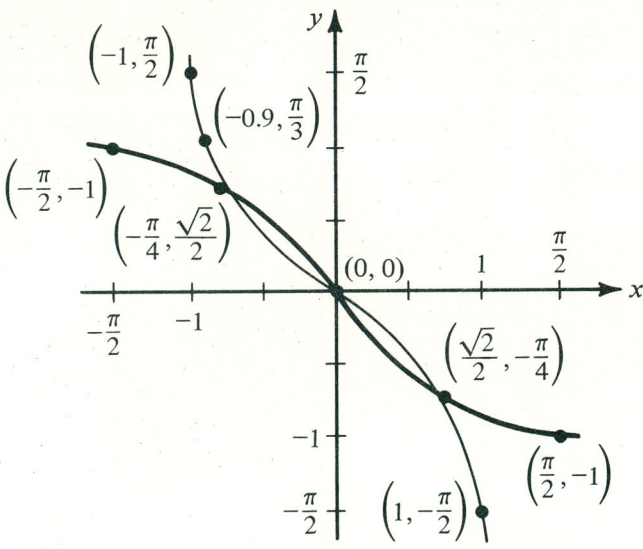

2.

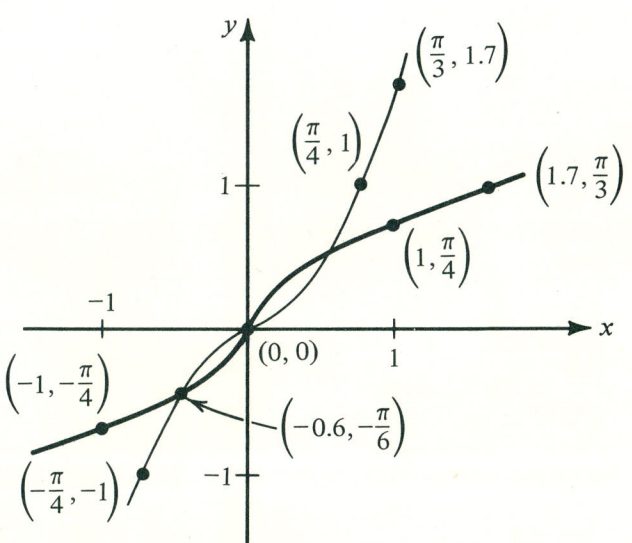

3.

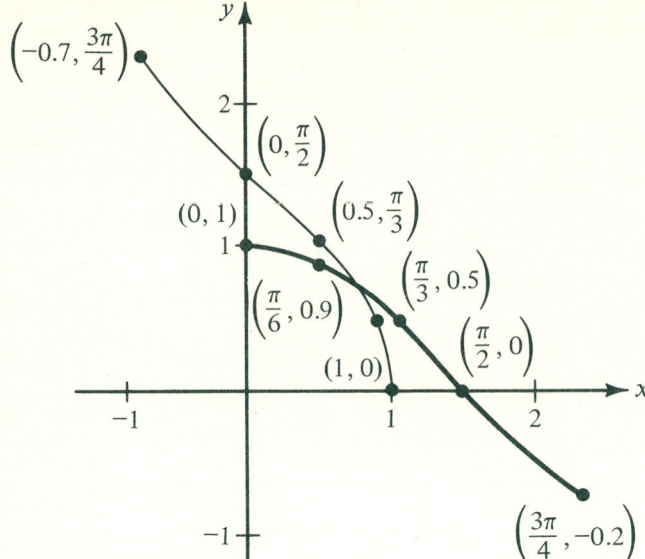

5. (a) $f^{-1}(x) = \cos^{-1} x$
   (b) $g^{-1}(x) = \cos^{-1} x - \pi/4$
   (c) $h^{-1}(x) = \cos^{-1} x + \pi/4$

7. (a) $f^{-1}(x) = 1/3 \sin^{-1} x$
   (b) $g^{-1}(x) = \sin^{-1} x + 2$
   (c) $h^{-1}(x) = 1/3 \sin^{-1} x + 2$

8. (a) This sine function, by convention, is restricted to $-\pi/2 \leqslant x \leqslant \pi/2$ to obtain a one-to-one function. In (b) and (c) we will use this same restriction on the arguments of the functions to find their domains.
   (b) $-\pi/2 \leqslant x - \pi/4 \leqslant \pi/2 \Rightarrow -\pi/4 \leqslant x \leqslant 3\pi/4$
   (c) $-\pi/2 \leqslant 2x + \pi/3 \leqslant \pi/2 \Rightarrow -5\pi/6 \leqslant 2x \leqslant \pi/6 \Rightarrow$
       $-5\pi/12 \leqslant x \leqslant \pi/12$

9. (a) $0 \leqslant x \leqslant \pi$
   (b) $-\pi/3 \leqslant x \leqslant 2\pi/3$
   (c) $\pi/4 \leqslant x \leqslant 5\pi/4$
   Notice that the change in amplitude in (c) did not influence the domain.

11. (a) ∞
    (b) ± 1
    (c) 1

13. (a) $\pi/3$
    (b) $\pi/3$ (Not $4\pi/3$, since domain of Cos x is $0 \leqslant x \leqslant \pi$.)
    (c) $\pi/3 + 2K\pi, 5\pi/3 + 2K\pi, K \in$ integers

15. $-\pi/4$

17. $\pi/4 + K\pi, K \in$ integers

19. $-\pi/4 + K\pi, K \in$ integers

21. $f(x) = 2 \text{ Sin } 3x$ means the argument, $3x$, is restricted to values between $-\pi/2$ and $\pi/2$. That is,
$$-\pi/2 \leqslant 3x \leqslant \pi/2 \Rightarrow -\pi/6 \leqslant x \leqslant \pi/6$$
The amplitude is 2. The equation of the inverse is found as follows by first solving for x,
$$y = 2 \sin 3x \Rightarrow \sin 3x = \frac{y}{2} \Rightarrow 3x = \sin^{-1}\left(\frac{y}{2}\right) \Rightarrow$$
$$x = \frac{1}{3} \sin^{-1}\left(\frac{y}{2}\right)$$

Then, renaming the variables,

$$y = \frac{1}{3}\sin^{-1}\left(\frac{x}{2}\right)$$

$$D_f = \left\{x \mid -\frac{\pi}{6} \leqslant x \leqslant \frac{\pi}{6}\right\} = R_{f^{-1}}$$

and

$$D_{f^{-1}} = \{x \mid -2 \leqslant x \leqslant 2\} = R_f$$

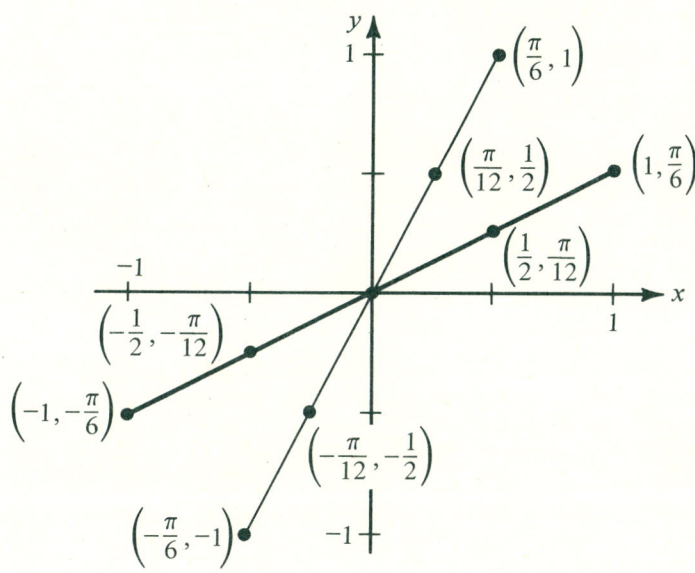

23.

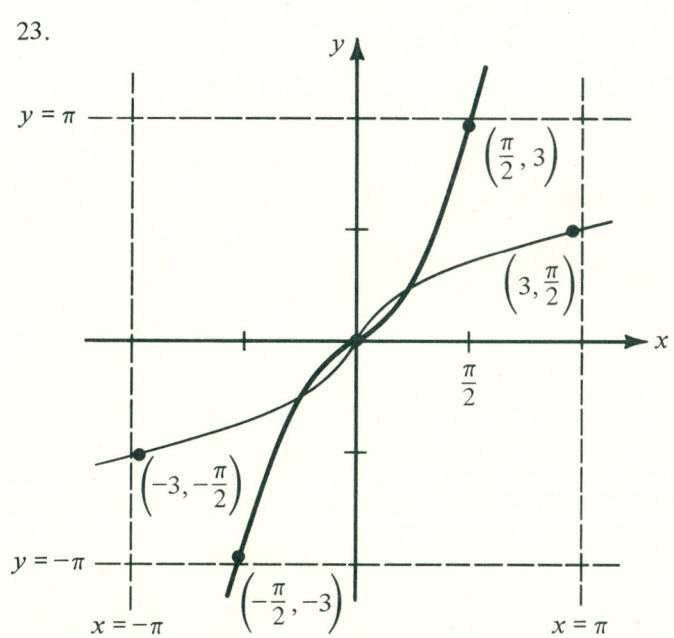

24. $y = 3 \cos(x - \pi/4)$ has an amplitude of 3. There is no restriction on this function's domain, but one complete period will occur when the argument, $x - \pi/4$, assumes all the values between 0 and 2π,

$$0 \leq x - \frac{\pi}{4} \leq 2\pi \Rightarrow \frac{\pi}{4} \leq x \leq \frac{9\pi}{4}$$

The equation of the inverse is found by first solving for x,

$$y = 3\cos\left(x - \frac{\pi}{4}\right) \Rightarrow \cos\left(x - \frac{\pi}{4}\right) = \frac{y}{3} \Rightarrow$$

$$x - \frac{\pi}{4} = \cos^{-1}\left(\frac{y}{3}\right) \Rightarrow x = \frac{\pi}{4} + \cos^{-1}\left(\frac{y}{3}\right)$$

Then, renaming the variables,

$$y = \frac{\pi}{4} + \cos^{-1}\left(\frac{x}{3}\right)$$

The $D_f = \{x | x \in \text{reals}\} = R_{f^{-1}}$
and $D_{f^{-1}} = \{y | -1 \leq y \leq 1\} = R_f$

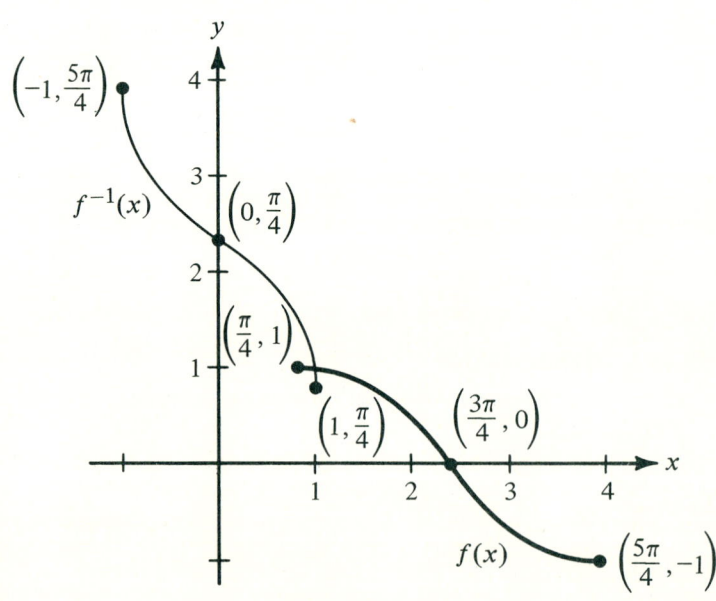

25.
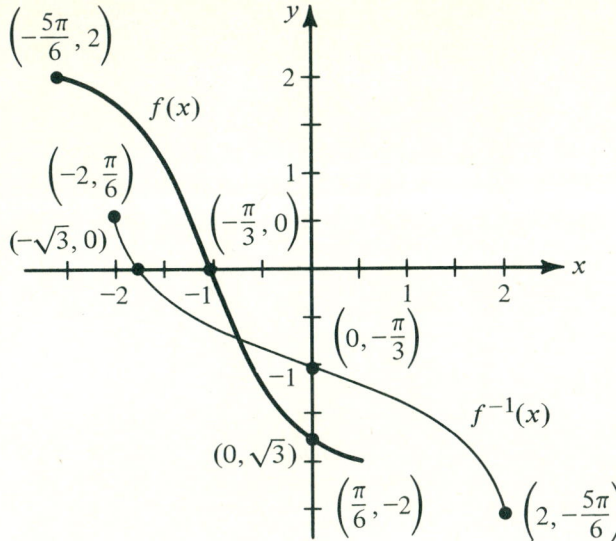

27.
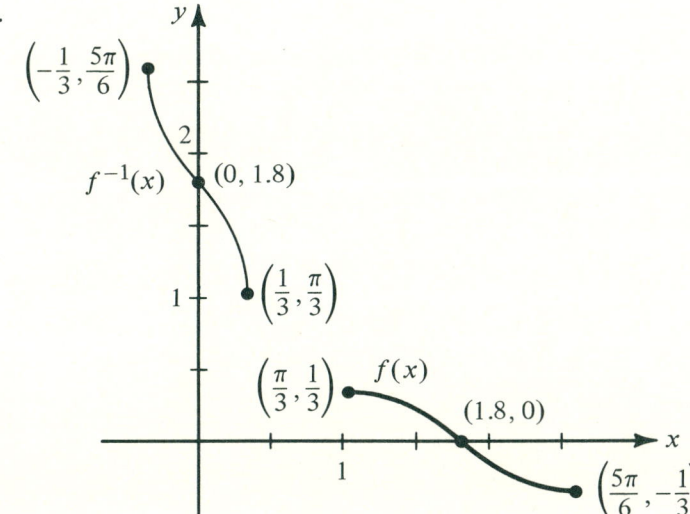

28. $1/2 \sin^{-1}\sqrt{3}/2 = 2x \Rightarrow 1/2(\pi/3) = 2x \Rightarrow x = \pi/12$
29. $x = (5\pi - 24)/24$
30. $1/3 \tan^{-1}(-\sqrt{3}) = x - 1 \Rightarrow x = 1 + 1/3(-\pi/3 + K\pi), K \in$ integers $\Rightarrow$
$$x = \frac{9 + (3K - 1)\pi}{9}$$

31. $\pi/6 + K\pi/4$ or $5\pi/24 + K\pi/4, K \in$ integers

32. $\sin^{-1}x = \tan^{-1}x \Rightarrow \sin(\sin^{-1}x) = \sin(\tan^{-1}x) \Rightarrow$
    $x = \sin\theta \Rightarrow x = x/\sqrt{1+x^2} \Rightarrow x = 0$

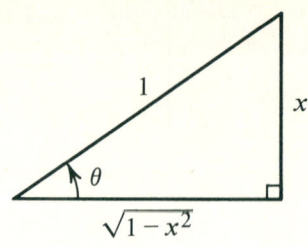

33. $x = \sqrt{1-4x^2} \Rightarrow x = \sqrt{5}/5$
35. $\sqrt{1-x^2}/x = 2x \Rightarrow x \approx 0.625$
41. $\pi \leq x \leq 2\pi$
43. $\pi \leq x \leq 2\pi$

CHAPTER 10

Exercise 10.2

1. (a) $\angle A = 30°$
       $b = \sqrt{3}$
   (b) $\angle A = 30°$
       $\angle B = 60°$
       $a = 1$
2. (a) $\angle A = \angle B = 45°$
       $a = b = 1$
   (b) $\angle A = \angle B = 45°$
       $c = 2\sqrt{2}$
3. (a) $\angle B = 60°$
       $c = 1/2$
       $b = \sqrt{3}$
   (b) $\angle B = 60°$
       $c = 8$
       $b = 4\sqrt{3}$
4. (a) $\angle B = 45°$
       $b = 1$
       $c = \sqrt{2}$
   (b) $\angle B = 45°$
       $b = 3$
       $c = 3\sqrt{2}$
5. (a) $\sqrt{2}/2$
   (b) $\sqrt{2}/2$
   (c) 1
   $\tan 45° = \sin 45°/\cos 45°$
6. (a) $1/2$
   (b) $1/2$
   (c) $-1/2$ (The reference angle is 30°.)
7. (a) $1/2$
   (b) $-1/2$
   (c) $1/2$ (Cosine is positive quadrant in I and II.)
8. (a) 1
   (b) 1
   (c) -1
9. (a) $\sqrt{2}/2$
   (b) $\sqrt{2}$
10. (a) $1/2$
    (b) 2
11. (a) $\sqrt{3}/3$
    (b) $\sqrt{3}$
12. (a) 1
    (b) 1
13. (a) $1/2$
    (b) $1/2$
14. (a) $\sqrt{3}$ (Problems 12, 13, and 14 show cofunctions of complementary angles.)
    (b) $\sqrt{3}$
15. (a) $1/4$ (Their sum is one.)
    (b) $3/4$
16. (a) 2
    (b) 2
17. (a) $\sqrt{3}/2$
    (b) $1/4$
18. (a) $\sqrt{3}/2$
    (b) 1

The following illustrate identities, relationships that hold for all values of the variable for which they are defined. Your answers for Problems 19–24 (a) and (b) should be equal.

19. 1   20. $-1/2$ 21. $-1/2$ 22. $\sqrt{3}/2$ 23. $1/2$ 24. 0

Exercise 10.3

1. (a) $-1/2$
   (b) $-\sqrt{3}/2$
   (c) $\sqrt{3}/3$
2. (a) $1/2$
   (b) $-\sqrt{3}/2$
   (c) $-\sqrt{3}/3$
3. (a) $-\sqrt{2}/2$
   (b) $\sqrt{3}/2$
   (c) $-\sqrt{3}$
4. (a) $-1/2$
   (b) $\sqrt{2}/2$
   (c) $-\sqrt{3}$
7. (a) $A = 150° \Rightarrow 1 + \cot^2 A = 1 + (\cot 150°)^2 = 1 + (-\cot 30°)^2 = 1 + (-\sqrt{3})^2 = 4$ and $\csc^2 A = \csc^2 150° = (\csc 30°)^2 = (2)^2 = 4$
9. $A = 45° \Rightarrow \sin A \cdot \cot A = \sin 45° \cdot \cot 45° = (\sqrt{2}/2)(1) = \sqrt{2}/2$ and $\cos A = \cos 45° = \sqrt{2}/2$
11. Satisfied by both.
13. Satisfied by both.
15.

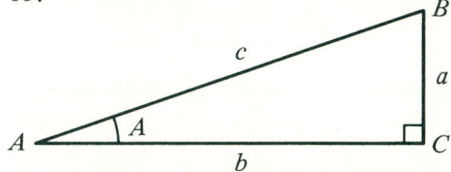

$90° - A = B \Rightarrow$
$\sec(90° - A) = \sec B = c/a = \csc A$

17.

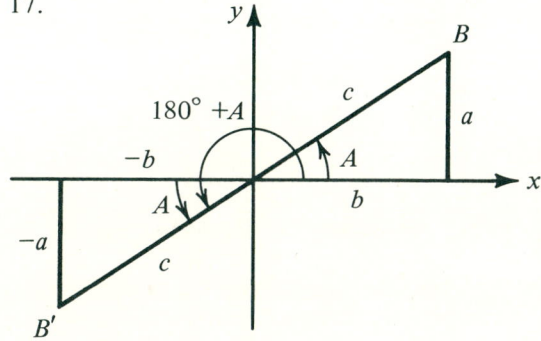

$\angle B'OB$ is $180°$, since $\angle B'OB = A + (180° - A)$
$\sin(180° + A) = -a/c$ and $\sin A = a/c \Rightarrow$
$\sin(180° + A) = -\sin A$
If $\angle A$ is not acute, the illustration is analogous to the one above.

Exercise 10.4

1. (a) $120°, 300°$
   (b) $60°, 240°$
2. (a) $30°, 150°$
   (b) $60°, 300°$

A-123

3. (a) $45°, 225°$
   (b) $135°, 315°$
4. (a) $45°, 225°$
   (b) $45°, 225°$
5. (a) $45°, 135°, 225°, 315°$
   (b) $45°, 225°$
6. (a) $0, 90°, 180°, 270°$
   (b) $0, 180°$
7. $\sin^2 A - \sin A = 0$
   Let $u = \sin A \Rightarrow u^2 - u = 0, u(u - 1) = 0, u = 0$ or $u = 1 \Rightarrow \sin A = 0$ or $\sin A = 1 \Rightarrow A = 0$ or π or $A = \pi/2$
9. $\pi/2, 3\pi/2$ or 0
10. $\sec^2 A - \tan A = 1 \Rightarrow (1 + \tan^2 A) - \tan A = 1$
    Let $u = \tan A \Rightarrow 1 + u^2 - u = 1 \Rightarrow u^2 - u = 0 \Rightarrow u(u - 1) = 0 \Rightarrow$
    $$u = 0 \text{ or } u = 1 \Rightarrow \tan A = 0 \text{ or } \tan A = 1 \Rightarrow$$
    $$A = 0 \text{ or } \pi \text{ or } A = \pi/4 \text{ or } 5\pi/4$$
11. 0 or π
13. $\pi/2, 3\pi/2$ (0 is not in the domain of $\cot x$)
15. (a) $\sin A \cdot \tan A + \cos A = \sin A \left(\dfrac{\sin A}{\cos A}\right) + \cos A \left(\dfrac{\cos A}{\cos A}\right)$

    $= \dfrac{\sin^2 A + \cos^2 A}{\cos A} = \dfrac{1}{\cos A} = \sec A$

    Therefore, $\sin A \cdot \tan A + \cos A = \sec A$ by the transitive law of equality ($a = b$ and $b = c \Rightarrow a = c$).

17. (HINT: Change to sine and cosine using $\tan B = \sin B/\cos B$ and $\cot B = \cos B/\sin B$. Remember that $1/\sin B = \csc B$.)

18. $\dfrac{1 - \cos B}{\sin B} \cdot \dfrac{1 + \cos B}{1 + \cos B} = \dfrac{1 - \cos^2 B}{\sin B(1 + \cos B)}$

    $= \dfrac{\sin^2 B}{\sin B(1 + \cos B)} = \dfrac{\sin B}{1 + \cos B}$

    Therefore,

    $\dfrac{1 - \cos B}{\sin B} = \dfrac{\sin B}{1 + \cos B}$

19. Multiply $1 - \cos^2 A = \sin^2 A$ by $\cos A/\cos A$ and regroup. It is easier to change the right side into the left.
21. Change everything to sine using $\cos^2 A = 1 - \sin^2 A$ and $\csc A = 1/\sin A$, then simply solve algebraically. Let $u = \sin A$. Your expression will be of the form $(1 - u^2)/u(1 + 1/u)$.

CHAPTER 11

Exercise 11.1

1. (a) $u = 1/2$
   (b) $\sin t = 1/2 \Rightarrow t = \pi/6$ or $t = 5\pi/6$
   (c) $\cos x = 1/2 \Rightarrow x = \pi/3$ or $5\pi/3$
2. (a) $u = 0$ or $u = 1$
   (b) $0, \pi,$ or $\pi/2$
   (c) $0, \pi,$ or $\pi/4, 5\pi/4$
3. (a) $u = 1$ or $u = -1/2$
   (b) $\pi/2$ or $7\pi/6, 11\pi/6$
   (c) 0 or $2\pi/3, 4\pi/3$
4. (a) $u = \pm\sqrt{3}$
   (b) $\pi/3, 2\pi/3, 4\pi/3, 5\pi/3$
   (c) $\pi/6, 5\pi/6, 7\pi/6, 11\pi/6$

5. (a) $u = \pm\sqrt{2}/2$
   (b) $\pi/4, 3\pi/4, 5\pi/4, 7\pi/4$
   (c) $\pi/4, 3\pi/4, 5\pi/4, 7\pi/4$
7. $[5(b)](\pi/4) + K\pi$ or $3\pi/4 + K\pi, K \in I$
   $[5(c)](\pi/4) + K\pi$ or $3\pi/4 + K\pi, K \in I$
9. (a) $1/2$ or -1
   (b) $\pi/6 + 2k\pi, 5\pi/6 + 2k\pi, 3\pi/2 + 2k\pi, K \in I$
   (c) $\pi/3 + 2k\pi, 5\pi/3 + 2k\pi, \pi + 2k\pi, K \in I$

6. (a) $u = \pm 1$
   (b) $\pi/2, 3\pi/2$
   (c) $\pi/4, 3\pi/4, 5\pi/4, 7\pi/4$
8. $[6(b)](\pi/2) + K\pi, K \in I$
   $[6(c)](\pi/4) + K\pi$ or $3\pi/4 + K\pi, K \in I$
11. $u = 1/2$ or $u = 1$
    (b) $\pi/2 + 2k\pi, K \in I$

12. $\tan^3 x - \tan^2 x = 3\tan x - 3$
    Let $u = \tan x \Rightarrow u^3 - u^2 - 3u + 3 = 0$

    $$\begin{array}{r|rrrr} 1 & 1 & -1 & -3 & 3 \\ & & 1 & 0 & -3 \\ \hline & 1 & 0 & -3 & 0 \end{array}$$

    $(u - 1)(u^2 - 3) = 0$
    $u = 1$ or $u = \pm\sqrt{3} \Rightarrow \tan x = 1$ or $\tan x = \pm\sqrt{3} \Rightarrow$
    $x = \pi/4 + k\pi, \pi/3 + k\pi$ or $2\pi/3 + k\pi, K \in I$

13. $\pi/4 + k\pi$ or $\pi/6 + k\pi$ or $7\pi, K \in I$

14. $16\cos^4 x = 11 - 8\sin^2 x$
    Let
    $u = \cos x$ and $\sin^2 x = 1 - \cos^2 x \Rightarrow$
    $16u^4 = 11 - 8(1 - u^2) \Rightarrow 16u^4 - 8u^2 - 3 = 0 \Rightarrow (4u^2 + 1)(4u^2 - 3) = 0 \Rightarrow$
    $4u^2 - 3 = 0 \Rightarrow u = \pm\sqrt{3}/2 \Rightarrow \cos x = \pm\sqrt{3}/2 \Rightarrow$
    $x = \pi/6 + 2k\pi$ or $5\pi/6 + 2k\pi, K \in I$

15. $\pi/3 + 2k\pi, 5\pi/3 + 2k\pi$, or $2k\pi, K \in$ integers

17. (a) $u = 1$
    (b) $\pi/4 + k\pi, K \in I$
    (c) $\pi/2 + 2k\pi, K \in I$

19. $2\sin x \cos x - \cos x = 0$
    $\cos x(2\sin x - 1) = 0$
    $\cos x = 0$ or $2\sin x - 1 = 0$
    $\sin x = \dfrac{1}{2}$
    $x = \pi/2, 3\pi/2, \pi/6,$ or $5\pi/6$

21. $x = 7\pi/6 + 2K\pi$ or $x = 11\pi/6 + 2K\pi$ or $x = \pi/4 + K\pi, K \in I$

23. The solution is the x-coordinate of the points of intersection of their graphs.

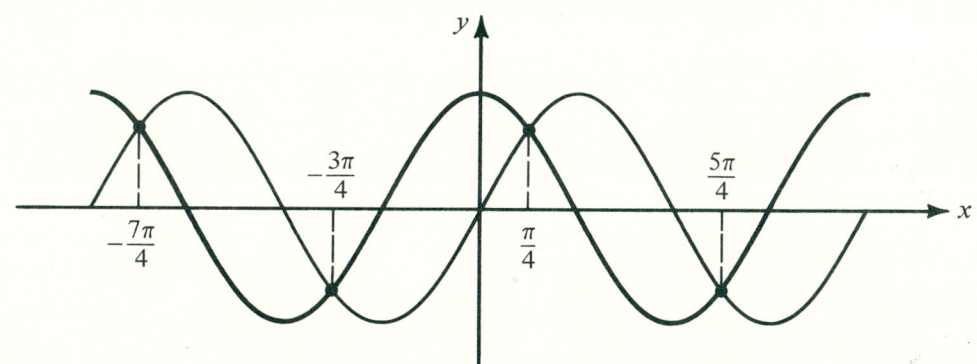

25. (a) The solutions are the points of intersections pictured below.

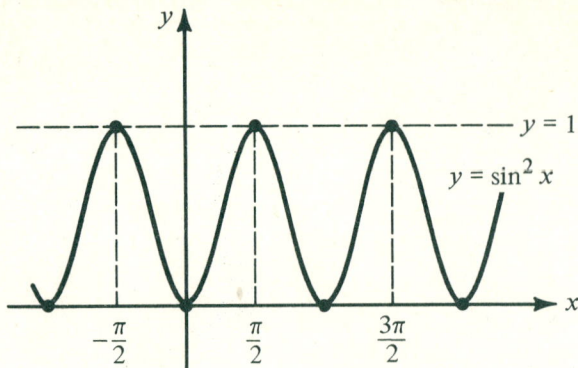

(b)

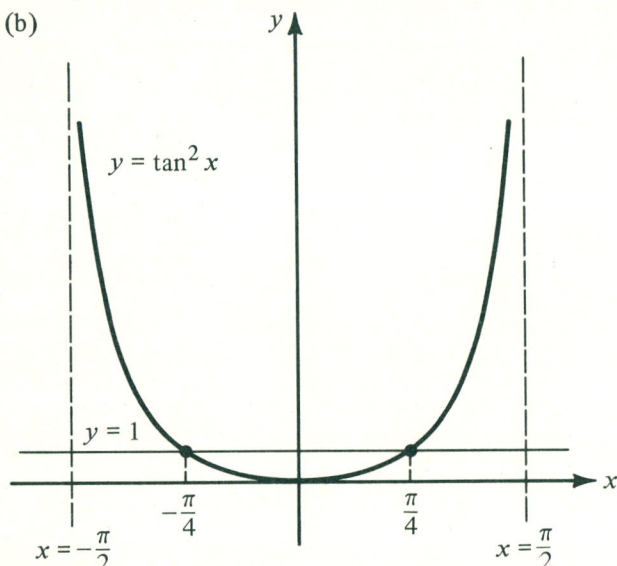

Exercise 11.2

1. (a) $(\sqrt{6} - \sqrt{2})/4$
   (b) $(\sqrt{6} + \sqrt{2})/4$
   (c) $(3 - \sqrt{3})^2/6$
2. (a) $(\sqrt{6} - \sqrt{2})/4$
   (b) $(\sqrt{6} + \sqrt{2})/4$
   (c) $(3 + \sqrt{3})^2/6$
3. (a) 1 ($u + v = \pi/2$)
   (b) 7/5
4. (a) 2
   (b) $-2/11$
5. (a) 1
   (b) 0
   (c) ∞ ($x + y = \pi/2$)
6. (a) ∞
   (b) $(x^2 - 1)/2x$ $\alpha + \beta = \pi/2$
   (c) 1
   (d) $2x/(1 + x^2)$
7. (a) Reflecting the graph of $f(\alpha) = \sin \alpha$ about y-axis, then translating it to the right $\pi/2$ yields the graph of $g(\alpha) = \cos \alpha$.

$$\sin(\pi/2 - \alpha) = \sin[-(\alpha - \pi/2)]$$

(c) Here the translation is to the left.
$$\sin(\pi/2 + \alpha) = \sin(\alpha + \pi/2)$$

9. (a) Reflecting the graph of $f(\alpha) = \sin \alpha$ about the y-axis, then translating to the right 2π yields the same graph as reflection $f(\alpha) = \sin \alpha$ about the x-axis.

10. (a) $\sin^{-1} \frac{3}{5} + \sin^{-1} \frac{12}{13} = \sin^{-1} x \Leftrightarrow \sin(\sin^{-1} \frac{3}{5} + \sin^{-1} \frac{12}{13}) = x$. If $\alpha = \sin^{-1} \frac{3}{5}$ and $\beta = \sin^{-1} \frac{12}{13}$, then $\sin \alpha = \frac{3}{5}$ and $\sin \beta = \frac{12}{13} \Rightarrow \cos \alpha = \frac{4}{5}$ and $\cos \beta = \frac{5}{13}$

Then,
$$x = \sin(\alpha + \beta) = \sin \alpha \cos \beta + \cos \alpha \sin \pi$$
$$= \left(\frac{3}{5}\right)\left(\frac{5}{13}\right) + \left(\frac{4}{5}\right)\left(\frac{12}{13}\right) = \frac{63}{65}$$

11. (a) $x = \pm\sqrt{3}/2$ (b) $x = -1/2$ or 2

12. $\cos 4x \cos 1 + \sin 4x \sin 1 = 1 \Leftrightarrow \cos(4x - 1) = 1 \Leftrightarrow 4x - 1 = \cos^{-1} 1 = 0 \Rightarrow x = 1/4$

13. $x = 0$ or 1 or $x = (1 \pm \sqrt{1 + 4\pi})/2$

14. $\sin(x + y) + \sin(x - y) = \cos y \Leftrightarrow \sin x \cos y + \cos x \sin y + \sin x \cos y - \cos x \sin y = \cos y \Rightarrow$
$2 \sin x \cos y - \cos y = 0 \Rightarrow \cos y(2 \sin x - 1) = 0 \Rightarrow \cos y = 0$ or $\sin x = 1/2 \Rightarrow$
$y = \pi/2$ or $3\pi/2$ or $x = \pi/6$ or $5\pi/6$

15. $x = \pi/6$ or $11\pi/6$ and $y = \pi/2$ or $3\pi/2$

16. We will prove this one is not an identity. Let $\theta = \pi/3$. Then, $\sin 3\theta = \sin \pi = 0$ but $2 \cos \pi/3 - \cos^3 \pi/3 = 2(1/2) - 1/8 \neq 0$. Hence, $\theta = \pi/3$ is one counterexample that proves the equation is not true for all values of θ.

17. It is an identity.   19. It is an identity.

21. It is an identity.

22. Let $\alpha = \tan^{-1} x$ and $\beta = \tan^{-1} y$. Then, $\tan x = \alpha$ and $\tan \beta = y$ and $\tan(\alpha - \beta)$
$= \tan \alpha - \tan \beta/1 + \tan \alpha \tan \beta = x - y/1 + xy \Rightarrow$
$\alpha - \beta = \tan^{-1}(x - y/1 + xy) \Rightarrow \tan^{-1} x - \tan^{-1} y = \tan^{-1}(x - y/1 + xy)$.

23. It is an identity.   25. This is a very important identity.

Exercise 11.3

1. (a) $24/25$
   (b) $-7/25$
   (c) $\sqrt{5}/5$
   (d) $2\sqrt{5}/5$

2. (a) $\sqrt{2 - \sqrt{3}}/2$
   (b) $\sqrt{2 + \sqrt{3}}/2$

3. (a) $\sin 2\alpha = 2(3/5)(4/5) = 24/25$ and $\sin 2\beta = 2(4/5)(3/5) = 24/25$
   (b) $\cos 2\alpha = (4/5)^2 - (3/5)^2 = 7/25$
   (c) $\cos 2\alpha + \cos 2\beta = [7/25] + [(3/5)^2 - (4/5)^2] = 0$
   (d) $\sin 3\alpha = \sin(2\alpha + \alpha)$
   $= \sin 2\alpha \cos \alpha + \cos 2\alpha \sin \alpha$
   $= (24/25)(4/5) + (7/25)(3/5) = 117/125$

4. $\sin 2x = 2 \sin x \Rightarrow 2 \sin x \cos x - 2 \sin x = 0 \Rightarrow 2 \sin x(\cos x - 1) = 0 \Rightarrow$
$\sin x = 0$ or $\cos x = 1 \Rightarrow x = 0$ or π or $x = 0 \Rightarrow x = 0$ or π

5. $x = 2\pi/3, 4\pi/3$, or 0 7. $t = 0, \pi$, or $\tan^{-1} 2$

9. $\theta = \pi/2, 3\pi/2, \pi/4, 3\pi/4, 5\pi/4$, or $7\pi/4$.

11. $\cos t \cos 2t = \sin 2t \sin t$
    $\cos t(1 - 2\sin^2 t) = 2\sin^2 t \cos t$
    $\cos t[1 - 2\sin^2 t - 2\sin^2 t] = 0$
    $\cos t = 0$ or $1 - 4\sin^2 t = 0 \Rightarrow t = \pi/2$, or $3\pi/2$
    or $(1 - 2\sin t) \cdot (1 + 2\sin t) = 0 \Rightarrow t = \pi/2$ or $3\pi 2$ or $\sin t = \pm 1/2 \Rightarrow$
    $t = \pi/2, 3\pi 2, \pi/6, 5\pi/6, 7\pi/6$, or $11\pi/6$.

12. $2\sin^{-1} x \cos^{-1} 1 \Rightarrow$ (by definition of the arccosine)
    $\cos 2\alpha = 1$
    where
    $2\alpha = 0$ or 2π
    $\alpha = 0$ or π

13. $x = 1/2$ or $x = 2$
15. $x = -1/2$

16. $\dfrac{\cos 2t}{1 - \sin 2t} = \dfrac{1 - 2\sin^2 t}{1 - 2\sin t \cos t} \cdot \dfrac{1/\cos^2 t}{1/\cos^2 t}$

$= \dfrac{\sec^2 t - 2\tan^2 t}{\sec^2 t - 2\tan t} = \dfrac{1 + \tan^2 t - 2\tan^2 t}{1 + \tan^2 t - 2\tan t} = \dfrac{1 - \tan^2 t}{(1 - \tan t)^2}$

$= \dfrac{(1 - \tan t)(1 + \tan t)}{(1 - \tan t)^2} = \dfrac{1 + \tan t}{1 - \tan t}$

20. $\dfrac{\sin 3x - \sin x}{\cos 3x + \cos x} = \dfrac{\sin 2x \cos x + \cos 2x \sin x - \sin x}{\cos 2x \cos x - \sin 2x \sin x + \cos x}$

$= \dfrac{2\sin x \cos^2 x + \sin x - 2\sin^3 x - \sin x}{2\cos^3 x - \cos x - 2\sin^2 x \cos x + \cos x}$

$= \dfrac{2\sin x(1 - \sin^2 x) - 2\sin^3 x}{2\cos^3 x - 2(1 - \cos^2 x)\cos x} = \dfrac{2\sin x}{2\cos x} = \tan x$

27. (a) $\cos(\alpha - \beta) + \cos(\alpha + \beta) = (\cos\alpha\cos\beta + \sin\alpha\sin\beta) + (\cos\alpha\cos\beta - \sin\alpha\sin\beta) = 2\cos\alpha\cos\beta$

(b) $\alpha = \dfrac{x+y}{2}$ and $\beta = \dfrac{x-y}{2} \Rightarrow \cos(\alpha - \beta) + \cos(\alpha + \beta)$

$= \cos\left(\dfrac{x+y}{2} - \dfrac{x-y}{2}\right) + \cos\left(\dfrac{x+y}{2} + \dfrac{x-y}{2}\right) = \cos y + \cos x$ and

$2\cos\alpha\cos\beta = 2\cos\left(\dfrac{x+y}{2}\right)\cos\left(\dfrac{x-y}{2}\right)$

(c) Let $x = u$ and $y = v - \pi$. Then,

$\dfrac{x+y}{2} = \dfrac{u+v}{2} - \dfrac{\pi}{2}$ and $\dfrac{x-y}{2} = \dfrac{u-v}{2} + \dfrac{\pi}{2} \Rightarrow$

$\cos x + \cos y = \cos u + \cos(v - \pi) = \cos u - \cos v$ and

$2\cos\left(\dfrac{x+y}{2}\right)\cos\left(\dfrac{x-y}{2}\right) = 2\cos\left(\dfrac{u+v}{2} - \dfrac{\pi}{2}\right)\cos\left(\dfrac{u-v}{2} + \dfrac{\pi}{2}\right)$

$= 2\sin\left(\dfrac{u+v}{2}\right)\sin\left(-\dfrac{u-v}{2}\right) = -2\sin\left(\dfrac{u+v}{2}\right)\sin\left(\dfrac{u-v}{2}\right)$

CHAPTER 12

Exercise 12.2

1. (a) $\angle A = 30°, \angle B = 60°, b = \sqrt{3}$
   (b) $\angle A = \angle B = 45°, c = \sqrt{2}$
   (c) $\angle B = 30°, b = \sqrt{3}/2, c = \sqrt{3}$
      [1(a) and (c) are similar triangles.]
2. (a) $a = 4, \angle A \approx 53°, \angle B \approx 37°$
   (b) $a = 12, \angle A \approx 67°, \angle B \approx 23°$
   (c) $c = 10, \angle A \approx 53°, \angle B \approx 37°$
      [2(a) and (c) are similar triangles.]
3. (a) $b \approx 38.1, c \approx 40.6, \angle B = 69°50'$
   (b) $a \approx 2.7, b \approx 9.6, \angle A = 16°40'$
   (c) $b \approx 117.1, c \approx 154, \angle B = 49°30'$
5. 30 meters
7. 132 meters
9. $d = 2.58$ centimeters, $h = 4.66$ centimeters
11. 5.25 meters
13. Since the satellite must complete one revolution in 24 hours and $d = r \cdot t$, where d = circumference of its "circular" orbit, we have $r \cdot t = c \Rightarrow$ (1694 kilometers per hour) 24 hours $= \pi d \Rightarrow d = 12{,}941$ kilometers, which, since the diameter of the earth is 12,742 kilometers, yields an altitude of $a = 12{,}941 - 6{,}371 = 6{,}570$ kilometers.

$$\cos\frac{\alpha}{2} = \frac{6{,}371 \text{ km}}{12{,}941 \text{ km}}, \qquad \frac{\alpha}{2} = \cos^{-1}(0{,}491), \qquad \alpha = 121°$$

Exercise 12.3

1.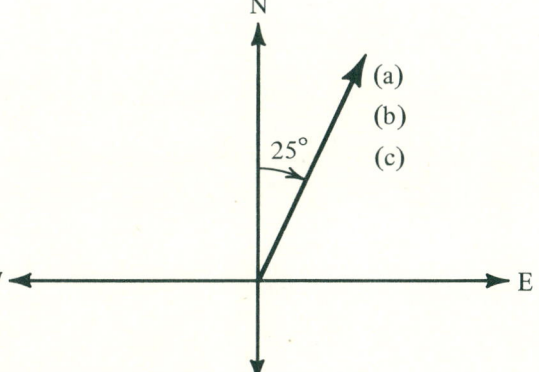

2.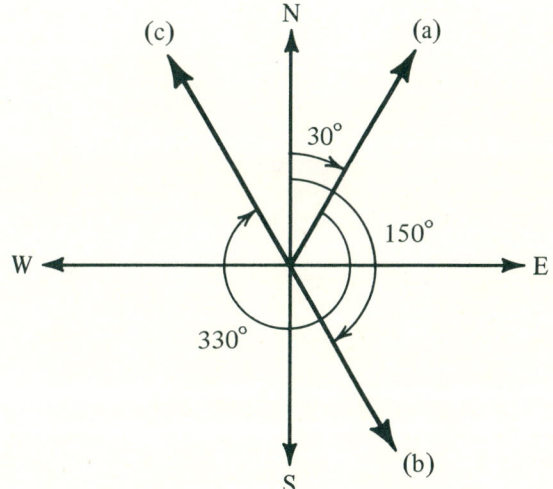

3.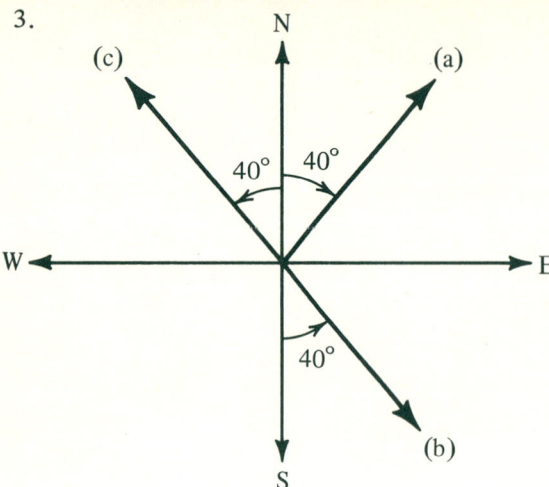

4.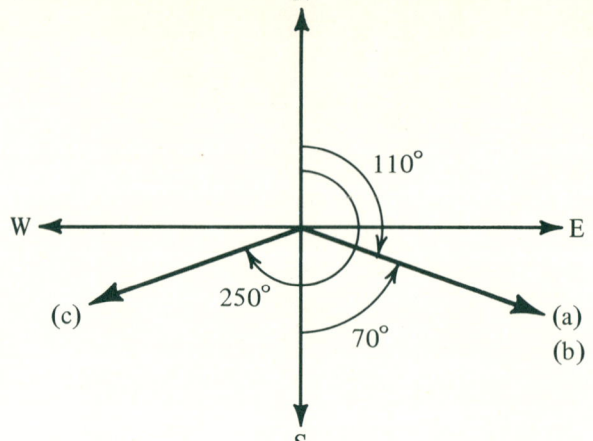

5. 300°
6. N40°E
7. Due South
8. Due West
9. 1.414 meters
11. They are about 25.6 nautical miles apart at a heading of about 83°40' from B to A.
13. The bearing from C to A is about S36°44'W at a distance of 128.65 meters.

Exercise 12.4

1.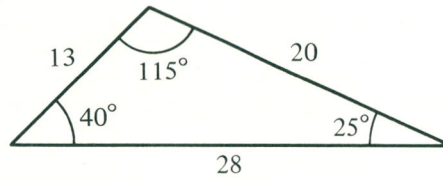

2. (a)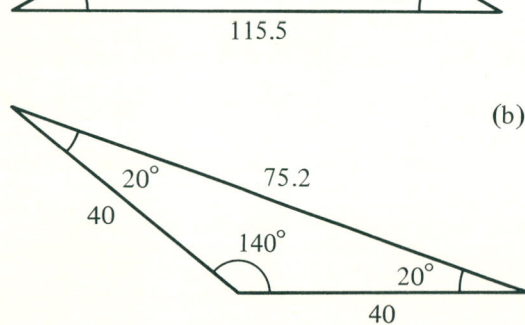

(c)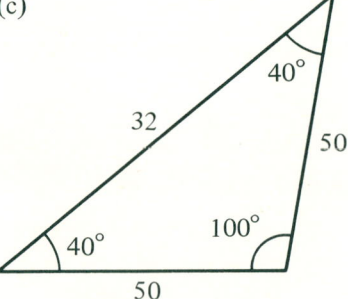

A-130

3.

(a)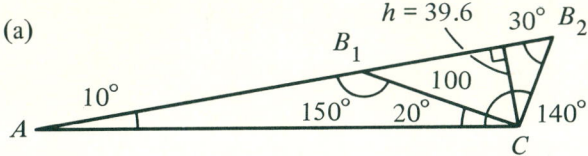

(b)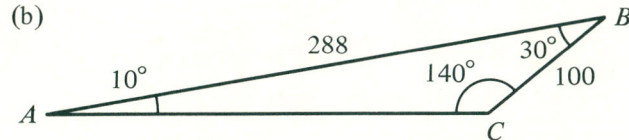

4.

(a)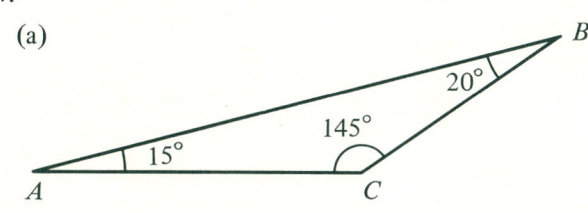

(b)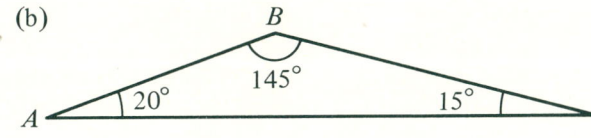

5.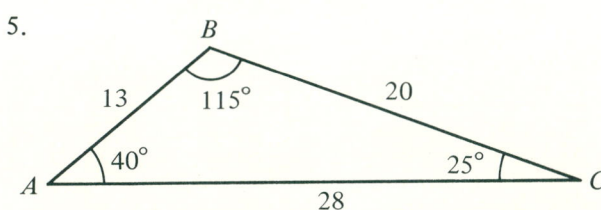

7.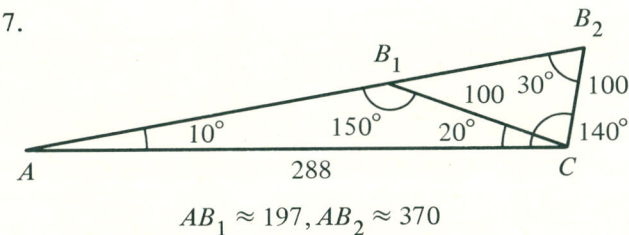

$AB_1 \approx 197, AB_2 \approx 370$

9. (a)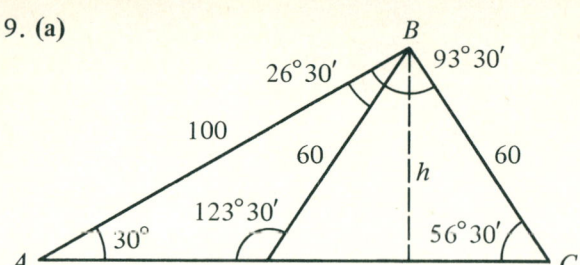

(b) No triangle, $h > a$
11. N83°54′E at a range of 12.5 kilometers
13. 2133 meters
15. Approximately 17.4 kilometers per hour

Exercise 12.5

1.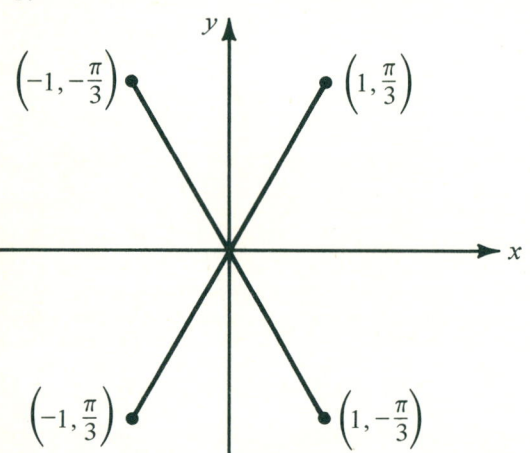

2.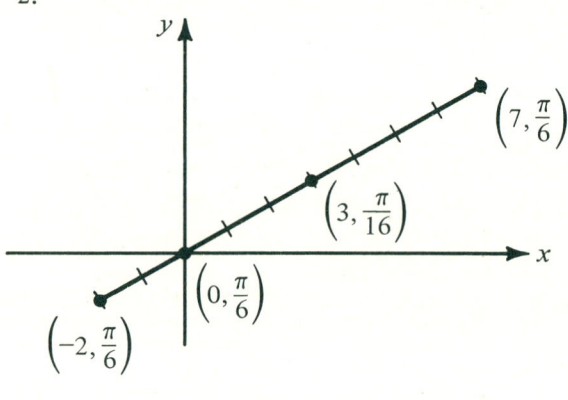

3.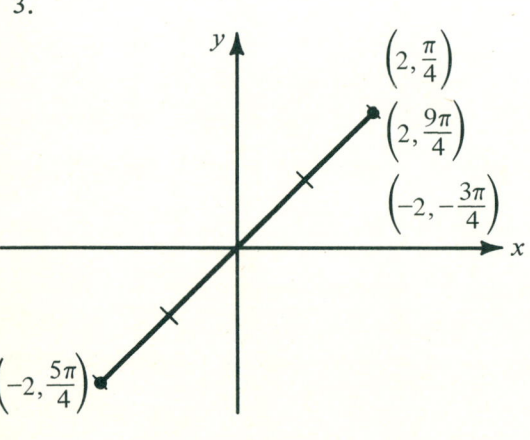

4.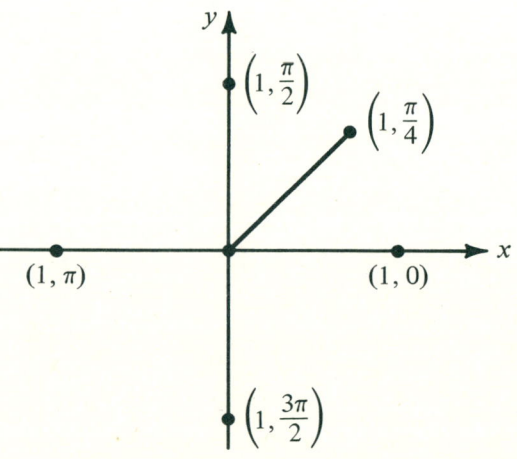

5.

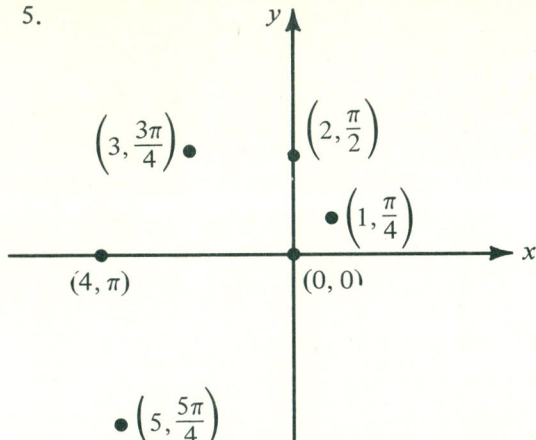

7.

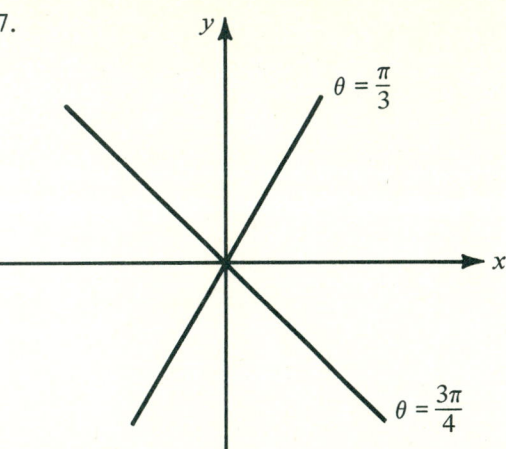

8.

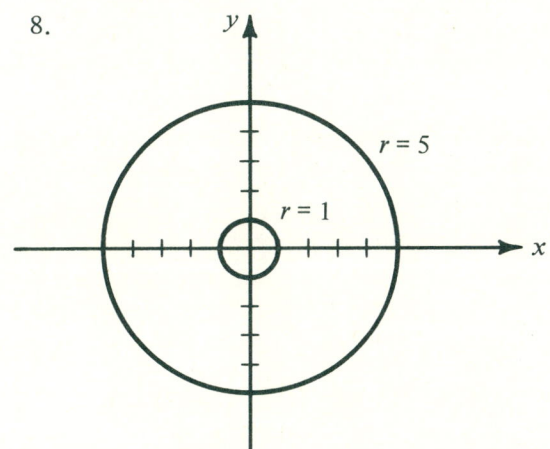

9.

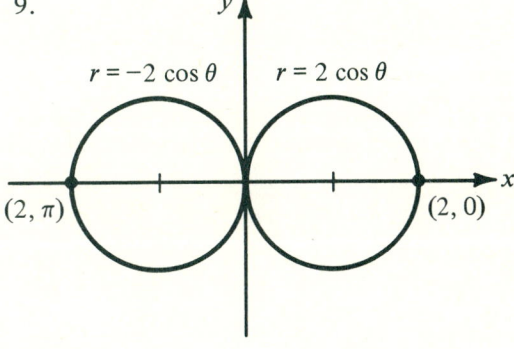

10.

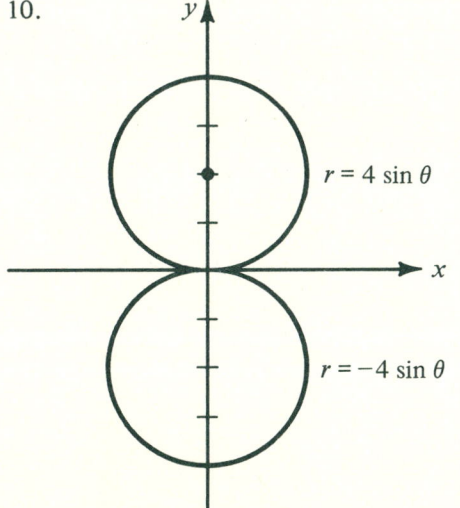

11. 7(a) $y = 1.7x$
    (b) $y = -x$
    9(a) $(x - 1)^2 + y^2 = 1$
    (b) $(x + 1)^2 + y^2 = 1$

    8(a) $x^2 + y^2 = 1$
    (b) $x^2 + y^2 = 25$
    10(a) $x^2 + (y - 2)^2 = 4$
    (b) $x^2 + (y + 2)^2 = 4$

13. (a) $r = 4$ (a circle)
    (b) $r = 4, 0 \leq \theta \leq \pi$ (semicircle)

15. (a) $r^2 = \sec 2\theta$
    (b) $r^2 = -\sec 2\theta$ (hyperbolas)

17. (a) $r = \sin \theta / \cos^2 \theta$
    (b) $r \sin \theta = (r \cos \theta - 1)^2$ (parabolas)
    Notice that the polar equations are less complicated for equations whose graphs are symmetric about a point, for example, those in Problems 12, 13, and 14.

18. (a) $r = 8 \cos \theta \Rightarrow r^2 = 8r \cos \theta \Rightarrow x^2 + y^2 = 8x \Rightarrow x^2 - 8x + y^2 = 0 \Rightarrow (x - 4)^2 + y^2 = 16$. This is a circle, centered at $(4, 0)$, with a radius of 4.

19. (a) $x^2 + (y - 5)^2 = 25$
    (b) $(x + 5)^2 + y^2 = 25$
    Circles of radius 5, centered at $(0, 5)$ and $(-5, 0)$.

21. (a)

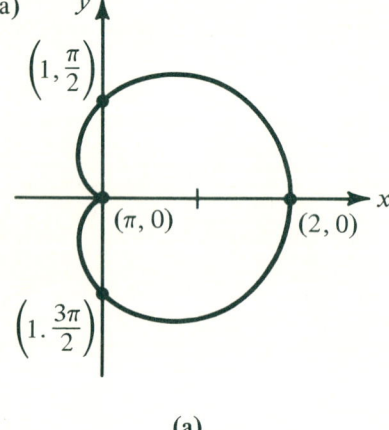

(a)

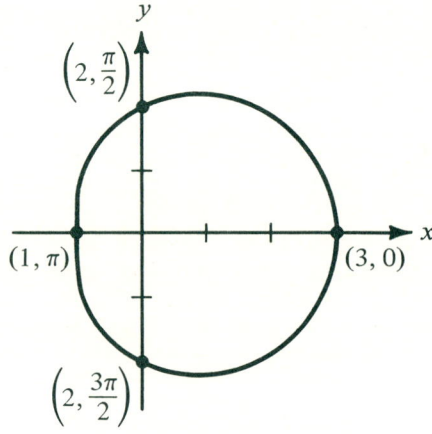

(b)

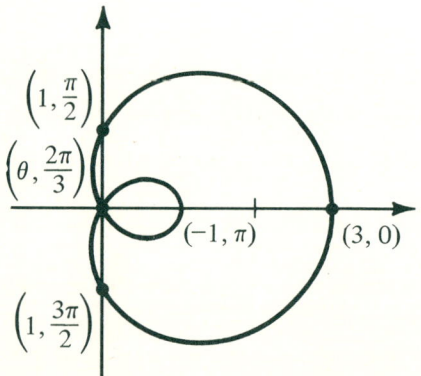

(c)

23. (a)

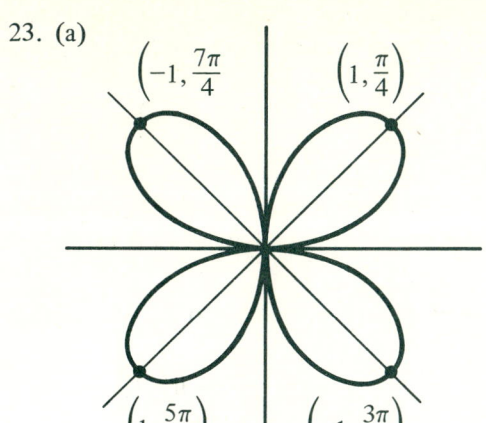

(a)

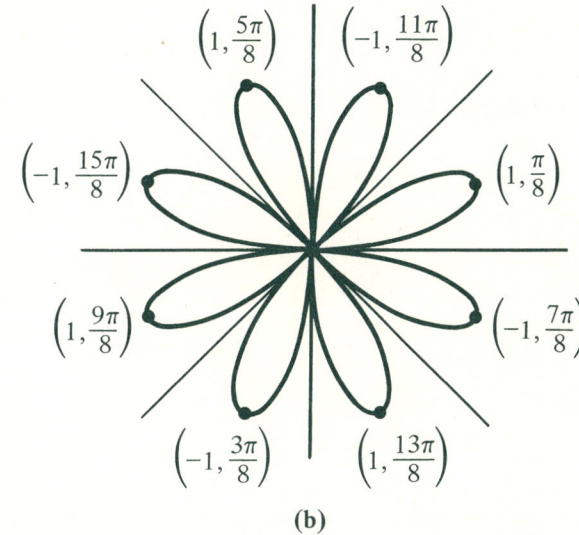

(b)

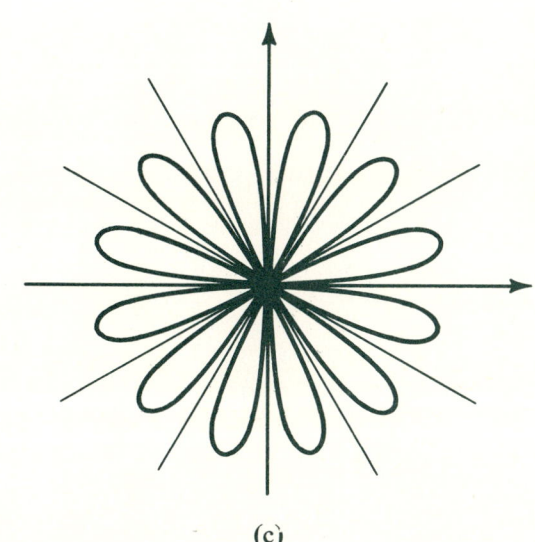

(c)

25. (a)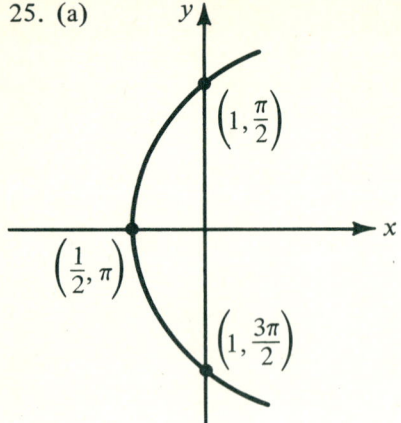

(b)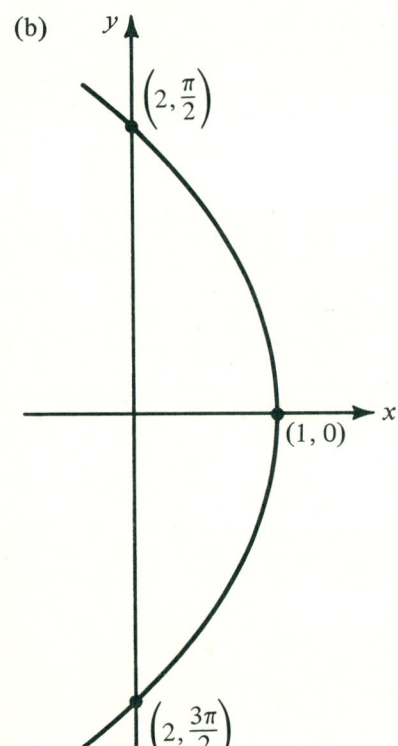

27.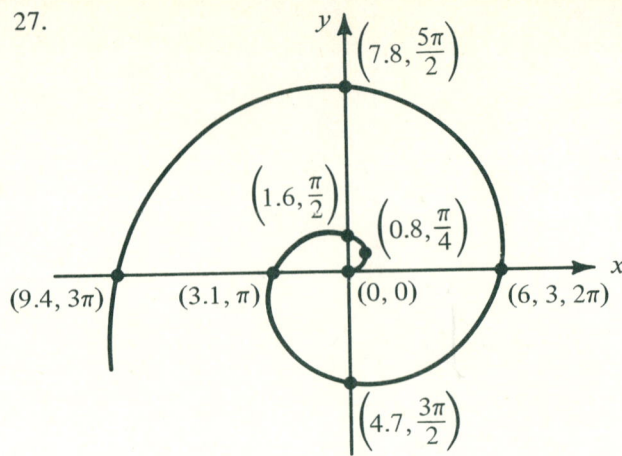

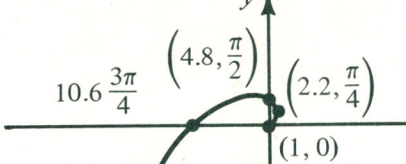

Exercise 12.6

1. (a) $\sin \pi/3 = \sqrt{3}/2$
   $\sin 2\pi/3 = \sqrt{3}/2$
   (b) $\sin \pi/4 = \sqrt{2}/2$
   $\sin 3\pi/4 = \sqrt{2}/2$
   (c) $\sin(-\pi/6) = -1/2$
   $\sin(7\pi/6) = -1/2$
2. (a) $\cos \pi/2 = 0$
   $\cos(-\pi/2) = 0$
   (b) $\cos(3\pi/4) = -\sqrt{2}/2$
   $\cos(5\pi/4) = -\sqrt{2}/2$
   (c) $\cos(-5\pi/6) = -\sqrt{3}/2$
   $\cos(5\pi/6) = -\sqrt{3}/2$

3. (a) $F(1, 0) = \cos 0 - 1 = 0$
       $F(-1, 0) = \cos 0 - 1 = 0$
   (b) $F(\sqrt{2}/2, \pi/3) = 0$
       $F(-\sqrt{2}/2, \pi/3) = 0$
   (c) $F(1/\sqrt[4]{2}, \pi/4) = 0$
       $F(-1/\sqrt[4]{2}, \pi/4) = 0$
4. x-axis
   $f(\theta) = f(-\theta)$
5. y-axis
   $f(\theta) = f(\pi - \theta)$
6. origin
   $F(r, \theta) = F(-r, \theta)$
7.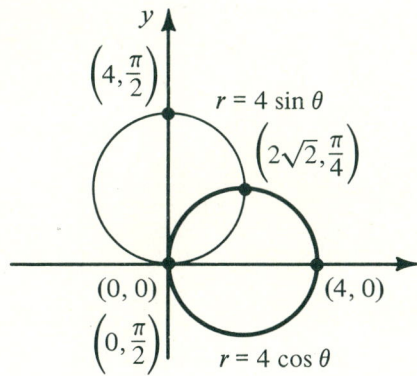
8.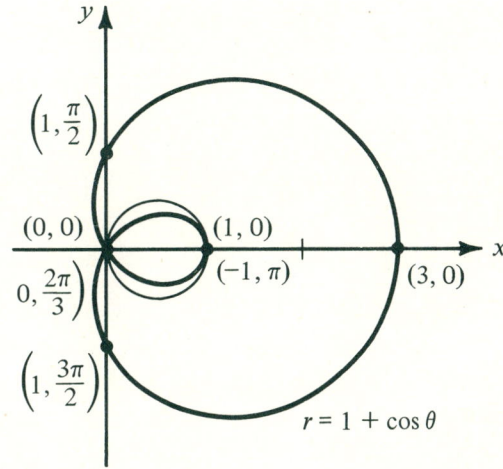
9.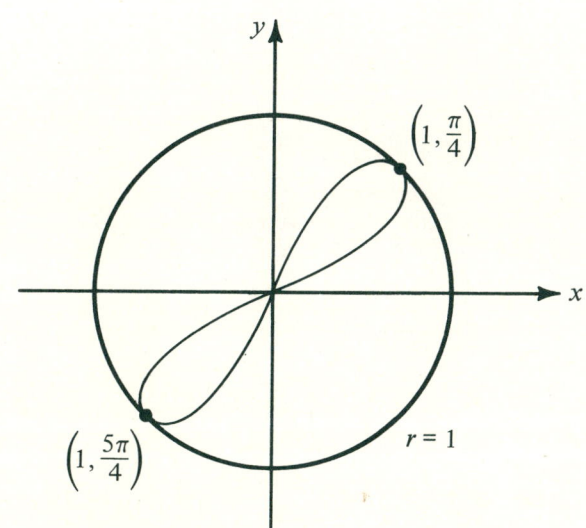

11. (a) Symmetric about the origin and y-axis. There are two leaves.

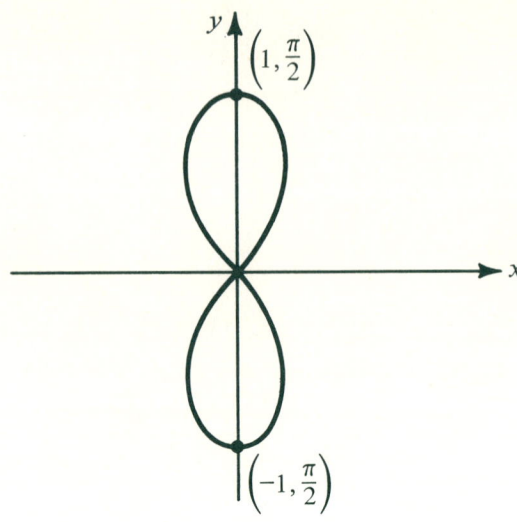

(b) Symmetric about the origin and (because of the period of $2\pi/4 = \pi/2$) the graph repeats itself every $\pi/2$.

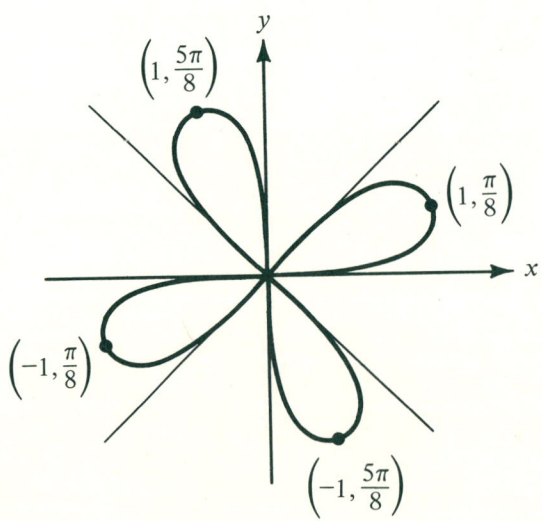

(c) Symmetric about the origin and (because of the period of $2\pi/8 = \pi/4$) the graph repeats itself every $\pi/4$.

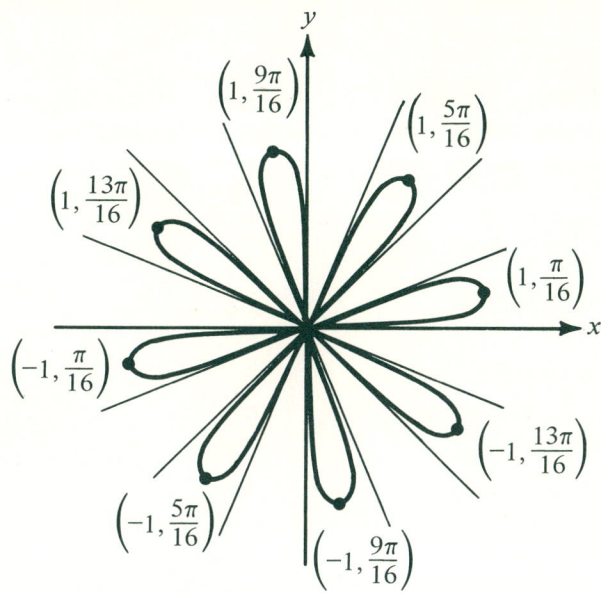

13. (a) Symmetric about x-axis

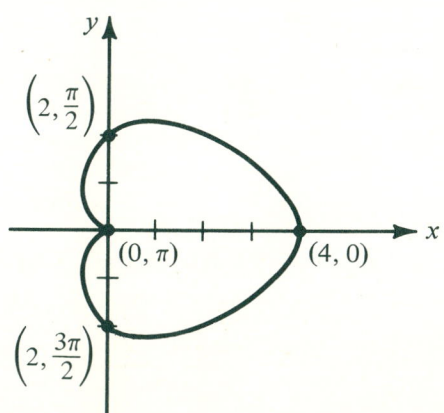

(b) Symmetric about y-axis

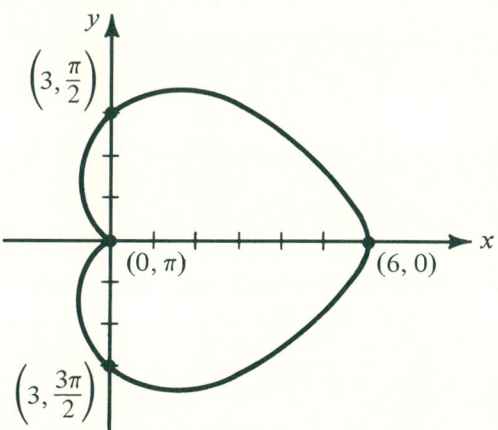

15. (a) *x*-axis symmetry   (b) *y*-axis symmetry

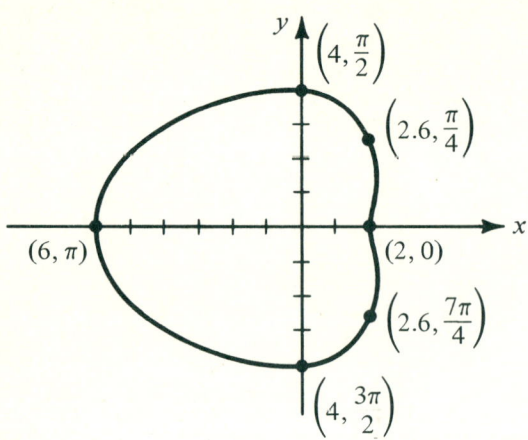
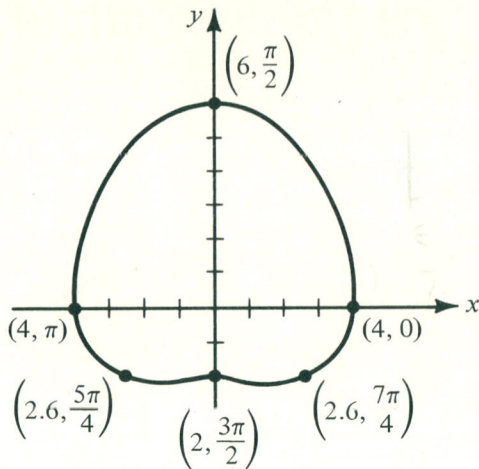

17. $r = \cos 2\theta$ and $r = -\cos \theta$ are both symmetric about the *x*-axis. They intersect at the origin, since $(0, \pi/4)$ and $(0, \pi/2)$ satisfy the first and second, respectively. Also,

$$\cos 2\theta = -\cos \theta \Rightarrow 2\cos^2\theta - 1 = -\cos \theta \Rightarrow 2\cos^2\theta + \cos \theta - 1 = 0 \Rightarrow$$

$$(2\cos\theta - 1)(\cos\theta + 1) = 0 \Rightarrow \cos\theta = \frac{1}{2} \text{ or } \cos\theta = -1 \Rightarrow$$

$$\theta = \frac{\pi}{3} \text{ or } \theta = \pi \Rightarrow r = -\frac{1}{2} \text{ or } r = 1$$

Because of the symmetry, if $(-\frac{1}{2}, \pi/3)$ is on both curves, then $(\frac{1}{2}, 2\pi/3)$ must also be on both curves.

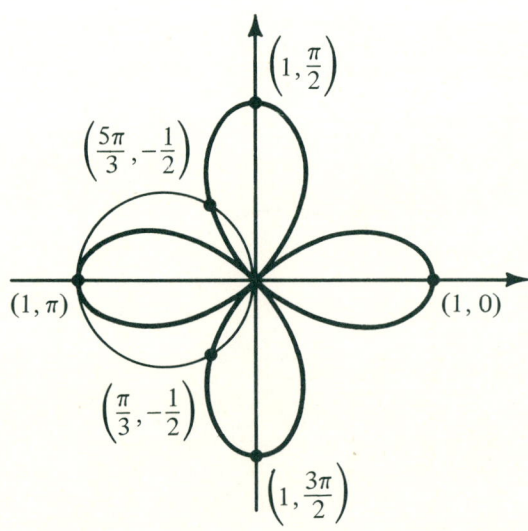

A–140

19. $(3, \pi)$

21. the origin $(2, 0)(2, \pi)$

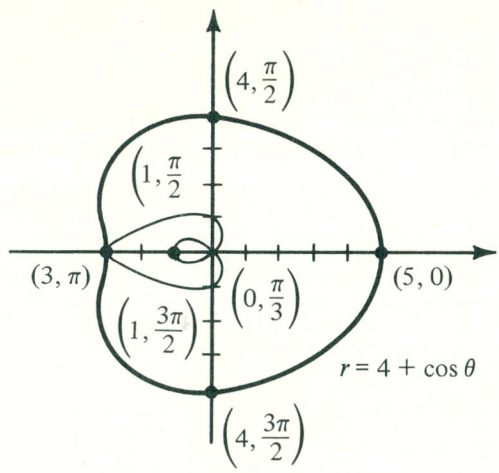
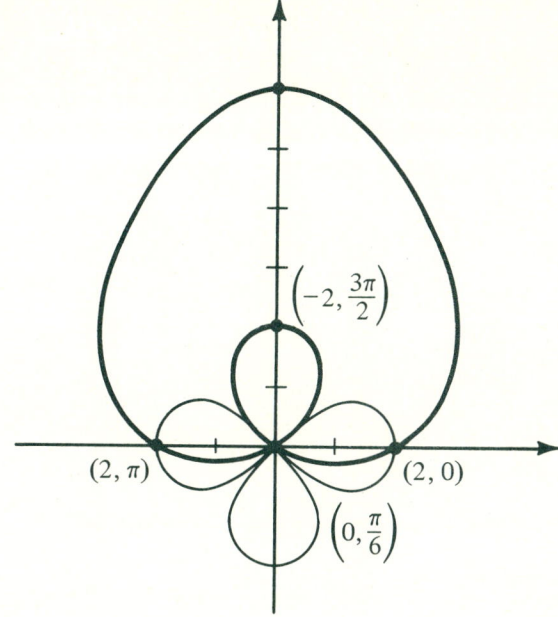

Exercise 12.7

1. (a) $z = \sqrt{2} \cos \pi/4 + (\sqrt{2} \sin \pi/4)i$
   (b) $z = \sqrt{2} \cos (-\pi/4) + [\sqrt{2} \sin (-\pi/4)]i$
2. (a) $z = \cos 0 + i \sin 0$
   (b) $z = \cos \pi/2 + i \sin \pi/2$
3. (a) $z = 2 \cos \pi/6 + (2 \sin \pi/6)i$
   (b) $z = 2 \cos (-\pi/3) + [2 \sin (-\pi/3)]i$
4. $z^2 = (4 + 4i)(4 + 4i) = (16 - 16) + (16i + 16i) = 32i$
   $z^2 = [4\sqrt{2} \cos \pi/4 + (4\sqrt{2} \sin \pi/4)i]^2 = 32 \cos \pi/2 + (32 \sin \pi/2)i$
   $z^3 = (4 + 4i)32i = -128 + 128i$
   $z^3 = [4\sqrt{2} (\cos \pi/4 + i \sin \pi/4)]^3 = 128\sqrt{2} (\cos 3\pi/4 + i \sin 3\pi/4)$
5. (a) 16

$$z_1 \cdot z_2 = \left[4\left(\cos \frac{\pi}{6} + i \sin \frac{\pi}{6}\right)\right]\left\{4\left[\cos\left(-\frac{\pi}{6}\right) + i \sin\left(-\frac{\pi}{6}\right)\right]\right\}$$

$$= 16(\cos 0 + i \sin 0)$$

(b) $z_1 \cdot z_2 = \{2[\cos(-\pi/6) + i \sin(-\pi/6)]\}[2(\cos \pi/3 + i \sin \pi/3)]$
   $z = 4(\cos \pi/6 + i \sin \pi/6) = (\sqrt{3} - i)(1 + \sqrt{3}i) = 2\sqrt{3} + 2i$
(c) $z_1 \cdot z_2 = (5\{\cos[\tan^{-1}(-4/3)]\} + i \sin[\tan^{-1}(-4/3)])[2\sqrt{2}(\cos \pi/4 + i \sin \pi/4)]$
   $\approx 10\sqrt{2}[\cos(-8°)] + i \sin(-8°) \approx 14 - 2i$
(d) $z_1 \cdot z_2 = 56 - 8i$
   $z_1 \cdot z_2 = (10\{\cos[\tan^{-1}(-8/6)]\} + i \sin[\tan^{-1}(-8/6)]) [4\sqrt{2}(\cos -\pi/4 + i \sin -\pi/4)]$
   $\approx 40\sqrt{2}[\cos(-8°)] + i \sin(-8°)$

Did you notice that those in (d) were multiples of those in (c)? Does this affect the answer in a predictable way?

A-141

7. $\pm 1, \pm i$

8. (e) $\quad z^3 = 1 + \sqrt{3}i = 2(\cos \pi/3 + i \sin \pi/3) \Rightarrow$

$$z = \sqrt[3]{2}\left[\cos\left(\frac{\pi/3 + 2\pi k}{3}\right)\right] + i \sin\left(\frac{\pi/3 + 2\pi k}{3}\right)$$

$k = 0 \Rightarrow z = \sqrt[3]{2}(\cos \pi/9 + i \sin \pi/9) = -\sqrt[3]{2}[0.94 + i(0.41)] = 1.093 + 0.55i$
$k = 1 \Rightarrow z = \sqrt[3]{2}(\cos 7\pi/9 + i \sin 7\pi/9) \approx -0.965 + 0.81i$
$k = 2 \Rightarrow z = \sqrt[3]{2}(\cos 13\pi/9 + i \sin 13\pi/9) \approx -0.219 - 1.24i$

(g) $z^{-3} = -27 - 27i \Rightarrow$
$\quad z = [27\sqrt{2}(\cos 5\pi/4 + i \sin 5\pi/4)]^3 \Rightarrow$
$\quad z = 39366\sqrt{2}(\cos 15\pi/4 + i \sin 15\pi/4)$
$\quad = 39366 - 39366i$

9. (a) $(1 + \sqrt{3}i)i = -\sqrt{3} + i$
This product is a rotation through $\pi/2$.
(b) This rotates z through $3\pi/2$.

11. (a) $z = \sqrt{2}e^{(\pi/4)i}$
(b) $z = re^{i\theta} \Rightarrow \ln z = \ln re^{i\theta}$
$\qquad = \ln r + \ln e^{i\theta}$
$\qquad = \ln r + i\theta \ln e$
$\qquad = \ln r + i\theta \cdot 1$
$\qquad = \ln r + i\theta$
(c) $\ln(1 + i) = \ln(\sqrt{2}e^{(\pi/4)i}) = \ln\sqrt{2} + (\pi/4)i \approx 0.35 + 0.78i$

13. $z = i^i \Rightarrow \ln z = i \ln i$, where $i = 1 \cdot e^{(\pi/2)i} \Rightarrow$
$\quad \ln z = i[\ln 1 + (\pi/2)i] = i[0 + (\pi/2)i] = -\pi/2$
Therefore, $z = e^{-\pi/2} \approx 0.208 \in$ reals.

Exercise 12.8

1. $\sin 2 \approx 0.909$ $\qquad$ $\sin 0.2 \approx 0.1987$ $\qquad$ $\sin 0.02 \approx 0.019998$
   $\tan 2 \approx -2.185$ $\qquad$ $\tan 0.2 \approx 0.2027$ $\qquad$ $\tan 0.02 \approx 0.020003$

2. $\sin 0.003 \approx 0.0029$ $\qquad$ $\sin 0.17 \approx 0.174$
   $\tan 0.003 \approx 0.0029$ $\qquad$ $\tan 0.17 \approx 0.176$

3. $\sin 0.105 \approx 0.1045$ $\qquad$ $\sin 0.1396 \approx 0.1392$ $\qquad$ $\sin 0.1745 \approx 0.1736$
   $\tan 0.105 \approx 0.1051$ $\qquad$ $\tan 0.1396 \approx 0.1405$ $\qquad$ $\tan 0.1745 \approx 0.1763$

5. 3,459 kilometers

7. approximately 0.77"

9. (a) 1 $\qquad$ (b) 2

11. (a) $\displaystyle\lim_{x \to 0} \frac{x}{\sin 3x} = \frac{1}{3} \lim_{x \to 0} \frac{3x}{\sin 3x} = \frac{1}{3} \cdot 1 = \frac{1}{3}$

(b) $\displaystyle\lim_{x \to 0} \frac{3x}{\tan 2x} = \frac{3}{2} \lim_{x \to 0} \frac{2x}{\tan 2x} = \frac{3}{2} \cdot 1 = \frac{3}{2}$

13. 2/3

CHAPTER 13

Exercise 13.2

1.

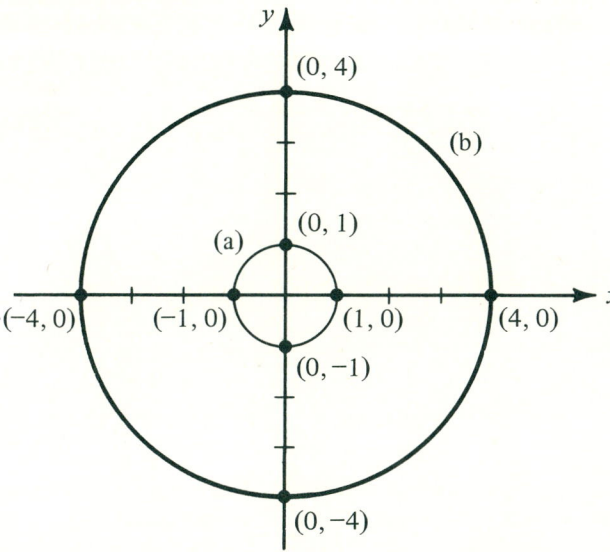

2.

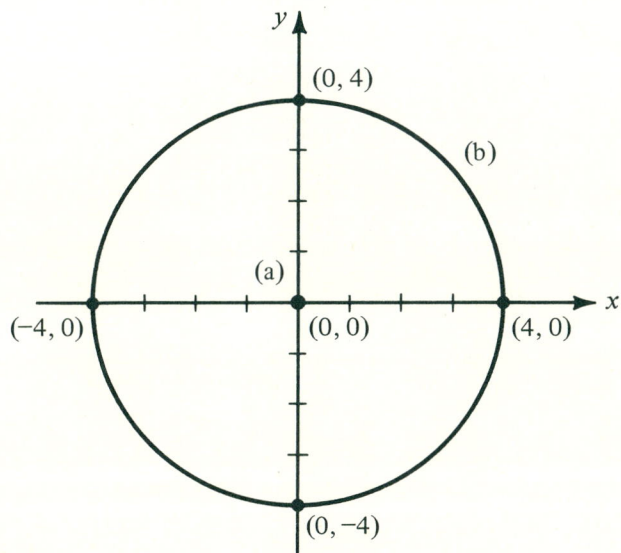

3.
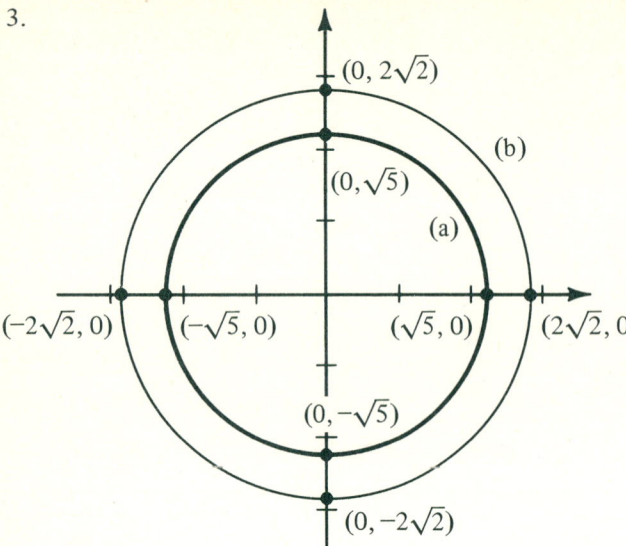

4.
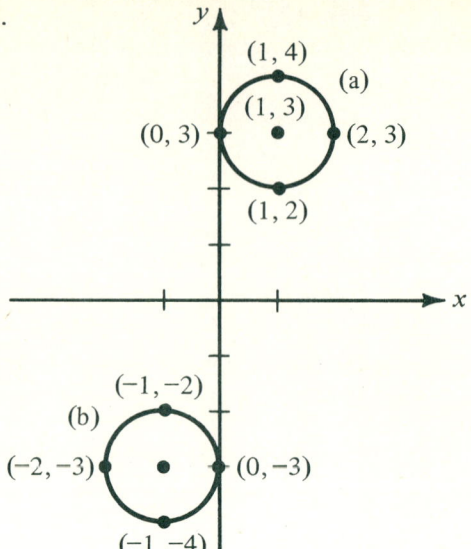

5.
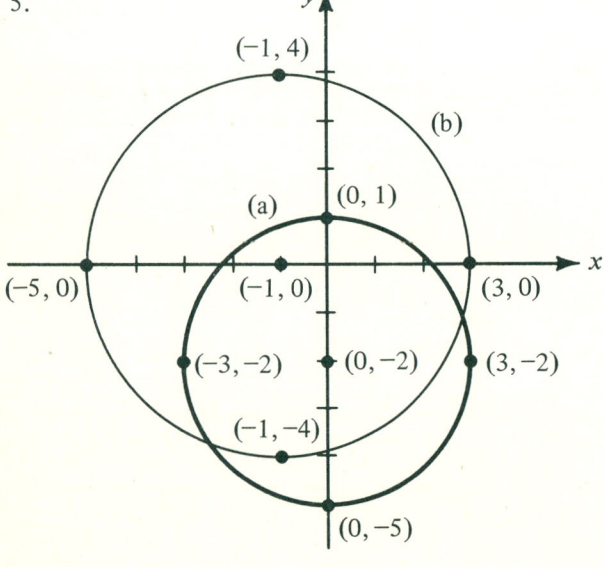

6.
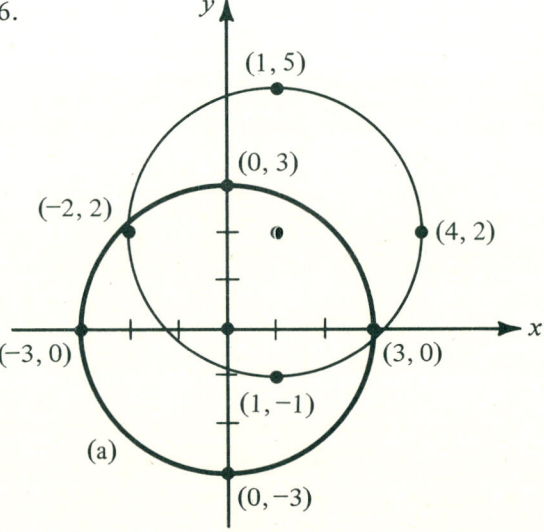

A-144

7.

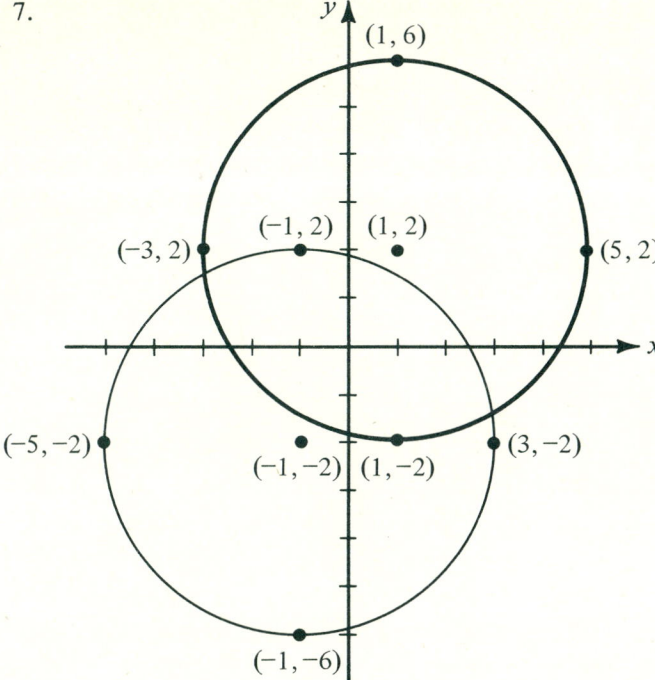

9.

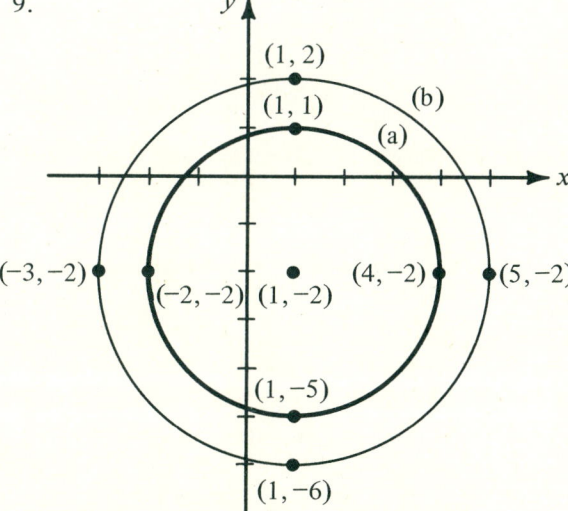

A-145

10. (a) $2x^2 + 2y^2 - 4x + 8y + 2 = 0 \Rightarrow 2(x^2 - 2x) + 2(y^2 + 4y) = -2 \Rightarrow$
$(x^2 - 2x + 1) + (y^2 + 4y + 4) = -2/2 + 1 + 4 \Rightarrow (x - 1)^2 + (y + 2)^2 = 4$
The graph will be a circle of radius 2 centered at $(1, -2)$.

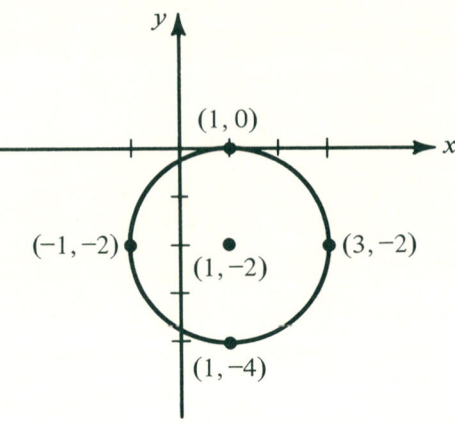

11.

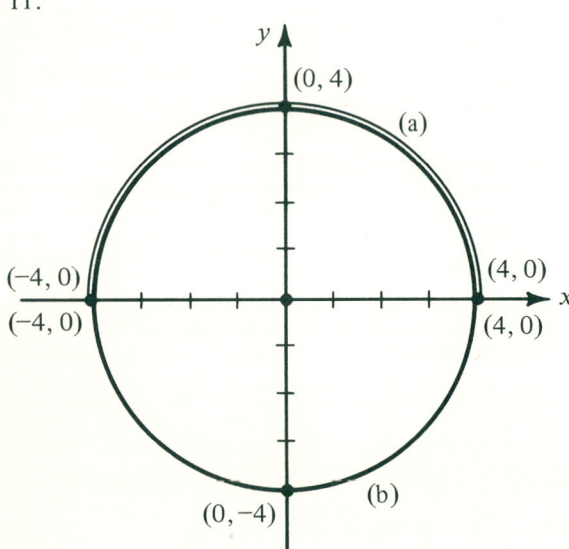

13. By examining the graphs in Problem 11, we see that both domains are $-4 \leq x \leq 4$. The range for 11(a) is $-4 \leq y \leq 4$, since that is the equation for the entire circle. 11(b), however, is only the top semicircle. Its range is therefore $0 \leq y \leq 4$.

Exercise 13.3

1.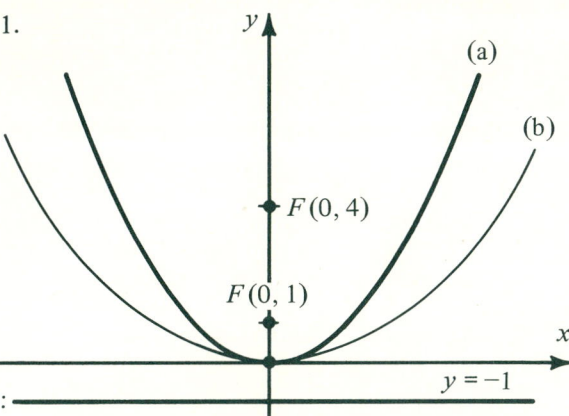

2.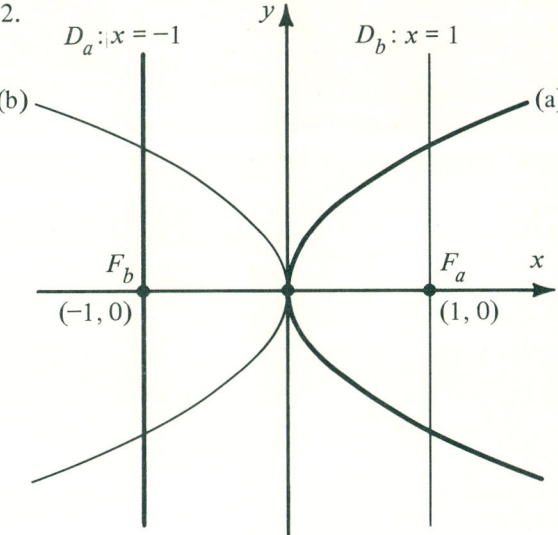

3.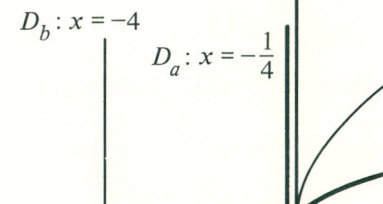

4.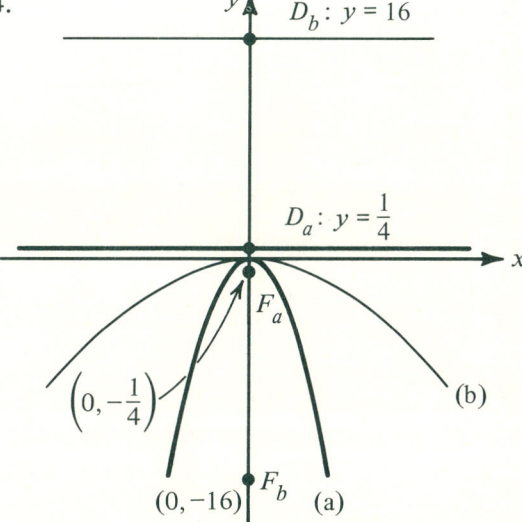

A–147

5.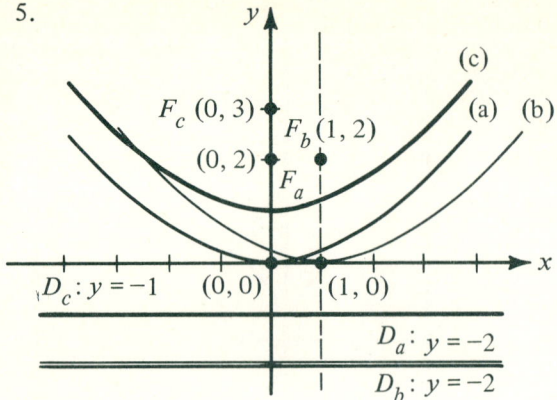

6.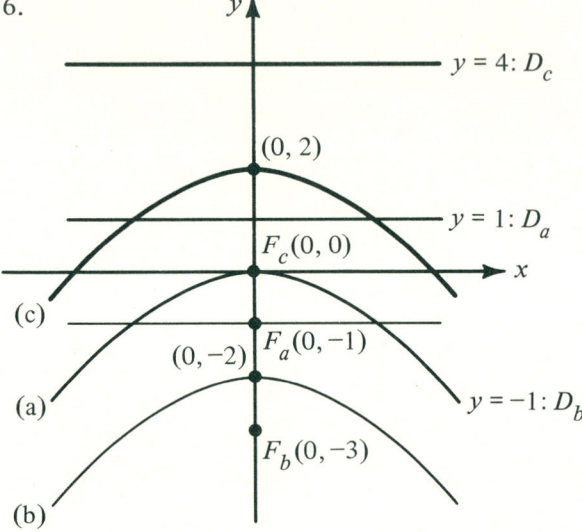

7.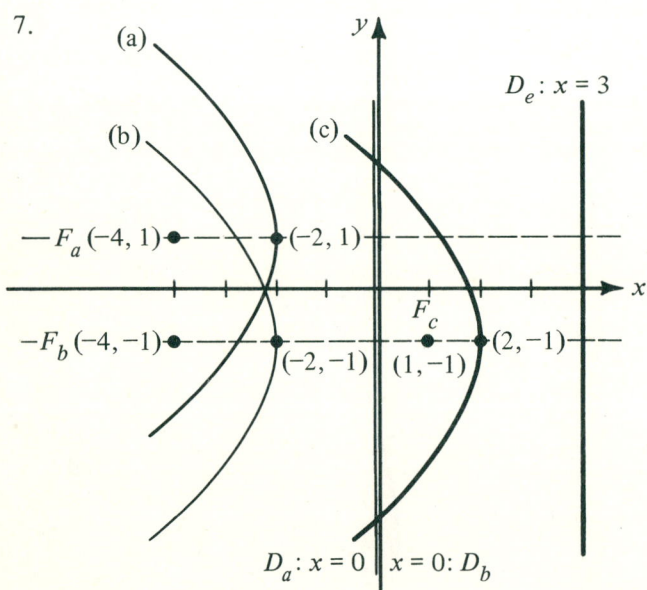

A-148

9.

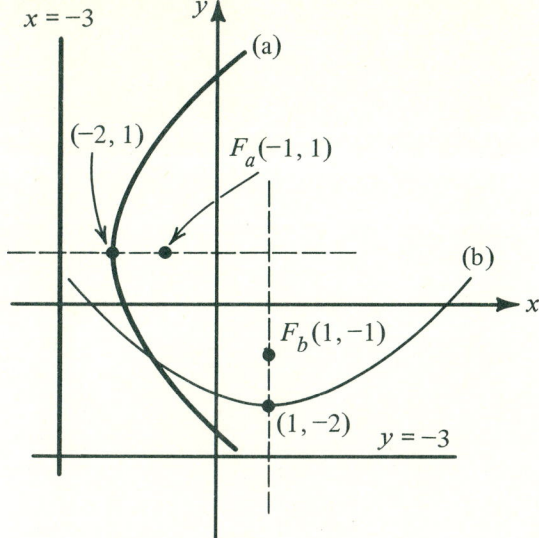

11.

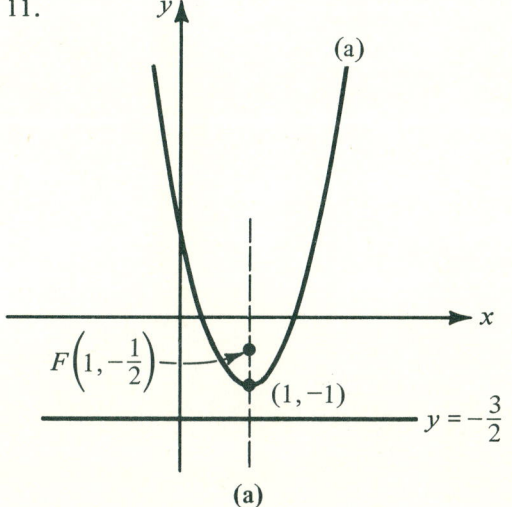

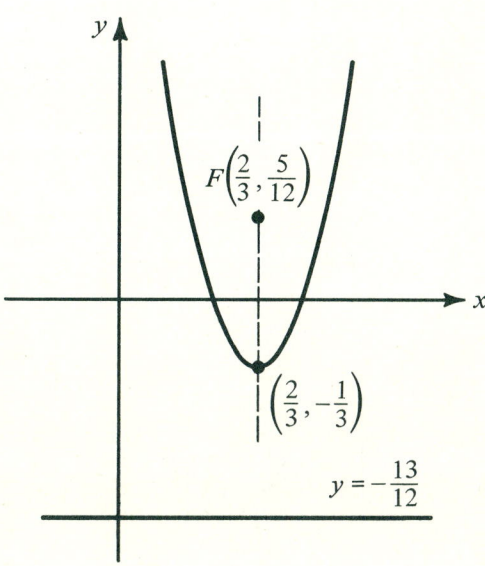

12. (a) If the directrix is $x = 4$ and the focus is $(-4, 0)$, then the equation is of the form $y^2 = -4px$, since the vertex must be at $(0, 0)$, half way between $x = 4$ and $x = -4$. Since $(-4, 0)$ is the focus, $p = 4$, and the equation is $y^2 = -16x$.

13. (a) $(x - 1)^2 = 14(y - 1/2)$ (b) $(y - 1)^2 = 10\left(x - \dfrac{1}{2}\right)$

15. $x^2 = 4py \Rightarrow$
    $a^2 = 4p(1)$ and $(100 + a)^2 = 4p(9)$

    $$\frac{a^2}{4} = \frac{(100 + a)^2}{36}$$

    $9a^2 = 10{,}000 + 200a + a^2$
    $8a^2 - 200a - 10{,}000 = 0$
    $a^2 - 25a - 1250 = 0 \Rightarrow a = 50$

Exercise 13.4

1.

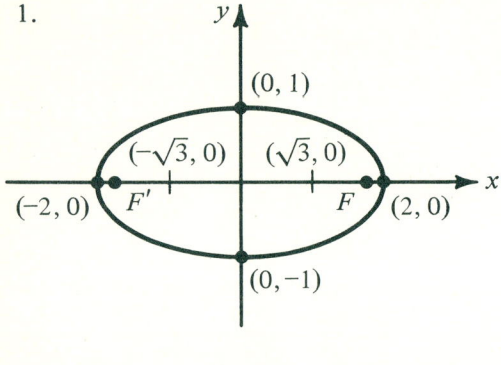

(a)

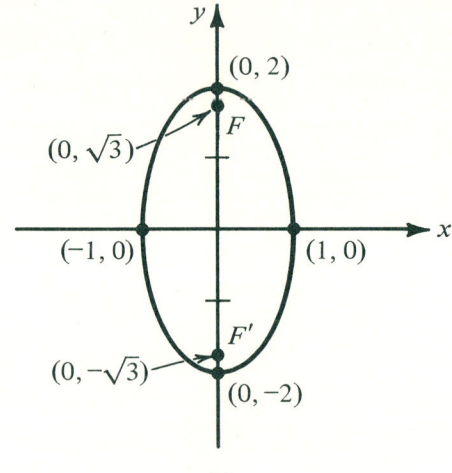

(b)

2.

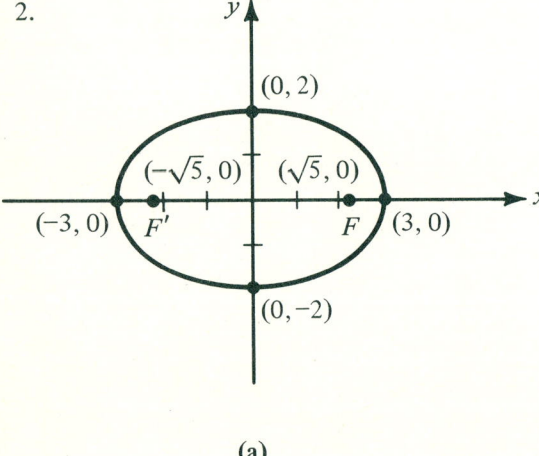

(a)

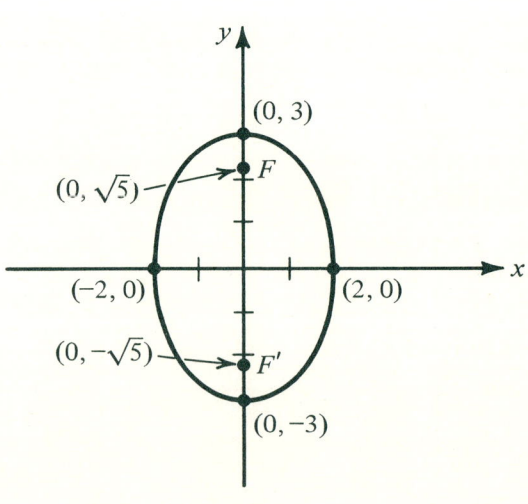

(b)

A–150

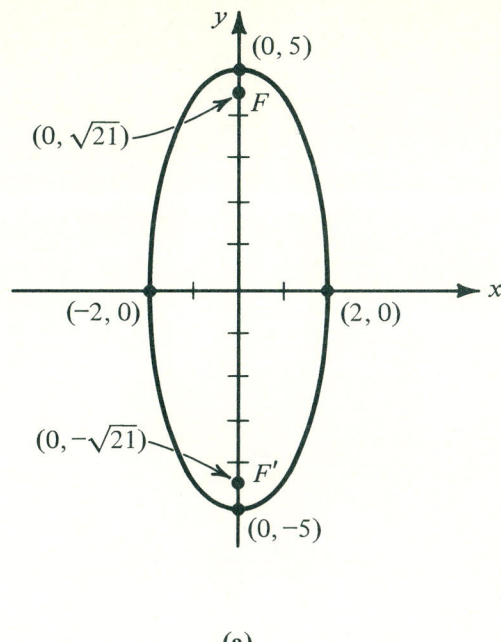

(a)

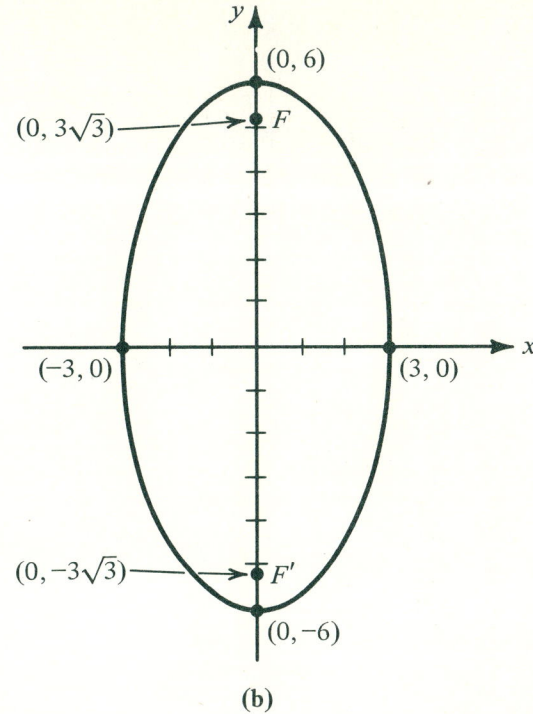

(b)

4.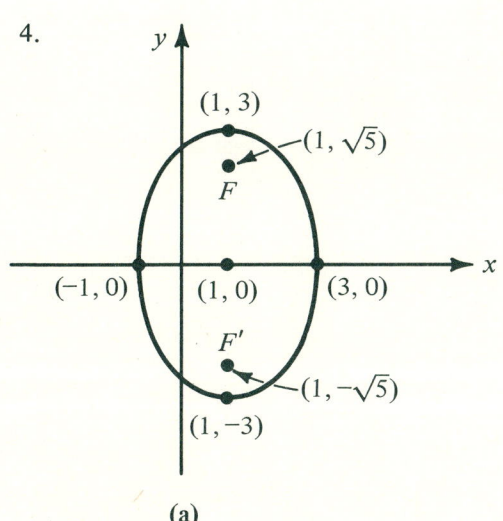

(a)

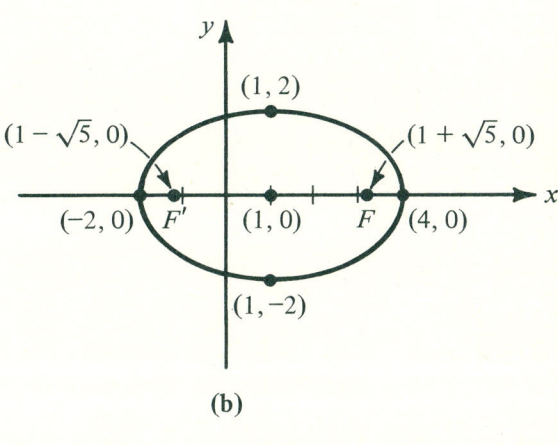

(b)

A-151

5.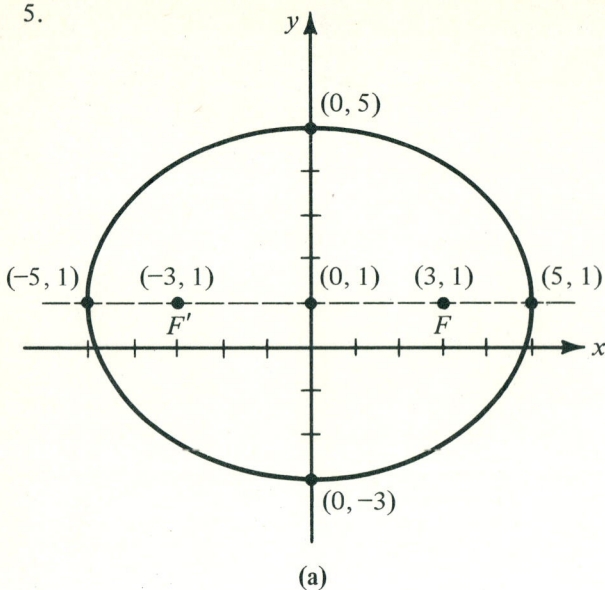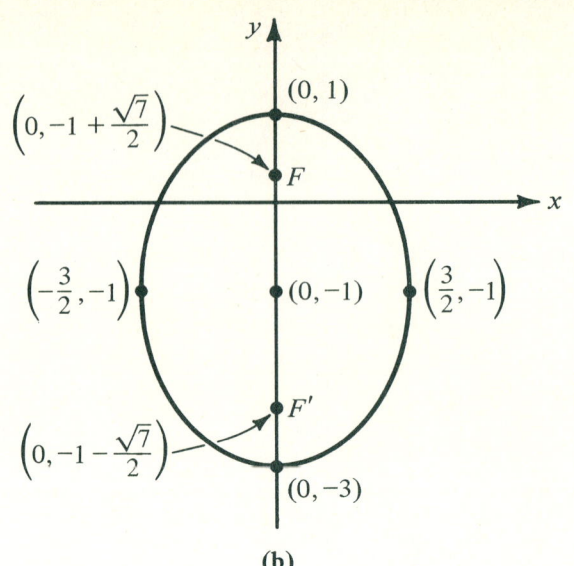

(a)   (b)

6.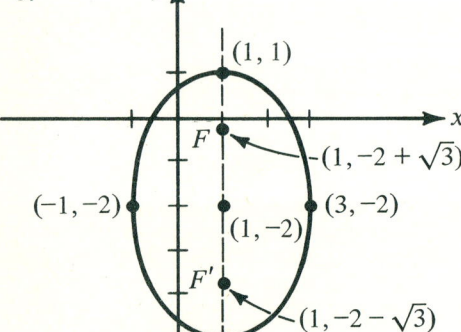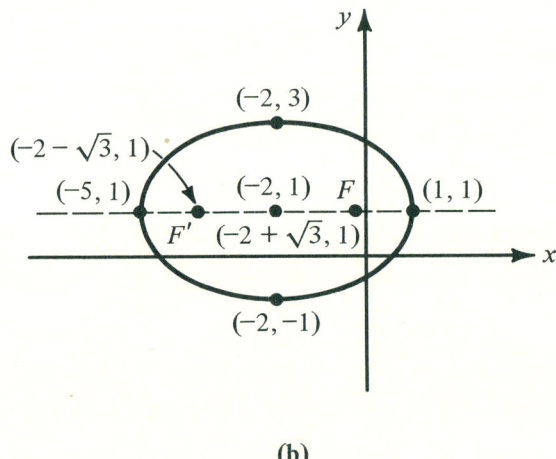

(a)   (b)

7. (a) $x^2 + 9y^2 = 9$

   (b) $\dfrac{(x-2)^2}{25} + \dfrac{(y+3)^2}{4} = 1$

8. (a) $9x^2 + 4y^2 + 18x - 27 = 0 \Rightarrow 9(x^2 + 2x + 1) + 4y^2 = 27 + 9$

$9(x + 1)^2 + 4y^2 = 36,$ $\quad \dfrac{(x + 1)^2}{4} + \dfrac{y^2}{9} = 1$

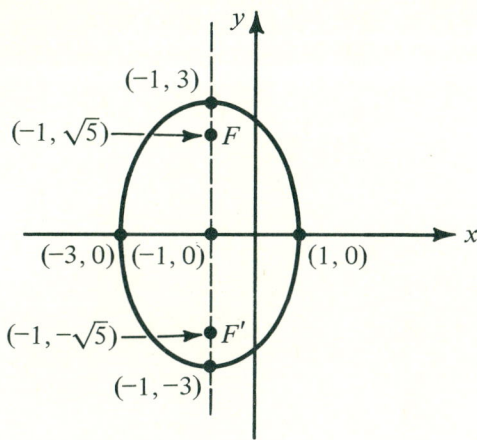

9. (a)

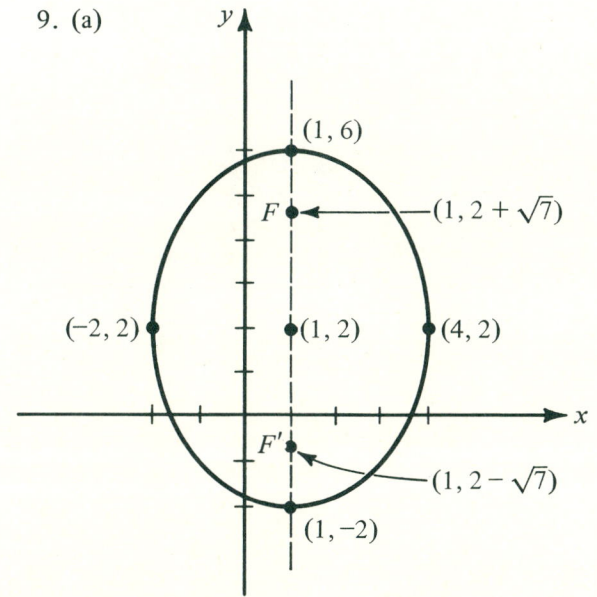

(b)

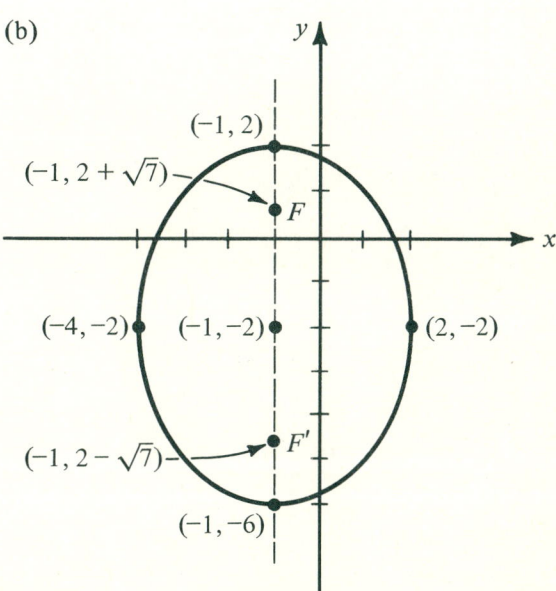

11.

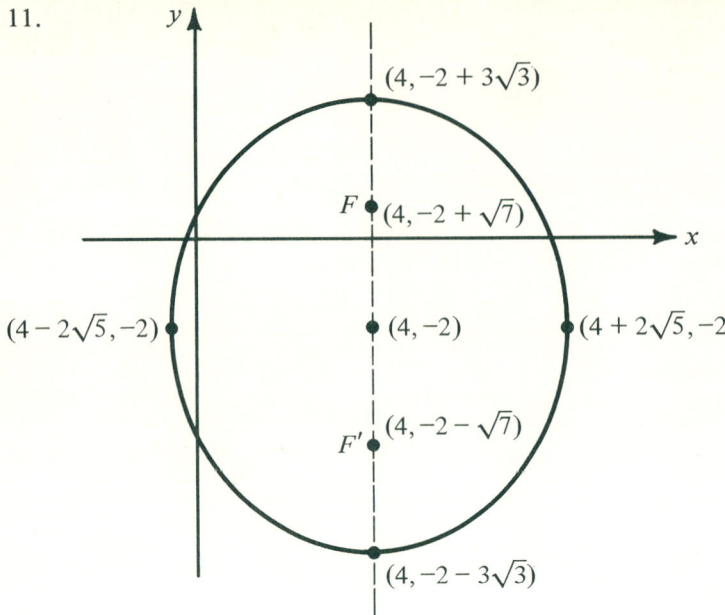

13. The foci are located about 6.3 meters from the center along the major axis in both directions. The speaker should be at one focus and your chair at the other.

15. $\dfrac{x^2}{2.238 \times 10^{16}} + \dfrac{y^2}{2.237 \times 10^{16}} = 1$ (Notice that it is almost circular.)

Exercise 13.5

1.

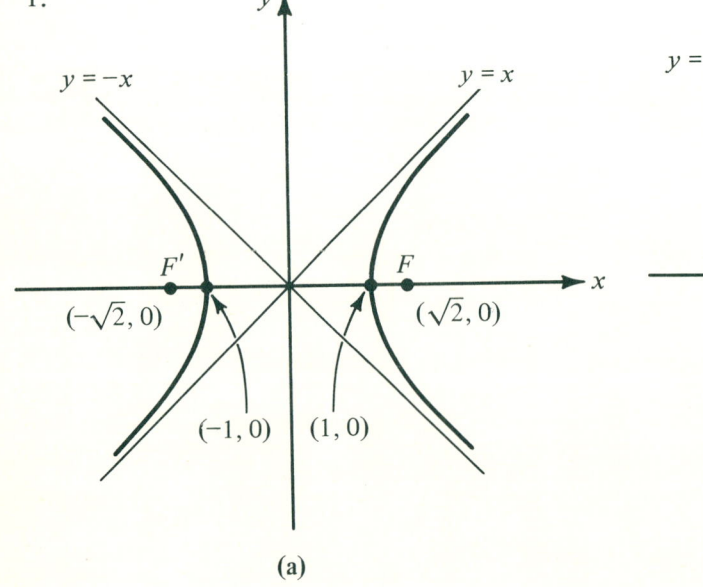

(a)

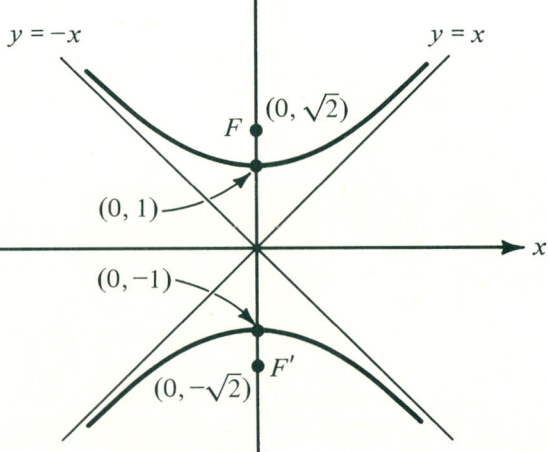

(b)

2.

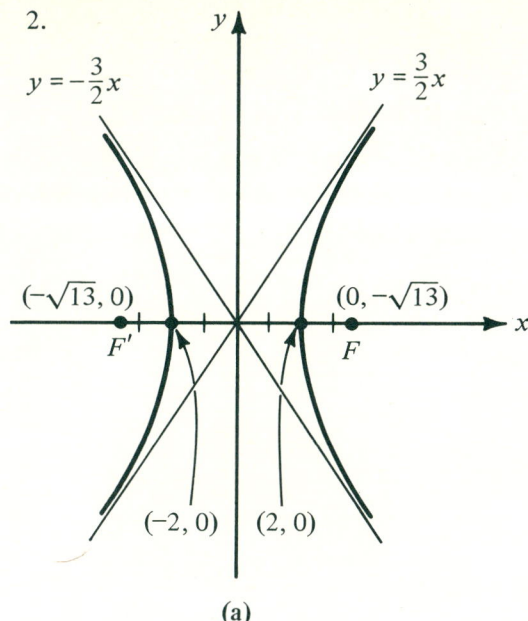

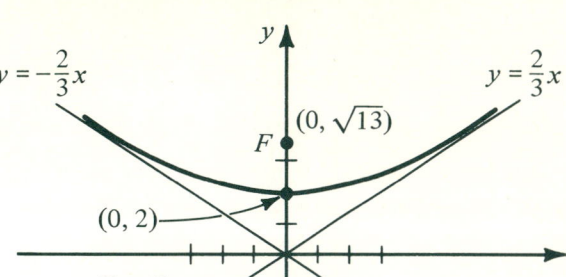

(a)

(b)

3.

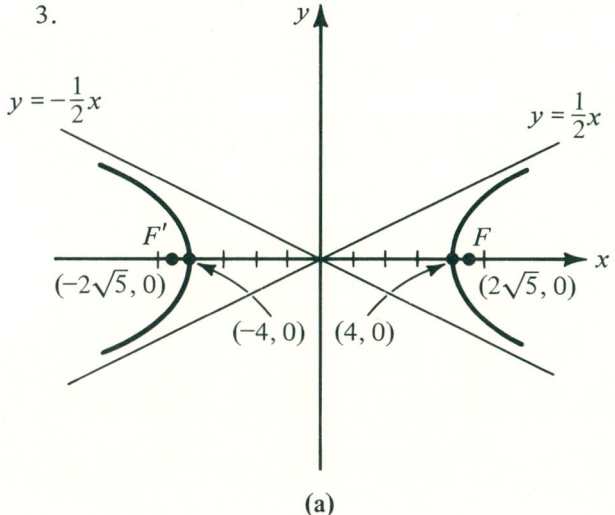

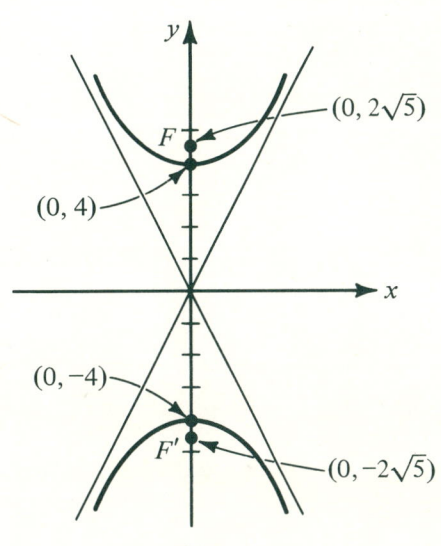

(a)

(b)

4. (a) $m = \dfrac{\Delta y}{\Delta x} = \dfrac{1-0}{2-1} = \dfrac{1}{1}$

$y - y_0 = m(x - x_0)$
$y - 0 = 1(x - 1)$
$y = x - 1$

(b) $y = -x + 1$

5. (a) $y = \dfrac{1}{2}x$

(b) $y = -\dfrac{1}{2}x + 2$

7.

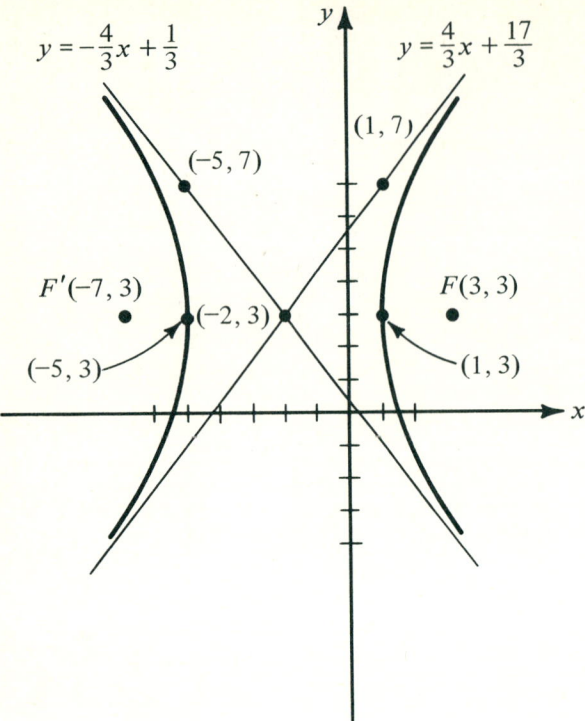

9.

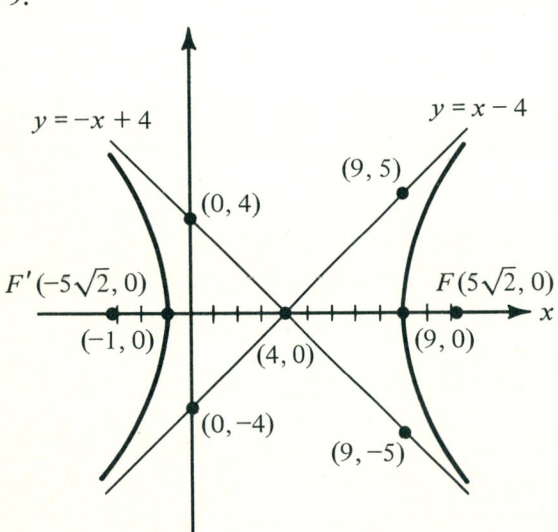
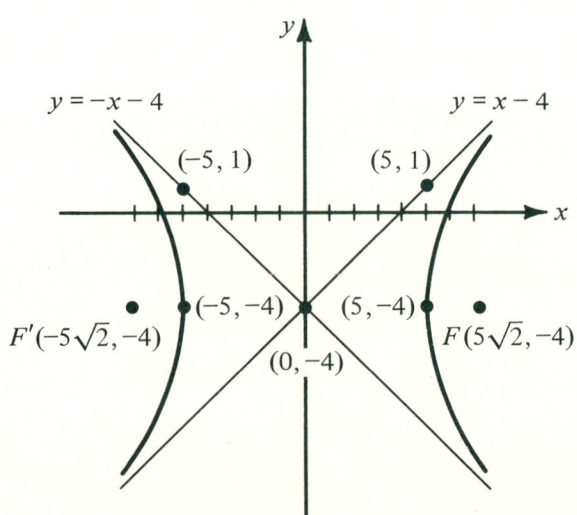

A-156

11.

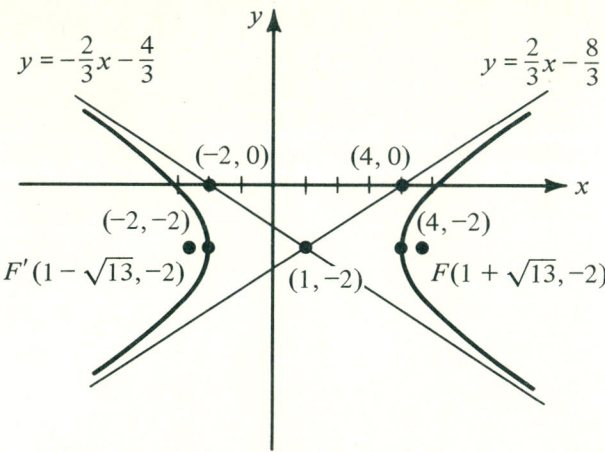

13. $\dfrac{x^2}{9} - \dfrac{y^2}{16} = 1$

15. $\dfrac{(x-1)^2}{9} - \dfrac{(y+2)^2}{16} = 1$

17. (a) Its graph is a hyperbola because 25 and −144 are of opposite signs.

$$25(x^2 - 4x + 4) - 144(y^2 - 2y + 1) = 3644 + 100 - 144$$
$$25(x-2)^2 - 144(y-1)^2 = 3600$$
$$\dfrac{(x-2)^2}{144} - \dfrac{(y-1)^2}{25} = 1$$

(b)

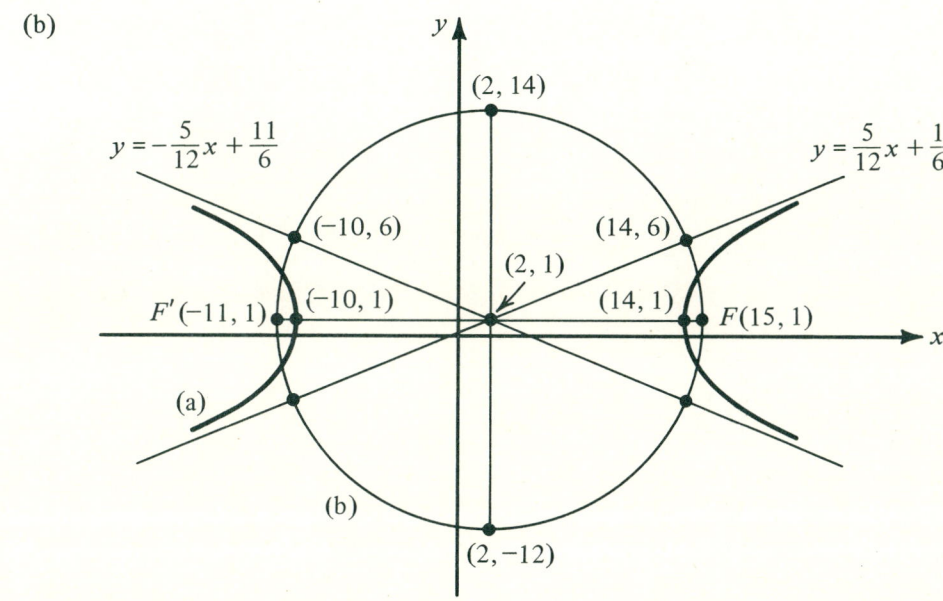

19.

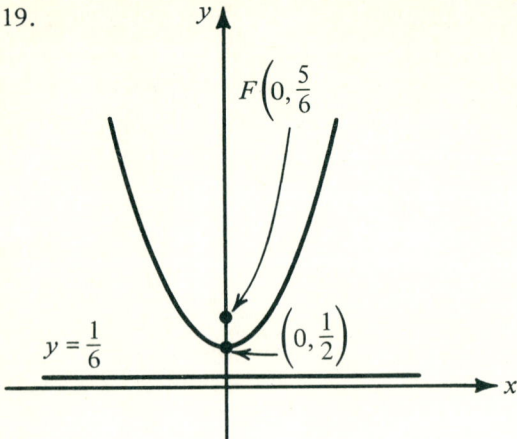

21.

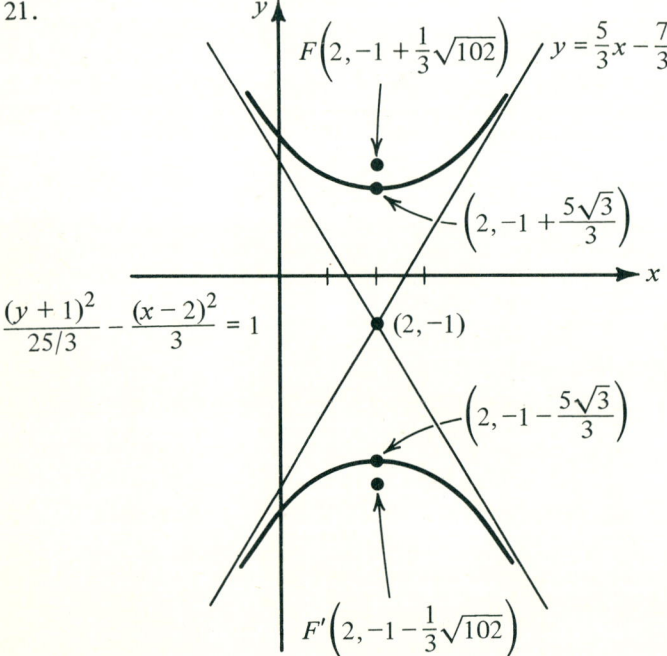

23.
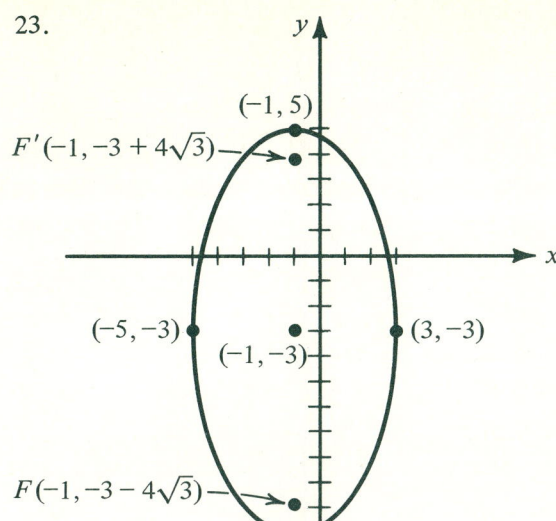

25.
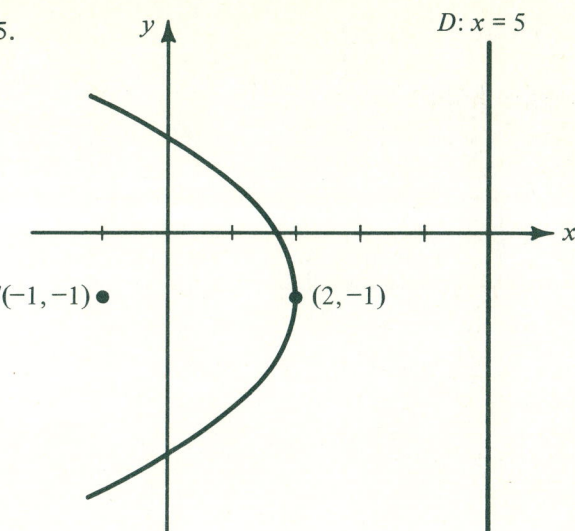

27.
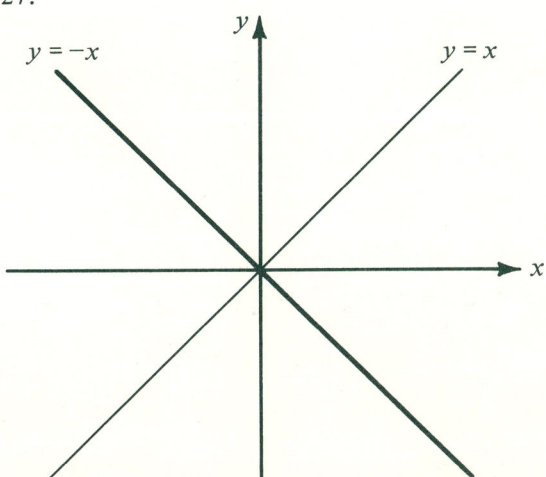

Exercise 13.6

1. parabola
3. hyperbola
5. $\pi/8$
7. $-\pi/12$
9. 0

2. hyperbola
4. ellipse
6. $\pi/6$
8. $\pi/4$

10.
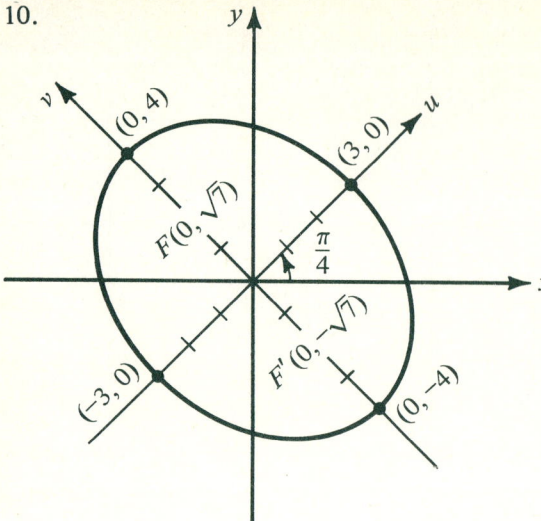

11.
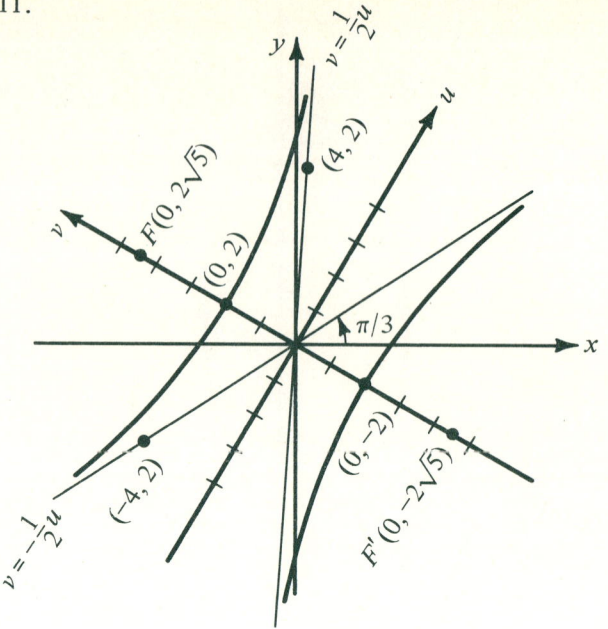

12.
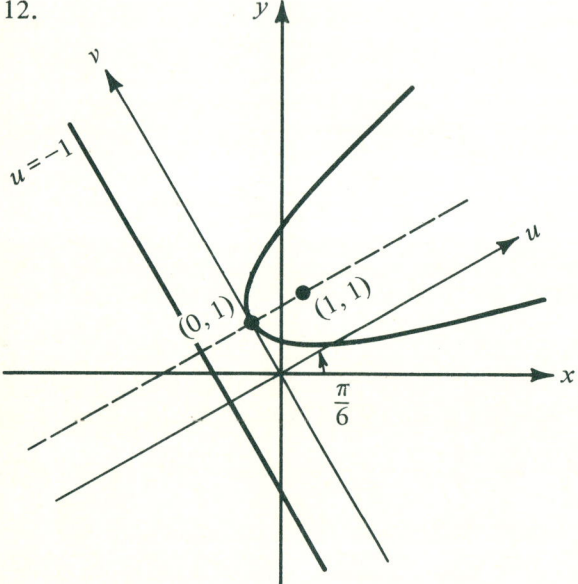

13.
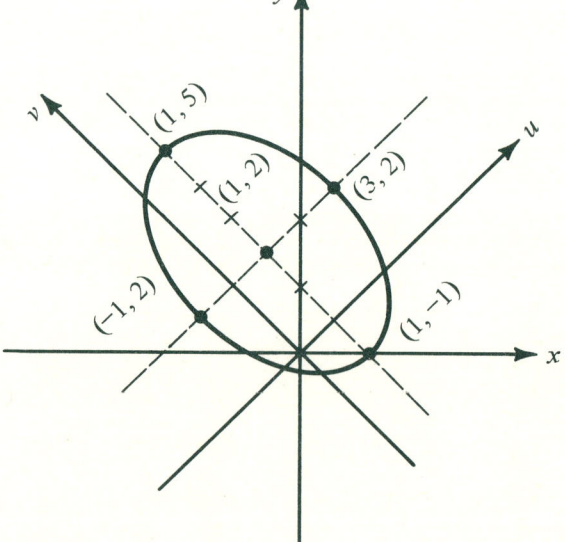

A-160

15. $11x^2 + 10\sqrt{3}xy - y^2 - 64 = 0$

17. $x^2 + 2xy + y^2 = \sqrt{2}y - \sqrt{2}x$

Since $A = C = 1$, $\theta = \pi/4$, and $x = u \cos \theta - v \sin \theta = \dfrac{\sqrt{2}}{2}u - \dfrac{\sqrt{2}}{2}v$

$$y = u \sin \theta + v \cos \theta = \dfrac{\sqrt{2}}{2}u + \dfrac{\sqrt{2}}{2}v$$

which, when substituted into our original equation, yields

$$\left(\dfrac{\sqrt{2}}{2}u - \dfrac{\sqrt{2}}{2}v\right)^2 + 2\left(\dfrac{\sqrt{2}}{2}u - \dfrac{\sqrt{2}}{2}v\right)\left(\dfrac{\sqrt{2}}{2}u + \dfrac{\sqrt{2}}{2}v\right) + \left(\dfrac{\sqrt{2}}{2}u + \dfrac{\sqrt{2}}{2}v\right)^2$$
$$= \sqrt{2}\left(\dfrac{\sqrt{2}}{2}u + \dfrac{\sqrt{2}}{2}v\right) - \sqrt{2}\left(\dfrac{\sqrt{2}}{2}u - \dfrac{\sqrt{2}}{2}v\right)$$

$$\dfrac{1}{2}u^2 - uv + \dfrac{1}{2}v^2 + u^2 - v^2 + \dfrac{1}{2}u^2 + uv + \dfrac{1}{2}v^2 = u + v - u + v$$

$$2u^2 = 2v \Rightarrow u^2 = v$$

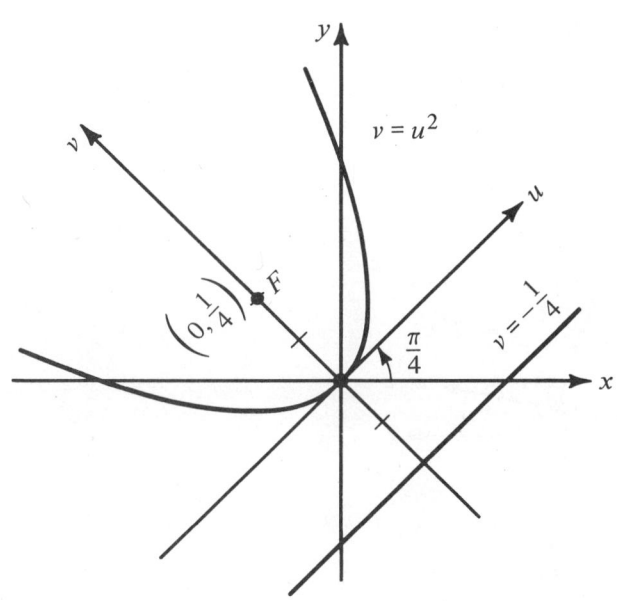

A–161

19.

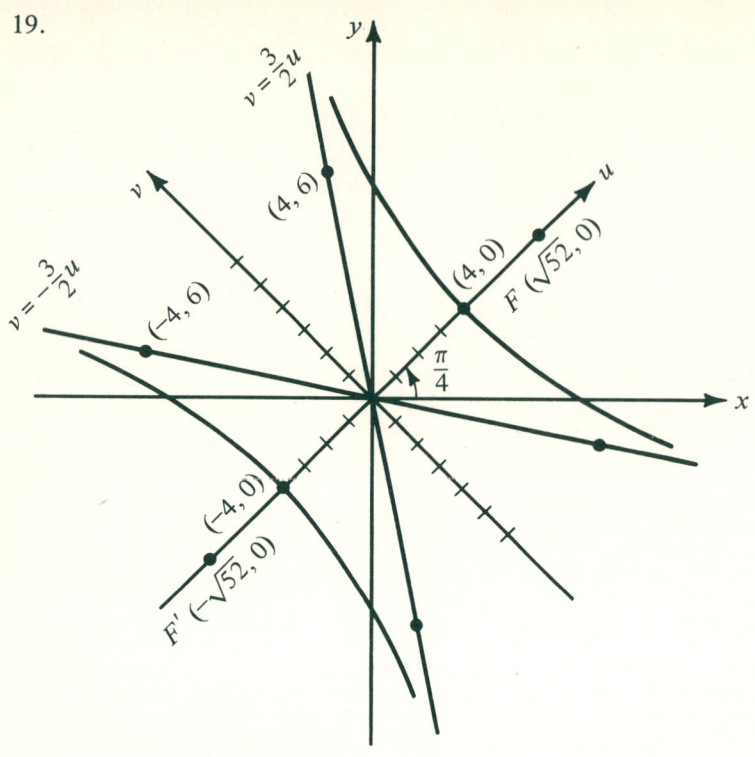

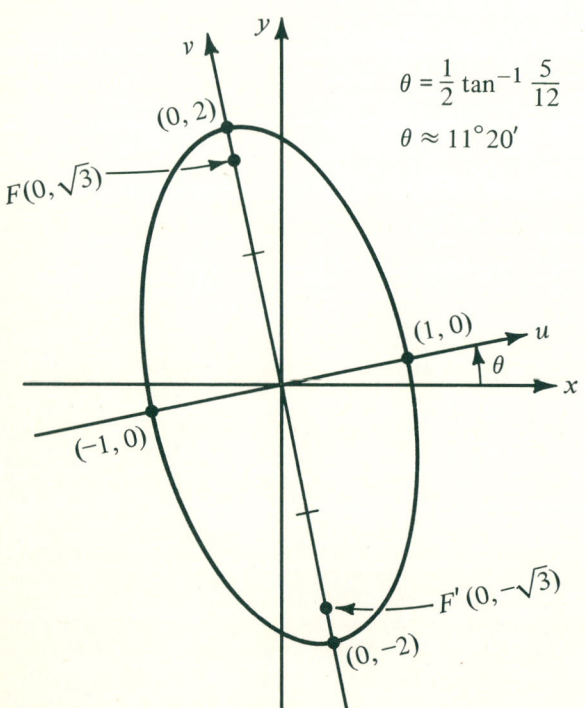

$$\theta = \frac{1}{2}\tan^{-1}\frac{5}{12}$$
$$\theta \approx 11°20'$$

23.

25.

26. $-6x^2 + 20xy - 20\sqrt{2}x - 6y^2 + 44\sqrt{2}y = 92$

$A = C \Rightarrow \theta = \dfrac{\pi}{4}$ $\quad x = u \cos \theta - v \sin \theta = \dfrac{1}{\sqrt{2}}(u - v)$

$\qquad\qquad\qquad y = u \sin \theta + v \cos \theta = \dfrac{1}{\sqrt{2}}(u + v)$

$-\dfrac{6}{2}(u - v)^2 + \dfrac{20}{2}(u - v)(u + v) - 20(u - v) - \dfrac{6}{2}(u + v)^2 + 44(u + v) = 92$

$-3(u^2 - 2uv + v^2) + 10(u^2 - v^2) - 20u + 20v - 3(u^2 + 2uv + v^2) + 44u + 44v = 92$

$4u^2 - 16v^2 + 24u + 64v = 92$

$4(u^2 + 6u + 9) - 16(v^2 - 4v + 4) = 92 + 36 - 64$

$\dfrac{(u + 3)^2}{16} - \dfrac{(v - 2)^2}{4} = 1$

Exercise 13.7

1.

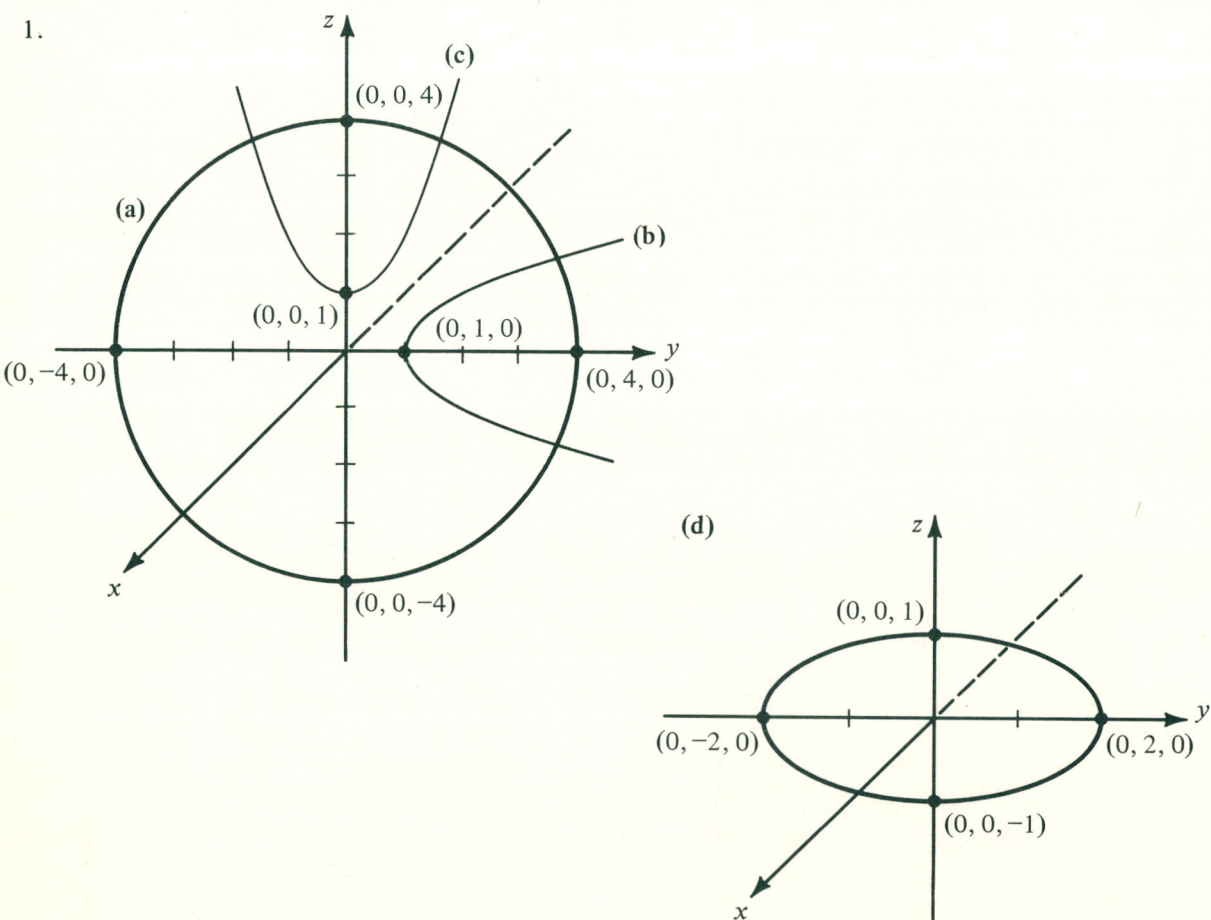

A-164

2. (a)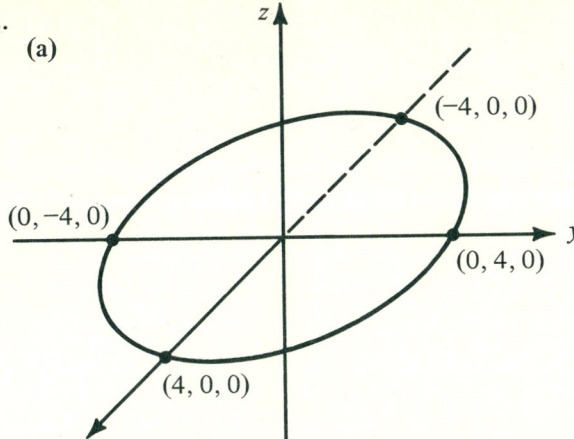

(b)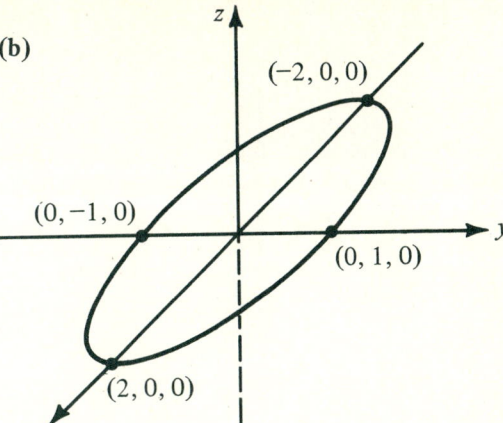

(c)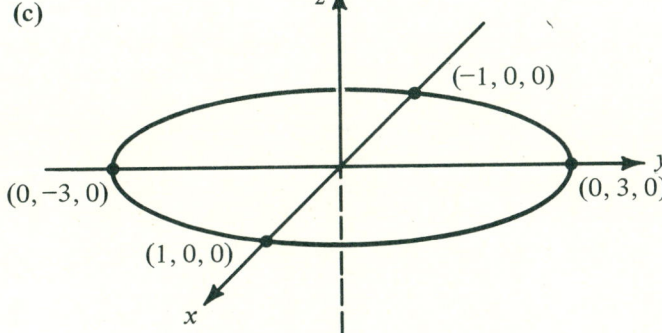

(d)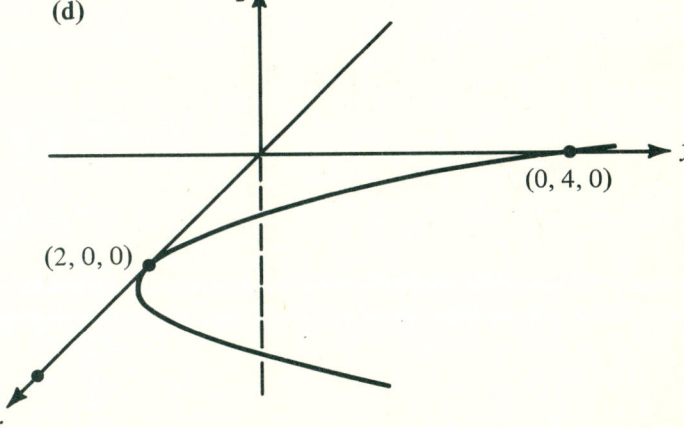

A-165

3.

(a)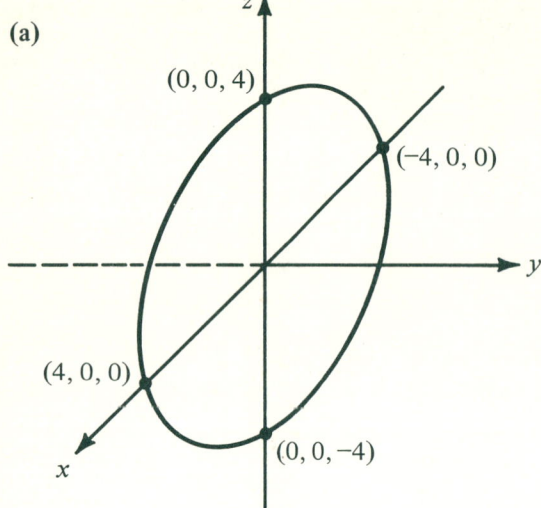

(b)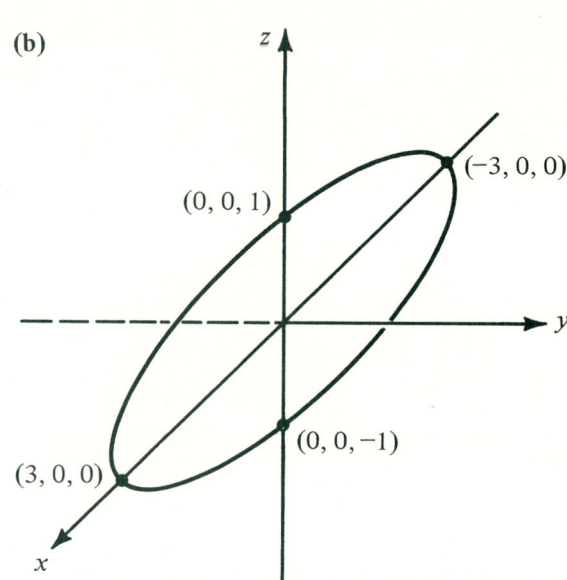

(c)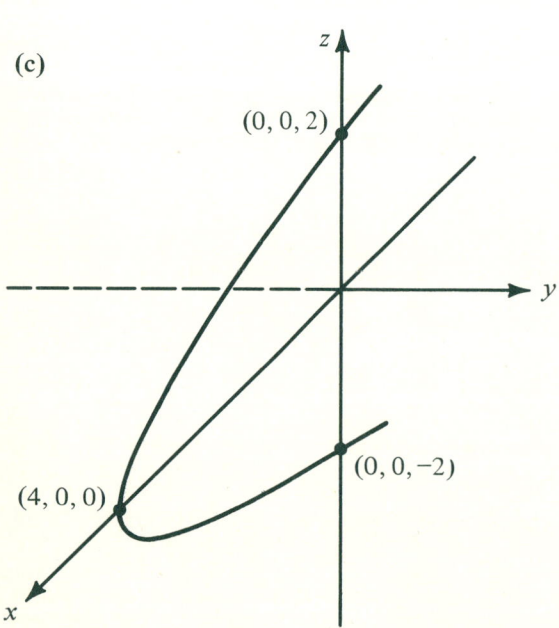

4.

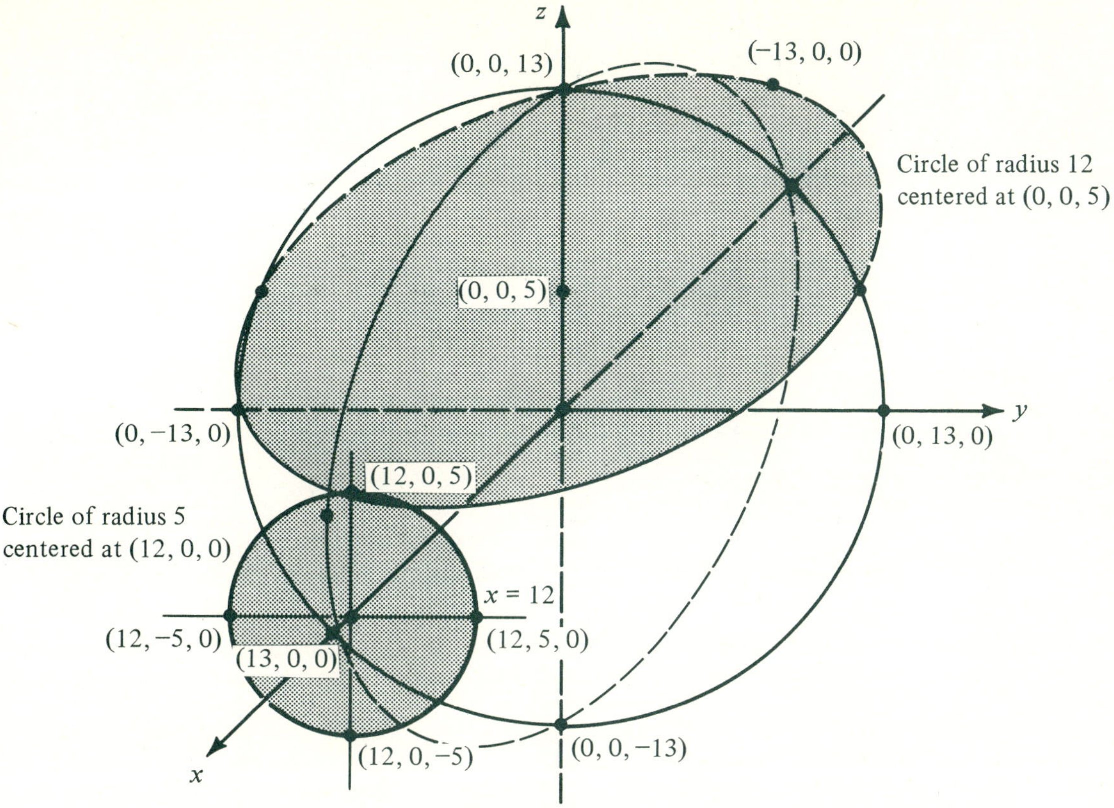

5.

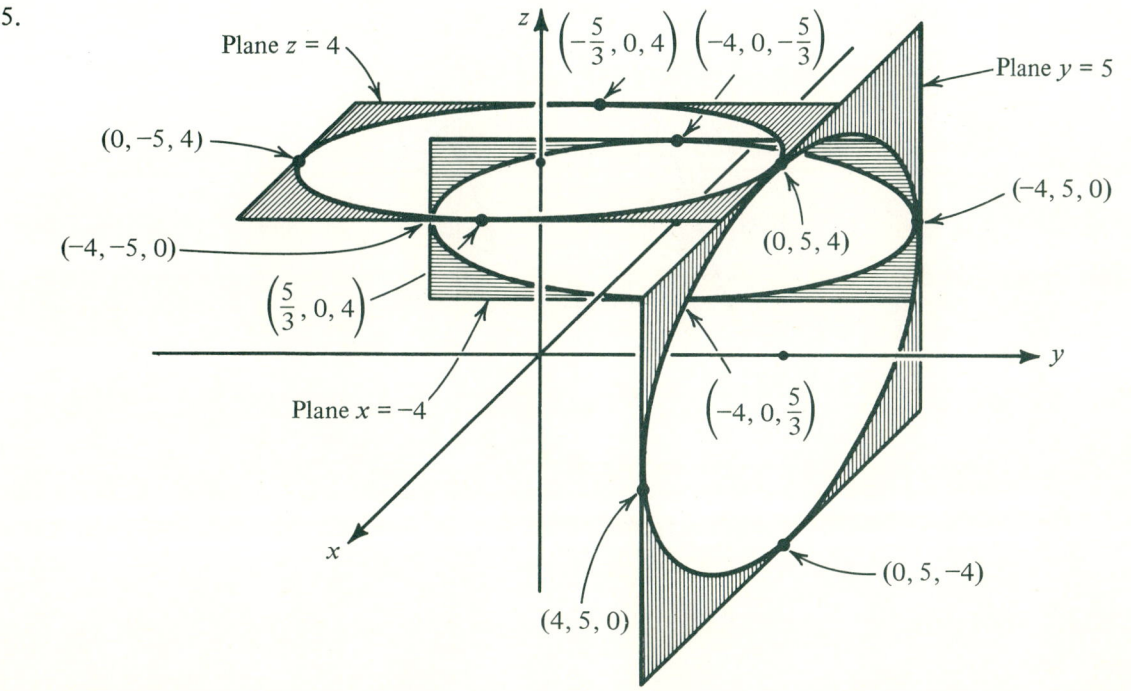

A-167

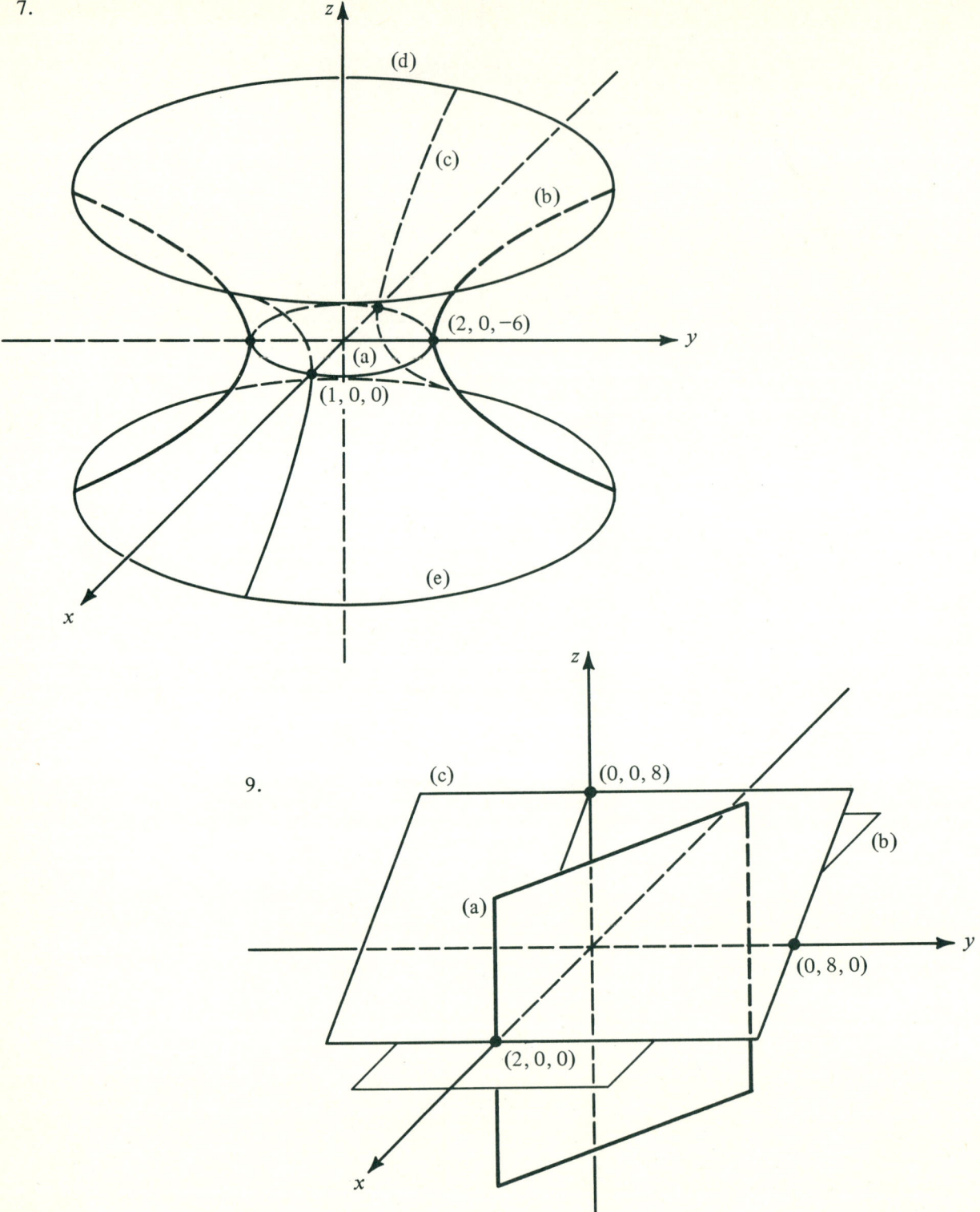

11.

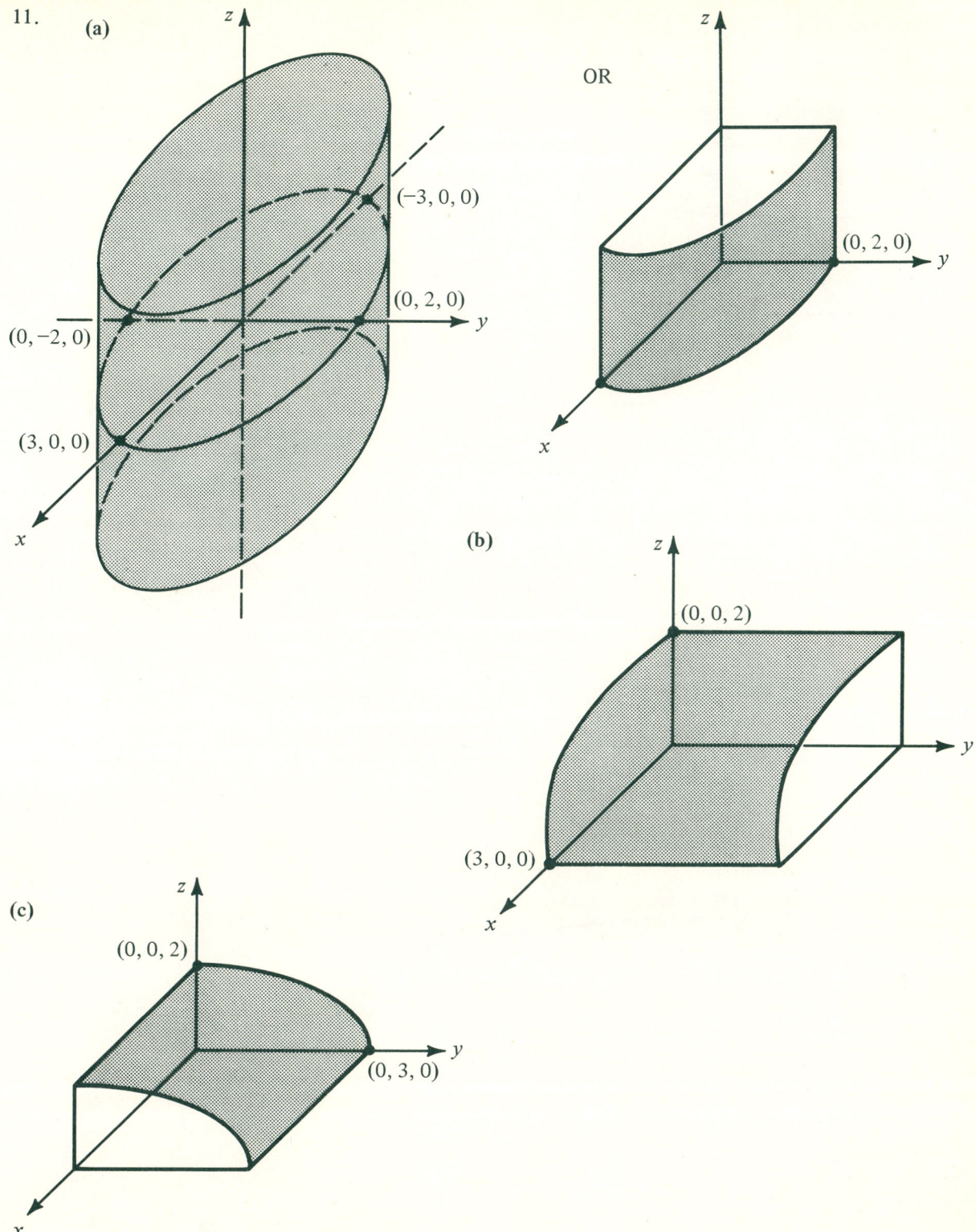

13. (a)

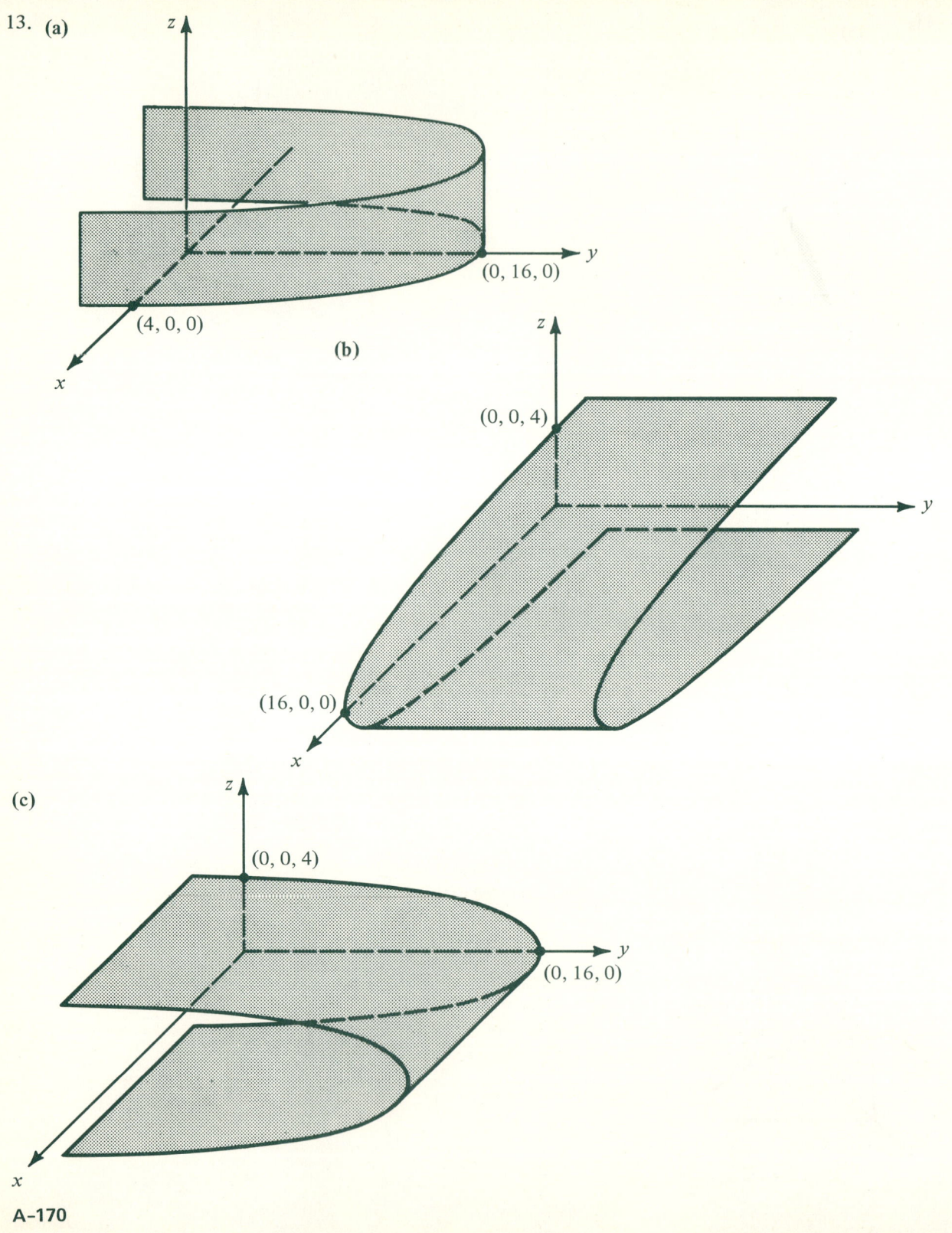

(b)

(c)

A-170

15.

(a)

(b)

(c)

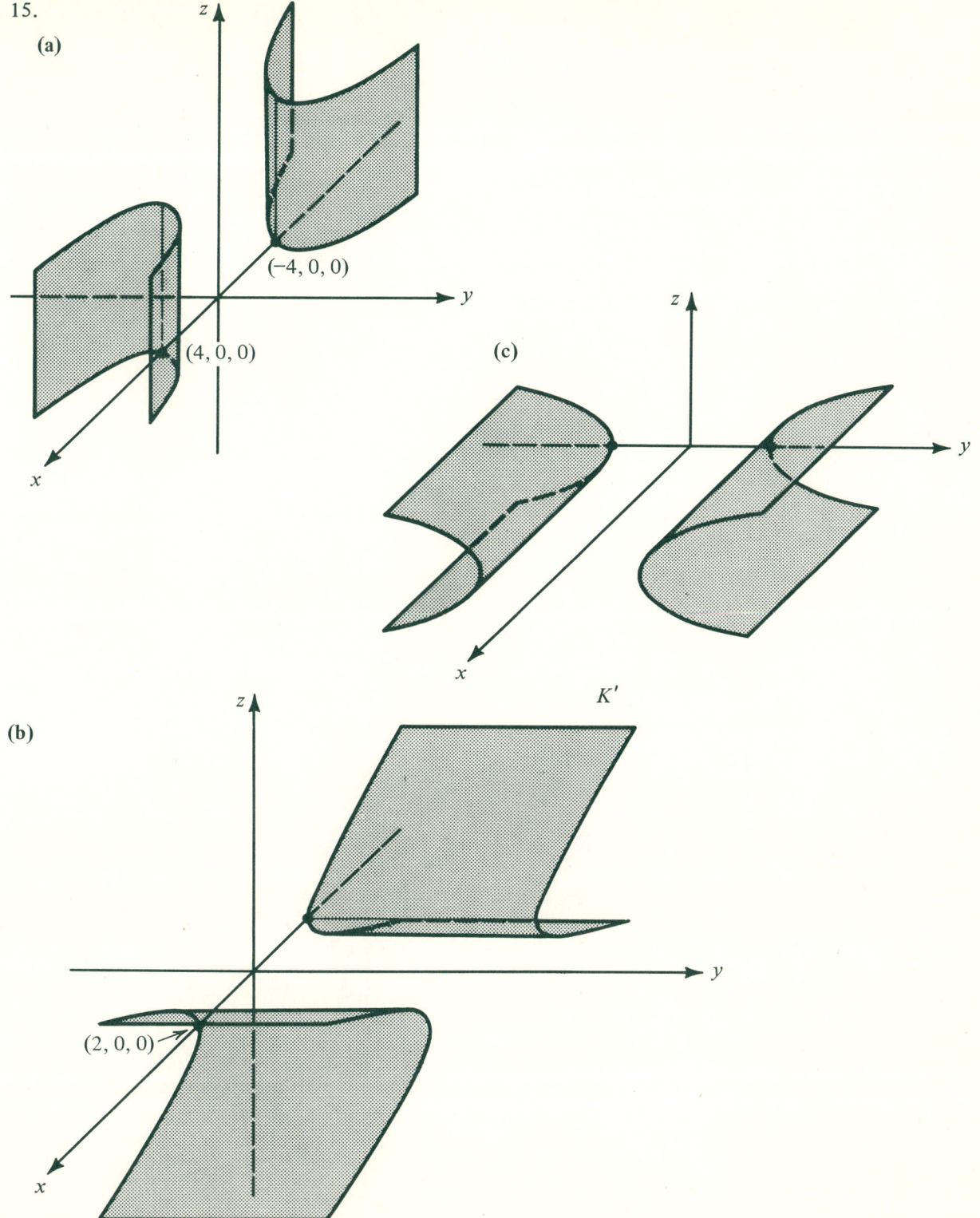

17.
(a)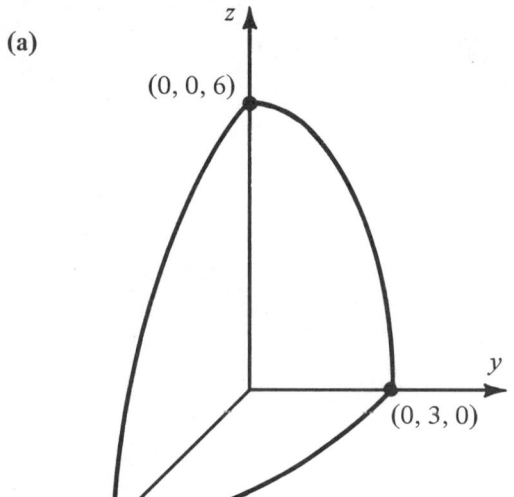
(b)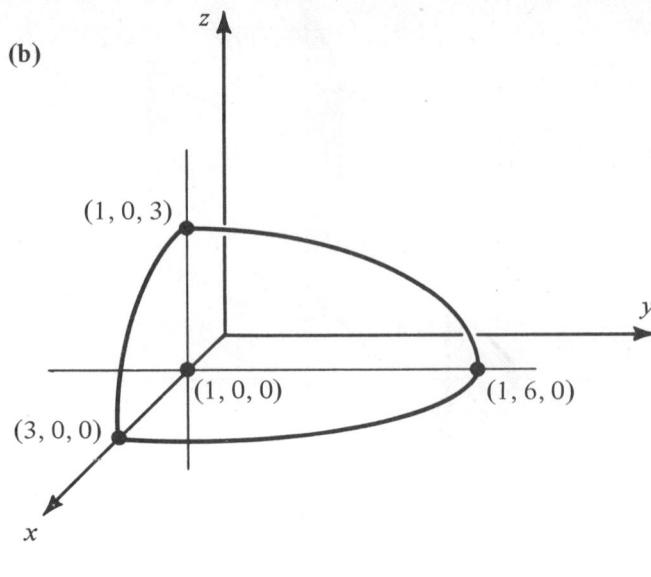

19.
(a)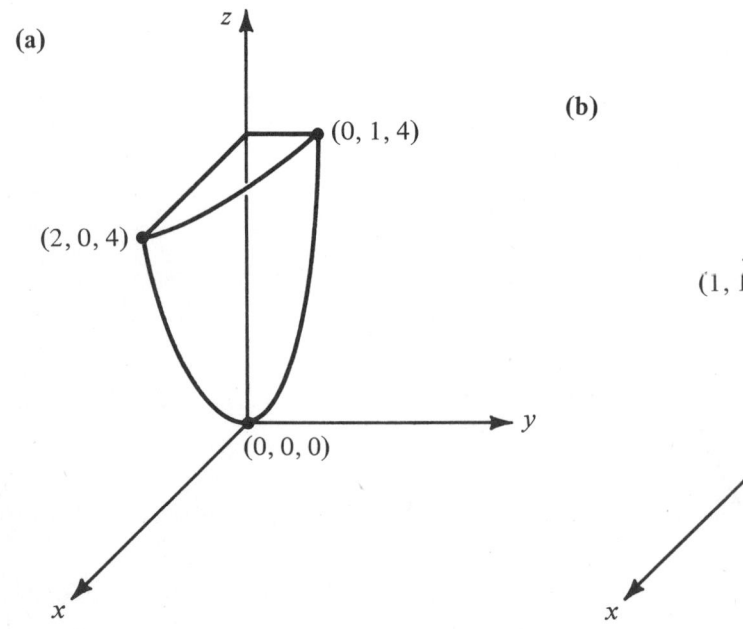
(b)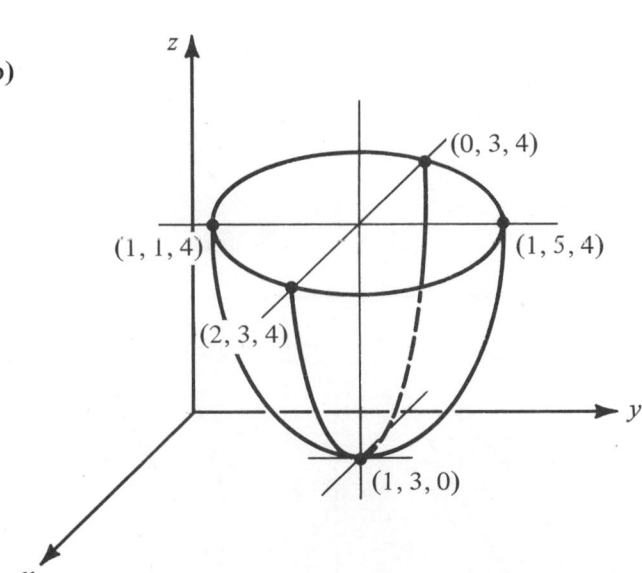

A–172

21.

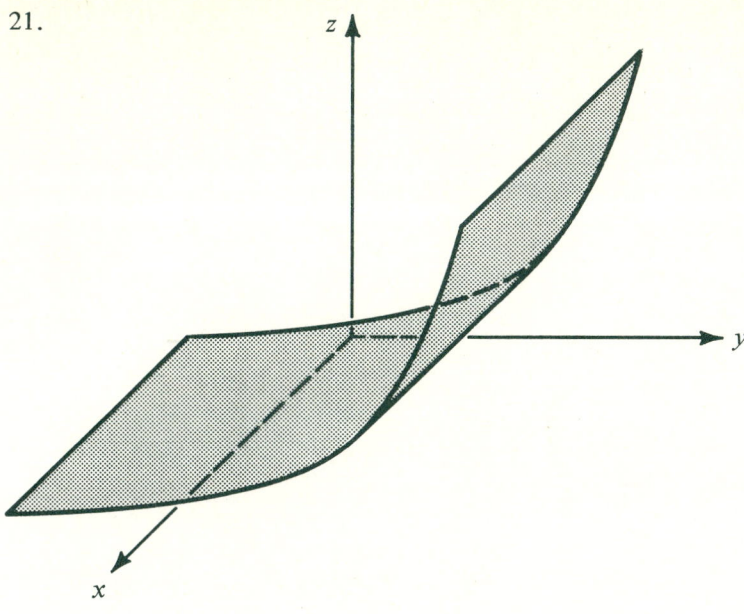

23.

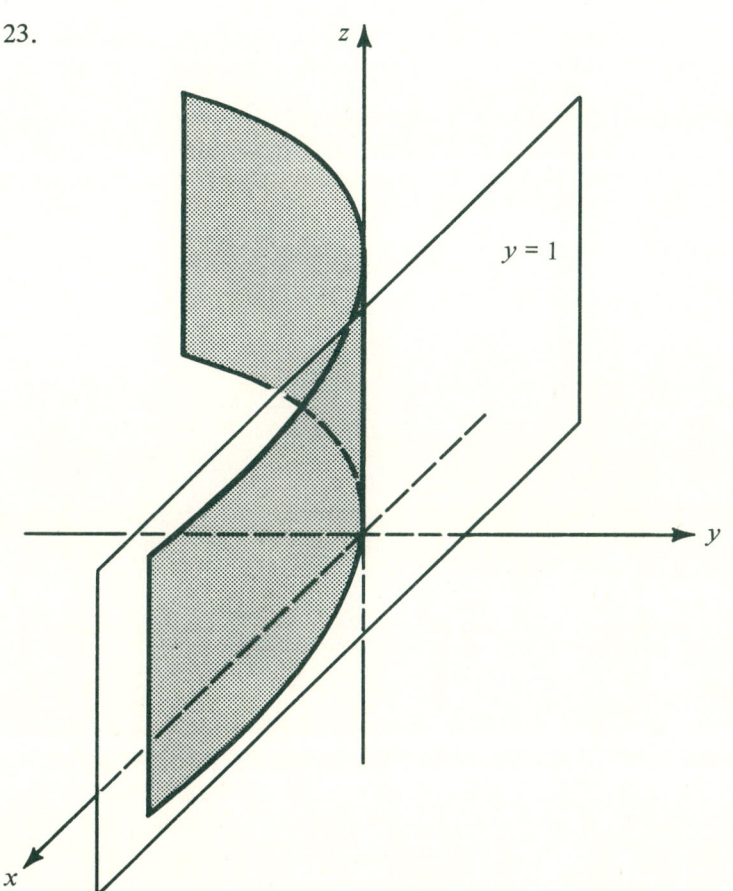

$y = 1$

25.

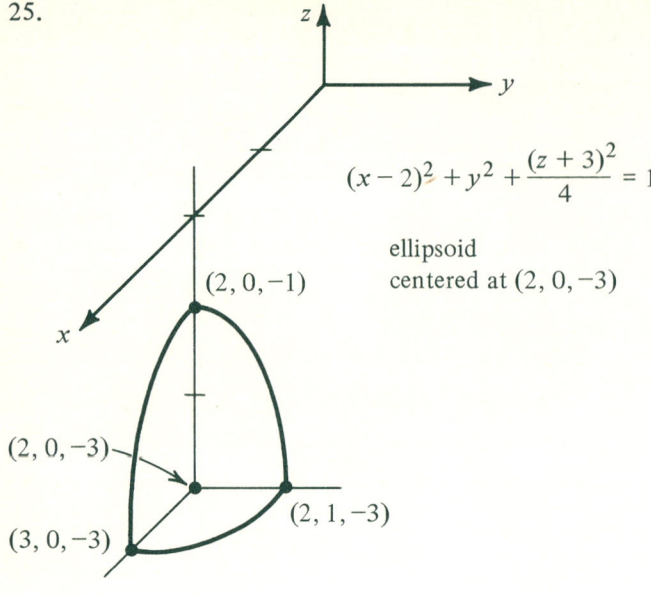

ellipsoid centered at $(2, 0, -3)$

25. $4x^2 - 16x + 4y^2 + z^2 + 6z = -24 \Rightarrow 4(x^2 - 4x + 4) + 4y^2 + (z^2 + 6z + 9) = -24 + 16 + 9$
$4(x - 2)^2 + 4y^2 + (z + 3)^2 = 1$. This is the equation of an ellipsoid centered at $(2, 0, -3)$.

26. $z = 11 - x^2 + 2x - y^2 - 4y \Rightarrow z - 11 - 1 - 4 = -(x^2 - 2x + 1) - (y^2 + 4y + 4)$
$z - 16 = -(x + 1)^2 - (y + 2)^2$. This is the equation of a concave-down paraboloid whose vertex is at $(-1, -2, 16)$ and whose $x - y$-trace is the circle $(x + 1)^2 + (y - 2)^2 = 16$, of radius 4, centered at $(-1, 2, 0)$.

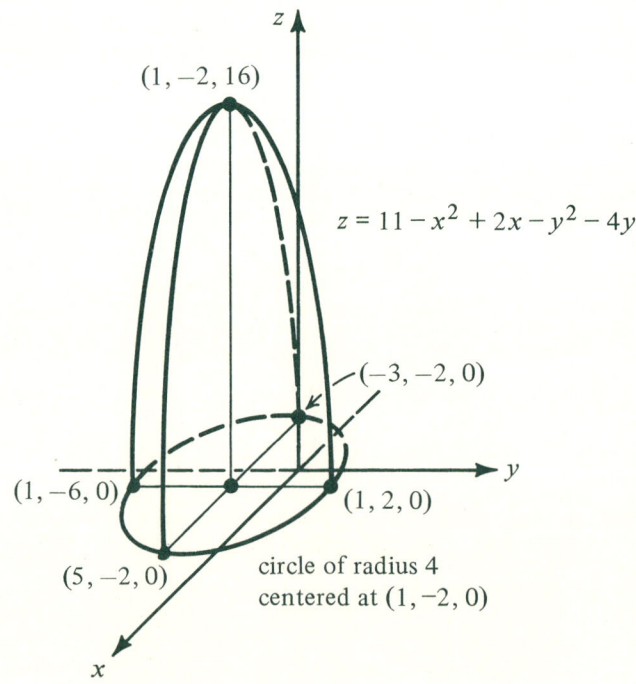

27.

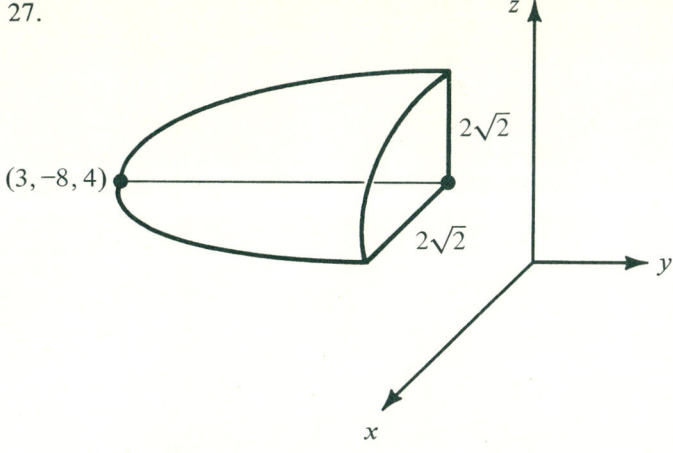

29.

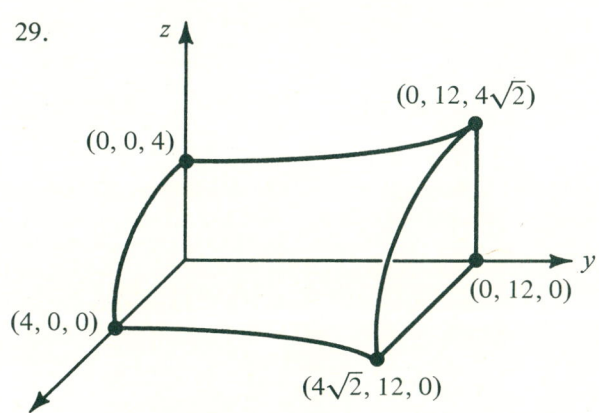

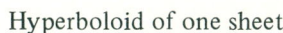

Hyperboloid of one sheet

31.

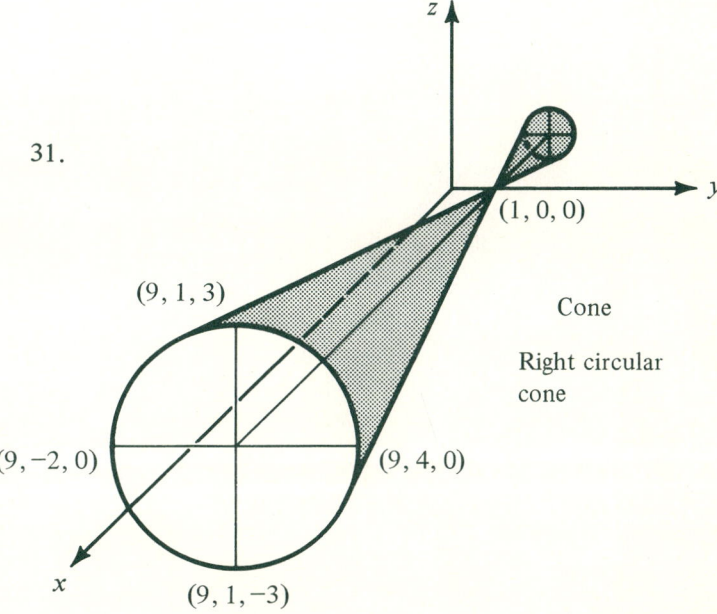

Cone

Right circular cone

A–175

33.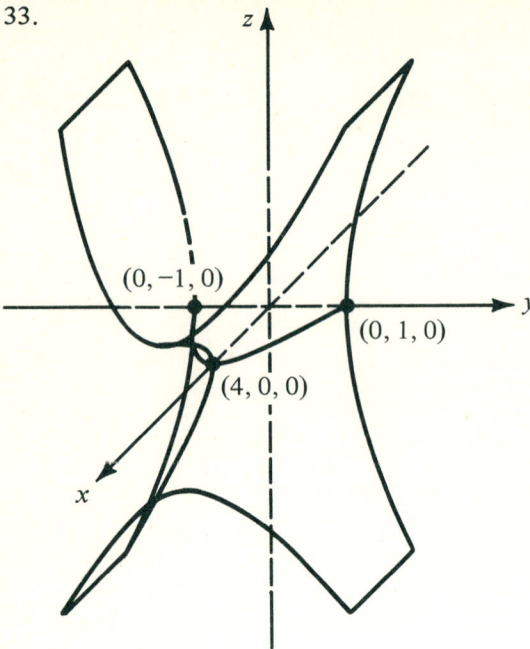

CHAPTER 14

Exercise 14.2

1. (a) and (b) are equal because matrix addition is commutative, $A + B = B + A$.

$$\begin{pmatrix} 3 & -4 \\ 5 & 11 \end{pmatrix}$$

2. (a) and (b) are equal illustrating the associative law for matrix addition $(A + B) + C = A + (B + C)$.

$$\begin{pmatrix} 4 & 13 & -1 \\ 5 & -4 & 15 \end{pmatrix}$$

3. $\begin{pmatrix} 5 \\ 7 \\ -10 \end{pmatrix}$

4. These matrices are not conformable to matrix addition because they are not of the same order.

5. (a) The "0 matrix" is the *additive identity* for 3×3 matrices because $A + 0 = A \; \forall \; A$.

$$\begin{pmatrix} 4 & 7 & 1 \\ 3 & 2 & -2 \\ 1 & 5 & 7 \end{pmatrix}$$

(b) These matrices are *additive inverses* of each other because their sum is the *additive identity*.

$$\begin{pmatrix} 0 & 0 & 0 \\ 0 & 0 & 0 \\ 0 & 0 & 0 \end{pmatrix}$$

6. $\begin{pmatrix} 1/3 & 1/6 \\ 1/2 & 1/2 \end{pmatrix}$

7. $\begin{pmatrix} -1 & -3/2 \\ 1 & 2 \end{pmatrix}$

8. $(a + b)A = aA + bA$ the distributive property for scalar multiplication over scalar addition. Both answers are

$$\begin{pmatrix} 5 & 10 & 15 \\ 20 & 25 & 30 \end{pmatrix}$$

9. $a(A + B) = aA + aB$ illustrates a different distributive property for scalar multiplication over matrix addition. Both answers are

$$\begin{pmatrix} 12 & -8 \\ 0 & 8 \end{pmatrix}$$

10. They are equal.

13. Yes $\begin{pmatrix} 0 & 0 \\ 0 & 0 \end{pmatrix}$

15. $A + B = B + A$ since the corresponding entries in their sums are equal, $a_1 + a_2 = a_2 + a_1, b_1 + b_2 = b_2 + b_1$, etc., because the real numbers $a_1, a_2, \ldots$ are commutative.

$$\begin{pmatrix} a_1 & b_1 \\ c_1 & d_1 \end{pmatrix} + \begin{pmatrix} a_2 & b_2 \\ c_2 & d_2 \end{pmatrix} = \begin{pmatrix} a_1 + a_2 & b_1 + b_2 \\ c_1 + c_2 & d_1 + d_2 \end{pmatrix}$$

$$\begin{pmatrix} a_2 & b_2 \\ c_2 & d_2 \end{pmatrix} + \begin{pmatrix} a_1 & b_1 \\ c_1 & d_1 \end{pmatrix} = \begin{pmatrix} a_2 + a_1 & b_2 + b_1 \\ c_2 + c_1 & d_2 + d_1 \end{pmatrix}$$

17. The set is closed. There is an additive identity, 0. Every 3 x 3 matrix, A, has an additive inverse, $-A$, such that $A + (-A) = (-A) + A = 0$. The associative property $(A + B) + C = A + (B + C)$ holds for all 3 x 3 matrices A, B, and C. By satisfying these properties, the system is called a group, but since the commutative property, $A + B = B + A$, also holds, it is a commutative or Abelian group.

$$\begin{pmatrix} -3 & -6 \\ -9 & -12 \end{pmatrix}$$

19. $(\alpha + \beta)A = (\alpha + \beta)\begin{pmatrix} a & b \\ c & d \end{pmatrix} = \begin{pmatrix} (\alpha + \beta)a & (\alpha + \beta)b \\ (\alpha + \beta)a & (\alpha + \beta)d \end{pmatrix}$

$\alpha A + \beta A = \alpha \begin{pmatrix} a & b \\ c & d \end{pmatrix} + \beta \begin{pmatrix} a & b \\ c & d \end{pmatrix} = \begin{pmatrix} \alpha a + \beta a & \alpha b + \beta b \\ \alpha c + \beta c & \alpha d + \beta d \end{pmatrix}$.

These results are equal because the corresponding entries $(\alpha + \beta)a = \alpha a + \beta a$, $(\alpha + \beta)b = \alpha b + \beta b$, etc., because in the real numbers the distributive properties hold. Therefore, $(\alpha + \beta)A = \alpha A + \beta A$

21. $\alpha A = \alpha \begin{pmatrix} a & b \\ c & d \end{pmatrix} = \begin{pmatrix} \alpha a & \alpha b \\ \alpha c & \alpha d \end{pmatrix}$ $A\alpha = \begin{pmatrix} a & b \\ c & d \end{pmatrix}\alpha = \begin{pmatrix} a\alpha & b\alpha \\ c\alpha & d\alpha \end{pmatrix}$

These results are equal because the real numbers are commutative under multiplication. Therefore, $\alpha A = A\alpha$.

24. Assume there are two identities and arrive at a contradiction proving that your original assumption that the identity was not unique was incorrect. Then, it must be unique.

Exercise 14.3

1. The commutative law does not hold for matrix multiplication.

   (a) $\begin{pmatrix} -10 & 9 \\ -10 & 7 \end{pmatrix}$ (b) $\begin{pmatrix} -1 & 6 \\ -3 & -2 \end{pmatrix}$

2. Every matrix does, however, commute with its inverse. Here, the equal answers are the multiplicative identity

   $\begin{pmatrix} 1 & 0 \\ 0 & 1 \end{pmatrix}$

3. (a) (-1) (b) not conformable

4. These illustrate an associative property involving scalar multiplication and matrix multiplication, $(\alpha A)B = \alpha(AB)$.

   $\begin{pmatrix} 6 & 8 \\ -2 & -6 \end{pmatrix}$

5. The associative property for matrix multiplication holds, as illustrated by these equal answers.

   $\begin{pmatrix} 11 & 54 & -16 \\ 13 & 3 & 7 \\ -23 & 17 & -19 \end{pmatrix}$

A-177

6. Here we are using the commutative property, $\alpha \cdot A = A \cdot \alpha$, and the associative property to regroup. The answers are equal.

$$\begin{pmatrix} -16 & -32 \\ 0 & -16 \end{pmatrix}$$

7. Here we see the multiplicative identities for 2 × 2 and for 3 × 3 matrices in action. Notice that $A \cdot I = A$ in both cases.

8. These pairs are multiplicative inverses since their products are the 2 × 2 and the 3 × 3 multiplicative identities, respectively. (See Problem 7.)

9. $\begin{pmatrix} 5 & 11 \\ 11 & 25 \end{pmatrix}$ 11. No. Try $A^T A$ from Problem 9.

13. The commutative law does not hold for matrix multiplication. We shall see in Section 14.6 that all nonzero matrices do not have multiplicative inverses, either.

15. (a) $A^2 = \begin{pmatrix} 4 & 0 & 0 \\ 0 & 36 & 0 \\ 0 & 0 & 9 \end{pmatrix}$ and $A^3 = \begin{pmatrix} 8 & 0 & 0 \\ 0 & 216 & 0 \\ 0 & 0 & 27 \end{pmatrix}$ (b) $A^n = \begin{pmatrix} 2^n & 0 & 0 \\ 0 & 6^n & 0 \\ 0 & 0 & 3^n \end{pmatrix}$

Exercise 14.4

1. (a) −2   (b) −4   [The first column of (a) is multiplied by 2 to get (b). The value of the determinant is also multiplied by 2.]
   (c) −6   (d) −12   (The first column again!)
2. (a) −10   (b) 10   (c) −10   [(c) is (a) multiplied by (−1) twice, first column and second row.]
3. (a) −3   (b) 3   (Rows 2 and 3 were interchanged.)
   (c) 3   [From (a), columns 2 and 3 were switched.]
4. (a) 4   (b) 4   (Here det A = det A^T.)
5. Here again det A = det A^T = 14.
6. These matrices are not equal, but the value of their determinants are all 10. The following manipulations apparently do not affect the value of the determinant. To change (a) to (b), 2 times column 1 was added to column 2. To change (a) to (c), 2 times row 1 was added to row 2. To change (a) to (d), 5 times column 2 was added to column 1.
7. The value of the determinants of all three matrices is 24. Notice that the diagonal entries, 3, 2, and 4, are the same for each. Matrix (a) is in lower triangular form (notice where the zeros are located), (b) is in upper triangular form, and (c) is a diagonal matrix. The value of each determinant is the product of its diagonal entries.
8. (a) 0   (b) 0   (c) 0   In (a) two rows are equal. In (b) we have a column of zeros. In (c) two columns are equal.
9. The value of the determinants of these three unequal matrices is −14. Using the second column in (c) yields the easiest calculation. To change (a) into (b), row 2 was added to row 1. To change (b) to (c), 3 times row 2 was added to row 3.
11. det (BC) = det B det C since,

$$\det B \det C = \begin{vmatrix} 4 & 0 & -2 \\ 1 & 3 & 1 \\ 2 & 1 & 0 \end{vmatrix} \cdot \begin{vmatrix} 3 & 1 & -2 \\ -10 & 0 & 4 \\ 2 & -1 & 2 \end{vmatrix} = (6)(20) = 120$$

and
$$\det BC = \det \begin{pmatrix} 4 & 0 & -2 \\ 1 & 3 & 1 \\ 2 & 1 & 0 \end{pmatrix} \begin{pmatrix} 3 & 1 & -2 \\ -10 & 0 & 4 \\ 2 & -1 & 2 \end{pmatrix}$$
$$= \det \begin{pmatrix} 8 & 6 & -12 \\ -25 & 0 & 12 \\ -4 & 2 & 0 \end{pmatrix} = \begin{vmatrix} 8 & 6 & -12 \\ -25 & 0 & 12 \\ -4 & 2 & 0 \end{vmatrix} = 120$$

13. (a) $\det 4A = 4^2 \det A = 16(-8) = -128$
    (b) $4 \det A = -32$

14. (a) $\det 2B = \det \begin{pmatrix} 8 & 0 & -4 \\ 2 & 6 & 2 \\ 4 & 2 & 0 \end{pmatrix} = 48$ and $2^3 \det B = 8 \cdot 6 = 48$

    (b) $\det 3B = 3^3 \det B = 27 \cdot 6 = 162$

15. (a) 3240 (b) 3240

17. It is singular for all values of x.

19. (a) $\det A = \det B = 1$ and $\det AB = \det \begin{pmatrix} 1 & 0 \\ 0 & 1 \end{pmatrix} = 1$

    (b) $\det A = -1$, $\det B = -1$ and $\det AB = 1$

21. $\det \begin{pmatrix} a & b & c \\ 0 & d & e \\ 0 & 0 & f \end{pmatrix} = a \begin{vmatrix} d & e \\ 0 & f \end{vmatrix} - 0 \begin{vmatrix} b & c \\ 0 & f \end{vmatrix} + 0 \begin{vmatrix} b & c \\ d & e \end{vmatrix}$

    $= a(df - 0 \cdot e) = adf$

22. (a) $A - \lambda I = \begin{pmatrix} 1 & 0 & 0 \\ 0 & 3 & 0 \\ 0 & 0 & 4 \end{pmatrix} - \lambda \begin{pmatrix} 1 & 0 & 0 \\ 0 & 1 & 0 \\ 0 & 0 & 1 \end{pmatrix} = \begin{pmatrix} 1-\lambda & 0 & 0 \\ 0 & 3-\lambda & 0 \\ 0 & 0 & 4-\lambda \end{pmatrix} \Rightarrow$

    $\det(A - \lambda I) = 0 \Leftrightarrow \begin{vmatrix} 1-\lambda & 0 & 0 \\ 0 & 3-\lambda & 0 \\ 0 & 0 & 4-\lambda \end{vmatrix} = 0 \Rightarrow$

    $(1-\lambda)(3-\lambda)(4-\lambda) = 0 \Rightarrow \lambda = 1, 3, \text{ or } 4$

    Here $d(A - \lambda I) = 0$ is a third-degree polynomial in λ.

23. (a) The equations for rotating the coordinate system through θ,
    $$x \cos \theta + y \sin \theta = u$$
    $$-x \sin \theta + y \cos \theta = v$$
    can be expressed in matrix form as
    $$\begin{pmatrix} \cos \theta & \sin \theta \\ -\sin \theta & \cos \theta \end{pmatrix} \begin{pmatrix} x \\ y \end{pmatrix} = \begin{pmatrix} u \\ v \end{pmatrix}$$
    Here $\theta = \pi/6$ yields
    $$\begin{pmatrix} \sqrt{3}/2 & 1/2 \\ -1/2 & \sqrt{3}/2 \end{pmatrix} \begin{pmatrix} 3 \\ 4 \end{pmatrix} = \begin{pmatrix} (3\sqrt{3}/2) + 2 \\ (-3/2) + 2\sqrt{3} \end{pmatrix} \approx \begin{pmatrix} 4.6 \\ 2 \end{pmatrix}$$

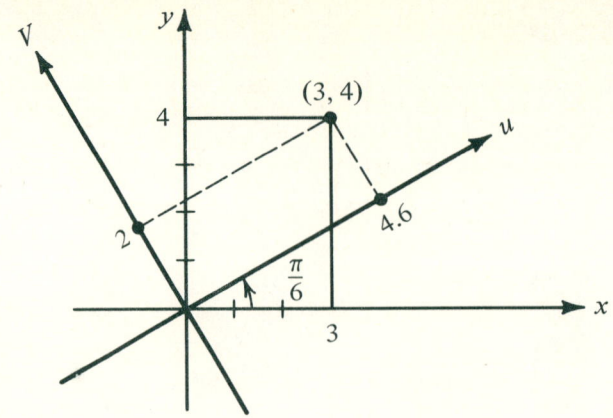

(b) This matrix is the inverse of the matrix in (a). Here (3, 4) is a coordinate in the u, v system, and the indicated product will yield its x, y coordinates. See Section 13.6.

Exercise 14.5

1. (a) 5   (b) 20   (Multiplying column 1 by 4 did it.)
2. (a) 11   (b) 33   (Multiplying row 1 by 3 did it.)
3. (a) 4   (b) 12   (Multiplying row 1 by 3 did it.)
4. (a) −21   (b) 42   [Multiplying row 3 by −2 did it. Did you use row 1 to evaluate (a) and (b)?]
5. 10   6. 0   7. 8
9. On the right side multiply column 2 by 4 and add it to column 1. Then multiply column 2 by −2 and add to column 3. Then multiply column 2 by 2 and divide column 3 by 2. The result equals the left side.
11. Add row 1 to row 2, then multiply column 4 by −2 and add to column 3.

$$\begin{vmatrix} 2 & 1 & 3 & 4 \\ 0 & 0 & 8 & 4 \\ 3 & 2 & -2 & 3 \\ 4 & 1 & 0 & 2 \end{vmatrix} = \begin{vmatrix} 2 & 1 & -5 & 4 \\ 0 & 0 & 0 & 4 \\ 2 & 2 & -8 & 3 \\ 4 & 1 & -4 & 2 \end{vmatrix} = 4 \begin{vmatrix} 2 & 1 & -5 \\ 3 & 2 & -8 \\ 4 & 1 & -4 \end{vmatrix}$$

Add column 3 to column 1, then add 4 times column 2 to column 3.

$$4 \begin{vmatrix} -3 & 1 & -5 \\ -5 & 2 & -8 \\ 0 & 1 & -4 \end{vmatrix} = 4 \begin{vmatrix} -3 & 1 & -17 \\ -5 & 2 & -28 \\ 0 & 1 & 0 \end{vmatrix} = 4(-1) \begin{vmatrix} -3 & -17 \\ -5 & -28 \end{vmatrix} = 4$$

13. −198   15. 0

Exercise 14.6

1. $\begin{pmatrix} 4/-2 & -2/-2 \\ -3/-2 & 1/-2 \end{pmatrix} = \begin{pmatrix} -2 & 1 \\ 3/2 & -1/2 \end{pmatrix}$ 2. $\begin{pmatrix} -1 & 0 \\ 2/3 & 1/3 \end{pmatrix}$

3. Singular. The determinant is zero.

4. $\begin{pmatrix} -7/8 & -5/8 \\ 3/8 & 1/8 \end{pmatrix}$ 5. $AA^{-1} = \begin{pmatrix} 1 & 0 \\ 0 & 1 \end{pmatrix}$ 6. $A^{-1}A = \begin{pmatrix} 1 & 0 \\ 0 & 1 \end{pmatrix}$

7. Singular.

8. $A^{-1} = \begin{pmatrix} 2 & -3 \\ -1 & 2 \end{pmatrix}$ and $B^{-1} = \begin{pmatrix} -2 & 3/2 \\ 1 & -1/2 \end{pmatrix}$

$(AB)^{-1} = \begin{pmatrix} 8 & 18 \\ 5 & 11 \end{pmatrix}^{-1} = \begin{pmatrix} 11/-2 & -18/-2 \\ -5/-2 & 8/-2 \end{pmatrix} = \begin{pmatrix} -11/2 & 9 \\ -5/2 & -4 \end{pmatrix}$

$A^{-1}B^{-1} = \begin{pmatrix} 2 & -3 \\ -1 & 2 \end{pmatrix} \cdot \begin{pmatrix} -2 & 3/2 \\ 1 & -1/2 \end{pmatrix} = \begin{pmatrix} -7 & 9/2 \\ 4 & -5/2 \end{pmatrix}$

$B^{-1}A^{-1} = \begin{pmatrix} -2 & 3/2 \\ 1 & -1/2 \end{pmatrix} \cdot \begin{pmatrix} 2 & -3 \\ -1 & 2 \end{pmatrix} = \begin{pmatrix} -11/2 & 9 \\ 5/2 & -4 \end{pmatrix}$

It appears that $(AB)^{-1} = B^{-1}A^{-1}$.

9. $(AB)^{-1} = B^{-1}A^{-1}$ again.

10. (a) $\det A = \begin{vmatrix} 3 & 1 \\ 2 & 1 \end{vmatrix} = 3 - 2 = 1$

$\det A^{-1} = \begin{vmatrix} 1 & -1 \\ -2 & 3 \end{vmatrix} = 3 - 2 = 1$

It appears that $\det A = \det A^{-1}$, but try another example.

11. They are equal! $\begin{pmatrix} -1 & 2/3 \\ 1/2 & -1/6 \end{pmatrix}$

16. We know that $(AB) \cdot (AB)^{-1} = I$ and that the inverse for (AB) is unique. If we can also prove that $(AB)(B^{-1}A^{-1}) = I$, then $(AB)^{-1}$ must equal $B^{-1}A^{-1}$ because there is only one inverse for (AB).
$(AB)(B^{-1}A^{-1}) = A(BB^{-1})A^{-1} = A(IA^{-1})$
$= AA^{-1} = I$

Exercise 14.7

1. (a) (8, 16) (b) (8, 16)
2. (a) 5 (b) $2\|v_1\| = 10$
3. (a) $5 + 13 = 18$ (b) $8\sqrt{5}$
4. (a) 27 (b) $3\sqrt{66}$
5. (1/3, 2/3, 2/3)
6. (a) $3 + \sqrt{66}$ (b) 11

7.
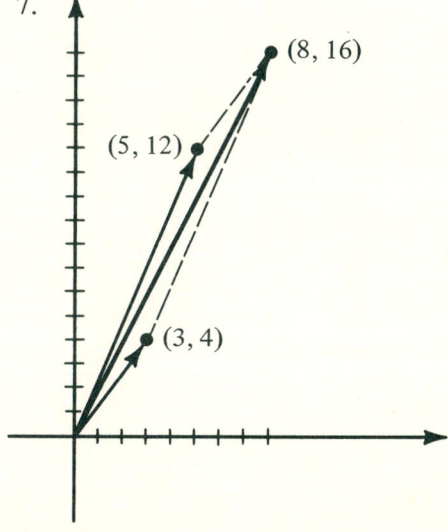

8.
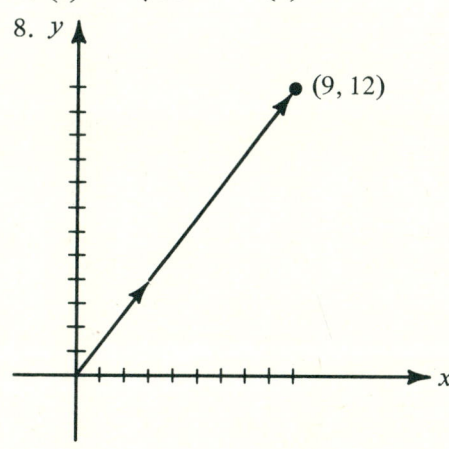

9.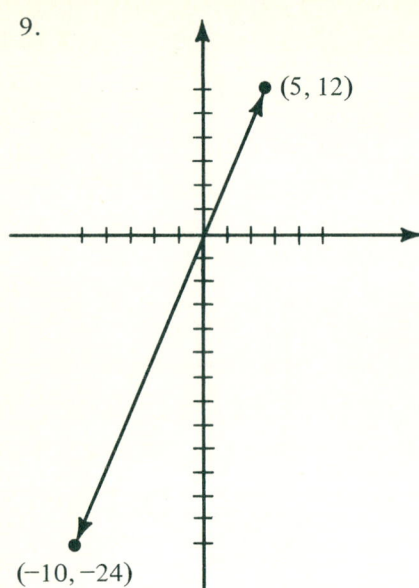

10.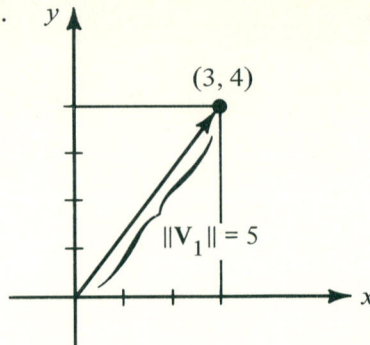

11. $v_1 \cdot v_2 = (3, 4) \cdot (2, 1) = 3 \cdot 2 + 4 \cdot 1 = 10$

13. 0. Draw these vectors. How are they related?

15. $v_1 \cdot v_2 = (3, 2) \cdot (4, -6) = 3 \cdot 4 + 2(-6) = 0$

$$\cos \theta = \frac{v_1 \cdot v_2}{\|v_1\| \|v_2\|} = 0 \Rightarrow \theta = \pi/2$$

17. $\theta = \pi/6$

18. (a)
$$v_1 \cdot v_2 = \begin{vmatrix} i & j & k \\ 3 & 2 & -1 \\ -1 & 3 & -2 \end{vmatrix} = -i + 7j + 11k$$

19. (a) $-13i - j + 5k$ (b) $-26i - 2j + 10k$

Since the first vectors are scalar multiples of each other and the second ones are equal, the cross-products must be in the same direction with magnitudes that are multiples.

21. $W = -18$

23. $\|F\| = 20$, applied at an angle of 30° with the distance, means
$$W = -F \cdot S = -\|F\| \|S\| \cos \theta = -(20)(10)\cos 150° = 100\sqrt{3}$$

The units are newton–meters (1kg-m/sec² is a Newton) called *joules*.

25. (a) $F = ma = -98$ kg-m/sec² $= -98$ Newtons
$$\cos 60° = \frac{F}{F_p} \Rightarrow F_p = F \cos 60° = -196 \text{ Newtons}$$

(b) $W = 1960$ Joules

27. (a) $\|v_x\| = 606.2$ m/sec
$\|v_y\| = 350$ m/sec

(b) 218.2 km

(c) $\|v_x\| = 606.2$
$\|v_y\| = -9.8(36) + 350 = -2.8$
$\|v\| = \|606.2i - 2.8j\| \approx 606.2$

CHAPTER 15

Exercise 15.1

1. Yes
2. Yes
3. No
4. Yes
5. No
6. Yes
7. (4, 1)
9. (3, −5)
11. (3, 2)
13. (3, 1)
15. (8, 13)
17. These equations are equivalent, hence, there are an infinite number of solutions — all points on the line.
19. These are parallel lines with $m = 2/3$.
21. (9, 3)
23. $\sqrt{5}$ km/hr
25. Cold 2/3 1/min and hot 4/3 1/min

Exercise 15.2

1. (a) 8
   (b) 39
2. (a) −4
   (b) 12
3. (a) 0
   (b) 0
4. (a) 0
   (b) 0
5. (a) −2
   (b) −8
7. (a) (4, 1)
   (b) (−10, −3)

$$x = \frac{\begin{vmatrix} 1 & -7 \\ t & -2 \end{vmatrix}}{17} = \frac{-2 + 7t}{17}, \quad y = \frac{\begin{vmatrix} 2 & 1 \\ 3 & t \end{vmatrix}}{17} = \frac{2t - 3}{17}$$

In (a) above, $t = 10$, and in (b), $t = -24$.

9. (a) Market equilibrium occurs when supply and demand are equal, which is the solution of the following system.

$$x + 2p = 151$$
$$3x - 4p = 448$$

$$x = \frac{\begin{vmatrix} 151 & 2 \\ 448 & -4 \end{vmatrix}}{\begin{vmatrix} 1 & 2 \\ 3 & -4 \end{vmatrix}} = \frac{-1500}{-10} = 150$$

$$p = \frac{\begin{vmatrix} 1 & 151 \\ 3 & 448 \end{vmatrix}}{-10} = \frac{1}{2}$$

Supply 150 items at 50¢ each.
  (b) $x = 2, p = 25$
11. State tax = \$241.45, Federal tax = \$1,951.71.

Exercise 15.3

1. (a) −16
   (b) −48
   (c) −80
2. (a) 77
   (b) −77
   (c) 77

3. (a) 9
   (b) 30
   (c) −4

5. (a) (−1, 1, 3)
   (b) (11/7, 20/7, −26/7)
   (c) (20/7, 17/7, −51/7)

7. $D = \begin{vmatrix} 2 & -3 & -1 \\ 1 & -2 & -4 \\ 3 & -1 & -2 \end{vmatrix} = 25$

$x = \dfrac{1}{25} \begin{vmatrix} 11 & -3 & -1 \\ -4 & -2 & -4 \\ 8 & -1 & -2 \end{vmatrix} = \dfrac{1}{25}(100) = 4$

$y = \dfrac{1}{25} \begin{vmatrix} 2 & 11 & -1 \\ 1 & -4 & -4 \\ 3 & 8 & -2 \end{vmatrix} = \dfrac{1}{25}(-50) = -2$

$z = 2x - 3y - 11 = 8 + 6 - 11 = 3$
(4, −2, 3)

9. (−1, 8, 11, 16)

11. N = 4, D = 3, Q = 2

13. $y = 2x^2 - 3x - 4$

14. G = 75, H = 20.24, B = 4.76

Exercise 15.4

1. (−5, −4)
2. (8/5, 14/5)
3. (12/67, −68/67)
4. (−1/5, 1/5)

Exercise 15.5

1. (2, 1)
2. (1, −1, 2)
3. (1, −2, −3)
4. (2, 3, −1)
5. (2, 1, 3, 1)
6. (0, 1, 2, −1)
7. (7t + 1, −4t, t)
9. (−2, 3, 2)
11. $\left(\dfrac{2t-3}{3}, \dfrac{3-2t}{2}, t \right)$

Exercise 15.6

1. $\begin{pmatrix} 1 & 2 & 1 & 2 \\ 0 & 1 & 3/4 & 1/4 \\ 0 & 0 & 1 & -5 \end{pmatrix}$

2. $\begin{pmatrix} 1 & 0 & 1 & 0 \\ 0 & 1 & -3 & 5 \\ 0 & 0 & 1 & -1 \end{pmatrix}$

3. $\begin{pmatrix} 1 & -2 & 1 & -8 \\ 0 & 1 & -4/7 & 26/7 \\ 0 & 0 & 1 & -63/25 \end{pmatrix}$

4. $\begin{pmatrix} 1 & 0 & 1 & -1 & 5 \\ 0 & 1 & -1 & 1 & -8 \\ 0 & 0 & 1 & -3/2 & 7/2 \\ 0 & 0 & 0 & 1 & 11 \end{pmatrix}$

5. $\begin{pmatrix} 1 & 2 & 0 & -1 & 6 \\ 0 & 1 & -1 & -1 & 8 \\ 0 & 0 & 1 & 1/7 & -26/7 \\ 0 & 0 & 0 & 1 & -9/4 \end{pmatrix}$

6. $\begin{pmatrix} 1 & 2 & -1 & 2 & 8 \\ 0 & 1 & -1/3 & 1 & 13/3 \\ 0 & 0 & 1 & 0 & 16/5 \\ 0 & 0 & 0 & 1 & 43/20 \end{pmatrix}$

7. $(2, 1, -3)$

8. $(1, 3, -4)$

9. $\left(\dfrac{t+5}{5}, \dfrac{7t}{5}, t\right)$

11. $(1, 1, -2, 3)$

Exercise 15.7

1. (a) circle, line $(0, 4)$
   (b) ellipse, line $(1, 0)$
2. (a) circle, line $(3, 4), (-3, -4)$
   (b) ellipse, line $(2, 1), (-2, 1)$
3. (a) $(2, 3)(-2, 3)(2, -3)(-2, -3)$
   (b) ϕ
4. (a) $(1, 1)(1, -1)(-1, 1)(-1, -1)$
   (b) $(2, 4)(2, -4)(-2, 4)(-2, -4)$
5. circle, parabola $(5, 2\sqrt{5})(5, -2\sqrt{5})$
7. hyperbola, ellipse $(2, 2)(2, -2)(-2, 2)(-2, -2)$
9. $(2, 3)(-2, -3)\left(\dfrac{2\sqrt{21}}{3}, \dfrac{\sqrt{21}}{3}\right)\left(\dfrac{-2\sqrt{21}}{3}, \dfrac{-\sqrt{21}}{3}\right)$
11. $(2, 0)(-2, 0)(\sqrt{2}, \sqrt{2})(-\sqrt{2}, -\sqrt{2})$
13. rate of the two-man boat is 2.45 km/hr

Exercise 15.8

1.

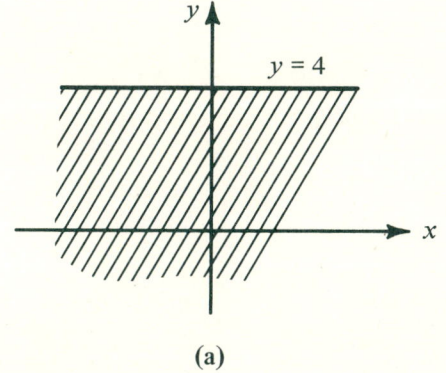
(a)

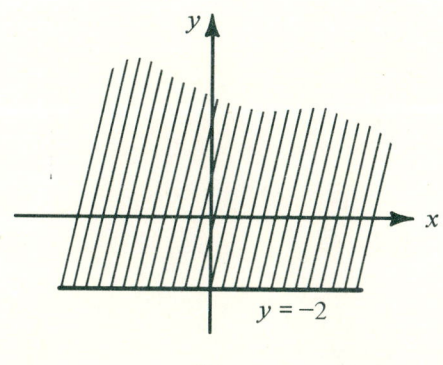
(b)

2.

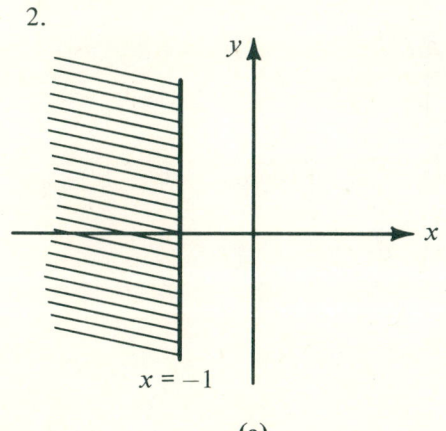
(a)

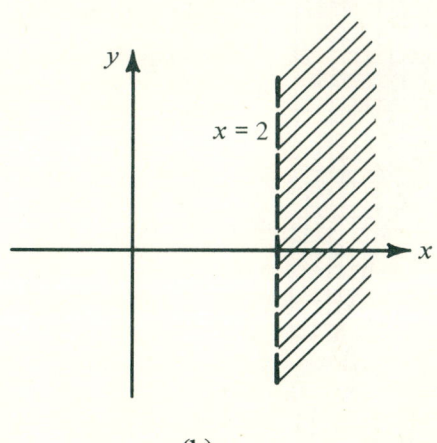
(b)

3.

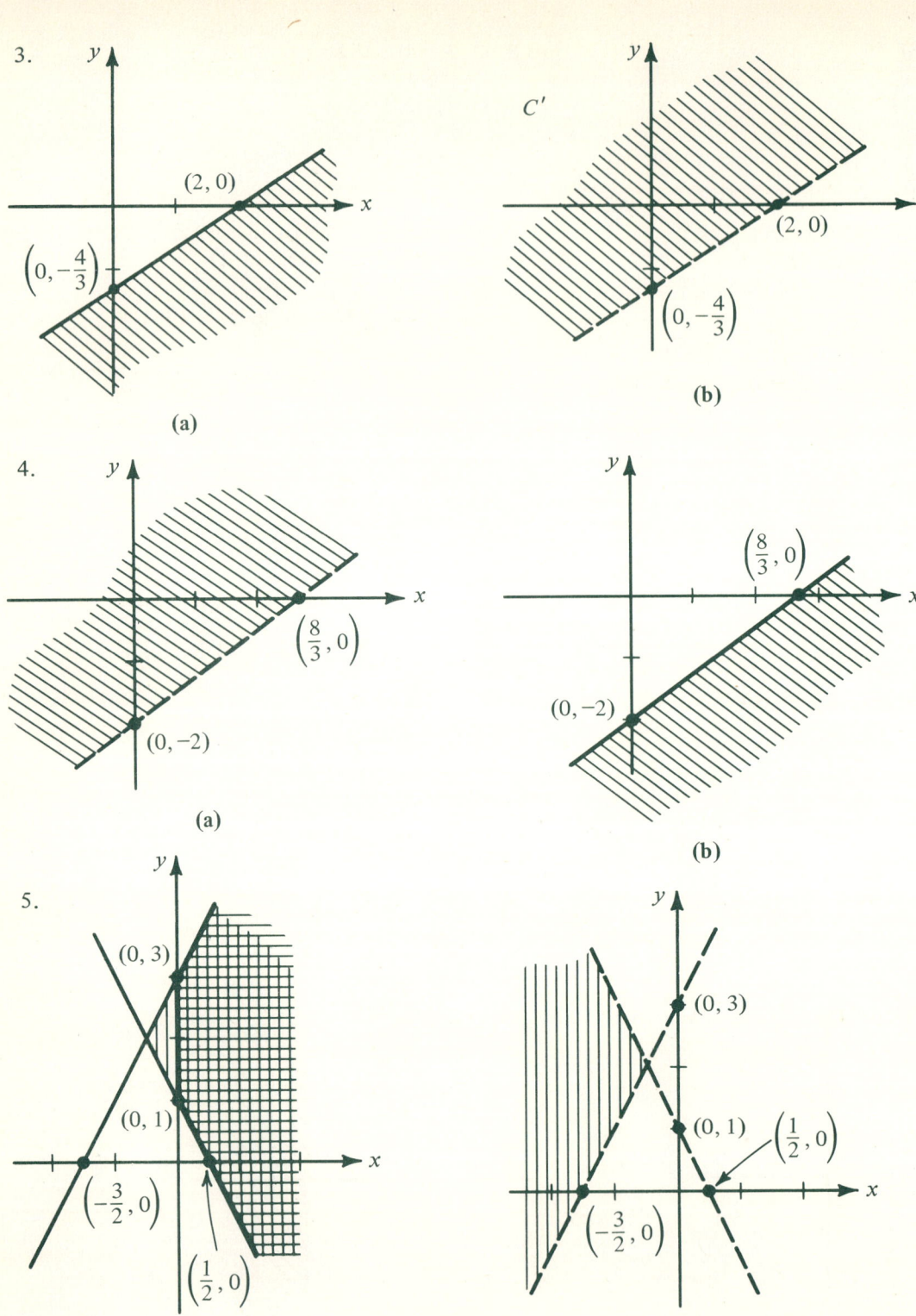

A-186

6.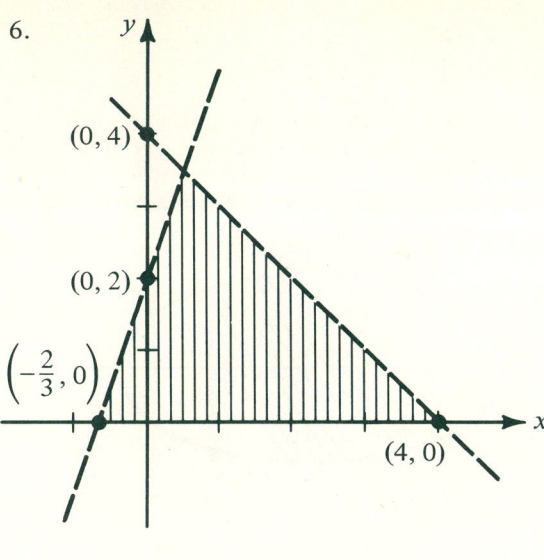

(a)

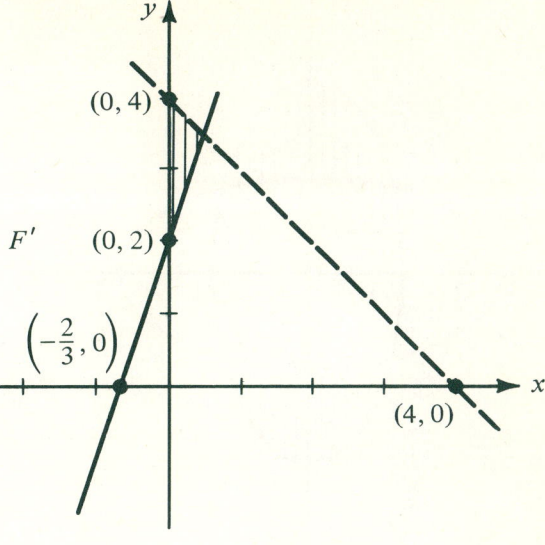

(b)

7.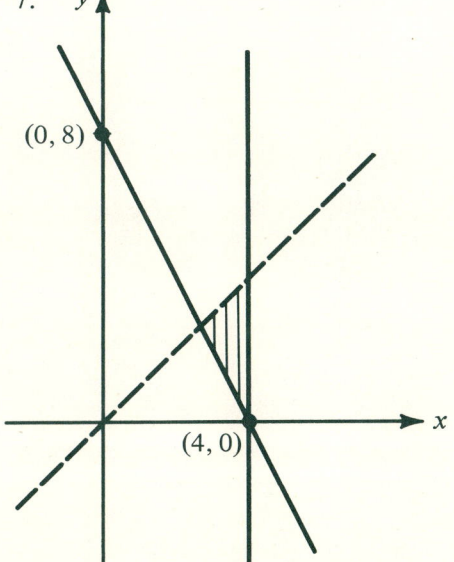

A-187

9.

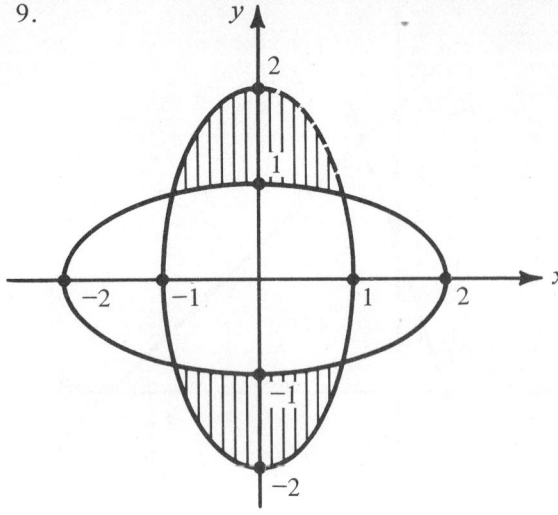

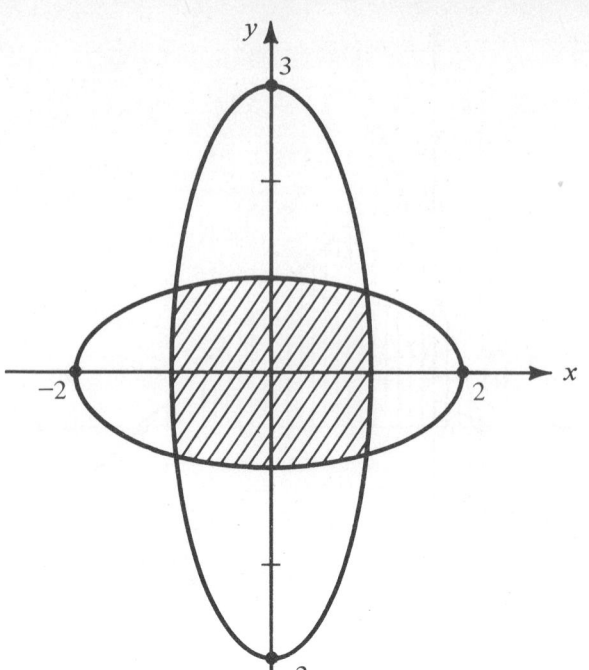

11.

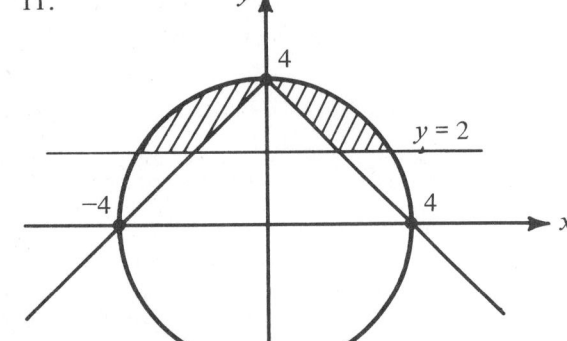

13.

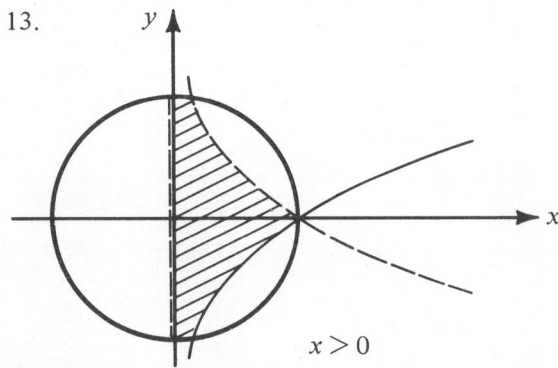

$x > 0$

Exercise 15.9

1.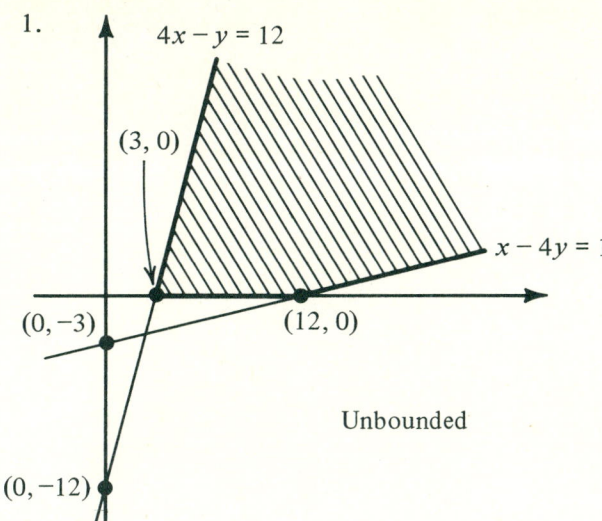

2.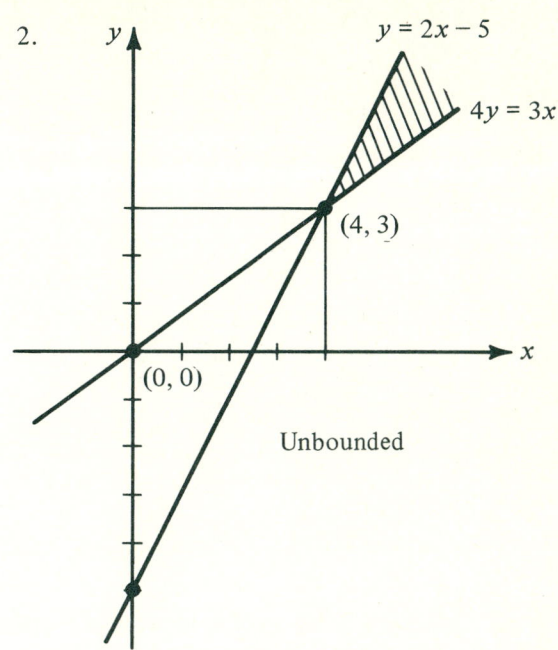

3.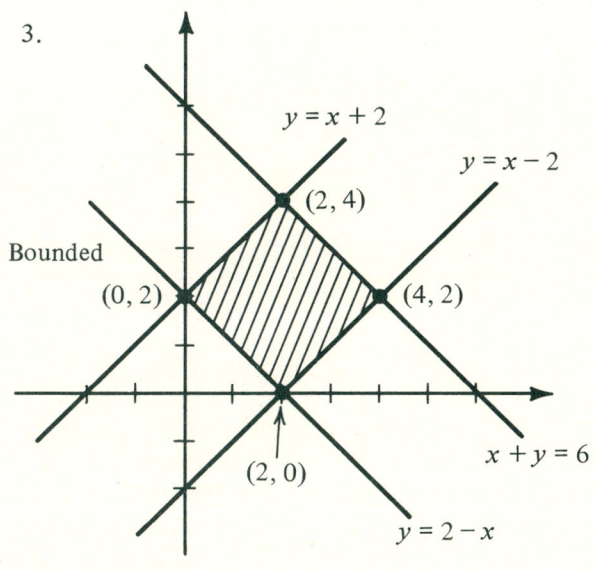

4.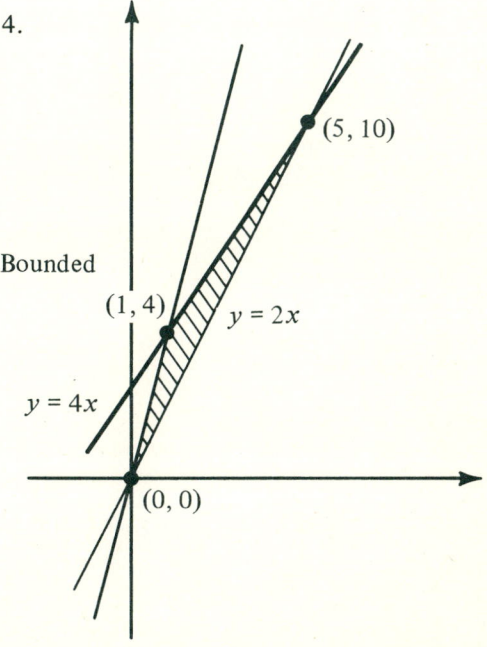

A-189

5.

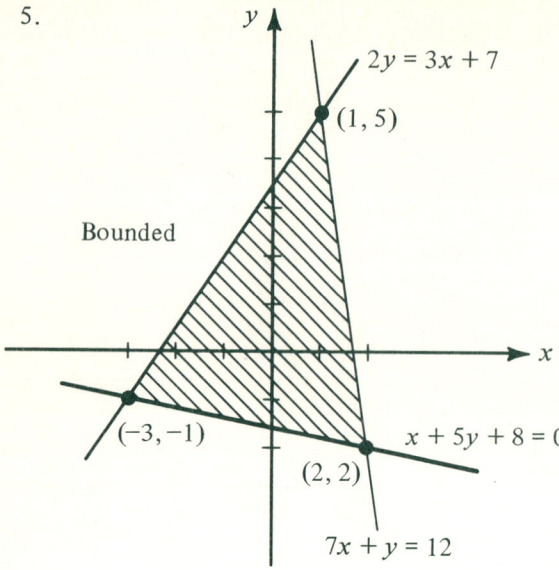

6.

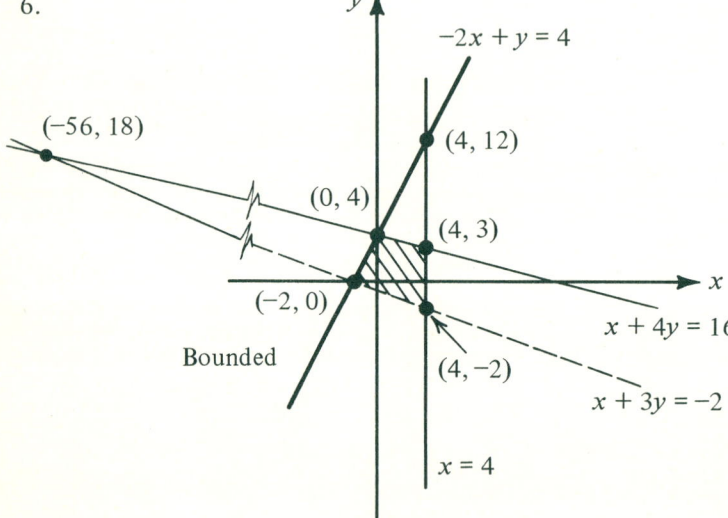

7. Minimum $z = 3$ at $(3, 0)$.   8. Minimum $z = 20$ at $(2, 0)$. Maximum $z = 100$ at $(2, 4)$.
9. Maximum along line joining $(0, 2)$ and $(2, 4)$. Minimum along line joining $(2, 0)$ and $(4, 2)$.
11. Minimum of 6 along $x + y = 3$ from $(3, 0)$ to $(0, 3)$ and maximum at $(8, 6, 17)$.
13. Maximum at $(2, 0, 6)$, minimum of -51 along $4y = x + 17$ from $(-5, 3)$ to $(-1, 4)$.
15. $C = 10x + 15y$, where x is the number of 707's (times a fixed distance from Seattle to New York) and y is the number of 747's. Number of seats at least 2000 means

$$100x + 200y \geq 2000$$
$$20 \text{ pilots: } x + y \leq 20$$
All can fly 707's but only 5 can fly 747's:
$$0 \leq x \leq 20 \quad \text{and} \quad 0 \leq y \leq 5$$

At $(15, 5), C = 225$
At $(10, 5), C = 175$
At $(20, 0), C = 200$
So, they will fly ten 707's and five 747's.

17. Use any point along the line $2x + 3y = 40$ between $(25/2, 5)$ and $(20, 0)$.

CHAPTER 16

Exercise 16.1

1. (a) A head. A tail.
   (b) Two heads, a head and a tail, a tail and a head, two tails.
   (c) HHH, HHT, HTH, HTT, THH, THT, TTH, TTT.
2. (a) True or false.
   (b) TTT, TTF, TFT, TFF, FTT, FTF, FFT, FFF.
   (c) TTTT, TTTF, TTFT, TTFF, TFTT, TFTF, TFFT, TFFF, FTTT, FTTF, FTFT, FTFF, FFTT, FFTF, FFFT, FFFF.
3. (a) {a, b}
   (b) {a, b}, {a, c}, {b, c}
   (c) {a, b}, {a, c}, {a, d}, {b, c}, {b, d}, {c, d}
5. (a) 2   (b) $2 \cdot 2 = 4$ (c) $2 \cdot 2 \cdot 2 = 8$
7. $3 \cdot 2 = 6$ 8. $3 \cdot 2 \cdot 1 = 6$

Exercise 16.2

1. (a) *ABC, ACB, BAC, BCA, CAB, CBA*
   (b) 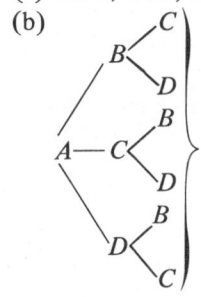 6 ways for *A* first, *B* first, *C* first, and *D* first, for a total of 4 groups of 6 or 24.

2. (a) *AB, AC, BA, BC, CA, CB*

   (b)
   ```
        B              A              A              A
   A ← C         B ← C         C ← B         D ← B      for a total of 4 groups
        D              D              D              C      of 3 or 12
   ```

3. (a) (i)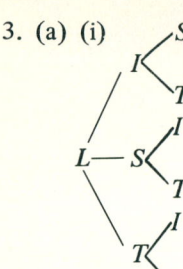

6 ways for L first, 6 for I first, 6 for S first and 6 for T first for 4 groups of 6 or 24

(ii) $_4P_2 = 12$
(b) (i) 24 (ii) 12
(c) (i) 2 (ii) 1

4. (a) $3 \cdot 2 \cdot 1 = 6$
   (b) $4 \cdot 3 \cdot 2 \cdot 1 = 24$

5. (a) $3 \cdot 2 = 6$
   (b) $4 \cdot 3 = 12$

6. (a) $4! = 24$, $4 \cdot 3 = 12$ (b) $6! = 720$, $6 \cdot 5 \cdot 4 = 120$ (c) $8! = 40320$, $8 \cdot 7 \cdot 6 \cdot 5 = 1680$

7. (a) $4! = 24$

(b) $6! = 720$

(c) $5! = 120$

9. (a) $10 \cdot 9 = 90$ (b) $10 \cdot 9 \cdot 8 = 720$

11. $\boxed{24}\ \boxed{24}\ \boxed{24}\ \boxed{10}\ \boxed{10}\ \boxed{10} = 13{,}824{,}000$

13. $\boxed{24}\ \boxed{24}\ \boxed{24} = 13{,}824$

15. (a) ABC, ACB, BAC, BCA, CAB, CBA. BCA and CAB are the only derangements. A is in the same place in ABC and ACB, C is in its old place in BAC and B in its place in CBA.
(b) BA (c) $BADC$, $BCDA$, $BDAC$
$CADB$, $CDAB$, $CDBA$
$DABC$, $DCAB$, $DCBA$
(c) There is 1 derangement for 2 items, 2 derangements for 3 items, and 9 derangements for 4 items.

Exercise 16.3

1. (a) 28
   (b) 28
2. (a) 161,700
   (b) 1
3. (a) 35
   (b) 1
4. (a) 19,900
   (b) 19,900
5. Combination
6. Permutation
7. Permutation
8. Permutation
9. Permutation

10. (a) *AB, BA, AC, CA, BC, CB*
    (b) *AB, AC, BC*
11. (5) 252   (6) 20   (7) 5040   (8) 720   (9) $3 \cdot 3 = 9$
13. $_{13}C_5 = \dfrac{13!}{5!(13-5)!} = 1287$ of these, 10 are straight flushes.

15. (a) 1260   (b) 180

(c) $\dfrac{11!}{4! \cdot 4! \cdot 2!} = 34{,}650$

17. $n = 4$

19. $_nC_r = \dfrac{n!}{r!(n-r)!}$ and $_nC_{(n-r)} = \dfrac{n!}{(n-r)![(n-r)-r!]} = \dfrac{n!}{(n-r)!r!}$

Therefore, $_nC_r = {_nC_{(n-r)}}$

Exercise 16.4

1. (a) 5/16   (b) $1 - (3/4) = 1/4$
2. (a) $4/52 = 1/13$ (b) $13/52 = 1/4$ (c) 1/52
3. (a) 16%   (b) 100%   (c) 0%
4. (a) 8/13   (b) 5/13
5. (a) 4/5   (b) 1/5
6. (a) $3/6 = 1/2$ (b) $1 - 1/2 = 1/2$
7. (a) $15/36 = 5/12$ (b) $1 - (5/12) = 7/12$
9. (a) $8/48 \approx 0.167$ (b) $4/48 \approx 0.0833$ 10. (a) 1/6   (b) 1/36
11. (a) $13/52 = 1/4$ (b) $4/52 = 1/13$ (c) $1/52 = (1/4)(1/13)$ (d) $16/52 = (13/52) + (4/52) - (1/52)$
13. $4/16 = 1/4$
15. The number of different ways of distributing r indistinguishable items among n cells is

$$\binom{n+r-1}{r} = {_{(n+r-1)}C_r}$$

We have $n = 3$ and $r = 6$, so there are 28 ways to line up the items.
    (a) 10/28   (b) 3/28   (c) 7/28
17. (a) $aA \times aA \Rightarrow aa, aA, Aa \Leftrightarrow aA, AA$ (b) 1/16   (c) aA, AA (d) aa, aA (e) 19/36

Exercise 16.5

1. (a) independent   (b) dependent
2. (a) independent   (b) dependent   (c) dependent
3. (a) independent (there is only one event)   (b) independent
4. (a) dependent   (b) dependent
5. (a) not mutually exclusive outcomes   (b) mutually exclusive

6. (a) dependent events (one usually must be a high school graduate before becoming a college graduate)
   (b) mutually exclusive, since $p(A \land B)$ is essentially zero
7. (a) not mutually exclusive, since, if both parents have bB genotypes, they can have a bb genotype child
   (b) not mutually exclusive
   (c) mutually exclusive (two bb genotypes — blue-eyed phenotype — cannot parent a brown-eyed child)
8. (a) 3/10   (b) 1/2
9. (a) $P(A \land B) = 0$, $P(A \lor B) = (3/10) + (5/10) = 4/5$
   (b) $P(A \land B) = 0$, $P(A \lor B) = (2/5) + (3/5) = 1$
   (c) 2/3   (d) 5/12   (e) 0
11. 5(a) 1/52   6(a) $6/36 = 1/6$ 13. (a) 1/3
    (b) 0   (b) $3/6 = 1/2$ (b) 1/2
    (c) 3/51
15. (a) 3/13   (b) 70/79
16. (a) Let B = {drawing a black marble}. Then, from Bayes' Theorem,

$$P(A_1|B) = \frac{P(A_1) \cdot P(B|A_2)}{P(A_1)P(B|A_1) + P(A_2) \cdot P(B|A_2)}$$

$$= \frac{(1/2)(4/6)}{(1/2)(4/6) + (1/2)(1/6)} = \frac{2/6}{5/12} = 4/5$$

   (b) 9/19

Exercise 16.6

1. (a) 6/15   (b) $13/15 = 87\%$ (c) practically zero 2/15
2. (a) 8/13   (b) 8/13   (c) 8:5, 5:8, 7:6
3. 1/8, 3/8, 3/8, 1/8
5. There are two categories, hearts ($K = 13$) and non-hearts ($N - K = 52 - 13 = 39$). The total number of ways of dealing the heart times the ways of dealing the non-hearts gives the total possible successful outcomes. $_NC_n = {}_{52}C_5 = 2{,}598{,}960$ represents the number of different five-card poker hands.
   (a) 1%
   (b) $p(0) = 0.222$ $p(1) = 0.4114$ $p(2) = 0.274$
       $p(3) = 0.082$ $p(4) = 0.011$ $p(5) = 0.0005$
   (c) $p(3) = 0.24$ $p(1) = 0.022$

Exercise 16.7

1. (a) $x^5 + 5x^4y + 10x^4y^2 + 10x^2y^3 + 5xy^4 + y^5$
   (b) $x^5 - 5x^4y + 10x^3y^2 - 10x^2y^3 + 5xy^4 - y^5$
2. (a) $A^6 + 6A^5B + 15A^4B^2 + 20A^3B^3 + 15A^2B^4 + 6AB^5 + B^6$
   (b) $64x^6 - 192x^5y + 240x^4y^2 - 160x^3y^2 + 60x^2y^4 - 12xy^5 + y^6$
3. (a) $x^4 + 4x^3 + 6x^2 + 4x + 1$
   (b) $x^8 + 4x^6 + 6x^4 + 4x^2 + 1$
4. (a) $16x^4 - 96x^3 + 216x^2 - 216x + 81$
   (b) $32x^5 - 240x^4y + 720x^3y^2 - 1080x^2y^3 + 810xy^4 + 243y^5$

5. (a) $8x^3 - 12x^2 + 6x - 1$
   (b) $8x^9 - 12x^6 + 6x^3 - 1$
6. (a) $(1/2)^7(2x-1)^7 = (1/128)(128x^7 - 448x^6 + 672x^5 - 560x^4 + 280x^3 - 84x^2 + 14x + 1)$
   (b) $2187x^7 - 5103x^6y + 5103x^5y^2 - 2835x^4y^3 + 945x^3y^4 - 63x^2y^5 + 3xy^6 - y^7$
7. $x^7 - 7x^6y + 21x^5y^2 - 35x^4y^3 + 35x^3y^4 - 21x^2y^5 + 7xy^6 - y^7$
8. $64x^6 + 192x^5y + 240x^4y^2 + 160x^3y^3 + 60x^2y^4 + 12xy^5 + y^6$
9. (a) x^{10} (b) y^{10} 10. (a) $-220a^3b^9$ (b) $66a^{10}b^2$
11. (a) $-960x^3y^7$ (b) $-5120x^9y$ 13. $-61{,}236$
15. 45   17. -120
19. (a) 1.4194
    (b) 12.4645 (This is not close to the exact answer, 32, because the fact that 0.41 is rather large requires using more terms.)
21. (a) 4   (b) $235 + 828i$
23. (a) $(1 + .4)^{1/2} = 1 + \dfrac{1/2(0.4)}{1!} + \dfrac{(1/2)(-1/2)}{2!}(.4)^2 + \dfrac{(1/2)(-1/2)(-3/2)}{3!}(.4)^3$

$+ \dfrac{(1/2)(-1/2)(-3/2)(-5/2)}{4!}(.4)^4 = 1.183 \approx 1.18322$

Index

Abel, Niels, 44, 385
Absolute value
  definition of, 9, 11
  function, 34
  geometric interpretation of, 9, 10
  theorems on, 10, 11, 12
Acceleration, 120
Addition formulas, 231
Adjoint, 321
Algebra, the fundamental theorem of, 141
Amplitude, 200
Angle
  degree measurement of, 216
  of depression, 239
  of elevation, 239
  radian measurement of, 191
  reference, 221
Antilogarithm, 177
Appollonius, 308, 383
Approximations of irrational zeros, 139
Arc-functions, 214
Archimedes, 242, 383
Arc length, 104
Argand, Jean-Robert, 40, 385
Aristotle, 60, 326, 383
Aristotelian logic, 47, 61
Arithmetic
  sequences, 36
  series, 37
  of signed numbers, 8
Associative property, 4
Asymptote
  horizontal, 148
  of a hyperbola, 292
  oblique, 154
  vertical, 146
Axis (axes)
  of Cartesian coordinate system
    in R^2, 31
    in R^3, 301
  of polar coordinate system, 252
Azimuth, 243

Babylonians, 22, 41, 216, 227
Barometric pressure, 168
Base of logarithms, 169
Bayes' theorem, 371
Bearing, 243
Beat phenomena, 276, 277
Bernoulli, Jacob, 41, 98, 278, 377, 384
Bernoulli, Johann, 41, 384
Bernoulli, trials *see also* Binomial trials.
Bessel, Frederick, 274
Bias, 203
Binomial
  distributions, 371
  theorem, 375
  trials, 373

BOD, 183
Bolzano, Bernhard, 60, 385
Boole, George, 61, 385
Bounds
  greatest lower, 134
  least upper, 134
  for irrational zeros, 140
  for rational zeros, 133
Briggs, Henry, 189, 384
Burgi, Joost, 189, 384

Cantor, Georg, 60, 385
Carbon dating, 166
Cardan, Girolamo, 40, 143, 384
Carrier wave, 278
Cartesian coordinate system
  for R^2, 31
  for R^3, 302
Cartesian product, 46, 47
Cauchy, Augustin, 144, 385
Cayley, Arthur, 327, 385
Change of variable, 23
Characteristic equation, 317
Circles
  definition of, 282
  equation of, 282
  unit, 193
Circular functions, 190–216
Closure, 4
Coefficient
  of polynomials, 126
  uniqueness of, 158
Comatrix, 319, 321
Combinations, 360–362
Commutative, 4
Completing the square, 20
Complex numbers, 15–19, 265–271
  absolute value of, 18
  addition and multiplication of, 16
  conjugate of, 18
  definition of, 16
  division of, 17
  equality of, 16
  geometric representation of, 18
  imaginary part of, 16
  polar form of, 265
  powers of, 267
  product of, 266
  rational powers of, 270
  real part of, 16
  zeros, 141
Component, vector, 326
Composite function, 92–96
Conclusion, 48
Conditional probability, 368
Conditional proposition, 52
Conic sections, 281–308
Conjunction
  definition of, 47
  negation of, 50
  truth table for, 48

Continuous, 148, 391
Contradiction, proof by, 55
Contrapositive, 50
Converse, 50
Conversion factors, 389
Coordinate systems
   Cartesian, 31, 302
   complex, 18
   polar, 252
Cosecant, 206, 218
Cosine, 195
   as a circular function, 193
   graph of, 200
   identities for, 222
   law of, 246
   as a trigonometric function, 218
Cost, minimize, 122
Cotangent, 206, 218
Counter example, proof by, 55
Counting, 256, 257
Cramer, Gabriel, 355
Cramer's rule, 331
Curve fitting
   linear, 107–110
   quadratic, 122–125
Cylinders, 303

Decibel scale, 182
Dedekind cut, 6
Dedekind, Richard, 6, 40, 385
Deductive reasoning, 56
Degree
   measure, 216, 217
   of a polynomial, 126
DeMoivre, Abraham, 384
DeMoivre's theorem, 270
DeMorgan, Augustus, 61
DeMorgan's Laws, 44, 50
Dependent events, 370
Dependent variable, 65
Derangements, 360
Descartes, Rene, 40, 41, 60, 143, 189, 384
Descartes' rule of signs, 133
Determinants, 314–319, 331
Directrix, 285
Discontinuity, point of, 148
Discriminant
   for conic sections, 300
   for quadratic equations, 21
Disjoint set, 43
Disjunction
   definition of, 47
   negation of, 50
   truth table for, 48
Distance formula, 105
Distributive law, 4
Divergence, 41
Division
   of polynomials, 126
   synthetic, 127

by zero, 145
Domain, 66
Double angle formulas, 235

e, 39, 164
Echelon form, 342–344
Elementary transformations on matrices, 342
Ellipse
   definition of, 289
   graph of, 289–291
Ellipsoid, 307
Elliptic paraboloid, 307
Empirical formula, 97
Empty set, 43
Equality
   of complex numbers, 16
   of matrices, 310
   of polynomials, 158
   properties of, 4
Equations
   dependent, 328
   exponential, 162
   inconsistent, 328
   independent, 328
   logarithmic, 174
   polar, 252
   quadratic, 19
   trigonometric, 225, 228
Equivalence, logical, 49
Euclid, 61, 227, 383
Eudoxus, 308, 383
Euler, Leonard, 40, 60, 98, 189, 384
Euler's formula, 265
Even functions, 76
Event, 363
   independent, 366
   mutually exclusive, 368
Exhaustion, method of, 242
Existance quantifier, 53
Exponential
   decay, 166
   function, 160–169
   growth, 164
   tables, 393
Exponents
   definition, 12, 14
   integer, 12
   rational, 13
   theorems on, 14
Extraneous roots, 30
Extrema. See Maximum

Factorial notation, 386
Factors, multiplicity of, 135
Feasible region, 350
Fermat, Pierre de, 41, 60, 377, 384
Fibonacci, 41, 384
Field properties, 4
Finite differences, 125, 139

Focus, 285
Force, 325
Fourier, Jean, 41, 144, 385
Fractions, 4
Fractions, partial, 157
Frege, Gottlog, 61
Frequency, 275
Functions
   absolute value, 34
   circular, 190–216
   composite, 92–95
   compressions of, 84
   continuity of, 148, 391
   definition of, 67
   domain of, 67
   even, 34
   expansions of, 84
   greatest integer, 35
   inverse of, 86–91
   linear, 99–111
   logarithmic, 169–189
   odd, 34
   one-to-one, 67
   periodic, 200
   polynomials, 126–144
   product of, 92
   quadratic, 112–125
   range of, 67
   rational, 145–159
   reflections of, 70–76
   sum of, 91
   symmetry of, 76–78
   translations of, 78–83
Fundamental Counting Principle, 358
Fundamental Theorem of Algebra, 141

Galileo, 326, 384
Galois, Everiste, 144, 385
Gaus, Karl, 40, 143, 189, 385
Gaussian reduction, 338–342
Genetics, 360
Geometric
   sequence, 37
   series, 38
Gibbs, J.W., 327
Godel, Kurt, 61, 385
Gomperz curves, 169
Graph
   of the absolute value function, 34
   of the circular functions, 200, 201, 206, 208
   of the conics, 282, 285, 289, 293
   of equations, 31–36
   of exponential functions, 161, 162
   of the greatest integer function, 35
   of inequalities, 32, 347–350
   of inverse relations, 89
   of linear functions, 108
   in polar coordinates, 252
   of polynomials, 134–139
   of quadratic functions, 113–114

Graph (*continued*)
   of rational functions, 148–157
   symmetry of, 76–78
   in three dimensions, 301–308
Greatest lower bound, 134
Gregory, James, 98

Half-angle formula, 235
Half-life, 166
Hamilton, William, 326, 385
Harmonic motion, 275
Heading, 243
Heaviside, Oliver, 327
Hilbert, David, 61, 385
Hippachus, 227, 383
Horizontal asymptote, 148
Huygens, Christian, 384
Hyperbola
   definition of, 292
   graphs of, 292–295
Hyperbolic paraboloid, 308
Hyperboloid, 306
Hypergeometric distribution, 373
Hypothesis, 48

i, 15
Identities, trigonometric, 222, 233, 235
Identity
   additive, 4
   for matrices, 320
   multiplicative, 4
   uniqueness of, 4
Imaginary
   axis, 18
   number, 15
Implication
   definition of, 47
   negation of, 50
   transitive law of, 51
   truth table for, 48
Inclusion, set, 42, 43
Inconsistent equations, 328
Independent
   equations, 328
   events, 366
   variables, 65
Indeterminant, 145, 332, 333
Indirect proof, 55
Induction, mathematical, 56–61, 267, 375
Inequalities
   linear, 26
   quadratic, 27
   systems, of, 347–350
Infinite series, 39, 40
Infinity, 145
Integers, 3, 4
Interest, 166
Interpolating polynomial, 123, 139
Interpolation, linear, 162

Intersection of sets, 43
Inverse
   additive, 4
   of circular functions, 212
   functions, 86–91
   linear functions, 110
   logical, 50
   multiplicative, 4
   of quadratic functions, 116
   relations, 87, 88
Irrational
   exponents, 163
   numbers, 3, 6
   zeros, 140
Isosceles triangle, 218

Kepler, Johann, 189

Lagrange, Joseph, 385
Laplace, Pierre Simon de, 385
Law
   of cosines, 246
   of the excluded middle, 47
   of sines, 248
Least common multiple, 24
Least squares, 122
Least upper bound, 134
Legendre, Adrien-Marie, 143
Leibniz, Gottfried, 40, 60, 98, 327, 377, 384
Lemniscates, 262
L'Huspital, Guillaume de, 384
Libby, Willard, 166
Limacon, 261
Limits
   definition of, 390
   of functions, 146
   notation, 149
   of $\sin x/x$, 271
   theorems on, 275
Linear
   algebra, 309
   combinations, 324
   inequalities, 26
   interpolation, 162
   programming, 350
   systems, 328–344
Linear functions, 99–111
   curve fitting, 107–110
   definition, 99
   graphing, 107–110
   inverses of, 110–111
   slope of, 101
Logarithm(s), 169–189
   applications of, 180–189
   base of, 169
   change of base of, 175, 176
   characteristic of, 177
   common, 170
   computations with, 177–180

   definitions of, 169
   equations involving, 174
   history of, 189
   interpolation of, 178
   mantissa of, 177
   Neperian (natural), 170
   properties of, 174
   scales for, 180
Logarithmic functions, 169–189
   definition of, 169
   graph of, 170–174
   tables, 394–396
Logic, 47–61
Log-log plots, 185
LORAN navigational system, 296

Maclaurin, Colin, 355, 384
Magnitude, celestial, 168
Marginal cost, 103
Marginal revenue, 103
Market equilibrium, 334
Mathematical induction, 56–61, 267, 375
Matrices, 309–314
   addition of, 310
   additive inverses of, 311
   adjoint of, 321
   augmented, 342
   column, 309
   comatrix, 319, 321
   definition of, 309
   determinant of, 314–319
   elementary row transformations of, 342
   equality of, 310
   identities of, 320
   inverses of, 319–322, 321
   multiplication of
      matrix, 312
      scalar, 310
   order of, 309
   row, 309
   transpose of, 314, 321
Maurolycus, Francesco, 61, 384
Maximum, 118–121
Maxwell, James, 327, 385
Menaechmus, 308, 383
Mendel, 366
Metric system, 389
Midpoint formula, 106
Minimum, 118–121
Minor, 315
Modulus, 17, 18
Modus ponens, 51
Modus tollens, 51
Muir, Thomas, 216
Multiplicity of zeros, 135
Mutually exclusive events, 368

$n!$, 386

Napier, John, 189, 384
Natural numbers, 3
Negation(s), 49–52
  of conjunctions, 50
  of disjunctions, 50
  of implications, 51
  of negations, 49
  of quantified propositions, 53
Negative
  angles, 192
  exponents, 12, 13
Newton, Isaac, 40, 41, 98, 143, 189, 377, 384
Newton's Law of Cooling, 187
Node, 350
Null set. See empty set
Numbers
  complex, 15–19, 265–271
  imaginary, 15
  integers, 3, 4
  irrational, 3, 6
  natural, 3, 4
  prime, 3
  rational, 3
  real, 3

Object function, 351
Odd function, 76
Odds, 365
One-to-one function, 67
Optimization, 350. See also Maximum
Ordered
  pair, 31
  triple, 301
Order of a matrix, 307
Order, properties of, 7
Oresme, Nicole, 189, 384
Origin, 31

p series, 39
Parabola
  as a conic section, 284–288
  definition of, 285
  as a quadratic function, 112–125
Paraboloid, 307
Parallel lines, 103
Parallelogram rule, 323
Pareto, Vilfredo, 154
Partial fractions, 157–159
Partial sums, 39, 40
Pascal, Blaise, 384
Pascal's triangle, 374
Period, 275
Periodic function, 199, 275
Periodic motion, 275
Permutations, 358–360
Perpendicular lines, 104
pH, 183
Phase shift, 202

Pi, 39
Planes, 302
Point slope formula, 107
Polar coordinates, 252
Polar form of complex numbers, 265
Polynomial(s), 126–144
  bounds for zeros of, 133–134
  coefficients of, 126
  complex zeros of, 141
  definition of, 126
  degree of, 126
  division of, 127
  equality of, 158
  factoring of, 129–134
  graph of, 134–139
  history of, 143
  irrational zeros of, 139
  rational zeros of, 129
  zeros of, 129
Population growth, 164
Power function, 184
Probability, 362–371
Product transformation equations, 154
Programming, linear, 350
Progression. See Sequence
Proof, methods of, 54–61
Proposition, 47
Pythagorean Theorem, 106, 227
Pythagoras, 111, 383

Quadrants, 31, 32
Quadratic
  equations, 19
  formula, 20
  inequalities, 27
Quadratic functions, 112–125
  applications of, 118–122
  curve fitting with, 122–125
  definition of, 112
  graphing of, 113, 114
  inverses of, 116–118
Quadratic in form, 23–25
Quadric surfaces, 301–308
Quantifier
  existential, 53
  universal, 53

Radian, 191
Radical, 30
Radioactive decay, 166
Range, 67
Rational
  exponents, 13, 14
  numbers, 3, 6
  zeros, 129
Rational functions, 145–159
  graph of, 148–157
Rationalize the denominator, 13, 14
Ray, 192

Real
  axis, 18
  number field, 3
  numbers, 4
Rectangular coordinate system. See Cartesian coordinate system
Recursive functions, 222
Reference triangle, 221
Reflections, 70–76
Relations, 67
Relatively prime, 3
Remainder theorem, 127
Resultant, 323
Richter scale, 180
Riemann, Georg, 385
Roots of complex numbers, 270
Roses, 261
Rotation of axes, 297–301
Row matrices, 309
Rules of signs, 133
Russell, Bertrand, 61, 385

Sample space, 363
Scales
  decibel, 182
  logarithmic, 180
  musical, 276
  Richter, 180
Secant, 69, 206, 218
Seki, Kowa, 327
Sequence, 36–41
  arithmetic, 36
  geometric, 37
  infinite, 39
  semilog plot, 184
Series, 36–41
Sets, 42–47
  Cartesian product of, 46
  compliment of, 44
  definition of, 42
  disjoint, 43
  element of, 42
  empty, 43
  intersection of, 43
  subsets of, 42, 43
  union of, 43
  universal, 43
Similar triangles, 218
Simultaneous
  linear equations, 329
  nonlinear equations, 345
Sine(s), 195
  as a circular function, 193
  graph of, 201
  identities for, 222
  Law of, 248
  as a trigonometric function, 218
Slope
  definition of, 99, 100
  of parallel lines, 103
  of perpendicular lines, 104

Slope (*continued*)
   of tangent line, 104
Slope intercept form, 108
Sphere, 307
Spirals, 262
Square, completing the, 20
Straight line, 107
Subsets, 42, 43
Sum formulas, 233
Supply and demand, 334
Symmetry, 76–78
   for polar coordinates, 260
Synthetic division, 127
Systems
   Cramer's rule for, 331
   dependent, 328
   echelon form for, 342–344
   equivalent, 330
   Gaussian reduction for, 338–342
   geometric interpretation of, 329, 335
   inconsistent, 328
   independent, 328
   of inequalities, 347–350
   of linear equations, 328–344
   of nonlinear equations, 344–347

Tables, 392–401
   for exponential functions, 393
   for logarithms base e, 369
   for logarithms base 10, 394–395
   for trigonometric functions, 397–401
Tangent
   as a circular function, 206
   graph of, 208
   to a graph, 110
   identities for, 222
   as a trigonometric function, 218
Tautology, 50
Taylor, Brook, 41, 384
Tchebycheff, Pafnuti, 385

Theorem
   binomial, 375
   DeMoivre's, 270
   DeMorgan's, 44, 50
   fundamental, of algebra, 141
   proof of, 54–56
   remainder, 127
Thompson, James T., 216
Three dimensions, 301, 334
Trace, 304
Transcendental functions, 160
Transitive law for implications, 51
Translations, 78–83
Triangle inequality, 106
Triangular form, 317
Trichotomy law, 7
Trigonometric
   equations, 225, 228
   functions, 217
   functions of small angles, 271
   identities, 222, 233, 235
   tables, 397–401
Truth set, 52
Truth table(s)
   for conjunction, 48
   definition of, 47, 48
   for disjunction, 48
   for implication, 48
   for logical equivalence, 49
   for negation, 49–51
Truth value, 47

Unbiased, 364
Union, 43
Unit circle, 193
Universal quantifier, 53
Universal set, 43
Upper bound, least, 134

Vapor pressure, 188
Variable
   change of, 23
   dependent, 65
   independent, 65
Vector(s), 322–327
   addition of, 323
   components, 326
   definition of, 322
   history of, 326
   magnitude of, 322, 323
   product of (cross product), 325
   resultant of, 323
   scalar product of, 325
   unit, 323, 324
Velocity, 120, 121, 326
Venn diagrams, 43, 44
Vertex
   of hyperbolas, 294
   of parabolas, 285
Vertical asymptote, 146
Vieta, Francois, 41, 384

Wallis, John, 384
Weierstrass, Karl, 385
Work, 325, 326

x-axis, 32
x-coordinate, 31

y-axis, 32
y-coordinate, 31

z-axis, 301
Zeno, 41, 385
Zero(s)
   complex, 141
   in division, 145
   exponent, 13
   factorial, 390
   multiple, 135
   of polynomials, 129